MasteringAstronomy® encourages continuous learning before, during, and after class

BEFORE CLASS

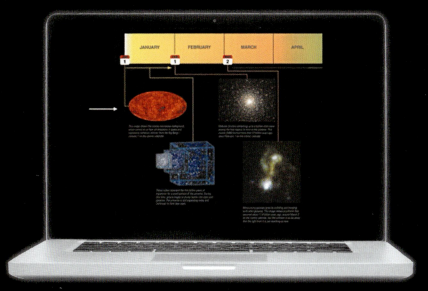

NEW! Interactive, book-specific prelecture videos provide subject overview for exposure to key concepts before class, opening the classroom time for active learning or deeper discussions of topics. These can be used for simple pre-class exposure or fully flipped classrooms.

NEW! e-Text 2.0

- **Full eReader functionality** includes page navigation, search, glossary, highlighting, note taking, annotations, and more.

- **A responsive design** allows the eText to reflow/resize to a device or screen. eText 2.0 now works on supported smartphones, tablets, and laptop/desktop computers.

- **In-context glossary** offers students instant access to definitions by simply hovering over key terms.

- **Seamlessly integrated videos and activities** allow students to watch and practice key concepts within the eText learning experience.

- **Accessible** (screen-reader ready)

- **Configurable reading settings,** including resizable type and night reading mode

DURING CLASS

NEW! Learning Catalytics™ generates class discussion, guides lecture, and promotes peer-to-peer learning with real-time analytics. MasteringAstronomy® with eText now provides Learning Catalytics—an interactive student response tool that uses students' smartphones, tablets, or laptops to engage them in more sophisticated tasks and thinking. Instructors can

- Pose a variety of open-ended questions that help students develop critical thinking skills

- Monitor responses to find out where students are struggling

- Use real-time data to adjust instructional strategy and try other ways of engaging students during class

- Manage student interactions by automatically grouping students for discussion, teamwork, and peer-to-peer learning

MasteringAstronomy®

AFTER CLASS

NEW! and enhanced **interactive figures and narrated figures** formatted for mobile use are now woven throughout the eText as well as assignable within MasteringAstronomy®.

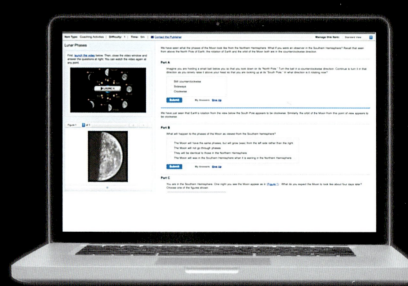

▶ NEW! **20 additional tutorials** focus on critical thinking and tie into the extraordinary claims boxes such as the one on martians in chapter 9, climate change in chapter 10, and neutron stars and black holes in chapter 18.

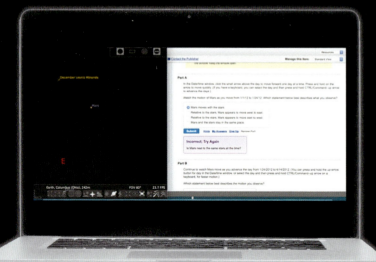

◀ **24 Virtual Astronomy Labs** are assignable online laboratory activities utilizing Stellarium open source planetarium software and interactive figures. The labs include night sky activities, data collection activities and inquiry based activities. These labs are housed in MasteringAstronomy and are gradable by Mastering. As a bonus, the Virtual Astronomy labs include 5 inquiry based labs based on education research principles (not auto graded by Mastering) in which students learn through exploring.

New features further reinforce the importance of critical thinking in astronomy

NEW! Icons call attention to the features that promote critical thinking.

► **NEW! Extraordinary Claims boxes** provide students with examples of extraordinary claims about the universe and how they were either supported or debunked as scientists collected more evidence. This engages student curiosity and promotes critical thinking.

extraordinary claims — The Death of the Dinosaurs Was Catastrophic, Not Gradual

By the mid-19th century, geologists recognized that Earth is shaped by both gradual and sudden changes. For example, sediments build up gradually as they are deposited, while volcanic eruptions can spew out vast amounts of rock in a very short time. But which type of process has been more important over the history of Earth? This question became a topic of such great debate among scientists that the two views were given names: *Uniformitarianism* was the view that today's Earth was shaped primarily by slow, gradual changes, while *catastrophism* held that the most important processes were dramatic events like global floods and gigantic volcanic eruptions.

A key issue in the debate was Earth's age. Geologists were already aware that major changes had occurred in the past. For example, the commonplace finding of seashells on mountaintops shows that seas can be transformed over time into mountains. If Earth was very old, then there had been plenty of time for such changes to occur gradually, but if Earth was young, catastrophic changes would have been necessary. During the latter half of the 19th century, evidence began to accumulate for a very old Earth, a case that was sealed in the mid-20th century when radiometric dating showed our planet to be more than 4 billion years old. There had clearly been enough time for uniformitarianism to explain most of Earth's features. But could it explain all of them?

The extinction of the dinosaurs posed at least a potential challenge. Dinosaur fossils are abundant in sediment layers dating to more than about 65 million years ago but absent from younger layers, indicating that the dinosaur extinction occurred in a "geological blink of the eye." But that still could have meant a period of a few million years, and most scientists assumed that the extinction occurred through relatively gradual change, such as changes in climate that were detrimental to dinosaur survival. So when Luis and Walter Alvarez proposed that the extinction had been a sudden event due to an asteroid impact, it was an extraordinary claim not only in terms of the dinosaur extinction but also in that it reintroduced the idea of catastrophism as an important part of Earth's history.

The initial evidence, which was from the iridium-rich sediment layer (see Figure 9.22), was strong enough to generate widespread scientific interest in the impact hypothesis, but many scientists (including the Alvarezes themselves) pointed out a major flaw: An impact large enough to deposit so much iridium should have left a noticeable crater, and no such crater was known at the time. That changed with the discovery of the *Chicxulub crater* in 1991 (see Figure 9.23). As discussed in this chapter, there is still scientific debate about whether or not the impact was the sole cause of the mass extinction, but at least part of the Alvarezes' extraordinary claim is not subject to any dispute at all: While most of Earth's history can be explained through the gradual processes of uniformitarianism, occasional catastrophic events can and do occur, sometimes with crucial consequences.

Verdict: Very likely.

---

the big picture — Putting Chapter 9 into Context

In this chapter we concluded the study of our own solar system by focusing on its smallest objects, finding that these objects can have big consequences. Keep in mind the following "big picture" ideas:

- Asteroids and comets may be small compared to planets, but they are very important scientifically, because they provide evidence that has helped us understand how the solar system formed.
- Pluto, once considered a "misfit" among the planets, is now recognized as just one of many moderately sized objects in the Kuiper belt. In terms of composition, the dwarf planets Pluto and Eris—along with other similar objects—are essentially comets of unusually large size.
- The small bodies are subject to the gravitational whims of the largest. The jovian planets shaped the asteroid belt, the Kuiper belt, and the Oort cloud, and they continue to nudge objects onto collision courses with the planets.
- Collisions not only bring meteorites and leave impact craters but also can profoundly affect life on Earth. An impact probably wiped out the dinosaurs, and future impacts pose a threat that we cannot ignore.

my cosmic perspective Perhaps more so than any other type of astronomical object, the small bodies of the solar system have the potential to touch our lives directly. We can hold pieces of them as meteorites, their impacts pose at least some threat to our survival, their compositions make them potentially useful, and, some 65 million years ago, one of them hit the Earth in an event that may have paved the way for our existence.

◄ **NEW! "My Cosmic Perspective" sections** focus on a personal connection between students and the cosmos, encouraging students

Scientific updates reflect new developments in the field of astronomy, engaging student curiosity

▶ **Recent results from major space missions and observatories,** including Hubble, *Kepler*, *New Horizons*, Mars *MAVEN*, and ALMA, bring the exhilaration of astronomy to students, showing them that it is happening right now in their lives.

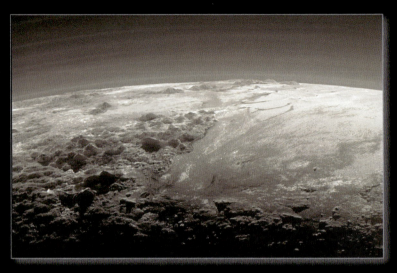

FIGURE 9.20
Pluto's jagged mountains, smooth plains, and hazy atmosphere, seen as *New Horizons* looked back toward the Sun after the encounter.

100,000 AU

FIGURE 6.15
The Orion Nebula, an interstellar cloud in which new star systems are forming. The insets show close-ups of six protoplanetary disks (nicknamed "proplyds")—disks of material in which planets can form—each surrounding a young star that is being born. The nebula will ultimately give birth to thousands of stars and planetary systems over the next few million years. (Images from the Hubble Space Telescope.)

◀ **Updated data and models** on topics such as the formation of planetary systems, global warming, and galaxy formation and evolution keep chapter content exciting and current.

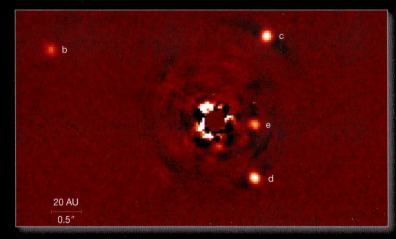

20 AU
0.5″

This infrared image from the Large Binocular Telescope shows direct detection of four planets (marked b, c, d, e) orbiting the star HR 8799. Light from the star itself (center) was mostly blocked out during the exposure, as indicated by the solid red circle.

▶ **Major new discoveries** and their implications on areas of study such as extrasolar planets and the timing and possible origins of life on Earth, including the recent finding of a potentially Earth-like planet around Proxima Centauri, elicit student curiosity and critical thinking about what we know.

THE ESSENTIAL Cosmic Perspective

EIGHTH EDITION

Jeffrey Bennett
University of Colorado at Boulder

Megan Donahue
Michigan State University

Nicholas Schneider
University of Colorado at Boulder

Mark Voit
Michigan State University

330 Hudson Street, NY NY 10013

Executive Editor: Nancy Whilton
Project Manager: Lifland et al., Bookmakers
Editor-in-Chief: Jeanne Zalesky
Media Producer: Jenny Moryan
Marketing Manager: Elizabeth Ellsworth Bell
Senior Content Developer: M. Amir Said
Director, Content Development, Science: Caroline Power
Program and Project Management Team Lead: Kristen Flathman
Production Service: Lifland et al., Bookmakers
Composition, Interior Design, Permissions: Cenveo® Publisher Services
Design Manager: Mark Ong
Cover Designer: John Walker
Photo Research: Julie Laffin
Photo Research Management: Laura Murray
Illustrations: Rolin Graphics, Jim Atherton
Manufacturing Buyer: Maura Zaldivar-Garcia
Printer and Binder: LSC Communications
Cover Image: Max Seigal
Cover Printer: LSC Communications

Library of Congress Cataloging-in-Publication Data

Names: Bennett, Jeffrey O. | Donahue, M. (Megan), 1962- | Schneider,
  Nicholas. | Voit, Mark.
Title: The essential cosmic perspective.
Other titles: Cosmic perspective
Description: Eighth edition / Jeffrey Bennett, University of Colorado at
  Boulder, Megan Donahue, Michigan State University, Nicholas Schneider,
  University of Colorado at Boulder, Mark Voit, Michigan State University. |
  New York, NY: Pearson, [2018] | Includes index.
Identifiers: LCCN 2016039737 (print) | LCCN 2016040278 (ebook) | ISBN
  9780134446431 | ISBN 9780134552101 ()
Subjects: LCSH: Astronomy—Textbooks.
Classification: LCC QB43.3 .B46 2018 (print) | LCC QB43.3 (ebook) | DDC
  520—dc23
LC record available at https://lccn.loc.gov/2016039737

ISBN 10-digit: 0-134-44643-7
ISBN 13-digit: 978-0-134-44643-1

www.pearsonhighered.com

5 18

Dedication

TO ALL WHO HAVE EVER WONDERED about the mysteries of the universe. We hope this book will answer some of your questions—and that it will also raise new questions in your mind that will keep you curious and interested in the ongoing human adventure of astronomy.

And, especially, to Michaela, Emily, Sebastian, Grant, Nathan, Brooke, and Angela. The study of the universe begins at birth, and we hope that you will grow up in a world with far less poverty, hatred, and war so that all people will have the opportunity to contemplate the mysteries of the universe into which they are born.

Brief Contents

KV 02.05.2019 0618

Contents

Preface

We humans have gazed into the sky for countless generations. We have wondered how our lives are connected to the Sun, Moon, planets, and stars that adorn the heavens. Today, through the science of astronomy, we know that these connections go far deeper than our ancestors ever imagined. This book tells the story of modern astronomy and the new perspective, *The Essential Cosmic Perspective*, that astronomy gives us on ourselves and our planet.

Who Is This Book For?

The Essential Cosmic Perspective is designed as a textbook for college courses in introductory astronomy, but is suitable for anyone who is curious about the universe. We assume no prior knowledge of astronomy or physics, and the book is especially written for students who do not intend to major in mathematics or science.

The Essential Cosmic Perspective is the mid-level of the three general astronomy textbooks we offer. Our longer book, *The Cosmic Perspective*, provides a comprehensive survey of modern astronomy with enough depth to fill a two-semester introductory astronomy sequence. This book, *The Essential Cosmic Perspective*, is trimmed down to fit what can realistically be covered in a one-semester survey of astronomy, though it may also be used with two-semester sequences. Our shortest textbook, *The Cosmic Perspective Fundamentals*, covers only the most fundamental topics in astronomy and is designed for courses that address a more limited set of topics.

New to This Edition

The underlying philosophy, goals, and structure of *The Essential Cosmic Perspective* remain the same as in past editions, but we have thoroughly updated the text and made a number of other improvements. Here, briefly, is a list of the significant changes you'll find in this eighth edition:

- **Major Chapter-Level Changes:** We have made numerous significant changes both to update the science and to improve the pedagogical flow in this edition. The full list is too long to put here, but major changes include the following:
  - **Chapter 7** has been significantly rewritten to reflect new results from *MESSENGER* at Mercury, *Curiosity* and *MAVEN* at Mars, and the latest data on global warming.
  - **Chapter 9** has been reorganized and rewritten to reflect recent developments in the study of small bodies, particularly the revolutionary new views provided by recent spacecraft including *Dawn, Rosetta,* and *New Horizons*.
  - **Chapter 10** has been heavily revised in light of thousands of new discoveries of extrasolar planets since the prior edition.
  - In **Chapter 14,** we have almost completely rewritten Section 14.4 to focus on events in which black holes can form and neutron star mergers.

- **Chapter 15** has been revised to reduce jargon and to include a new full-page figure showing the Milky Way in different wavelengths. In addition, Section 15.4 on the galactic center has been rewritten and features a new two-page Cosmic Context spread.
  - **Chapter 16** has been significantly revised in light of new research into galactic evolution.
  - **Chapter 19** has been significantly rewritten, particularly in Sections 19.2 and 19.3 (which has also been completely reorganized), thanks to new understanding of the potential habitability of Mars, Titan, and extrasolar planets.

- **Fully Updated Science:** Astronomy is a fast-moving field, and numerous new developments have occurred since the prior edition was published. In addition to the major chapter-level changes above, other scientific updates in this edition include the following:
  - New results and images from spacecraft exploring our solar system, including *Curiosity* and *MAVEN* at Mars, *Cassini* at Saturn, *MESSENGER* at Mercury, *Dawn* at Ceres, *New Horizons* at Pluto, and more
  - Recent results from major space observatories, including *Hubble* and *Kepler*, and from powerful ground-based observatories such as ALMA
  - Updated data and models on topics including the formation of planetary systems, global warming, and galaxy formation and evolution
  - Major new discoveries and statistics relating to the study of extrasolar planets, new research on the timing and possible origin of life on Earth, and much more

- **Reinforced Focus on Critical Thinking:** We have always placed a strong emphasis on helping students develop critical thinking skills, both by showing students the process through which we have acquired our current understanding of the universe and through features that encourage critical thinking, such as our Think About It and See It for Yourself questions and many of our exercises. To further reinforce the importance of critical thinking, we have added the following new features to this edition:
  - **New Feature—Extraordinary Claims boxes:** Carl Sagan made famous the statement "extraordinary claims require extraordinary evidence." With this new feature, we provide students with examples of extraordinary claims about the universe and how they were either supported or debunked as scientists collected more evidence. The first of these features appears in Chapter 3, where the context of Sagan's dictum is also explained. Nine additional Extraordinary Claims boxes are sprinkled throughout the rest of the text. Instructors will find assignable tutorials based on these boxes on the Mastering Astronomy® site.
  - **New Feature—My Cosmic Perspective:** As in prior editions, every chapter ends with a feature titled "The

Big Picture," designed to help students put the chapter content into the context of a larger cosmic perspective. For this edition, we have added in each of these sections a final entry entitled "My Cosmic Perspective," which aims to focus on a more personal connection between students and the cosmos. We believe that such a personal connection encourages students to think more critically about the meaning of all that they learn in their astronomy course.

- **New Icons:** You'll see a new icon designed to call attention to a few of the features that promote critical thinking: ○ (a "C" for "critical"). While we believe the entire structure of the book promotes critical thinking, you can use the features identified by these icons for special assignments to help students with these skills.

- **New Content in MasteringAstronomy®:** We have reached the point where *The Essential Cosmic Perspective* is no longer just a textbook; rather, it is a "learning package" that combines a printed book with deeply integrated, interactive media developed to support every chapter of our book. For students, the MasteringAstronomy® Study Area provides a wealth of tutorials and activities to build understanding, while quizzes and exercises allow them to test what they've learned. For instructors, the MasteringAstronomy® Item Library provides the unprecedented ability to quickly build, post, and automatically grade pre- and post-lecture diagnostic tests, weekly homework assignments, and exams of appropriate difficulty, duration, and content coverage. It also provides the ability to record detailed information on the step-by-step work of every student directly into a powerful and easy-to-use gradebook, and to evaluate results with a sophisticated suite of diagnostics. Among the changes you'll find to the MasteringAstronomy® site for this edition are numerous new narrated video tours of key figures from the textbook; numerous new tutorials in the Item Library; and a fully updated set of reading, concept, and visual quizzes in both the Study Area and the Item Library.

Themes of *The Essential Cosmic Perspective*

The Essential Cosmic Perspective offers a broad survey of our modern understanding of the cosmos and of how we have built that understanding. Such a survey can be presented in a number of different ways. We have chosen to interweave a few key themes throughout the book, each selected to help make the subject more appealing to students who may never have taken any formal science courses and who may begin the course with little understanding of how science works. Our book is built around the following five key themes:

- **Theme 1:** *We are a part of the universe and thus can learn about our origins by studying the universe.* This is the overarching theme of *The Essential Cosmic Perspective,* as we continually emphasize that learning about the universe helps us understand ourselves. Studying the intimate connections between human life and the cosmos gives students a

reason to care about astronomy and also deepens their appreciation of the unique and fragile nature of our planet.

- **Theme 2:** *The universe is comprehensible through scientific principles that anyone can understand.* We can understand the universe because the same physical laws appear to be at work in every aspect, on every scale, and in every age of the universe. Moreover, while professional scientists generally have discovered the laws, anyone can understand their fundamental features. Students can learn enough in one or two terms of astronomy to comprehend the basic reasons for many phenomena that they see around them—ranging from seasonal changes and phases of the Moon to the most esoteric astronomical images that appear in the news.

- **Theme 3:** *Science is not a body of facts but rather a process through which we seek to understand the world around us.* Many students assume that science is just a laundry list of facts. The long history of astronomy shows that science is a process through which we learn about our universe—a process that is not always a straight line to the truth. That is why our ideas about the cosmos sometimes change as we learn more, as they did dramatically when we first recognized that Earth is a planet going around the Sun rather than the center of the universe. In this book, we continually emphasize the nature of science so that students can understand how and why modern theories have gained acceptance and why these theories may change in the future.

- **Theme 4:** *A course in astronomy is the beginning of a lifelong learning experience.* Building on the prior themes, we emphasize that what students learn in their astronomy course is not an end but a beginning. By remembering a few key physical principles and understanding the nature of science, students can follow astronomical developments for the rest of their lives. We therefore seek to motivate students to continue to participate in the ongoing human adventure of astronomical discovery.

- **Theme 5:** *Astronomy affects each of us personally with the new perspectives it offers.* We all conduct the daily business of our lives with reference to some "world view"—a set of personal beliefs about our place and purpose in the universe that we have developed through a combination of schooling, religious training, and personal thought. This world view shapes our beliefs and many of our actions. Although astronomy does not mandate a particular set of beliefs, it does provide perspectives on the architecture of the universe that can influence how we view ourselves and our world, which can potentially affect our behavior. In many respects, the role of astronomy in shaping world views may represent the deepest connection between the universe and the everyday lives of humans.

Pedagogical Principles of *The Essential Cosmic Perspective*

No matter how an astronomy course is taught, it is very important to present material according to a clear set of pedagogical principles. The following list briefly summarizes the major pedagogical principles that we apply throughout this book. (The Instructor Guide describes these principles in more detail.)

- **Stay focused on the big picture.** Astronomy is filled with interesting facts and details, but they are meaningless unless they fit into a big picture view of the universe. We therefore take care to stay focused on the big picture (essentially the themes discussed above) at all times. A major benefit of this approach is that although students may forget individual facts and details after the course is over, the big picture framework should stay with them for life.

- **Always provide context first.** We all learn new material more easily when we understand why we are learning it. We therefore begin the book (in Chapter 1) with a broad overview of modern understanding of the cosmos so that students know what they will be studying in the rest of the book. We maintain this "context first" approach throughout the book by always telling students what they will be learning, and why, before diving into the details.

- **Make the material relevant.** It's human nature to be more interested in subjects that seem relevant to our lives. Fortunately, astronomy is filled with ideas that touch each of us personally. By emphasizing our personal connections to the cosmos, we make the material more meaningful, inspiring students to put in the effort necessary to learn it.

- **Emphasize conceptual understanding over the "stamp collecting" of facts.** If we are not careful, astronomy can appear to be an overwhelming collection of facts that are easily forgotten when the course ends. We therefore emphasize a few key concepts that we use over and over again. For example, the laws of conservation of energy and conservation of angular momentum (introduced in Section 4.3) reappear throughout the book, and we find that the wide variety of features found on the terrestrial planets can be understood through just a few basic geological processes. Research shows that, long after the course is over, students are far more likely to retain such conceptual ideas than individual facts or details.

- **Emphasize critical thinking and understanding of the process of science.** One of the major problems in public understanding of science is that too many people don't understand the difference between evidence-based science and opinion or faith. For that reason, we place particular focus on making sure students can think critically about scientific evidence and how we have arrived at current scientific understanding. For example, we discuss the nature of science in detail in Chapter 3 and then continue to show examples of how science has progressed throughout the book, while also encouraging students to think critically about the evidence that has led to this progression.

- **Proceed from the more familiar and concrete to the less familiar and abstract.** It's well known that children learn best by starting with concrete ideas and then generalizing to abstractions. The same is true for many adults. We therefore always try to "build bridges to the familiar"—that is, to begin with concrete or familiar ideas and then gradually develop more general principles from them.

- **Use plain language.** Surveys have found that the number of new terms in many introductory astronomy books is larger than the number of words taught in many first-year foreign language courses. This means that most books are teaching astronomy in what looks to students like a foreign language! It is much easier for students to understand key astronomical concepts if they are explained in plain English without resorting to unnecessary jargon. We have gone to great lengths to eliminate jargon as much as possible or, at minimum, to replace standard jargon with terms that are easier to remember in the context of the subject matter.

- **Recognize and address student misconceptions.** Students do not arrive as blank slates. Most students enter our courses not only lacking the knowledge we hope to teach but often holding misconceptions about astronomical ideas. Therefore, to teach correct ideas, we must also help students recognize the paradoxes in their prior misconceptions. We address this issue in a number of ways, most overtly with Common Misconceptions boxes. These summarize commonly held misconceptions and explain why they cannot be correct.

The Topical (Part) Structure of *The Essential Cosmic Perspective*

The Essential Cosmic Perspective is organized into six broad topical areas (the six parts in the table of contents), each approached in a distinctive way designed to help maintain the focus on the themes discussed earlier. Here, we summarize the guiding philosophy through which we have approached each topic. Every part concludes with a two-page Cosmic Context figure, which ties together into a coherent whole the diverse ideas covered in the individual chapters.

PART I Developing Perspective (Chapters 1–3)

Guiding Philosophy Introduce the big picture, the process of science, and the historical context of astronomy.

The basic goal of these chapters is to give students a big picture overview and context for the rest of the book and to help them develop an appreciation for the process of science and how science has developed through history. Chapter 1 outlines our modern understanding of the cosmos, so that students gain perspective on the entire universe before diving into its details. Chapter 2 introduces basic sky phenomena, including seasons and phases of the Moon, and provides perspective on how phenomena we experience every day are tied to the broader cosmos. Chapter 3 discusses the nature of science, offering a historical perspective on the development of science and giving students perspective on how science works and how it differs from nonscience.

The Cosmic Context for Part I appears on pp. 80–81.

PART II Key Concepts for Astronomy (Chapters 4–5)

Guiding Philosophy Connect the physics of the cosmos to everyday experiences.

These chapters lay the groundwork for understanding astronomy through what is sometimes called the "universality of physics"—the idea that a few key principles governing matter, energy, light, and motion explain both the phenomena of our daily lives and the mysteries of the cosmos. Chapter 4 covers the laws of motion, the crucial conservation laws of angular momentum and energy, and the universal law of gravitation. Chapter 5 covers the nature of light and matter, spectra, and telescopes.

The Cosmic Context for Part II appears on pp. 134–135.

PART III **Learning from Other Worlds (Chapters 6–10)**

Guiding Philosophy Learn about Earth by studying other planets in our solar system and beyond.

This set of chapters begins in Chapter 6 with a broad overview of the solar system and its formation, including a 10-page tour that highlights some of the most important and interesting features of the Sun and each of the planets. Chapters 7 to 9 focus, respectively, on the terrestrial planets, the jovian planets, and the small bodies of the solar system. Finally, Chapter 10 turns to the exciting topic of other planetary systems that have been discovered in recent years. Note that Part III is essentially independent of Parts IV and V, and can be covered either before or after them.

The Cosmic Context for Part III appears on pp. 286–287.

PART IV **Stars (Chapters 11–14)**

Guiding Philosophy We are intimately connected to the stars.

These are our chapters on stars and stellar life cycles. Chapter 11 covers the Sun in depth, so that it can serve as a concrete model for building an understanding of other stars. Chapter 12 describes the general properties of stars, how we measure these properties, and how we classify stars using the H-R diagram. Chapter 13 covers stellar evolution, tracing the birth-to-death lives of both low- and high-mass stars. Chapter 14 covers the end points of stellar evolution: white dwarfs, neutron stars, and black holes.

The Cosmic Context for Part IV appears on pp. 384–385.

PART V **Galaxies and Beyond (Chapters 15–18)**

Guiding Philosophy Present galaxy evolution and cosmology together as intimately related topics.

These chapters cover galaxies and cosmology. Chapter 15 presents the Milky Way as a paradigm for galaxies in much the same way that Chapter 11 uses the Sun as a paradigm for stars. Chapter 16 presents the variety of galaxies, how we determine key parameters such as galactic distances and age, and current understanding of galaxy evolution. Chapter 17 then presents the Big Bang theory and the evidence supporting it, setting the stage for Chapter 18, which explores dark matter and its role in galaxy formation, as well as dark energy and its implications for the fate of the universe.

The Cosmic Context for Part V appears on pp. 494–495.

PART VI **Life on Earth and Beyond (Chapter 19)**

Guiding Philosophy The study of life on Earth helps us understand the search for life in the universe.

This part consists of a single chapter. It may be considered optional, to be used as time allows. Those who wish to teach a more detailed course on astrobiology may consider the text *Life in the Universe*, by Jeffrey Bennett and Seth Shostak.

The Cosmic Context for Part VI appears on pp. 532–533.

Pedagogical Features of *The Essential Cosmic Perspective*

Alongside the main narrative, *The Essential Cosmic Perspective* includes a number of pedagogical devices designed to enhance student learning:

- **Basic Chapter Structure** Each chapter is carefully structured to ensure that students understand the goals up front, learn the details, and pull together all the ideas at the end. In particular, note the following key structural elements:
  - **Chapter Learning Goals** Each chapter opens with a page offering an enticing image and a brief overview of the chapter, including a list of the section titles and associated learning goals. The learning goals are presented as key questions designed to help students both understand what they will be learning about and stay focused on these key goals as they work through the chapter.
  - **Introduction** The first page of the main chapter text begins with a two- to three-paragraph introduction to the chapter material.
  - **Section Structure** Chapters are divided into numbered sections, each addressing one key aspect of the chapter material. Each section begins with a short introduction that leads into a set of learning goals relevant to the section—the same learning goals listed at the beginning of the chapter.
  - **The Big Picture** Every chapter narrative ends with this feature, designed to help students put what they've learned in the chapter into the context of the overall goal of gaining a broader perspective on ourselves, our planet, and our place in the universe.
  - **Chapter Summary** The end-of-chapter summary offers a concise review of the learning goal questions, helping to reinforce student understanding of key concepts from the chapter. Thumbnail figures are included to remind students of key illustrations and photos in the chapter.
- **End-of-Chapter Exercises** Each chapter includes an extensive set of exercises that can be used for study, discussion, or assignment. All of the end-of-chapter exercises are organized into the following subsets:
  - **Visual Skills Check** A set of questions designed to help students build their skills at interpreting the many types of visual information used in astronomy.
  - **Review Questions** Questions that students should be able to answer from the reading alone.
  - **Does It Make Sense? (or similar title)** A set of short statements, each of which students are expected to evaluate critically so that they can explain why it does or does not make sense. These exercises are generally easy once students understand a particular concept, but very difficult otherwise; this makes them an excellent probe of comprehension.

- **Quick Quiz** A short multiple-choice quiz that allows students to check their basic understanding.

- **Process of Science Questions** Essay or discussion questions that help students focus on how science progresses over time.

- **Group Work Exercise** A suggested activity designed for collaborative learning in class.

- **Short-Answer/Essay Questions** Questions that go beyond the Review Questions in asking for conceptual interpretation.

- **Quantitative Problems** Problems that require some mathematics, usually based on topics covered in the Cosmic Calculations boxes.

- **Discussion Questions** Open-ended questions for class discussions.

- **Web Projects** A few suggestions for additional Web-based research.

- **Additional Features** You'll find a number of other features designed to increase student understanding, both within individual chapters and at the end of the book, including the following:

  - **Annotated Figures** Key figures in each chapter incorporate the research-proven technique of "annotation"—carefully crafted text placed on the figure (in blue) to guide students through interpreting graphs, following process figures, and translating between different representations.

  - **Cosmic Context Two-Page Figures** These two-page spreads provide visual summaries of key processes and concepts.

  - **Wavelength/Observatory Icons** For astronomical images, simple icons indicate whether the image is a photo, artist's impression, or computer simulation; whether a photo came from ground-based or space-based observations; and the wavelength band used to take the photo.

  - **Think About It** This feature, which appears throughout the book as short questions integrated into the narrative, gives students the opportunity to reflect on important new concepts. It also serves as an excellent starting point for classroom discussions.

  - **See It for Yourself** This feature also occurs throughout the book, integrated into the narrative, and gives students the opportunity to conduct simple observations or experiments that will help them understand key concepts.

  - **Common Misconceptions** These boxes address popularly held but incorrect ideas related to the chapter material.

  - **Special Topic Boxes** These boxes contain supplementary discussion topics related to the chapter material but not prerequisite to the continuing discussion.

  - **Extraordinary Claims Boxes** Carl Sagan made famous the statement "extraordinary claims require extraordinary evidence." These boxes provide students with examples of extraordinary claims about the universe and how they were either supported or debunked as scientists collected more evidence.

  - **Cosmic Calculations Boxes** These boxes contain most of the mathematics used in the book and can be covered or skipped depending on the level of mathematics that you wish to include in your course.

  - **Cross-References** When a concept is covered in greater detail elsewhere in the book, we include a cross-reference, printed in blue and surrounded by brackets, to the relevant section (e.g., [Section 5.2]).

  - **Glossary** A detailed glossary makes it easy for students to look up important terms.

  - **Appendixes** The appendixes include a number of useful references and tables, including key constants (Appendix A), key formulas (Appendix B), key mathematical skills (Appendix C), and numerous data tables and star charts (Appendixes D–I).

About MasteringAstronomy®

What is the single most important factor in student success in astronomy? Both research and common sense reveal the same answer: *study time*. No matter how good the teacher or how good the textbook, students learn only when they spend adequate time studying. Unfortunately, limitations on resources for grading have prevented most instructors from assigning much homework despite its obvious benefits to student learning. And limitations on help and office hours have made it difficult for students to make sure they use self-study time effectively. That, in a nutshell, is why we created MasteringAstronomy®. For students, it provides adaptive learning designed to coach them *individually*— responding to their errors with specific, targeted feedback and providing optional hints for those who need additional guidance. For professors, MasteringAstronomy® provides the unprecedented ability to automatically monitor and record students' step-by-step work and evaluate the effectiveness of assignments and exams. As a result, we believe that MasteringAstronomy® can change the way astronomy courses are taught: It is now possible, even in large classes, to ensure that each student spends his or her study time on optimal learning activities outside of class.

MasteringAstronomy® provides students with a wealth of self-study resources, including interactive tutorials targeting the most difficult concepts of the course, interactive or narrated versions of key figures and photos, and quizzes and other activities for self-assessment covering every chapter and every week. For professors, MasteringAstronomy® provides a library of tutoring activities that is periodically updated based on the performance of students nationwide. You can create assignments tailored to your specific class goals from among hundreds of activities and problems including pre- and post-lecture diagnostic quizzes, tutoring activities, end-of-chapter problems from this textbook, and test bank questions. MasteringAstronomy® now also includes Learning Catalytics, which provides additional capabilities for in-class learning. Visit MasteringAstronomy® to learn more.

Finally, in a world where everyone claims to have the best website, we'd like to point out three reasons why you'll discover that MasteringAstronomy® really does stand out from the crowd:

- MasteringAstronomy® has been built specifically to support the structure and pedagogy of *The Essential Cosmic Perspective*. You'll find the same concepts emphasized in the book and on the website, using the same terminology and the same pedagogical approaches. This type of consistency ensures that students focus on the concepts, without the risk of becoming confused by different presentations.

- Nearly all MasteringAstronomy® content has been developed either directly by *The Essential Cosmic Perspective* author team or in close collaboration with outstanding educators including Jim Dove, Jim Cooney, Jonathan Williams, Richard Gelderman, Ed Prather, Tim Slater, Daniel Lorenz, and Lauren Jones. The direct involvement of book authors ensures consistency from our website to the textbook, resulting in an effective, high-quality learning program.

- The MasteringAstronomy® platform uses the same unique student-driven engine as the highly successful MasteringPhysics® product (the most widely adopted physics tutorial and assessment system), developed by a group led by MIT physicist David Pritchard. This robust platform gives instructors unprecedented power not only to tailor content to their own courses, but also to evaluate the effectiveness of assignments and exams.

Additional Supplements for *The Essential Cosmic Perspective*

The Essential Cosmic Perspective is much more than just a textbook. It is a complete package of teaching, learning, and assessment resources designed to help both teachers and students. In addition to MasteringAstronomy®, the following supplements are available with this book:

- **SkyGazer v5.0**: Based on *Voyager V*, SkyGazer, one of the world's most popular planetarium programs now available for download, makes it easy for students to learn constellations and explore the wonders of the sky through interactive exercises and demonstrations. Accompanying activities are available in LoPresto's Astronomy Media Workbook, Seventh Edition, available both on the MasteringAstronomy® study area and on the SkyGazer site. Ask your Pearson sales representative for details.

- **Starry Night™ College** (ISBN 0-321-71295-1): Now available as an additional option with *The Essential Cosmic Perspective*, Starry Night has been acclaimed as the world's most realistic desktop planetarium software. This special version has an easy-to-use point-and-click interface and is available as an additional bundle. The Starry Night Activity Workbook, consisting of thirty-five worksheets for homework or lab, based on Starry Night planetarium software, is available for download in the MasteringAstronomy® study area or with a Starry Night College access code. Ask your Pearson sales representative for details.

- **Lecture Tutorials for Introductory Astronomy** (ISBN 0-321-82046-0) by Edward E. Prather, Timothy F. Slater, Jeffrey P. Adams, and Gina Brissenden: The forty-four lecture tutorials included are designed to engage students in critical reasoning and spark classroom discussion.

- **Sky and Telescope: Special Student Supplement** (ISBN 0-321-70620-X): The nine articles, each with an assessment following, provide a general review as well as covering such topics as the process of science, the scale of the universe, and our place in the universe. The supplement is available for bundling; ask your Pearson sales representative for details.

- **Observation Exercises in Astronomy** (ISBN 0-321-63812-3): This manual includes fifteen observation activities that can be used with a number of different planetarium software packages.

- **McCrady/Rice Astronomy Labs: A Concept Oriented Approach** (ISBN: 0-321-86177-9): This customizable lab is available in the Pearson Custom Library. It consists of 40 conceptually oriented introductory astronomy labs that focus on the mid to higher levels of Bloom's taxonomy: application, synthesis, and analysis. The labs are all written to minimize equipment requirements and are largely created to maximize the use of inexpensive everyday objects such as flashlights, construction paper, and theater gels.

Instructor-Only Supplements

Several additional supplements are available for instructors only. Contact your local Pearson sales representative to find out more about the following supplements:

- **The Instructor Resources tab** in MasteringAstronomy® provides a wealth of lecture and teaching resources, including high-resolution JPEGs of all images from the book for in-class projection, Narrated Figures, based on figures from the book, pre-built PowerPoint® Lecture Outlines, answers to SkyGazer and Starry Night workbooks, and PRS-enabled Clicker Quizzes based on the book and book-specific interactive media.

- **Instructor Guide** (ISBN 0-134-53247-3): This guide contains a detailed overview of the text, sample syllabi for courses of different emphasis and duration, suggestions for teaching strategies, answers or discussion points for all Think About It and See It for Yourself questions in the text, solutions to end-of-chapter problems, and a detailed reference guide summarizing media resources available for every chapter and section in the book. Word files can be downloaded from the instructor resource section of MasteringAstronomy®.

- **Test Bank** (ISBN 0-134-60203-X): The Test Bank includes hundreds of multiple-choice, true/false, and short-answer questions, plus Process of Science questions for each chapter. TestGen® and Word files can be downloaded from the instructor resource section of the study area in MasteringAstronomy®.

Acknowledgments

Our textbook carries only four author names, but in fact it is the result of hard work by a long list of committed individuals. We could not possibly list everyone who has helped, but we would like to call attention to a few people who have played particularly important roles. First, we thank our editors and friends at Pearson, who have stuck with us through thick and thin, including Adam Black, Nancy Whilton, Jim Smith, Michael Gillespie, Mary Ripley, Chandrika Madhavan, and Corinne Benson. Special thanks to our production teams, especially Sally Lifland, and our art and design team.

We've also been fortunate to have an outstanding group of reviewers, whose extensive comments and suggestions helped us shape the book. We thank all those who have reviewed drafts of the book in various stages, including

Marilyn Akins, *Broome Community College*
Christopher M. Anderson, *University of Wisconsin*
John Anderson, *University of North Florida*
Peter S. Anderson, *Oakland Community College*
Keith Ashman
Simon P. Balm, *Santa Monica College*
Reba Bandyopadhyay, *University of Florida*
Nadine Barlow, *Northern Arizona University*
John Beaver, *University of Wisconsin at Fox Valley*
Peter A. Becker, *George Mason University*
Timothy C. Beers, *National Optical Astronomy Observatory*
Jim Bell, *Arizona State University*
Priscilla J. Benson, *Wellesley College*
Philip Blanco, *Grossmont College*
Jeff R. Bodart, *Chipola College*
Bernard W. Bopp, *University of Toledo*
Sukanta Bose, *Washington State University*
David Brain, *University of Colorado*
David Branch, *University of Oklahoma*
John C. Brandt, *University of New Mexico*
James E. Brau, *University of Oregon*
Jean P. Brodie, *UCO/Lick Observatory, University of California, Santa Cruz*
Erik Brogt, *University of Canterbury*
James Brooks, *Florida State University*
Daniel Bruton, *Stephen F. Austin State University*
Debra Burris, *University of Central Arkansas*
Scott Calvin, *Sarah Lawrence College*
Amy Campbell, *Louisiana State University*
Eugene R. Capriotti, *Michigan State University*
Eric Carlson, *Wake Forest University*
David A. Cebula, *Pacific University*
Supriya Chakrabarti, *University of Massachusetts, Lowell*
Kwang-Ping Cheng, *California State University, Fullerton*
Dipak Chowdhury, *Indiana University—Purdue University Fort Wayne*
Chris Churchill, *New Mexico State University*
Josh Colwell, *University of Central Florida*
James Cooney, *University of Central Florida*
Anita B. Corn, *Colorado School of Mines*
Philip E. Corn, *Red Rocks Community College*
Kelli Corrado, *Montgomery County Community College*
Peter Cottrell, *University of Canterbury*
John Cowan, *University of Oklahoma*
Kevin Crosby, *Carthage College*
Christopher Crow, *Indiana University—Purdue University Fort Wayne*
Manfred Cuntz, *University of Texas at Arlington*
Christopher De Vries, *California State University, Stanislaus*
John M. Dickey, *University of Minnesota*

Matthias Dietrich, *Worcester State University*
Bryan Dunne, *University of Illinois, Urbana-Champaign*
Suzan Edwards, *Smith College*
Robert Egler, *North Carolina State University at Raleigh*
Paul Eskridge, *Minnesota State University*
David Falk, *Los Angeles Valley College*
Timothy Farris, *Vanderbilt University*
Robert A. Fesen, *Dartmouth College*
Tom Fleming, *University of Arizona*
Douglas Franklin, *Western Illinois University*
Sidney Freudenstein, *Metropolitan State College of Denver*
Martin Gaskell, *University of Nebraska*
Richard Gelderman, *Western Kentucky University*
Harold A. Geller, *George Mason University*
Donna Gifford, *Pima Community College*
Mitch Gillam, *Marion L. Steele High School*
Bernard Gilroy, *The Hun School of Princeton*
Owen Gingerich, *Harvard–Smithsonian* (Historical Accuracy Reviewer)
David Graff, *U.S. Merchant Marine Academy*
Richard Gray, *Appalachian State University*
Kevin Grazier, *Jet Propulsion Laboratory*
Robert Greeney, *Holyoke Community College*
Henry Greenside, *Duke University*
Alan Greer, *Gonzaga University*
John Griffith, *Lin-Benton Community College*
David Griffiths, *Oregon State University*
David Grinspoon, *Planetary Science Institute*
John Gris, *University of Delaware*
Bruce Gronich, *University of Texas at El Paso*
Thomasana Hail, *Parkland University*
Jim Hamm, *Big Bend Community College*
Charles Hartley, *Hartwick College*
J. Hasbun, *University of West Georgia*
Joe Heafner, *Catawba Valley Community College*
David Herrick, *Maysville Community College*
Scott Hildreth, *Chabot College*
Tracy Hodge, *Berea College*
Mark Hollabaugh, *Normandale Community College*
Richard Holland, *Southern Illinois University, Carbondale*
Joseph Howard, *Salisbury University*
James Christopher Hunt, *Prince George's Community College*
Richard Ignace, *University of Wisconsin*
James Imamura, *University of Oregon*
Douglas R. Ingram, *Texas Christian University*
Assad Istephan, *Madonna University*
Bruce Jakosky, *University of Colorado*
Adam G. Jensen, *University of Colorado*
Adam Johnston, *Weber State University*
Lauren Jones, *Gettysburg College*
Kishor T. Kapale, *Western Illinois University*
William Keel, *University of Alabama*
Julia Kennefick, *University of Arkansas*
Steve Kipp, *University of Minnesota, Mankato*
Kurtis Koll, *Cameron University*
Ichishiro Konno, *University of Texas at San Antonio*
John Kormendy, *University of Texas at Austin*
Eric Korpela, *University of California, Berkeley*
Arthur Kosowsky, *University of Pittsburgh*
Kevin Krisciunas, *Texas A&M*
David Lamp, *Texas Technical University*

Ted La Rosa, *Kennesaw State University*
Kristine Larsen, *Central Connecticut State University*
Ana Marie Larson, *University of Washington*
Stephen Lattanzio, *Orange Coast College*
Chris Laws, *University of Washington*
Larry Lebofsky, *University of Arizona*
Patrick Lestrade, *Mississippi State University*
Nancy Levenson, *University of Kentucky*
David M. Lind, *Florida State University*
Abraham Loeb, *Harvard University*
Michael LoPresto, *Henry Ford Community College*
William R. Luebke, *Modesto Junior College*
Ihor Luhach, *Valencia Community College*
Darrell Jack MacConnell, *Community College of Baltimore City*
Marie Machacek, *Massachusetts Institute of Technology*
Loris Magnani, *University of Georgia*
Steven Majewski, *University of Virginia*
Phil Matheson, *Salt Lake Community College*
John Mattox, *Fayetteville State University*
Marles McCurdy, *Tarrant County College*
Stacy McGaugh, *Case Western University*
Barry Metz, *Delaware County Community College*
William Millar, *Grand Rapids Community College*
Dinah Moche, *Queensborough Community College of City University, New York*
Stephen Murray, *University of California, Santa Cruz*
Zdzislaw E. Musielak, *University of Texas at Arlington*
Charles Nelson, *Drake University*
Gerald H. Newsom, *Ohio State University*
Lauren Novatne, *Reedley College*
Brian Oetiker, *Sam Houston State University*
Richard Olenick, *University of Dallas*
John P. Oliver, *University of Florida*
Stacy Palen, *Weber State University*
Russell L. Palma, *Sam Houston State University*
Mark Pecaut, *Rockhurst University*
Jon Pedicino, *College of the Redwoods*
Bryan Penprase, *Pomona College*
Eric S. Perlman, *Florida Institute of Technology*
Peggy Perozzo, *Mary Baldwin College*
Greg Perugini, *Burlington County College*
Charles Peterson, *University of Missouri, Columbia*
Cynthia W. Peterson, *University of Connecticut*
Jorge Piekarewicz, *Florida State University*
Lawrence Pinsky, *University of Houston*
Stephanie Plante, *Grossmont College*
Jascha Polet, *California State Polytechnic University, Pomona*
Matthew Price, *Oregon State University*
Harrison B. Prosper, *Florida State University*
Monica Ramirez, *Aims College, Colorado*
Christina Reeves-Shull, *Richland College*
Todd M. Rigg, *City College of San Francisco*
Elizabeth Roettger, *DePaul University*
Roy Rubins, *University of Texas at Arlington*
April Russell, *Siena College*
Carl Rutledge, *East Central University*
Bob Sackett, *Saddleback College*
Rex Saffer, *Villanova University*
John Safko, *University of South Carolina*
James A. Scarborough, *Delta State University*
Britt Scharringhausen, *Ithaca College*

Ann Schmiedekamp, *Pennsylvania State University, Abington*
Joslyn Schoemer, *Denver Museum of Nature and Science*
James Schombert, *University of Oregon*
Gregory Seab, *University of New Orleans*
Larry Sessions, *Metropolitan State College of Denver*
Anwar Shiekh, *Colorado Mesa University*
Ralph Siegel, *Montgomery College, Germantown Campus*
Philip I. Siemens, *Oregon State University*
Caroline Simpson, *Florida International University*
Paul Sipiera, *William Harper Rainey College*
Earl F. Skelton, *George Washington University*
Evan Skillman, *University of Minnesota*
Michael Skrutskie, *University of Virginia*
Mark H. Slovak, *Louisiana State University*
Norma Small-Warren, *Howard University*
Jessica Smay, *San Jose City College*
Dale Smith, *Bowling Green State University*
Brent Sorenson, *Southern Utah University*
James R. Sowell, *Georgia Technical University*
Kelli Spangler, *Montgomery County Community College*
John Spencer, *Southwest Research Institute*
Darryl Stanford, *City College of San Francisco*
George R. Stanley, *San Antonio College*
Peter Stein, *Bloomsburg University of Pennsylvania*
Adriane Steinacker, *University of California, Santa Cruz*
John Stolar, *West Chester University*
Irina Struganova, *Valencia Community College*
Jack Sulentic, *University of Alabama*
C. Sean Sutton, *Mount Holyoke College*
Beverley A. P. Taylor, *Miami University*
Brett Taylor, *Radford University*
Donald M. Terndrup, *Ohio State University*
Frank Timmes, *Arizona State University*
David Trott, *Metro State College*
David Vakil, *El Camino College*
Trina Van Ausdal, *Salt Lake Community College*
Licia Verde, *Institute of Cosmological Studies, Barcelona*
Nicole Vogt, *New Mexico State University*
Darryl Walke, *Rariton Valley Community College*
Fred Walter, *State University of New York, Stony Brook*
James Webb, *Florida International University*
Mark Whittle, *University of Virginia*
Paul J. Wiita, *The College of New Jersey*
Lisa M. Will, *Mesa Community College*
Jonathan Williams, *University of Hawaii*
Terry Willis, *Chesapeake College*
Grant Wilson, *University of Massachusetts, Amherst*
J. Wayne Wooten, *Pensacola Junior College*
Guy Worthey, *Washington State University, Pullman*
Scott Yager, *Brevard College*
Andrew Young, *Casper College*
Arthur Young, *San Diego State University*
Tim Young, *University of North Dakota*
Min S. Yun, *University of Massachusetts, Amherst*
Dennis Zaritsky, *University of Arizona*
Robert L. Zimmerman, *University of Oregon*

In addition, we thank the following colleagues who helped us clarify technical points or checked the accuracy of technical discussions in the book:

Caspar Amman, *NCAR*
Nahum Arav, *Virginia Technical University*
Phil Armitage, *University of Colorado*
Thomas Ayres, *University of Colorado*
Cecilia Barnbaum, *Valdosta State University*
Rick Binzel, *Massachusetts Institute of Technology*
Howard Bond, *Space Telescope Science Institute*
David Brain, *University of Colorado*
Humberto Campins, *University of Central Florida*
Robin Canup, *Southwest Research Institute*
Clark Chapman, *Southwest Research Institute*
Kelly Cline, *Carroll College*
Josh Colwell, *University of Central Florida*
James Cooney, *University of Central Florida*
Mark Dickinson, *National Optical Astronomy Observatory*
Jim Dove, *Metropolitan State College of Denver*
Doug Duncan, *University of Colorado*
Dan Fabrycky, *University of Chicago*
Harry Ferguson, *Space Telescope Science Institute*
Andrew Hamilton, *University of Colorado*
Todd Henry, *Georgia State University*
Dennis Hibbert, *Everett Community College*
Seth Hornstein, *University of Colorado*
Dave Jewitt, *University of California, Los Angeles*
Julia Kregenow, *Penn State University*
Emily Lakdawalla, *The Planetary Society*
Hal Levison, *Southwest Research Institute*
Mario Livio, *Space Telescope Science Institute*
J. McKim Malville, *University of Colorado*
Geoff Marcy
Mark Marley, *Ames Research Center*
Linda Martel, *University of Hawaii*
Kevin McLin, *University of Colorado*
Michael Mendillo, *Boston University*

Steve Mojzsis, *University of Colorado*
Francis Nimmo, *University of California, Santa Cruz*
Tyler Nordgren, *University of Redlands*
Rachel Osten, *Space Telescope Science Institute*
Bob Pappalardo, *Jet Propulsion Laboratory*
Bennett Seidenstein, *Arundel High School*
Michael Shara, *American Museum of Natural History*
Evan Skillman, *University of Minnesota*
Brad Snowder, *Western Washington University*
Bob Stein, *Michigan State University*
Glen Stewart, *University of Colorado*
John Stolar, *West Chester University*
Jeff Taylor, *University of Hawaii*
Dave Tholen, *University of Hawaii*
Nick Thomas, *University of Bern*
Dimitri Veras, *Cambridge University*
John Weiss, *Carleton College*
Francis Wilkin, *Union College*
Jeremy Wood, *Hazard Community College*
Jason Wright, *Penn State University*
Don Yeomans, *Jet Propulsion Laboratory*

Finally, we thank the many people who have greatly influenced our outlook on education and our perspective on the universe over the years, including Tom Ayres, Fran Bagenal, Forrest Boley, Robert A. Brown, George Dulk, Erica Ellingson, Katy Garmany, Jeff Goldstein, David Grinspoon, Robin Heyden, Don Hunten, Geoffrey Marcy, Joan Marsh, Catherine McCord, Dick McCray, Dee Mook, Cherilynn Morrow, Charlie Pellerin, Carl Sagan, Mike Shull, John Spencer, and John Stocke.

Jeff Bennett
Megan Donahue
Nick Schneider
Mark Voit

About the Authors

Jeffrey Bennett

Jeffrey Bennett, a recipient of the American Institute of Physics Science Communication Award, holds a B.A. in biophysics (UC San Diego), and an M.S. and Ph.D. in astrophysics (University of Colorado). He specializes in science and math education and has taught at every level from preschool through graduate school. Career highlights include serving 2 years as a visiting senior scientist at NASA headquarters, where he developed programs to build stronger links between research and education, and proposing and helping to develop the Voyage scale model solar system on the National Mall (Washington, DC). He is the lead author of textbooks in astronomy, astrobiology, mathematics, and statistics, and of critically acclaimed books for the public including *Beyond UFOs* (Princeton University Press, 2008/2011), *Math for Life* (Bid Kid Science, 2014), *What Is Relativity?* (Columbia University Press, 2014), *On Teaching Science* (Big Kid Science, 2014), and *A Global Warming Primer* (Big Kid Science, 2016). He is also the author of six science picture books for children, including *Max Goes to the Moon*, *The Wizard Who Saved the World*, and *I, Humanity*; all six have been launched to the International Space Station and read aloud by astronauts for NASA's Story Time From Space program. Jeff lives in Boulder, CO with his wife, children, and dog. His personal website is www.jeffreybennett.com.

Megan Donahue

Megan Donahue is a full professor in the Department of Physics and Astronomy at Michigan State University (MSU) and a Fellow of the American Association for the Advancement of Science. Her current research is mainly about using X-ray, UV, infrared, and visible light to study galaxies and clusters of galaxies: their contents—dark matter, hot gas, galaxies, active galactic nuclei—and what they reveal about the contents of the universe and how galaxies form and evolve. She grew up on a farm in Nebraska and received an S.B. in physics from MIT, where she began her research career as an X-ray astronomer. She has a Ph.D. in astrophysics from the University of Colorado. Her Ph.D. thesis on theory and optical observations of intergalactic and intracluster gas won the 1993 Trumpler Award from the Astronomical Society for the Pacific for an outstanding astrophysics doctoral dissertation in North America. She continued postdoctoral research as a Carnegie Fellow at Carnegie Observatories in Pasadena, California, and later as an STScI Institute Fellow at Space Telescope. Megan was a staff astronomer at the Space Telescope Science Institute until 2003, when she joined the MSU faculty. She is also actively involved in advising national and international astronomical facilities and NASA, including planning future NASA missions. Megan is married to Mark Voit, and they collaborate on many projects, including this textbook, over 50 peer-reviewed astrophysics papers, and the raising of their children, Michaela, Sebastian, and Angela. Megan has run three full marathons, including Boston. These days she does trail running, orienteers, and plays piano and bass guitar for fun and no profit.

Nicholas Schneider

Nicholas Schneider is an associate professor in the Department of Astrophysical and Planetary Sciences at the University of Colorado and a researcher in the Laboratory for Atmospheric and Space Physics. He received his B.A. in physics and astronomy from Dartmouth College in 1979 and his Ph.D. in planetary science from the University of Arizona in 1988. In 1991, he received the National Science Foundation's Presidential Young Investigator Award. His research interests include planetary atmospheres and planetary astronomy. One research focus is the odd case of Jupiter's moon Io. Another is the mystery of Mars's lost atmosphere, which he hopes to answer by leading the Imaging UV Spectrograph team on NASA's *MAVEN* mission now orbiting Mars. Nick enjoys teaching at all levels and is active in efforts to improve undergraduate astronomy education. In 2010 he received the Boulder Faculty Assembly's Teaching Excellence Award. Off the job, Nick enjoys exploring the outdoors with his family and figuring out how things work.

Mark Voit

Mark Voit is a professor in the Department of Physics and Astronomy and Associate Dean for Undergraduate Studies at Michigan State University. He earned his A.B. in astrophysical sciences at Princeton University and his Ph.D. in astrophysics at the University of Colorado in 1990. He continued his studies at the California Institute of Technology, where he was a research fellow in theoretical astrophysics, and then moved on to Johns Hopkins University as a Hubble Fellow. Before going to Michigan State, Mark worked in the Office of Public Outreach at the Space Telescope, where he developed museum exhibitions about the Hubble Space Telescope and helped design NASA's award-winning HubbleSite. His research interests range from interstellar processes in our own galaxy to the clustering of galaxies in the early universe, and he is a Fellow of the American Association for the Advancement of Science. He is married to coauthor Megan Donahue, and cooks terrific meals for her and their three children. Mark likes getting outdoors whenever possible and particularly enjoys running, mountain biking, canoeing, orienteering, and adventure racing. He is also author of the popular book *Hubble Space Telescope: New Views of the Universe*.

How to Succeed
in Your Astronomy Course

If Your Course Is	Times for Reading the Assigned Text (per week)	Times for Homework Assignments (per week)	Times for Review and Test Preparation (average per week)	Total Study Time (per week)
3 credits	2 to 4 hours	2 to 3 hours	2 hours	6 to 9 hours
4 credits	3 to 5 hours	2 to 4 hours	3 hours	8 to 12 hours
5 credits	3 to 5 hours	3 to 6 hours	4 hours	10 to 15 hours

The Key to Success: Study Time

The single most important key to success in any college course is to spend enough time studying. A general rule of thumb for college classes is that you should expect to study about 2 to 3 hours per week *outside* of class for each unit of credit. For example, based on this rule of thumb, a student taking 15 credit hours should expect to spend 30 to 45 hours each week studying outside of class. Combined with time in class, this works out to a total of 45 to 60 hours spent on academic work—not much more than the time a typical job requires, and you get to choose your own hours. Of course, if you are working or have family obligations while you attend school, you will need to budget your time carefully.

As a rough guideline, your study time might be divided as shown in the table above. If you find that you are spending fewer hours than these guidelines suggest, you can probably improve your grade by studying longer. If you are spending more hours than these guidelines suggest, you may be studying inefficiently; in that case, you should talk to your instructor about how to study more effectively.

Using This Book

Each chapter in this book is designed to help you to study effectively and efficiently. To get the most out of each chapter, you might wish to use the following study plan.

- A textbook is not a novel, and you'll learn best by reading the elements of this text in the following order:
  1. Start by reading the Learning Goals and the introductory paragraphs at the beginning of the chapter so that you'll know what you are trying to learn.
  2. Get an overview of key concepts by studying the illustrations and their captions and annotations. The illustrations highlight most major concepts, so this "illustrations first" strategy gives you an opportunity to survey the concepts before you read about them in depth. You will find the two-page Cosmic Context figures especially useful.
  3. Read the chapter narrative, trying the Think About It questions and the See It for Yourself activities as you go along, but save the boxed features (e.g., Common Misconceptions, Special Topics) to read later. As you read, make notes on the pages to remind yourself of ideas you'll want to review later. Take notes as you read, but avoid using a highlight pen (or a highlighting tool if you are using an e-book), which makes it too easy to highlight mindlessly.
  4. After reading the chapter once, go back through and read the boxed features.
  5. Review the Chapter Summary, ideally by trying to answer the Learning Goal questions for yourself before reading the given answers.

- After completing the reading as outlined above, test your understanding with the end-of-chapter exercises. A good way to begin is to make sure you can answer all of the Review and Quick Quiz Questions; if you don't know an answer, look back through the chapter until you figure it out.
- Visit the MasteringAstronomy® site and make use of resources that will help you further build your understanding. These resources have been developed specifically to help you learn the most important ideas in your course, and they have been extensively tested to make sure they are effective. They really do work, and the only way you'll gain their benefits is by going to the website and using them.

General Strategies for Studying

- Budget your time effectively. Studying 1 or 2 hours each day is more effective, and far less painful, than studying all night before homework is due or before exams.
- Engage your brain. Learning is an active process, not a passive experience. Whether you are reading, listening to a lecture, or working on assignments, always make sure that your mind is actively engaged. If you find your mind drifting or find yourself falling asleep, make a conscious effort to revive yourself, or take a break if necessary.
- Don't miss class, and come prepared. Listening to lectures and participating in discussions is much more effective than reading someone else's notes or watching a video later. Active participation will help you retain what you are learning. Also, be sure to complete any assigned reading *before* the class in which it will be discussed. This

is crucial, since class lectures and discussions are designed to reinforce key ideas from the reading.

- Take advantage of resources offered by your professor, whether it be email, office hours, review sessions, online chats, or other opportunities to talk to and get to know your professor. Most professors will go out of their way to help you learn in any way that they can.
- Start your homework early. The more time you allow yourself, the easier it is to get help if you need it. If a concept gives you trouble, do additional reading or studying beyond what has been assigned. And if you still have trouble, ask for help: You surely can find friends, peers, or teachers who will be glad to help you learn.
- Working together with friends can be valuable in helping you understand difficult concepts, but be sure that you learn *with* your friends and do not become dependent on them.
- Don't try to multitask. Research shows that human beings simply are not good at multitasking: When we attempt it, we do more poorly at all of the individual tasks. And in case you think you are an exception, research has also found that those people who believe they are best at multitasking are often the worst! So when it is time to study, turn off your electronic devices, find a quiet spot, and concentrate on your work. (If you *must* use a device to study, as with an e-book or online homework, turn off email, text, and other alerts so that they will not interrupt your concentration; some apps will do this for you.)

Preparing for Exams

- Study the Review Questions, and rework problems and other assignments; try additional questions to be sure you understand the concepts. Study your performance on assignments, quizzes, or exams from earlier in the term.
- Work through the relevant chapter quizzes and other study resources available at the MasteringAstronomy® site.
- Study your notes from lectures and discussions. Pay attention to what your instructor expects you to know for an exam.
- Reread the relevant sections in the textbook, paying special attention to notes you have made on the pages.
- Study individually *before* joining a study group with friends. Study groups are effective only if every individual comes prepared to contribute.
- Don't stay up too late before an exam. Don't eat a big meal within an hour of the exam (thinking is more difficult when blood is being diverted to the digestive system).
- Try to relax before and during the exam. If you have studied effectively, you are capable of doing well. Staying relaxed will help you think clearly.

Presenting Homework and Writing Assignments

All work that you turn in should be of *collegiate quality:* neat and easy to read, well organized, and demonstrating mastery of the subject matter. Future employers and teachers will expect this quality of work. Moreover, although submitting homework of collegiate quality requires "extra" effort, it serves two important purposes directly related to learning:

1. The effort you expend in clearly explaining your work solidifies your learning. In particular, research has shown that writing (or typing) and speaking trigger different areas of your brain. Writing something down—even when you think you already understand it—reinforces your learning by involving other areas of your brain.
2. If you make your work clear and self-contained (that is, make it a document that you can read without referring to the questions in the text), you will have a much more useful study guide when you review for a quiz or exam.

The following guidelines will help ensure that your assignments meet the standards of collegiate quality:

- Always use proper grammar, proper sentence and paragraph structure, and proper spelling. Do not use texting shorthand.
- Make all answers and other writing fully self-contained. A good test is to imagine that a friend will be reading your work and to ask yourself whether the friend will understand exactly what you are trying to say. It is also helpful to read your work out loud to yourself, making sure that it sounds clear and coherent.
- In problems that require calculation:
  1. Be sure to *show your work* clearly so that both you and your instructor can follow the process you used to obtain an answer. Also, use standard mathematical symbols, rather than "calculator-ese." For example, show multiplication with the $\times$ symbol (not with an asterisk), and write 10^5, not 10^5 or 10E5.
  2. *Check that word problems have word answers.* That is, after you have completed any necessary calculations, make sure that any problem stated in words is answered with one or more *complete sentences* that describe the point of the problem and the meaning of your solution.
  3. Express your word answers in a way that would be *meaningful* to most people. For example, most people would find it more meaningful if you expressed a result of 720 hours as 1 month. Similarly, if a precise calculation yields an answer of 9,745,600 years, it may be more meaningfully expressed in words as "nearly 10 million years."

- Include illustrations whenever they help explain your answer, and make sure your illustrations are neat and clear. For example, if you graph by hand, use a ruler to make straight lines. If you use software to make illustrations, be careful not to make them overly cluttered with unnecessary features.
- If you study with friends, be sure that you turn in your own work stated in your own words—you should avoid anything that might give even the *appearance* of possible academic dishonesty.

Foreword

The Meaning of *The Cosmic Perspective*

© Neil deGrasse Tyson

by Neil deGrasse Tyson

Astrophysicist Neil deGrasse Tyson is the Frederick P. Rose Director of New York City's Hayden Planetarium at the American Museum of Natural History. He has written numerous books and articles, has hosted the PBS series NOVA scienceNOW *and the globally popular* Cosmos: A Spacetime Odyssey, *and was named one of the "Time 100"—*Time *magazine's list of the 100 most influential people in the world. He contributed this essay about the meaning of "The Cosmic Perspective," abridged from his 100th essay written for* Natural History *magazine.*

Of all the sciences cultivated by mankind, Astronomy is acknowledged to be, and undoubtedly is, the most sublime, the most interesting, and the most useful. For, by knowledge derived from this science, not only the bulk of the Earth is discovered ... ; but our very faculties are enlarged with the grandeur of the ideas it conveys, our minds exalted above [their] low contracted prejudices.

James Ferguson, *Astronomy Explained Upon Sir Isaac Newton's Principles, and Made Easy To Those Who Have Not Studied Mathematics* (1757)

LONG BEFORE ANYONE knew that the universe had a beginning, before we knew that the nearest large galaxy lies two and a half million light-years from Earth, before we knew how stars work or whether atoms exist, James Ferguson's enthusiastic introduction to his favorite science rang true.

But who gets to think that way? Who gets to celebrate this cosmic view of life? Not the migrant farm worker. Not the sweatshop worker. Certainly not the homeless person rummaging through the trash for food. You need the luxury of time not spent on mere survival. You need to live in a nation whose government values the search to understand humanity's place in the universe. You need a society in which intellectual pursuit can take you to the frontiers of discovery, and in which news of your discoveries can be routinely disseminated.

When I pause and reflect on our expanding universe, with its galaxies hurtling away from one another, embedded with the ever-stretching, four-dimensional fabric of space and time, sometimes I forget that uncounted people walk this Earth without food or shelter, and that children are disproportionately represented among them.

When I pore over the data that establish the mysterious presence of dark matter and dark energy throughout the universe, sometimes I forget that every day—every twenty-four-hour rotation of Earth—people are killing and being killed. In the name of someone's ideology.

When I track the orbits of asteroids, comets, and planets, each one a pirouetting dancer in a cosmic ballet choreographed by the forces of gravity, sometimes I forget that too many people act in wanton disregard for the delicate interplay of Earth's atmosphere, oceans, and land, with consequences that our children and our children's children will witness and pay for with their health and well-being.

And sometimes I forget that powerful people rarely do all they can to help those who cannot help themselves.

I occasionally forget those things because, however big the world is—in our hearts, our minds, and our outsize atlases—the universe is even bigger. A depressing thought to some, but a liberating thought to me.

Consider an adult who tends to the traumas of a child: a broken toy, a scraped knee, a schoolyard bully. Adults know that kids have no clue what constitutes a genuine problem, because inexperience greatly limits their childhood perspective.

As grown-ups, dare we admit to ourselves that we, too, have a collective immaturity of view? Dare we admit that our thoughts and behaviors spring from a belief that the world revolves around us? Part the curtains of society's racial, ethnic, religious, national, and cultural conflicts, and you find the human ego turning the knobs and pulling the levers.

Now imagine a world in which everyone, but especially people with power and influence, holds an expanded view of our place in the cosmos. With that perspective, our problems would shrink—or never arise at all—and we could celebrate our earthly differences while shunning the behavior of our predecessors who slaughtered each other because of them.

* * *

Back in February 2000, the newly rebuilt Hayden Planetarium featured a space show called "Passport to the Universe," which took visitors on a virtual zoom from New York City to the edge of the cosmos. En route the audience saw Earth, then the solar system, then the 100 billion stars of the Milky Way galaxy shrink to barely visible dots on the planetarium dome.

I soon received a letter from an Ivy League professor of psychology who wanted to administer a questionnaire to visitors, assessing the depth of their depression after viewing the show. Our show, he wrote, elicited the most dramatic feelings of smallness he had ever experienced.

How could that be? Every time I see the show, I feel alive and spirited and connected. I also feel large, knowing that the goings-on within the three-pound human brain are what enabled us to figure out our place in the universe.

Allow me to suggest that it's the professor, not I, who has misread nature. His ego was too big to begin with, inflated by delusions of significance and fed by cultural assumptions that human beings are more important than everything else in the universe.

In all fairness to the fellow, powerful forces in society leave most of us susceptible. As was I . . . until the day I learned in biology class that more bacteria live and work in one centimeter of my colon than the number of people who have ever existed in the world. That kind of information makes you think twice about who—or what—is actually in charge.

From that day on, I began to think of people not as the masters of space and time but as participants in a great cosmic chain of being, with a direct genetic link across species both living and extinct, extending back nearly 4 billion years to the earliest single-celled organisms on Earth.

* * *

Need more ego softeners? Simple comparisons of quantity, size, and scale do the job well.

Take water. It's simple, common, and vital. There are more molecules of water in an eight-ounce cup of the stuff than there are cups of water in all the world's oceans. Every cup that passes through a single person and eventually rejoins the world's water supply holds enough molecules to mix 1,500 of them into every other cup of water in the world. No way around it: some of the water you just drank passed through the kidneys of Socrates, Genghis Khan, and Joan of Arc.

How about air? Also vital. A single breathful draws in more air molecules than there are breathfuls of air in Earth's entire atmosphere. That means some of the air you just breathed passed through the lungs of Napoleon, Beethoven, Lincoln, and Billy the Kid.

Time to get cosmic. There are more stars in the universe than grains of sand on any beach, more stars than seconds have passed since Earth formed, more stars than words and sounds ever uttered by all the humans who ever lived.

Want a sweeping view of the past? Our unfolding cosmic perspective takes you there. Light takes time to reach Earth's observatories from the depths of space, and so you see objects and phenomena not as they are but as they once were. That means the universe acts like a giant time machine: The farther away you look, the further back in time you see—back almost to the beginning of time itself. Within that horizon of reckoning, cosmic evolution unfolds continuously, in full view.

Want to know what we're made of? Again, the cosmic perspective offers a bigger answer than you might expect. The chemical elements of the universe are forged in the fires of high-mass stars that end their lives in stupendous explosions, enriching their host galaxies with the chemical arsenal of life as we know it. We are not simply in the universe. The universe is in us. Yes, we are stardust.

* * *

Again and again across the centuries, cosmic discoveries have demoted our self-image. Earth was once assumed to be astronomically unique, until astronomers learned that Earth is just another planet orbiting the Sun. Then we presumed the Sun was unique, until we learned that the countless stars of the night sky are suns themselves. Then we presumed our galaxy, the Milky Way, was the entire known universe, until we established that the countless fuzzy things in the sky are other galaxies, dotting the landscape of our known universe.

The cosmic perspective flows from fundamental knowledge. But it's more than just what you know. It's also about having the wisdom and insight to apply that knowledge to assessing our place in the universe. And its attributes are clear:

- The cosmic perspective comes from the frontiers of science, yet is not solely the provenance of the scientist. It belongs to everyone.

- The cosmic perspective is humble.

- The cosmic perspective is spiritual—even redemptive—but is not religious.

- The cosmic perspective enables us to grasp, in the same thought, the large and the small.

- The cosmic perspective opens our minds to extraordinary ideas but does not leave them so open that our brains spill out, making us susceptible to believing anything we're told.

- The cosmic perspective opens our eyes to the universe, not as a benevolent cradle designed to nurture life but as a cold, lonely, hazardous place.

- The cosmic perspective shows Earth to be a mote, but a precious mote and, for the moment, the only home we have.

- The cosmic perspective finds beauty in the images of planets, moons, stars, and nebulae but also celebrates the laws of physics that shape them.

- The cosmic perspective enables us to see beyond our circumstances, allowing us to transcend the primal search for food, shelter, and sex.

- The cosmic perspective reminds us that in space, where there is no air, a flag will not wave—an indication that perhaps flag waving and space exploration do not mix.

- The cosmic perspective not only embraces our genetic kinship with all life on Earth but also values our chemical kinship with any yet-to-be discovered life in the universe, as well as our atomic kinship with the universe itself.

* * *

At least once a week, if not once a day, we might each ponder what cosmic truths lie undiscovered before us, perhaps awaiting the arrival of a clever thinker, an ingenious experiment, or an innovative space mission to reveal them. We might further ponder how those discoveries may one day transform life on Earth.

Absent such curiosity, we are no different from the provincial farmer who expresses no need to venture beyond the county line, because his forty acres meet all his needs. Yet if all our predecessors had felt that way, the farmer would instead be a cave dweller, chasing down his dinner with a stick and a rock.

During our brief stay on planet Earth, we owe ourselves and our descendants the opportunity to explore—in part because it's fun to do. But there's a far nobler reason. The day our knowledge of the cosmos ceases to expand, we risk regressing to the childish view that the universe figuratively and literally revolves around us. In that bleak world, arms-bearing, resource-hungry people and nations would be prone to act on their "low contracted prejudices." And that would be the last gasp of human enlightenment—until the rise of a visionary new culture that could once again embrace the cosmic perspective.

▲ Astronauts get a unique opportunity to experience a cosmic perspective. Here, astronaut John Grunsfeld has a CD of *The Cosmic Perspective* floating in front of him while orbiting Earth during the Space Shuttle's final servicing mission to the Hubble Space Telescope (May 2009).

1 A Modern View of the Universe

This Hubble Space Telescope photo shows thousands of galaxies in a region of the sky so small you could cover it with a grain of sand held at arm's length.

LEARNING GOALS

1.1 The Scale of the Universe
- What is our place in the universe?
- How big is the universe?

1.2 The History of the Universe
- How did we come to be?
- How do our lifetimes compare to the age of the universe?

1.3 Spaceship Earth
- How is Earth moving through space?
- How do galaxies move within the universe?

Far from city lights on a clear night, you can gaze upward at a sky filled with stars. Lie back and watch for a few hours, and you will observe the stars marching steadily across the sky. Confronted by the seemingly infinite heavens, you might wonder how Earth and the universe came to be. If you do, you will be sharing an experience common to humans around the world and in thousands of generations past.

Modern science offers answers to many of our fundamental questions about the universe and our place within it. We now know the basic content and scale of the universe. We know the ages of Earth and the universe. And, although much remains to be discovered, we are rapidly learning how the simple ingredients of the early universe developed into the incredible diversity of life on Earth—and, perhaps, of life on other worlds as well.

In this first chapter, we will survey the scale, history, and motion of the universe. This "big picture" perspective on our universe will provide a base on which you'll be able to build a deeper understanding in the rest of the book.

1.1 The Scale of the Universe

For most of human history, our ancestors imagined Earth to be stationary at the center of a relatively small universe. This idea made sense at a time when understanding was built upon everyday experience. After all, we cannot feel the constant motion of Earth as it rotates on its axis and orbits the Sun, and if you observe the sky you'll see that the Sun, Moon, planets, and stars all appear to revolve around us each day. Nevertheless, we now know that Earth is a planet orbiting a rather average star in a rather typical galaxy in a vast universe.

The historical path to this knowledge was long and complex. In later chapters, we'll see that the ancient belief in an Earth-centered (or *geocentric*) universe changed only when people were confronted by strong evidence to the contrary, and we'll explore how the method of learning that we call *science* enabled us to acquire this evidence. First, however, it's useful to have a general picture of the universe as we know it today.

◆ What is our place in the universe?

Take a look at the remarkable photo that opens this chapter (on page 1). This photo, taken by the Hubble Space Telescope, shows a piece of the sky so small that you could block your view of it with a grain of sand held at arm's length. Yet it covers an almost unimaginable expanse of both space and time: Nearly every object within it is a galaxy containing billions of stars, and some of the smaller smudges are galaxies so far away that their light has taken billions of years to reach us. Let's begin our study of astronomy by exploring what a photo like this one tells us about our own place in the universe.

Our Cosmic Address   The galaxies that we see in the Hubble Space Telescope photo make up just one of several key levels of structure in our universe, all illustrated as our "cosmic address" in Figure 1.1.

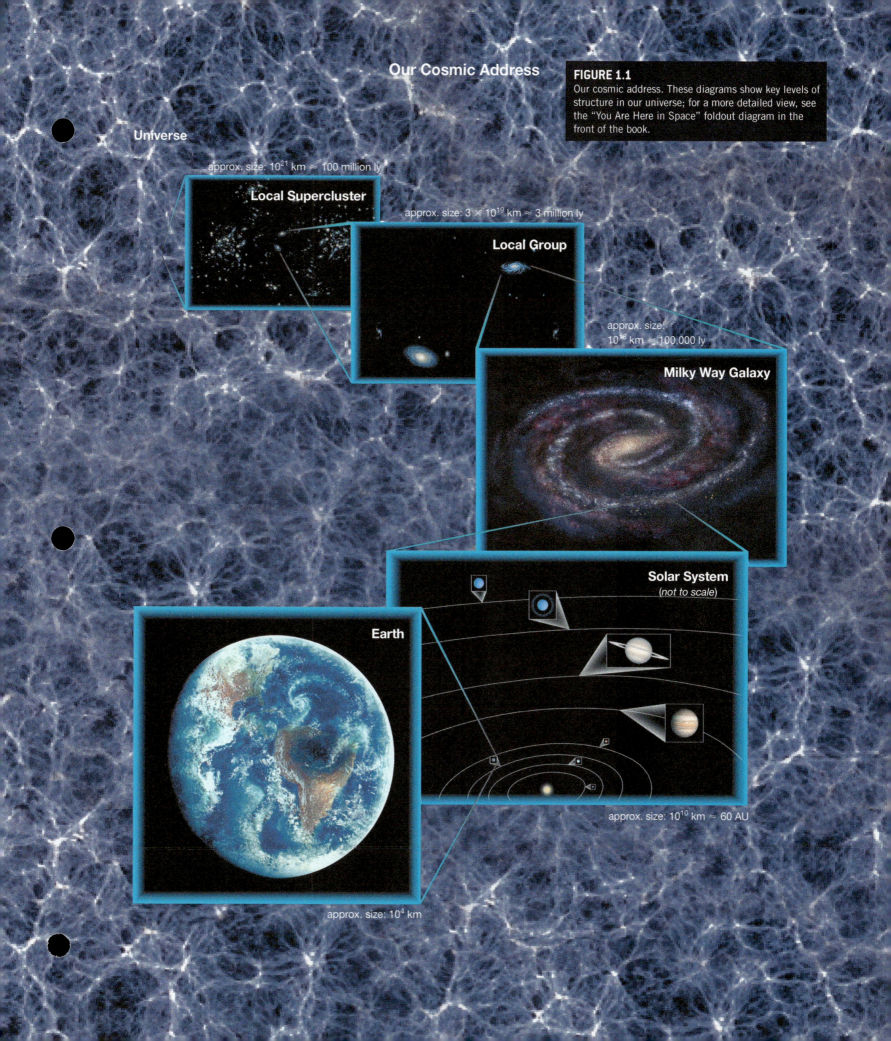

Our Cosmic Address

Universe

approx. size: 10^{21} km ≈ 100 million ly

Local Supercluster

approx. size: 3×10^{19} km ≈ 3 million ly

Local Group

approx. size: 10^{18} km ≈ 100,000 ly

Milky Way Galaxy

Solar System
(*not to scale*)

Earth

approx. size: 10^{10} km ≈ 60 AU

approx. size: 10^4 km

Earth is a planet in our **solar system,** which consists of the Sun, the planets and their moons, and countless smaller objects that include rocky *asteroids* and icy *comets.* Keep in mind that our Sun is a *star*, just like the stars we see in our night sky.

Our solar system belongs to the huge, disk-shaped collection of stars called the **Milky Way Galaxy.** A **galaxy** is a great island of stars in space, all held together by gravity and orbiting a common center. The Milky Way is a relatively large galaxy, containing more than 100 billion stars, and many of these stars are orbited by planets. Our solar system is located a little over halfway from the galactic center to the edge of the galactic disk.

Billions of other galaxies are scattered throughout space. Some galaxies are fairly isolated, but most are found in groups. Our Milky Way, for example, is one of the two largest among more than 70 galaxies (most relatively small) in the **Local Group.** Groups of galaxies with many more large members are often called **galaxy clusters.**

> We live on one planet orbiting one star among more than 100 billion stars in the Milky Way Galaxy, which in turn is one of billions of galaxies in the universe.

On a very large scale, galaxies and galaxy clusters appear to be arranged in giant chains and sheets with huge voids between them; the background of Figure 1.1 represents this large-scale structure. The regions in which galaxies and galaxy clusters are most tightly packed are called **superclusters,** which are essentially clusters of galaxy clusters. Our Local Group is located in the outskirts of the Local Supercluster (which was recently named *Laniakea*, Hawaiian for "immense heaven").

Together, all these structures make up our **universe.** In other words, the universe is the sum total of all matter and energy, encompassing the superclusters and voids and everything within them.

think about it Some people think that our tiny physical size in the vast universe makes us insignificant. Others think that our ability to learn about the wonders of the universe gives us significance despite our small size. What do *you* think?

Astronomical Distance Measurements The labels in Figure 1.1 give approximate sizes for the various structures in kilometers (recall that 1 kilometer $\approx$ 0.6 mile), but many distances in astronomy are so large that kilometers are not the most convenient unit. Instead, we often use two other units:

- One **astronomical unit (AU)** is Earth's average distance from the Sun, which is about 150 million kilometers (93 million miles). We commonly describe distances within our solar system in AU.

- One **light-year (ly)** is the distance that light can travel in 1 year, which is about 10 trillion kilometers (6 trillion miles). We generally use light-years to describe the distances of stars and galaxies.

Be sure to note that a light-year is a unit of *distance*, not of time. Light travels at the speed of light, which is about 300,000 kilometers per second. We therefore say that one *light-second* is about 300,000 kilometers, because that is the distance light travels in 1 second. Similarly, one light-minute is the distance that light travels in 1 minute, one light-hour is the distance that light travels in 1 hour, and so on. Cosmic Calculations 1.1 shows that light travels about 10 trillion kilometers in 1 year, so that distance represents a light-year.

Looking Back in Time The speed of light is extremely fast by earthly standards. It is so fast that if you could make light go in circles, it could circle Earth nearly eight times in a single second. Nevertheless, even light

cosmic calculations 1.1

How Far Is a Light-Year?

We can calculate the distance represented by a light-year by recalling that

$$\text{distance} = \text{speed} \times \text{time}$$

For example, at a speed of 50 km/hr, in 2 hours you travel 50 km/hr $\times$ 2 hr = 100 km. To find the distance represented by 1 light-year, we multiply the speed of light by 1 year. Because we are given the speed of light in kilometers per *second* but the time as 1 *year*, we must carry out the multiplication while converting 1 year into seconds. (See Appendix C for a review of unit conversions.) The result is

$$1 \text{ light-year} = (\text{speed of light}) \times (1 \text{ yr})$$

$$= \left(300,00 \, \frac{\text{km}}{\text{s}}\right) \times (1 \, \text{yr}) \times \frac{365 \, \text{days}}{1 \, \text{yr}}$$

$$\times \frac{24 \, \text{hr}}{1 \, \text{day}} \times \frac{60 \, \text{min}}{1 \, \text{hr}} \times \frac{60 \, \text{s}}{1 \, \text{min}}$$

$$= 9,460,000,000,000 \text{ km}$$

$$= 9.46 \text{ trillion km}$$

That is, 1 light-year is about 9.46 trillion kilometers, which we can approximate as 10 trillion kilometers. This can be easier to write with powers of 10 (see Appendix C.1 for a review); recall that 1 trillion is a 1 followed by 12 zeros, or 10^{12}, so 10 trillion can be written as 10^{13}.

takes time to travel the vast distances in space. Light takes a little more than 1 second to reach Earth from the Moon, and about 8 minutes to reach Earth from the Sun. Stars are so far away that their light takes years to reach us, which is why we measure their distances in light-years.

Because light takes time to travel through space, we are led to a remarkable fact: **The farther away we look in distance, the further back we look in time.** For example, the brightest star in the night sky, Sirius, is about 8 light-years away, which means its light takes about 8 years to reach us. When we look at Sirius, we are seeing it not as it is today but as it was about 8 years ago.

Light takes time to travel the vast distances in space. When we look deep into space, we also look far into the past.

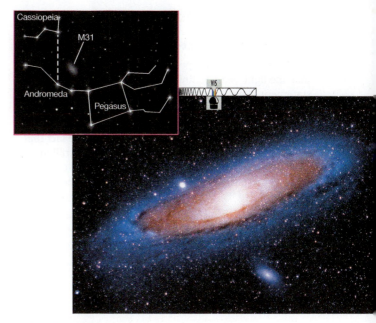

The effect is more dramatic at greater distances. The Andromeda Galaxy (Figure 1.2) lies about 2.5 million light-years from Earth, which means we see it as it looked about 2.5 million years ago. We see more distant galaxies as they were even further in the past. Some of the galaxies in the Hubble Space Telescope photo that opens the chapter are more than 12 billion light-years away, meaning we see them as they were more than 12 billion years ago.

It's also amazing to realize that any "snapshot" of a distant galaxy is a picture of both space and time. For example, because the Andromeda Galaxy is about 100,000 light-years in diameter, the light we see from the far side of the galaxy must have left on its journey to us 100,000 years

▲ **FIGURE 1.2**
The Andromeda Galaxy (M31). When we look at this galaxy, we see light that has been traveling through space for 2.5 million years. The inset shows the galaxy's location in the constellation Andromeda.

Basic Astronomical Definitions

Basic Astronomical Objects

star A large, glowing ball of gas that generates heat and light through nuclear fusion in its core. Our Sun is a star.

planet A moderately large object that orbits a star and shines primarily by reflecting light from its star. According to the current definition, an object can be considered a planet only if it (1) orbits a star, (2) is large enough for its own gravity to make it round, and (3) has cleared most other objects from its orbital path. An object that meets the first two criteria but has not cleared its orbital path, like Pluto, is designated a **dwarf planet.**

moon (or **satellite**) An object that orbits a planet. The term *satellite* is also used more generally to refer to any object orbiting another object.

asteroid A relatively small and rocky object that orbits a star.

comet A relatively small and ice-rich object that orbits a star.

small solar system body An asteroid, comet, or other object that orbits a star but is too small to qualify as a planet or dwarf planet.

Collections of Astronomical Objects

solar system The Sun and all the material that orbits it, including planets, dwarf planets, and small solar system bodies. Although the term *solar system* technically refers only to our own star system (*solar* means "of the Sun"), it is often applied to other star systems as well.

star system A star (sometimes more than one star) and any planets and other materials that orbit it.

galaxy A great island of stars in space, all held together by gravity and orbiting a common center, with a total mass equivalent to that of millions, billions, or even trillions of stars.

cluster (or group) of galaxies A collection of galaxies bound together by gravity. Small collections of galaxies are generally called *groups*, while larger collections are called *clusters*.

supercluster A gigantic region of space in which many groups and clusters of galaxies are packed more closely together than elsewhere in the universe.

universe (or cosmos) The sum total of all matter and energy—that is, all galaxies and everything between them.

observable universe The portion of the entire universe that can be seen from Earth, at least in principle. The observable universe is probably only a tiny portion of the entire universe.

Astronomical Distance Units

astronomical unit (AU) The average distance between Earth and the Sun, which is about 150 million kilometers. More technically, 1 AU is the length of the semimajor axis of Earth's orbit.

light-year The distance that light can travel in 1 year, which is about 10 trillion kilometers (more precisely, 9.46 trillion km).

Terms Relating to Motion

rotation The spinning of an object around its axis. For example, Earth rotates once each day around its axis, which is an imaginary line connecting the North and South Poles.

orbit (revolution) The orbital motion of one object around another due to gravity. For example, Earth orbits the Sun once each year.

expansion (of the universe) The increase in the average distance between galaxies as time progresses.

The Meaning of a Light-Year

You've probably heard people say things like "It will take me light-years to finish this homework!" But a statement like this one doesn't make sense, because a light-year is a unit of *distance*, not time. If you are unsure whether the term *light-year* is being used correctly, try testing the statement by using the fact that 1 light-year is about 10 trillion kilometers, or 6 trillion miles. The statement then reads "It will take me 6 trillion miles to finish this homework," which clearly does not make sense.

before the light from the near side. Figure 1.2 therefore shows different parts of the galaxy spread over a time period of 100,000 years. When we study the universe, it is impossible to separate space and time.

see it for yourself The central region of the Andromeda Galaxy is faintly visible to the naked eye and easy to see with binoculars. Use a star chart to find it in the night sky and remember that you are seeing light that spent 2.5 million years in space before reaching your eyes. If students on a planet in the Andromeda Galaxy were looking at the Milky Way, what would they see? Could they know that we exist here on Earth?

The Observable Universe

As we'll discuss in Section 1.2, the measured the age of the universe is about 14 billion years. This fact, combined with the fact that looking deep into space means looking far back in time, places a limit on the portion of the universe that we can see, even in principle.

Figure 1.3 shows the idea. If we look at a galaxy that is 7 billion light-years away, we see it as it looked 7 billion years ago*—which means we see it as it was when the universe was half its current age. If

Because the universe is about 14 billion years old, we cannot observe light coming from anything more than 14 billion light-years away.

we look at a galaxy that is 12 billion light-years away (like the most distant ones in the Hubble Space Telescope photo), we see it as it was 12 billion years ago, when the universe was only 2 billion years old.

If we tried to look beyond 14 billion light-years, we'd be looking to a time more than 14 billion years ago—which is before the universe existed and therefore means that there is nothing to see. This distance of 14 billion light-years therefore marks the boundary (or *horizon*) of our **observable universe**—the portion of the entire universe that we can potentially observe. Note that this fact does not put any limit on the size of the *entire* universe, which we assume to be far larger than our observable universe. We simply cannot see or study anything beyond the bounds of our observable universe, because the light from such distances has not yet had time to reach us in a 14-billion-year-old universe.

▼ **FIGURE 1.3**
The farther away we look in space, the further back we look in time. The age of the universe therefore puts a limit on the size of the observable universe—the portion of the entire universe that we could observe in principle.

*As we'll see in Chapter 16, distances to faraway galaxies must be defined carefully in an expanding universe; in this book, we use distances based on the light-travel time from a distant object (called the *lookback time*).

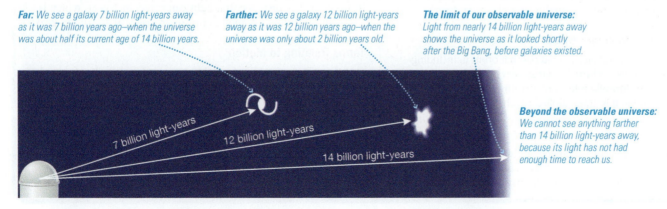

Far: We see a galaxy 7 billion light-years away as it was 7 billion years ago—when the universe was about half its current age of 14 billion years.

Farther: We see a galaxy 12 billion light-years away as it was 12 billion years ago—when the universe was only about 2 billion years old.

The limit of our observable universe: Light from nearly 14 billion light-years away shows the universe as it looked shortly after the Big Bang, before galaxies existed.

Beyond the observable universe: We cannot see anything farther than 14 billion light-years away, because its light has not had enough time to reach us.

7 billion light-years

12 billion light-years

14 billion light-years

◆ How big is the universe?

Figure 1.1 put numbers on the sizes of different structures in the universe, but these numbers have little meaning for most people—after all, they are literally astronomical. To help you develop a greater appreciation of our modern view of the universe, we'll discuss a few ways of putting these numbers into perspective.

The Scale of the Solar System   One of the best ways to develop perspective on cosmic sizes and distances is to imagine our solar system shrunk down to a scale that would allow you to walk through it. The Voyage scale model solar system (Figure 1.4) makes such a walk possible by showing the Sun, the planets, and the distances between them at *one ten-billionth* of their actual sizes and distances.

Figure 1.5a shows the Sun and planets at their correct sizes (but not distances) on the Voyage scale. The model Sun is about the size of a large

> On a scale in which the Sun is the size of a grapefruit, Earth is the size of a ballpoint from a pen, orbiting the Sun at a distance of 15 meters.

grapefruit, Jupiter is about the size of a marble, and Earth is about the size of the ballpoint in a pen. You can immediately see some key facts about our solar system. For example, the Sun is far larger than any of the planets; in mass, the Sun outweighs all the planets combined by a factor of nearly 1000. The planets also vary considerably in size: The storm on Jupiter known as the Great Red Spot (visible near Jupiter's lower left in the painting) could swallow up the entire Earth.

The scale of the solar system is even more remarkable when you combine the sizes shown in Figure 1.5a with the distances illustrated by the map of the Voyage model in Figure 1.5b. For example, the ballpoint-size Earth is located about 15 meters (16.5 yards) from the grapefruit-size Sun, which means you can picture Earth's orbit as a circle of radius 15 meters around a grapefruit.

Perhaps the most striking feature of our solar system when we view it to scale is its emptiness. The Voyage model shows the planets along a straight path, so we'd need to draw each planet's orbit around the model Sun to show the full extent of our planetary system. Fitting all these orbits would require an area measuring more than a kilometer on a side—an area equivalent to more than 300 football fields arranged in a grid. Spread over this large area, only the grapefruit-size Sun, the planets,

▲ **FIGURE 1.4**
This photo shows the pedestals housing the Sun (the gold sphere on the nearest pedestal) and the inner planets in the Voyage scale model solar system (Washington, D.C.). The model planets are encased in the sidewalk-facing disks visible at about eye level on the planet pedestals. To the left is the National Air and Space Museum.

---

special topic   **How Many Planets Are in Our Solar System?**

Until recently, children were taught that our solar system had nine planets. However, in 2006 astronomers voted to demote Pluto to a *dwarf planet*, leaving our solar system with only eight official planets. Why the change?

When Pluto was discovered in 1930, it was assumed to be similar to other planets. But as we'll discuss in Chapter 9, we now know that Pluto is much smaller than any of the first eight planets and that it shares the outer solar system with thousands of other icy objects. Still, as long as Pluto was the largest known of these objects, most astronomers were content to leave the planetary status quo. Change was forced by the 2005 discovery of an object called Eris. Because Eris is slightly larger in mass than Pluto, astronomers could no longer avoid the question of what objects should count as planets.

Official decisions on astronomical names and definitions rest with the International Astronomical Union (IAU), an organization made up of professional astronomers from around the world. In 2006, an IAU vote defined "planet" in a way that left out Pluto and Eris (see Basic Astronomical Definitions on page 5) but added the "dwarf planet" category to accommodate them. Three smaller solar system objects are also now considered dwarf planets (the asteroid Ceres and the Kuiper belt objects Makemake and Haumea). More than a half dozen other objects are still being studied to determine if they meet the dwarf planet definition.

Some astronomers object to these definitions, which may yet be revisited. Pluto and other objects will remain the same either way. Indeed, much as there are no well-defined distinctions between the flowing waterways that we call creeks, streams, or rivers, this case offers a good example of the difference between the fuzzy boundaries of nature and the human preference for categories.

a The scaled sizes (but not distances) of the Sun, the planets, and the two largest known dwarf planets.

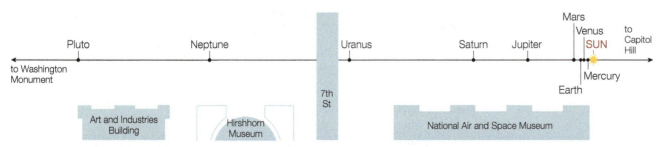

b Locations of the major objects in the Voyage model (Washington, D.C.); the distance from the Sun to Pluto is about 600 meters (1/3 mile). Planets are lined up in the model, but in reality each planet orbits the Sun independently and a perfect alignment never occurs.

▲ FIGURE 1.5
The Voyage scale model represents the solar system at one *ten-billionth* of its actual size. Pluto is included in the Voyage model for context.

common misconceptions

Confusing Very Different Things

Most people are familiar with the terms *solar system* and *galaxy*, but few realize how incredibly different they are. Our solar system is a single star system, while our galaxy is a collection of more than 100 billion star systems—so many that it would take thousands of years just to count them. Moreover, if you look at the sizes in Figure 1.1, you'll see that our galaxy is about 100 million times larger in diameter than our solar system. So be careful; numerically speaking, mixing up *solar system* and *galaxy* is a gigantic mistake!

and a few moons would be big enough to see. The rest of it would look virtually empty (that's why we call it *space!*).

Seeing our solar system to scale also helps put space exploration into perspective. The Moon, the only other world on which humans have ever stepped (Figure 1.6), lies only about 4 centimeters (1½ inches) from Earth in the Voyage model. On this scale, the palm of your hand can cover the entire region of the universe in which humans have so far traveled. The trip to Mars is more than 150 times as far as the trip to the Moon, even when Mars is on the same side of its orbit as Earth. And while you can walk from Earth to Pluto in a few minutes on the Voyage scale, the *New Horizons* spacecraft, which flew past Pluto in 2015, took more than 9 years to make the real journey, despite traveling at a speed nearly 100 times as fast as that of a commercial jet.

Distances to the Stars If you visit the Voyage model in Washington, D.C., you can walk the roughly 600-meter distance from the Sun to Pluto in just a few minutes. How much farther would you have to walk to reach the next star on this scale?

Amazingly, you would need to walk to California. If this answer seems hard to believe, you can check it for yourself. A light-year is about 10 trillion kilometers, which becomes 1000 kilometers on the 1-to-10-billion scale (because 10 trillion ÷ 10 billion = 1000). The nearest star

On the same scale on which Pluto is a few minutes' walk from Earth, you'd have to walk across the United States to reach the nearest stars. system to our own, a three-star system called Alpha Centauri (Figure 1.7), is about 4.4 light-years away. That distance is about 4400 kilometers (2700 miles) on the 1-to-10-billion scale, or roughly equivalent to the distance across the United States.

The tremendous distances to the stars give us some perspective on the technological challenge of astronomy. For example, because the largest star of the Alpha Centauri system is roughly the same size and brightness as our Sun, viewing it in the night sky is somewhat like being in Washington, D.C., and seeing a very bright grapefruit in San Francisco (neglecting the problems introduced by the curvature of Earth). It may seem remarkable that we can see the star at all, but the blackness of the night sky allows the naked eye to see it as a faint dot of light. It looks much brighter through powerful telescopes, but we still cannot see features of the star's surface.

Now, consider the difficulty of detecting *planets* orbiting nearby stars, which is equivalent to looking from Washington, D.C., and trying to find ballpoints or marbles orbiting grapefruits in California or beyond. When you consider this challenge, it is all the more remarkable to realize that we now have technology capable of finding such planets [Section 10.1].

The vast distances to the stars also offer a sobering lesson about interstellar travel. Although science fiction shows like *Star Trek* and *Star Wars* make such travel look easy, the reality is far different. Consider the *Voyager 2* spacecraft. Launched in 1977, *Voyager 2* flew by Jupiter in 1979, Saturn in 1981, Uranus in 1986, and Neptune in 1989. It is now bound for the stars at a speed of close to 50,000 kilometers per hour—about 100 times as fast as a speeding bullet. But even at this speed, *Voyager 2* would take about 100,000 years to reach Alpha Centauri if it were headed in that direction (which it's not). Convenient interstellar travel remains well beyond our present technology.

The Size of the Milky Way Galaxy The vast separation between our solar system and Alpha Centauri is typical of the separations between star systems in our region of the Milky Way Galaxy. We therefore cannot use the 1-to-10-billion scale for thinking about distances beyond the nearest stars, because more distant stars would not fit on Earth with this scale. To visualize the galaxy, let's reduce our scale by another factor of 1 billion (making it a scale of 1 to 10^{19}).

On this new scale, each light-year becomes 1 millimeter, and the 100,000-light-year diameter of the Milky Way Galaxy becomes 100 meters, or about the length of a football field. Visualize a football field with a scale model of our galaxy centered over midfield. Our entire solar system is a microscopic dot located around the 20-yard line. The 4.4-light-year separation between our solar system and Alpha Centauri becomes just 4.4 millimeters on this scale—smaller than the width of your little finger. If you stood at the position of our solar system in this model, millions of star systems would lie within reach of your arms.

Another way to put the galaxy into perspective is to consider its number of stars—more than 100 billion. Imagine that tonight you are having difficulty falling asleep (perhaps because you are contemplating the scale of the universe). Instead of counting sheep, you decide to count stars. If you are able to count about one star each second, how long would it take you to count 100 billion stars in the Milky Way? Clearly, the answer is 100 billion (10^{11}) seconds, but how long is that?

Amazingly, 100 billion seconds is more than 3000 years. (You can confirm this by dividing 100 billion by the number of seconds in 1 year.) You

▲ **FIGURE 1.6**

This famous photograph from the first Moon landing (*Apollo 11* in July 1969) shows astronaut Buzz Aldrin, with Neil Armstrong reflected in his visor. Armstrong was the first to step onto the Moon's surface, saying, "That's one small step for a man, one giant leap for mankind."

▲ **FIGURE 1.7**

On the same 1-to-10-billion scale on which you can walk from the Sun to Pluto in just a few minutes, you'd need to cross the United States to reach Alpha Centauri, the nearest other star system. The inset shows the location and appearance of Alpha Centauri in the night sky.

▲ **FIGURE 1.8**
The number of stars in the observable universe is comparable to the number of grains of dry sand on all the beaches on Earth.

It would take thousands of years just to count out loud the number of stars in the Milky Way Galaxy. would need thousands of years just to *count* the stars in the Milky Way Galaxy, and this assumes you never take a break—no sleeping, no eating, and absolutely no dying!

The Observable Universe   As incredible as the scale of our galaxy may seem, the Milky Way is only one of roughly 100 billion galaxies in the observable universe. Just as it would take thousands of years to count the stars in the Milky Way, it would take thousands of years to count all the galaxies.

Think for a moment about the total number of stars in all these galaxies. If we assume 100 billion stars per galaxy, the total number of stars in the observable universe is roughly 100 billion × 100 billion, or 10,000,000,000,000,000,000,000 (10^{22}).

How big is this number? Visit a beach. Run your hands through the fine-grained sand. Imagine counting each tiny grain of sand as it slips *Roughly speaking, there are as many stars in the observable universe as there are grains of sand on all the beaches on Earth.* through your fingers. Then imagine counting every grain of sand on the beach and continuing to count *every* grain of dry sand on *every* beach on Earth. If you could actually complete this task, you would find that the number of grains of sand is comparable to the number of stars in the observable universe (Figure 1.8).

think about it   Contemplate the incredible numbers of stars in our galaxy and in the universe, and the fact that each star is a potential sun for a system of planets. How does this perspective affect your thoughts about the possibilities for finding life—or intelligent life—beyond Earth? Explain.

1.2 The History of the Universe

Our universe is vast not only in space, but also in time. In this section, we will briefly discuss the history of the universe as we understand it today.

Before we begin, you may wonder how we can claim to know anything about what the universe was like in the distant past. We'll devote much of this textbook to understanding how science enables us to do this, but you already know part of the answer: Because looking farther into space means looking further back in time, we can actually *see* parts of the universe as they were long ago, simply by looking far enough away. In other words, telescopes are somewhat like time machines, enabling us to observe the history of the universe.

◆ How did we come to be?

Figure 1.9 (pp. 12–13) summarizes the history of the universe according to modern science. Let's start at the upper left of the figure, and discuss the key events and what they mean.

The Big Bang, Expansion, and the Age of the Universe   Telescopic observations of distant galaxies show that the entire universe is *expanding*, meaning that average distances between galaxies are increasing with time. This fact implies that galaxies must have been closer together in the past, and if we go back far enough, we must reach the point at which the expansion began. We call this beginning the **Big Bang**, and

The rate at which galaxies are moving apart suggests that the universe was born about 14 billion years ago, in the event we call the Big Bang.

scientists use the observed rate of expansion to calculate that it occurred about 14 billion years ago. The three cubes in the upper left portion of Figure 1.9 represent the expansion of a small piece of the universe through time.

The universe as a whole has continued to expand ever since the Big Bang, but on smaller size scales, the force of gravity has drawn matter together. Structures such as galaxies and galaxy clusters occupy regions where gravity has won out against the overall expansion. That is, while the universe as a whole continues to expand, individual galaxies and galaxy clusters (and objects within them such as stars and planets) do *not* expand. This idea is also illustrated by the three cubes in Figure 1.9. Notice that as the cube as a whole grew larger, the matter within it clumped into galaxies and galaxy clusters. Most galaxies, including our own Milky Way, formed within a few billion years after the Big Bang.

Stellar Lives and Galactic Recycling

Within galaxies like the Milky Way, gravity drives the collapse of clouds of gas and dust to form stars and planets. Stars are not living organisms, but they nonetheless go through "life cycles." A star is born when gravity compresses the material in a cloud to the point at which the center becomes dense enough and

Stars are born in interstellar clouds, produce energy and new elements through nuclear fusion, and release those new elements in interstellar space when they die.

hot enough to generate energy by **nuclear fusion,** the process in which lightweight atomic nuclei smash together and stick (or fuse) to make heavier nuclei. The star "lives" as long as it can generate energy from fusion and "dies" when it exhausts its usable fuel.

In its final death throes, a star blows much of its content back out into space. The most massive stars die in titanic explosions called *supernovae*. The returned matter mixes with other matter floating between the stars in the galaxy, eventually becoming part of new clouds of gas and dust from which future generations of stars can be born. Galaxies therefore function as cosmic recycling plants, recycling material expelled from dying stars into new generations of stars and planets. This cycle is illustrated in the lower right of Figure 1.9. Our own solar system is a product of many generations of such recycling.

Star Stuff

The recycling of stellar material is connected to our existence in an even deeper way. By studying stars of different ages, we have learned that the early universe contained only the simplest chemical elements: hydrogen and helium (and a trace of lithium). We and Earth are made primarily of other elements, such as carbon,

We are "star stuff"—made of material that was manufactured in stars from the simple elements born in the Big Bang.

nitrogen, oxygen, and iron. Where did these other elements come from? Evidence shows that they were manufactured by stars, some through the nuclear fusion that makes stars shine, and others through nuclear reactions accompanying the explosions that end stellar lives.

By the time our solar system formed, about 4½ billion years ago, earlier generations of stars had already converted up to 2% of our galaxy's original hydrogen and helium into heavier elements. Therefore, the cloud that gave birth to our solar system was made of roughly 98% hydrogen and helium and 2% other elements. This 2% may sound small, but it was more than enough to make the small rocky planets of

Throughout this book we will see that human life is intimately connected with the development of the universe as a whole. This illustration presents an overview of our cosmic origins, showing some of the crucial steps that made our existence possible.

① Birth of the Universe: The expansion of the universe began with the hot and dense Big Bang. The cubes show how one region of the universe has expanded with time. The universe continues to expand, but on smaller scales gravity has pulled matter together to make galaxies.

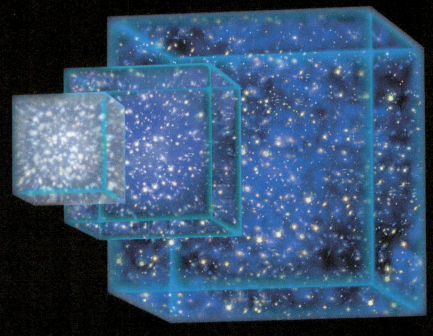

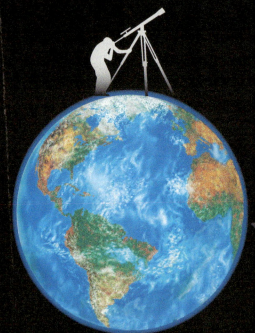

④ Earth and Life: By the time our solar system was born, 4½ billion years ago, about 2% of the original hydrogen and helium had been converted into heavier elements. We are therefore "star stuff," because we and our planet are made from elements manufactured in stars that lived and died long ago.

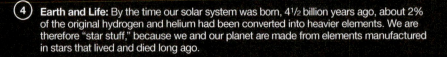

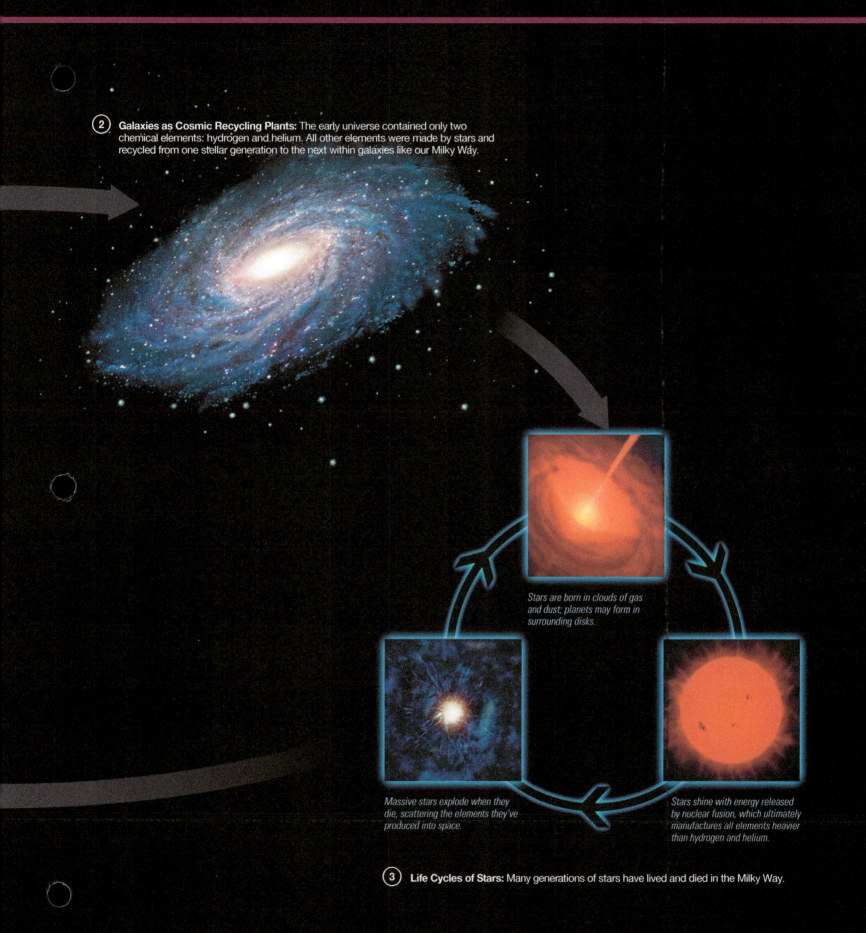

② **Galaxies as Cosmic Recycling Plants:** The early universe contained only two chemical elements: hydrogen and helium. All other elements were made by stars and recycled from one stellar generation to the next within galaxies like our Milky Way.

Stars are born in clouds of gas and dust; planets may form in surrounding disks.

Massive stars explode when they die, scattering the elements they've produced into space.

Stars shine with energy released by nuclear fusion, which ultimately manufactures all elements heavier than hydrogen and helium.

③ **Life Cycles of Stars:** Many generations of stars have lived and died in the Milky Way.

THE HISTORY OF THE UNIVERSE IN 1 YEAR

January 1: The Big Bang
February: The Milky Way forms
September 3: Earth forms
September 22: Early life on Earth
December 17: Cambrian explosion
December 26: Rise of the dinosaurs
December 30: Extinction of the dinosaurs

▲ **FIGURE 1.10**

The cosmic calendar compresses the 14-billion-year history of the universe into 1 year, so that each month represents a little more than 1 billion years. *Adapted from the cosmic calendar created by Carl Sagan.* (For a more detailed version, see the "You Are Here in Time" foldout diagram in the front of the book.)

our solar system, including Earth. On Earth, some of these elements became the raw ingredients of life, which ultimately blossomed into the great diversity of life on Earth today.

In summary, most of the material from which we and our planet are made was created inside stars that lived and died before the birth of our Sun. As astronomer Carl Sagan (1934–1996) said, we are "star stuff."

◆ How do our lifetimes compare to the age of the universe?

We can put the 14-billion-year age of the universe into perspective by imagining this time compressed into a single year, so each month represents a little more than 1 billion years. On this *cosmic calendar*, the Big Bang occurred at the first instant of January 1 and the present is the stroke of midnight on December 31 (Figure 1.10).

On this time scale, the Milky Way Galaxy probably formed in February. Many generations of stars lived and died in the subsequent cosmic months, enriching the galaxy with the "star stuff" from which we and our planet are made.

Our solar system and our planet did not form until early September on this scale, or 4½ billion years ago in real time. By late September, life on Earth was flourishing. However, for most of Earth's history, living organisms remained relatively primitive and microscopic. On the scale of the cosmic calendar, recognizable animals became prominent only in mid-December. Early dinosaurs appeared on the day after Christmas. Then, in a cosmic instant, the dinosaurs disappeared forever—probably because of the impact of an asteroid or a comet [Section 9.5]. In real time, the death of the dinosaurs occurred some 65 million years ago, but on the cosmic calendar it was only yesterday. With the dinosaurs gone, small furry mammals inherited Earth. Some 60 million years later, or around 9 p.m. on December 31 of the cosmic calendar, early hominids (human ancestors) began to walk upright.

> If we imagine the 14-billion-year history of the universe compressed into 1 year, a human lifetime lasts only a fraction of a second.

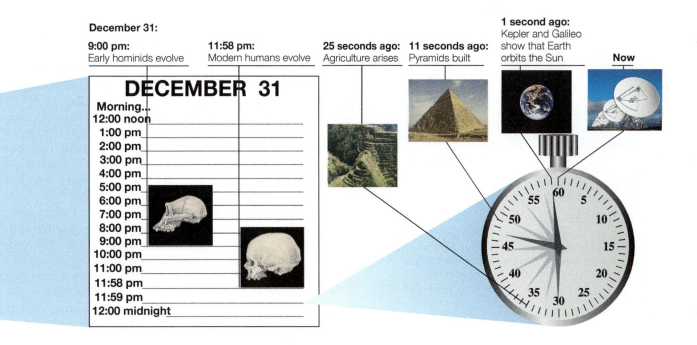

December 31:

9:00 pm:
Early hominids evolve

11:58 pm:
Modern humans evolve

25 seconds ago:
Agriculture arises

11 seconds ago:
Pyramids built

1 second ago:
Kepler and Galileo
show that Earth
orbits the Sun

Now

DECEMBER 31

Morning...
12:00 noon
1:00 pm
2:00 pm
3:00 pm
4:00 pm
5:00 pm
6:00 pm
7:00 pm
8:00 pm
9:00 pm
10:00 pm
11:00 pm
11:58 pm
11:59 pm
12:00 midnight

Perhaps the most astonishing fact about the cosmic calendar is that the entire history of human civilization falls into just the last half-minute. The ancient Egyptians built the pyramids only about 11 seconds ago on this scale. About 1 second ago, Kepler and Galileo provided the key evidence that led us to understand that Earth orbits the Sun rather than vice versa. The average college student was born about 0.05 second ago, around 11:59:59.95 p.m. on the cosmic calendar. On the scale of cosmic time, the human species is the youngest of infants, and a human lifetime is a mere blink of an eye.

think about it Study the more detailed cosmic calendar found on the foldout in the front of this book. How does an understanding of the scale of time affect your view of human civilization? Explain.

1.3 Spaceship Earth

Wherever you are as you read this book, you probably have the feeling that you're "just sitting here." Nothing could be further from the truth. As we'll discuss in this section, all of us are moving through space in so many ways that noted inventor and philosopher R. Buckminster Fuller (1895–1983) described us as travelers on *spaceship Earth*.

◆ How is Earth moving through space?

Rotation and Orbit The most basic motions of Earth are its daily **rotation** (spin) and its yearly **orbit** (or *revolution*) around the Sun.

Earth rotates once each day around its axis (Figure 1.11), which is the imaginary line connecting the North Pole to the South Pole. Earth rotates from west to east—counterclockwise as viewed from above the North Pole—which is why the Sun and stars appear to rise in the east and set in the west each day. Although the physical effects of rotation

Earth rotates once each day and orbits the Sun once each year. Its average orbital distance, called an astronomical unit (AU), is about 150 million kilometers.

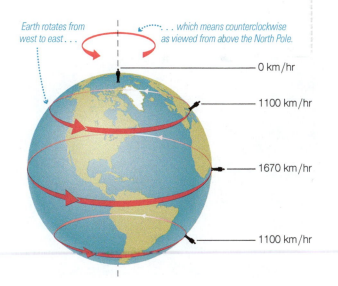

Earth rotates from west to east . . .

. . . which means counterclockwise as viewed from above the North Pole.

0 km/hr

1100 km/hr

1670 km/hr

1100 km/hr

▲ **FIGURE 1.11**

As Earth rotates, your speed around Earth's axis depends on your location: The closer you are to the equator, the faster you travel with rotation.

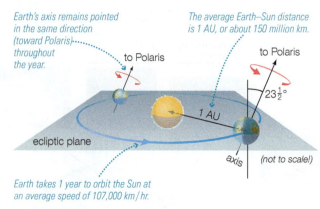

Earth's axis remains pointed in the same direction (toward Polaris) throughout the year.

The average Earth–Sun distance is 1 AU, or about 150 million km.

to Polaris

to Polaris

23½°

1 AU

ecliptic plane

axis

(not to scale!)

Earth takes 1 year to orbit the Sun at an average speed of 107,000 km/hr.

▲ **FIGURE 1.12**
Earth takes a year to complete an orbit of the Sun, but its orbital speed is still surprisingly fast. Notice that Earth both rotates and orbits counterclockwise as viewed from above the North Pole.

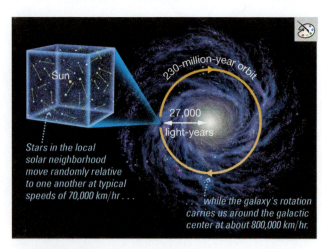

Sun

230-million-year orbit

27,000 light-years

Stars in the local solar neighborhood move randomly relative to one another at typical speeds of 70,000 km/hr . . .

. . . while the galaxy's rotation carries us around the galactic center at about 800,000 km/hr.

▲ **FIGURE 1.13**
This painting illustrates the motion of our solar system within our local solar neighborhood and around the center of the Milky Way Galaxy.

are so subtle that our ancestors assumed the heavens revolved around us, the rotation speed is substantial: Unless you live quite far north or south, you are whirling around Earth's axis at a speed of more than 1000 kilometers per hour (600 miles per hour)—faster than most airplanes travel.

At the same time as it is rotating, Earth also orbits the Sun, completing one orbit each year (Figure 1.12). Earth's orbital distance varies slightly over the course of each year, but as we discussed earlier, the average distance is one astronomical unit (AU), which is about 150 million kilometers. Again, even though we don't feel this motion, the speed is impressive: We are racing around the Sun at a speed in excess of 100,000 kilometers per hour (60,000 miles per hour), which is faster than any spacecraft yet launched.

As you study Figure 1.12, notice that Earth's orbital path defines a flat plane that we call the **ecliptic plane.** Earth's axis is tilted by 23½° from a line *perpendicular* to the ecliptic plane. This **axis tilt** happens to be oriented so that the axis points almost directly at a star called *Polaris,* or the *North Star*. Keep in mind that the idea of axis tilt makes sense only in relation to the ecliptic plane. That is, the idea of "tilt" by itself has no meaning in space, where there is no absolute up or down. In space, "up" and "down" mean only "away from the center of Earth (or another planet)" and "toward the center of Earth," respectively.

think about it If there is no up or down in space, why do you think most globes have the North Pole on top? Would it be equally correct to have the South Pole on top or to turn the globe sideways? Explain.

Notice also that Earth orbits the Sun in the same direction that it rotates on its axis: counterclockwise as viewed from above the North Pole. This is not a coincidence but a consequence of the way our planet was born. As we'll discuss in Chapter 6, strong evidence indicates that Earth and the other planets were born in a spinning disk of gas that surrounded our Sun as it formed, and Earth rotates and orbits in the same direction as the disk was spinning.

Motion Within the Milky Way Galaxy Rotation and orbit are only a small part of the travels of spaceship Earth. Our entire solar system is on a great journey within the Milky Way Galaxy. There are two major components to this motion, both shown in Figure 1.13.

First, our solar system is moving relative to nearby stars in our *local solar neighborhood*, the region of the Sun and nearby stars. The small box in Figure 1.13 shows that stars within the local solar neighborhood (like the stars of any other small region of the galaxy) move essentially at random relative to one another. The speeds are quite fast: On average, our Sun is moving relative to nearby stars at a speed of about 70,000 kilometers per hour (40,000 miles per hour), almost three times as fast as the international Space Station orbits Earth. Given these high speeds, you may wonder why we don't see stars racing around the sky. The answer lies in their vast distances from us. You've probably noticed that a distant airplane appears to move through the sky more slowly than one flying close overhead. Stars are so far away that even at speeds of 70,000 kilometers per hour, their motions would be noticeable to the naked eye only if we watched them for thousands of years. That is why the patterns in the constellations seem to remain fixed. Nevertheless, in 10,000 years the constellations will be noticeably different from those we see today. In 500,000 years they will be unrecognizable. If you could watch a time-lapse movie made over millions of years, you *would* see stars racing across the sky.

Most of the galaxy's light comes from stars and gas in the galactic disk and central bulge . . .

. . . but measurements suggest that most of the mass lies unseen in the spherical halo that surrounds the entire disk.

◀ **FIGURE 1.14**
This painting shows an edge-on view of the Milky Way Galaxy. Study of galactic rotation shows that although most visible stars lie in the disk and central bulge, most of the mass lies in the halo that surrounds and encompasses the disk. Because this mass emits no light that we have detected, we call it *dark matter*.

think about it Despite the chaos of motion in the local solar neighborhood over millions and billions of years, collisions between star systems are extremely rare. Explain why. (*Hint:* Consider the sizes of star systems, such as the solar system, relative to the distances between them.)

The second motion shown in Figure 1.13 is much more organized. If you look closely at leaves floating in a stream, their motions relative to one another might appear random, just like the motions of stars in the local solar neighborhood. As you widen your view, you see that all the leaves are being carried in the same general direction by the downstream current. In the same way, as we widen our view beyond the local solar neighborhood, the seemingly random motions of its stars give way to a simpler and even faster motion: rotation of the Milky Way Galaxy. Our solar system, located about 27,000 light-years from the galactic center, completes one orbit of the galaxy in about 230 million years. Even if you could watch from outside our galaxy, this motion would be unnoticeable to your naked eye. However, if you calculate the speed of our solar system as we orbit the center of the galaxy, you will find that it is close to 800,000 kilometers (500,000 miles) per hour.

Galactic rotation carries us around the center of the galaxy once every 230 million years.

Careful study of the galaxy's rotation reveals one of the greatest mysteries in science. Stars at different distances from the galactic center orbit at different speeds, and we can learn how mass is distributed in the galaxy by measuring these speeds. Such studies indicate that the stars in the disk of the galaxy represent only the "tip of the iceberg" compared to the mass of the entire galaxy (Figure 1.14). Most of the mass of the galaxy seems to be located outside the visible disk (occupying the galactic *halo* that surrounds and encompasses the disk), but the matter that makes up this mass is completely invisible to our telescopes. We therefore know very little about the nature of this matter, which we refer to as *dark matter* (because of the lack of light from it). Studies of other galaxies indicate that they also are made mostly of dark matter. These and other observations imply that dark matter significantly outweighs the ordinary matter that makes up planets and stars, making it the dominant source of gravity that has led to the formation of galaxies, clusters, and superclusters. We know even less about the mysterious **dark energy** that astronomers first recognized when they discovered that the expansion of the universe is actually getting faster with time, and that scientists have since found to make up the majority of the total energy content of the universe. We'll discuss the mysteries of dark matter and dark energy in Chapter 18.

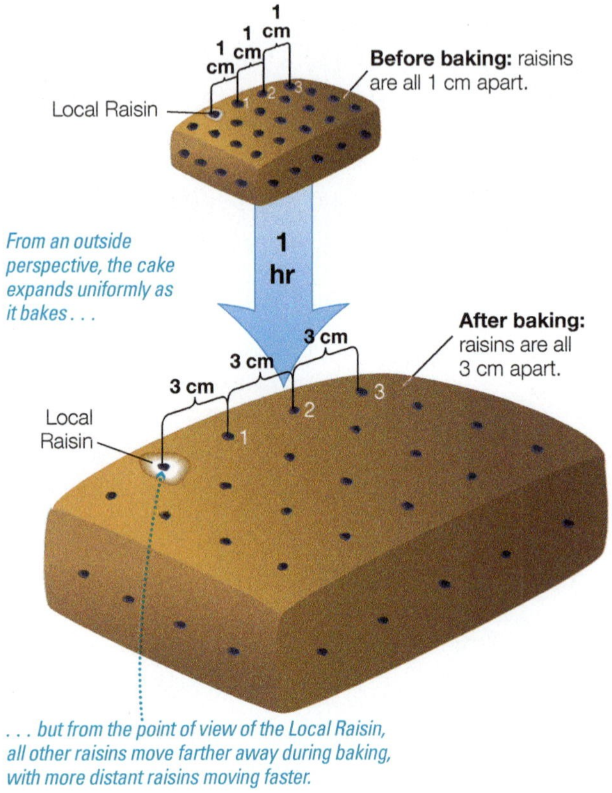

◆ How do galaxies move within the universe?

The billions of galaxies in the universe also move relative to one another. Within the Local Group (see Figure 1.1), some of the galaxies move toward us, some move away from us, and numerous small galaxies (including the Large and Small Magellanic Clouds) apparently orbit our Milky Way Galaxy. Again, the speeds are enormous by earthly standards. For example, the Milky Way and Andromeda galaxies are moving toward each other at about 300,000 kilometers (180,000 miles) per hour. Despite this high speed, we needn't worry about a collision anytime soon. Even if the Milky Way and Andromeda galaxies are approaching each other head-on, it will be billions of years before any collision begins.

When we look outside the Local Group, however, we find two astonishing facts recognized in the 1920s by Edwin Hubble, for whom the Hubble Space Telescope was named:

1. Virtually every galaxy outside the Local Group is moving *away* from us.

2. The more distant the galaxy, the faster it appears to be racing away.

These facts might make it sound as if we suffer from a cosmic case of chicken pox, but there is a much more natural explanation: *The entire universe is expanding.* We'll save the details for later in the book, but you can understand the basic idea by thinking about a raisin cake baking in an oven.

The Raisin Cake Analogy Imagine that you make a raisin cake in which the distance between adjacent raisins is 1 centimeter. You place the cake in the oven, where it expands as it bakes. After 1 hour, you remove the cake, which has expanded so that the distance between adjacent raisins has increased to 3 centimeters (Figure 1.15). The expansion of the cake seems fairly obvious. But what would you see if you lived *in* the cake, as we live in the universe?

Pick any raisin (it doesn't matter which one) and call it the Local Raisin. Figure 1.15 shows one possible choice, with three nearby raisins also labeled. The accompanying table summarizes what you would see if you lived within the Local Raisin. Notice, for example, that Raisin 1 starts out at a distance of 1 centimeter before baking and ends up at a distance of 3 centimeters after baking, which means it moves a distance of 2 centimeters farther away from the Local Raisin during the hour of baking. Hence, its speed as seen from the Local Raisin is 2 centimeters per hour. Raisin 2 moves from a distance of 2 centimeters before baking to a distance of 6 centimeters after baking, which means it moves a distance of 4 centimeters farther away from the Local Raisin during the hour. Hence, its speed is 4 centimeters per hour, or twice the speed of Raisin 1. Generalizing, the fact that the cake is expanding means that all the raisins are moving away from the Local Raisin, with more distant raisins moving away faster.

Hubble's discovery that galaxies are moving in much the same way as the raisins in the cake, with most moving away from us and more distant ones moving away faster, implies that the universe is expanding much like the raisin cake. If you now imagine the Local Raisin as representing our Local Group of galaxies and the other raisins as representing more distant galaxies or clusters of galaxies, you have a

Distant galaxies are all moving away from us, with more distant ones moving faster, indicating that we live in an expanding universe.

From an outside perspective, the cake expands uniformly as it bakes...

Before baking: raisins are all 1 cm apart.

Local Raisin

1 hr

After baking: raisins are all 3 cm apart.

Local Raisin

... but from the point of view of the Local Raisin, all other raisins move farther away during baking, with more distant raisins moving faster.

Distances and Speeds as Seen from the Local Raisin

Raisin Number	Distance Before Baking	Distance After Baking (1 hour later)	Speed
1	1 cm	3 cm	2 cm/hr
2	2 cm	6 cm	4 cm/hr
3	3 cm	9 cm	6 cm/hr
:	:	:	:

▲ **FIGURE 1.15**

An expanding raisin cake offers an analogy to the expanding universe. Someone living in one of the raisins inside the cake could figure out that the cake is expanding by noticing that all other raisins are moving away, with more distant raisins moving away faster. In the same way, we know that we live in an expanding universe because all galaxies outside our Local Group are moving away from us, with more distant ones moving faster.

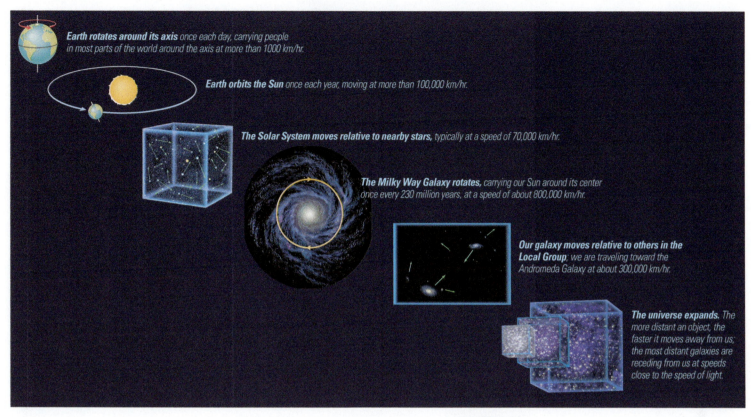

Earth rotates around its axis once each day, carrying people in most parts of the world around the axis at more than 1000 km/hr.

Earth orbits the Sun once each year, moving at more than 100,000 km/hr.

The Solar System moves relative to nearby stars, typically at a speed of 70,000 km/hr.

The Milky Way Galaxy rotates, carrying our Sun around its center once every 230 million years, at a speed of about 800,000 km/hr.

Our galaxy moves relative to others in the Local Group; we are traveling toward the Andromeda Galaxy at about 300,000 km/hr.

The universe expands. The more distant an object, the faster it moves away from us; the most distant galaxies are receding from us at speeds close to the speed of light.

▲ **FIGURE 1.16**

This figure summarizes the basic motions of Earth in the universe, along with their associated speeds.

basic picture of the expansion of the universe. Like the expanding dough between the raisins in the cake, *space* itself is growing between galaxies. More distant galaxies move away from us faster because they are carried along with this expansion like the raisins in the expanding cake. You can also now see how observations of expansion allow us to measure the age of the universe: The faster the rate of expansion, the more quickly the galaxies reached their current positions, and therefore the younger the universe must be. It is by precisely measuring the expansion rate that astronomers have learned that the universe is approximately 14 billion years old.

The Real Universe   There's at least one important distinction between the raisin cake and the universe: A cake has a center and edges, but we do not think the same is true of the entire universe. Anyone living in any galaxy in an expanding universe sees just what we see—other galaxies moving away, with more distant ones moving away faster. Because the view from each point in the universe is about the same, no place can claim to be more "central" than any other place.

It's also important to realize that, unlike the case with a raisin cake, we can't actually *see* galaxies moving apart with time—the distances are too vast for any motion to be noticeable on the time scale of a human life. Instead, we measure the speeds of galaxies by spreading their light into spectra and observing what we call *Doppler shifts* [Section 5.2]. This illustrates how modern astronomy depends both on careful observations and on using current understanding of the laws of nature to explain what we see.

Motion Summary   Figure 1.16 summarizes the motions we have discussed. As we have seen, we are never truly sitting still. We spin around Earth's axis at more than 1000 kilometers per hour, while our planet orbits the Sun at more than 100,000 kilometers per hour. Our solar system moves among the stars of the local solar neighborhood at typical speeds of 70,000 kilometers per hour, while also orbiting the center of the Milky Way Galaxy at a speed of about 800,000 kilometers per hour. Our galaxy moves among the other galaxies of the Local Group, while all other galaxies move away from us at speeds that grow greater with distance in our expanding universe. Spaceship Earth is carrying us on a remarkable journey.

the big picture — Putting Chapter 1 into Perspective

In this first chapter, we developed a broad overview of our place in the universe. As we consider the universe in more depth in the rest of the book, remember the following "big picture" ideas:

- Earth is not the center of the universe but instead is a planet orbiting a rather ordinary star in the Milky Way Galaxy. The Milky Way Galaxy, in turn, is one of billions of galaxies in our observable universe.

- Cosmic distances are literally astronomical, but we can put them in perspective with the aid of scale models and other scaling techniques. When you think about these enormous scales, don't forget that every star is a sun and every planet is a unique world.

- We are "star stuff." The atoms from which we are made began as hydrogen and helium in the Big Bang and

were later fused into heavier elements by massive stars. Stellar deaths released these atoms into space, where our galaxy recycled them into new stars and planets. Our solar system formed from such recycled matter some 4½ billion years ago.

- We are latecomers on the scale of cosmic time. The universe was already more than half its current age when our solar system formed, and it took billions of years more before humans arrived on the scene.

- All of us are being carried through the cosmos on spaceship Earth. Although we cannot feel this motion, the associated speeds are surprisingly high. Learning about the motions of spaceship Earth gives us a new perspective on the cosmos and helps us understand its nature and history.

my cosmic perspective The science of astronomy affects all of us on many levels. In particular, it helps us understand how we as humans fit into the universe as a whole, and the history of astronomy has been deeply intertwined with the development of civilization.

summary of key concepts

1.1 The Scale of the Universe

◆ What is our place in the universe?

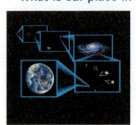

Earth is a planet orbiting the Sun. Our Sun is one of more than 100 billion stars in the **Milky Way Galaxy.** Our galaxy is one of more than 70 galaxies in the **Local Group.** The Local Group is one small part of the **Local Supercluster,** which is one small part of the **universe.**

◆ How big is the universe?

If we imagine our Sun as a large grapefruit, Earth is a ballpoint that orbits 15 meters away; the nearest stars are thousands of kilometers away on the same scale. Our galaxy contains more than 100 billion stars—so many that it would take thousands of years just to count them out loud. The **observable universe** contains roughly 100 billion galaxies, and the total number of stars is comparable to the number of grains of dry sand on all the beaches on Earth.

1.2 The History of the Universe

◆ How did we come to be?

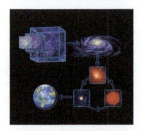

The universe began in the **Big Bang** and has been expanding ever since, except in localized regions where gravity has caused matter to collapse into galaxies and stars. The Big Bang essentially produced only two chemical elements: hydrogen and helium. The rest have been produced by stars and recycled within galaxies from one generation of stars to the next, which is why we are "star stuff."

◆ How do our lifetimes compare to the age of the universe?

On a cosmic calendar that compresses the history of the universe into 1 year, human civilization is just a few seconds old, and a human lifetime lasts only a fraction of a second.

1.3 Spaceship Earth

◆ How is Earth moving through space?

Earth **rotates** on its axis once each day and **orbits** the Sun once each year. At the same time, we move with our Sun in random directions relative to other stars in our local solar neighborhood, while the galaxy's rotation carries us around the center of the galaxy every 230 million years.

◆ How do galaxies move within the universe?

Galaxies move essentially at random within the Local Group, but all galaxies beyond the Local Group are moving away from us. More distant galaxies are moving faster, which tells us that we live in an expanding universe.

---

visual skills check   Check your understanding of some of the many types of visual information used in astronomy. For additional practice, try the Chapter 1 Visual Quiz at MasteringAstronomy®.

The figure at right shows the sizes of Earth and the Moon to scale; the scale used is 1 cm = 4000 km. Using what you've learned about astronomical scale in this chapter, answer the following questions. *Hint:* If you are unsure of the answers, you can calculate them using the given data.

Earth–Sun distance = 150,000,000 km

Diameter of Sun = 1,400,000 km

Earth–Moon distance = 384,000 km

Diameter of Earth = 12,800 km

1. If you wanted to show the distance between Earth and the Moon on the same scale, about how far apart would you need to place the two photos above?
   a. 10 centimeters (about the width of your hand)
   b. 1 meter (about the length of your arm)
   c. 100 meters (about the length of a football field)
   d. 1 kilometer (a little more than a half mile)

2. Suppose you wanted to show the Sun on the same scale. About how big would it need to be?
   a. 3.5 centimeters in diameter (the size of a golf ball)
   b. 35 centimeters in diameter (a little bigger than a basketball)
   c. 3.5 meters in diameter (about 11½ feet across)
   d. 3.5 kilometers in diameter (the size of a small town)

3. About how far away from Earth would the Sun be located on this scale?
   a. 3.75 meters (about 12 feet)
   b. 37.5 meters (about the height of a 12-story building)
   c. 375 meters (about the length of four football fields)
   d. 37.5 kilometers (the size of a large city)

4. Could you use the same scale to represent the distances to nearby stars? Why or why not?

---

exercises and problems

MasteringAstronomy®   For instructor-assigned homework and other learning materials, go to MasteringAstronomy®.

Review Questions

1. Briefly describe the major levels of structure (such as planet, star, galaxy) in the universe.
2. Define *astronomical unit* and *light-year*.
3. Explain the statement *The farther away we look in distance, the further back we look in time.*
4. What do we mean by the *observable universe*? Is it the same thing as the entire universe?
5. Using techniques described in the chapter, put the following into perspective: the size of our solar system; the distance to nearby stars; the size and number of stars in the Milky Way Galaxy; the number of stars in the observable universe.
6. What do we mean when we say that the universe is *expanding*, and how does expansion lead to the idea of the *Big Bang* and our current estimate of the age of the universe?
7. In what sense are we "star stuff"?
8. Use the cosmic calendar to describe how the human race fits into the scale of time.
9. Briefly explain Earth's daily rotation and annual orbit, defining the terms *ecliptic plane* and *axis tilt*.
10. Briefly describe our solar system's location and motion within the Milky Way Galaxy.
11. What is *dark matter*? Where does it reside in our galaxy? What makes dark matter and *dark energy* so mysterious and so important?

12. What key observations lead us to conclude that the universe is expanding? Use the raisin cake model to explain how these observations imply expansion.

13. How does the expansion rate of the universe allow us to determine the age of the universe? Would a faster expansion rate imply an older or a younger age for the universe? Explain.

Test Your Understanding

Does It Make Sense?

Decide whether the statement makes sense (or is clearly true) or does not make sense (or is clearly false). Explain clearly; not all of these have definitive answers, so your explanation is more important than your chosen answer.

Example: I walked east from our base camp at the North Pole.

Solution: The statement does not make sense because east has no meaning at the North Pole—all directions are south from the North Pole.

14. Our solar system is bigger than some galaxies.
15. The universe is billions of light-years in age.
16. It will take me light-years to complete this homework assignment!
17. Someday we may build spaceships capable of traveling a light-year in only a decade.
18. Astronomers discovered a moon that does not orbit a planet.
19. NASA will soon launch a spaceship that will photograph our Milky Way Galaxy from beyond its halo.
20. The observable universe is the same size today as it was a few billion years ago.
21. Photographs of distant galaxies show them as they were when they were much younger than they are today.
22. At a nearby park, I built a scale model of our solar system in which I used a basketball to represent Earth.
23. Because nearly all galaxies are moving away from us, we must be located at the center of the universe.

Quick Quiz

Choose the best answer to each of the following. Explain your reasoning with one or more complete sentences.

24. Which of the following correctly lists our "cosmic address" from small to large? (a) Earth, solar system, Milky Way Galaxy, Local Group, Local Supercluster, universe (b) Earth, solar system, Local Group, Local Supercluster, Milky Way Galaxy, universe (c) Earth, Milky Way Galaxy, solar system, Local Group, Local Supercluster, universe

25. An *astronomical unit* is (a) any planet's average distance from the Sun. (b) Earth's average distance from the Sun. (c) any large astronomical distance.

26. The star Betelgeuse is about 600 light-years away. If it explodes tonight, (a) we'll know because it will be brighter than the full Moon in the sky. (b) we'll know because debris from the explosion will rain down on us from space. (c) we won't know about it until about 600 years from now.

27. If we represent the solar system on a scale that allows us to walk from the Sun to Pluto in a few minutes, then (a) the planets are the size of basketballs and the nearest stars are a few miles away. (b) the planets are marble-size or smaller and the nearest stars are thousands of miles away. (c) the planets are microscopic and the stars are light-years away.

28. The total number of stars in the observable universe is roughly equivalent to (a) the number of grains of sand on all the beaches on Earth. (b) the number of grains of sand on Miami Beach. (c) infinity.

29. When we say the universe is *expanding*, we mean that (a) everything in the universe is growing in size. (b) the average distance between galaxies is growing with time. (c) the universe is getting older.

30. If stars existed but galaxies did not, (a) we would probably still exist anyway. (b) we would not exist because life on Earth depends on the light of galaxies. (c) we would not exist because we are made of material that was recycled in galaxies.

31. Could we see a galaxy that is 50 billion light-years away? (a) Yes, if we had a big enough telescope. (b) No, because it would be beyond the bounds of our observable universe. (c) No, because a galaxy could not possibly be that far away.

32. The age of our solar system is about (a) one-third of the age of the universe. (b) three-fourths of the age of the universe. (c) two billion years less than the age of the universe.

33. The fact that nearly all galaxies are moving away from us, with more distant ones moving faster, helped us to conclude that (a) the universe is expanding. (b) galaxies repel each other like magnets. (c) our galaxy lies near the center of the universe.

Process of Science

34. *Earth as a Planet.* For most of human history, scholars assumed Earth was the center of the universe. Today, we know that Earth is just one planet orbiting the Sun, and the Sun is just one star in a vast universe. How did science make it possible for us to learn these facts about Earth?

35. *Thinking About Scale.* One key to success in science is finding a simple way to evaluate new ideas, and making a simple scale model is often helpful. Suppose someone tells you that the reason it is warmer during the day than at night is that the day side of Earth is closer to the Sun than the night side. Evaluate this idea by thinking about the size of Earth and its distance from the Sun in a scale model of the solar system.

36. *Looking for Evidence.* In this first chapter, we have discussed the scientific story of the universe but have not yet discussed most of the evidence that backs it up. Choose one idea presented in this chapter—such as the idea that there are billions of galaxies in the universe, or that the universe was born in the Big Bang, or that the galaxy contains more dark matter than ordinary matter—and briefly discuss the type of evidence you would want to see before accepting the idea. (*Hint:* It's okay to look ahead in the book to see the evidence presented in later chapters.)

Group Work Exercise

37. *Counting the Milky Way's Stars.* **Roles:** *Scribe* (takes notes on the group's activities), *Proposer* (proposes explanations to the group), *Skeptic* (points out weaknesses in proposed explanations), *Moderator* (leads group discussion and makes sure everyone contributes). **Activity:** Work as a group to answer each part.
    a. Estimate the number of stars in the Milky Way from two facts: (1) the number of stars within 12 light-years of the Sun, which you can count in Appendix F; (2) the total volume of the Milky Way's disk (100,000 light-years in diameter and 1000 light-years thick) is about 1 billion times the volume of the region of your star count.
    b. Compare your value from part (a) to the value given in this chapter. Write down a list of possible reasons why your technique may have given you an underestimate or overestimate of the actual number.

Investigate Further

Short-Answer/Essay Questions

38. *Alien Technology.* Some people believe that Earth is regularly visited by aliens who travel here from other star systems. For this to be true, how much more advanced than our own technology would the aliens' technology have to be? Write one to two paragraphs to give a sense of the technological difference. (*Hint:* The ideas of scale in this chapter can help you contrast the distance the aliens would have to travel with the distances we currently are capable of traveling.)

39. *Stellar Collisions.* Is there any danger that another star will come crashing through our solar system in the near future? Explain.

40. *Raisin Cake Universe.* Suppose that all the raisins in a cake are 1 centimeter apart before baking and 4 centimeters apart after baking.
    a. Draw diagrams to represent the cake before and after baking.
    b. Identify one raisin as the Local Raisin on your diagrams. Construct a table showing the distances and speeds of other raisins as seen from the Local Raisin.
    c. Briefly explain how your expanding cake is similar to the expansion of the universe.

41. *The Cosmic Perspective.* Write a short essay describing how the ideas presented in this chapter affect your perspectives on your own life and on human civilization.

Quantitative Problems

Be sure to show all calculations clearly and state your final answers in complete sentences.

42. *Distances by Light.* Just as a light-year is the distance that light can travel in 1 year, we define a light-second as the distance that light can travel in 1 second, a light-minute as the distance that light can travel in 1 minute, and so on. Calculate the distance in both kilometers and miles represented by each of the following:
    a. 1 light-second
    b. 1 light-minute
    c. 1 light-hour
    d. 1 light-day

43. *Moonlight and Sunlight.* How long does it take light to travel from
    a. the Moon to Earth?
    b. the Sun to Earth?

44. *Saturn vs. the Milky Way.* Photos of Saturn and photos of galaxies can look so similar that children often think the photos show similar objects. In reality, a galaxy is far larger than any planet. About how many times larger is the diameter of the Milky Way Galaxy than the diameter of Saturn's rings? (*Data:* Saturn's rings are about 270,000 km in diameter; the Milky Way is 100,000 light-years in diameter.)

45. *Driving Trips.* Imagine that you could drive your car at a constant speed of 100 km/hr (62 mi/hr), even across oceans and in space. How long would it take to drive
    a. around Earth's equator? (Earth's circumference ≈ 40,000 km)
    b. from the Sun to Earth?
    c. from the Sun to Pluto? (Pluto distance ≈ 5.9×10^9 km)
    d. to Alpha Centauri (4.4 light-years)?

46. *Faster Trip.* Suppose you wanted to reach Alpha Centauri in 100 years.
    a. How fast would you have to go, in km/hr?
    b. How many times faster is the speed you found in (a) than the speeds of our fastest current spacecraft (around 50,000 km/hr)?

47. *Age of the Universe.* Suppose we did not yet know the expansion rate of the universe, and two astronomers came up with two different measurements: Allen measured an expansion rate for the universe that was 50% faster than the expansion rate Wendy measured. Is the age of the universe that Allen inferred older or younger than the age that Wendy inferred? By how much? Explain.

Discussion Questions

48. *Vast Orbs.* Dutch astronomer Christiaan Huygens may have been the first person to truly understand both the large sizes of other planets and the great distances to other stars. In 1690, he wrote, "How vast those Orbs must be, and how inconsiderable this Earth, the Theatre upon which all our mighty Designs, all our Navigations, and all our Wars are transacted, is when compared to them. A very fit consideration, and matter of Reflection, for those Kings and Princes who sacrifice the Lives of so many People, only to flatter their Ambition in being Masters of some pitiful corner of this small Spot." What do you think he meant? Explain.

49. *Infant Species.* In the last few tenths of a second before midnight on December 31 of the cosmic calendar, we have developed an incredible civilization and learned a great deal about the universe, but we also have developed technology through which we could destroy ourselves. The midnight bell is striking, and the choice for the future is ours. How far into the next cosmic year do you think our civilization will survive? Defend your opinion.

50. *A Human Adventure.* Astronomical discoveries clearly are important to science, but are they also important to our personal lives? Defend your opinion.

Web Projects

51. *NASA Missions.* Visit the NASA website to learn about upcoming astronomy missions. Write a one-page summary of the mission you feel is most likely to provide new astronomical information during the time you are enrolled in this astronomy course.

52. *The Hubble Extreme Deep Field.* The photo that opens this chapter is called the Hubble *Extreme* Deep Field. Find the photo on the Hubble Space Telescope website. Learn how it was taken, what it shows, and what we've learned from it. Write a short summary of your findings.

53. *Dwarf Planets.* The 2006 decision to call Pluto a "dwarf planet" still generates controversy. Gather and summarize information about the results from the *New Horizons* mission to Pluto and the *Dawn* mission to the dwarf planet (and large asteroid) Ceres. Where did you get your information? How do you know it is reliable? Has this information shed any light on the classification of Pluto, Ceres, and other solar system bodies? Overall, what is your opinion about the appropriate classification of these bodies?

2 Discovering the Universe for Yourself

This time-exposure photograph shows star paths at Arches National Park, Utah.

LEARNING GOALS

2.1 Patterns in the Night Sky
- ◆ What does the universe look like from Earth?
- ◆ Why do stars rise and set?
- ◆ Why do the constellations we see depend on latitude and time of year?

2.2 The Reason for Seasons
- ◆ What causes the seasons?
- ◆ How does the orientation of Earth's axis change with time?

2.3 The Moon, Our Constant Companion
- ◆ Why do we see phases of the Moon?
- ◆ What causes eclipses?

2.4 The Ancient Mystery of the Planets
- ◆ Why was planetary motion so hard to explain?
- ◆ Why did the ancient Greeks reject the real explanation for planetary motion?

We live in an exciting time in the history of astronomy. New and powerful telescopes are scanning the depths of the universe. Sophisticated space probes are exploring our solar system. Rapid advances in computing technology are allowing scientists to analyze the vast amount of new data and to model the processes that occur in planets, stars, galaxies, and the universe.

One goal of this book is to help *you* share in the ongoing adventure of astronomical discovery. One of the best ways to become a part of this adventure is to do what other humans have done for thousands of generations: Go outside, observe the sky around you, and contemplate the awe-inspiring universe of which you are a part. In this chapter, we'll discuss a few key ideas that will help you understand what you see in the sky.

ESSENTIAL PREPARATION

1. What is our place in the universe? [Section 1.1]
2. How big is the universe? [Section 1.1]
3. How is Earth moving through space? [Section 1.3]

2.1 Patterns in the Night Sky

Today we take for granted that we live on a small planet orbiting an ordinary star in one of many galaxies in the universe. But this fact is not obvious from a casual glance at the night sky, and we've learned about our place in the cosmos only through a long history of careful observations. In this section, we'll discuss major features of the night sky and how we understand them in light of our current knowledge of the universe.

◆ What does the universe look like from Earth?

Shortly after sunset, as daylight fades to darkness, the sky appears to fill slowly with stars. On clear, moonless nights far from city lights, more than 2000 stars may be visible to your naked eye, along with the whitish band of light that we call the *Milky Way* (Figure 2.1). As you look at the stars, your mind may group them into patterns that look like familiar shapes or objects. If you observe the sky night after night or year after year, you will recognize the same patterns of stars. These patterns have not changed noticeably in the past few thousand years.

Constellations People of nearly every culture gave names to patterns they saw in the sky. We usually refer to such patterns as constellations, but to astronomers the term has a more precise meaning: A **constellation** is a *region* of the sky with well-defined borders; the familiar patterns of stars merely help us locate these constellations.

The names and borders of the 88 official constellations [Appendix H] were chosen in 1928 by members of the International Astronomical Union. Note that, just as every spot of land in the continental United States is part of some

Bright stars help us identify constellations, which officially are regions of the sky.

state, every point in the sky belongs to some constellation. For example, Figure 2.2 shows the borders of the constellation Orion and several of its neighbors.

Recognizing the patterns of just 20 or so constellations is enough to make the sky seem as familiar as your own neighborhood. The best way to learn the constellations is to go out and view them, guided by a few visits to a planetarium, star charts [Appendix I], or sky-viewing apps.

▲ **FIGURE 2.1**
This photo shows the Milky Way over Haleakala crater on the island of Maui, Hawaii. The bright spot just below (and slightly left of) the center of the band is the planet Jupiter.

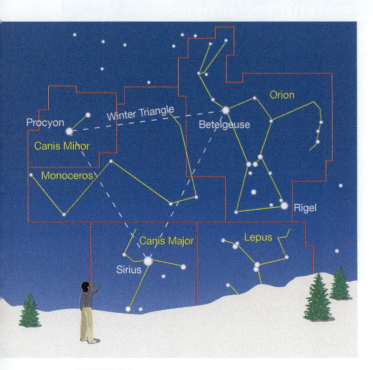

▲ FIGURE 2.2

Red lines mark official borders of several constellations near Orion. Yellow lines connect recognizable patterns of stars. Sirius, Procyon, and Betelgeuse form the *Winter Triangle*, which spans several constellations. This view shows how it appears (looking south) on winter evenings from the Northern Hemisphere.

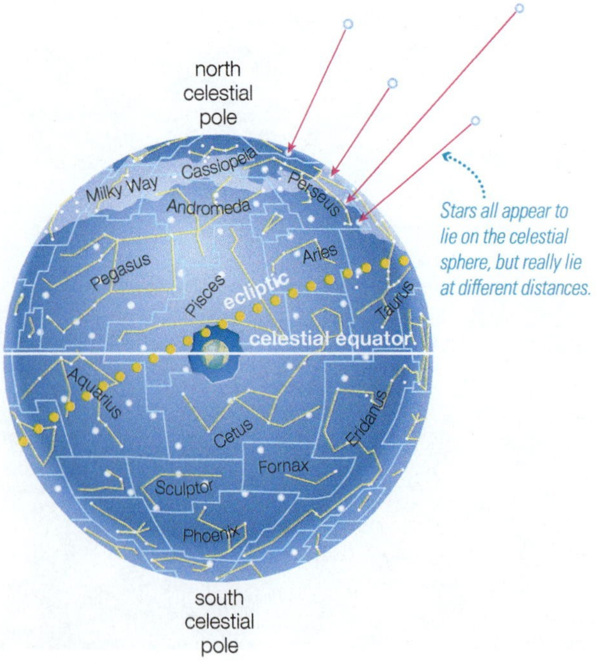

The Celestial Sphere   The stars in a particular constellation appear to lie close to one another but may be quite far apart in reality, because they may lie at very different distances from Earth. This illusion occurs because we lack depth perception when we look into space, a consequence of the fact that the stars are so far away [Section 1.1]. The ancient Greeks mistook this illusion for reality, imagining the stars and constellations as lying on a great **celestial sphere** that surrounds Earth (Figure 2.3).

> **All stars appear to lie on the *celestial sphere*, but in reality they lie at different distances from Earth.**

We now know that Earth seems to be in the center of the celestial sphere only because it is where we are located as we look into space. Nevertheless, the celestial sphere is a useful illusion, because it allows us to map the sky as seen from Earth. For reference, we identify two special points and two special circles on the celestial sphere (Figure 2.4).

- The **north celestial pole** is the point directly over Earth's North Pole.
- The **south celestial pole** is the point directly over Earth's South Pole.
- The **celestial equator,** which is a projection of Earth's equator into space, makes a complete circle around the celestial sphere.
- The **ecliptic** is the path the Sun follows as it appears to circle around the celestial sphere once each year. It crosses the celestial equator at a 23½° angle, because that is the tilt of Earth's axis.

The Milky Way   The band of light that we call the *Milky Way* circles all the way around the celestial sphere, passing through more than a dozen constellations, and bears an important relationship to the Milky Way Galaxy: *It traces our galaxy's disk of stars—the galactic plane—as it appears from our location within the galaxy.*

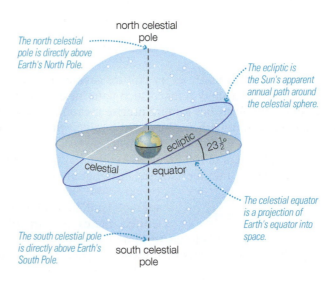

▲ FIGURE 2.3

The stars appear to lie on a great celestial sphere that surrounds Earth. This is an illusion created by our lack of depth perception in space, but it is useful for mapping the sky.

▲ FIGURE 2.4

This schematic diagram shows key features of the celestial sphere.

Figure 2.5 shows the idea. Our galaxy is shaped like a thin pancake with a bulge in the middle. We view the universe from our location a little more than halfway out from the center of this "pancake." In all directions that we look within the pancake, we see the countless stars and vast interstellar clouds that make up the Milky Way in the night sky; that is why the band of light makes a full circle around our sky. The Milky Way appears somewhat wider in the direction of the constellation Sagittarius, because that is the direction in which we are looking toward the galaxy's central bulge. We have a clear view to the distant universe only when we look *away* from the galactic plane, along directions that have relatively few stars and clouds to block our view.

The Milky Way in the night sky is our view in all directions into the disk of our galaxy.

The dark lanes that run down the center of the Milky Way contain the densest clouds, obscuring our view of stars behind them. In most directions, these clouds prevent us from seeing more than a few thousand light-years into our galaxy's disk. As a result, much of our own galaxy remained hidden from view until just a few decades ago, when new technologies allowed us to peer through the clouds by observing forms of light that are invisible to our eyes (such as radio waves, infrared light, and X rays [Section 5.1]).

The Local Sky

The celestial sphere provides a useful way of thinking about the appearance of the universe from Earth. But it is not what we actually see when we go outside. Instead, your **local sky**—the sky as seen from wherever you happen to be standing—appears to take the shape of a hemisphere or dome. The dome shape arises from the fact that we see only half of the celestial sphere at any particular moment from any particular location, while the ground blocks the other half from view.

Figure 2.6 shows key reference features of the local sky. The boundary between Earth and sky defines the **horizon.** The point directly overhead is the **zenith.** The **meridian** is an imaginary half-circle stretching from the horizon due south, through the zenith, to the horizon due north.

We pinpoint an object in the local sky by stating its altitude above the horizon and direction along the horizon.

We can pinpoint the position of any object in the local sky by stating its **direction** along the horizon (sometimes stated as *azimuth*, which is degrees clockwise from due north) and its **altitude** above the horizon. For example, Figure 2.6 shows a person pointing to a star located in the southeast direction at an altitude of 60°. Note that the zenith has altitude 90° but no direction, because it is straight overhead.

Angular Sizes and Distances

Our lack of depth perception on the celestial sphere makes it difficult to judge the true sizes or separations of the objects we see in the sky. However, we can describe the *angular* sizes or separations of objects without knowing how far away they are.

The farther away an object is, the smaller its angular size.

The **angular size** of an object is the angle it appears to span in your field of view. For example, the angular sizes of the Sun and the Moon are each about ½° (Figure 2.7a). Note that angular size does not by itself tell us an object's true size, because angular size also depends on distance. The Sun is about 400 times as large in diameter as the Moon, but it has the same angular size in our sky because it is also about 400 times as far away.

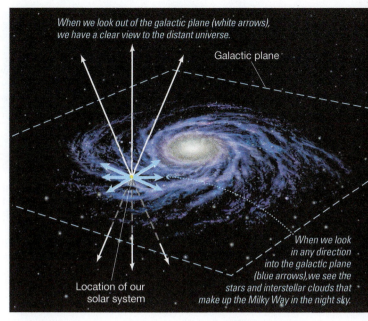

When we look out of the galactic plane (white arrows), we have a clear view to the distant universe.

Galactic plane

When we look in any direction into the galactic plane (blue arrows), we see the stars and interstellar clouds that make up the Milky Way in the night sky.

Location of our solar system

▲ **FIGURE 2.5**

This painting shows how our galaxy's structure affects our view from Earth.

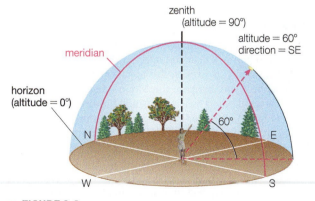

zenith (altitude = 90°)

altitude = 60° direction = SE

meridian

horizon (altitude = 0°)

60°

N

E

W

S

▲ **FIGURE 2.6**

From any place on Earth, the local sky looks like a dome (hemisphere). This diagram shows key reference points in the local sky. It also shows how we can describe any position in the local sky by its altitude and direction.

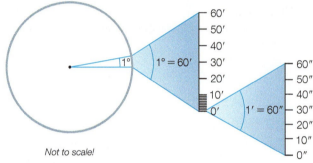

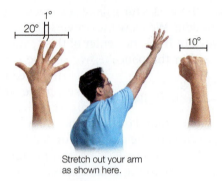

a The angular sizes of the Sun and the Moon are about 1/2°.

b The angular distance between the "pointer stars" of the Big Dipper is about 5°, and the angular length of the Southern Cross is about 6°.

c You can estimate angular sizes or distances with your outstretched hand.

Stretch out your arm as shown here.

▲ **FIGURE 2.7**
We measure *angular sizes* or *angular distances*, rather than actual sizes or distances, when we look at objects in the sky.

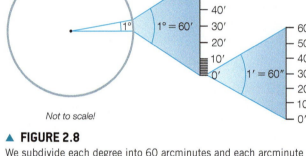

Not to scale!

▲ **FIGURE 2.8**
We subdivide each degree into 60 arcminutes and each arcminute into 60 arcseconds.

---

think about it Children often try to describe the sizes of objects in the sky (such as the Moon or an airplane) in inches or miles, or by holding their fingers apart and saying, "It was THIS big." Can we really describe objects in the sky in this way? Why or why not?

The **angular distance** between a pair of objects in the sky is the angle that appears to separate them. For example, the angular distance between the "pointer stars" at the end of the Big Dipper's bowl is about 5° and the angular length of the Southern Cross is about 6° (Figure 2.7b). You can use your outstretched hand to make rough estimates of angles in the sky (Figure 2.7c).

For greater precision, we subdivide each degree into 60 **arcminutes** (symbolized by ') and each arcminute into 60 arcseconds (symbolized by "), as shown in Figure 2.8. For example, we read 35°27'15" as *35 degrees, 27 arcminutes, 15 arcseconds*.

◆ Why do stars rise and set?

If you spend a few hours out under a starry sky, you'll notice that the universe seems to be circling around us, with stars moving gradually across the sky from east to west. Many ancient people took this appearance at face value, concluding that we lie in the center of a universe that rotates around us each day. Today we know that the ancients had it backward: It is Earth that rotates daily, not the rest of the universe.

We can picture the movement of the sky by imagining the celestial sphere rotating around Earth (Figure 2.9). From this perspective you can see how the universe seems to turn around us: Every object on the celestial sphere appears to make a simple daily circle around Earth. However, the motion can look a little more complex in the local sky, because the horizon cuts the celestial sphere in half. Figure 2.10 shows the idea for a typical Northern Hemisphere location (latitude 40°N). If you study the figure carefully, you'll notice the following key facts about the paths of various stars through the local sky:

- Stars near the north celestial pole are **circumpolar,** meaning that they remain perpetually above the horizon, circling (counterclockwise) around the north celestial pole each day.

- Stars near the south celestial pole never rise above the horizon at all.

- All other stars have daily circles that are partly above the horizon and partly below, which means they appear to rise in the east and set in the west.

---

common **misconceptions**

The Moon Illusion

You've probably noticed that the full moon appears to be larger when it is near the horizon than when it is high in your sky. However, this apparent size change is an illusion: If you compare the Moon's angular size to that of a small object (such as a small button) held at arm's length, you'll see that it remains essentially the same throughout the night. The reason is that the Moon's angular size depends on its true size and distance, and while the latter varies over the course of the Moon's monthly orbit, it does not change enough to cause a noticeable effect on a single night. The Moon illusion clearly occurs within the human brain, though its precise cause is still hotly debated. Interestingly, you may be able to make the illusion go away by viewing the Moon upside down between your legs.

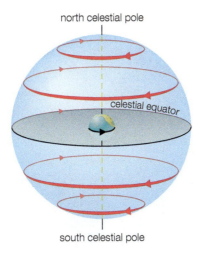

▲ FIGURE 2.9
Earth rotates from west to east (black arrow), making the celestial sphere *appear* to rotate around us from east to west (red arrows).

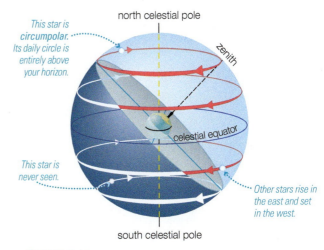

▲ FIGURE 2.10
The local sky for a location at latitude 40°N. The horizon slices through the celestial sphere at an angle to the celestial equator, causing the daily circles of stars to appear tilted in the local sky. Note: It is easier to follow the star paths if you rotate the page so that the zenith points up.

The time-exposure photograph that opens this chapter (p. 24) shows a part of the daily paths of stars. Paths of circumpolar stars are visible within the arch; notice that the complete daily circles for these stars are above the horizon, although the photo shows only a portion of each circle. The north celestial pole lies at the center of these circles. The circles grow larger for stars farther from the north celestial pole.

Earth's west-to-east rotation makes stars appear to move from east to west through the sky as they circle around the celestial poles.

If they are large enough, the circles cross the horizon, so that the stars rise in the east and set in the west. The same ideas apply in the Southern Hemisphere, except that circumpolar stars are those near the south celestial pole and they circle clockwise rather than counterclockwise.

think about it Do distant galaxies also rise and set like the stars in our sky? Why or why not?

◆ Why do the constellations we see depend on latitude and time of year?

If you stay in one place, the basic patterns of motion in the sky will stay the same from one night to the next. However, if you travel far north or south, you'll see a different set of constellations than you see at home. And even if you stay in one place, you'll see different constellations at different times of year. Let's explore why.

Variation with Latitude **Latitude** measures north-south position on Earth and **longitude** measures east-west position (Figure 2.11). Latitude is defined to be 0° at the equator, increasing to 90°N at the North Pole and 90°S at the South Pole. By international treaty, longitude is defined to be 0° along the **prime meridian,** which passes through Greenwich, England. Stating a latitude and a longitude pinpoints a location on Earth. For example, Miami lies at about 26°N latitude and 80°W longitude.

---

cosmic calculations 2.1

Angular Size, Physical Size, and Distance

If you hold a coin in front of your eye, it can block your entire field of view. But as you move it farther away, it appears to get smaller and it blocks less of your view. As long as a coin or any other object is far enough away so that its angular size is relatively small (less than a few degrees), the following formula relates the object's angular size (in degrees), physical size, and distance:

$$\frac{\text{angular size}}{360°} = \frac{\text{physical size}}{2\pi \times \text{distance}}$$

Example: The Moon's angular diameter is about 0.5° and its distance is about 380,000 km. What is the Moon's physical diameter?

Solution: To solve the formula for physical size, we multiply both sides by $2\pi \times$ distance and rearrange:

$$\text{physical size} = \text{angular size} \times \frac{2\pi \times \text{distance}}{360°}$$

We now plug in the given values of the Moon's angular size and distance:

$$\text{physical size} = 0.5° \times \frac{2\pi \times 380,000 \text{ km}}{360°}$$

$$\approx 3300 \text{ km}$$

The Moon's diameter is about 3300 km. We could find a more exact value (3476 km) by using more precise values for the angular diameter and distance.

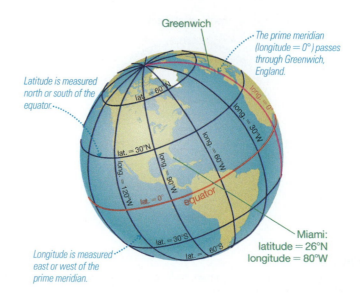

▶ FIGURE 2.11

We can locate any place on Earth's surface by its latitude and longitude.

common misconceptions

Stars in the Daytime

Stars may appear to vanish in the daytime and "come out" at night, but in reality the stars are always present. The reason you don't see stars in the daytime is that their dim light is overwhelmed by the bright daytime sky. You *can* see bright stars in the daytime with the aid of a telescope, or if you are fortunate enough to observe a total solar eclipse. Astronauts can also see stars in the daytime. Above Earth's atmosphere, where there is no air to scatter sunlight, the Sun is a bright disk against a dark sky filled with stars. (However, the Sun is so bright that astronauts must block its light if they wish to see the stars.)

Latitude affects the constellations we see because it affects the locations of the horizon and zenith relative to the celestial sphere. Figure 2.12 shows how this works for the latitudes of the North Pole (90°N) and Sydney, Australia (34°S). Note that although the sky varies with latitude, it does *not* vary with longitude. For example, Charleston (South Carolina) and San Diego (California) are at about the same latitude, so people in both cities see the same set of constellations at night.

> The constellations you see depend on your latitude, but not on your longitude.

You can learn more about how the sky varies with latitude by studying diagrams like those in Figures 2.10 and 2.12. For example, at the North Pole, you can see only objects that lie on the northern half of the celestial sphere, and they are all circumpolar. That is why the Sun remains above the horizon for 6 months at the North Pole: The Sun lies north of the celestial equator for half of each year (see Figure 2.4), so during these 6 months it circles the sky at the North Pole just like a circumpolar star.

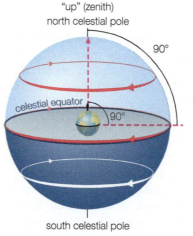

a The local sky at the North Pole (latitude 90°N).

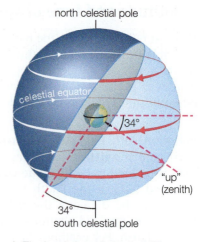

b The local sky at latitude 34°S.

▶ FIGURE 2.12

The sky varies with latitude. Notice that the altitude of the celestial pole that is visible in your sky is always equal to your latitude.

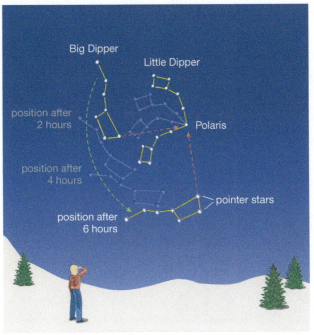

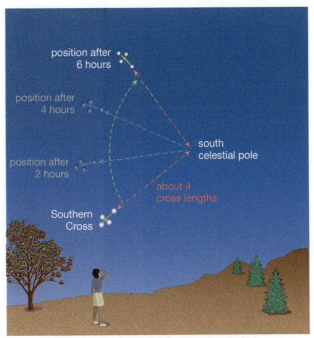

looking northward in the Northern Hemisphere

looking southward in the Southern Hemisphere

a The pointer stars of the Big Dipper point to the North Star, Polaris, which lies within 1° of the north celestial pole. The sky appears to turn *counterclockwise* around the north celestial pole.

b The Southern Cross points to the south celestial pole, which is not marked by any bright star. The sky appears to turn *clockwise* around the south celestial pole.

▲ **FIGURE 2.13**

You can determine your latitude by measuring the altitude of the celestial pole in your sky.

The diagrams also show a fact that is very important to navigation: *The altitude of the celestial pole in your sky is equal to your latitude.* For example, if you see the north celestial pole at an altitude of 40° above your

> **The altitude of the celestial pole in your sky is equal to your latitude.**

north horizon, your latitude is 40°N. Similarly, if you see the south celestial pole at an altitude of 34° above your south horizon, your latitude is 34°S. Finding the north celestial pole is fairly easy, because it lies very close to the star Polaris, also known as the North Star (Figure 2.13a). In the Southern Hemisphere, you can find the south celestial pole with the aid of the Southern Cross (Figure 2.13b).

see it for yourself What is *your* latitude? Use Figure 2.13 to find the celestial pole in your sky, and estimate its altitude with your hand as shown in Figure 2.7c. Is its altitude what you expect?

Variation with Time of Year The night sky changes throughout the year because of Earth's changing position in its orbit around the Sun. Figure 2.14 shows the idea. From our vantage point on Earth, the annual orbit of Earth around the Sun makes the Sun *appear* to move steadily eastward along the ecliptic, with the stars of different constellations in the background at different times of year. The constellations along the ecliptic make up what we call the **zodiac;** tradition places 12 constellations along the zodiac, but the official borders include a thirteenth constellation, Ophiuchus.

> **The constellations visible at a particular time of night change as we orbit the Sun.**

The Sun's apparent location along the ecliptic determines which constellations we see at night. For example, Figure 2.14 shows that the Sun appears to be in Leo

common misconceptions

What Makes the North Star Special?

Most people are aware that the North Star, Polaris, is a special star. Contrary to a relatively common belief, however, it is *not* the brightest star in the sky. More than 50 other stars are just as bright or brighter. Polaris is special not because of its brightness, but because it is so close to the north celestial pole and therefore very useful in navigation.

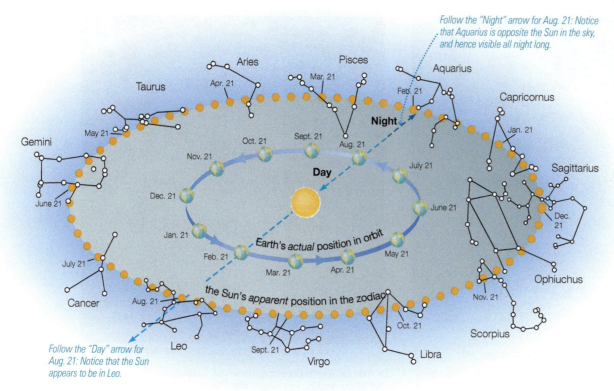

Follow the "Night" arrow for Aug. 21: Notice that Aquarius is opposite the Sun in the sky, and hence visible all night long.

Follow the "Day" arrow for Aug. 21: Notice that the Sun appears to be in Leo.

▲ **FIGURE 2.14**

The Sun appears to move steadily eastward along the ecliptic as Earth orbits the Sun, so we see the Sun against the background of different zodiac constellations at different times of year. For example, on August 21 the Sun appears to be in Leo, because it is between us and the much more distant stars that make up Leo.

in late August. We therefore cannot see Leo at this time (because it is in our daytime sky), but we can see Aquarius all night long because of its location opposite Leo on the celestial sphere. Six months later, in February, we see Leo at night while Aquarius is above the horizon only in the daytime.

see it for yourself Based on Figure 2.14 and today's date, in what constellation does the Sun currently appear? What constellation of the zodiac will be on your meridian at midnight? What constellation of the zodiac will you see in the west shortly after sunset? Go outside at night to confirm your answers.

2.2 The Reason for Seasons

We have seen how Earth's rotation makes the sky appear to circle us daily and how the night sky changes as Earth orbits the Sun each year. The combination of Earth's rotation and orbit also leads to the progression of the seasons.

◆ What causes the seasons?

You know that we have seasonal changes, such as longer and warmer days in summer and shorter and cooler days in winter. But why do the seasons occur? The answer is that the tilt of Earth's axis causes sunlight to fall differently on Earth at different times of year.

Figure 2.15 illustrates the key ideas. Step 1 illustrates the tilt of Earth's axis, which remains pointed in the same direction in space (toward Polaris) throughout the year. As a result, the orientation of the axis *relative to the Sun* changes over the course of each orbit: The Northern Hemisphere is tipped toward the Sun in June and away from the

Sun in December, while the reverse is true for the Southern Hemisphere. That is why the two hemispheres experience opposite seasons. The rest of the figure shows how the changing angle of sunlight on the two hemispheres leads directly to seasons.

Step 2 shows Earth in June, when the axis tilt causes sunlight to strike the Northern Hemisphere at a steeper angle and the Southern Hemisphere at a shallower angle. The steeper sunlight angle makes it summer in the Northern Hemisphere for two reasons. First, as shown in the zoom-out, the steeper angle means more concentrated sunlight, which tends to make it warmer. Second, if you visualize what happens as Earth rotates each day, you'll see that the steeper angle also means the Sun follows a longer and higher path through the sky, giving the Northern Hemisphere more hours of daylight during which it is warmed by the Sun. The opposite is true for the Southern Hemisphere at this time: The shallower sunlight angle makes it winter there because sunlight is less concentrated and the Sun follows a shorter, lower path through the sky.

Earth's axis points in the same direction all year round, which means its orientation relative to the Sun changes as Earth orbits the Sun.

The sunlight angle gradually changes as Earth orbits the Sun. At the opposite side of Earth's orbit, Step 4 shows that it has become winter for the Northern Hemisphere and summer for the Southern Hemisphere. In between these two extremes, Step 3 shows that both hemispheres are illuminated equally in March and September. It is therefore spring for the hemisphere that is on the way from winter to summer, and fall for the hemisphere on the way from summer to winter.

Notice that the seasons on Earth are caused only by the axis tilt and *not* by any change in Earth's distance from the Sun. Although Earth's orbital

common misconceptions

The Cause of Seasons

Many people guess that seasons are caused by variations in Earth's distance from the Sun. But if this were true, the whole Earth would have summer or winter at the same time, and it doesn't: The seasons are opposite in the Northern and Southern Hemispheres. In fact, Earth's slightly varying orbital distance has virtually no effect on the weather. The real cause of seasons is Earth's axis tilt, which causes the two hemispheres to take turns being tipped toward the Sun over the course of each year.

special topic   How Long Is a Day?

We usually associate our 24-hour day with Earth's rotation, but if you measure the rotation period, you'll find that it is about 23 hours and 56 minutes (more precisely $23^h56^m4.09^s$)—or about 4 minutes short of 24 hours. What's going on?

Astronomically, we define two different types of day. Earth's 23 hour and 56 minute rotation period, which we measure by timing how long it takes any star to make one full circuit through our sky, is called a **sidereal day;** *sidereal* (pronounced *sy-dear-ee-al*) means "related to the stars." Our 24-hour day, which we call a **solar day,** is the average time it takes *the Sun* to make one circuit through the sky.

A simple demonstration shows why the solar day is about 4 minutes longer than the sidereal day. Set an object representing the Sun on a table, and stand a few steps away to represent Earth. Point at the Sun and imagine that you also happen to be pointing toward a distant star that lies in the same direction. If you rotate (counterclockwise) while standing in place, you'll again be pointing at both the Sun and the star after one full rotation. However, to show that Earth also orbits the Sun, you should take a couple of steps around the Sun (counterclockwise) as you rotate (see figure). After one full rotation, you will again be pointing in the direction of the distant star, so this rotation represents a sidereal day. But notice that you need to rotate a little extra to point back at the Sun. This "extra" bit of rotation makes a solar day longer than a sidereal day. To figure out how long this extra rotation takes, note

that Earth completes a full 360° orbit around the Sun in about 365 days (1 year), which is a rate of about 1° per day. This extra rotation therefore takes about $\frac{1}{360}$ of Earth's rotation period—which is about 4 minutes.

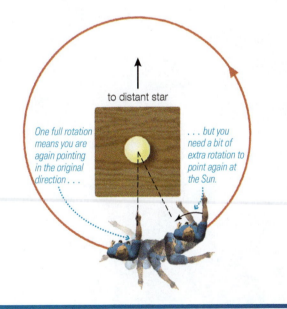

to distant star

One full rotation means you are again pointing in the original direction . . .

. . . but you need a bit of extra rotation to point again at the Sun.

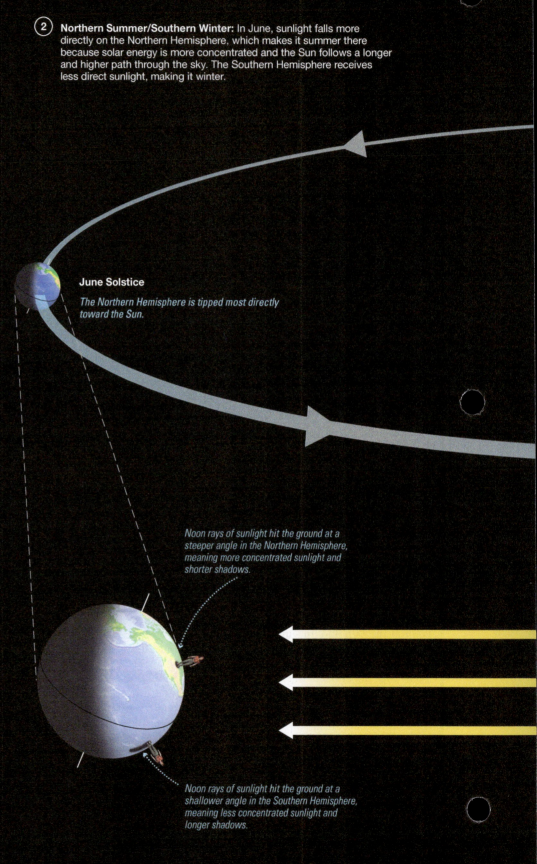

Earth's seasons are caused by the tilt of its rotation axis, which is why the seasons are opposite in the two hemispheres. The seasons do *not* depend on Earth's distance from the Sun, which varies only slightly throughout the year.

① **Axis Tilt:** Earth's axis points in the same direction throughout the year, which causes changes in Earth's orientation *relative to the Sun*.

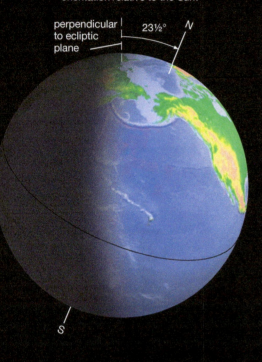

perpendicular to ecliptic plane

23½°

N

S

② **Northern Summer/Southern Winter:** In June, sunlight falls more directly on the Northern Hemisphere, which makes it summer there because solar energy is more concentrated and the Sun follows a longer and higher path through the sky. The Southern Hemisphere receives less direct sunlight, making it winter.

June Solstice

The Northern Hemisphere is tipped most directly toward the Sun.

Noon rays of sunlight hit the ground at a steeper angle in the Northern Hemisphere, meaning more concentrated sunlight and shorter shadows.

Noon rays of sunlight hit the ground at a shallower angle in the Southern Hemisphere, meaning less concentrated sunlight and longer shadows.

Interpreting the Diagram

To interpret the seasons diagram properly, keep in mind:

1. Earth's size relative to its orbit would be microscopic on this scale, meaning that both hemispheres are at essentially the same distance from the Sun.

2. The diagram is a side view of Earth's orbit. A top-down view (below) shows that Earth orbits in a nearly perfect circle and comes closest to the Sun in January.

March Equinox

147.1 million km

January 3

152.1 million km

July 4

September Equinox

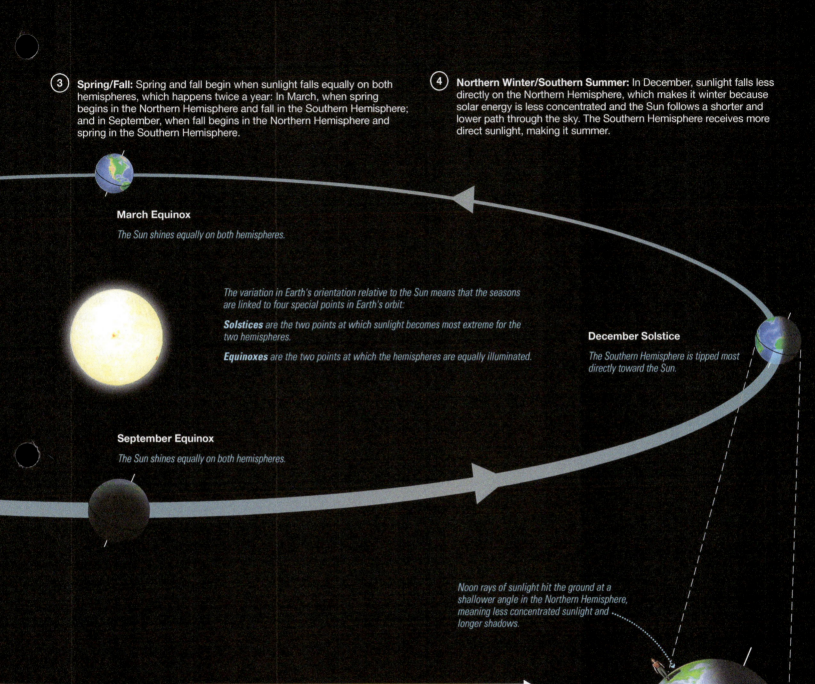

③ Spring/Fall: Spring and fall begin when sunlight falls equally on both hemispheres, which happens twice a year: In March, when spring begins in the Northern Hemisphere and fall in the Southern Hemisphere; and in September, when fall begins in the Northern Hemisphere and spring in the Southern Hemisphere.

④ Northern Winter/Southern Summer: In December, sunlight falls less directly on the Northern Hemisphere, which makes it winter because solar energy is less concentrated and the Sun follows a shorter and lower path through the sky. The Southern Hemisphere receives more direct sunlight, making it summer.

March Equinox

The Sun shines equally on both hemispheres.

The variation in Earth's orientation relative to the Sun means that the seasons are linked to four special points in Earth's orbit:

Solstices *are the two points at which sunlight becomes most extreme for the two hemispheres.*

Equinoxes *are the two points at which the hemispheres are equally illuminated.*

December Solstice

The Southern Hemisphere is tipped most directly toward the Sun.

September Equinox

The Sun shines equally on both hemispheres.

Noon rays of sunlight hit the ground at a shallower angle in the Northern Hemisphere, meaning less concentrated sunlight and longer shadows.

Noon rays of sunlight hit the ground at a steeper angle in the Southern Hemisphere, meaning more concentrated sunlight and shorter shadows.

distance varies over the course of each year, the variation is fairly small: Earth is only about 3% farther from the Sun at its farthest point (which is in July) than at its nearest (in January). The difference in the strength of sunlight due to this small change in distance is overwhelmed by the effects caused by the axis tilt. If Earth did not have an axis tilt, we would not have seasons.

think about it Jupiter has an axis tilt of about 3°, while Saturn has an axis tilt of about 27°. Both planets have nearly circular orbits around the Sun. Do you expect Jupiter to have seasons? Do you expect Saturn to have seasons? Explain.

Solstices and Equinoxes   To help us mark the changing seasons, we define four special moments in the year, each of which corresponds to one of the four special positions in Earth's orbit shown in Figure 2.15.

- The **June solstice,** called the *summer solstice* by people in the Northern Hemisphere, occurs around June 21 and is the moment when the Northern Hemisphere is tipped most directly toward the Sun and receives the most direct sunlight.

- The **December solstice,** called the *winter solstice* by people in the Northern Hemisphere, occurs around December 21 and is the moment when the Northern Hemisphere receives the least direct sunlight.

- The **March equinox,** called the *spring equinox* (or *vernal equinox*) by people in the Northern Hemisphere, occurs around March 21 and is the moment when the Northern Hemisphere goes from being tipped slightly away from the Sun to being tipped slightly toward the Sun.

- The **September equinox,** called the *fall equinox* (or *autumnal equinox*) by people in the Northern Hemisphere, occurs around September 22 and is the moment when the Northern Hemisphere first starts to be tipped away from the Sun.

The exact dates and times of the solstices and equinoxes can vary by up to a couple days from the dates given above, depending on where we are in the leap year cycle. In fact, our modern calendar includes leap years (usually adding one day—February 29—every fourth year) specifically to keep the solstices and equinoxes around the same dates; the leap year pattern is based on the fact that the true length of the year* is very close to 365¼ days.

We use the equinoxes and solstices to mark the progression of the seasons.

We can mark the dates of the equinoxes and solstices by observing changes in the Sun's path through our sky (Figure 2.16). The equinoxes occur on the only two days of the year on which the Sun rises precisely due east and sets precisely due west; these are also the two days when the Sun is above and below the horizon for equal times of 12 hours (*equinox* means "equal night"). The June solstice occurs on the day on which the Sun follows its longest and highest path through the Northern Hemisphere sky (and its shortest and lowest path

The Sun rises precisely due east and sets precisely due west only on the days of the March and September equinoxes.

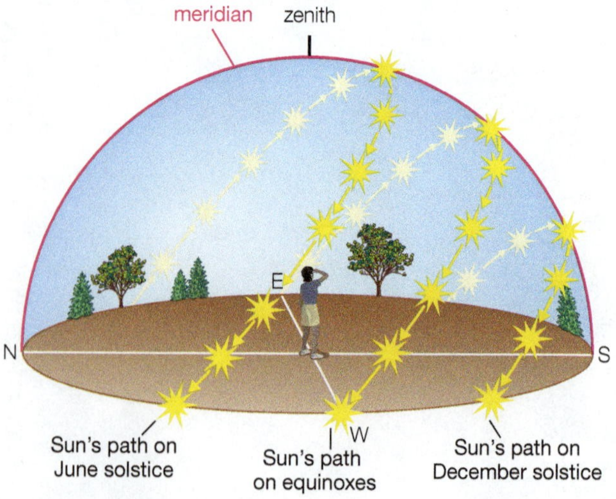

▲ **FIGURE 2.16**
This diagram shows the Sun's path on the solstices and equinoxes for a Northern Hemisphere sky (latitude 40°N). The precise paths are different for other latitudes; for example, at latitude 40°S, the paths look similar except tilted to the north rather than to the south. Notice that the Sun rises exactly due east and sets exactly due west only on the equinoxes.

---

*Technically, we are referring here to the *tropical year*—the time from one March equinox to the next. Axis precession (discussed later in this section) causes the tropical year to be slightly shorter (by about 20 minutes) than Earth's orbital period, called the *sidereal year*.

◀ **FIGURE 2.17**
This composite photograph shows images of the Sun taken at the same time of morning (technically, at the same "mean solar time") and from the same spot (over a large sundial in Carefree, Arizona) at 7- to 11-day intervals over the course of a year; the photo looks eastward, so north is to the left and south is to the right. Because this location is in the Northern Hemisphere, the Sun images that are high and to the north represent times near the June solstice and the images that are low and south represent times near the December solstice. The "figure 8" shape (called an *analemma*) arises from a combination of Earth's axis tilt and Earth's varying speed as it orbits the Sun.

through the Southern Hemisphere sky). It is therefore the day on which the Sun rises and sets farthest to the north of due east and due west; it is also the day on which the Northern Hemisphere has its longest hours of daylight and the Sun rises highest in the midday sky. The opposite is true on the day of the December solstice, when the Sun rises and sets farthest to the south and the Northern Hemisphere has its shortest hours of daylight and lowest midday Sun. Figure 2.17 shows how the Sun's position in the sky varies over the course of the year.

First Days of Seasons We usually say that each equinox and solstice marks the first day of a season. For example, the day of the June solstice is usually called the "first day of summer" in the Northern Hemisphere. Notice, however, that the Northern Hemisphere has its *maximum* tilt toward the Sun at this time. You might then wonder why we consider the solstice to be the beginning rather than the midpoint of summer.

The choice is somewhat arbitrary, but it makes sense in at least two ways. First, it was much easier for ancient people to identify the days on which the Sun reached extreme positions in the sky—such as when it reached its highest point on the summer solstice—than other days in between. Second, we usually think of the seasons in terms of weather, and the warmest summer weather tends to come 1 to 2 months after the solstice. To understand why, think about what happens when you heat a pot of cold soup. Even though you may have the stove turned on high from the start, it takes a while for the soup to warm up. In the same way, it takes some time for sunlight to heat the ground and oceans from the cold of winter to the warmth of summer. "Midsummer" in terms of weather therefore comes in late July or early August, which makes the June solstice a pretty good choice for the "first day of summer." Similar logic applies to the starting times for spring, fall, and winter.

Seasons Around the World The seasons have different characteristics in different parts of the world. High latitudes have more extreme seasons. For example, Vermont has much longer summer days and much longer winter nights than Florida. At the Arctic Circle (latitude 66½°), the Sun remains above the horizon all day long on the June solstice (Figure 2.18), and never rises on the December solstice (although

| Approximate time: | Midnight | 6:00 A.M. | Noon | 6:00 P.M. |
| Direction: | due north | due east | due south | due west |

▲ **FIGURE 2.18**

This sequence of photos shows the progression of the Sun around the horizon on the June solstice at the Arctic Circle. Notice that the Sun skims the northern horizon at midnight, then gradually rises higher, reaching its highest point when it is due south at noon.

bending of light by the atmosphere makes the Sun *appear* to be about a half-degree higher than it really is). The most extreme cases occur at the North and South Poles, where the Sun remains above the horizon for 6 months in summer and below the horizon for 6 months in winter.

Seasons also differ in equatorial regions, because the equator gets its most direct sunlight on the two equinoxes and its least direct sunlight on the solstices. As a result, instead of the four seasons experienced at higher latitudes, equatorial regions tend to have rainy and dry seasons, with the rainy seasons coming when the Sun is higher in the sky.

> **At very high latitudes, the summer Sun remains above the horizon all day long.**

◆ How does the orientation of Earth's axis change with time?

Our calendar keeps the solstices and equinoxes around the same dates each year, but the constellations associated with them change gradually over time. The reason is **precession,** a gradual wobble that alters the orientation of Earth's axis in space.

Precession occurs with many rotating objects. You can see it easily by spinning a top (Figure 2.19a). As the top spins rapidly, you'll notice that its axis also sweeps out a circle at a slower rate. We say that the top's axis *precesses.* Earth's axis precesses in much the same way, but far more slowly (Figure 2.19b). Each cycle of Earth's precession takes about 26,000 years. This gradually changes the direction in which the axis points in space.

think about it   Was Polaris the North Star in ancient times? Explain.

Note that precession does not change the *amount* of the axis tilt (which stays close to 23½°) and therefore does not affect the pattern of the seasons. However, it changes the points in Earth's orbit at which the solstices and equinoxes occur, and therefore changes the constellations that

> **The tilt of Earth's axis remains close to 23½°, but the direction the axis points in space changes slowly with the 26,000-year cycle of precession.**

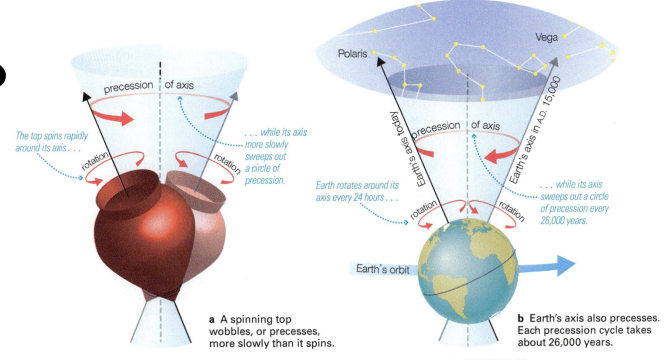

a A spinning top wobbles, or precesses, more slowly than it spins.

precession of axis

The top spins rapidly around its axis . . .

rotation

rotation

. . . while its axis more slowly sweeps out a circle of precession.

Polaris

Vega

Earth's axis today

precession of axis

Earth's axis in A.D. 15,000

Earth rotates around its axis every 24 hours . . .

rotation

rotation

. . . while its axis sweeps out a circle of precession every 26,000 years.

Earth's orbit

b Earth's axis also precesses. Each precession cycle takes about 26,000 years.

▲ **FIGURE 2.19**
Precession affects the orientation of a spinning object's axis, but not the amount of its tilt.

we see at those times. For example, a couple thousand years ago the June solstice occurred when the Sun appeared in the constellation Cancer, but it now occurs when the Sun appears in Gemini. This explains something you can see on any world map: The latitude at which the Sun is directly overhead on the June solstice (23½°N) is called the *Tropic of Cancer*, telling us that it was named back when the Sun appeared in Cancer on this solstice.

Precession is caused by gravity's effect on a tilted, rotating object. A spinning top precesses because Earth's gravity tries to pull over its lopsided, tilted spin axis. Gravity does not succeed in pulling it over—at least until friction slows the rate of spin—but instead causes the axis to precess. The spinning Earth precesses because gravitational tugs from the Sun and Moon try to "straighten out" our planet's bulging equator, which has the same tilt as the axis. Again, gravity does not succeed in straightening out the tilt but only causes the axis to precess.

2.3 The Moon, Our Constant Companion

Aside from the Sun, the Moon is the brightest and most noticeable object in our sky. The Moon is our constant companion in space, traveling with us as we orbit the Sun.

◆ Why do we see phases of the Moon?

As the Moon orbits Earth, it returns to the same position relative to the Sun in our sky (such as along the Earth–Sun line) about every 29½ days. This time period marks the cycle of **lunar phases,** in which the Moon's appearance in our sky changes as its position relative to the Sun

common misconceptions

Sun Signs

You probably know your astrological "Sun sign." When astrology began a few thousand years ago, your Sun sign was supposed to represent the constellation in which the Sun appeared on your birth date. However, because of precession, this is no longer the case for most people. For example, if your birthday is March 21, your Sun sign is Aries even though the Sun now appears in Pisces on that date. The problem is that astrological Sun signs are based on the positions of the Sun among the stars as they were almost 2000 years ago. Because Earth's axis has moved about 1/13 of the way through its 26,000-year precession cycle since that time, astrological Sun signs are off by nearly a month from the actual positions of the Sun among the constellations today.

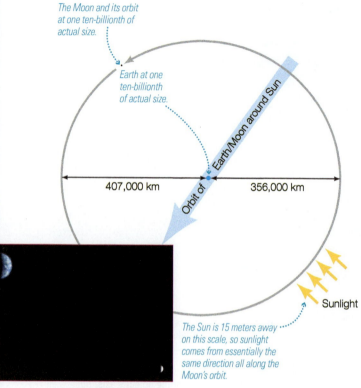

▲ **FIGURE 2.20**

The Moon's orbit on the 1-to-10-billion scale introduced in Chapter 1 (see Figure 1.5); black labels indicate the Moon's actual distances when it is nearest and farthest from Earth. The orbit is so small compared to the distance to the Sun that sunlight strikes the entire orbit from the same direction. You can see this in the inset photo, which shows the Moon and Earth photographed from Mars by the *Mars Reconnaissance Orbiter*.

common misconceptions

Shadows and the Moon

Many people guess that the Moon's phases are caused by Earth's shadow falling on its surface, but this is not the case. As we've seen, the Moon's phases are caused by the fact that we see different portions of its day and night sides at different times as it orbits around Earth. The only time Earth's shadow falls on the Moon is during the relatively rare event of a lunar eclipse.

changes. This 29½-day period is also the origin of the word *month* (think "moonth").*

Understanding Phases The first step in understanding phases is to recognize that sunlight essentially comes at both Earth and the Moon from the same direction. You can see why by studying Figure 2.20, which shows the Moon's orbit on the same scale we used for the model solar system in Chapter 1. Recall that the Sun is located 15 meters away from Earth and the Moon on this scale, which is far enough that the Sun would seem to be in almost precisely the same direction no matter whether you looked at it from Earth or from the Moon.

You can now understand the lunar phases with the simple demonstration illustrated in Figure 2.21. Take a ball outside on a sunny day. (If it's dark or cloudy, you can use a flashlight instead of the Sun; put the flashlight on a table a few meters away and shine it toward you.) Hold the ball at arm's length to represent the Moon while your head represents Earth. Slowly spin counterclockwise so that the ball goes around you the way the Moon orbits Earth. (If you live in the Southern Hemisphere, spin clockwise because you view the sky "upside down" compared to people in the Northern Hemisphere.) As you turn, you'll see the ball go through phases just like the Moon. If you think about what's happening, you'll realize that the phases of the ball result from just two basic facts:

The phase of the Moon depends on its position relative to the Sun as it orbits Earth.

1. Half the ball always faces the Sun (or flashlight) and therefore is bright, while the other half faces away from the Sun and is dark.

2. As you look at the ball at different positions in its "orbit" around your head, you see different combinations of its bright and dark faces.

For example, when you hold the ball directly opposite the Sun, you see only the bright portion of the ball, which represents the "full" phase. When you hold the ball at its "first-quarter" position, half the face you see is dark and the other half is bright.

We see lunar phases for the same reason. Half the Moon is always illuminated by the Sun, but the amount of this illuminated half that we see from Earth depends on the Moon's position in its orbit. The photographs in Figure 2.21 show how the phases look. (The new moon photo shows blue sky, because a new moon is nearly in line with the Sun and therefore hidden from view in the bright daytime sky.)

The Moon's phase also determines the times of day at which we see it in the sky. For example, the full moon must rise around sunset, because it occurs when the Moon is opposite the Sun in the sky. It therefore reaches its highest point in the sky at midnight and sets around sunrise. Similarly, a first-quarter moon must rise around noon, reach its highest point around sunset, and set around midnight, because it occurs when the Moon is about 90° east of the Sun in our sky.

The Moon's phase affects not only its appearance, but also its rise and set times.

think about it Suppose you go outside in the morning and notice that the visible face of the Moon is half-light and half-dark. Is this a first-quarter or third-quarter moon? How do you know?

---

*The cycle of phases is about two days longer than the Moon's actual orbital period (27⅓ days) because of Earth's motion around the Sun during this time; this reason is analogous to the reason why the solar day is longer than the sidereal day (see Special Topic, page 33).

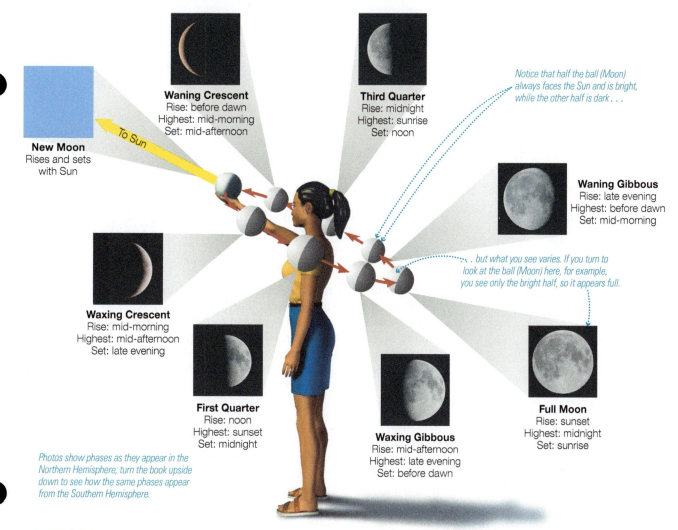

Notice that half the ball (Moon) always faces the Sun and is bright, while the other half is dark...

...but what you see varies. If you turn to look at the ball (Moon) here, for example, you see only the bright half, so it appears full.

New Moon
Rises and sets
with Sun

Waning Crescent
Rise: before dawn
Highest: mid-morning
Set: mid-afternoon

Third Quarter
Rise: midnight
Highest: sunrise
Set: noon

Waning Gibbous
Rise: late evening
Highest: before dawn
Set: mid-morning

Waxing Crescent
Rise: mid-morning
Highest: mid-afternoon
Set: late evening

First Quarter
Rise: noon
Highest: sunset
Set: midnight

Waxing Gibbous
Rise: mid-afternoon
Highest: late evening
Set: before dawn

Full Moon
Rise: sunset
Highest: midnight
Set: sunrise

To Sun

Photos show phases as they appear in the Northern Hemisphere; turn the book upside down to see how the same phases appear from the Southern Hemisphere.

▲ **FIGURE 2.21**
A simple demonstration illustrates the phases of the Moon. The half of the ball (Moon) facing the Sun is always illuminated while the half facing away is always dark, but you see the ball go through phases as it orbits around your head (Earth). The figure also shows the approximate times at which we see each phase as it moves across the sky; the exact times depend on your location, the time of year, and details of the Moon's orbit.

Notice that the phases from new to full are said to be *waxing*, which means "increasing." Phases from full to new are *waning*, or "decreasing." Also notice that no phase is called a "half moon." Instead, we see half the Moon's face at first-quarter and third-quarter phases; these phases mark the times when the Moon is one-quarter or three-quarters of the way through its monthly cycle (which begins at new moon). The phases just before and after new moon are called *crescent*, while those just before and after full moon are called *gibbous* (pronounced with a hard *g* as in "gift").

The Moon's Synchronous Rotation Although we see many *phases* of the Moon, we do not see many *faces*. From Earth we always see (nearly) the same face of the Moon. This happens because the Moon rotates on its axis in the same amount of time it takes to orbit Earth, a trait called

common misconceptions

Moon in the Daytime

Night is so closely associated with the Moon in traditions and stories that many people mistakenly believe that the Moon is visible only in the nighttime sky. In fact, the Moon is above the horizon as often in the daytime as at night, though it is easily visible only when its light is not drowned out by sunlight. For example, a first-quarter moon is easy to spot in the late afternoon as it rises through the eastern sky, and a third-quarter moon is visible in the morning as it heads toward the western horizon.

a If you do not rotate while walking around the model, you will not always face it.

b You will face the model at all times only if you rotate exactly once during each orbit.

▲ **FIGURE 2.22**

The fact that we always see the same face of the Moon means that the Moon must rotate once in the same amount of time it takes to orbit Earth once. You can see why by walking around a model of Earth while imagining that you are the Moon.

synchronous rotation. A simple demonstration shows the idea. Place a ball on a table to represent Earth while you represent the Moon (Figure 2.22). The only way you can face the ball at all times is by completing exactly one rotation while you complete one orbit. Note that the Moon's synchronous rotation is *not* a coincidence; it is a consequence of Earth's gravity affecting the Moon in much the same way the Moon's gravity causes tides on Earth.

The View from the Moon A good way to solidify your understanding of the lunar phases is to imagine that you live on the side of the Moon that faces Earth. For example, what would you see if you looked at Earth when people on Earth saw a new moon? By remembering that a new moon occurs when the Moon is between the Sun and Earth, you'll realize that from the Moon you'd be looking at Earth's daytime side and hence would see a *full earth*. Similarly, at full moon you would be facing the night side of Earth and would see a *new earth*. In general, you'd always see Earth in a phase opposite the phase of the Moon seen by people on Earth at the same time. Moreover, because the Moon always shows nearly the same face to Earth, Earth would appear to hang nearly stationary in your sky as it went through its cycle of phases.

think about it About how long would each day and night last if you lived on the Moon? Explain.

common misconceptions

The "Dark Side" of the Moon

Some people refer to the far side of the Moon—meaning the side that we never see from Earth—as the *dark side*. But this is not correct, because the far side is not always dark. For example, during new moon the far side faces the Sun and hence is completely sunlit. In fact, because the Moon rotates with a period of approximately one month (the same time it takes to orbit Earth), points on both the near and the far side have two weeks of daylight alternating with two weeks of darkness. The only time the far side is completely dark is at full moon, when it faces away from both the Sun and Earth.

◆ **What causes eclipses?**

Occasionally, the Moon's orbit around Earth causes events much more dramatic than lunar phases. The Moon and Earth both cast shadows in sunlight, and these shadows can create **eclipses** when the Sun, Earth, and Moon fall into a straight line. Eclipses come in two basic types:

- A **lunar eclipse** occurs when Earth lies directly between the Sun and the Moon, so Earth's shadow falls on the Moon.
- A **solar eclipse** occurs when the Moon lies directly between the Sun and Earth, so the Moon's shadow falls on Earth.

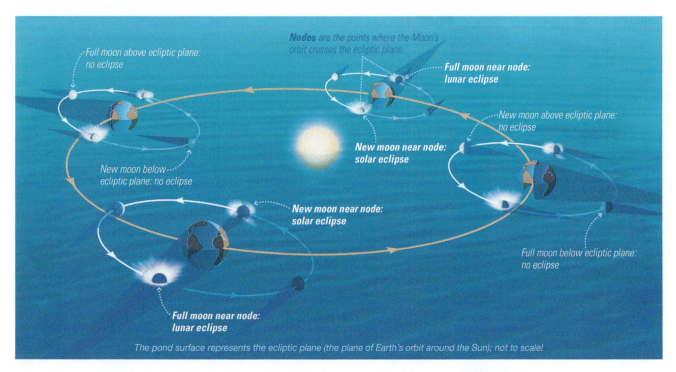

Nodes are the points where the Moon's orbit crosses the ecliptic plane.

Full moon above ecliptic plane: no eclipse

Full moon near node: lunar eclipse

New moon above ecliptic plane: no eclipse

New moon near node: solar eclipse

New moon below ecliptic plane: no eclipse

New moon near node: solar eclipse

Full moon below ecliptic plane: no eclipse

Full moon near node: lunar eclipse

The pond surface represents the ecliptic plane (the plane of Earth's orbit around the Sun); not to scale!

▲ **FIGURE 2.23**

This illustration represents the ecliptic plane as the surface of a pond. The Moon's orbit is tilted by about 5° to the ecliptic plane, so the Moon spends half of each orbit above the plane (the pond surface) and half below it. Eclipses occur only when the Moon is near a node (passing through the pond surface) *and* in a phase of either new moon (for a solar eclipse) or full moon (for a lunar eclipse)—as is the case with the lower left and top right orbits shown.

Conditions for Eclipses Look again at Figure 2.21. The figure makes it look as if the Sun, Earth, and Moon line up with every new and full moon. If this figure told the whole story of the Moon's orbit, we would have both a lunar and a solar eclipse every month—but we don't.

The missing piece of the story in Figure 2.21 is that the Moon's orbit is slightly inclined (by about 5°) to the ecliptic plane (the plane of Earth's orbit around the Sun). To visualize this inclination, imagine the ecliptic plane as the surface of a pond, as shown in Figure 2.23.

> We see a lunar eclipse when Earth's shadow falls on the Moon, and a solar eclipse when the Moon blocks our view of the Sun.

Because of the inclination of its orbit, the Moon spends most of its time either above or below this surface. It crosses *through* this surface only twice during each orbit: once coming out and once going back in. The two points in each orbit at which the Moon crosses the surface are called the **nodes** of the Moon's orbit.

Notice that the nodes are aligned approximately the same way throughout the year (diagonally in Figure 2.23), which means they lie along a nearly straight line with the Sun and Earth about twice each year. Eclipses can occur only during these periods, because these are the only times when the Moon can be directly in line with the Sun and Earth. In other words, eclipses can occur only when

1. the phase of the Moon is full (for a lunar eclipse) or new (for a solar eclipse) *and*

2. the new or full moon occurs at a time when the Moon is very close to a node.

Note also that while Figure 2.23 shows the Moon and Earth casting only simple "shadow cones" (extending away from the Sun), a closer look at the geometry shows that each shadow consists of two distinct regions (Figure 2.24): a central **umbra,** where sunlight is

> We see an eclipse only when a full or new moon occurs near one of the points where the Moon's orbit crosses the ecliptic plane.

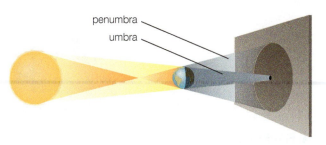

penumbra
umbra

▲ **FIGURE 2.24**

The shadow cast by an object in sunlight. Sunlight is fully blocked in the umbra and partially blocked in the penumbra.

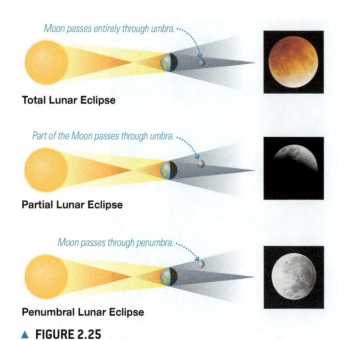

Total Lunar Eclipse

Moon passes entirely through umbra.

Partial Lunar Eclipse

Part of the Moon passes through umbra.

Penumbral Lunar Eclipse

Moon passes through penumbra.

▲ **FIGURE 2.25**

The three types of lunar eclipse.

A total solar eclipse occurs in the small central region.

Moon

path of total eclipse

A partial solar eclipse occurs in the lighter area surrounding the area of totality.

Moon

path of annular eclipse

If the Moon's umbral shadow does not reach Earth, an annular eclipse occurs in the small central region.

Total Solar Eclipse

Partial Solar Eclipse

Annular Solar Eclipse

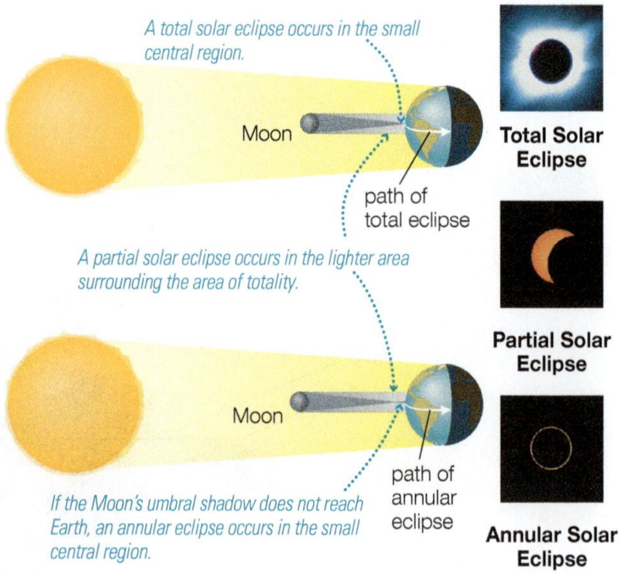

▲ **FIGURE 2.26**

The three types of solar eclipse. The diagrams show the Moon's shadow falling on Earth; note the dark central umbra surrounded by the much lighter penumbra.

completely blocked, and a surrounding **penumbra,** where sunlight is only partially blocked. Lunar and solar eclipses vary in appearance depending on which part of the shadow is involved.

Lunar Eclipses A lunar eclipse begins at the moment when the Moon's orbit first carries it into Earth's penumbra. After that, we will see one of three types of lunar eclipse (Figure 2.25). If the Sun, Earth, and Moon are nearly perfectly aligned, the Moon will pass through Earth's umbra and we see a **total lunar eclipse.** If the alignment is somewhat less perfect, only part of the full moon will pass through the umbra (with the rest in the penumbra) and we see a **partial lunar eclipse.** If the Moon passes *only* through Earth's penumbra, we see a **penumbral lunar eclipse.** Penumbral eclipses are the most common, but they are the least visually impressive because the full moon darkens only slightly.

Total lunar eclipses are the most spectacular. The Moon becomes dark and eerily red during **totality,** when the Moon is entirely engulfed in the umbra. Totality usually lasts about an hour, with partial phases both before and after. The curvature of Earth's shadow during partial phases shows that Earth is round. To understand the redness during totality, consider the view of an observer on the eclipsed Moon, who would see Earth's night side surrounded by the reddish glow of all the sunrises and sunsets occurring on the Earth at that moment. It is this reddish light that illuminates the Moon during total eclipse.

Solar Eclipses We can also see three types of solar eclipse (Figure 2.26). If a solar eclipse occurs when the Moon is in a part of its orbit where it is relatively close to Earth (see Figure 2.20), the Moon's umbra can cover a small area of Earth's surface (up to about 270 kilometers in diameter). Within this area you will see a **total solar eclipse.** If the eclipse occurs when the Moon is in a part of its orbit that puts it farther from Earth, the umbra may not reach Earth's surface, leading to an **annular eclipse**—a ring of sunlight surrounding the Moon—in the small region of Earth directly behind the umbra. In either case, the region of totality or annularity will be surrounded by a much larger region (typically about 7000 kilometers in diameter) that falls within the Moon's penumbral shadow. Here you will see a **partial solar eclipse,** in which only part of the Sun is blocked from view. (Some solar eclipses are partial only, meaning that no locations on Earth see a total or annular eclipse.) The combination of Earth's rotation and the Moon's orbital motion causes the Moon's shadow to race across the face of Earth at a typical speed of about 1700 kilometers per hour. As a result, the umbral shadow traces a narrow path across Earth, and totality never lasts more than a few minutes in any particular place.

A total solar eclipse is a spectacular sight. It begins when the disk of the Moon first appears to touch the Sun. Over the next couple of hours, the Moon appears to take a larger and larger "bite" out of the Sun. As totality approaches, the sky darkens and temperatures fall. Birds head back to their nests, and crickets begin their nighttime chirping. During the few minutes of totality, the Moon completely blocks the normally visible disk of the Sun, allowing the faint *corona* to be seen (Figure 2.27). The surrounding sky takes on a twilight glow, and planets and bright stars become visible in the daytime. As totality ends, the Sun slowly emerges from behind the Moon over the next couple of hours. However, because your eyes have adapted to the darkness, totality appears to end far more abruptly than it began.

A total solar eclipse is visible only within the narrow path that the Moon's umbral shadow makes across Earth's surface.

Predicting Eclipses Few phenomena have so inspired and humbled humans throughout the ages as eclipses. For many cultures, eclipses were mystical events associated with fate or the gods, and countless stories and legends surround them.

Much of the mystery of eclipses probably stems from the relative difficulty of predicting them. Look once more at Figure 2.23, focusing on the two periods—called *eclipse seasons*—in which the nodes of the Moon's orbit are closely aligned with the Sun. If the figure showed the whole story, these periods would always occur 6 months apart and predicting eclipses would be easy. For example, if the eclipse seasons occurred in January and July, eclipses would always occur on the dates of new and full moons in those months. Actual eclipse prediction is more difficult because of something the figure does not show: The nodes slowly move around the Moon's orbit, so the eclipse seasons occur slightly *less* than 6 months apart (about 173 days apart).

The combination of the changing dates of eclipse seasons and the 29½-day cycle of lunar phases makes eclipses recur in a cycle of about 18 years, 11⅓ days, called the **saros cycle.** Astronomers in many ancient cultures identified the saros cycle and used it to predict *when* eclipses would occur. However, they still could not predict eclipses in every detail (such as exactly where an eclipse would be visible and whether it would be total or partial).

Today, we can predict eclipses because we know the precise details of the orbits of Earth and the Moon. Table 2.1 lists upcoming lunar eclipses; notice that, as we expect,

The general pattern of eclipses repeats with the roughly 18-year saros cycle.

eclipses come a little less than 6 months apart. (On rare occasions, two lunar eclipses may occur in the same eclipse season and therefore come just a month apart.) Figure 2.28 shows paths of totality for upcoming

▲ **FIGURE 2.27**
This multiple-exposure photograph shows the progression of a total solar eclipse above a thorny acacia tree near Chisamba, Zambia. Totality (central image) lasts only a few minutes, during which time we can see the faint corona around the outline of the Sun.

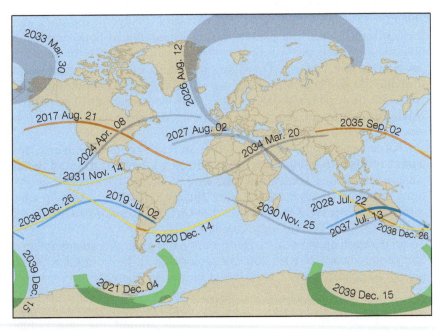

▲ **FIGURE 2.28**
This map shows the paths of totality for solar eclipses from 2017 through 2040. Paths of the same color represent eclipses occurring in successive saros cycles, separated by 18 years 11 days. For example, the 2039 eclipse occurs 18 years 11 days after the 2021 eclipse, both shown in green. Eclipse predictions by Fred Espenak, NASA GSFC. **Note:** Students in the United States should pay special attention to the eclipses of 2017 and 2024, the first total eclipses in the continental United States since 1979. It is worth traveling to make sure you are *in* the path of totality when the eclipse occurs.

TABLE 2.1	Lunar Eclipses 2017–2020*	
Date	**Type**	**Where You Can See It**
Feb. 11, 2017	penumbral	Americas, Europe, Africa, Asia
Aug. 7, 2017	partial	Europe, Africa, Asia, Australia
Jan. 31, 2018	total	Asia, Australia, Pacific, western N. America
July 27, 2018	total	S. America, Europe, Africa, Asia, Australia
Jan. 21, 2019	total	Pacific, Americas, Europe, Africa
Jul. 16, 2019	partial	S. America, Europe, Africa, Asia, Australia
Jan. 10, 2020	penumbral	Americas, Europe, Africa
Jun. 5, 2020	penumbral	Asia, Australia, Pacific, Americas

*Dates are based on Universal Time and hence are those in Greenwich, England, at the time of the eclipse; check a news source for the local time and date. Eclipse predictions by Fred Espenak, NASA GSFC.

total solar eclipses (but not for partial or annular eclipses), using color coding to show eclipses that repeat with the saros cycle.

2.4 The Ancient Mystery of the Planets

We've now covered the appearance and motion of the stars, Sun, and Moon in the sky. That leaves us with the planets to discuss. As you'll see, planetary motion posed an ancient mystery that played a critical role in the development of modern civilization.

Five planets are easy to find with the naked eye: Mercury, Venus, Mars, Jupiter, and Saturn. Mercury is visible infrequently, and only just after sunset or just before sunrise because it is so close to the Sun. Venus often shines brightly in the early evening in the west or before dawn in the east. If you see a very bright "star" in the early evening or early morning, it is probably Venus. Jupiter, when it is visible at night, is the brightest object in the sky besides the Moon and Venus. Mars is often recognizable by its reddish color, though you should check a star chart to make sure you aren't looking at a bright red star. Saturn is also easy to see with the naked eye, but because many stars are just as bright as Saturn, it helps to know where to look. (It also helps to know that planets tend not to twinkle as much as stars.)

◆ Why was planetary motion so hard to explain?

Over the course of a single night, planets behave like all other objects in the sky: Earth's rotation makes them appear to rise in the east and set in the west. But if you continue to watch the planets night after night, you will notice that their movements among the constellations are quite complex. Instead of moving steadily eastward relative to the stars, like the Sun and Moon, the planets vary substantially in both speed and brightness; in fact, the word *planet* comes from a Greek term meaning "wandering star." Moreover, while the planets *usually* move eastward through the constellations, they occasionally reverse course, moving westward through the zodiac (Figure 2.29). These periods of **apparent retrograde motion** (*retrograde* means "backward") last from a few weeks to a few months, depending on the planet.

For ancient people who believed in an Earth-centered universe, apparent retrograde motion was very difficult to explain. After all, what could make planets sometimes turn around and go backward if everything moves in circles around Earth? The ancient Greeks came up with some very clever ways to explain it, but their explanations (which we'll study in Chapter 3) were quite complex.

In contrast, apparent retrograde motion has a simple explanation in a Sun-centered solar system. You can demonstrate it for yourself with the help of a friend (Figure 2.30a). Pick a spot in an open area to represent the Sun. You can represent Earth by walking counterclockwise around the Sun, while your friend represents a more distant planet (such as Mars or Jupiter) by walking in the same direction around the Sun at a greater distance. Your friend should walk more slowly than you, because more distant planets orbit the Sun more slowly. As you walk, watch how your friend appears to move relative to buildings or trees in the distance. Although both of you always walk the same way around the Sun, your friend will appear to move backward against

A planet appears to move backward relative to the stars during the period when Earth passes it in its orbit.

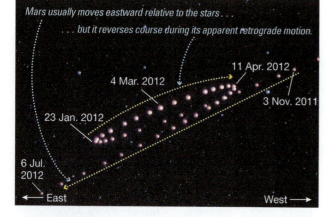

▲ **FIGURE 2.29**

This composite of images (taken at 5- to 7-day intervals in 2011 and 2012) shows a retrograde loop of Mars. Note that Mars is biggest and brightest in the middle of the retrograde loop, because that is where it is closest to Earth in its orbit.

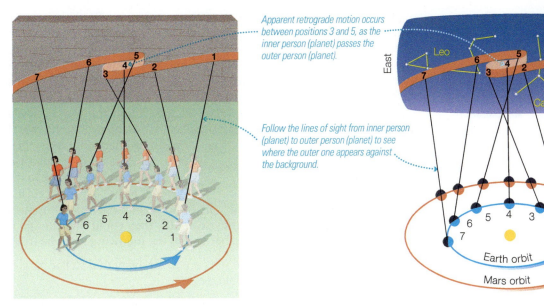

Apparent retrograde motion occurs between positions 3 and 5, as the inner person (planet) passes the outer person (planet).

Follow the lines of sight from inner person (planet) to outer person (planet) to see where the outer one appears against the background.

a The retrograde motion demonstration: Watch how your friend (in red) usually appears to move forward against the background of the building in the distance but appears to move backward as you (in blue) catch up to and pass her in your "orbit."

b This diagram shows the same idea applied to a planet. Follow the lines of sight from Earth to Mars in numerical order. Notice that Mars appears to move westward relative to the distant stars (from points 3 to 5) as Earth passes it by in its orbit.

▲ **FIGURE 2.30**
Apparent retrograde motion—the occasional "backward" motion of the planets relative to the stars—has a simple explanation in a Sun-centered solar system.

the background during the part of your "orbit" at which you catch up to and pass him or her. Figure 2.30b shows how the same idea applies to Mars. Note that Mars never actually changes direction; it only *appears* to go backward as Earth passes Mars in its orbit. (To understand the apparent retrograde motions of Mercury and Venus, which are closer to the Sun than is Earth, simply switch places with your friend and repeat the demonstration.)

◆ Why did the ancient Greeks reject the real explanation for planetary motion?

If the apparent retrograde motion of the planets is so readily explained by recognizing that Earth orbits the Sun, why wasn't this idea accepted in ancient times? In fact, the idea that Earth goes around the Sun was suggested as early as 260 B.C. by the Greek astronomer Aristarchus. No one knows why Aristarchus proposed a Sun-centered solar system, but the fact that it so naturally explains planetary motion probably played a role. Nevertheless, Aristarchus's contemporaries rejected his idea, and the Sun-centered solar system did not gain wide acceptance until almost 2000 years later.

Although there were many reasons why the Greeks were reluctant to abandon the idea of an Earth-centered universe, one of the most important was their inability to detect what we call **stellar parallax.** Extend your arm and hold up one finger. If you keep your finger still and alternately close your left eye and right eye, your finger will appear to jump back and forth against the background. This apparent shifting, called *parallax*, occurs because your two eyes view your finger from opposite sides of your nose. If you move your finger closer to your face, the parallax increases. If you look at a distant tree or flagpole instead of your finger, you may not notice any parallax at all. In other words, parallax depends on distance, with nearer objects exhibiting greater parallax than more distant objects.

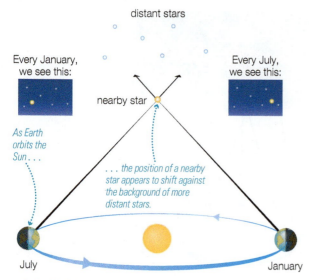

distant stars

Every January,
we see this:

Every July,
we see this:

nearby star

As Earth orbits the Sun . . .

. . . the position of a nearby star appears to shift against the background of more distant stars.

July

January

▲ **FIGURE 2.31**

Stellar parallax is an apparent shift in the position of a nearby star as we look at it from different places in Earth's orbit. This figure is greatly exaggerated; in reality, the amount of shift is far too small to detect with the naked eye.

If you now imagine that your two eyes represent Earth at opposite sides of its orbit around the Sun and that the tip of your finger represents a relatively nearby star, you have the idea of stellar parallax. Because we view the stars from different places in our orbit at different times of year, nearby stars should *appear* to shift back and forth against the background of more distant stars (Figure 2.31).

Because the Greeks believed that all stars lie on the same celestial sphere, they expected to see stellar parallax in a slightly different way. If Earth orbited the Sun, they reasoned, at different times of year we would be closer to different parts of the celestial sphere and would notice changes in the angular separations of stars. However, no matter how hard they searched, they could find no sign of stellar parallax. They concluded that one of the following must be true:

The Greeks knew that stellar parallax should occur if Earth orbits the Sun, but they could not detect it.

1. Earth orbits the Sun, but the stars are so far away that stellar parallax is not detectable to the naked eye.

2. There is no stellar parallax because Earth remains stationary at the center of the universe.

Aside from notable exceptions, such as Aristarchus, the Greeks rejected the correct answer (the first one) because they could not imagine that the stars could be *that* far away. Today, we can detect stellar parallax with the aid of telescopes, providing direct proof that Earth really does orbit the Sun. Careful measurements of stellar parallax also provide the most reliable means of measuring distances to nearby stars [Section 12.1].

think about it How far apart are opposite sides of Earth's orbit? How far away are the nearest stars? Using the 1-to-10-billion scale from Chapter 1, describe the challenge of detecting stellar parallax.

The ancient mystery of the planets drove much of the historical debate over Earth's place in the universe. In many ways, the modern technological society we take for granted today can be traced directly to the scientific revolution that began in the quest to explain the strange wanderings of the planets among the stars in our sky. We will turn our attention to this revolution in the next chapter.

the big picture Putting Chapter 2 into Perspective

In this chapter, we surveyed the phenomena of our sky. Keep the following "big picture" ideas in mind as you continue your study of astronomy:

- You can enhance your enjoyment of astronomy by observing the sky. The more you learn about the appearance and apparent motions of objects in the sky, the more you will appreciate what you can see in the universe.

- From our vantage point on Earth, it is convenient to imagine that we are at the center of a great celestial sphere—even though we really are on a planet orbiting a star in a vast universe. We can then understand what we see in the local sky by thinking about how the celestial sphere appears from our latitude.

- Most of the phenomena of the sky are relatively easy to observe and understand. The more complex phenomena—particularly eclipses and apparent retrograde motion of the planets—challenged our ancestors for thousands of years. The desire to understand these phenomena helped drive the development of science and technology.

my cosmic perspective Astronomy begins with an understanding of and appreciation for the motions of the Sun, Moon, planets, and stars in our sky. Studies of these patterns ultimately helped ignite the scientific and technological revolution upon which our modern civilization is based.

2.1 Patterns in the Night Sky

◆ What does the universe look like from Earth?

Stars and other celestial objects appear to lie on a great **celestial sphere** surrounding Earth. We divide the celestial sphere into **constellations** with well-defined borders. From any location on Earth, we see half the celestial sphere at any given time as the dome of our **local sky,** in which the **horizon** is the boundary between Earth and sky, the **zenith** is the point directly overhead, and the **meridian** runs from due south to due north through the zenith.

◆ Why do stars rise and set?

Earth's rotation makes stars appear to circle around Earth each day. A star whose complete circle lies above our horizon is said to be **circumpolar.** Other stars have circles that cross the horizon, so they rise in the east and set in the west each day.

◆ Why do the constellations we see depend on latitude and time of year?

The visible constellations vary with time of year because our night sky lies in different directions in space as we orbit the Sun. The constellations vary with **latitude** because your latitude determines the orientation of your horizon relative to the celestial sphere. The sky does not vary with **longitude.**

2.2 The Reason for Seasons

◆ What causes the seasons?

The tilt of Earth's axis causes the seasons. The axis points in the same direction throughout the year, so as Earth orbits the Sun, sunlight hits different parts of Earth more directly at different times of year.

◆ How does the orientation of Earth's axis change with time?

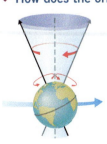

Earth's 26,000-year cycle of **precession** changes the orientation of its axis in space, although the tilt remains about 23½°. The changing orientation of the axis does not affect the pattern of seasons, but it changes the identity of the North Star and shifts the locations of the solstices and equinoxes in Earth's orbit.

2.3 The Moon, Our Constant Companion

◆ Why do we see phases of the Moon?

The phase of the Moon depends on its position relative to the Sun as it orbits Earth. The half of the Moon facing the Sun is always illuminated while the other half is dark, but from Earth we see varying combinations of the illuminated and dark halves.

◆ What causes eclipses?

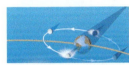

We see a **lunar eclipse** when Earth's shadow falls on the Moon and a **solar eclipse** when the Moon blocks our view of the Sun. We do not see an eclipse at every new and full moon because the Moon's orbit is slightly inclined to the ecliptic plane.

2.4 The Ancient Mystery of the Planets

◆ Why was planetary motion so hard to explain?

Planets generally appear to move eastward relative to the stars over the course of the year, but for weeks or months they reverse course during periods of **apparent retrograde motion.** This motion occurs when Earth passes by (or is passed by) another planet in its orbit, but it posed a major mystery to ancient people who assumed Earth to be at the center of the universe.

◆ Why did the ancient Greeks reject the real explanation for planetary motion?

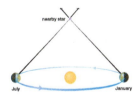

The Greeks rejected the idea that Earth goes around the Sun in part because they could not detect **stellar parallax**—slight apparent shifts in stellar positions over the course of the year. To most Greeks, it seemed unlikely that the stars could be so far away as to make parallax undetectable to the naked eye, even though that is, in fact, the case.

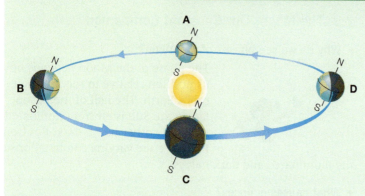

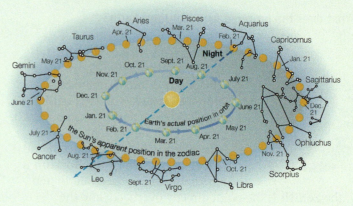

Use the diagram above to answer questions 1–5.

1. Which of the four labeled points (A through D) represents the day with the most hours of daylight for the Northern Hemisphere?

2. Which of the four labeled points represents the day with the most hours of daylight for the Southern Hemisphere?

3. Which of the four labeled points represents the beginning of spring for the Southern Hemisphere?

4. The diagram exaggerates the sizes of Earth and the Sun relative to the orbit. If Earth were correctly scaled relative to the orbit in the figure, how big would it be?
   a. about half the size shown
   b. about 2 millimeters across
   c. about 0.1 millimeter across
   d. microscopic

5. Given that Earth's actual distance from the Sun varies by less than 3% over the course of a year, why does the diagram look so elliptical?
   a. It correctly shows that Earth is closest to the Sun at points A and C and farthest at points B and D.
   b. The elliptical shape is an effect of perspective, since the diagram shows an almost edge-on view of a nearly circular orbit.
   c. The shape of the diagram is meaningless and is only for artistic effect.

Use the diagram above to answer questions 6–8.

6. As viewed from Earth, in which zodiac constellation does the Sun appear to be located on April 21?
   a. Leo   b. Aquarius
   c. Libra   d. Aries

7. If the date is April 21, what zodiac constellation will be visible on your meridian at midnight?
   a. Leo   b. Aquarius
   c. Libra   d. Aries

8. If the date is April 21, what zodiac constellation will you see setting in the west shortly after sunset?
   a. Scorpius   b. Pisces
   c. Taurus   d. Virgo

exercises and problems

MasteringAstronomy® For instructor-assigned homework and other learning materials, go to MasteringAstronomy®.

Review Questions

1. What are *constellations*? How did they get their names?
2. Suppose you were making a model of the celestial sphere with a ball. Briefly describe all the things you would need to mark on your celestial sphere.
3. On a clear, dark night, the sky may appear to be "full" of stars. Does this appearance accurately reflect the way stars are distributed in space? Explain.
4. Why does the *local sky* look like a dome? Define *horizon*, *zenith*, and *meridian*. How do we describe the location of an object in the local sky?
5. Explain why we can measure only *angular sizes* and *angular distances* for objects in the sky. What are *arcminutes* and *arcseconds*?
6. What are *circumpolar stars*? Are more stars circumpolar at the North Pole or in the United States? Explain.
7. What are *latitude* and *longitude*? Does the local sky vary with latitude? Does it vary with longitude? Explain.
8. What is the *zodiac*, and why do we see different parts of it at different times of year?
9. Suppose Earth's axis had no tilt. Would we still have seasons? Why or why not?
10. Briefly describe key facts about the solstices and equinoxes.

11. What is *precession*? How does it affect our view of the sky?
12. Briefly describe the Moon's cycle of *phases*. Can you ever see a full moon at noon? Explain.
13. Why do we always see the same face of the Moon?
14. Why don't we see an *eclipse* at every new and full moon? Describe the conditions needed for a *solar* or *lunar eclipse*.
15. What do we mean by the *apparent retrograde motion* of the planets? Why was it difficult for ancient astronomers to explain but easy for us to explain?
16. What is *stellar parallax*? How did an inability to detect it support the ancient belief in an Earth-centered universe?

Test Your Understanding

◐ Does It Make Sense?

Decide whether the statement makes sense (or is clearly true) or does not make sense (or is clearly false). Explain clearly; not all of these have definitive answers, so your explanation is more important than your chosen answer.

17. The constellation Orion didn't exist when my grandfather was a child.
18. When I looked into the dark lanes of the Milky Way with my binoculars, I saw a cluster of distant galaxies.
19. Last night the Moon was so big that it stretched for a mile across the sky.
20. I live in the United States, and during a trip to Argentina I saw many constellations that I'd never seen before.
21. Last night I saw Jupiter in the middle of the Big Dipper. (*Hint*: Is the Big Dipper part of the zodiac?)
22. Last night I saw Mars move westward through the sky in its apparent retrograde motion.
23. Although all the known stars rise in the east and set in the west, we might someday discover a star that will rise in the west and set in the east.
24. If Earth's orbit were a perfect circle, we would not have seasons.
25. Because of precession, someday it will be summer everywhere on Earth at the same time.
26. This morning I saw the full moon setting at about the same time the Sun was rising.

Quick Quiz

Choose the best answer to each of the following. Explain your reasoning with one or more complete sentences.

27. Two stars that are in the same constellation (a) must both be part of the same cluster of stars in space. (b) must both have been discovered at about the same time. (c) may actually be very far away from each other.
28. The north celestial pole is 35° above your northern horizon. This tells you that you are at (a) latitude 35°N. (b) longitude 35°E. (c) latitude 35°S.
29. Beijing and Philadelphia have about the same latitude but different longitudes. Therefore, tonight's night sky in these two places will (a) look about the same. (b) have completely different sets of constellations. (c) have partially different sets of constellations.
30. In winter, Earth's axis points toward the star Polaris. In spring, the axis points toward (a) Polaris. (b) Vega. (c) the Sun.
31. When it is summer in Australia, the season in the United States is (a) winter. (b) summer. (c) spring.

32. If the Sun rises precisely due east, (a) you must be located at Earth's equator. (b) it must be the day of either the March or the September equinox. (c) it must be the day of the June solstice.
33. A week after full moon, the Moon's phase is (a) first quarter. (b) third quarter. (c) new.
34. The fact that we always see the same face of the Moon tells us that the Moon (a) does not rotate. (b) rotates with the same period that it orbits Earth. (c) looks the same on both sides.
35. If there is going to be a total lunar eclipse tonight, then you know that (a) the Moon's phase is full. (b) the Moon's phase is new. (c) the Moon is unusually close to Earth.
36. When we see Saturn going through a period of apparent retrograde motion, it means (a) Saturn is temporarily moving backward in its orbit of the Sun. (b) Earth is passing Saturn in its orbit, with both planets on the same side of the Sun. (c) Saturn and Earth must be on opposite sides of the Sun.

◐ Process of Science

37. *Earth-Centered or Sun-Centered?* For each of the following, decide whether the phenomenon is consistent or inconsistent with a belief in an Earth-centered system. If consistent, describe how. If inconsistent, explain why, and also explain why the inconsistency did not immediately lead people to abandon the Earth-centered model.
    a. The daily paths of stars through the sky
    b. Seasons
    c. Phases of the Moon
    d. Eclipses
    e. Apparent retrograde motion of the planets
38. *Shadow Phases.* Many people incorrectly guess that the phases of the Moon are caused by Earth's shadow falling on the Moon. How would you convince a friend that the phases of the Moon have nothing to do with Earth's shadow? Describe the observations you would use to show that Earth's shadow isn't the cause of phases.

Group Work Exercise

39. *Lunar Phases and Time of Day.* **Roles:** *Scribe* (takes notes on the group's activities), *Proposer* (proposes explanations to the group), *Skeptic* (points out weaknesses in proposed explanations), *Moderator* (leads group discussion and makes sure everyone contributes). **Activity:** The diagram below represents the Moon's orbit as seen from above Earth's North Pole (not to scale). Each group member should draw a copy of the diagram and label it as you work together on the following questions.

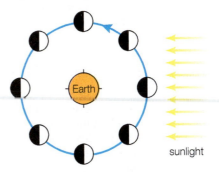

sunlight

a. How would the Moon appear from Earth at each of the eight Moon positions? Label each one with the corresponding phase.

b. What time of day corresponds to each of the four tick marks on Earth? Label each tick mark accordingly.

c. Why doesn't the Moon's phase change during the course of one night? Explain your reasoning.

d. At what times of day would a full moon be visible to someone standing on Earth? Write down when a full moon rises and explain why it appears to rise at that time.

e. At what times of day would a third-quarter moon be visible to someone standing on Earth? Write down when a third-quarter moon sets and explain why it appears to set at that time.

f. At what times of day would a waxing crescent moon be visible to someone standing on Earth? Write down when a waxing crescent moon rises and explain why it appears to rise at that time.

Investigate Further

Short-Answer/Essay Questions

40. *New Planet.* A planet in another solar system has a circular orbit and an axis tilt of 35°. Would you expect this planet to have seasons? If so, would you expect them to be more extreme than the seasons on Earth? If not, why not?

41. *Your View of the Sky.*
    a. What are your latitude and longitude?
    b. Where does the north (or south) celestial pole appear in your sky?
    c. Is Polaris a circumpolar star in your sky? Explain.

42. *View from the Moon.* Suppose you lived on the Moon, in which case you would see Earth going through phases in your sky. Assume you live near the center of the face that looks toward Earth.
    a. Suppose you see a full Earth in your sky. What phase of the Moon would people on Earth see? Explain.
    b. Suppose people on Earth see a full moon. What phase would you see for Earth? Explain.
    c. Suppose people on Earth see a waxing gibbous moon. What phase would you see for Earth? Explain.
    d. Suppose people on Earth are viewing a total lunar eclipse. What would you see from your home on the Moon? Explain.

43. *View from the Sun.* Suppose you lived on the Sun (and could ignore the heat). Would you still see the Moon go through phases as it orbits Earth? Why or why not?

44. *Farther Moon.* Suppose the distance to the Moon were twice its actual value. Would it still be possible to have a total solar eclipse? Why or why not?

45. *Smaller Earth.* Suppose Earth were smaller. Would solar eclipses be any different? What about lunar eclipses? Explain.

46. *Observing Planetary Motion.* Find out what planets are currently visible in your evening sky. At least once a week, observe these planets and draw a diagram showing their positions relative to stars in a zodiac constellation. How long does it take to notice the "wandering" features of planetary motion? Explain.

Quantitative Problems

Be sure to show all calculations clearly and state your final answers in complete sentences.

47. *Arcminutes and Arcseconds.* There are 360° in a full circle.
    a. How many arcminutes are in a full circle?
    b. How many arcseconds are in a full circle?
    c. The Moon's angular size is about ½°. What is this in arcminutes? In arcseconds?

48. *Find the Sun's Diameter.* The Sun has an angular diameter of about 0.5° and an average distance from Earth of about 150 million km. What is the Sun's approximate physical diameter? Compare your answer to the actual value of 1,390,000 km.

49. *Find a Star's Diameter.* Estimate the diameter of the supergiant star Betelgeuse, using its measured angular diameter of about 0.5 arcsecond and distance of about 600 light-years. Compare your answer to the size of our Sun and the Earth–Sun distance.

50. *Eclipse Conditions.* The Moon's precise equatorial diameter is 3476 km, and its orbital distance from Earth varies between 356,400 km and 406,700 km. The Sun's diameter is 1,390,000 km, and its distance from Earth ranges between 147.5 and 152.6 million km.
    a. Find the Moon's angular size at its minimum and maximum distances from Earth.
    b. Find the Sun's angular size at its minimum and maximum distances from Earth.
    c. Based on your answers to (a) and (b), is it possible to have a total solar eclipse when the Moon and Sun are both at their maximum distances? Explain.

Discussion Questions

51. *Earth-Centered Language.* Many common phrases reflect the ancient Earth-centered view of our universe. For example, the phrase "the Sun rises each day" implies that the Sun is really moving over Earth. We know that the Sun only *appears* to rise as the rotation of Earth carries us to a place where we can see the Sun in our sky. Identify other common phrases that imply an Earth-centered viewpoint.

52. *Flat Earth Society.* Believe it or not, there is an organization called the Flat Earth Society. Its members hold that Earth is flat and that all indications to the contrary (such as pictures of Earth from space) are fabrications made as part of a conspiracy to hide the truth from the public. Discuss the evidence for a round Earth and how you can check it for yourself. In light of the evidence, is it possible that the Flat Earth Society is correct? Defend your opinion.

Web Projects

53. *Sky Information.* Search the Web for sources of daily information about sky phenomena (such as lunar phases, times of sunrise and sunset, or dates of equinoxes and solstices). Identify and briefly describe your favorite source.

54. *Constellations.* Search the Web for information about the constellations and their mythology. Write a one- to three-page report about one or more constellations.

55. *Eclipse Trip.* Find details about an upcoming total solar eclipse that you may be able to observe. Create a plan for a trip to see the eclipse, including details of where you will view it and how you will get there, and describe what you should expect to see. Bonus: Describe how you could photograph the eclipse.

3 The Science of Astronomy

Astronaut Bruce McCandless orbits Earth as if he were a tiny moon during Space Shuttle mission STS-41-B.

LEARNING GOALS

3.1 The Ancient Roots of Sciences
- In what ways do all humans use scientific thinking?
- How is modern science rooted in ancient astronomy?

3.2 Ancient Greek Science
- Why does modern science trace its roots to the Greeks?
- How did the Greeks explain planetary motion?

3.3 The Copernican Revolution
- How did Copernicus, Tycho, and Kepler challenge the Earth-centered model?
- What are Kepler's three laws of planetary motion?
- How did Galileo solidify the Copernican revolution?

3.4 The Nature of Science
- How can we distinguish science from nonscience?
- What is a scientific theory?

Today we know that Earth is a planet orbiting a rather ordinary star, in a galaxy of more than a hundred billion stars, in an incredibly vast universe. We know that Earth, along with the entire cosmos, is in constant motion. We know that, on the scale of cosmic time, human civilization has existed for only the briefest moment. How did we manage to learn these things?

It wasn't easy. In this chapter, we will trace how modern astronomy grew from its roots in ancient observations, including those of the Greeks. We'll discuss the Copernican revolution, which overturned the ancient belief in an Earth-centered universe and laid the foundation for the rise of our technological civilization. Finally, we'll explore the nature of modern science and how science can be distinguished from nonscience.

3.1 The Ancient Roots of Science

The rigorous methods of modern science have proven to be one of the most valuable inventions in human history. These methods have enabled us to discover almost everything we now know about nature and the universe, and they also have made our modern technology possible. In this section, we will explore the ancient roots of science, which grew out of experiences common to nearly all people and all cultures.

◆ In what ways do all humans use scientific thinking?

Scientific thinking comes naturally to us. By about a year of age, a baby notices that objects fall to the ground when she drops them. She lets go of a ball—it falls. She pushes a plate of food from her high chair—it falls, too. She continues to drop all kinds of objects, and they all plummet to Earth. Through her powers of observation, the baby learns about the physical world, finding that things fall when they are unsupported. Eventually, she becomes so certain of this fact that, to her parents' delight, she no longer needs to test it continually.

One day someone gives the baby a helium balloon. She releases it, and to her surprise it rises to the ceiling! Her understanding of nature must be revised. She now knows that the principle "all things fall" does not represent the whole truth, although it still serves her quite well in most situations. It will be years before she learns enough about the atmosphere, the force of gravity, and the concept of density to understand *why* the balloon rises when most other objects fall. For now, she is delighted to observe something new and unexpected.

The baby's experience with falling objects and balloons exemplifies scientific thinking. In essence, science is a way of learning about nature **Scientific thinking is based on everyday observations and trial-and-error experiments.** through careful observation and trial-and-error experiments. Rather than thinking differently than other people, modern scientists simply are trained to organize everyday thinking in a way that makes it easier for them to share their discoveries and use their collective wisdom.

think about it Describe a few cases where you have learned by trial and error—while cooking, participating in sports, fixing something, learning on the job, or in any other situation.

Just as learning to communicate through language, art, or music is a gradual process for a child, the development of science has been a gradual process for humanity. Science in its modern form requires painstaking attention to detail, relentless testing of each piece of information to ensure its reliability, and a willingness to give up old beliefs that are not consistent with observed facts about the physical world. For professional scientists, these demands are the "hard work" part of the job. At heart, professional scientists are like the baby with the balloon, delighted by the unexpected and motivated by those rare moments when they—and all of us—learn something new about the universe.

◆ How is modern science rooted in ancient astronomy?

Astronomy has been called the oldest of the sciences, because its roots stretch deepest into antiquity. Ancient civilizations did not always practice astronomy in the same ways or for the same reasons that we study it today, but they nonetheless had some amazing achievements. Understanding this ancient astronomy can give us a greater appreciation of how and why science developed through time.

Practical Benefits of Astronomy Humans have been making careful observations of the sky for many thousands of years. Part of the reason for this interest in astronomy probably comes from our inherent curiosity as humans, but ancient cultures also discovered that astronomy had practical benefits for timekeeping, keeping track of seasonal changes, and navigation.

One amazing example comes from people of central Africa. Although we do not know exactly when they developed the skill, people in some regions learned to predict rainfall patterns through observations of the Moon. Figure 3.1 shows

Ancient people used observations of the sky to keep track of the time and seasons and as an aid in navigation.

how the method works. The orientation of the "horns" of a waxing crescent moon (relative to the horizon) varies over the course of the year, primarily because the angle at which the ecliptic intersects the horizon changes during the

▼ **FIGURE 3.1**

In central Nigeria, the orientation of the "horns" of a waxing crescent moon (shown along the top) correlates with the average amount of rainfall at different times of year. Local people could use this fact to predict the weather with reasonable accuracy. (Adapted from *Ancient Astronomers* by Anthony F. Aveni.)

TABLE 3.1 The Seven Days of the Week and the Astronomical Objects They Honor

The seven days were originally linked directly to the seven objects. The correspondence is no longer perfect, but the pattern is clear in many languages; some English names come from Germanic gods.

Object	Germanic God	English	French	Spanish
Sun	—	Sunday	dimanche	domingo
Moon	—	Monday	lundi	lunes
Mars	Tiw	Tuesday	mardi	martes
Mercury	Woden	Wednesday	mercredi	miércoles
Jupiter	Thor	Thursday	jeudi	jueves
Venus	Fria	Friday	vendredi	viernes
Saturn	—	Saturday	samedi	sábado

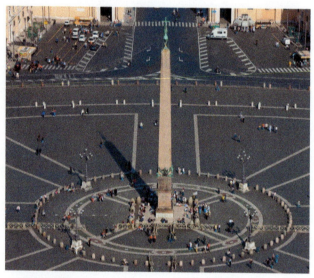

▲ **FIGURE 3.2**

This ancient Egyptian obelisk resides in St. Peter's Square at the Vatican in Rome; it is one of 21 surviving Egyptian obelisks. Shadows cast by the obelisks may have been used to tell time.

▲ **FIGURE 3.3**

The remains of Stonehenge, which was built in stages from about 2750 B.C. to about 1550 B.C.

year. (The orientation also depends on latitude.) In tropical regions in which there are distinct rainy and dry seasons—rather than the four seasons familiar at temperate latitudes—the orientation of the crescent moon can be used to predict how much rainfall should be expected over coming days and weeks.

Astronomy and Measures of Time The impact of ancient astronomical observations is still with us in our modern measures of time. The length of our day is the time it takes the Sun to make one full circuit of the sky. The length of a month comes from the Moon's cycle of phases [Section 2.3], and our year is based on the cycle of the seasons [Section 2.2]. The seven days of the week were named after the seven "planets" of ancient times (Table 3.1), which were the Sun, the Moon, and the five planets that are easily visible to the naked eye: Mercury, Venus, Mars, Jupiter, Saturn. Note that the ancient definition of *planet* (which meant "wandering star") applied to any object that appeared to wander among the fixed stars. That is why the Sun and Moon were on the list while Earth was not, because we don't see our own planet moving in the sky.

think about it Uranus is faintly visible to the naked eye, but it was not recognized as a planet in ancient times. If Uranus had been brighter, would we now have eight days in a week? Defend your opinion.

Because timekeeping was so important and required precise observations, many ancient cultures built structures or created special devices to help with it. Let's briefly investigate a few of the ways that ancient cultures learned to keep track of time.

The seven days of the week are named for the seven "planets" known in ancient times.

Determining the Time of Day In the daytime, ancient peoples could tell time by observing the Sun's path through the sky. Many cultures probably used the shadows cast by sticks as simple sundials. The ancient Egyptians built huge obelisks, often decorated in homage to the Sun, which probably also served as simple clocks (Figure 3.2). At night, ancient people could estimate the time from the position and phase of the Moon (see Figure 2.21) or by observing the constellations visible at a particular time (see Figure 2.14).

We also trace the origins of our modern clock to ancient Egypt. Some 4000 years ago, the Egyptians divided daytime and nighttime into 12 equal parts each, which is how we got our 12 hours each of a.m. and p.m. The abbreviations a.m. and p.m. stand for the Latin terms *ante meridiem* and *post meridiem*, respectively, which mean "before the middle of the day" and "after the middle of the day."

Marking the Seasons Many ancient cultures built structures to help them mark the seasons. Stonehenge (Figure 3.3) is a well-known example that served both as an astronomical device and as a social and religious gathering place. In the Americas, one of the most spectacular structures was the Templo Mayor (Figure 3.4) in the Aztec city of Tenochtitlán (in modern-day Mexico City), which featured twin temples on a flat-topped pyramid. From the vantage point of a royal observer watching from the opposite side of the plaza, the Sun rose through the notch between the temples on the equinoxes.

Many cultures aligned buildings and streets with the cardinal directions (north, south, east, and west), which made it easier to keep track of the rise and set positions of the Sun over the course of the year. Other structures marked special dates. For example, the Ancestral Pueblo People carved a spiral—known as the *Sun Dagger*—on a cliff face in Chaco Canyon, New Mexico (Figure 3.5). The Sun's rays formed a dagger of sunlight that pierced the center of the carved spiral only at noon on the summer solstice.

Solar and Lunar Calendars   The tracking of the seasons eventually led to the advent of written calendars. Today, we use a *solar calendar*, meaning a calendar that is synchronized with the seasons so that seasonal events such as the solstices and equinoxes occur on approximately the same dates each year. However, recall that the length of our month comes from the Moon's 29½-day cycle of phases. Some cultures therefore created *lunar calendars* that aimed to stay synchronized with the lunar cycle, so that the Moon's phase was always the same on the first day of each month.

A basic lunar calendar has 12 months, with some months lasting 29 days and others lasting 30 days; the lengths are chosen to make the average agree with the approximately 29½-day lunar cycle. A 12-month lunar calendar therefore has 354 or 355 days, or about 11 days fewer than a calendar based on the Sun. Such a calendar is still used in the Muslim religion. That is why the month-long fast of Ramadan (the ninth month) begins about 11 days earlier with each subsequent year.

A lunar calendar always has the same moon phase on the first day of each month.

It's possible to keep lunar calendars roughly synchronized with solar calendars by taking advantage of a timing coincidence: 19 years on a solar calendar is almost precisely 235 months on a lunar calendar. As a result, the lunar phases repeat on the same solar dates about every 19 years (a pattern known as the *Metonic cycle*). For example, there was a full moon on February 11, 2017, and there will be a full moon 19 years later, on February 11, 2036. Because an ordinary lunar calendar has only $19 \times 12 = 228$ months in a 19-year period, adding 7 extra months (to make 235) can keep the lunar calendar roughly synchronized to the seasons. The Jewish calendar does this by adding a thirteenth month in the third, sixth, eighth, eleventh, fourteenth, seventeenth, and nineteenth years of each 19-year cycle.

Learning About Ancient Achievements   The study of ancient astronomical achievements is a rich field of research. Many ancient cultures made careful observations of planets and stars, and some left remarkably detailed records. The Chinese, for example, began recording astronomical observations at least 5000 years ago, allowing ancient Chinese astronomers to make many important discoveries.

Other cultures either did not leave clear written records or had records that were lost or destroyed, so we must piece together their astronomical achievements by studying the physical evidence they left behind. This type of study is usually called *archaeoastronomy*, a word that combines archaeology and astronomy.

Archaeoastronomy is the study of astronomical uses of ancient structures.

The cases we've discussed to this point have been fairly straightforward for archaeoastronomers to interpret, but many other cases are more ambiguous. For example, ancient people in what is now Peru etched hundreds of lines and patterns in the sand of the Nazca desert. Many of the lines point to places where the Sun or bright stars rise at particular times of year, but that doesn't prove anything: With hundreds of lines, random chance ensures that many will have astronomical alignments no matter how or why they were made. The patterns, many of which are large figures of animals (Figure 3.6), have evoked even more debate. Some people think they may be representations of constellations

▲ **FIGURE 3.4**
This scale model shows the Templo Mayor and the surrounding plaza as they are thought to have looked five centuries ago.

▲ **FIGURE 3.5**
The Sun Dagger. Three large slabs of rock are arranged so that a dagger of sunlight pierced the carved spiral only at noon on the summer solstice. (Unfortunately, within just 12 years of the site's 1977 discovery, erosion caused the rocks to shift so that the effect no longer occurs.)

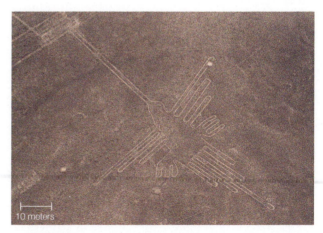

▲ **FIGURE 3.6**
Hundreds of lines and patterns are etched in the sand of the Nazca desert in Peru. This aerial photo shows a large figure of a hummingbird.

▲ **FIGURE 3.7**
The World Heritage Site of Machu Picchu has structures aligned with sunrise at the winter and summer solstices.

▲ **FIGURE 3.8**
Polynesian Navigators used astronomy and ocean swell patterns to navigate among Pacific islands. Here, we see a Micronesian stick chart, an instrument used to represent swell patterns around islands.

recognized by the people who lived in the region, but we do not know for sure.

In some cases, scientists can use other clues to establish the intentions of ancient builders. For example, traditions of the Inca Empire of South America held that its rulers were descendants of the Sun and therefore demanded that movements of the Sun be watched closely. This fact supports the idea that astronomical alignments in Inca cities and ceremonial centers, such as the World Heritage Site of Machu Picchu (Figure 3.7), were deliberate rather than accidental.

A different type of evidence makes a convincing case for the astronomical sophistication of ancient Polynesians, who lived and traveled among the many islands of the mid- and South Pacific. Navigation was crucial to survival, because the next island in a journey usually was too distant to be seen. The most esteemed position in Polynesian culture was that of the Navigator, a person who had acquired the knowledge necessary to navigate great distances among the islands. Navigators used a detailed knowledge of astronomy for their broad navigational sense, and a deep understanding of wave and swell patterns to locate precise landing points (Figure 3.8).

3.2 Ancient Greek Science

Before a structure such as Stonehenge or the Templo Mayor could be built, careful observations had to be made and repeated over and over to ensure their accuracy. Careful, repeatable observations also underlie modern science. Elements of modern science were therefore present in many early human cultures. If the circumstances of history had been different, almost any culture might have been the first to develop what we consider to be modern science. In the end, however, history takes only one of countless possible paths. The path that led to modern science emerged from the ancient civilizations of the Mediterranean and the Middle East—especially from ancient Greece.

◆ Why does modern science trace its roots to the Greeks?

Greece gradually rose as a power in the Middle East beginning around 800 B.C. and was well established by about 500 B.C. Its geographical location placed it at a crossroads for travelers, merchants, and armies from northern Africa, Asia, and Europe. Building on the diverse ideas brought forth by the meeting of these many cultures, ancient Greek philosophers soon began their efforts to move human understanding of nature from the mythological to the rational.

Three Philosophical Innovations Greek philosophers developed at least three major innovations that helped pave the way for modern science. First, they developed a tradition of trying to understand nature without relying on supernatural explanations, and of working communally to debate and challenge each other's ideas. Second, the Greeks used mathematics to give precision to their ideas, which allowed them to explore the implications of new ideas in much greater depth than would have otherwise been possible. Third, while much of their philosophical activity consisted of subtle debates grounded only in thought and was not scientific in the modern sense, the Greeks also saw the power of reasoning from observations. They understood that an explanation could not be right if it disagreed with observed facts.

Models of Nature Perhaps the greatest Greek contribution to science came from the way they synthesized all three of the above innovations in creating models of nature, a practice that is central to modern science. Scientific models differ somewhat from the models you may be familiar with in everyday life. In our daily lives, we tend to think of models as miniature physical representations, such as model cars or airplanes. In contrast, a scientific **model** is a conceptual representation created to explain and predict observed phenomena.

For example, a scientific model of Earth's climate uses logic and mathematics to represent what we know about how the climate works. Its purpose is to explain and predict climate changes, such as the changes that may occur with global warming. Just as a model airplane does not faithfully represent every aspect of a real airplane, a scientific model may not fully explain our observations of nature. Nevertheless, even the failings of a scientific model can be useful, because they often point the way toward building a better model.

> The Greeks developed models of nature that aimed to explain and predict observed phenomena.

In astronomy, the Greeks constructed conceptual models of the universe in an attempt to explain what they observed in the sky, an effort that quickly led them past simplistic ideas of a flat Earth under a dome-shaped sky. We do not know precisely when other Greeks began to think that Earth is round, but this idea was being taught by about 500 B.C. by the famous mathematician Pythagoras (c. 560–480 B.C.). He and his followers envisioned Earth as a sphere floating at the center of the celestial sphere. More than a century later, Aristotle cited observations of Earth's curved shadow on the Moon during lunar eclipses as evidence for a spherical Earth. Greek philosophers therefore adopted a **geocentric model** of the universe with a spherical Earth at the center of a great celestial sphere.

From Greece to the Renaissance

Before we discuss the Greek geocentric model, it's worth briefly discussing how ancient Greek philosophy was passed to Europe, where it ultimately grew into the principles of modern science. Greek philosophy began to spread widely with the conquests of Alexander the Great (356–323 B.C.). Alexander had a deep interest in science, perhaps in part because Aristotle had been his personal tutor. Alexander founded the city of Alexandria in Egypt, and his successors founded the renowned Library of Alexandria (Figure 3.9). Though it is sometimes difficult to distinguish fact from legend in stories of this great library, there is little doubt that it was once the world's preeminent center of research, housing up to a half million books written on papyrus scrolls. Most were ultimately burned, their contents lost forever.

> **think about it** Estimate the number of books you're likely to read in your lifetime and compare this number to the half million books once housed in the Library of Alexandria. Can you think of other ways to put into perspective the loss of ancient wisdom resulting from the destruction of the Library of Alexandria?

The relatively few books from the library that survive today were preserved primarily thanks to scholars of the new religion of Islam. Around A.D. 800, the Islamic leader Al-Mamun (A.D. 786–833) established a "House of Wisdom" in Baghdad (in present-day Iraq), where Islamic scholars—often working together with colleagues from other

---

cosmic calculations 3.1

Eratosthenes Measures Earth

The first accurate estimate of Earth's circumference was made by the Greek scientist Eratosthenes in about 240 B.C. Eratosthenes knew that the Sun passed directly overhead in the Egyptian city of Syene (modern-day Aswan) on the summer solstice and that on the same day the Sun came only within 7° of the zenith in Alexandria. He concluded that Alexandria must be 7° of latitude north of Syene (see figure), making the north-south distance between the two cities $\frac{7}{360}$ of Earth's circumference.

Eratosthenes estimated the north-south distance between Syene and Alexandria to be 5000 stadia (the *stadium* was a Greek unit of distance), which meant

$$\frac{7}{360} \times \text{Earth's circumference} = 5000 \text{ stadia}$$

Multiplying both sides by $\frac{360}{7}$ gives us

$$\text{Earth's circumference} = \frac{360}{7} \times 5000 \text{ stadia}$$
$$\approx 250,000 \text{ stadia}$$

Based on the actual sizes of Greek stadiums, we estimate that stadia must have been about $\frac{1}{6}$ km each, making Eratosthenes' estimate about $\frac{250,000}{6} = 42,000$ km— remarkably close to the actual value of just over 40,000 km.

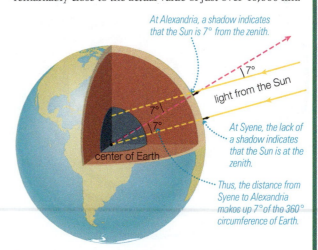

At Alexandria, a shadow indicates that the Sun is 7° from the zenith.

7°

light from the Sun

7°

7°

At Syene, the lack of a shadow indicates that the Sun is at the zenith.

center of Earth

Thus, the distance from Syene to Alexandria makes up 7° of the 360° circumference of Earth.

This diagram shows how Eratosthenes concluded that the north-south distance from Syene to Alexandria is $\frac{7}{360}$ of Earth's circumference.

a This painting shows a courtyard of the ancient Library of Alexandria, as it may have looked shortly after its completion.

b The New Library of Alexandria in Egypt, which opened in 2003.

▲ **FIGURE 3.9**
The ancient Library of Alexandria thrived for centuries, starting some time after about 300 B.C.

religions—translated and thereby saved many ancient Greek works. They were also in frequent contact with Hindu scholars from India, who in turn brought knowledge of ideas and discoveries from China, which allowed the Islamic scholars to achieve a synthesis of the surviving work of the ancient Greeks along with that of the Indians and the Chinese. Using all these ideas as building blocks, scholars in the House of Wisdom developed the mathematics of algebra and many new instruments and techniques for astronomical observation. The latter explains why many official constellation and star names come from Arabic; for example, the names of many bright stars begin with *al* (e.g., Aldebaran, Algol), which simply means "the" in Arabic.

The accumulated knowledge of the Baghdad scholars spread throughout the Byzantine empire (part of the former Roman Empire).

Islamic scholars preserved and extended ancient Greek scholarship, and their work helped ignite the European Renaissance.

When the Byzantine capital of Constantinople (modern-day Istanbul) fell to the Turks in 1453, many scholars headed west to Europe, carrying with them the knowledge that helped ignite the European Renaissance.

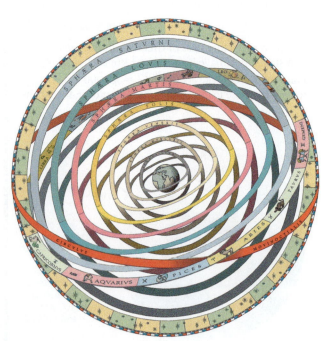

▲ **FIGURE 3.10**
This model represents the Greek idea of the heavenly spheres (c. 400 B.C.). Earth is a sphere that rests in the center. The Moon, the Sun, and the planets each have their own spheres. The outermost sphere holds the stars.

◆ How did the Greeks explain planetary motion?

Greek philosophers quickly realized that there had to be more to the heavens than just a single sphere surrounding Earth. To account for the fact that the Sun and Moon each move gradually eastward through the constellations, the Greeks added separate spheres for them, with these spheres turning at different rates from the sphere of the stars. The planets also move relative to the stars, so the Greeks added an additional sphere for each planet (Figure 3.10).

The difficulty with this model was that it made it hard to explain the apparent retrograde motion of the planets [Section 2.4]. You might guess that the Greeks would simply have allowed the planetary spheres to sometimes turn forward and sometimes turn backward relative to the sphere of the stars, but they did not because it would have violated their deeply held belief in "heavenly perfection." According to this idea, enunciated most clearly by Plato, heavenly objects could move only in perfect circles. But how could the planets sometimes go backward in our sky if they were moving in perfect circles?

One potential answer would have been to discard the geocentric model and replace it with a Sun-centered model, which gives a simple and natural explanation for apparent retrograde motion (see Figure 2.30). However, while such a model had been proposed by Aristarchus in about 260 B.C., it never gained much support in ancient times (in part because of the lack of detectable stellar parallax [Section 2.4]). Instead, the Greeks found ingenious ways to explain planetary motion while preserving Earth's central position and motion in perfect circles. The final synthesis of these ideas came with the work of Claudius Ptolemy (c. A.D. 100–170; pronounced *tol-e-mee*). We refer to Ptolemy's model as the **Ptolemaic model** to distinguish it from earlier geocentric models.

The essence of the Ptolemaic model was that each planet moves on a small circle whose center moves around Earth on a larger circle (Figure 3.11).

In the Ptolemaic model, each planet moved on a small circle whose center moved around Earth on a larger circle.

(The small circle is called an *epicycle*, and the larger circle is called a *deferent*.) A planet following this circle-upon-circle motion traces a loop as seen from Earth, with the backward portion of the loop mimicking apparent retrograde motion. However, to make his model agree well with observations, Ptolemy had to include a number of other complexities, such as positioning some of the large circles slightly off-center from Earth. As a result, the full Ptolemaic model was mathematically complex and tedious. Many centuries later, while supervising computations based on the Ptolemaic model, the Spanish monarch Alphonso X (1221–1284) is said to have complained, "If I had been present at the creation, I would have recommended a simpler design for the universe."

Despite its complexity, the Ptolemaic model worked remarkably well: It could correctly forecast future planetary positions to within a few degrees of arc, which is about the angular size of your hand held at arm's length against the sky. This was sufficiently accurate to keep the model in use for the next 1500 years. When Ptolemy's book describing the model was translated by Arabic scholars around A.D. 800, they gave it the title *Almagest*, derived from words meaning "the greatest work."

3.3 The Copernican Revolution

The Greeks and other ancient peoples developed many important scientific ideas, but what we now think of as science arose during the European Renaissance. Within a half century of the fall of Constantinople, Polish scientist Nicholas Copernicus began the work that ultimately overturned the Earth-centered Ptolemaic model.

◆ How did Copernicus, Tycho, and Kepler challenge the Earth-centered model?

The ideas introduced by Copernicus fundamentally changed the way we perceive our place in the universe. The story of this dramatic change, known as the **Copernican revolution,** is in many ways the story of the origin of modern science. It is also the story of several key personalities, beginning with Copernicus himself.

Copernicus Copernicus was born in Torún, Poland, on February 19, 1473. He began studying astronomy in his late teens, and soon learned

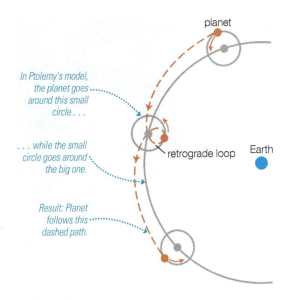

In Ptolemy's model, the planet goes around this small circle . . .

. . . while the small circle goes around the big one.

retrograde loop

Earth

planet

Result: Planet follows this dashed path.

▲ **FIGURE 3.11**

This diagram shows how the Ptolemaic model accounted for apparent retrograde motion. Each planet is assumed to move around a small circle that turns upon a larger circle. The resulting path (dashed) includes a loop in which the planet goes backward as seen from Earth.

common **misconceptions**

Columbus and a Flat Earth

A widespread myth gives credit to Columbus for learning that Earth is round, but knowledge of Earth's shape predated Columbus by nearly 2000 years. Not only were scholars of Columbus's time well aware that Earth is round, they even knew its approximate size: Earth's circumference was first measured in about 240 B.C. by the Greek scientist Eratosthenes. In fact, a likely reason Columbus had so much difficulty finding a sponsor for his voyages was that he tried to argue a point on which he was wrong: He claimed the distance by sea from western Europe to eastern Asia to be much less than scholars knew it to be. When he finally found a patron in Spain and left on his journey, he was so woefully underprepared that the voyage would almost certainly have ended in disaster if the Americas hadn't stood in his way.

Copernicus (1473–1543)

that tables of planetary motion based on the Ptolemaic model had been growing increasingly inaccurate. He began a quest to find a better way to predict planetary positions.

Copernicus was aware of and adopted Aristarchus's ancient Sun-centered idea, perhaps because it offered such a simple explanation for the apparent retrograde motion of the planets. But he went beyond Aristarchus in working out mathematical details of the model. In the process, Copernicus discovered simple geometric relationships that strengthened his belief in the Sun-centered idea, because they allowed him to calculate each planet's orbital period and relative distance (compared to Earth's distance) from the Sun.

Copernicus was nonetheless hesitant to publish his work, fearing that the idea of a moving Earth would be considered absurd. However, he discussed his system with other scholars, including high-ranking officials of the Church, who urged him to publish a book. Copernicus saw the first printed copy of his book, *De Revolutionibus Orbium Coelestium* ("On the Revolutions of the Heavenly Spheres"), on the day he died—May 24, 1543.

Publication of the book spread the Sun-centered idea widely, and many scholars were drawn to its aesthetic advantages. However, the Copernican model gained relatively few converts over the next 50 years,

Copernicus's Sun-centered model was based on the right general ideas, but its predictions were not substantially better than those of Ptolemy's Earth-centered model.

for a good reason: It didn't work all that well. The primary problem was that while Copernicus had been willing to overturn Earth's central place in the cosmos, he held fast to the ancient belief that heavenly motion must occur in perfect circles. This incorrect assumption forced him to add numerous complexities to his system (including circles on circles much like those used by Ptolemy) to get it to make decent predictions. In the end, his complete model was no more accurate and no less complex than the Ptolemaic model, and few people were willing to throw out thousands of years of tradition for a new model that worked just as poorly as the old one.

Tycho Part of the difficulty faced by astronomers who sought to improve either the Ptolemaic or the Copernican models was a lack of quality data. The telescope had not yet been invented, and existing naked-eye observations were not very accurate. Better data were needed, and they were provided by the Danish nobleman Tycho Brahe (1546–1601), usually known simply as Tycho (pronounced "tie-koe").

Tycho was an eccentric genius who once lost part of his nose in a sword fight with another student over who was the better mathematician. In 1563, Tycho decided to observe a widely anticipated alignment of Jupiter and Saturn. To his surprise, the alignment occurred nearly 2 days later than the date Copernicus had predicted. Resolving to improve the state of astronomical prediction, he set about compiling careful observations of stellar and planetary positions in the sky.

Tycho's fame grew after he observed what he called a *nova*, meaning "new star," in 1572 and showed that it was much farther away than the Moon. (Today, we know that Tycho saw a *supernova*—the explosion of a distant star [Section 13.3].) In 1577, Tycho observed a comet and showed that it too lay in the realm of the heavens. Others, including Aristotle, had argued that comets were phenomena of Earth's atmosphere. King Frederick II of Denmark decided to sponsor Tycho's ongoing work, providing him with money to build an unparalleled observatory for

Tycho Brahe (1546–1601)

naked-eye observations (Figure 3.12). After Frederick II died in 1588, Tycho moved to Prague, where his work was supported by German emperor Rudolf II.

Over a period of three decades, Tycho and his assistants compiled naked-eye observations accurate to within less than 1 arcminute—less

Tycho's accurate naked-eye observations provided the data needed to improve the Copernican system.

than the thickness of a fingernail viewed at arm's length. Despite the quality of his observations, Tycho never succeeded in coming up with a satisfying explanation for planetary motion. He was convinced that the *planets* must orbit the Sun, but his inability to detect stellar parallax [Section 2.4] led him to conclude that Earth must remain stationary. He therefore advocated a model in which the Sun orbits Earth while all other planets orbit the Sun. Few people took this model seriously.

Kepler   Tycho failed to explain the motions of the planets satisfactorily, but he succeeded in finding someone who could: In 1600, he hired the young German astronomer Johannes Kepler (1571–1630). Kepler and Tycho had a strained relationship, but Tycho recognized the talent of his young apprentice. In 1601, as he lay on his deathbed, Tycho begged Kepler to find a system that would make sense of his observations so "that it may not appear I have lived in vain."

Kepler was deeply religious and believed that understanding the geometry of the heavens would bring him closer to God. Like Copernicus, he believed that planetary orbits should be perfect circles, so he worked diligently to match circular motions to Tycho's data. After years of effort, he found a set of circular orbits that matched most of Tycho's observations quite well. Even in the worst cases, which were for the planet Mars, Kepler's predicted positions differed from Tycho's observations by only about 8 arcminutes.

Kepler surely was tempted to attribute these discrepancies to errors by Tycho. After all, 8 arcminutes is barely one-fourth the angular diameter of the full moon. But Kepler trusted Tycho's work. The small discrepancies finally led Kepler to abandon the idea of circular orbits—and to find the correct solution to the ancient riddle of planetary motion. About this event, Kepler wrote:

If I had believed that we could ignore these eight minutes [of arc], I would have patched up my hypothesis accordingly. But, since it was not permissible to ignore, those eight minutes pointed the road to a complete reformation in astronomy.

Kepler's key discovery was that planetary orbits are not circles but instead are a special type of oval called an **ellipse.** You can draw a circle by putting a pencil on the end of a string, tacking the string to a board, and pulling the pencil around (Figure 3.13a). Drawing an ellipse is similar, except that you must stretch the string around *two* tacks (Figure 3.13b). The locations of the two tacks are called the **foci** (singular, **focus**) of the ellipse. The long axis of the ellipse is called its *major axis,* each half of which is called a **semimajor axis;** as we'll see shortly, the length of the semimajor axis is particularly important in astronomy. The short axis is called the *minor axis.* By altering the distance between the two foci while keeping the length of string the same, you can draw ellipses of varying **eccentricity,** a quantity that describes how much an ellipse is stretched out compared to a perfect circle (Figure 3.13c). A circle is an ellipse with zero eccentricity, and greater eccentricity means a more elongated ellipse.

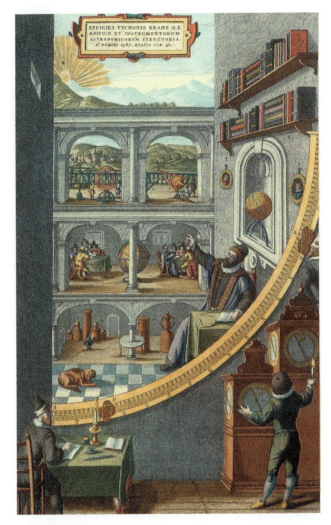

▲ **FIGURE 3.12**

Tycho Brahe in his naked-eye observatory, which worked much like a giant protractor. He could sit and observe a planet through the rectangular hole in the wall as an assistant used a sliding marker to measure the angle on the protractor.

Johannes Kepler (1571–1630)

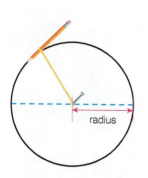

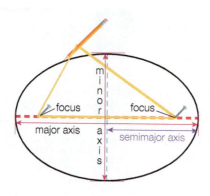

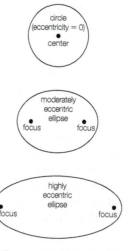

a Drawing a circle with a string of fixed length.

b Drawing an ellipse with a string of fixed length.

c Eccentricity describes how much an ellipse deviates from a perfect circle.

▲ FIGURE 3.13

An ellipse is a special type of oval. These diagrams show how an ellipse differs from a circle and how different ellipses vary in their eccentricity.

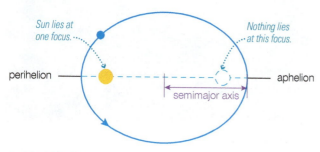

▲ FIGURE 3.14

Kepler's first law: The orbit of each planet about the Sun is an ellipse with the Sun at one focus. (The eccentricity shown here is exaggerated compared to the actual eccentricities of the planets.)

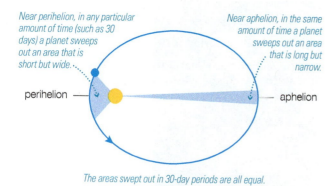

The areas swept out in 30-day periods are all equal.

▲ FIGURE 3.15

Kepler's second law: As a planet moves around its orbit, an imaginary line connecting it to the Sun sweeps out equal areas (the shaded regions) in equal times.

Kepler's decision to trust the data over his preconceived beliefs marked an important transition point in the history of science. Once he abandoned perfect circles in favor of ellipses, Kepler soon came up with a model that could predict planetary positions with far greater accuracy than Ptolemy's Earth-centered model. Kepler's model withstood the test of time and became accepted not only as a model of nature but also as a deep, underlying truth about planetary motion.

By using elliptical orbits, Kepler created a Sun-centered model that predicted planetary positions with outstanding accuracy.

◆ What are Kepler's three laws of planetary motion?

Kepler summarized his discoveries with three simple laws that we now call **Kepler's laws of planetary motion.** He published the first two laws in 1609 and the third in 1619.

Kepler's first law tells us that the orbit of each planet about the Sun is an ellipse with the Sun at one focus (Figure 3.14). This law tells us that a planet's distance from the Sun varies during its orbit. It is closest at the point called **perihelion** (from the Greek for "near the Sun") and farthest at the point called **aphelion** ("away from the Sun"). The *average* of a planet's perihelion and aphelion distances is the length of its *semimajor axis*. We will refer to this simply as the planet's average distance from the Sun.

Kepler's first law: The orbit of each planet about the Sun is an ellipse with the Sun at one focus.

Kepler's second law states that as a planet moves around its elliptical orbit, it moves faster when it is nearer the Sun and slower when it is farther from the Sun, sweeping out equal areas in equal times. As shown in Figure 3.15, the "sweeping" refers to an imaginary line connecting the planet to the Sun, and keeping the areas equal means that the planet moves a greater distance (and hence is moving faster) when it is near perihelion than it does in the same amount of time near aphelion.

Kepler's second law: A planet moves faster in the part of its orbit nearer the Sun and slower when farther from the Sun, sweeping out equal areas in equal times.

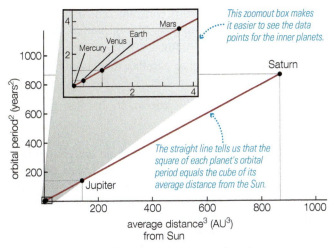

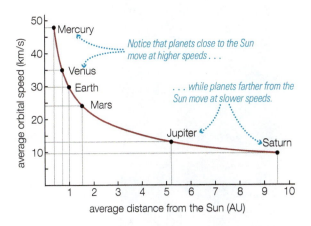

a This graph shows that Kepler's third law ($p^2 = a^3$) holds true; the graph shows only the planets known in Kepler's time.

b This graph, based on Kepler's third law and modern values of planetary distances, shows that more distant planets orbit the Sun more slowly.

▲ **FIGURE 3.16**

Graphs based on Kepler's third law.

Kepler's third law tells us that more distant planets orbit the Sun at slower average speeds. Mathematically, it states that planets obey the relationship $p^2 = a^3$, where p is the planet's orbital period in years and a is its average distance from the Sun in astronomical units. Figure 3.16a shows the $p^2 = a^3$ law graphically; notice that the square of each planet's orbital period (p^2) is indeed equal to the cube of its average distance from the Sun (a^3). Because Kepler's third law relates orbital distance to orbital time (period), we can use the law to calculate a planet's average orbital speed. Figure 3.16b shows the result, confirming that more distant planets orbit the Sun more slowly.

Kepler's third law: More distant planets orbit the Sun at slower average speeds, obeying the relationship $p^2 = a^3$.

think about it Suppose a comet has an orbit that brings it quite close to the Sun at its perihelion and beyond Mars at its aphelion, but with an average distance (semimajor axis) of 1 AU. How long would the comet take to complete each orbit of the Sun? Would it spend most of its time close to the Sun, far from the Sun, or somewhere in between? Explain.

The fact that more distant planets move more slowly led Kepler to suggest that planetary motion might be the result of a force from the Sun. He did not know the nature of the force, which was later identified as gravity by Isaac Newton.

◆ How did Galileo solidify the Copernican revolution?

The success of Kepler's laws in matching Tycho's data provided strong evidence in favor of Copernicus's placement of the Sun at the center of the solar system. Nevertheless, many scientists still voiced objections to the Copernican view. There were three basic objections, all rooted in the 2000-year-old beliefs of Aristotle and other ancient Greeks.

- First, Aristotle had held that Earth could not be moving because, if it were, objects such as birds, falling stones, and clouds would be left behind as Earth moved along its way.

cosmic calculations 3.2

Kepler's Third Law

Kepler's third law in its original form, $p^2 = a^3$, applies to any orbiting object that meets the following two conditions:

1. it orbits the Sun *or* a star of the same mass, and
2. we measure *period in years* and *distance in AU*.

(Newton extended the law to *all* orbiting objects; see Cosmic Calculations 4.1.)

Example 1: Find the orbital period of the dwarf planet (and largest asteroid) Ceres, which orbits the Sun at an average distance (semimajor axis) of 2.77 AU.

Solution: Both conditions are met, so we solve Kepler's third law for the orbital period p and substitute the given orbital distance, $a = 2.77$ AU:

$$p^2 = a^3 \Rightarrow p = \sqrt{a^3} = \sqrt{2.77^3} = 4.6$$

Ceres has an orbital period of 4.6 years.

Example 2: Find the orbital distance of a planet that orbits a star of the same mass as our Sun with a period of 3 months.

Solution: The first condition is met, and we can satisfy the second by converting the orbital period from months to years: $p = 3$ months $= 0.25$ year. We now solve Kepler's third law for the average distance a:

$$p^2 = a^3 \Rightarrow a = \sqrt[3]{p^2} = \sqrt[3]{0.25^2} = 0.40$$

The planet orbits its star at an average distance of 0.40 AU, which is slightly greater than Mercury's orbital distance from the Sun.

Galileo (1564–1642)

- Second, the idea of noncircular orbits contradicted Aristotle's claim that the heavens—the realm of the Sun, Moon, planets, and stars—must be perfect and unchanging.

- Third, no one had detected the stellar parallax that should occur if Earth orbits the Sun [Section 2.4].

Galileo Galilei (1564–1642), usually known by his first name, answered all three objections.

Galileo's Evidence Galileo defused the first objection with experiments that almost single-handedly overturned the Aristotelian view of physics. In particular, he used experiments with rolling balls to demonstrate that a moving object remains in motion *unless* a force acts to stop it (an idea now codified in Newton's first law of motion [Section 4.2]). This insight explained why objects that share Earth's motion through space—such as birds, falling stones, and clouds—should *stay* with Earth rather than falling behind as Aristotle had argued. This same idea explains why passengers stay with a moving airplane even when they leave their seats.

The second objection had already been challenged by Tycho's supernova and comet observations, which demonstrated that the heavens could change. Galileo shattered the idea of heavenly perfection after he built a telescope in late 1609. (Galileo did not invent the telescope, but his innovations made it much more powerful.) Through his telescope, Galileo saw sunspots on the Sun, which were considered "imperfections" at the time. He also used his telescope to observe that the Moon has mountains and valleys like the "imperfect" Earth by noticing the shadows cast near the dividing line between the light and dark portions of the lunar face (Figure 3.17). If the heavens were in fact not perfect, then the idea of elliptical orbits (as opposed to "perfect" circles) was not so objectionable.

The third objection—the absence of observable stellar parallax—had been of particular concern to Tycho. Based on his estimates of the distances of stars, Tycho believed that his naked-eye observations were sufficiently precise to detect stellar parallax if Earth did in fact orbit the Sun. Refuting Tycho's argument required showing that the stars were more distant than Tycho had thought and therefore too distant for him to have

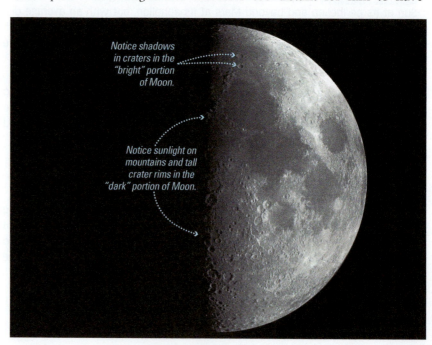

Notice shadows in craters in the "bright" portion of Moon.

Notice sunlight on mountains and tall crater rims in the "dark" portion of Moon.

▶ **FIGURE 3.17**

The shadows cast by mountains and crater rims near the dividing line between the light and dark portions of the lunar face show that the Moon's surface is not perfectly smooth.

observed stellar parallax. Although Galileo didn't actually prove this fact, he provided strong evidence in its favor. For example, he saw with his telescope that the Milky Way resolved into countless individual stars. This discovery helped him argue that the stars were far more numerous and more distant than Tycho had believed.

In hindsight, the final nails in the coffin of the Earth-centered model came with two of Galileo's earliest discoveries through the tele-

Galileo's experiments and telescopic observations overcame remaining scientific objections to the Copernican idea, sealing the case for the Sun-centered solar system.

scope. First, he observed four moons clearly orbiting Jupiter, not Earth (Figure 3.18). Soon thereafter, he observed that Venus goes through phases in a way that meant that it must orbit the Sun and not Earth (Figure 3.19).

Galileo and the Church
Although we now recognize that Galileo won the day, the story was more complex in his own time, when Catholic Church doctrine still held Earth to be the center of the universe. On June 22, 1633, Galileo was brought before a Church inquisition in Rome and ordered to recant his claim that Earth orbits the Sun. Nearly 70 years old and fearing for his life, Galileo did as ordered. His life was spared. However, legend has it that as he rose from his knees he whispered under his breath, *Eppur si muove*—Italian for "And yet it moves." (Given the likely consequences if Church officials had heard him say this, most historians doubt the legend.)

The Church did not formally vindicate Galileo until 1992, but Church officials gave up the argument long before that: In 1757, all works backing the idea of a Sun-centered solar system were removed from the Church's Index of banned books. Moreover, Catholic scientists have long worked at the forefront of much astronomical research, and today's official Church teachings are compatible not only with Earth's planetary status but also with the theories of the Big Bang and the subsequent evolution of the cosmos and of life.

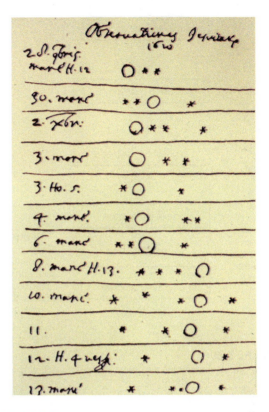

▲ **FIGURE 3.18**

A page from Galileo's notebook written in 1610. His sketches show four "stars" near Jupiter (the circle) but in different positions at different times (and sometimes hidden from view). Galileo soon realized that the "stars" were actually moons orbiting Jupiter.

▼ **FIGURE 3.19**

Galileo's telescopic observations of Venus showed that it orbits the Sun rather than Earth.

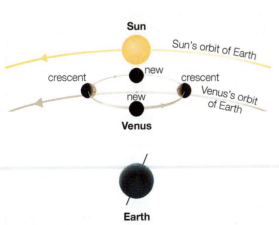

Ptolemaic View of Venus

a In the Ptolemaic system, Venus orbits Earth, moving around a smaller circle on its larger orbital circle; the center of the smaller circle lies on the Earth–Sun line. If this view were correct, Venus's phases would range only from new to crescent.

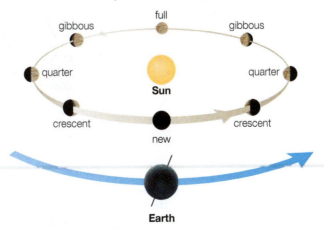

Copernican View of Venus

b In reality, Venus orbits the Sun, so from Earth we can see it in many different phases. This is just what Galileo observed, which convinced him that Venus orbits the Sun.

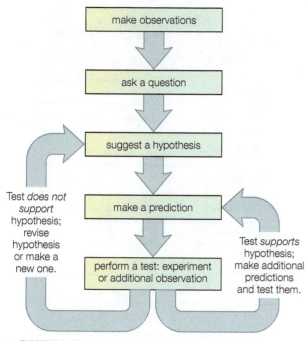

▲ **FIGURE 3.20**
This diagram illustrates what we often call the *scientific method*.

3.4 The Nature of Science

The story of how our ancestors gradually figured out the basic architecture of the cosmos exhibits many features of what we now consider "good science." For example, we have seen how models were formulated and tested against observations and modified or replaced when they failed those tests. The story also illustrates some classic mistakes, such as the apparent failure of anyone before Kepler to question the belief that orbits must be circles. The ultimate success of the Copernican revolution led scientists, philosophers, and theologians to reassess the various modes of thinking that played a role in the 2000-year process of discovering Earth's place in the universe. Let's examine how the principles of modern science emerged from the lessons of the Copernican revolution.

◆ How can we distinguish science from nonscience?

It's surprisingly difficult to define the term *science* precisely. The word comes from the Latin *scientia*, meaning "knowledge," but not all knowledge is science. For example, you may know what music you like best, but your musical taste is not a result of scientific study.

Approaches to Science One reason science is difficult to define is that not all science works in the same way. For example, you've probably heard that science is supposed to proceed according to something called the "scientific method." As an idealized illustration of this method, consider what you would do if your flashlight suddenly stopped working. You might *hypothesize* that the flashlight's batteries have died. This type of tentative explanation, or **hypothesis,** is sometimes called an *educated guess*—in this case, it is "educated" because you already know that flashlights need batteries. Your hypothesis allows you to make a simple prediction: If you replace the batteries with new ones, the flashlight should work. You can test this prediction by replacing the batteries. If the flashlight now works, you've confirmed your hypothesis. If it doesn't, you must revise or discard your hypothesis, perhaps in favor of some other one that you can also test (such as that the bulb is burned out). Figure 3.20 illustrates the basic flow of this process.

The scientific method can be a useful idealization, but real science rarely progresses in such an orderly way. Scientific progress often begins

---

extraordinary claims — Earth Orbits the Sun

In the 21st century, claiming that Earth orbits the Sun will not raise any eyebrows, but it was quite an extraordinary claim in the 3rd century B.C., when Greek astronomer Aristarchus of Samos put it forward [Section 2.4]. To almost everyone else of his time, the idea that the Sun moves while Earth remains stationary seemed like plain common sense. However, Aristarchus was also a mathematician and he used mathematical reasoning to conclude that observations of the sky made more sense if the Sun, and not Earth, were at the center of the solar system.

In this and similar boxes elsewhere in the book, we will look at scientific claims that seemed extraordinary in their time. As astronomer Carl Sagan was fond of saying, "extraordinary claims require extraordinary evidence," and we will discuss how scientific evidence ended up supporting or debunking those claims, or in some cases leaving them still unanswered. Each case will illustrate the self-correcting nature of

science: Mistaken ideas are eventually disproved, while a few ideas that once appeared extraordinary end up gaining widespread acceptance.

In the case of Aristarchus, the evidence proving his claim did not become strong enough to convince most other scholars until almost two millennia after his death. Nevertheless, Aristarchus's Sun-centered idea remained alive throughout this time, and apparently influenced Copernicus when he proposed his own, more detailed Sun-centered model. As discussed in this chapter, others including Tycho, Kepler, and Galileo then collected the evidence that ultimately led to widespread acceptance of Aristarchus's extraordinary claim. The case was later sealed after Newton provided a physical understanding of *why* Kepler's laws hold and after astronomers collected direct evidence, including measurements of stellar parallax, that proved beyond a shadow of doubt that Earth orbits the Sun.

Verdict: Clearly correct.

---

with someone going out and looking at nature in a general way, rather than by conducting a careful set of experiments. For example, Galileo wasn't looking for anything in particular when he pointed his telescope at the sky and made his first startling discoveries.

The scientific method is a useful idealization of scientific thinking, but science rarely progresses in such an orderly way.

Furthermore, scientists are human beings, and their intuition and personal beliefs inevitably influence their work. Copernicus, for example, adopted the idea that Earth orbits the Sun not because he had carefully tested it but because he believed it made more sense than the prevailing view of an Earth-centered universe. While his intuition guided him to the right general idea, he erred in the specifics because he still held Plato's ancient belief that heavenly motion must be in perfect circles.

Given that the idealized scientific method is an overly simplistic characterization of science, how can we tell what is science and what is not? To answer this question, we must look a little deeper at the distinguishing characteristics of scientific thinking.

Hallmarks of Science

One way to define scientific thinking is to list the criteria that scientists use when they judge competing models of nature. Historians and philosophers of science have examined (and continue to examine) this issue in great depth, and different experts express different viewpoints on the details. Nevertheless, everything we now consider to be science shares the following three basic characteristics, which we will refer to as the "hallmarks" of science (Figure 3.21):

- Modern science seeks explanations for observed phenomena that rely solely on natural causes.
- Science progresses through the creation and testing of models of nature that explain the observations as simply as possible.
- A scientific model must make testable predictions about natural phenomena that would force us to revise or abandon the model if the predictions did not agree with observations.

Each of these hallmarks is evident in the story of the Copernican revolution. The first shows up in the way Tycho's careful measurements of planetary motion motivated Kepler to come up with a better explanation for those motions. The second is evident in the way several competing models were compared and tested, most notably those of Ptolemy, Copernicus, and Kepler. We see the third in the fact that each model

Science seeks to explain observed phenomena using testable models of nature that explain the observations as simply as possible.

could make precise predictions about the future motions of the Sun, Moon, planets, and stars in our sky. Kepler's model gained acceptance because it worked, while the competing models lost favor because their predictions failed to match the observations. Figure 3.22 summarizes the Copernican revolution and how it illustrates the hallmarks of science.

Occam's Razor

The criterion of simplicity in the second hallmark deserves additional explanation. Remember that Copernicus's original model did *not* match the data noticeably better than Ptolemy's model. If scientists had judged this model solely on the accuracy of its predictions, they might have rejected it immediately. However, many scientists found elements of the Copernican model appealing, such as its simple explanation for apparent retrograde motion. They therefore kept the model alive until Kepler found a way to make it work.

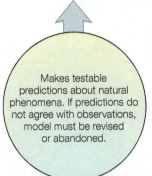

Hallmarks of Science

Seeks explanations for observed phenomena that rely solely on natural causes.

Progresses through creation and testing of models of nature that explain the observations as simply as possible.

Science

Makes testable predictions about natural phenomena. If predictions do not agree with observations, model must be revised or abandoned.

▲ **FIGURE 3.21**
Hallmarks of science.

Ancient Earth-centered models of the universe easily explained the simple motions of the Sun and Moon through our sky, but had difficulty explaining the more complicated motions of the planets. The quest to understand planetary motions ultimately led to a revolution in our thinking about Earth's place in the universe that illustrates the process of science. This figure summarizes the major steps in that process.

(1) Night by night, planets usually move from west to east relative to the stars. However, during periods of *apparent retrograde motion,* they reverse direction for a few weeks to months [Section 2.4]. The ancient Greeks knew that any credible model of the solar system had to explain these observations.

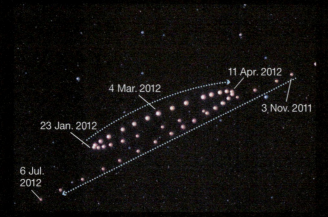

11 Apr. 2012
4 Mar. 2012
3 Nov. 2011
23 Jan. 2012
6 Jul. 2012

This composite photo shows the apparent retrograde motion of Mars.

(2) Most ancient Greek thinkers assumed that Earth remained fixed at the center of the solar system. To explain retrograde motion, they therefore added a complicated scheme of circles moving upon circles to their Earth-centered model. However, at least some Greeks, such as Aristarchus, preferred a Sun-centered model, which offered a simpler explanation for retrograde motion.

The Greek geocentric model explained apparent retrograde motion by having planets move around Earth on small circles that turned on larger circles.

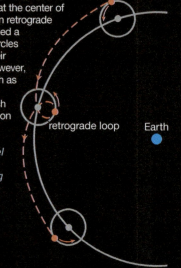

planet

retrograde loop   Earth

HALLMARK OF SCIENCE   **A scientific model must seek explanations for observed phenomena that rely solely on natural causes.** The ancient Greeks used geometry to explain their observations of planetary motion.

(Left page)
A schematic map of the universe from 1539 with Earth at the center and the Sun (Solis) orbiting it between Venus (Veneris) and Mars (Martis).

(Right page)
A page from Copernicus's De Revolutionibus, published in 1543, showing the Sun (Sol) at the center and Earth (Terra) orbiting between Venus and Mars.

3 By the time of Copernicus (1473–1543), predictions based on the Earth-centered model had become noticeably inaccurate. Hoping for improvement, Copernicus revived the Sun-centered idea. He did not succeed in making substantially better predictions because he retained the ancient belief that planets must move in perfect circles, but he inspired a revolution continued over the next century by Tycho, Kepler, and Galileo.

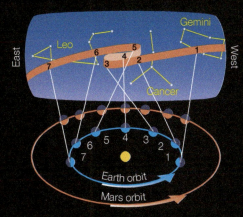

Apparent retrograde motion is simply explained in a Sun-centered system. Notice how Mars appears to change direction as Earth moves past it.

HALLMARK OF SCIENCE **Science progresses through creation and testing of models of nature that explain the observations as simply as possible.** Copernicus developed a Sun-centered model in hopes of explaining observations better than the more complicated Earth-centered model.

4 Tycho exposed flaws in both the ancient Greek and Copernican models by observing planetary motions with unprecedented accuracy. His observations led to Kepler's breakthrough insight that planetary orbits are elliptical, not circular, and enabled Kepler to develop his three laws of planetary motion.

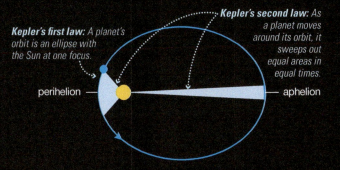

Kepler's first law: *A planet's orbit is an ellipse with the Sun at one focus.*

Kepler's second law: *As a planet moves around its orbit, it sweeps out equal areas in equal times.*

perihelion   aphelion

Kepler's third law: *More distant planets orbit at slower average speeds, obeying $p^2 = a^3$.*

HALLMARK OF SCIENCE **A scientific model makes testable predictions about natural phenomena. If predictions do not agree with observations, the model must be revised or abandoned.** Kepler could not make his model agree with observations until he abandoned the belief that planets move in perfect circles.

5 Galileo's experiments and telescopic observations overcame remaining scientific objections to the Sun-centered model. Together, Galileo's discoveries and the success of Kepler's laws in predicting planetary motion overthrew the Earth-centered model once and for all.

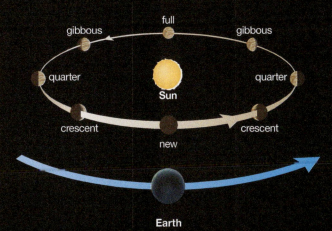

gibbous   full   gibbous

quarter   **Sun**   quarter

crescent   new   crescent

Earth

With his telescope, Galileo saw phases of Venus that are consistent only with the idea that Venus orbits the Sun rather than Earth.

If agreement with data were the sole criterion for judgment, we could imagine a modern-day Ptolemy adding millions or billions of additional circles to the geocentric model in an effort to improve its agreement with observations. A sufficiently complex geocentric model could in principle reproduce the observations with almost perfect accuracy—but it still would not convince us that Earth is the center of the universe. We would still choose the Copernican view over the geocentric view because its predictions would be just as accurate but follow a much simpler model of nature. The idea that scientists should prefer the simpler of two models that agree equally well with observations is called *Occam's razor*, after the medieval scholar William of Occam (1285–1349).

Verifiable Observations   The third hallmark of science forces us to face the question of what counts as an "observation" against which a prediction can be tested. Consider the claim that aliens are visiting Earth in UFOs. Proponents of this claim say that thousands of eyewitness reports of UFO encounters provide evidence that it is true. But do these personal testimonials count as *scientific* evidence? On the surface, the answer isn't obvious, because all scientific studies involve eyewitness accounts on some level. For example, only a handful of scientists have personally made detailed tests of Einstein's theory of relativity, and it is their personal reports of the results that have convinced other scientists of the theory's validity. However, there's an important difference between personal testimony about a scientific test and a UFO: The first can be verified by anyone, at least in principle, while the second cannot.

Understanding this difference is crucial to understanding what counts as science and what does not. Even though you may never have conducted a test of Einstein's theory of relativity yourself, there's nothing stopping you from doing so. It might require several years of study before you had the necessary background to conduct the test, but you could then confirm the results reported by other scientists. In other words, while you may currently be trusting the eyewitness testimony of scientists, you always have the option of verifying their testimony for yourself.

In contrast, there is no way for you to verify someone's eyewitness account of a UFO. Moreover, scientific studies of eyewitness testimony show it to be notoriously unreliable, because different eyewitnesses often disagree on what they saw even immediately after an event has occurred. As time passes, memories of the event may change further. In some cases in which memory has been checked against reality, people have reported vivid memories of events that never happened at all. This explains something that virtually all of us have experienced: disagreements with a friend about who did what and when. Since both people cannot be right in such cases, at least one person must have a memory that differs from reality.

The demonstrated unreliability of eyewitness testimony explains why it is generally considered insufficient for a conviction in criminal court; at least some other evidence is required. For the same reason, we cannot accept eyewitness testimony by itself as sufficient evidence in science, no matter who reports it or how many people offer similar testimony.

Objectivity in Science   It's important to realize that science is not the only valid way of seeking knowledge. For example, suppose you are shopping for a car, learning to play drums, or pondering the meaning of life. In each case, you might make observations, exercise logic, and test hypotheses. Yet these pursuits are not science, because they are not directed at developing testable explanations for observed natural phenomena. As long as nonscientific searches for knowledge make no claims about how the natural world works, they do not conflict with science. Unfortunately, the boundaries between science and nonscience are not always so clear cut.

common misconceptions

Eggs on the Equinox

One of the hallmarks of science holds that you needn't take scientific claims on faith. In principle, at least, you can always test them for yourself. Consider the claim, repeated in news reports every year, that the spring equinox is the only day on which you can balance an egg on its end. Many people believe this claim, but you'll be immediately skeptical if you think about the nature of the spring equinox. The equinox is merely a point in time at which sunlight strikes both hemispheres equally (see Figure 2.15). It's difficult to see how sunlight could affect an attempt to balance eggs (especially if the eggs are indoors), and there's nothing special about either Earth's or the Sun's gravity on that day.

More important, you can test this claim directly. It's not easy to balance an egg on its end, but with practice you can do it on any day of the year, not just on the spring equinox. Not all scientific claims are so easy to test for yourself, but the basic lesson should be clear: Before you accept any scientific claim, you should demand at least a reasonable explanation of the evidence that backs it up.

We generally think of science as being objective, meaning that all people should be able to find the same scientific results. However, there is a difference between the overall objectivity of science and the objectivity of individual scientists. Science is practiced by human beings, and individual scientists may bring their personal biases and beliefs to their scientific work. For example, most scientists choose their research projects based on personal interests rather than on some objective formula. In extreme cases, scientists have been known to cheat—either deliberately or subconsciously—to obtain a result they desire. For example, in the late 19th century, astronomer Percival Lowell claimed to see a network of artificial canals in blurry telescopic images of Mars, leading him to conclude that there was a great Martian civilization. But no such canals exist, so Lowell must have allowed his beliefs about extraterrestrial life to influence the way he interpreted what he saw—in essence, a form of cheating, though almost certainly not intentional.

Bias can occasionally show up even in the thinking of the scientific community as a whole. Some valid ideas may not be considered by any scientist because they fall too far outside the general patterns of thought, or **paradigm,** of the time. Einstein's theory of relativity is an example. Many scientists in the decades before Einstein had gleaned hints of the theory but did not investigate them, at least in part because they seemed too outlandish.

The beauty of science is that it encourages continued testing by many people. Even if personal biases affect some results, tests by others should eventually uncover the mistakes. Similarly, if a new idea is correct but falls outside the accepted paradigm, sufficient testing and verification of the idea should eventually force a paradigm shift. In that sense, *science ultimately provides a means of bringing people to agreement*, at least on topics that can be subjected to scientific study.

Individual scientists inevitably carry personal biases into their work, but the collective action of many scientists should ultimately make science objective.

---

special topic | Astrology

The terms *astrology* and *astronomy* sound very similar, but today they describe very different practices. In ancient times, however, astrology and astronomy often went hand in hand, and astrology played an important role in the historical development of astronomy. Indeed, astronomers and astrologers were usually one and the same.

The basic tenet of astrology is that human events are influenced by the apparent positions of the Sun, Moon, and planets among the stars in our sky. The origins of this idea are easy to understand. The position of the Sun in the sky clearly influences our lives—it determines the seasons and hence the times of planting and harvesting, of warmth and cold, and of daylight and darkness. Similarly, the Moon determines the tides, and the cycle of lunar phases coincides with many biological cycles. Because the planets also appear to move among the stars, it seemed reasonable to imagine that planets also influence our lives, even if these influences were much more difficult to discover.

Ancient astrologers hoped that they might learn *how* the positions of the Sun, Moon, and planets influence our lives. They charted the skies, seeking correlations with events on Earth. For example, if an earthquake occurred when Saturn was entering the constellation of Leo, might Saturn's position have caused the earthquake? If the king became ill when Mars was in Gemini and the first-quarter moon was in Scorpio, might it mean another misfortune for the king when this particular alignment of the Moon and Mars next recurred? Ancient astrologers thought that the patterns of influence eventually would become clear and they would then be able to forecast human events with the same reliability with which observations of the Sun could forecast the coming of spring.

This hope was never realized. Although many astrologers still attempt to predict future events, scientific tests have shown that their predictions come true no more often than would be expected by pure chance. Moreover, in light of our current understanding of the universe, the original ideas behind astrology no longer make sense. For example, today we use ideas of gravity and energy to explain the influences of the Sun and the Moon, and these same ideas tell us that the planets are too far from Earth to have a similar influence.

Of course, many people continue to practice astrology, perhaps because of its ancient and rich traditions. Scientifically, we cannot say anything about such traditions, because traditions are not testable predictions. But if you want to understand the latest discoveries about the cosmos, you'll need a science that can be tested and refined—and astrology fails to meet these requirements.

◆ What is a scientific theory?

The most successful scientific models explain a wide variety of observations in terms of just a few general principles. When a powerful yet simple model makes predictions that survive repeated and varied testing, scientists elevate its status and call it a **theory.** Some famous examples are Isaac Newton's theory of gravity, Charles Darwin's theory of evolution, and Albert Einstein's theory of relativity.

Note that the scientific meaning of the word *theory* is quite different from its everyday meaning, in which we equate a theory more closely with speculation or a hypothesis. For example, someone might say, "I have a new theory about why people enjoy the beach." Without the support of a broad range of evidence that others have tested and confirmed, this "theory" is really only a guess. In contrast, Newton's theory of gravity qualifies as a scientific theory because it uses simple physical principles to explain many observations and experiments. *Theory* is just one of many terms that are used with different meaning in science than in everyday life. Table 3.2 summarizes a few of the most common of these terms.

Despite its success in explaining observed phenomena, a scientific theory can never be proved true beyond all doubt, because future

TABLE 3.2 Scientific Terminology

This table lists some words you will encounter in the media that have a different meaning in science than in everyday life. (Adapted from a table published by Richard Somerville and Susan Joy Hassol in Physics Today, Oct. 2011.)

Term	Everyday Meaning	Scientific Meaning	Example
model	something you build, like a model airplane	a representation of nature, sometimes using mathematics or computer simulations, that is intended to explain or predict observed phenomena	A model of planetary motion can be used to calculate exactly where planets should appear in our sky.
hypothesis	a guess or assumption of almost any type	a model that has been proposed to explain some observations, but which has not yet been rigorously confirmed	Scientists hypothesize that the Moon was formed by a giant impact, but there is not enough evidence to be fully confident in this model.
theory	speculation	a particularly powerful model that has been so extensively tested and verified that we have extremely high confidence in its validity	Einstein's theory of relativity successfully explains a broad range of natural phenomena and has passed a great many tests of its validity.
bias	distortion, political motive	tendency toward a particular result	Current techniques for detecting extrasolar planets are biased toward detecting large planets.
critical	really important; involving criticism, often negative	right on the edge	A boiling point is a "critical value" because above that temperature, a liquid will boil away.
deviation	strangeness or unacceptable behavior	change or difference	The recent deviation in global temperatures from their long-term average implies that something is heating the planet.
enhance/enrich	improve	increase or add more, but not necessarily to make something "better"	"Enhanced colors" means colors that have been brightened. "Enriched with iron" means containing more iron.
error	mistake	range of uncertainty	The "margin of error" tells us how closely measured values are likely to reflect true values.
feedback	a response	a self-regulating (negative feedback) or self-reinforcing (positive feedback) cycle	Gravity can provide positive feedback to a forming planet: Adding mass leads to stronger gravity, which leads to more added mass, and so on.
state (as a noun)	a place or location	a description of current condition	The Sun is in a state of balance, and so it shines steadily.
uncertainty	ignorance	a range of possible values around some central value	The measured age of our solar system is 4.55 billion years with an uncertainty of 0.02 billion years.
values	ethics, monetary value	numbers or quantities	The speed of light has a measured value of 300,000 km/s.

observations may disagree with its predictions. However, anything that qualifies as a scientific theory must be supported by a large, compelling body of evidence.

A scientific theory is a simple yet powerful model whose predictions have been borne out by repeated and varied testing.

In this sense, a scientific theory is not at all like a hypothesis or any other type of guess. We are free to change a hypothesis at any time, because it has not yet been carefully tested. In contrast, we can discard or replace a scientific theory only if we have an alternative way of explaining the evidence that supports it.

Again, the theories of Newton and Einstein offer good examples. A vast body of evidence supports Newton's theory of gravity, but by the late 19th century scientists had begun to discover cases where its predictions did not perfectly match observations. These discrepancies were explained only when Einstein developed his general theory of relativity, which was able to match the observations. Still, the many successes of Newton's theory could not be ignored, and Einstein's theory would not have gained acceptance if it had not been able to explain these successes equally well. It did, and that is why we now view Einstein's theory as a broader theory of gravity than Newton's theory. Some scientists today are seeking a theory of gravity that will go beyond Einstein's. If any new theory ever gains acceptance, it will have to match all the successes of Einstein's theory as well as apply to new realms where Einstein's theory does not.

think about it When people claim that something is "only a theory," what do you think they mean? Does this meaning of "theory" agree with the definition of a theory in science? Do scientists always use the word *theory* in its "scientific" sense? Explain.

the big picture Putting Chapter 3 into Perspective

In this chapter, we focused on the scientific principles through which we have learned so much about the universe. Key "big picture" concepts from this chapter include the following:

- The basic ingredients of scientific thinking—careful observation and trial-and-error testing—are a part of everyone's experience. Modern science simply provides a way of organizing this thinking to facilitate the learning and sharing of new knowledge.

- Although our understanding of the universe is growing rapidly today, each new piece of knowledge builds on ideas that came before.

- The Copernican revolution, which overthrew the ancient Greek belief in an Earth-centered universe, unfolded over a period of more than a century. Many of the characteristics of modern science first appeared during this time.

- Science exhibits several key features that distinguish it from nonscience and that in principle allow anyone to come to the same conclusions when studying a scientific question.

my cosmic perspective All of us are connected to ancient astronomy through our ancestors, and to modern astronomy through the fact that it was driven by the Copernican revolution, which also drove the development of virtually all modern science and technology.

3.1 The Ancient Roots of Science

◆ In what ways do all humans use scientific thinking?
Scientific thinking relies on the same type of trial-and-error thinking that we use in our everyday lives, but done in a carefully organized way.

◆ How is modern science rooted in ancient astronomy?

Ancient astronomers were accomplished observers who learned to tell the time of day and the time of year, to track cycles of the Moon, and to observe planets and stars. The care and effort that went into these observations helped set the stage for modern science.

3.2 Ancient Greek Science

◆ Why does modern science trace its roots to the Greeks?
The Greeks developed **models** of nature and emphasized the importance of agreement between the predictions of those models and observations of nature.

◆ How did the Greeks explain planetary motion?

retrograde loop

The Greek **geocentric model** reached its culmination with the **Ptolemaic model,** which explained apparent retrograde motion by having each planet move on a small circle whose center moves around Earth on a larger circle.

3.3 The Copernican Revolution

◆ How did Copernicus, Tycho, and Kepler challenge the Earth-centered model?
Copernicus created a Sun-centered model of the solar system designed to replace the Ptolemaic model, but it was no more accurate than Ptolemy's because Copernicus still used perfect circles. Tycho's accurate, naked-eye observations provided the data needed to improve on Copernicus's model. Kepler developed a model of planetary motion that fit Tycho's data.

◆ What are Kepler's three laws of planetary motion?
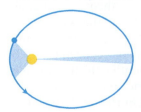
(1) The orbit of each planet is an ellipse with the Sun at one focus. (2) A planet moves faster in the part of its orbit nearer the Sun and slower when farther from the Sun, sweeping out equal areas in equal times. (3) More distant planets orbit the Sun at slower average speeds, obeying the precise mathematical relationship $p^2 = a^3$.

◆ How did Galileo solidify the Copernican revolution?

Venus

Earth

Galileo's experiments and telescopic observations overcame remaining objections to the Copernican idea of Earth as a planet orbiting the Sun. Although not everyone accepted his results immediately, in hindsight we see that Galileo sealed the case for the Sun-centered solar system.

3.4 The Nature of Science

◆ How can we distinguish science from nonscience?
Science generally exhibits three hallmarks: (1) Modern science seeks explanations for observed phenomena that rely solely on natural causes. (2) Science progresses through the creation and testing of models of nature that explain the observations as simply as possible. (3) A scientific model must make testable predictions about natural phenomena that would force us to revise or abandon the model if the predictions did not agree with observations.

◆ What is a scientific theory?
A scientific **theory** is a simple yet powerful model that explains a wide variety of observations using just a few general principles, and that has survived repeated and varied testing.

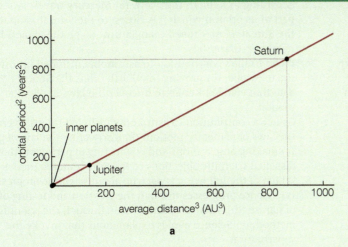

a

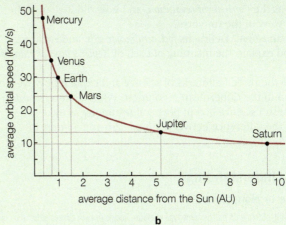

b

Use the information in the graphs to answer the following questions.

1. Approximately how fast is Jupiter orbiting the Sun?
   a. cannot be determined from the information provided
   b. 20 km/s
   c. 10 km/s
   d. a little less than 15 km/s

2. An asteroid with an average orbital distance of 2 AU will orbit the Sun at an average speed that is _____.
   a. a little slower than the orbital speed of Mars
   b. a little faster than the orbital speed of Mars
   c. the same as the orbital speed of Mars

3. Uranus, not shown on the graph, orbits about 19 AU from the Sun. Based on the graph, its approximate orbital speed is between about _____.
   a. 20 and 25 km/s
   b. 15 and 20 km/s
   c. 10 and 15 km/s
   d. 5 and 10 km/s

4. Kepler's third law is often stated as $p^2 = a^3$. The value a^3 for a planet is shown on _____.
   a. the horizontal axis of Figure a
   b. the vertical axis of Figure a
   c. the horizontal axis of Figure b
   d. the vertical axis of Figure b

5. On Figure a, you can see Kepler's third law ($p^2 = a^3$) from the fact that _____.
   a. the data fall on a straight line
   b. the axes are labeled with values for p^2 and a^3
   c. the planet names are labeled on the graph

6. Suppose Figure a showed a planet on the red line directly above a value of 1000 AU^3 along the horizontal axis. On the vertical axis, this planet would be at _____.
   a. 1000 years^2
   b. 1000^2 years^2
   c. $\sqrt{1000}$ years^2
   d. 100 years

7. How far does the planet in question 6 orbit from the Sun?
   a. 10 AU
   b. 100 AU
   c. 1000 AU
   d. $\sqrt{1000}$ AU

MasteringAstronomy® For instructor-assigned homework and other learning materials, go to MasteringAstronomy®.

Review Questions

1. In what way is scientific thinking natural to all of us, and how does modern science build upon this everyday type of thinking?
2. Why did ancient peoples study astronomy? Describe the astronomical origins of our day, week, month, and year.
3. What is a lunar calendar? How can it be kept roughly synchronized with a solar calendar?
4. What do we mean by a *model* in science? Briefly summarize the Greek *geocentric model*.

5. What do we mean by the *Ptolemaic model*? How did this model account for the apparent retrograde motion of planets in our sky?
6. What was the *Copernican revolution*, and how did it change the human view of the universe?
7. What is an *ellipse*? Define its *foci, semimajor axis*, and *eccentricity*.
8. State and explain the meaning of each of *Kepler's laws of planetary motion*.
9. Describe the three hallmarks of science and explain how we can see them in the Copernican revolution. What is *Occam's razor*? Why doesn't science accept personal testimony as evidence?
10. What is the difference between a *hypothesis* and a *theory* in science?

Test Your Understanding

Science or Nonscience?

Each of the following statements makes some type of claim. Decide in each case whether the claim could be evaluated scientifically or whether it falls into the realm of nonscience. Explain clearly; not all of these have definitive answers, so your explanation is more important than your chosen answer.

11. The Yankees are the best baseball team of all time.
12. Several kilometers below its surface, Jupiter's moon Europa has an ocean of liquid water.
13. My house is haunted by ghosts who make the creaking noises I hear each night.
14. There are no lakes or seas on the surface of Mars today.
15. Dogs are smarter than cats.
16. Children born when Jupiter is in the constellation Taurus are more likely to be musicians than other children.
17. Aliens can manipulate time so that they can abduct and perform experiments on people who never realize they were taken.
18. Newton's law of gravity works as well for explaining the orbits of planets around other stars as it does for explaining the orbits of planets in our own solar system.
19. God created the laws of motion that were discovered by Newton.
20. A huge fleet of alien spacecraft will land on Earth and introduce an era of peace and prosperity on January 1, 2035.

Quick Quiz

Choose the best answer to each of the following. Explain your reasoning with one or more complete sentences.

21. In the Greek geocentric model, the retrograde motion of a planet occurs when (a) Earth is about to pass the planet in its orbit around the Sun. (b) the planet actually goes backward in its orbit around Earth. (c) the planet is aligned with the Moon in our sky.
22. Which of the following was *not* a major advantage of Copernicus's Sun-centered model over the Ptolemaic model? (a) It made significantly better predictions of planetary positions in our sky. (b) It offered a more natural explanation for the apparent retrograde motion of planets in our sky. (c) It allowed calculation of the orbital periods and distances of the planets.
23. When we say that a planet has a highly *eccentric* orbit, we mean that (a) it is spiraling in toward the Sun. (b) its orbit is an ellipse with the Sun at one focus. (c) in some parts of its orbit it is much closer to the Sun than in other parts.
24. Earth is closer to the Sun in January than in July. Therefore, in accord with Kepler's second law, (a) Earth travels faster in its orbit around the Sun in July than in January. (b) Earth travels faster in its orbit around the Sun in January than in July. (c) it is summer in January and winter in July.
25. According to Kepler's third law, (a) Mercury travels fastest in the part of its orbit in which it is closest to the Sun. (b) Jupiter orbits the Sun at a faster speed than Saturn. (c) all the planets have nearly circular orbits.
26. Tycho Brahe's contribution to astronomy included (a) inventing the telescope. (b) proving that Earth orbits the Sun. (c) collecting data that enabled Kepler to discover the laws of planetary motion.
27. Galileo's contribution to astronomy included (a) discovering the laws of planetary motion. (b) discovering the law of gravity. (c) making observations and conducting experiments that dispelled scientific objections to the Sun-centered model.
28. Which of the following is *not* true about scientific progress? (a) Science progresses through the creation and testing of models of nature. (b) Science advances only through the scientific method. (c) Science avoids explanations that invoke the supernatural.
29. Which of the following is *not* true about a scientific theory? (a) A theory must explain a wide range of observations or experiments. (b) Even the strongest theories can never be proved true beyond all doubt. (c) A theory is essentially an educated guess.
30. When Einstein's theory of gravity (general relativity) gained acceptance, it demonstrated that Newton's theory had been (a) wrong. (b) incomplete. (c) really only a guess.

Process of Science

31. *What Makes It Science?* Choose a single idea in the modern view of the cosmos discussed in Chapter 1, such as "The universe is expanding," or "We are made from elements manufactured by stars," or "The Sun orbits the center of the Milky Way Galaxy once every 230 million years."
    a. Describe how this idea reflects each of the three hallmarks of science, discussing how it is based on observations, how our understanding of it depends on a model, and how that model is testable.
    b. Describe a hypothetical observation that, if it were actually made, might cause us to call the idea into question. Then briefly discuss whether you think that, overall, the idea is likely or unlikely to hold up to future observations.
32. *Earth's Shape.* It took thousands of years for humans to deduce that Earth is spherical. For each of the following alternative models of Earth's shape, identify one or more observations that you could make for yourself that would invalidate the model.
    a. A flat Earth
    b. A cylindrical Earth [which was actually proposed by the Greek philosopher Anaximander (c. 610–546 B.C.)]
    c. A football-shaped Earth

Group Work Exercise

33. *Galileo on Trial.* **Roles:** *Scribe* (takes notes on the group's activities), *Galileo* (argues in favor of the idea that Earth orbits the Sun), *Prosecutor* (argues against the idea that Earth orbits the Sun), and *Moderator* (leads group discussion and makes sure the debate remains civil). **Activity:** Conduct a mock trial in which you consider the following three pieces of evidence: (1) observations of mountains and valleys on the Moon; (2) observations of moons orbiting Jupiter; (3) observations of

the phases of Venus. *Galileo* should explain why the evidence indicates that Earth orbits the Sun. The *Prosecutor* should present a rebuttal. The *Scribe* and *Moderator* should serve as jury to decide whether the evidence is convincing beyond a reasonable doubt, somewhat convincing, or not convincing. The group should record the final verdict along with an explanation of their reasoning.

Investigate Further

Short-Answer/Essay Questions

34. *Copernican Players.* Using a bulleted list format, make a one-page "executive summary" of the major roles that Copernicus, Tycho, Kepler, and Galileo played in overturning the ancient belief in an Earth-centered universe.

35. *Influence on History.* Based on what you have learned about the Copernican revolution, write a one- to two-page essay about how you believe it altered the course of human history.

36. *Cultural Astronomy.* Choose a particular culture of interest to you, and research the astronomical knowledge and accomplishments of that culture. Write a two- to three-page summary of your findings.

Quantitative Problems

Be sure to show all calculations clearly and state your final answers in complete sentences.

37. *Method of Eratosthenes.* You are an astronomer on planet Nearth, which orbits a distant star. It has recently been accepted that Nearth is spherical in shape, though no one knows its size. One day, while studying in the library of Alectown, you learn that on the equinox your sun is directly overhead in the city of Nyene, located 1000 km due north of you. On the equinox, you go outside in Alectown and observe that the altitude of your sun is 80°. What is the circumference of Nearth? (*Hint:* Apply the technique used by Eratosthenes to measure Earth's circumference.)

38. *Eris Orbit.* The dwarf planet Eris orbits the Sun every 557 years. What is its average distance (semimajor axis) from the Sun? How does its average distance compare to that of Pluto?

39. *Halley Orbit.* Halley's comet orbits the Sun every 76.0 years and has an orbital eccentricity of 0.97.
    a. Find its average distance (semimajor axis).
    b. Halley's perihelion distance is approximately 90 million km. Approximately what is its aphelion distance?

Discussion Questions

40. *The Impact of Science.* The modern world is filled with ideas, knowledge, and technology that developed through science and application of the scientific method. Discuss some of these things and how they affect our lives. Which of these impacts do you think are positive? Which are negative? Overall, do you think science has benefited the human race? Defend your opinion.

41. *The Importance of Ancient Astronomy.* Why was astronomy important to people in ancient times? Discuss both the practical importance of astronomy and the importance it may have had for religious or other traditions. Which do you think was more important in the development of ancient astronomy: its practical or its philosophical role? Defend your opinion.

42. *Astronomy and Astrology.* Why do you think astrology remains so popular around the world even though it has failed all scientific tests of its validity? Do you think this popularity has any social consequences? Defend your opinions.

Web Projects

43. *The Ptolemaic Model.* This chapter gives only a very brief description of Ptolemy's model of the universe. Investigate this model in greater depth. Using diagrams and text as needed, create a two- to three-page description of the model.

44. *The Galileo Affair.* In recent years, the Roman Catholic Church has devoted a lot of resources to learning more about the trial of Galileo and to understanding past actions of the Church in the Galilean case. Learn more about these studies, and write a two- to three-page report about the current Vatican view of the case.

45. *Astrology.* Find out about at least one scientific test of the validity of astrology. Write a short summary of the methods and results of the test.

46. *Your Test of Astrology.* Collect the 12 astrological predictions for the previous day or month from any astrology site. Remove the names of the astrological signs from the predictions, and then ask other students to identify which prediction best fits what happened to them over the prior day or month. (Be sure to randomize the order in which you show your list.) Also ask students their astrological sign, so that you can check whether the prediction they picked matches their sign. What fraction of students choose their own sign? What does your test suggest about the validity of astrological predictions? (Note: You can do a simplified version of the test by showing each student just three predictions: one that corresponds to his or her "correct" sign and the other two randomly drawn from the rest.)

Our perspective on the universe has changed dramatically throughout human history. This timeline summarizes some of the key discoveries that have shaped our modern perspective.

Stonehenge

Earth-centered model of the universe

Galileo's telescope

| <500 B.C. | 400 B.C. –170 A.D. | 1543–1648 A.D. |

(1) Ancient civilizations recognized patterns in the motion of the Sun, Moon, planets, and stars through our sky. They also noticed connections between what they saw in the sky and our lives on Earth, such as the cycles of seasons and of tides [Section 3.1].

(2) The ancient Greeks tried to explain observed motions of the Sun, Moon, and planets using a model with Earth at the center, surrounded by spheres in the heavens. The model explained many phenomena well, but could explain the apparent retrograde motion of the planets only with the addition of many complex features— and even then, its predictions were not especially accurate [Section 3.2].

(3) Copernicus suggested that Earth is a planet orbiting the Sun. The Sun-centered model explained apparent retrograde motion simply, though it made accurate predictions only after Kepler discovered his three laws of planetary motion. Galileo's telescopic observations confirmed the Sun-centered model, and revealed that the universe contains far more stars than had been previously imagined [Section 3.3].

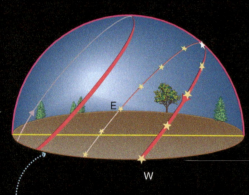

Earth's rotation around its axis leads to the daily east-to-west motions of objects in the sky.

The tilt of Earth's rotation axis leads to seasons as Earth orbits the Sun.

Planets are much smaller than the Sun. At a scale of 1 to 10 billion, the Sun is the size of a grapefruit, Earth is the size of a ball point of a pen, and the distance between them is about 15 meters.

Yerkes Observatory

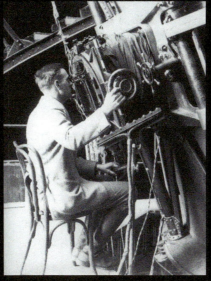
Edwin Hubble at the Mt. Wilson telescope

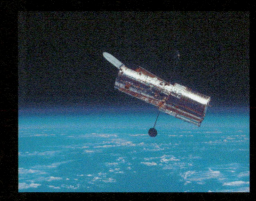
Hubble Space Telescope

1838–1920 A.D.   1924–1929 A.D.   1990 A.D.–present

④ Larger telescopes and photography made it possible to measure the parallax of stars, offering direct proof that Earth really does orbit the Sun and showing that even the nearest stars are light-years away. We learned that our Sun is a fairly ordinary star in the Milky Way [Sections 2.4, 12.1].

⑤ Edwin Hubble measured the distances of galaxies, showing that they lay far beyond the bounds of the Milky Way and proving that the universe is far larger than our own galaxy. He also discovered that more distant galaxies are moving away from us faster, telling us that the entire universe is expanding and suggesting that it began in an event we call the Big Bang [Sections 1.3, 16.2].

⑥ Improved measurements of galactic distances and the rate of expansion have shown that the universe is about 14 billion years old. These measurements have also revealed still unexplained surprises, including evidence for the existence of mysterious dark matter and dark energy [Sections 1.3, 18.1].

Distances between stars are enormous. At a scale of 1 to 10 billion, you can hold the Sun in your hand, but the nearest stars are thousands of kilometers away.

Our solar system is located about 27,000 light-years from the center of the Milky Way Galaxy.

The Milky Way Galaxy contains over 100 billion stars.

The observable universe contains over 100 billion galaxies.

4 Making Sense of the Universe
Understanding Motion, Energy, and Gravity

The same laws that govern motion on Earth also govern gargantuan collisions between galaxies.

LEARNING GOALS

4.1 Describing Motion: Examples from Daily Life
- How do we describe motion?
- How is mass different from weight?

4.2 Newton's Laws of Motion
- How did Newton change our view of the universe?
- What are Newton's three laws of motion?

4.3 Conservation Laws in Astronomy
- What keeps a planet rotating and orbiting the Sun?
- Where do objects get their energy?

4.4 The Force of Gravity
- What determines the strength of gravity?
- How does Newton's law of gravity extend Kepler's laws?
- How do gravity and energy allow us to understand orbits?
- How does gravity cause tides?

The history of the universe is essentially a story about the interplay between matter and energy. This interplay began in the Big Bang and continues today in everything from the microscopic jiggling of atoms to gargantuan collisions of galaxies. Understanding the universe therefore depends on becoming familiar with how matter responds to the ebb and flow of energy.

You might guess that it would be difficult to understand the many interactions that shape the universe, but we now know that just a few physical laws govern the movements of everything from atoms to galaxies. The Copernican revolution spurred the discovery of these laws, and Galileo deduced some of them from his experiments. But it was Sir Isaac Newton who put all the pieces together into a simple system of laws describing both motion and gravity.

In this chapter, we'll discuss Newton's laws of motion, the laws of conservation of angular momentum and of energy, and the universal law of gravitation. Understanding these laws will enable you to make sense of many of the wide-ranging phenomena you will encounter as you study astronomy.

ESSENTIAL PREPARATION

1. How is Earth moving through space? [Section 1.3]
2. How did Copernicus, Tycho, and Kepler challenge the Earth-centered model? [Section 3.3]
3. What are Kepler's three laws of planetary motion? [Section 3.3]

4.1 Describing Motion: Examples from Daily Life

We all have experience with motion and a natural intuition as to what motion is, but in science we need to define our ideas and terms precisely. In this section, we'll use examples from everyday life to explore some of the fundamental ideas of motion.

◆ How do we describe motion?

You are probably familiar with common terms used to describe motion in science, such as *velocity, acceleration,* and *momentum*. However, their scientific definitions may differ subtly from those you use in casual conversation. Let's investigate the precise meanings of these terms.

Speed, Velocity, and Acceleration   A car provides a good illustration of the three basic terms that we use to describe motion:

- The **speed** of the car tells us how far it will go in a certain amount of time. For example, "100 kilometers per hour" (about 60 miles per hour) is a speed, and it tells us that the car will cover a distance of 100 kilometers if it is driven at this speed for an hour.
- The **velocity** of the car tells us both its speed and its direction. For example, "100 kilometers per hour going due north" describes a velocity.
- The car has an **acceleration** if its velocity is changing in any way, whether in speed or direction or both.

Note that while we normally think of *acceleration* as an increase in speed, in science we also say that you are accelerating when you slow down or turn (Figure 4.1). Slowing represents a negative acceleration,

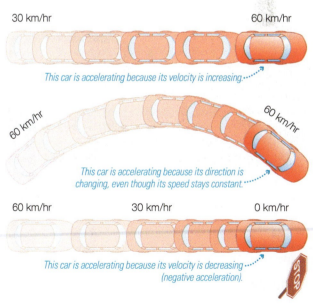

30 km/hr · 60 km/hr

This car is accelerating because its velocity is increasing.

60 km/hr · 60 km/hr

This car is accelerating because its direction is changing, even though its speed stays constant.

60 km/hr · 30 km/hr · 0 km/hr

This car is accelerating because its velocity is decreasing (negative acceleration).

▲ **FIGURE 4.1**
Speeding up, turning, and slowing down are all examples of acceleration.

causing your velocity to decrease. Turning means a change in direction—which therefore means a change in velocity—so turning is a form of acceleration even if your speed remains constant.

You can often feel the effects of acceleration. For example, as you speed up in a car, you feel yourself being pushed back into your seat. As you slow down, you feel yourself being pulled forward. As you drive around a curve, you feel yourself being pushed away from the direction of your turn. In contrast, you don't feel such effects when moving at *constant velocity*. That is why you don't feel any sensation of motion when you're traveling in an airplane on a smooth flight.

The Acceleration of Gravity   One of the most important types of acceleration is the acceleration caused by gravity. In a legendary experiment in which he supposedly dropped weights from the Leaning Tower of Pisa, Galileo demonstrated that gravity accelerates all objects by the same amount, regardless of their mass. This fact may be surprising because it seems to contradict everyday experience: A feather floats gently to the ground, while a rock plummets. However, air resistance causes this difference in acceleration. If you dropped a feather and a rock on the Moon, where there is no air, both would fall at exactly the same rate.

see it for yourself   Find a piece of paper and a small rock. Hold both at the same height and let them go at the same instant. The rock, of course, hits the ground first. Next, crumple the paper into a small ball and repeat the experiment. What happens? Explain how this experiment suggests that gravity accelerates all objects by the same amount.

The acceleration of a falling object is called the **acceleration of gravity,** abbreviated g. On Earth, the acceleration of gravity causes falling objects to fall faster by 9.8 meters per second (m/s), or about 10 m/s, with each passing second. For example, suppose you drop a rock from a tall building. At the moment you let it go, its speed is 0 m/s. After 1 second, the rock will be falling downward at about 10 m/s. After 2 seconds, it will be falling at about 20 m/s. In the absence of air resistance, its speed will continue to increase by about 10 m/s each second until it hits the ground (Figure 4.2). We therefore say that the acceleration of gravity is about 10 *meters per second per second,* or 10 *meters per second squared,* which we write as 10 m/s^2 (more precisely, $g = 9.8 \text{ m/s}^2$).

Momentum and Force   The concepts of speed, velocity, and acceleration describe how an individual object moves, but most of the interesting phenomena we see in the universe result from *interactions* between objects. We need two additional concepts to describe these interactions:

- An object's **momentum** is the product of its mass and its velocity; that is, momentum = mass × velocity.
- The only way to change an object's momentum is to apply a **force** to it.

We can understand these concepts by considering the effects of collisions. Imagine that you're stopped in your car at a red light when a bug flying at a velocity of 30 km/hr due south slams into your windshield. What will happen to your car? Not much, except perhaps a bit of a mess on your windshield. Next, imagine that a 2-ton truck runs the red light and hits you head-on with the same velocity as the bug. Clearly, the truck will cause far more damage. We can understand why by considering the momentum and force in each collision.

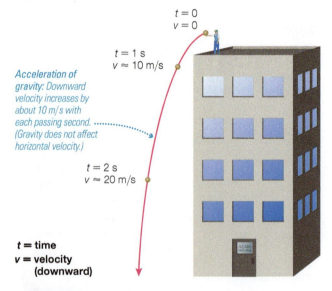

Acceleration of gravity: Downward velocity increases by about 10 m/s with each passing second. (Gravity does not affect horizontal velocity.)

$t = 0$
$v = 0$

$t = 1 \text{ s}$
$v \approx 10 \text{ m/s}$

$t = 2 \text{ s}$
$v \approx 20 \text{ m/s}$

$t = $ time
$v = $ velocity
(downward)

▲ **FIGURE 4.2**
On Earth, gravity causes an unsupported object to accelerate downward at about 10 m/s^2, which means its downward velocity increases by about 10 m/s with each passing second. (Gravity does not affect horizontal velocity.)

Before the collisions, the truck's much greater mass means it has far more momentum than the bug, even though both the truck and the bug are moving with the same velocity. During the collisions, the bug and the truck each transfer some of their momentum to your car. The bug has very little momentum to give to your car, so it does not exert much of a force. In contrast, the truck imparts enough of its momentum to cause a dramatic and sudden change in your car's momentum. You feel this sudden change in momentum as a force, and it can do great damage to you and your car.

The mere presence of a force does not always cause a change in momentum. For example, a moving car is always affected by forces of air resistance and friction with the road—forces that will slow your car if you take your foot off the gas pedal. However, you can maintain a constant velocity, and hence constant momentum, if you step on the gas pedal hard enough to overcome the slowing effects of these forces.

In fact, forces of some kind are always present, such as the force of gravity or the electromagnetic forces acting between atoms. The **net force** (or *overall force*) acting on an object represents the combined effect of all the individual forces put together. There is no net force on your car when you are driving at constant velocity, because the force generated by the engine to turn the wheels precisely offsets the forces of air resistance and road friction. A change in momentum occurs only when the net force is not zero. (The car is experiencing a net force in all three examples in Figure 4.1.)

Changing an object's momentum means changing its velocity, as long as its mass remains constant. A net force that is not zero therefore causes an object to accelerate. Conversely, whenever an object accelerates, a net force must be causing the acceleration. That is why you feel forces (pushing you forward, backward, or to the side) when you accelerate in your car. We can use the same ideas to understand many astronomical processes. For example, planets are always accelerating as they orbit the Sun, because their direction of travel constantly changes as they go around their orbits. We can therefore conclude that some force must be causing this acceleration. As we'll discuss shortly, Isaac Newton identified this force as gravity.

An object must accelerate whenever a net force acts on it.

◆ How is mass different from weight?

In daily life, we usually think of *mass* as something you can measure with a bathroom scale, but technically the scale measures your *weight*, not your mass. The distinction between mass and weight rarely matters when we are talking about objects on Earth, but it is very important in astronomy:

- Your **mass** is the amount of matter in your body.
- Your **weight** (or **apparent weight***) is the *force* that a scale measures when you stand on it; that is, weight depends both on your mass and on the forces (including gravity) acting on your mass.

To understand the difference between mass and weight, imagine standing on a scale in an elevator (Figure 4.3). Your mass will be the same no matter how the elevator moves, but your weight can vary. When the elevator is stationary or moving at constant velocity, the scale

---

*Some physics texts distinguish between "true weight" due only to gravity and "apparent weight" that also depends on other forces (as in an elevator).

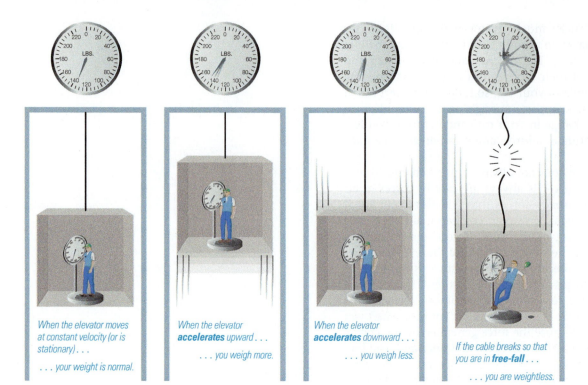

When the elevator moves at constant velocity (or is stationary) . . .
. . . your weight is normal.

When the elevator **accelerates** upward . . .
. . . you weigh more.

When the elevator **accelerates** downward . . .
. . . you weigh less.

If the cable breaks so that you are in **free-fall** . . .
. . . you are weightless.

▲ **FIGURE 4.3**

Mass is not the same as weight. In an elevator, your mass never changes, but your weight is different when the elevator accelerates.

reads your "normal" weight. When the elevator accelerates upward, the floor exerts a greater force than it does when you are at rest. You feel heavier, and the scale verifies your greater weight. When the elevator accelerates downward, the floor and the scale exert a weaker force on you, so the scale registers less weight. Note that the scale shows a weight different from your "normal" weight only when the elevator is *accelerating*, not when it is going up or down at constant speed.

see it for yourself   Take a small bathroom scale on an elevator and stand on it. How does your weight change when the elevator accelerates upward or downward? Does it change when the elevator is moving at constant speed? Explain your observations.

Your mass is the same no matter where you are, but your weight can vary.   Your mass therefore depends only on the amount of matter in your body and is the same anywhere, but your weight can vary because the forces acting on you can vary. For example, your mass would be the same on the Moon as on Earth, but you would weigh less on the Moon because of its weaker gravity.

common misconceptions

No Gravity in Space?

If you ask people why astronauts are weightless in space, one of the most common answers is "There is no gravity in space." But you can usually convince people that this answer is wrong by following up with another simple question: Why does the Moon orbit Earth? Most people know that the Moon orbits Earth because of gravity, proving that there is gravity in space. In fact, at the altitude of the International Space Station's orbit, the acceleration of gravity is only about 10% less than it is on Earth's surface.

The real reason astronauts are weightless is that they are in a constant state of free-fall. Imagine being an astronaut. You'd have the sensation of free-fall—just as when you jump from a diving board—the entire time you were in orbit. This constant falling sensation makes many astronauts sick to their stomachs when they first experience weightlessness. Fortunately, they quickly get used to the sensation, which allows them to work hard and enjoy the view.

Free-Fall and Weightlessness   Now consider what happens if the elevator cable breaks (see the last frame in Figure 4.3). The elevator and you are suddenly in **free-fall**—falling without any resistance to slow you down. The floor drops away at the same rate that you fall, allowing you to "float" freely above it, and the scale reads zero because you are no longer held to it. In other words, your free-fall has made you **weightless.**

In fact, you are in free-fall whenever there's nothing to *prevent* you from falling. For example, you are in free-fall when you jump off a chair or spring from a diving board or trampoline. Surprising as it may seem, you have therefore experienced weightlessness many times in your life.

You can experience it right now simply by jumping off your chair—though your weightlessness lasts for only a very short time until you hit the ground.

Weightlessness in Space You've probably seen videos of astronauts floating weightlessly in the International Space Station. But why are they weightless? Many people guess that there's no gravity in space, but that's not true. After all, it is gravity that makes the Space Station orbit Earth. Astronauts are weightless for the same reason that you are weightless when you jump off a chair: They are in free-fall.

People or objects are weightless whenever they are falling freely, and astronauts in orbit are weightless because they are in a constant state of free-fall.

Astronauts are weightless the entire time they orbit Earth because they are in a *constant state of free-fall*. To understand this idea, imagine a tower that reaches all the way to the Space Station's orbit, about 350 kilometers above Earth (Figure 4.4). If you stepped off the tower, you would fall downward, remaining weightless until you hit the ground (or until air resistance had a noticeable effect on you). Now, imagine that instead of stepping off the tower, you ran and jumped out of the tower. You'd still fall to the ground, but because of your forward motion you'd land a short distance away from the base of the tower.

The faster you ran out of the tower, the farther you'd go before landing. If you could somehow run fast enough—about 28,000 km/hr (17,000 mi/hr) at the orbital altitude of the Space Station—a very interesting thing would happen: By the time gravity had pulled you downward as far as the length of the tower, you'd already have moved far enough around Earth that you'd no longer be going down at all. Instead, you'd be just as high above Earth as you'd been all along, but a good portion of the way around the world. In other words, you'd be orbiting Earth.

The Space Station and all other orbiting objects stay in orbit because they are constantly "falling around" Earth. Their constant state of free-fall makes spacecraft and everything in them weightless.

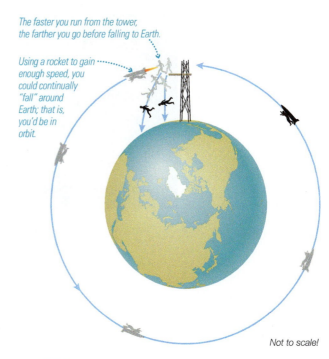

The faster you run from the tower, the farther you go before falling to Earth.

Using a rocket to gain enough speed, you could continually "fall" around Earth; that is, you'd be in orbit.

Not to scale!

▲ **FIGURE 4.4**

This figure explains why astronauts are weightless and float freely in space. If you could leap from a tall tower with enough speed, you could travel forward so fast that you'd orbit Earth. You'd then be in a constant state of free-fall, which means you'd be weightless. *Note:* On the scale shown here, the tower extends far higher than the Space Station's orbit; the rocket's orientation shows it rotating once with each orbit. The rocket has also been given a slight amount of rotation so that the driver remains facing Earth. (Adapted from *Space Station Science* by Marianne Dyson.)

4.2 Newton's Laws of Motion

The complexity of motion in daily life might lead you to guess that the laws governing motion would also be complex. For example, if you watch a falling piece of paper waft lazily to the ground, you'll see it rock back and forth in a seemingly unpredictable pattern. However, the complexity of this motion arises because the paper is affected by a variety of forces, including gravity and the changing forces caused by air currents. If you could analyze the forces individually, you'd find that each force affects the paper's motion in a simple, predictable way. Sir Isaac Newton (1642–1727) discovered the remarkably simple laws that govern motion.

◆ **How did Newton change our view of the universe?**

Newton was born in Lincolnshire, England, on Christmas Day in 1642. He had a difficult childhood and showed few signs of unusual talent. He attended Trinity College at Cambridge, where he earned his keep by

Sir Isaac Newton (1642–1727)

performing menial labor, such as cleaning the boots and bathrooms of wealthier students and waiting on their tables.

The plague hit Cambridge shortly after Newton graduated, and he returned home. By his own account, he experienced a moment of inspiration in 1666 when he saw an apple fall to the ground. He suddenly realized that the gravity making the apple fall was the same force that held the Moon in orbit around Earth. In that moment, Newton shattered the remaining vestiges of the Aristotelian view of the world, which for centuries had been accepted as unquestioned truth.

Aristotle had made many claims about the physics of motion, using his ideas to support his belief in an Earth-centered cosmos. He had also maintained that the heavens were totally distinct from Earth, so physical laws on Earth did not apply to heavenly motion. By the time Newton saw the apple fall, the Copernican revolution had displaced Earth from a central position, and Galileo's experiments had shown that the laws of physics were not what Aristotle had believed.

Newton's sudden insight delivered the final blow to Aristotle's view. By recognizing that gravity operated in the heavens as well as on Earth, Newton eliminated Aristotle's distinction between the two realms and brought the heavens and Earth together as one *universe*.

Newton showed that the same physical laws that operate on Earth also operate in the heavens.

This insight also marked the birth of the modern science of *astrophysics* (although the term wasn't coined until much later), which applies physical laws discovered on Earth to phenomena throughout the cosmos.

Over the next 20 years, Newton's work completely revolutionized mathematics and science. He quantified the laws of motion and gravity, conducted crucial experiments regarding the nature of light, built the first reflecting telescopes, and invented the mathematics of calculus. We'll discuss his laws of motion in the rest of this section, and later in the chapter we'll turn our attention to Newton's discoveries about gravity.

◆ What are Newton's three laws of motion?

Newton published the laws of motion and gravity in 1687, in his book *Philosophiae Naturalis Principia Mathematica* ("Mathematical Principles of Natural Philosophy"), usually called *Principia*. He enumerated three laws that apply to all motion, which we now call **Newton's laws of motion.** These laws govern the motion of everything from our daily movements on Earth to the movements of planets, stars, and galaxies throughout the universe. Figure 4.5 summarizes the three laws.

Newton's First Law Newton's first law of motion states that in the absence of a net force, an object will move with constant velocity. Objects at rest (velocity = 0) tend to remain at rest, and objects in motion tend to remain in motion with no change in either their speed or their direction.

Newton's first law: An object moves at constant velocity if there is no net force acting upon it.

The idea that an object at rest should remain at rest is rather obvious: A car parked on a flat street won't suddenly start moving for no reason. But what if the car is traveling along a flat, straight road? Newton's first law says that the car should keep going at the same speed forever *unless* a force acts to slow it down. You know that the car eventually will come to a stop if you take your foot off the gas pedal, so one or more forces must be stopping the car—in this case, forces arising from friction and air resistance. If the car were in space, and therefore unaffected by friction or

Newton's first law of motion:
An object moves at constant velocity unless a net force acts to change its speed or direction.

Example: A spaceship needs no fuel to keep moving in space.

Newton's second law of motion:
Force = mass × acceleration

Example: A baseball accelerates as the pitcher applies a force by moving his arm. (Once the ball is released, the force from the pitcher's arm ceases, and the ball's path changes only because of the forces of gravity and air resistance.)

Newton's third law of motion:
For any force, there is always an equal and opposite reaction force.

Example: A rocket is propelled upward by a force equal and opposite to the force with which gas is expelled out its back.

▲ **FIGURE 4.5**
Newton's three laws of motion.

air, it would keep moving forever (though gravity would gradually alter its speed and direction). That is why interplanetary spacecraft need no fuel to keep going after they are launched into space, and why astronomical objects don't need fuel to travel through the universe.

Newton's first law also explains why you don't feel any sensation of motion when you're traveling in an airplane on a smooth flight. As long as the plane is traveling at constant velocity, no net force is acting on it or on you. Therefore, you feel no different from the way you would feel at rest. You can walk around the cabin, play catch with someone, or relax and go to sleep just as though you were "at rest" on the ground.

Newton's Second Law   Newton's second law of motion tells us what happens to an object when a net force *is* present. We have already seen that a net force will change an object's momentum, accelerating it in the direction of the force. Newton's second law quantifies this relationship, and is most commonly written as force = mass × acceleration, or $F = ma$ for short.

This law explains why you can throw a baseball farther than you can throw a shot in the shot put. The force your arm delivers to both the baseball and the shot equals the product of mass and acceleration. Because the mass of the shot is greater than that of the baseball, the same force from your arm gives the shot a smaller acceleration.

Newton's second law:
Force = mass × acceleration ($F = ma$).

Because of its smaller acceleration, the shot leaves your hand with less speed than the baseball and therefore travels a shorter distance before hitting the ground. Astronomically, Newton's second law explains why large planets such as Jupiter have a greater effect on asteroids and comets than small planets such as Earth [Section 9.5]. Because Jupiter is much more massive than Earth, it exerts a stronger gravitational force on passing asteroids and comets, and therefore sends them scattering with greater acceleration.

Newton's Third Law   Think for a moment about standing still on the ground. Your weight exerts a downward force; if this force were acting alone, Newton's second law would demand that you accelerate downward. The fact that you are not falling means there must be no *net* force acting on you, which is possible only if the ground is exerting an upward force on you that precisely offsets your weight. The fact that the ground pushes up on you with a force equal and opposite to the one your feet apply to the ground is one example of Newton's third law of motion,

What Makes a Rocket Launch?

If you watch a rocket launch, it's easy to see why many people believe that the rocket "pushes off" the ground. However, the ground has nothing to do with the rocket launch, which is actually explained by Newton's third law of motion. To balance the force driving gas out the back of the rocket, an equal and opposite force must propel the rocket forward. Rockets can be launched horizontally as well as vertically, and a rocket can be "launched" in space (for example, from a space station) with no need for a solid surface to push off from.

which tells us that every force is always paired with an equal and opposite reaction force.

This law is very important in astronomy, because it tells us that objects always attract *each other* through gravity. For example, your body always exerts a gravitational force on Earth identical to the force that Earth exerts on you, except that it acts in the opposite direction. Of course, the same force means a much greater acceleration for you than for Earth (because your mass is so much smaller than Earth's), which is why you fall toward Earth when you jump off a chair, rather than Earth falling toward you.

Newton's third law: For any force, there is always an equal and opposite reaction force.

Newton's third law also explains how a rocket works: A rocket engine generates a force that drives hot gas out the back, which creates an equal and opposite force that propels the rocket forward.

4.3 Conservation Laws in Astronomy

Newton's laws of motion are easy to state, but they may seem a bit arbitrary. Why, for example, should every force be opposed by an equal and opposite reaction force? In the centuries since Newton first stated his laws, we have learned that they are not arbitrary at all, but instead reflect deeper aspects of nature known as *conservation laws*.

Consider what happens when two objects collide. Newton's second law tells us that object 1 exerts a force that will change the momentum of object 2. At the same time, Newton's third law tells us that object 2 exerts an equal and opposite force on object 1—which means that object 1's momentum changes by precisely the same amount as object 2's momentum, but in the opposite direction. The total combined momentum of objects 1 and 2 remains the same both before and after the collision. We say that the total momentum of the colliding objects is conserved, reflecting a principle that we call *conservation of momentum*. In essence, the law of conservation of momentum tells us that the total momentum of all interacting objects always stays the same. An individual object can gain or lose momentum only when a force causes it to exchange momentum with another object.

Conservation of momentum is one of several important conservation laws that underlie Newton's laws of motion and other physical laws in the universe. Two other conservation laws—one for *angular momentum* and one for *energy*—are especially important in astronomy. Let's see how these important laws work.

◆ What keeps a planet rotating and orbiting the Sun?

Perhaps you've wondered how Earth manages to keep rotating and going around the Sun day after day and year after year. The answer relies on a special type of momentum that we use to describe objects turning in circles or going around curves. This special type of "circling momentum" is called **angular momentum.** (The term *angular* arises because an object moving in a circle turns through an *angle* of 360°.)

Conservation of angular momentum: An object's angular momentum cannot change unless it transfers angular momentum to or from another object.

The **law of conservation of angular momentum** tells us that total angular momentum can never change. An individual object can change its angular momentum only by transferring some angular momentum to or from another object.

Orbital Angular Momentum   Consider Earth's orbit around the Sun. A simple formula tells us Earth's angular momentum at any point in its orbit:

$$\text{angular momentum} = m \times v \times r$$

where *m* is Earth's mass, *v* is its orbital velocity (or, more technically, the component of velocity perpendicular to *r*), and *r* is the "radius" of the orbit, by which we mean Earth's distance from the Sun (Figure 4.6). Because there are no objects around to give or take angular momentum from Earth as it orbits the Sun, Earth's orbital angular momentum must always stay the same. This explains two key facts about Earth's orbit:

1. Earth needs no fuel or push of any kind to keep orbiting the Sun—it will keep orbiting as long as nothing comes along to take angular momentum away.

2. Because Earth's angular momentum at any point in its orbit depends on the product of its speed and orbital radius (distance from the Sun), Earth's orbital speed must be faster when it is nearer to the Sun (and the radius is smaller) and slower when it is farther from the Sun (and the radius is larger).

The second fact is just what Kepler's second law of planetary motion states [Section 3.3]. That is, the law of conservation of angular momentum tells us *why* Kepler's law is true.

Rotational Angular Momentum   The same idea explains why Earth keeps rotating. As long as Earth isn't transferring any of the angular momentum of its rotation to another object, it will keep rotating at the same rate. (In fact, Earth is very gradually transferring some of its rotational angular momentum to the Moon, and as a result Earth's rotation is gradually slowing down; see Special Topic, page 99.)

Conservation of angular momentum also explains why we see so many spinning disks in the universe, such as the disks of galaxies like the Milky Way and disks of material orbiting young stars. The idea is easy to illustrate with an ice skater spinning in place (Figure 4.7). Because there is so little friction on ice, the angular momentum

Earth is not exchanging substantial angular momentum with any other object, so its rotation rate and orbit must stay about the same.

of the ice skater remains essentially constant. When she pulls in her extended arms, she decreases her radius—which means her velocity of rotation must increase. Stars and galaxies are both born from clouds of gas that start out much larger in size. These clouds almost inevitably have some small net rotation, though it may be imperceptible. Like the spinning skater as she pulls in her arms, these clouds must spin faster as gravity makes them shrink in size. (We'll discuss why the clouds also flatten into disks in Chapter 6.)

> **think about it**   How does conservation of angular momentum explain the spiraling of water going down a drain?

◆ Where do objects get their energy?

The **law of conservation of energy** tells us that, like momentum and angular momentum, energy cannot appear out of nowhere or disappear into nothingness. Objects can gain or lose energy only by exchanging energy with other objects. Because of this law, the story of the universe is a story of

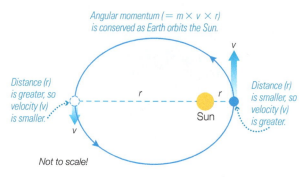

Angular momentum ($= m \times v \times r$) is conserved as Earth orbits the Sun.

Distance (r) is greater, so velocity (v) is smaller.

Distance (r) is smaller, so velocity (v) is greater.

Sun

Not to scale!

▲ **FIGURE 4.6**
Earth's angular momentum always stays the same as it orbits the Sun, so it moves faster when it is closer to the Sun and slower when it is farther from the Sun. It needs no fuel to keep orbiting because no forces are acting in a way that could change its angular momentum.

In the product $m \times v \times r$, extended arms mean larger radius and smaller velocity of rotation.

Bringing in her arms decreases her radius and therefore increases her rotational velocity.

▲ **FIGURE 4.7**
A spinning skater conserves angular momentum.

Energy can be converted from one form to another.

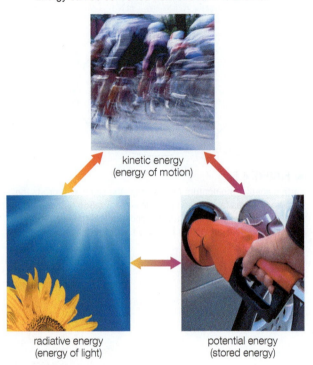

kinetic energy
(energy of motion)

radiative energy
(energy of light)

potential energy
(stored energy)

▲ **FIGURE 4.8**

The three basic categories of energy. Energy can be converted from one form to another, but it can never be created or destroyed, an idea embodied in the law of conservation of energy.

TABLE 4.1 Energy Comparisons

Item	Energy (joules)
Energy of sunlight at Earth (per square meter per second)	1.3×10^3
Energy from metabolism of a candy bar	1×10^6
Energy needed to walk for 1 hour	1×10^6
Kinetic energy of a car going 60 mi/hr	1×10^6
Daily food energy need of average adult	1×10^7
Energy released by burning 1 liter of oil	1.2×10^7
Thermal energy of parked car	1×10^8
Energy released by fission of 1 kilogram of uranium-235	5.6×10^{13}
Energy released by fusion of hydrogen in 1 liter of water	7×10^{13}
Energy released by 1-megaton H-bomb	4×10^{15}
Energy released by magnitude 8 earthquake	2.5×10^{16}
Annual U.S. energy consumption	10^{20}
Annual energy generation of Sun	10^{34}
Energy released by a supernova	10^{44}–10^{46}

the interplay of energy and matter: All actions involve exchanges of energy or the conversion of energy from one form to another.

Throughout the rest of this book, we'll see numerous cases in which we can understand astronomical processes simply by studying how energy is transformed and exchanged. For example, we'll see that planetary interiors cool with time because they radiate energy into space, and that the Sun became hot because of energy released by the gas that formed it. By applying the laws of conservation of angular momentum and conservation of energy, we can understand almost every major process that occurs in the universe.

Conservation of energy: Energy can be transferred from one object to another or transformed from one type to another, but the total amount of energy is always conserved.

Basic Types of Energy   Before we can fully understand the law of conservation of energy, we need to know what energy is. In essence, energy is what makes matter move. Because this statement is so broad, we often distinguish between different types of energy. For example, we talk about the energy we get from the food we eat, the energy that makes our cars go, and the energy a light bulb emits. Fortunately, we can classify nearly all types of energy into just three major categories (Figure 4.8):

- Energy of motion, or **kinetic energy** (*kinetic* comes from a Greek word meaning "motion"). Falling rocks, orbiting planets, and the molecules moving in the air are all examples of objects with kinetic energy.

- Energy carried by light, or **radiative energy** (the word *radiation* is often used as a synonym for *light*). All light carries energy, which is why light can cause changes in matter. For example, light can alter molecules in our eyes—thereby allowing us to see—or warm the surface of a planet.

- Stored energy, or **potential energy,** which might later be converted into kinetic or radiative energy. For example, a rock perched on a ledge has *gravitational* potential energy because it will fall if it slips off the edge, and gasoline contains *chemical* potential energy that can be converted into the kinetic energy of a moving car.

Regardless of which type of energy we are dealing with, we can measure the amount of energy with the same standard units. For Americans, the most familiar units of energy are *Calories*, which are shown on food labels to tell us how much energy our bodies can draw from the food. A typical adult needs about 2500 Calories of energy from food each day. In science, the standard unit of energy is the **joule.** One food Calorie is equivalent to about 4184 joules, so the 2500 Calories used daily by a typical adult is equivalent to about 10 million joules. Table 4.1 compares various energies in joules.

There are three basic categories of energy: energy of motion (kinetic), energy of light (radiative), and stored energy (potential).

Thermal Energy—The Kinetic Energy of Many Particles   Although there are only three major categories of energy, we sometimes divide them into various subcategories. In astronomy, the most important subcategory of kinetic energy is **thermal energy,** which represents the collective kinetic energy of the many individual particles (atoms and molecules) moving randomly within a substance like a rock or the air or the gas within a distant star. In such cases, it is much easier to talk about the thermal energy of the object rather than about the kinetic energies of its billions upon billions of individual particles.

Thermal energy gets its name because it is related to temperature, but temperature and thermal energy are not quite the same thing. Thermal energy measures the *total* kinetic energy of all the randomly moving particles in a substance, while **temperature** measures the *average* kinetic energy of the particles. For a particular object, a higher temperature simply means that the particles on average have more kinetic energy and hence are moving faster (Figure 4.9). You're probably familiar with temperatures measured in *Fahrenheit* or *Celsius*, but in science we often use the **Kelvin** temperature scale (Figure 4.10). The Kelvin scale does not have negative temperatures, because it starts from the coldest possible temperature, known as *absolute zero* (0 K).

Thermal energy depends on temperature, because a higher average kinetic energy for the particles in a substance means a higher total energy. But thermal energy also depends on the number and density of the particles, as you

Thermal energy is the total kinetic energy of many individual particles.

can see by imagining that you quickly thrust your arm in and out of a hot oven and a pot of boiling water (don't try this!). The air in a hot oven is much higher in temperature than the water boiling in a pot (Figure 4.11). However, the boiling water would scald your arm almost instantly, while you can safely put your arm into the oven air for a few seconds. The reason for this difference is density. In both cases, because the air or water is hotter than your body, molecules striking your skin transfer thermal energy to molecules in your arm. The higher temperature in the oven means that the air molecules strike your skin harder, on average, than the molecules in the boiling water. However, because the *density* of water is so much higher than the density of air (meaning water has far more molecules in the same amount of space), many more molecules strike your skin each second in the water. While each individual molecule that strikes your skin transfers a little less energy in the boiling water than in the oven, the sheer number of molecules hitting you in the water means that more thermal energy is transferred to your arm. That is why the boiling water causes a burn almost instantly.

think about it In air or water that is colder than your body temperature, thermal energy is transferred from you to the surrounding cold air or water. Use this fact to explain why falling into a 32°F (0°C) lake is much more dangerous than standing naked outside on a 32°F day.

Potential Energy in Astronomy Many types of potential energy are important in astronomy, but two are particularly important: *gravitational potential energy* and the potential energy of mass itself, or *mass-energy*.

An object's **gravitational potential energy** depends on its mass and how far it can fall as a result of gravity. An object has more gravita-

An object's gravitational potential energy increases when it moves higher and decreases when it moves lower.

tional potential energy when it is higher and less when it is lower. For example, if you throw a ball up into the air, it has more potential energy when it is high up than it does near the ground. Because energy must be conserved during the ball's flight, the ball's kinetic energy increases when its gravitational potential energy decreases, and vice versa (Figure 4.12a). That is why the ball travels fastest (has the most kinetic energy) when it is closest to the ground, where it has the least gravitational potential energy. The higher the ball is, the more gravitational potential energy it has and the slower the ball travels (less kinetic energy).

lower temperature   higher temperature

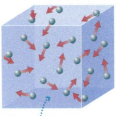
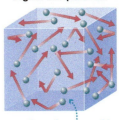

These particles are moving relatively slowly, which means low temperature . . .

. . . and now the same particles are moving faster, which means higher temperature.

▲ **FIGURE 4.9**
Temperature is a measure of the average kinetic energy of the particles (atoms and molecules) in a substance. Longer arrows represent faster speeds.

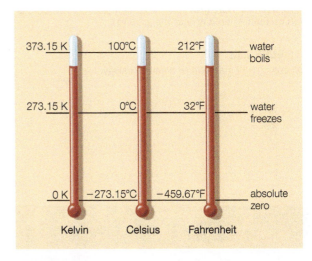

▲ **FIGURE 4.10**
Three common temperature scales: Kelvin, Celsius, and Fahrenheit. Scientists generally prefer the Kelvin scale.
(The degree symbol ° is not usually used with the Kelvin scale.)

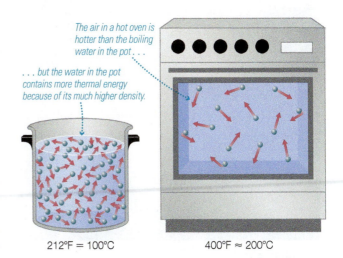

The air in a hot oven is hotter than the boiling water in the pot . . .

. . . but the water in the pot contains more thermal energy because of its much higher density.

212°F = 100°C   400°F ≈ 200°C

▲ **FIGURE 4.11**
Thermal energy depends on both the temperature and the density of a substance.

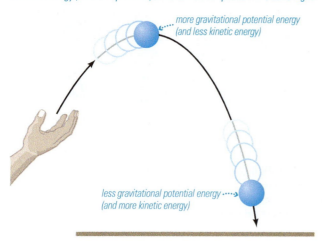

The total energy (kinetic + potential) is the same at all points in the ball's flight.

more gravitational potential energy
(and less kinetic energy)

less gravitational potential energy
(and more kinetic energy)

a The ball has more gravitational potential energy when it is high up than when it is near the ground.

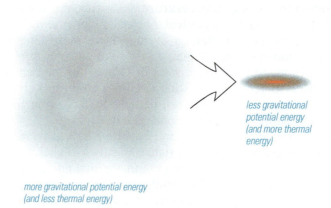

Energy is conserved: As the cloud contracts, gravitational potential energy is converted to thermal energy and radiation.

less gravitational
potential energy
(and more thermal
energy)

more gravitational potential energy
(and less thermal energy)

b A cloud of interstellar gas contracting because of its own gravity has more gravitational potential energy when it is spread out than when it shrinks in size.

▲ **FIGURE 4.12**
Two examples of gravitational potential energy.

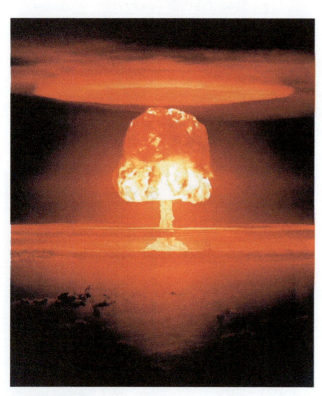

▲ **FIGURE 4.13**
The energy released by this H-bomb comes from converting only about 0.1 kilogram of mass into energy in accordance with the formula $E = mc^2$.

The same general idea explains how stars become hot (Figure 4.12b). Before a star forms, its matter is spread out in a large, cold cloud of gas. Most of the individual gas particles are far from the center of this large cloud and therefore have a lot of gravitational potential energy. The particles lose gravitational potential energy as the cloud contracts under its own gravity, and this "lost" potential energy ultimately gets converted into thermal energy, making the center of the cloud hot.

Einstein discovered that mass itself is a form of potential energy, often called **mass-energy.** The amount of potential energy contained in mass is described by Einstein's famous equation

$$E = mc^2$$

where E is the amount of potential energy, m is the mass of the object, and c is the speed of light. This equation tells us that a small amount of mass contains a huge amount of energy. For example, the energy released by a 1-megaton H-bomb comes from converting only about 0.05 kilogram of mass

Mass itself is a form of potential energy, as described by Einstein's equation $E = mc^2$.

(about 1½ ounces—an eighth of a can of soda) into energy (Figure 4.13). The Sun generates energy by converting a tiny fraction of its mass into energy through a similar process of nuclear fusion [Section 11.2].

Just as Einstein's formula tells us that mass can be converted into other forms of energy, it also tells us that energy can be transformed into mass. This process is especially important in understanding what we think happened during the early moments in the history of the universe, when some of the energy of the Big Bang turned into the mass from which all objects, including us, are made [Section 17.1]. Scientists also use this idea to search for undiscovered particles of matter, using large machines called *particle accelerators* to create subatomic particles from energy.

Conservation of Energy We have seen that energy comes in three basic categories—kinetic, radiative, and potential—and explored several subcategories that are especially important in astronomy: thermal energy, gravitational potential energy, and mass-energy. Now we are ready to return to the question of where objects get their energy. Because energy cannot be created or destroyed, objects always get their energy from other objects. Ultimately, we can always trace an object's energy back to the Big Bang [Section 1.2], the beginning of the universe in which all matter and energy is thought to have come into existence.

For example, imagine that you've thrown a baseball. It is moving, so it has kinetic energy. Where did this kinetic energy come from? The baseball got its kinetic energy from the motion of your arm as you threw it. Your arm,

The energy of any object can be traced back to the origin of the universe in the Big Bang.

in turn, got its kinetic energy from the release of chemical potential energy stored in your muscle tissues. Your muscles got this energy from the chemical potential energy stored in the foods you ate. The energy stored in the foods came from sunlight, which plants convert into chemical potential energy through photosynthesis. The radiative energy of the Sun was generated through the process of nuclear fusion, which releases some of the mass-energy stored in the Sun's supply of hydrogen. The mass-energy stored in the hydrogen came from the birth of the universe in the Big Bang. After you throw the ball, its kinetic energy will ultimately be transferred to molecules in the air or ground. It may be difficult to trace after this point, but it will never disappear.

4.4 The Force of Gravity

Newton's laws of motion describe how objects in the universe move in response to forces. The laws of conservation of momentum, angular momentum, and energy offer an alternative and often simpler way of thinking about what happens when a force causes some change in the motion of one or more objects. However, we cannot fully understand motion unless we also understand the forces that lead to changes in motion. In astronomy, the most important force is gravity, which governs virtually all large-scale motion in the universe.

◆ What determines the strength of gravity?

Isaac Newton discovered the basic law that describes how gravity works. Newton expressed the force of gravity mathematically with his **universal law of gravitation.** Three simple statements summarize this law:

- Every mass attracts every other mass through the force called *gravity.*
- The strength of the gravitational force attracting any two objects is *directly proportional* to the product of their masses. For example, doubling the mass of *one* object doubles the force of gravity between the two objects.
- The strength of gravity between two objects decreases with the *square* of the distance between their centers. We therefore say that the gravitational force follows an **inverse square law.** For example, doubling the distance between two objects weakens the force of gravity by a factor of 2^2, or 4.

Doubling the distance between two objects weakens the force of gravity by a factor of 2^2, or 4.

These three statements tell us everything we need to know about Newton's universal law of gravitation. Mathematically, all three statements can be combined into a single equation, usually written like this:

$$F_g = G \frac{M_1 M_2}{d^2}$$

where F_g is the force of gravitational attraction, M_1 and M_2 are the masses of the two objects, and d is the distance between their centers (Figure 4.14).

The **universal law of gravitation** *tells us the strength of the gravitational attraction between the two objects.*

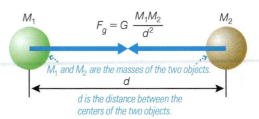

M_1 and M_2 *are the masses of the two objects.*

d *is the distance between the centers of the two objects.*

▲ **FIGURE 4.14**

The universal law of gravitation is an *inverse square law*, which means that the force of gravity declines with the *square* of the distance d between two objects.

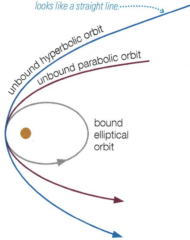

► **FIGURE 4.15**
Newton showed that ellipses are not the only possible orbital paths. Orbits can also be unbound, taking the mathematical shapes of either parabolas or hyperbolas.

Far from the focus, a hyperbolic orbit looks like a straight line.

unbound hyperbolic orbit

unbound parabolic orbit

bound elliptical orbit

The symbol G is a constant called the **gravitational constant,** and its numerical value has been measured to be $G = 6.67 \times 10^{-11}$ m^3/(kg $\times$ s^2).

think about it How does the gravitational force between two objects change if the distance between them triples? If the distance between them drops by half?

◆ How does Newton's law of gravity extend Kepler's laws?

By the time Newton published *Principia* in 1687, Kepler's laws of planetary motion [Section 3.3] had already been known for some 70 years and had proven so successful that there was little doubt about their validity. However, there was great debate among scientists about *why* Kepler's laws hold true, a debate resolved only when Newton showed mathematically that Kepler's laws are consequences of the laws of motion and the universal law of gravitation. In doing so, Newton discovered that he could generalize Kepler's laws in several ways, three of which are particularly important for our purposes.

First, Newton discovered that Kepler's first two laws apply to all orbiting objects, not just to planets orbiting the Sun. For example, the orbits of satellites around Earth, of moons around planets, and of asteroids around the Sun are all ellipses in which the orbiting object moves faster at the nearer points in its orbit and slower at the farther points.

Second, Newton found that ellipses are not the only possible orbital paths (Figure 4.15). Ellipses (which include circles) are the only possible shapes for **bound orbits**—orbits in which an object goes around another object over and over again. (The term *bound orbit* comes from the idea that gravity creates a *bond* that holds the objects together.) However, Newton discovered that objects can also follow **unbound orbits**—paths that bring an object close to another object just once. For example, some comets that enter the inner solar system follow unbound orbits. They come in from afar just once, loop around the Sun, and never return.

Third, and perhaps most important, Newton generalized Kepler's third law in a way that allows us to calculate the masses of distant objects. Recall that Kepler's third law states that $p^2 = a^3$, where p is a planet's orbital period in years and a is its average distance from the Sun in AU. Newton found that this statement is a special case of a more general equation that we call **Newton's version of Kepler's third law** (see Cosmic Calculations 4.1). This equation allows us to calculate the mass of a distant object if we measure the orbital period and distance of another object orbiting around it. For example, we can calculate the mass of the Sun from Earth's orbital period (1 year) and its average distance (1 AU); we can calculate Jupiter's mass from the orbital period and average distance of one of its moons; and we can determine the masses of distant stars if they are members of binary star systems, in which two stars orbit one another. In fact, Newton's version of Kepler's third law is the primary means by which we determine masses throughout the universe.

Newton's version of Kepler's third law allows us to calculate the masses of distant objects.

---

cosmic calculations 4.1

Newton's Version of Kepler's Third Law

For an object of mass M_1 orbiting another object of mass M_2, Newton's version of Kepler's third law states

$$p^2 = \frac{4\pi^2}{G(M_1 + M_2)}a^3$$

$$\left(G = 6.67 \times 10^{-11} \frac{m^3}{kg \times s^2}\right.$$

is the gravitational constant.)

This equation allows us to calculate the sum $M_1 + M_2$ if we know the orbital period p and average distance (semimajor axis) a. The equation is especially useful when one object is much more massive than the other.

Example: Use the fact that Earth orbits the Sun in 1 year at an average distance of 1 AU to calculate the Sun's mass.

Solution: Newton's version of Kepler's third law becomes

$$p_{Earth}^2 = \frac{4\pi^2}{G(M_{Sun} + M_{Earth})}a_{Earth}^3$$

Because the Sun is much more massive than Earth, the sum of their masses is nearly the mass of the Sun alone: $M_{Sun} + M_{Earth} \approx M_{Sun}$. Using this approximation, we find

$$p_{Earth}^2 \approx \frac{4\pi^2}{GM_{Sun}}a_{Earth}^3$$

We now solve for the mass of the Sun and plug in Earth's orbital period ($p_{Earth} = 1$ year $\approx 3.15 \times 10^7$ seconds) and average orbital distance ($a_{Earth} = 1$ AU $\approx 1.5 \times 10^{11}$ m):

$$M_{Sun} \approx \frac{4\pi^2 a_{Earth}^3}{Gp_{Earth}^2} \approx \frac{4\pi^2(1.5 \times 10^{11} \text{ m})^3}{\left(6.67 \times 10^{-11} \dfrac{m^3}{kg \times s^2}\right)(3.15 \times 10^7 \text{ s})^2}$$

$$= 2.0 \times 10^{30} \text{ kg}$$

The Sun's mass is about 2×10^{30} kilograms.

◆ How do gravity and energy allow us to understand orbits?

Newton's universal law of gravitation explains Kepler's laws of planetary motion, which describe the stable orbits of the planets, and Newton's extensions of Kepler's laws explain other stable orbits, such as the orbit of a satellite around Earth. But orbits do not always stay the same. For example, you've probably heard of satellites crashing to Earth from orbit, proving that orbits can sometimes change dramatically. To understand how and why orbits sometimes change, we need to consider the role of energy in orbits.

Orbital Energy A planet orbiting the Sun has both kinetic energy (because it is moving around the Sun) and gravitational potential energy (because it would fall toward the Sun if it stopped orbiting). The amount of kinetic energy depends on orbital speed, and the amount of gravitational potential energy depends on orbital distance. Because the planet's distance and speed both vary as it orbits the Sun, its gravitational potential energy and kinetic energy also vary (Figure 4.16). However, the planet's total **orbital energy**—the sum of its kinetic and gravitational potential energies—stays the same. This fact is a consequence of the law of conservation of energy. As long as no other object causes the planet to gain or lose orbital energy, its orbital energy cannot change and its orbit must remain the same.

> **Orbits cannot change spontaneously—an object's orbit can change only if it gains or loses orbital energy.**

Generalizing from planets to other objects leads to an important idea about motion throughout the cosmos: *Orbits cannot change spontaneously.* Left undisturbed, planets would forever keep the same orbits around the Sun, moons would keep the same orbits around planets, and stars would keep the same orbits in their galaxies.

Gravitational Encounters Although orbits cannot change spontaneously, they can change through exchanges of energy. One way that two objects can exchange orbital energy is through a **gravitational encounter,** in which they pass near enough that each can feel the effects of the other's gravity. For example, in the rare cases in which a comet happens to pass near a planet, the comet's orbit can change dramatically. Figure 4.17 shows a comet headed toward the Sun on an unbound orbit. The comet's close passage by Jupiter allows the comet and Jupiter to exchange energy. In this case, the comet loses so much orbital energy that its orbit changes from unbound to bound and elliptical. Jupiter gains exactly as much energy as the comet loses, but the effect on Jupiter is unnoticeable because of its much greater mass.

Spacecraft engineers can use the same basic idea in reverse. For example, on its way to Pluto, the *New Horizons* spacecraft was deliberately sent past Jupiter on a path that allowed it to gain orbital energy at Jupiter's expense. This extra orbital energy boosted the spacecraft's speed; without this boost, it would have needed four extra years to reach Pluto.

A similar dynamic can occur naturally and may explain why most comets orbit far from the Sun. Comets probably once orbited in the same region of the solar system as the large outer planets [Section 9.3]. Gravitational encounters with the planets then caused some of these comets to be "kicked out" into much more distant orbits around the Sun.

Total orbital energy = gravitational potential energy + kinetic energy

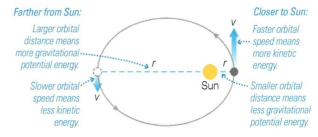

▲ **FIGURE 4.16**
The total orbital energy of a planet stays the same throughout its orbit, because its gravitational potential energy increases when its kinetic energy decreases, and vice versa.

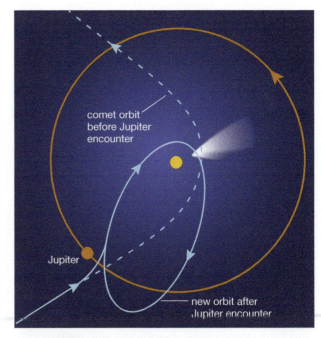

▲ **FIGURE 4.17**
This diagram shows a comet in an unbound orbit of the Sun that happens to pass near Jupiter. The comet loses orbital energy to Jupiter, changing its unbound orbit to a bound orbit around the Sun.

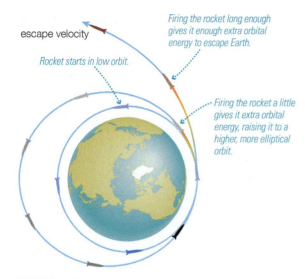

escape velocity

Firing the rocket long enough gives it enough extra orbital energy to escape Earth.

Rocket starts in low orbit.

Firing the rocket a little gives it extra orbital energy, raising it to a higher, more elliptical orbit.

▲ **FIGURE 4.18**

An object with escape velocity has enough orbital energy to escape Earth completely.

The Origin of Tides

Many people believe that tides arise because the Moon pulls Earth's oceans toward it. But if that were the whole story, there would be a bulge only on the side of Earth facing the Moon, and hence only one high tide each day. The correct explanation for tides must account for why Earth has two tidal bulges.

Only one explanation works: Earth must be stretching from its center in both directions (toward and away from the Moon). This stretching force, or tidal force, arises from the difference between the force of gravity attracting different parts of Earth to the Moon. In fact, stretching due to tides affects many objects, not just Earth. Many moons are stretched into slightly oblong shapes by tidal forces caused by their parent planets, and mutual tidal forces stretch close binary stars into teardrop shapes. In regions where gravity is extremely strong, such as near a black hole, tides can have even more dramatic effects.

Atmospheric Drag Friction can cause objects to lose orbital energy. A satellite in low-Earth orbit (a few hundred kilometers above Earth's surface) experiences a bit of drag from Earth's thin upper atmosphere. This drag gradually causes the satellite to lose orbital energy until it finally plummets to Earth. The satellite's lost orbital energy is converted to thermal energy in the atmosphere, which is why a falling satellite usually burns up.

Friction may also help explain why the outer planets have so many small moons. These moons may once have orbited the Sun independently, and their orbits could not have changed spontaneously. However, the outer planets probably once were surrounded by clouds of gas [Section 6.3], and friction would have slowed objects passing through this gas. Some of these small objects may have lost just enough energy to friction to allow them to be "captured" as moons. Mars may have captured its two small moons in a similar way.

Escape Velocity An object that gains orbital energy moves into an orbit with a higher average altitude. For example, if we want to boost the orbital altitude of a spacecraft, we can give it more orbital energy by firing a rocket. The chemical potential energy released by the rocket fuel is converted to orbital energy for the spacecraft.

If we give a spacecraft enough orbital energy, it may end up in an unbound orbit that allows it to *escape* Earth completely (Figure 4.18).

A spacecraft that achieves escape velocity can escape Earth completely.

For example, when we send a space probe to Mars, we must use a large rocket that gives the probe enough energy to leave Earth orbit. Although it would probably make more sense to say that the probe achieves "escape energy," we instead say that it achieves **escape velocity.** The escape velocity from Earth's surface is about 40,000 km/hr, or 11 km/s; this is the minimum velocity required to escape Earth's gravity for a spacecraft that starts near the surface.

Note that the escape velocity does not depend on the mass of the escaping object—*any* object must travel at a velocity of 11 km/s to escape from Earth, whether it is an individual atom or molecule escaping from the atmosphere, a spacecraft being launched into deep space, or a rock blasted into the sky by a large impact. Escape velocity *does* depend on whether you start from the surface or from someplace high above the surface. Because gravity weakens with distance, it takes less energy—and hence a lower velocity—to escape from a point high above Earth than from Earth's surface.

◆ How does gravity cause tides?

Newton's universal law of gravitation has applications that go far beyond explaining Kepler's laws and orbits. For our purposes, however, there is just one more topic we need to cover: how gravity causes tides.

If you've spent time near an ocean, you've probably observed the rising and falling of the tides. In most places, tides rise and fall twice each day. Tides arise because gravity attracts Earth and the Moon toward each other (with the Moon staying in orbit as it "falls around" Earth), but it affects different parts of Earth slightly differently: Because the strength of gravity declines with distance, the gravitational attraction of each part of Earth to the Moon becomes weaker as we go from the side of Earth facing the Moon to the side facing away from the Moon. This difference in attraction creates a "stretching force," or **tidal force,** that stretches the entire Earth to create two tidal bulges, one facing the Moon and one opposite the Moon (Figure 4.19). If you are still unclear about why there are *two* tidal bulges, think about a rubber band: If you pull on a rubber band, it will stretch in both directions relative to its center, even if you pull on only one side (while holding the other side still). In the same way, Earth stretches on both sides even though the Moon is tugging harder on only one side.

Tides affect both land and ocean, but we generally notice only the ocean tides because water flows much more readily than land. Earth's

Tidal forces cause the entire Earth to stretch along the Earth–Moon line, creating two tidal bulges.

rotation carries any location through each of the two bulges each day, creating two high tides. Low tides occur when the location is at the points halfway between the two tidal bulges. The height and timing of ocean tides varies considerably from place to place on Earth. For example, while the tide rises gradually in most locations,

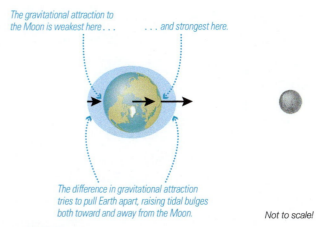

The gravitational attraction to the Moon is weakest here . . . . . . and strongest here.

The difference in gravitational attraction tries to pull Earth apart, raising tidal bulges both toward and away from the Moon.

Not to scale!

▲ **FIGURE 4.19**

Tides are created by the difference in the force of attraction between different parts of Earth and the Moon. The two daily high tides occur as a location on Earth rotates through the two tidal bulges. (The diagram highly exaggerates the tidal bulges, which raise the oceans only about 2 meters and the land only about a centimeter.)

---

special topic   Why Does the Moon Always Show the Same Face to Earth?

Tidal effects explain the Moon's *synchronous rotation,* in which the Moon keeps the same face toward Earth because its orbital period is the same as its rotation period. To understand why, let's start by considering the Moon's tidal effect on Earth.

By itself, the Moon's gravity would keep Earth's two tidal bulges on the Earth–Moon line. However, because the tidal bulges stretch Earth itself, Earth's rotation generates friction with the bulges that tries to pull them around with it. The resulting "compromise" keeps the bulges just ahead of the Earth–Moon line at all times (see the figure), which causes two important effects. First, the Moon's gravity always pulls back on the bulges, slowing Earth's rotation. Second, the gravity of the bulges pulls the Moon slightly ahead in its orbit, adding orbital energy that causes the Moon to move farther from Earth. These effects are barely noticeable on human time scales, but they add up over billions of years. Early in Earth's history, a day may have been only 5 or 6 hours long and the Moon may have been one-tenth or less of its current distance from Earth. These changes also provide a great example of conservation of angular momentum: The Moon's growing orbit gains the angular momentum that Earth loses as its rotation slows.

Now consider Earth's tidal force on the Moon, which must be greater than the Moon's tidal force on Earth because of Earth's greater mass. This tidal force gives the Moon two tidal bulges along the Earth–Moon line, much like the two tidal bulges that the Moon creates on Earth. (The Moon's tidal bulges are not visible but can be measured in terms of excess mass along the Earth–Moon line.) If the Moon rotated relative to its tidal bulges in the same way as

If Earth didn't rotate, tidal bulges would be oriented along the Earth–Moon line.

Friction with the rotating Earth pulls the tidal bulges slightly ahead of the Earth–Moon line.

The Moon's gravity tries to pull the bulges back into line, slowing Earth's rotation.

The gravity of the bulges pulls the Moon ahead, increasing its orbital distance.

Moon

Not to scale!

Earth's rotation pulls its tidal bulges slightly ahead of the Earth–Moon line, leading to gravitational effects that gradually slow Earth's rotation and increase the Moon's orbital distance.

Earth, the resulting tidal friction would cause the Moon's rotation to slow down. This is exactly what we think happened long ago.

The Moon probably once rotated much faster than it does today. As a result, it *did* rotate through its tidal bulges, and its rotation gradually slowed. Once the Moon's rotation slowed to the point at which the Moon and its bulges rotated at the same rate—that is, synchronously with the orbital period—there was no further source for tidal friction. The Moon's synchronous rotation was therefore a natural outcome of Earth's tidal effects on the Moon.

Similar tidal friction has led to synchronous rotation in many other cases. For example, Jupiter's four large moons (Io, Europa, Ganymede, and Callisto) keep nearly the same face toward Jupiter at all times, as do many other moons. Pluto and its moon Charon *both* rotate synchronously: Like two dancers, they always keep the same face toward each other. Many binary star systems also rotate in this way. Tidal forces may be most familiar because of their effects on our oceans, but they are important throughout the universe.

▲ **FIGURE 4.20**

Photographs of high and low tide at the abbey of Mont-Saint-Michel, France. Here the tide rushes in much faster than a person can swim. Before a causeway was built (visible at the far left), the Mont was accessible by land only at low tide. At high tide, it became an island.

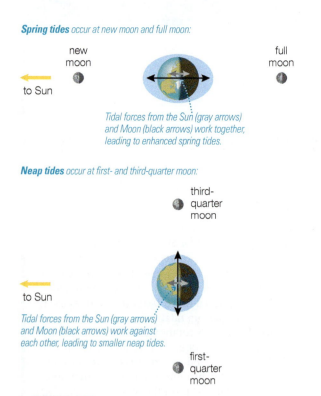

Spring tides occur at new moon and full moon:

Tidal forces from the Sun (gray arrows) and Moon (black arrows) work together, leading to enhanced spring tides.

Neap tides occur at first- and third-quarter moon:

Tidal forces from the Sun (gray arrows) and Moon (black arrows) work against each other, leading to smaller neap tides.

▲ **FIGURE 4.21**

The Sun exerts a tidal force on Earth less than half as strong as that from the Moon. When the tidal forces from the Sun and Moon work together at new moon and full moon, we get enhanced *spring tides*. When they work against each other, at first- and third-quarter moons, we get smaller *neap tides*.

the incoming tide near the famous abbey on Mont-Saint-Michel, France, moves much faster than a person can swim (Figure 4.20). In centuries past, the Mont was an island twice a day at high tide but was connected to the mainland at low tide. Many pilgrims drowned when they were caught unprepared by the tide rushing in. Another unusual tidal pattern occurs in coastal states along the northern shore of the Gulf of Mexico, where topography and other factors combine to make only one noticeable high tide and low tide each day.

The Sun also affects the tides. Although the Sun is much more massive than the Moon, its tidal effect on Earth is smaller because its much greater distance means that the *difference* in the Sun's pull on the near and far sides of Earth is relatively small. The overall tidal force caused by the Sun is a little less than half that caused by the Moon (Figure 4.21). When the tidal forces of the Sun and the Moon work together, as is the case at both new moon and full moon, we get the especially pronounced *spring tides* (so named because the water tends to "spring up" from Earth). When the tidal forces of the Sun and the Moon counteract each other, as is the case at first- and third-quarter moon, we get the relatively small tides known as *neap tides*.

think about it Explain why any tidal effects on Earth caused by the other planets would be unnoticeably small.

Tidal forces affect many objects in the solar system and universe. For example, Earth exerts tidal forces on the Moon that explain why the Moon always shows the same face to Earth (see Special Topic on page 99), and in Chapter 8 we'll see how tidal forces have led to the astonishing volcanic activity of Jupiter's moon Io and the possibility of a subsurface ocean on its moon Europa.

the big picture — Putting Chapter 4 into Perspective

We've covered a lot of ground in this chapter, from the scientific terminology of motion to the overarching principles that govern motion throughout the universe. Be sure you understand the following "big picture" ideas:

- Understanding the universe requires understanding motion. Motion may seem complex, but it can be described simply using Newton's three laws of motion.

- Today, we know that Newton's laws of motion stem from deeper physical principles, including the laws of conservation of angular momentum and of energy.

These principles enable us to understand a wide range of astronomical phenomena.

- Newton also discovered the universal law of gravitation, which explains how gravity holds planets in their orbits and much more—including how satellites can reach and stay in orbit, the nature of tides, and why the Moon rotates synchronously around Earth.

- Newton's discoveries showed that the same physical laws we observe on Earth apply throughout the universe. The universality of physics opens up the entire cosmos as a possible realm of human study.

my cosmic perspective The laws of physics that we use today in everything from building bridges to sending spacecraft to distant objects were first inspired and tested by astronomical observations.

summary of key concepts

4.1 Describing Motion: Examples from Daily Life

◆ How do we describe motion?

Speed is the rate at which an object is moving. **Velocity** is speed in a certain direction. **Acceleration** is a change in velocity, meaning a change in either speed or direction. **Momentum** is mass × velocity. A **force** can change an object's momentum, causing it to accelerate.

◆ How is mass different from weight?

An object's **mass** is the same no matter where it is located, but its **weight** varies with the strength of gravity or other forces acting on the object. An object becomes **weightless** when it is in **free-fall,** even though its mass is unchanged.

4.2 Newton's Laws of Motion

◆ How did Newton change our view of the universe?

Newton showed that the same physical laws that operate on Earth also operate in the heavens, making it possible to learn about the universe by studying physical laws on Earth.

◆ What are Newton's three laws of motion?

(1) An object moves at constant velocity if there is no net force acting upon it. (2) Force = mass × acceleration ($F = ma$). (3) For any force, there is always an equal and opposite reaction force.

4.3 Conservation Laws in Astronomy

◆ What keeps a planet rotating and orbiting the Sun?

Conservation of angular momentum means that a planet's rotation and orbit cannot change unless it transfers angular momentum to another object. The planets in our solar system do not exchange substantial angular momentum with each other or anything else, so their orbits and rotation rates remain fairly steady.

◆ Where do objects get their energy?

kinetic energy

radiative energy   potential energy

Energy is always conserved—it can be neither created nor destroyed. Objects received whatever energy they now have from exchanges of energy with other objects. Energy comes in three basic categories—**kinetic, radiative, and potential.**

4.4 The Force of Gravity

◆ What determines the strength of gravity?

The **universal law of gravitation** states that every object attracts every other object with a gravitational force that is directly proportional to the product of the objects' masses and declines with the square of the distance between their centers:

$$F_g = G \frac{M_1 M_2}{d^2}$$

How does Newton's law of gravity extend Kepler's laws?

(1) Newton showed that Kepler's first two laws apply to all orbiting objects, not just planets. (2) He showed that elliptical **bound orbits** are not the only possible orbital shape—orbits can also be **unbound** (taking the shape of a parabola or a hyperbola). (3) **Newton's version of Kepler's third law** allows us to calculate the masses of orbiting objects from their orbital periods and distances.

How do gravity and energy allow us to understand orbits?

Gravity determines orbits, and an object cannot change its orbit unless it gains or loses **orbital energy**—the sum of its kinetic and gravitational potential energy—through energy exchange with other objects. If an object gains enough orbital energy, it may achieve **escape velocity** and leave the gravitational influence of the object it was orbiting.

How does gravity cause tides?

The Moon's gravity creates a **tidal force** that stretches Earth along the Earth–Moon line, causing Earth to bulge both toward and away from the Moon. Earth's rotation carries us through the two bulges each day, giving us two daily high tides and two daily low tides.

---

visual skills check

Check your understanding of some of the many types of visual information used in astronomy. For additional practice, try the Chapter 4 Visual Quiz at MasteringAstronomy.

The figure represents how the Moon causes tides on Earth; the diagram looks down from above the North Pole, so the numbers 1 through 4 label points along Earth's equator.

1. What do the three black arrows represent?
   a. the tidal force Earth exerts on the Moon
   b. the Moon's gravitational force at different points on Earth
   c. the direction in which Earth's water is flowing
   d. Earth's orbital motion

2. Where is it high tide?
   a. point 1 only   b. point 2 only
   c. points 1 and 3   d. points 2 and 4

3. Where is it low tide?
   a. point 1 only   b. point 2 only
   c. points 1 and 3   d. points 2 and 4

4. What time is it at point 1?
   a. noon
   b. midnight
   c. 6 a.m.
   d. cannot be determined from the information in the figure

5. The light blue region represents tidal bulges. In what way are these bulges drawn inaccurately?
   a. There should be only one bulge rather than two.
   b. They should be aligned with the Sun rather than the Moon.
   c. They should be much smaller compared to Earth.
   d. They should be more pointy in shape.

---

exercises and problems

MasteringAstronomy® For instructor-assigned homework and other learning materials, go to MasteringAstronomy®.

Review Questions

1. Define *speed*, *velocity*, and *acceleration*. What are the units of acceleration? What is the *acceleration of gravity*?
2. Define *momentum* and *force*. What do we mean when we say that momentum can be changed only by a *net force*?
3. What is *free-fall*, and why does it make you *weightless*? Briefly describe why astronauts are weightless in the International Space Station.
4. State *Newton's three laws of motion*. For each law, give an example of its application.
5. Describe the laws of *conservation of angular momentum* and *conservation of energy*. Give an example of how each is important in astronomy.
6. Define *kinetic energy*, *radiative energy*, and *potential energy*, with at least two examples for each.
7. Define and distinguish *temperature* and *thermal energy*.

8. What is *mass-energy*? Explain the formula $E = mc^2$.
9. Summarize the *universal law of gravitation* both in words and with an equation.
10. What is the difference between a *bound* and an *unbound orbit*?
11. Under what conditions can we use *Newton's version of Kepler's third law* to calculate an object's mass, and what quantities must we measure in order to complete the calculation?
12. Explain why orbits cannot change spontaneously, and how a *gravitational encounter* can cause a change. How can an object achieve *escape velocity*?
13. Explain how the Moon creates tides on Earth. Why do we have two high and low tides each day?
14. How do the tides vary with the phase of the Moon? Why?

Test Your Understanding

Does It Make Sense?

Decide whether the statement makes sense (or is clearly true) or does not make sense (or is clearly false). Explain clearly; not all of these have definitive answers, so your explanation is more important than your chosen answer.

15. I've never been to space, so I've never experienced weightlessness.
16. Suppose you could enter a vacuum chamber (a chamber with no air in it) on Earth. Inside this chamber, a feather would fall at the same rate as a rock.
17. If an astronaut goes on a space walk outside the Space Station, she will quickly float away from the station unless she has a tether holding her to the station.
18. I used Newton's version of Kepler's third law to calculate Saturn's mass from orbital characteristics of its moon Titan.
19. If the Sun were magically replaced with a giant rock that had precisely the same mass, Earth's orbit would not change.
20. The fact that the Moon rotates once in precisely the time it takes to orbit Earth once is such an astonishing coincidence that scientists probably never will be able to explain it.
21. Venus has no oceans, so it could not have tides even if it had a moon (which it doesn't).
22. If an asteroid passed by Earth at just the right distance, Earth's gravity would capture it and make it our second moon.
23. When I drive my car at 30 miles per hour, it has more kinetic energy than it does at 10 miles per hour.
24. Someday soon, scientists are likely to build an engine that produces more energy than it consumes.

Quick Quiz

Choose the best answer to each of the following. Explain your reasoning with one or more complete sentences.

25. A car is accelerating when it is (a) traveling on a straight, flat road at 50 miles per hour. (b) traveling on a straight uphill road at 30 miles per hour. (c) going around a circular track at a steady 100 miles per hour.
26. Compared to their values on Earth, on another planet your (a) mass and weight would both be the same. (b) mass would be the same but your weight would be different. (c) weight would be the same but your mass would be different.
27. Which person is weightless? (a) a child in the air as she plays on a trampoline (b) a scuba diver exploring a deep-sea wreck (c) an astronaut on the Moon
28. Consider the statement "There's no gravity in space." This statement is (a) completely false. (b) false if you are close to a planet or moon, but true in between the planets. (c) completely true.

29. To make a rocket turn left, you need to (a) fire an engine that shoots out gas to the left. (b) fire an engine that shoots out gas to the right. (c) spin the rocket clockwise.
30. Compared to its angular momentum when it is farthest from the Sun, Earth's angular momentum when it is nearest to the Sun is (a) greater. (b) less. (c) the same.
31. The gravitational potential energy of a contracting interstellar cloud (a) stays the same at all times. (b) gradually transforms into other forms of energy. (c) gradually grows larger.
32. If Earth were twice as far from the Sun, the force of gravity attracting Earth to the Sun would be (a) twice as strong. (b) half as strong. (c) one-quarter as strong.
33. According to the law of universal gravitation, what would happen to Earth if the Sun were somehow replaced by a black hole of the same mass? (a) Earth would be quickly sucked into the black hole. (b) Earth would slowly spiral into the black hole. (c) Earth's orbit would not change.
34. If the Moon were closer to Earth, high tides would (a) be higher than they are now. (b) be lower than they are now. (c) occur three or more times a day rather than twice a day.

Process of Science

35. *Testing Gravity.* Scientists are constantly trying to learn whether our current understanding of gravity is complete or must be modified. Describe how the observed motion of spacecraft headed out of the solar system (such as the *Voyager* spacecraft) can be used to test the accuracy of our current theory of gravity.
36. *How Does the Table Know?* Thinking deeply about seemingly simple observations sometimes reveals underlying truths that we might otherwise miss. For example, think about holding a golf ball in one hand and a bowling ball in the other. To keep them motionless, you must actively adjust the tension in your arm muscles so that each arm exerts a different upward force that exactly balances the weight of each ball. Now, think about what happens when you set the balls on a table. Somehow, the table exerts exactly the right amount of upward force to keep the balls motionless, even though their weights are very different. How does a table "know" to make the same type of adjustment that you make consciously when you hold the balls motionless in your hands? (*Hint:* Think about the origin of the force pushing upward on the objects.)

Group Work Exercise

37. *Your Ultimate Energy Source.* **Roles:** *Scribe* (takes notes on the group's activities), *Proposer* (proposes explanations to the group), *Skeptic* (points out weaknesses in proposed explanations), *Moderator* (leads group discussion and makes sure the group works as a team). **Activity:** According to the law of conservation of energy, the energy your body is using right now had to come from somewhere else. Make a list going backwards in time describing how the energy you are using right now has proceeded through time. For each item on the list, identify the energy as kinetic energy, gravitational potential energy, chemical potential energy, electrical potential energy, mass-energy, or radiative energy.

Investigate Further

Short-Answer/Essay Questions

38. *Weightlessness.* Astronauts are weightless when in orbit in the Space Station. Are they also weightless during launch to the station? How about during their return to Earth? Explain.

39. *Einstein's Famous Formula.*
    a. What is the meaning of the formula $E = mc^2$? Define each variable.
    b. How does this formula explain the generation of energy by the Sun?
    c. How does this formula explain the destructive power of nuclear bombs?
40. *The Gravitational Law.*
    a. How does quadrupling the distance between two objects affect the gravitational force between them?
    b. Suppose the Sun were somehow replaced by a star with twice as much mass. What would happen to the gravitational force between Earth and the Sun?
    c. Suppose Earth were moved to one-third of its current distance from the Sun. What would happen to the gravitational force between Earth and the Sun?
41. *Allowable Orbits?*
    a. Suppose the Sun were replaced by a star with twice as much mass. Could Earth's orbit stay the same? Why or why not?
    b. Suppose Earth doubled in mass (but the Sun stayed the same as it is now). Could Earth's orbit stay the same? Why or why not?
42. *Head-to-Foot Tides.* You and Earth attract each other gravitationally, so you should also be subject to a tidal force resulting from the difference between the gravitational attraction felt by your feet and that felt by your head (at least when you are standing). Explain why you can't feel this tidal force.

Quantitative Problems

Be sure to show all calculations clearly and state your final answers in complete sentences.

43. *Energy Comparisons.* Use the data in Table 4.1 to answer each of the following questions.
    a. Compare the energy of a 1-megaton H-bomb to the energy released by a major earthquake.
    b. If the United States obtained all its energy from oil, how much oil would be needed each year?
    c. Compare the Sun's annual energy output to the energy released by a supernova.
44. *Fusion Power.* No one has yet succeeded in creating a commercially viable way to produce energy through nuclear fusion. However, suppose we could build fusion power plants using the hydrogen in water as a fuel. Based on the data in Table 4.1, how much water would we need each minute to meet U.S. energy needs? Could such a reactor power the entire United States with the water flowing from your kitchen sink? Explain. (*Hint:* Use the annual U.S. energy consumption to find the energy consumption per minute, and then divide by the energy yield from fusing 1 liter of water to figure out how many liters would be needed each minute.)
45. *Understanding Newton's Version of Kepler's Third Law.* Find the orbital period for the planet in each case. (*Hint:* The calculations for this problem are so simple that you will not need a calculator.)
    a. A planet with twice Earth's mass orbiting at a distance of 1 AU from a star with the same mass as the Sun.
    b. A planet with the same mass as Earth orbiting at a distance of 1 AU from a star with four times the Sun's mass.

46. *Using Newton's Version of Kepler's Third Law.*
    a. Find Earth's approximate mass from the fact that the Moon orbits Earth in an average time of 27.3 days at an average distance of 384,000 kilometers. (*Hint:* The Moon's mass is only about $\frac{1}{80}$ of Earth's.)
    b. Find Jupiter's mass from the fact that its moon Io orbits every 42.5 hours at an average distance of 422,000 kilometers.
    c. You discover a planet orbiting a distant star that has about the same mass as the Sun, with an orbital period of 63 days. What is the planet's orbital distance?
    d. Pluto's moon Charon orbits Pluto every 6.4 days with a semimajor axis of 19,600 kilometers. Calculate the *combined* mass of Pluto and Charon.
    e. Calculate the orbital period of a spacecraft in an orbit 300 kilometers above Earth's surface.
    f. Estimate the mass of the Milky Way Galaxy from the fact that the Sun orbits the galactic center every 230 million years at a distance of 27,000 light-years. (As we'll discuss in Chapter 15, this calculation actually tells us only the mass of the galaxy *within* the Sun's orbit.)

Discussion Questions

47. *Knowledge of Mass-Energy.* Einstein's discovery that energy and mass are equivalent has led to technological developments that are both beneficial and dangerous. Discuss some of these developments. Overall, do you think the human race would be better or worse off if we had never discovered that mass is a form of energy? Defend your opinion.
48. *Perpetual Motion Machines.* Every so often, someone claims to have built a machine that can generate energy perpetually from nothing. Why isn't this possible according to the known laws of nature? Why do you think claims of perpetual motion machines sometimes receive substantial media attention?

Web Projects

49. *Space Station.* Visit a NASA site with pictures from the International Space Station. Choose two photos that illustrate some facet of Newton's laws. Explain how Newton's laws apply to each photo.
50. *Nuclear Power.* There are two basic ways to generate energy from atomic nuclei: through nuclear fission (splitting nuclei) and through nuclear fusion (combining nuclei). All current nuclear reactors are based on fission, but using fusion would have many advantages if we could develop the technology. Research some of the advantages of fusion and some of the obstacles to developing fusion power. Do you think fusion power will be a reality in your lifetime? Explain.
51. *Space Elevator.* Some people have proposed using a "space elevator" to reach orbit high above Earth. Learn about the concept and write a short report on how it works, what advantages it would have over rockets, and whether it is feasible.

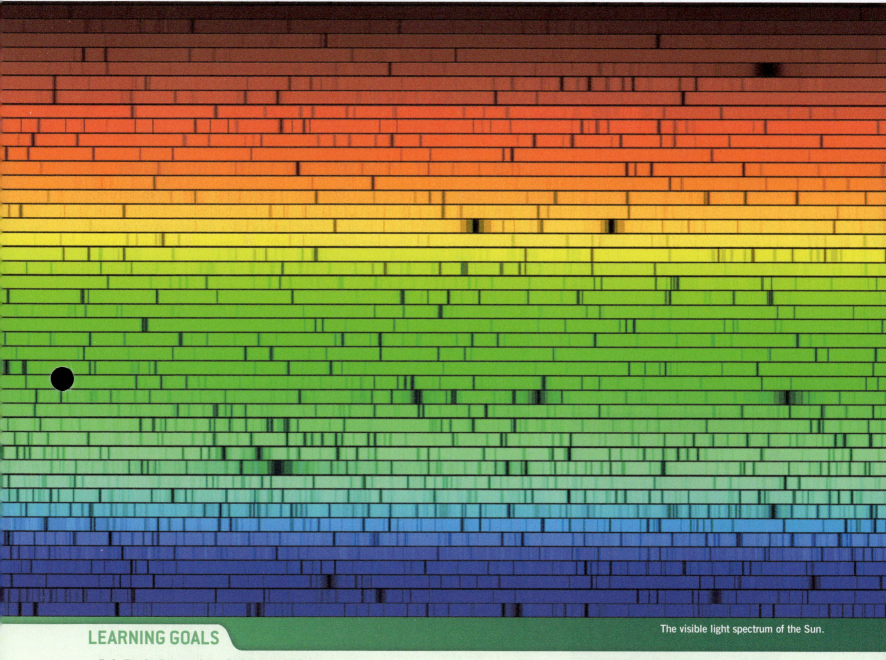

5 Light
The Cosmic Messenger

The visible light spectrum of the Sun.

LEARNING GOALS

5.1 Basic Properties of Light and Matter
- What is light?
- What is matter?
- How do light and matter interact?

5.2 Learning from Light
- What are the three basic types of spectra?
- How does light tell us what things are made of?
- How does light tell us the temperatures of planets and stars?
- How does light tell us the speed of a distant object?

5.3 Collecting Light with Telescopes
- How do telescopes help us learn about the universe?
- Why do we put telescopes in space?

ESSENTIAL PREPARATION

1. How did Galileo solidify the Copernican revolution? [Section 3.3]

2. What are Newton's three laws of motion? [Section 4.2]

3. Where do objects get their energy? [Section 4.3]

Ancient observers could discern only the most basic features of the light that they saw, such as color and brightness, but we now know that light carries far more information. Today, we can analyze the light of distant objects to learn what they are made of, how hot they are, how fast they are moving, and much more. Light is truly the cosmic messenger, bringing the stories of distant objects to Earth.

In this chapter, we'll explore the basic properties of light and matter that allow us to understand the messages encoded in light. We'll also discuss telescopes and technologies that allow us to collect and study light.

5.1 Basic Properties of Light and Matter

The chapter opening photo shows a detailed view of the Sun's **spectrum,** in which the colors of sunlight are spread out in horizontal rows from the upper left to the lower right. The rainbow of color probably reminds you of what we see when we pass white light through a prism (Figure 5.1). However, notice that the Sun's spectrum contains hundreds of dark lines, representing places where a small piece of the rainbow is missing. We see similar dark or bright lines when we look at almost any spectrum in detail, whether it is the spectrum of the flame from a backyard gas grill or the spectrum of a distant galaxy whose light we collect with a gigantic telescope.

The features in an object's spectrum are created by interactions between light and matter in the object. As a result, careful study of a spectrum can tell us a great deal about the object from which it comes—but only if we first understand the basic properties of light and matter.

◆ What is light?

Light is familiar to all of us, but its nature remained a mystery for most of human history. Experiments performed by Isaac Newton in the 1660s provided the first real insights into the nature of light. It was already known that passing white light through a prism produced a rainbow of color, but many people thought the colors came from the prism rather than from the light itself. Newton demonstrated that the colors came from the light by placing a second prism in front of the light of just one color, such as red, from the first prism. If the rainbow of color came from the prism itself, the second prism would have produced a rainbow just like the first. But it did not: When only red light entered the second prism, only red light emerged, proving that the color was a property of the light and not of the prism.

Newton's experiment showed that *white* light is a mix of all the colors in the rainbow. Later scientists found that just as there are sounds that our ears cannot hear (such as the sound of a dog whistle), there is light "beyond the rainbow" that our eyes cannot see. In fact, the **visible light** that splits into the rainbow of color is only a tiny part of the complete spectrum of light, usually called the **electromagnetic spectrum** for reasons that will soon be clear (Figure 5.2). Light itself is often called **electromagnetic radiation.**

Light is also known as electromagnetic radiation.

▲ **FIGURE 5.1**
When we pass white light through a prism, It disperses into a rainbow of color that we call a *spectrum*.

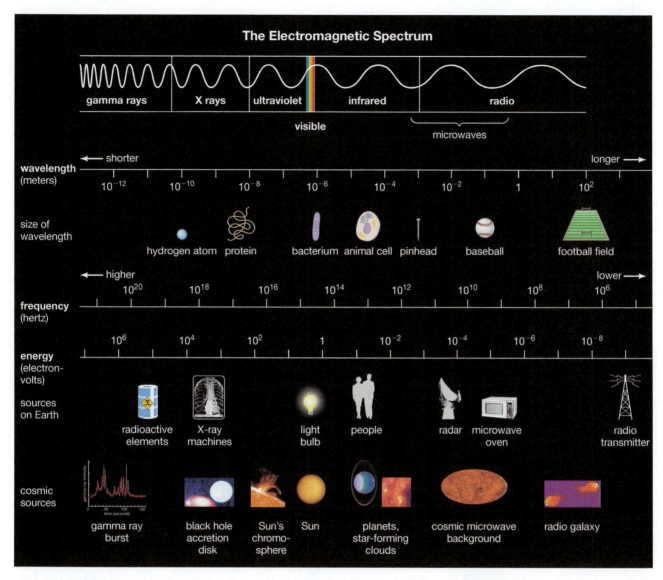

▲ FIGURE 5.2

The electromagnetic spectrum. Notice that wavelength increases as we go from gamma rays to radio waves, while frequency and energy increase in the opposite direction. (Energy is given in units of electron-volts [eV]: 1 eV = 1.60 × 10⁻¹⁹ joule.)

Wave Properties of Light   You've probably heard that light is a wave, but what does that mean? In general, a wave is something that can transmit energy without carrying material along with it. For example, shaking the end of a rope makes every piece of the rope bob up and down, creating a wave of peaks and troughs that move along the rope even though the rope itself doesn't go anywhere (Figure 5.3a). The **wavelength** is the distance between adjacent peaks and the **frequency** is the number of times that any piece of the rope moves up and down each second. For example, if a piece of the rope moves up and down three times each second, the wave has a frequency of three cycles per second, or three *hertz* for short.

Light is different from waves on a rope because we cannot see anything moving up and down as it travels. However, we can tell that light is a wave from its effect on matter. If you could set up a row of electrically charged particles such as electrons, it would wriggle like a snake as a wave of light passed by (Figure 5.3b). The distance between adjacent peaks would tell us the wavelength and the number of times each electron bobbed up and down would tell us the frequency (Figure 5.3c). Because light can interact with both electrically charged particles and magnetic

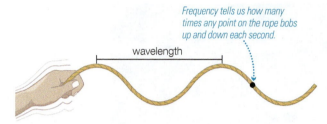

Frequency tells us how many times any point on the rope bobs up and down each second.

wavelength

a Shaking one end of a rope up and down generates waves moving along it.

b If you could line up electrons, they would wriggle up and down as light passes by, demonstrating that light is a wave.

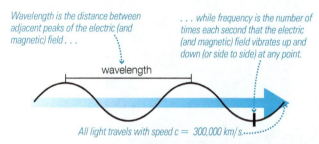

Wavelength is the distance between adjacent peaks of the electric (and magnetic) field . . .

. . . while frequency is the number of times each second that the electric (and magnetic) field vibrates up and down (or side to side) at any point.

wavelength

All light travels with speed c = 300,000 km/s.

c Light can affect both electrically charged particles and magnets, so we say that light is an *electromagnetic wave.*

▲ **FIGURE 5.3**

These diagrams explain the wave properties of light.

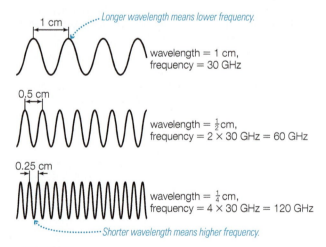

1 cm

Longer wavelength means lower frequency.

wavelength = 1 cm, frequency = 30 GHz

0.5 cm

wavelength = $\frac{1}{2}$ cm, frequency = 2 × 30 GHz = 60 GHz

0.25 cm

wavelength = $\frac{1}{4}$ cm, frequency = 4 × 30 GHz = 120 GHz

Shorter wavelength means higher frequency.

▲ **FIGURE 5.4**

Because all light travels through space at the same speed, light of longer wavelength must have lower frequency, and vice versa. (GHz stands for gigahertz, or 10^9 hertz.)

fields, we say that light is an **electromagnetic wave** (which explains the origin of the terms *electromagnetic radiation* and *electromagnetic spectrum*).

All light travels through empty space at the same speed—the **speed of light**—which is about 300,000 kilometers per second. The speed of any wave is its wavelength times its frequency, and this fact leads to an important relationship between wavelength and frequency for light: *The longer the wavelength, the lower the frequency, and vice versa* (Figure 5.4).

The longer the wavelength of light, the lower its frequency and energy.

Particle Properties of Light In everyday life, waves seem very different from particles. A wave exists only as a pattern of motion with a wavelength and a frequency, while a particle is a "thing" such as a marble, a baseball, or an individual atom. However, experiments show that light can behave both as a wave and as a particle.

The idea that light can be both a wave and a particle may seem quite strange, but it is fundamental to our modern understanding of physics. We think of light as consisting of many individual "pieces," or **photons.** Like baseballs, photons of light can be counted individually and can hit a wall one at a time. Like waves, each photon travels at the speed of light and is characterized by a wavelength and frequency. Moreover, each photon carries a particular amount of energy that depends on its frequency: The higher the frequency of the photon, the more energy it carries. That is why energy increases in the same direction as frequency in Figure 5.2.

Light comes in "pieces" called photons, each with a precise wavelength, frequency, and energy.

think about it Does the energy of a photon also depend on its wavelength? Explain.

The Many Forms of Light Figure 5.2 also shows the names we give to different portions of the electromagnetic spectrum. Visible light has wavelengths ranging from about 400 nm at the blue or violet end of the rainbow to about 700 nm at the red end. (A nanometer [nm] is a billionth of a meter.) Light with wavelengths somewhat longer than red light is called **infrared,** because it lies beyond the red end of the rainbow. **Radio waves** are the longest-wavelength light. The region near the border between infrared and radio waves, where wavelengths range from micrometers to centimeters, is often called **microwaves.** In astronomy, microwaves are sometimes divided further: Wavelengths from about one to a few millimeters are called *millimeter waves,* while wavelengths of tenths of a millimeter are called *submillimeter waves.*

On the other side of the spectrum, light with wavelengths somewhat shorter than blue light is called **ultraviolet,** because it lies beyond the blue (or violet) end of the rainbow. Light with even shorter wavelengths is called **X rays,** and the shortest-wavelength light is called **gamma rays.** Notice that visible light is an extremely small part of the entire electromagnetic spectrum: The reddest red that our eyes can see has only about twice the wavelength of the bluest blue, but the radio waves from your favorite radio station are a billion times as long as the X rays used in a doctor's office.

Radio waves, microwaves, infrared, visible light, ultraviolet, X rays, and gamma rays are all forms of light.

The various energies of light explain many familiar effects in everyday life. Radio waves carry so little energy that they have no noticeable effect on our bodies, but they can make electrons move up and down in an antenna, making them useful for radio communication. Molecules moving in a warm object emit infrared light, which is why we sometimes associate infrared light with heat. Receptors in our eyes respond to visible-light photons, making vision possible. Ultraviolet photons carry enough energy to damage skin cells, causing sunburn or cancer. X-ray photons have enough energy to penetrate skin and muscle but can be blocked by bones or teeth, which is why they can be used to make images of bone or tooth structures.

◆ What is matter?

We are usually more interested in the matter that light is coming from—such as planets, stars, and galaxies—than we are in the light itself. We must therefore explore the nature of matter if we are to decode the messages carried by light.

The ancient Greeks imagined that all material was made of four elements: fire, water, earth, and air. Some Greeks, beginning with the philosopher Democritus (c. 470–380 B.C.), further imagined that these four elements came in the form of tiny particles they called *atoms*, a Greek term meaning "indivisible." Modern ideas are similar in spirit but differ in many details. For example, we now know of more than 100 chemical **elements,** each composed of a different type of atom. Some of the most familiar elements are hydrogen, helium, carbon, oxygen, silicon, iron, gold, silver, lead, and uranium. (See Appendix D for a complete list.) Moreover, the atoms that make up the elements are composed of even smaller particles.

Atomic Structure   Atoms are made of particles that we call **protons, neutrons,** and **electrons** (Figure 5.5). Protons and neutrons are found in the tiny **nucleus** at the center of the atom. The rest of the atom's volume contains the electrons that surround the nucleus. Although

The chemical elements are made of atoms, which in turn are made of protons, neutrons, and electrons.

the nucleus is very small compared to the atom as a whole, it contains most of the atom's mass, because protons and neutrons are each about 2000 times as massive as an electron. Note that atoms are incredibly small: Millions could fit end to end across the period at the end of this sentence. The number of atoms in a single drop of water (typically, 10^{22} to 10^{23} atoms) may exceed the number of stars in the observable universe.

The properties of an atom depend mainly on the **electrical charge** in its nucleus. Electrical charge is a fundamental physical property that describes how strongly an object will interact with electromagnetic fields; total electrical charge is always conserved, just as energy is always conserved. We define the electrical charge of a proton as the basic unit of positive charge, which we write as +1. An electron has an electrical charge that is precisely opposite that of a proton, so we say it has negative charge (−1). Neutrons are electrically neutral, meaning that they have no charge.

Oppositely charged particles attract, and similarly charged particles repel. The attraction between the positively charged protons in the nucleus and the negatively charged electrons that surround it is what holds an atom together. Ordinary atoms have identical numbers of electrons and protons, making them electrically neutral overall. (You may wonder why electrical repulsion doesn't cause the positively charged protons in

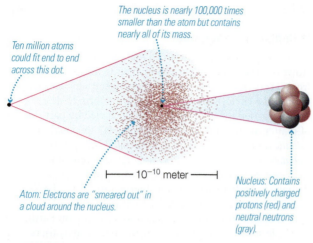

The nucleus is nearly 100,000 times smaller than the atom but contains nearly all of its mass.

Ten million atoms could fit end to end across this dot.

|— 10^{-10} meter —|

Atom: Electrons are "smeared out" in a cloud around the nucleus.

Nucleus: Contains positively charged protons (red) and neutral neutrons (gray).

▲ **FIGURE 5.5**
The structure of a typical atom. Note that atoms are extremely tiny: The atom shown in the middle is magnified to about 1 billion times its actual size, and the nucleus on the right is magnified to about 100 trillion times its actual size.

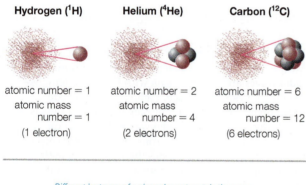

atomic number = number of protons
atomic mass number = number of protons + neutrons
(A neutral atom has the same number of electrons as protons.)

Hydrogen (^{1}H)

Helium (^{4}He)

Carbon (^{12}C)

atomic number = 1
atomic mass number = 1
(1 electron)

atomic number = 2
atomic mass number = 4
(2 electrons)

atomic number = 6
atomic mass number = 12
(6 electrons)

Different **isotopes** of a given element contain the same number of protons, but different numbers of neutrons.

Isotopes of Carbon

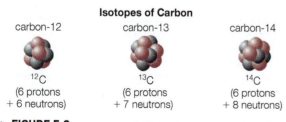

carbon-12

carbon-13

carbon-14

^{12}C
(6 protons + 6 neutrons)

^{13}C
(6 protons + 7 neutrons)

^{14}C
(6 protons + 8 neutrons)

▲ **FIGURE 5.6**
Terminology of atoms.

a nucleus to fly apart from one another. The answer is that an even stronger force, called the **strong force,** overcomes electrical repulsion and holds the nucleus together [Section 11.2].)

Although we can think of electrons as tiny particles, they are not quite like tiny grains of sand and they don't orbit the nucleus the way planets orbit the Sun. Instead, the electrons in an atom form a kind of "smeared out" cloud that surrounds the nucleus and gives the atom its apparent size. The electrons aren't really cloudy, but it is impossible to pinpoint their positions in the atom. The electrons therefore give the atom a size far larger than its nucleus even though they represent only a tiny portion of the atom's mass. If we imagine an atom on a scale that makes its nucleus the size of your fist, its electron cloud would be many kilometers wide.

Atomic Terminology   You've probably learned the basic terminology of atoms in past science classes, but let's review it just to be sure. Figure 5.6 summarizes the key terminology we will use in this book.

Each different chemical element contains a different number of protons in its nucleus. This number is its **atomic number.** For example, a hydrogen nucleus contains just one proton, so its atomic number is 1. A helium nucleus contains two protons, so its atomic number is 2. The *combined* number of protons and neutrons in an atom is called its **atomic mass number.** The atomic mass number of ordinary hydrogen is 1 because its nucleus is just a single proton. Helium usually has two neutrons in addition to its two protons, giving it an atomic mass number of 4. Carbon usually has six protons and six neutrons, giving it an atomic mass number of 12.

Atoms of different chemical elements have different numbers of protons.

Every atom of a given element contains exactly the same number of protons, but the number of neutrons can vary. For example, all carbon atoms have six protons, but they may have six, seven, or eight neutrons. Versions of an element with different numbers of neutrons are called **isotopes** of that element. Isotopes are named with their element name and atomic mass number. For example, the most common isotope of carbon has six protons and six neutrons, giving it atomic mass number $6 + 6 = 12$, so we call it *carbon-12*. Other isotopes of carbon are carbon-13 (six protons and seven neutrons) and carbon-14 (six protons and eight neutrons). We sometimes write the atomic mass number as a superscript to the left of the element symbol: ^{12}C, ^{13}C, ^{14}C. We read ^{12}C as "carbon-12."

Isotopes of a particular chemical element all have the same number of protons but different numbers of neutrons.

think about it   The symbol ^{4}He represents helium with an atomic mass number of 4. ^{4}He is the most common form of helium, containing two protons and two neutrons. What does the symbol ^{3}He represent?

The number of different material substances is far greater than the number of chemical elements because atoms can combine to form **molecules.** Some molecules consist of two or more atoms of the same element. For example, we breathe O_2, oxygen molecules made of two oxygen atoms. Other molecules, such as water, are made up of atoms of two or more different elements. The symbol H_2O tells us that a water molecule contains two hydrogen atoms and one oxygen atom. The chemical properties of a molecule are different from those of its individual atoms. For example, water behaves very differently than pure hydrogen or pure oxygen.

◆ How do light and matter interact?

Now that we have discussed the nature of light and of matter individually, we are ready to explore how light and matter interact. Energy carried by light can interact with matter in four general ways:

- **Emission:** A light bulb *emits* visible light; the energy of the light comes from electrical potential energy supplied to the bulb.
- **Absorption:** When you place your hand near an incandescent light bulb, your hand *absorbs* some of the light, and this absorbed energy warms your hand.
- **Transmission:** Some forms of matter, such as glass and air, *transmit* light, which means allowing it to pass through.
- **Reflection/scattering:** Light can bounce off matter, leading to what we call *reflection* (when the bouncing is all in the same general direction) or *scattering* (when the bouncing is more random).

Materials that transmit light are said to be *transparent*, and materials that absorb light are called *opaque*. Many materials are neither perfectly transparent nor perfectly opaque. For example, dark sunglasses and

Matter can emit, absorb, transmit, or reflect light.

clear eyeglasses are both partially transparent, but the dark glasses absorb more light and transmit less. Materials often interact differently with different colors of light. For example, red glass transmits red light but absorbs other colors, while a green lawn reflects (scatters) green light but absorbs all other colors.

Let's put these ideas together to understand what happens when you walk into a room and turn on the light switch (Figure 5.7). The light bulb begins to emit white light, which is a mix of all the colors in the spectrum. Some of this light exits the room, transmitted through the windows. The rest of the light strikes the surfaces of objects inside the room, and the material properties of each object determine the colors it absorbs or reflects. The light coming from each object therefore carries an enormous amount of information about the object's location, shape and structure, and composition. You acquire this information

▼ **FIGURE 5.7**

This diagram shows examples of the four basic interactions between light and matter: emission, absorption, transmission, and reflection (or scattering).

The Sun and the lamp both **emit** light.

The mirror **reflects** all colors of visible light.

Special cells in the eye **absorb** light, leading to vision.

The chair is red because it **scatters** red light but **absorbs** all other colors.

The snow **absorbs** some light, which aids melting . . .

. . . but **scatters** most light, so it looks bright.

The glass **transmits** all colors of visible light.

when light enters your eyes, where special cells in your retina absorb it and send signals to your brain. Your brain interprets the messages that light carries, recognizing materials and objects in the process we call *vision*.

5.2 Learning from Light

Light carries much more information than our naked eyes can recognize. Modern instruments can reveal otherwise hidden details in spectra, and special telescopes can record forms of light that are invisible to our eyes. In this section, we'll explore how detailed studies of light help us unlock the secrets of the universe.

◆ What are the three basic types of spectra?

Laboratory studies show that spectra come in three basic types,* summarized in Figure 5.8:

1. The spectrum of a traditional, or incandescent, light bulb is a rainbow of color. Because the rainbow spans a broad range of wavelengths without interruption, we call it a **continuous spectrum.**

2. A thin or low-density cloud of gas emits light only at specific wavelengths that depend on its composition and temperature. The spectrum therefore consists of bright **emission lines** against a black background and is called an **emission line spectrum.**

3. If the cloud of gas lies between us and a light bulb (and the cloud is cooler than the light bulb or other light source), we still see most of the continuous spectrum of the light bulb. However, the cloud absorbs light of specific wavelengths, so the spectrum shows dark **absorption lines** over the background rainbow. We call this an **absorption line spectrum.**

Notice that each of the spectra in Figure 5.8 is shown both as a band of light and as a graph. The band of light is essentially what you would see if you projected the light that passes through the prism onto a wall. The graph shows the amount, or **intensity,** of the light at each wavelength; the intensity is high at wavelengths where there is a lot of light and low where there is little light. For example, the graph of the absorption line spectrum shows dips in intensity at the wavelengths where the band of light shows dark lines. Astronomers usually display spectra as graphs because they make it easier to see how intensity varies across the spectrum.

The three basic types of spectra are continuous, emission line, and absorption line.

We can apply these ideas to the solar spectrum that opens this chapter. The many dark absorption lines over a background rainbow of color tell us that we are essentially looking at a hot light source through a cooler gas, much like the situation in Figure 5.8c. For the solar spectrum, the hot light source is the hot interior of the Sun, while the "cloud" is the relatively cool and low-density gas that makes up the Sun's visible surface, or *photosphere* [Section 11.1].

---

*The rules that specify the conditions producing each type are often called *Kirchhoff's laws*.

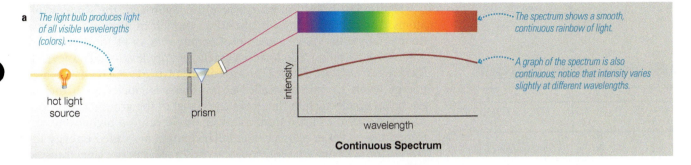

a *The light bulb produces light of all visible wavelengths (colors).* ⋯⋯

hot light source

prism

The spectrum shows a smooth, continuous rainbow of light.

A graph of the spectrum is also continuous; notice that intensity varies slightly at different wavelengths.

intensity

wavelength

Continuous Spectrum

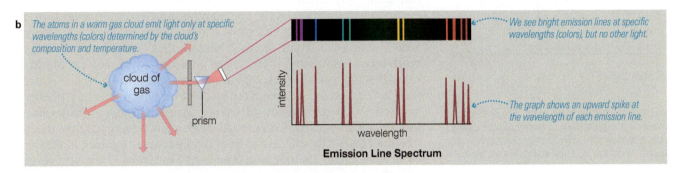

b *The atoms in a warm gas cloud emit light only at specific wavelengths (colors) determined by the cloud's composition and temperature.*

cloud of gas

prism

We see bright emission lines at specific wavelengths (colors), but no other light.

The graph shows an upward spike at the wavelength of each emission line.

intensity

wavelength

Emission Line Spectrum

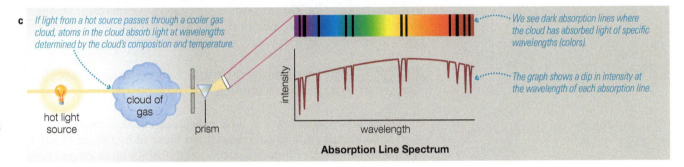

c *If light from a hot source passes through a cooler gas cloud, atoms in the cloud absorb light at wavelengths determined by the cloud's composition and temperature.*

hot light source

cloud of gas

prism

We see dark absorption lines where the cloud has absorbed light of specific wavelengths (colors).

The graph shows a dip in intensity at the wavelength of each absorption line.

intensity

wavelength

Absorption Line Spectrum

▲ **FIGURE 5.8**
These diagrams show examples of the conditions under which we see the three basic types of spectra.

◆ How does light tell us what things are made of?

We have just seen *how* different viewing conditions lead to different types of spectra, so it is time to discuss *why*. Let's start with absorption and emission line spectra. As we'll see, the lines in these spectra can tell us what distant objects are made of.

Energy Levels in Atoms   To understand why we sometimes see emission and absorption lines, we must first discuss a strange fact about electrons in atoms: The electrons can have only particular amounts of energy, and not other energies in between. As an analogy, suppose you're washing windows on a building. If you use an adjustable platform to reach high windows, you can stop the platform at any height above the ground. But if you use a ladder, you can stand only at particular heights—the heights of the rungs of the ladder—and not at other heights in between. The possible energies of electrons in atoms are like the possible heights on a ladder. Only a few particular energies are possible, and energies between these special few are not possible. The possible energies are known as the **energy levels** of an atom.

> Electrons in atoms can have only particular amounts of energy, and not other energies in between.

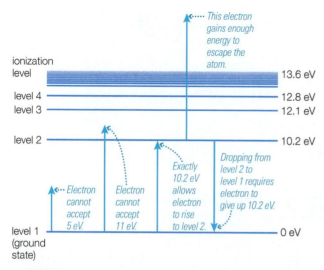

This electron gains enough energy to escape the atom.

ionization level — 13.6 eV

level 4 — 12.8 eV
level 3 — 12.1 eV

level 2 — 10.2 eV

Electron cannot accept 5 eV.

Electron cannot accept 11 eV.

Exactly 10.2 eV allows electron to rise to level 2.

Dropping from level 2 to level 1 requires electron to give up 10.2 eV.

level 1 (ground state) — 0 eV

▲ **FIGURE 5.9**

Energy levels for the electron in a hydrogen atom. The electron can change energy levels only if it gains or loses the amount of energy separating the levels. If the electron gains enough energy to reach the ionization level, it can escape from the atom, leaving behind a positively charged ion.

Figure 5.9 shows the energy levels in hydrogen, the simplest of all atoms. The energy levels are labeled on the left in numerical order and on the right with energies in units of *electron-volts*, or *eV* for short. (1 eV = 1.60×10^{-19} joule.) The lowest possible energy level—called level 1 or the *ground state*—is defined as an energy of 0 eV. Each of the higher energy levels (sometimes called *excited states*) is labeled with the extra energy of an electron in that level compared to the ground state.

An electron can rise from a low energy level to a higher one or fall from a high level to a lower one. Such changes are called **energy level transitions.** Because energy must be conserved, energy level transitions can occur only when an electron gains or loses the specific amount of energy separating two levels. For example, an electron in level 1 can rise to level 2 only if it gains 10.2 eV of energy. If you try to give the electron 5 eV of energy, it won't accept it because that is not enough energy to reach level 2. Similarly, if you try to give it 11 eV, it won't accept it because it is too much for level 2 but not enough to reach level 3. Once in level 2, the electron can return to level 1 by giving up 10.2 eV of energy.

Notice that the amount of energy separating the various levels gets smaller at higher levels. For example, it takes more energy to raise the electron from level 1 to level 2 than from level 2 to level 3, which in turn takes more energy than the transition from level 3 to level 4. If the electron gains enough energy to reach the *ionization level,* it escapes the atom completely. Because the escaping electron carries away negative electrical charge, the atom is left with positive electrical charge. Electrically charged atoms are called **ions,** so we say that the escape of the electron *ionizes* the atom.

think about it Are there any circumstances under which an electron in a hydrogen atom can gain 2.6 eV of energy? Explain.

Other atoms also have distinct energy levels, but the levels correspond to different amounts of energy than those of hydrogen. Every type of ion and every type of molecule also has a distinct set of energy levels.

Emission and Absorption Lines The fact that each type of atom, ion, or molecule possesses a unique set of energy levels is what causes emission and absorption lines to appear at specific wavelengths in spectra. It is also what allows us to learn the compositions of distant objects in the universe. To see how, let's consider what happens in a cloud of gas consisting solely of hydrogen atoms.

The atoms in any cloud of gas are constantly colliding with one another, exchanging energy in each collision. Most of the collisions simply send the atoms careening off in new directions. However, a few of the collisions transfer the right amount of energy to bump an electron from a low energy level to a higher energy level. Electrons can't stay in higher energy levels for long. They always fall back down to level 1, usually in a tiny fraction of a second. The en-

The photons that produce emission lines are created when electrons fall to lower energy levels.

ergy the electron loses when it falls to a lower energy level must go somewhere, and often it goes to *emitting* a photon of light. The emitted photon must have the same amount of energy that the electron loses, which means that it has a specific wavelength and frequency. Figure 5.10a shows the energy levels in hydrogen that we saw in Figure 5.9, but it is also labeled with the wavelengths of the photons emitted by various downward transitions of an electron from a higher energy level to a lower one. For example, the transition from level 2 to level 1 emits an ultraviolet photon of wavelength 121.6 nm, and the transition from level 3 to level 2 emits a red visible-light photon of wavelength 656.3 nm.

As long as the gas remains moderately warm, collisions will continually bump some electrons into higher levels from which they fall back down and emit photons with some of the wavelengths shown in Figure 5.10a. The gas therefore emits light with these specific wavelengths. That is why a warm gas cloud produces an emission line spectrum (Figure 5.10b). The bright emission lines appear at the wavelengths that correspond to downward transitions of electrons, and the rest of the spectrum is dark (black). The specific set of lines that we see depends on the cloud's temperature as well as its composition: At higher temperatures, electrons are more likely to be bumped to higher energy levels.

think about it If nothing continues to heat the hydrogen gas, all the electrons eventually will end up in the lowest energy level (the ground state, or level 1). Use this fact to explain why we should *not* expect to see an emission line spectrum from a very cold cloud of hydrogen gas.

Absorption lines occur when photons cause electrons to rise to higher energy levels.

Now, suppose a light bulb illuminates the hydrogen gas from behind (as in Figure 5.8c). The light bulb emits light of all wavelengths, producing a spectrum that looks like a rainbow of color. However, the hydrogen atoms can absorb those photons that have the right amount of energy needed to raise an electron from a low energy level to a higher one.* Figure 5.10c shows the result. It is an absorption line spectrum, because the light bulb produces a continuous rainbow of color while the hydrogen atoms absorb light at specific wavelengths.

You should now understand why the dark absorption lines in Figure 5.10c occur at the same wavelengths as the emission lines in Figure 5.10b: Both types of lines represent the same energy level transitions, except in opposite directions. For example, electrons moving downward from level 3 to level 2 in hydrogen can emit photons of wavelength 656.3 nm (producing an emission line at this wavelength), while electrons absorbing photons with this wavelength can rise up from level 2 to level 3 (producing an absorption line at this wavelength).

Chemical Fingerprints The fact that hydrogen emits and absorbs at specific wavelengths makes it possible to detect its presence in distant objects. For example, imagine that you look through a telescope at an interstellar gas cloud, and its spectrum looks like that shown in Figure 5.10b. Because only hydrogen produces this particular set of lines, you can conclude that the cloud is made of hydrogen. In essence, the spectrum contains a "fingerprint" left by hydrogen atoms.

Every kind of atom, ion, and molecule produces a unique spectral "fingerprint."

Real interstellar clouds are not made solely of hydrogen. However, the other chemical constituents in the cloud leave fingerprints on the spectrum in much the same way. Every type of atom, ion, and molecule has its own unique spectral fingerprint, because it has its own unique set of energy levels. Over the past century, scientists have done laboratory experiments to

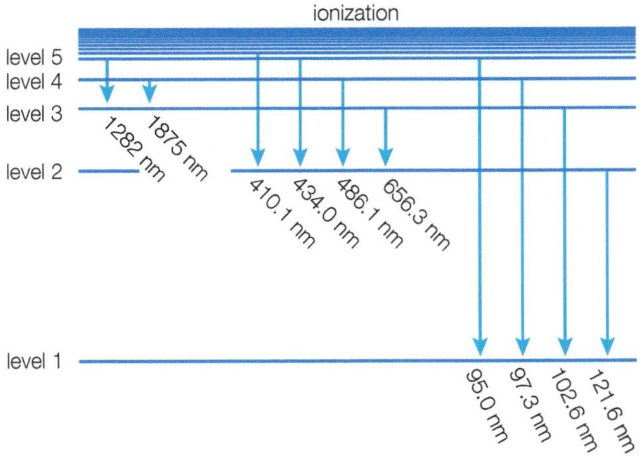

---

*Of course, the electrons quickly fall back down, which means they can emit photons of the same wavelength they absorbed. However, these photons are emitted in random directions, so we still see absorption lines because photons that were originally coming toward us have been redirected away from our line of sight.

a Energy level transitions in hydrogen correspond to photons with specific wavelengths. Only a few of the many possible transitions are labeled.

b This spectrum shows emission lines produced by downward transitions between higher levels and level 2 in hydrogen.

c This spectrum shows absorption lines produced by upward transitions between level 2 and higher levels in hydrogen.

▲ **FIGURE 5.10**
An atom emits or absorbs light only at specific wavelengths that correspond to changes in the atom's energy as an electron undergoes transitions between its allowed energy levels.

identify the spectral lines of every chemical element and many ions and molecules. When we see any of those lines in the spectrum of a distant object, we can determine what chemicals produced them. For example, if we see spectral lines of hydrogen, helium, and carbon in the spectrum of a distant star, we know that all three elements are present in the star. With more detailed analysis, we can determine the relative proportions of the various elements. That is how we have learned the chemical compositions of objects throughout the universe.

◆ How does light tell us the temperatures of planets and stars?

We next turn our attention to continuous spectra. Although continuous spectra can be produced in more than one way, light bulbs, planets, and stars produce a particular kind of continuous spectrum that can help us determine their temperatures.

Thermal Radiation: Every Body Does It In a cloud of gas that produces a simple emission or absorption line spectrum, the individual atoms or molecules are essentially independent of one another. Most photons pass easily through such a gas, except those that cause energy level transitions in the atoms or molecules of the gas. However, the atoms and molecules within most of the objects we encounter in everyday life—such as rocks, light bulb filaments, and people—cannot be considered independent and therefore have much more complex sets of energy levels. These objects tend to absorb light across a broad range of wavelengths, which means light cannot easily pass through them and light emitted inside them cannot easily escape. The same is true of almost any large or dense object, including planets and stars.

In order to understand the spectra of such objects, let's consider an idealized case in which an object absorbs all photons that strike it and does not allow photons inside it to escape easily. Photons tend to bounce around randomly inside such an object, constantly exchanging energy with its atoms or molecules. By the time the photons finally escape the object, their radiative energies have become randomized so that they are spread over a wide range of wavelengths. The wide wavelength range of the photons explains why the spectrum of light from such an object is smooth, or *continuous*, like a pure rainbow without any absorption or emission lines.

Most important, the spectrum from such an object depends on only one thing: the object's *temperature*. To understand why, remember that temperature represents the average kinetic energy of the atoms or molecules in an object [Section 4.3]. Because the randomly bouncing photons interact so many times with those atoms or molecules, they end up with energies that match the kinetic energies of the object's atoms or molecules—which means the photon energies depend only on the object's temperature, regardless of what the object is made of. The temperature dependence of this light explains why we call it **thermal radiation** (sometimes known as *blackbody* radiation) and why its spectrum is called a **thermal radiation spectrum.**

No real object emits a perfect thermal radiation spectrum, but almost all familiar objects—including the Sun, the planets, rocks, and even you—

Planets, stars, rocks, and people emit thermal radiation that depends only on temperature.

emit light that approximates thermal radiation. Figure 5.11 shows a graph of the idealized thermal radiation spectra of three stars and a human, each with its temperature given on the Kelvin scale (see Figure 4.10). Be sure to notice that these spectra show the intensity of light *per unit surface area*, not the total

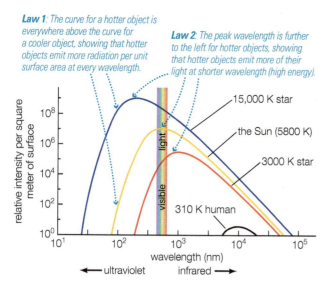

Law 1: *The curve for a hotter object is everywhere above the curve for a cooler object, showing that hotter objects emit more radiation per unit surface area at every wavelength.*

Law 2: *The peak wavelength is further to the left for hotter objects, showing that hotter objects emit more of their light at shorter wavelength (high energy).*

15,000 K star

the Sun (5800 K)

3000 K star

310 K human

▲ **FIGURE 5.11**

Graphs of idealized thermal radiation spectra demonstrate the two laws of thermal radiation: (1) Each square meter of a hotter object's surface emits more light at all wavelengths; (2) hotter objects emit photons with a higher average energy. Notice that the graph uses power-of-10 scales on both axes, so that we can see all the curves even though the differences between them are quite large.

amount of light emitted by the object. For example, a very large 3000 K star can emit more total light than a small 15,000 K star, even though the hotter star emits much more light per unit area of its surface.

The Two Laws of Thermal Radiation

If you compare the spectra in Figure 5.11, you'll see that they obey two laws of thermal radiation:

- **Law 1** (Stefan-Boltzmann law): *Each square meter of a hotter object's surface emits more light at all wavelengths.* For example, each square meter on the surface of the 15,000 K star emits a lot more light at every wavelength than each square meter of the 3000 K star, and the hotter star emits light at some ultraviolet wavelengths that the cooler star does not emit at all.

- **Law 2** (Wien's [pronounced *veen's*] law): *Hotter objects emit photons with a higher average energy,* which means a shorter average wavelength. That is why the peaks of the spectra are at shorter wavelengths for hotter objects. For example, the peak for the 15,000 K star is in ultraviolet light, the peak for the 5800 K Sun is in visible light, and the peak for the 3000 K star is in the infrared.

You can see these laws in action with a fireplace poker (Figure 5.12). While the poker is still relatively cool, it emits only infrared light, which we cannot see. As it gets hot (above about 1500 K), it begins to glow with visible light, and it glows more brightly as it gets hotter, demonstrating the first law. Its color demonstrates the second law. At first it glows "red hot," because red light has the longest wavelengths of visible light. As it gets even hotter, the average wavelength of the emitted photons moves toward the blue (short wavelength) end of the visible spectrum. The mix of colors emitted at this higher temperature makes the poker look white to your eyes, which is why "white hot" is hotter than "red hot."

see it for yourself Find an incandescent light that has a dimmer switch. What happens to the bulb temperature (which you can check by placing your hand near it) as you turn the switch up? How does the light change color? Explain how these observations demonstrate the two laws of thermal radiation.

Because thermal radiation spectra depend only on temperature, we can use them to measure the temperatures of distant objects. In many cases we can estimate temperatures simply from the object's colors. Notice that while hotter objects emit more light at *all* wavelengths, the biggest difference appears at the shortest wavelengths. With a human body temperature of about 310 K, *Hotter objects emit more total light per unit surface area and emit photons with a higher average energy.* people emit infrared light but not visible light—which explains why we don't glow in the dark! A relatively cool star, with a 3000 K surface temperature, emits mostly red light. That is why some bright stars in our sky, such as Betelgeuse (in Orion) and Antares (in Scorpius), appear reddish in color. The Sun's 5800 K surface emits most strongly in green light (around 500 nm), but the Sun looks yellow or white to our eyes because it also emits other colors throughout the visible spectrum. Hotter stars emit mostly in the ultraviolet but appear blue-white in color because our eyes cannot see their ultraviolet light. If an object were heated to a temperature of millions of degrees, it would radiate mostly X rays. Some astronomical objects are indeed hot enough to emit X rays, such as the Sun's *corona* and hot *accretion disks* around black holes.

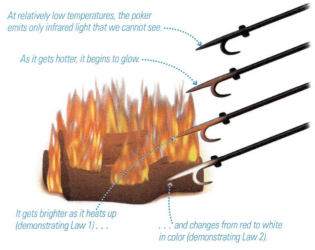

At relatively low temperatures, the poker emits only infrared light that we cannot see.

As it gets hotter, it begins to glow.

It gets brighter as it heats up (demonstrating Law 1) . . .

. . . and changes from red to white in color (demonstrating Law 2).

▲ **FIGURE 5.12**
A fireplace poker shows the two laws of thermal radiation in action.

---

cosmic calculations 5.1

Laws of Thermal Radiation

The two laws of thermal radiation have simple mathematical formulas.

Law 1 (Stefan-Boltzmann law):
emitted power (per square meter of surface) $= \sigma T^4$

where T is temperature (in Kelvin), σ (Greek letter *sigma*) is a constant with measured value $\sigma = 5.7 \times 10^{-8} \frac{\text{watt}}{(\text{m}^2 \times \text{K}^4)}$, and a *watt* is a unit of power equal to 1 joule per second.

Law 2 (Wien's law): $\lambda_{\max} \approx \dfrac{2,900,000}{T \text{ (in Kelvin)}} \text{ nm}$

where $\lambda_{\max}$ (read as "lambda max") is the wavelength (in nanometers) of maximum intensity, which is located at the peak in a thermal radiation spectrum.

Example: Find the emitted power per square meter and the wavelength of peak intensity for a 10,000 K object that emits thermal radiation.

Solution: We use the first law to calculate the emitted power per square meter for an object with $T = 10,000$ K:

$$\sigma T^4 = 5.7 \times 10^{-8} \frac{\text{watt}}{\text{m}^2 \times \text{K}^4} \times (10,000 \text{ K})^4$$
$$= 5.7 \times 10^8 \text{ watt} / \text{m}^2$$

The second law gives the wavelength of maximum intensity:

$$\lambda_{\max} \approx \frac{2,900,000}{10,000 \text{ (in Kelvin)}} \text{ nm} \approx 290 \text{ nm}$$

A 10,000 K object emits a total power of 570 million watts per square meter of surface. Its wavelength of maximum intensity is 290 nm, which is in the ultraviolet.

train stationary

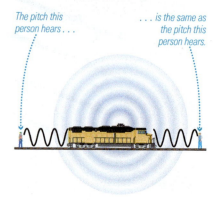

The pitch this person hears . . .

. . . is the same as the pitch this person hears.

train moving to right

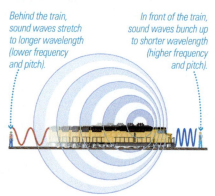

Behind the train, sound waves stretch to longer wavelength (lower frequency and pitch).

In front of the train, sound waves bunch up to shorter wavelength (higher frequency and pitch).

light source moving to right

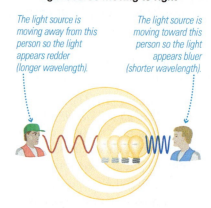

The light source is moving away from this person so the light appears redder (longer wavelength).

The light source is moving toward this person so the light appears bluer (shorter wavelength).

a The whistle sounds the same no matter where you stand near a stationary train.

b For a moving train, the sound you hear depends on whether the train is moving toward you or away from you.

c We get the same basic effect from a moving light source (although the shifts are usually too small to notice with our eyes).

▲ **FIGURE 5.13**

The Doppler effect. Each circle represents the crests of sound (or light) waves going in all directions from the source. For example, the circles from the train might represent waves emitted 0.001 second apart.

cosmic calculations 5.2

The Doppler Shift

We can calculate an object's radial velocity from its Doppler shift. For velocities that are small compared to the speed of light (less than a few percent of c), the formula is

$$\frac{v_{rad}}{c} = \frac{\lambda_{shift} - \lambda_{rest}}{\lambda_{rest}}$$

where v_{rad} is the object's radial velocity, λ_{rest} is the rest wavelength of a particular spectral line, and λ_{shift} is the shifted wavelength of the same line. (As always, c is the speed of light.) A positive answer means the object is redshifted and moving away from us; a negative answer means it is blueshifted and moving toward us.

Example: One of the visible lines of hydrogen has a rest wavelength of 656.285 nm, but it appears in the spectrum of the star Vega at 656.255 nm. How is Vega moving relative to us?

Solution: We use the rest wavelength $\lambda_{rest} = 656.285$ nm and the shifted wavelength $\lambda_{shift} = 656.255$ nm:

$$\frac{v_{rad}}{c} = \frac{\lambda_{shift} - \lambda_{rest}}{\lambda_{rest}}$$

$$= \frac{656.255 \text{ nm} - 656.285 \text{ nm}}{656.285 \text{ nm}}$$

$$\approx -4.6 \times 10^{-5}$$

The negative answer tells us that Vega is moving *toward* us. Its speed is about 4.6×10^{-5} of the speed of light c. Because $c = 300{,}000$ km/s, this is equivalent to about $4.6 \times 10^{-5} \times (3 \times 10^5 \text{ km/s}) \approx 13.8$ km/s.

How does light tell us the speed of a distant object?

There is still more that we can learn from light: We can use light to learn about the motion of distant objects (relative to us) from changes in their spectra caused by the **Doppler effect.**

The Doppler Effect You've probably noticed the Doppler effect on the *sound* of a train whistle near train tracks. If the train is stationary, the pitch of its whistle sounds the same no matter where you stand (Figure 5.13a). But if the train is moving, the pitch sounds higher when the train is coming toward you and lower when it's moving away from you. Just as the train passes by, you can hear the dramatic change from high to low pitch—a sort of "weeeeeeee–ooooooooooh" sound. To understand why, we have to think about what happens to the sound waves coming from the train (Figure 5.13b). When the train is moving toward you, each pulse of a sound wave is emitted a little closer to you. The result is that waves are bunched up between you and the train, giving them a shorter wavelength and higher frequency (pitch). After the train passes you by, each pulse comes from farther away, stretching out the wavelengths and giving the sound a lower frequency.

The Doppler effect causes similar shifts in the wavelengths of light (Figure 5.13c). If an object is moving toward us, the light waves bunch up between us and the object, so that its entire spectrum is shifted to shorter wavelengths. Because shorter wavelengths of visible light are bluer, the Doppler shift of an object coming toward us is called a **blueshift.** If an object is moving away from us, its light is shifted to longer wavelengths. We call this a **redshift** because longer wavelengths of visible light are redder. For convenience, astronomers use the terms *blueshift* and *redshift* even when they aren't talking about visible light.

Spectral lines shift to shorter wavelengths when an object is moving toward us, and to longer wavelengths when an object is moving away from us.

Spectral lines provide the reference points we use to identify and measure Doppler shifts (Figure 5.14). For example, suppose we recognize the pattern of hydrogen lines in the spectrum of a distant object. We know the **rest wavelengths** of the hydrogen lines—that is, their wavelengths in stationary clouds of hydrogen gas—from laboratory experiments in which a tube of hydrogen gas is heated so that the wavelengths of the spectral lines can be measured. If the hydrogen lines from the object appear at longer wavelengths, then we know they are redshifted and the object is moving away from us. The larger the shift, the faster the object is moving. If the lines appear at shorter wavelengths, then we know they are blueshifted and the object is moving toward us.

think about it Suppose the hydrogen emission line with a rest wavelength of 121.6 nm (the transition from level 2 to level 1) appears at a wavelength of 120.5 nm in the spectrum of a particular star. Given that these wavelengths are in the ultraviolet, is the shifted wavelength closer to or farther from blue visible light? Why, then, do we say that this spectral line is *blueshifted*?

Note that a Doppler shift tells us only the part of an object's full motion that is directed toward or away from us (the *radial* component of motion). Doppler shifts do not give us any information about how fast an object is moving across our line of sight (the *tangential* component of motion). For example, consider three stars moving at the same speed: one moving directly away from us, one moving across our line of sight, and one moving diagonally away from us (Figure 5.15). The Doppler shift will tell us the full speed only of the first star. It will not indicate any speed for the second star, and for the third star it will tell us only the part of the star's speed that is directed away from us. To measure how fast an object is moving across our line of sight, we must observe it long enough to notice how its position gradually shifts across our sky.

Spectral Summary We've covered the major ways in which we can learn from an object's spectrum, discussing how we learn about an object's composition, temperature, and motion. Figure 5.16 (pages 120–121) summarizes the ways in which we learn from spectra.

5.3 Collecting Light with Telescopes

Light carries a great deal of information, but we can obtain only a little of that information with our naked eyes. We need telescopes to learn more, and most great advances in astronomy are directly tied to advances in telescope technology.

◆ How do telescopes help us learn about the universe?

Telescopes are essentially giant eyes that can collect far more light than our own eyes. By combining this light-collecting capacity with cameras and other instruments—such as *spectrographs* that disperse the light into spectra—we can record and analyze light in detail.

Two Key Telescope Properties The first key property of a telescope is its **light-collecting area,** which determines how much total light the telescope can collect at one time. Telescopes are generally round, so we usually characterize a telescope's "size" as the *diameter* of its light-collecting area. For example, a "10-meter telescope" has a light-collecting area 10 meters in diameter.

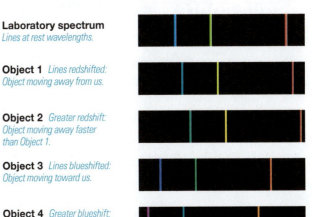

Laboratory spectrum
Lines at rest wavelengths.

Object 1 *Lines redshifted:*
Object moving away from us.

Object 2 *Greater redshift:*
Object moving away faster than Object 1.

Object 3 *Lines blueshifted:*
Object moving toward us.

Object 4 *Greater blueshift:*
Object moving toward us faster than Object 3.

▲ **FIGURE 5.14**
Spectral lines provide the crucial reference points for measuring Doppler shifts.

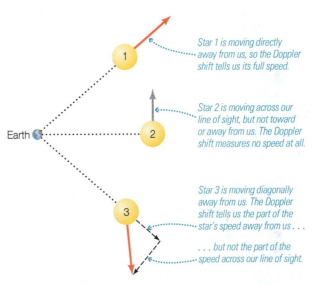

Earth

Star 1 is moving directly away from us, so the Doppler shift tells us its full speed.

Star 2 is moving across our line of sight, but not toward or away from us. The Doppler shift measures no speed at all.

Star 3 is moving diagonally away from us. The Doppler shift tells us the part of the star's speed away from us . . .

. . . but not the part of the speed across our line of sight.

▲ **FIGURE 5.15**
The Doppler shift tells us only the portion of an object's speed that is directed toward or away from us. It does not give us any information about how fast an object is moving across our line of sight.

An astronomical spectrum contains an enormous amount of information. This figure describes what we can learn from a schematic spectrum of Mars.

① Continuous Spectrum: The visible light we see from Mars is actually reflected sunlight. The Sun produces a nearly continuous spectrum of light, which includes the full rainbow of color.

hot light source

prism

Like the Sun, a light bulb produces light of all visible wavelengths (colors).

② Scattered/Reflected Light: Mars is red because it absorbs most of the blue light from the Sun but reflects (scatters) most of the red light. This pattern of absorption and reflection helps us learn the chemical composition of the surface.

Like Mars, a red chair looks red because it absorbs blue light and scatters red light.

intensity

The dashed curve is the continuous spectrum of the sunlight shining on Mars.

The graph and the "rainbow" contain the same information. The graph makes it easier to read the intensity at each wavelength of light . . .

Mars reflects relatively little of the blue sunlight . . .

. . . but a lot of the red sunlight.

. . . while the "rainbow" shows how the spectrum appears to the eye (for visible light) or instruments (for non-visible light).

ultraviolet   blue   green   red

wavelength

④ Emission Lines: Ultraviolet emission lines in the spectrum of Mars tell us that the atmosphere of Mars contains hot gas at high altitudes.

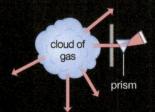

cloud of gas

prism

We see bright emission lines from gases in which collisions raise electrons in atoms to higher energy levels. The atoms emit photons at specific wavelengths as the electrons drop to lower energy levels.

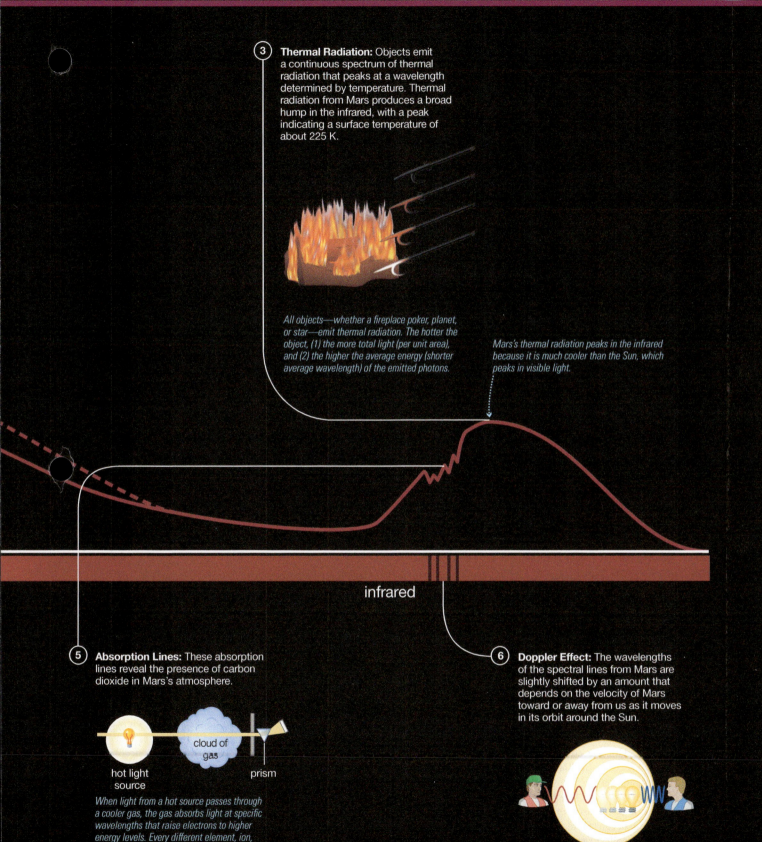

③ Thermal Radiation: Objects emit a continuous spectrum of thermal radiation that peaks at a wavelength determined by temperature. Thermal radiation from Mars produces a broad hump in the infrared, with a peak indicating a surface temperature of about 225 K.

All objects—whether a fireplace poker, planet, or star—emit thermal radiation. The hotter the object, (1) the more total light (per unit area), and (2) the higher the average energy (shorter average wavelength) of the emitted photons.

Mars's thermal radiation peaks in the infrared because it is much cooler than the Sun, which peaks in visible light.

infrared

⑤ Absorption Lines: These absorption lines reveal the presence of carbon dioxide in Mars's atmosphere.

hot light
source

cloud of
gas

prism

When light from a hot source passes through a cooler gas, the gas absorbs light at specific wavelengths that raise electrons to higher energy levels. Every different element, ion, and molecule has unique energy levels and hence its own spectral "fingerprint."

⑥ Doppler Effect: The wavelengths of the spectral lines from Mars are slightly shifted by an amount that depends on the velocity of Mars toward or away from us as it moves in its orbit around the Sun.

A Doppler shift toward the red side of the spectrum tells us the object is moving away from us. A shift toward the blue side of the spectrum

lens

starlight

eyepiece

focus

▲ **FIGURE 5.17**

A refracting telescope collects light with a large transparent lens (see diagram). The photo shows the just-over-1-meter (40-inch) refractor at the University of Chicago's Yerkes Observatory, the world's largest refracting telescope.

Because area is proportional to the *square* of diameter, a small increase in diameter can mean a big increase in light-collecting area. A 10-meter telescope has five times the diameter of a 2-meter telescope, which means $5^2 = 25$ times the light-collecting area.

think about it How does the light-collecting area of a 10-meter telescope compare to that of a human eye, which collects light through an opening (the pupil) about 5 millimeters in diameter?

A second key property of a telescope is the amount of detail it allows us to see, which we characterize by its **angular resolution**—the smallest

Telescopes collect far more light and allow us to see far more detail than does the naked eye.

angle over which we can tell that two dots (or two stars) are distinct. The human eye has an angular resolution of about 1 arcminute ($\frac{1}{60}$ degree), meaning that two stars can appear distinct only if they have at least this much angular separation in the sky. If the stars are separated by less than 1 arcminute, our eyes will not be able to distinguish them individually and they will look like a single star.

Large telescopes can have amazing angular resolution. The 2.4-meter Hubble Space Telescope has an angular resolution of about 0.05 arcsecond (for visible light), which would allow you to read this book from a distance of almost 1 kilometer. Larger telescopes can have even better (smaller) angular resolution, though Earth's atmosphere usually prevents ground-based telescopes from achieving their theoretical limits.

Basic Telescope Design Telescopes come in two basic designs: *refracting* and *reflecting*. A **refracting telescope** operates similar to an eye, using transparent glass lenses to collect and focus light (Figure 5.17). The earliest telescopes, including Galileo's, were refracting telescopes. The world's largest refracting telescope, completed in 1897, has a lens that is 1 meter in diameter and a telescope tube that is 19.5 meters long.

---

extraordinary claims We Can Never Learn the Composition of Stars

The limits of science have long been a topic of great debate among both scientists and philosophers. One of the most famous claims about such limits came in 1835, when the prominent French philosopher Auguste Comte declared that science could never allow us to learn the composition of stars. His rationale was based primarily on the idea that while we could observe things like the sizes of distant objects, we could not learn what they were made of without collecting physical samples to study here on Earth.

As you know from this chapter, Comte's extraordinary claim was proven wrong a few decades later with the advent of spectroscopy. Less than a century after he made his claim, Cecilia Payne-Gaposchkin used spectroscopy and an understanding of quantum mechanics to unlock the secret to the composition of stars [Section 12.1]. Moreover, even if spectroscopy had not been discovered, Comte's claim might still have been proven wrong, because it also presumed that we would never travel out into space to collect samples of distant objects. While stars besides the Sun remain far beyond our reach, sample

collection within our solar system is now well within our abilities. It's also worth noting that the possibility of space travel was already being considered seriously by Comte's time. For example, in 1593, Johannes Kepler wrote a letter to Galileo in which he made an extraordinary claim of his own: "Provide ships or sails adapted to the heavenly breezes, and there will be some who will not fear even that void . . . ." Of course, that didn't stop many people from claiming that space travel would prove impossible, some of them holding to that view almost right up until we achieved it.

There may well be limits to what we can learn through science, but the history of science teaches us to be wary of claims about what we may or may not do or learn in the future. Some questions that seem unanswerable today, such as why the universe exists or what (if anything) came before the Big Bang, may remain unanswerable, but we should be open to the possibility that science may yet open doors to knowledge that we cannot yet envision.

Verdict: Rejected.

A **reflecting telescope** uses a precisely curved *primary mirror* to gather light (Figure 5.18). This mirror reflects the gathered light to a *secondary mirror* that lies in front of it. The secondary mirror then reflects the light to a focus at a place where the eye or instruments can observe it—sometimes through a hole in the primary mirror and sometimes through the side of the telescope (often with the aid of additional small mirrors). The fact that the secondary mirror prevents some light from reaching the primary mirror might seem like a drawback, but in practice it is not a problem because only a small fraction of the incoming light is blocked.

Nearly all telescopes used in current astronomical research are reflectors. For a long time, the main factor limiting the size of reflecting telescopes was the sheer weight of the glass needed for their primary mirrors.

The world's largest reflecting telescopes have primary mirrors 10 meters or more in diameter. Recent technological innovations have made it possible to build lighter-weight mirrors, such as the 8-meter mirror in the Gemini telescope (see Figure 5.18), or to make many small mirrors work together as one large one, as in the twin 10-meter Keck telescopes (Figure 5.19). Several other large telescopes are currently in planning or construction, including the Large Synoptic Survey Telescope (8.4 meters, but with a very wide field of view), the Giant Magellan Telescope (effective size of 21 meters), the Thirty Meter Telescope (30 meters), and the European Extremely Large Telescope (39 meters).

Telescopes Across the Spectrum

If we studied only visible light, we'd be missing much of the picture. Planets are relatively cool and emit primarily infrared light. The hot upper layers of stars emit ultraviolet and X-ray light. Some violent events produce gamma rays. In fact, most objects emit light over a broad range of wavelengths, so astronomers seek to study light across the entire spectrum.

Telescopes specialized to observe different wavelengths of light allow us to learn far more than we could learn from visible light alone. The basic idea behind nearly all telescopes is the same: Light is collected by a primary mirror (sometimes more than one) and ultimately focused on cameras or other instruments. However, different wavelengths pose different challenges for telescope design.

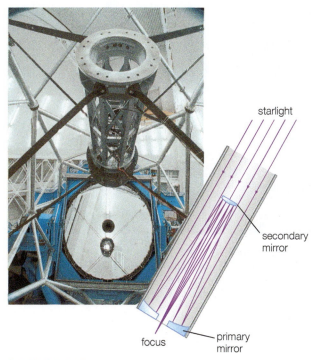

▲ **FIGURE 5.18**

A reflecting telescope collects light with a precisely curved primary mirror (see diagram). The photo shows the Gemini North telescope (Mauna Kea, Hawaii), which has an 8-meter-diameter primary mirror. The secondary mirror, located in the smaller central lattice, reflects light back down through the hole visible in the center of the primary mirror.

▲ **FIGURE 5.19**

The primary mirror of one of the Keck telescopes (Mauna Kea, Hawaii), with a man in the center for scale. It is made up of 36 hexagonal mirrors arranged in a honeycomb pattern. The inset shows the two Keck telescopes from above.

common misconceptions

Magnification and Telescopes

Many people guess that magnification is the most important function of a telescope. However, even though telescopes can magnify images—much like telephoto camera lenses or binoculars—the amount of magnification a telescope can provide is *not* one of its crucial properties. No matter how much a telescope image is magnified, you cannot see details if the telescope does not collect enough light to show them, or if they are smaller than the angular resolution of the telescope. Magnifying an image too much just makes it look blurry, which is why a telescope's light-collecting area and angular resolution are much more important than its magnification.

▲ FIGURE 5.20

China's Five-hundred-meter Aperture Spherical Telescope (FAST) is a radio dish with a diameter of 500 meters. It is nestled in a natural valley in southwest Guizhou province.

Some portions of the infrared and ultraviolet lie near enough to visible wavelengths that the light behaves similarly to visible light. This light can therefore be focused by visible-light telescopes, which is why the Hubble Space Telescope can be used to study infrared and ultraviolet light as well as visible light.

At longer infrared wavelengths, the heat of a telescope itself can create problems, because the telescope may emit thermal radiation at the same wavelengths it is trying to observe. To alleviate this problem, infrared telescopes in space generally have shielding to block the Sun, which allows them to cool to temperatures of 30–50 K. Some, like the Spitzer Space Telescope during its first 5 years (until the coolant ran out), are cooled further with liquid helium, which can take them down to temperatures only a few degrees above absolute zero.

Radio waves present a different challenge. For a telescope of a particular size, the angular resolution is poorer (larger) for longer wavelengths of light. For example, the Hubble Space Telescope has lower angular resolution for infrared light than for visible light. The long wavelengths of radio waves therefore mean that very large telescopes are necessary to achieve reasonable angular resolution. The world's largest single telescope is China's Five-hundred-meter Aperture Spherical Telescope, or FAST, a 500-meter-diameter radio dish nestled in a natural valley (Figure 5.20). Completed in 2016, FAST surpassed the previous record holder, Puerto Rico's 305-meter Arecibo radio telescope. Despite their large sizes, these telescopes have an angular resolution of only about 1 arcminute at commonly observed radio wavelengths—about 1000 times worse than the visible-light resolution of the Hubble Space Telescope.

---

special topic   Would You Like Your Own Telescope?

Just a couple decades ago, a decent telescope would have set you back a few thousand dollars and taken weeks of practice to learn to use. Today, you can get a good-quality telescope for a few hundred dollars, and built-in computer drives can make it very easy to use.

Before you consider buying a telescope, you should understand what a personal telescope can and cannot do. A telescope will allow you to look for yourself at light that has traveled vast distances through space to reach your eyes. This can be a rewarding experience, but the images in your telescope will *not* look like the beautiful photographs in this book, which were obtained with much larger telescopes and sophisticated cameras. In addition, while your telescope can in principle let you see many distant objects, including star clusters, nebulae, and galaxies, it won't allow you to find anything unless you first set it up properly. Even computer-driven telescopes (sometimes called "go to" telescopes) typically take 15 minutes to a half-hour to set up for each use, and longer when you are first learning.

If your goal is just to see the Moon and a few other objects with relatively little effort, you may want to skip the telescope in favor of a good pair of binoculars, which is usually less expensive. Binoculars are generally described by two numbers, such as 7 × 35 or 12 × 50. The first number is the magnification; for example, "7×" means that objects will look 7 times closer through the binoculars than to your eye. The second number is the diameter of each lens in millimeters. As with telescopes, larger lenses mean more light and better views. However, larger lenses also tend to be heavier and more difficult to hold steady, which means you may need a tripod.

If you decide to get a telescope, the first rule to remember is that magnification is *not* the key factor, and telescopes advertised only by their magnification (such as "650 power") are rarely high quality. Instead, focus on three factors when choosing your telescope:

1. *The light-collecting area (also called aperture).* Most personal telescopes are reflectors, so a "6-inch" telescope means it has a primary mirror that is 6 inches in diameter.
2. *Optical quality.* A poorly made telescope won't do you much good. If you cannot do side-by-side comparisons, stick with a major telescope manufacturer (such as Meade, Celestron, or Orion).
3. *Portability.* A large, bulky telescope can be great if you plan to keep it on a deck, but it will be difficult to carry on camping trips. Depending on how you plan to use your telescope, you'll need to make trade-offs between size and portability.

Most important, remember that a telescope is an investment that you will keep for many years. As with any investment, learn all you can before you settle on a particular model. Read buyers' guides and reviews of telescopes from sources such as *Astronomy*, *Sky and Telescope*, or the Astronomical Society of the Pacific. Talk to knowledgeable salespeople at stores that specialize in telescopes. And find a nearby astronomy club that holds observing sessions so that you can try out some telescopes and learn from experienced telescope owners.

Fortunately, astronomers have developed a technology known as **interferometry,** which can allow multiple telescopes to work together to achieve an angular resolution equivalent to that of a much larger single telescope. The Karl G. Jansky Very Large Array (JVLA) in New Mexico links 27 individual radio dishes laid out in the shape of a Y (Figure 5.21); when spaced as widely as possible, the JVLA can achieve an angular resolution that otherwise would require a single radio telescope with a diameter of almost 40 kilometers. Astronomers can achieve even higher angular resolution by linking radio telescopes around the world.

Interferometry is more difficult for shorter-wavelength (higher-frequency) light, but astronomers have achieved many successes. One spectacular example is the Atacama Large Millimeter/submillimeter Array (ALMA), in Chile, which combines light from 66 individual telescopes working at millimeter and submillimeter wavelengths (Figure 5.22).

Interferometry allows small telescopes to work together to obtain the angular resolution of a much larger telescope

Perhaps even more impressively, the Event Horizon Telescope project seeks to link ALMA and other radio and submillimeter telescopes around the world, with the goal of achieving the angular resolution needed to obtain an image of the Milky Way Galaxy's central black hole [Section 15.4]. Interferometry is also being used at infrared and visible wavelengths with telescopes built in pairs (such as the Keck and Magellan telescope pairs) or with more than one telescope on a common mount (such as the Large Binocular Telescope).

At the short-wavelength end of the spectrum, astronomers face a different challenge in building telescopes. Trying to focus high-energy ultraviolet or X-ray photons is somewhat like trying to focus a stream of bullets. If the bullets are fired directly at a metal sheet, they will puncture or damage the sheet. However, if the metal sheet is angled so that the bullets barely graze its surface, then it will slightly deflect the bullets. The mirrors of X-ray telescopes, such as NASA's Chandra X-Ray Observatory and the NuSTAR mission, are designed to deflect X rays in much the same way (Figure 5.23). Gamma rays have even more energy and

▲ **FIGURE 5.21**

The Karl G. Jansky Very Large Array (JVLA) in New Mexico consists of 27 telescopes that can be moved along train tracks. The telescopes work together through interferometry and can achieve an angular resolution equivalent to that of a single radio telescope almost 40 kilometers across.

◀ **FIGURE 5.22**

The Atacama Large Millimeter/submillimeter Array (ALMA) is located in a high desert (altitude 5000 meters) in Chile.

a Artist's illustration of the Chandra X-Ray Observatory, which orbits Earth.

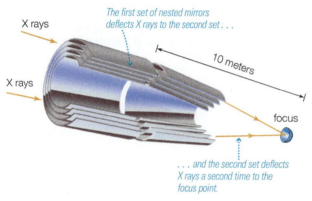

X rays

X rays

The first set of nested mirrors deflects X rays to the second set . . .

10 meters

focus

. . . and the second set deflects X rays a second time to the focus point.

b This diagram shows the arrangement of Chandra's nested, cylindrical X-ray mirrors. Each mirror is 0.8 meter long and between 0.6 and 1.2 meters in diameter.

▲ **FIGURE 5.23**

The Chandra X-Ray Observatory focuses X rays that enter the front of the telescope by deflecting them twice so that they end up focused at the back of the telescope.

▲ **FIGURE 5.24**

This photo shows an aerial view of a detector in Hanford, Washington, for the Advanced Laser Interferometer Gravitational-Wave Observatory (LIGO). A second detector is located in Livingston, Louisiana.

cannot generally be focused by mirrors, so gamma-ray telescopes (such as the Fermi Gamma-Ray Space Telescope) typically use massive detectors to capture photons and determine the direction they came from.

Looking Beyond Light Light is not the only form of information that travels through the universe, and astronomers have begun to build and use telescopes designed to observe at least three other types of "cosmic messengers." First, there's an extremely lightweight type of subatomic particle known as the *neutrino* [Section 11.2] that is produced by nuclear reactions, including nuclear fusion in the Sun and the reactions that accompany the explosions of distant stars. Astronomers have used "neutrino telescopes"—typically located in deep mines or under water or ice—to gain valuable insights about the Sun and stellar explosions. Second, Earth is continually bombarded by very high-energy subatomic particles from space known as *cosmic rays* [Section 15.2]. We still know relatively little about the origin of cosmic rays, but astronomers are now using both satellites and ground-based detectors to catch and study them. Third, Einstein's general theory of relativity (see Special Topic, page 373) predicts the existence of something called *gravitational waves*, which are different in nature from light but travel at the speed of light. For decades, we've had indirect evidence that gravitational waves really exist, but until recently, direct detection of them was beyond our technological capabilities. Today, the first gravitational wave telescopes are up and running (Figure 5.24), and they have already successfully detected gravitational waves from the mergers of pairs of black holes [Section 14.4].

◆ Why do we put telescopes in space?

Many telescopes are now placed in space, including the Hubble Space Telescope, the Chandra X-Ray Observatory, and the James Webb Space Telescope slated for launch in 2018. To understand why we do this despite the high cost of launching a telescope into space, we must understand the ways in which Earth's atmosphere hinders observations from the ground.

Atmospheric Effects on Visible Light Some of the problems created by Earth's atmosphere are obvious. The brightness of the daytime sky limits visible-light observations to nighttime, and clouds can block night observations. In addition, our atmosphere scatters the bright lights of cities, creating **light pollution** that can obscure the view even for the best telescopes (Figure 5.25). For example, the 2.5-meter telescope at Mount Wilson, the world's largest when it was built in 1917, would be much more useful today if it weren't located so close to the lights of what was once the small town of Los Angeles.

A somewhat less obvious problem is the blurring of light by the atmosphere. The ever-changing motion, or **turbulence,** of air in the atmosphere bends light in constantly shifting patterns. This turbulence causes the familiar twinkling of stars. Twinkling may be beautiful to the naked eye, but it causes problems for astronomers because it blurs astronomical images.

Telescopes in space are above the distorting effects of Earth's atmosphere.

see it for yourself Put a coin at the bottom of a cup of water. If you stir the water, the coin will appear to move around, even if it remains stationary on the bottom. What makes the coin appear to move? How is this similar to the way that our atmosphere makes stars appear to twinkle?

▲ **FIGURE 5.25**
Earth at night: It's pretty, but to astronomers it's light pollution. This image, a composite made from hundreds of satellite photos, shows the bright lights of cities as they appear from Earth orbit at night.

▲ **FIGURE 5.26**
Observatories on the summit of Mauna Kea in Hawaii. Mauna Kea meets all the key criteria for an observing site: It is far from big-city lights, high in altitude, and located in an area where the air tends to be calm and dry.

Astronomers can partially mitigate effects of weather, light pollution, and atmospheric blurring by choosing observing sites that are *dark* (limiting light pollution), *dry* (limiting rain and clouds), *calm* (limiting turbulence), and *high* (placing them above at least part of the atmosphere). Three particularly important sites are the 4300-meter (14,000-foot) summit of Mauna Kea on the Big Island of Hawaii (Figure 5.26), a 2400-meter-high site on the island of La Palma in Spain's Canary Islands, and, for the southern hemisphere, the 2600-meter-high Paranal Observatory site in Chile.

Of course, the ultimate solution to atmospheric distortion is to put telescopes in space, above the atmosphere. That is one reason why the Hubble Space Telescope (Figure 5.27) was built and why it has been so successful despite the relatively small size (2.4 meters) of its primary mirror.

Atmospheric Absorption and Emission of Light As we'll discuss shortly, new technologies sometimes make it possible for ground-based observatories to equal or better the visible-light observations of the Hubble Space Telescope. However, Earth's atmosphere poses one major problem that no Earth-bound technology can overcome: *Our atmosphere prevents most forms of light from reaching the ground at all* (Figure 5.28). Moreover, the atmosphere itself glows at many infrared wavelengths, generating a background glare for most infrared observations from the ground. Only radio waves, visible light (and the very longest wavelengths of ultraviolet light), and small parts of the infrared spectrum can be observed from the ground.

The most important reason for putting telescopes in space is to allow us to observe light that does not penetrate Earth's atmosphere. That is why all X-ray telescopes are in space—an X-ray telescope would be completely useless on the ground—and the same is true for telescopes observing in most portions of the infrared, ultraviolet, and gamma rays. Indeed, the Hubble Space Telescope often observes ultraviolet or infrared wavelengths that do not reach the ground (or that would be

Much of the electromagnetic spectrum can be observed only from space and not from the ground.

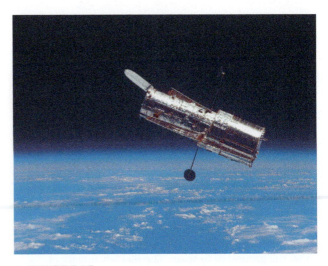

▲ **FIGURE 5.27**
The Hubble Space Telescope orbits Earth. Its position above the atmosphere allows it an undistorted view of space. Hubble can observe infrared and ultraviolet light as well as visible light.

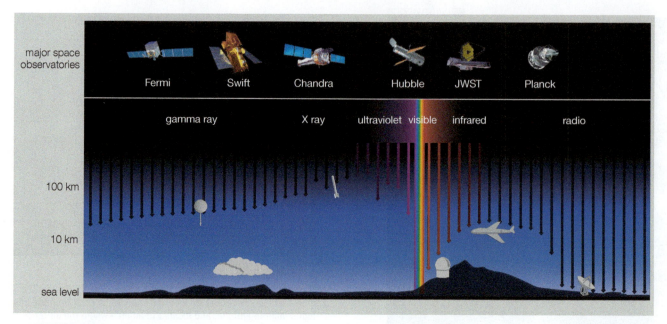

▲ **FIGURE 5.28**
This diagram shows the approximate depths to which different wavelengths of light penetrate Earth's atmosphere. Note that most of the electromagnetic spectrum can be observed only from very high altitudes or from space. Major space observatories for different wavelengths are also shown.

interfered with by atmospheric infrared emissions), which is why it would remain a valuable observatory even if ground-based telescopes matched its visible-light capabilities. The James Webb Space Telescope is being optimized for infrared observations, which will allow it to study the highly redshifted light of the most distant galaxies in the universe. Because we see distant objects as they were long ago, this will enable us to observe galaxies as they first began to form in the early universe. The telescope will be kept far from Earth's heat by being put in an orbit around the Sun at a greater distance than Earth, while an attached Sun shield will prevent sunlight from heating the telescope (Figure 5.29).

think about it Find the current status of the James Webb Space Telescope. Is it on track for launch or already launched? List at least three science objectives of the telescope.

common misconceptions

Closer to the Stars?

Many people mistakenly believe that space telescopes are advantageous because their locations above Earth put them closer to the stars. You can see why this is wrong by thinking about scale. On the scale of the Voyage model solar system (see Section 1.1), the Hubble Space Telescope is so close to the surface of the millimeter-diameter Earth that you would need a microscope to resolve its altitude, while the nearest stars are thousands of kilometers away. The distances to the stars are effectively the same whether a telescope is on the ground or in space. The real advantages of space telescopes arise from their being above Earth's atmosphere and the observational problems it presents.

▲ **FIGURE 5.29**
This photo shows a full-size model of the James Webb Space Telescope, with the sunshield at the bottom, on display in Austin, Texas, during the South by Southwest festival.

Adaptive Optics Despite the many advantages of putting telescopes in space, it is still much less expensive to build large telescopes on the ground. However, while these telescopes can obviously collect a lot of light, until recently they were unable to achieve the angular resolution that would in principle be allowed by their sizes, because air turbulence inevitably blurred their images.

Today, an amazing technology known as **adaptive optics** can eliminate much of the blurring caused by our atmosphere. The tech-

Adaptive optics allows ground-based telescopes to overcome atmospheric distortions.

nology works like this: Turbulence causes rays of light from a star to dance around as they reach a telescope. Adaptive optics essentially makes a telescope's mirrors do an opposite dance, canceling out the atmospheric distortions (Figure 5.30). The shape of a mirror (often the secondary or even a third or fourth mirror) is changed slightly many times each second to compensate for the rapidly changing atmospheric distortions. A computer calculates the necessary changes by monitoring distortions in the image of a bright star near the object under study. If there is no bright star near the object of interest, the observatory may shine a laser into the sky to create an *artificial star* (a point of light in Earth's atmosphere) that it can monitor for distortions.

Together, the combination of space telescopes, adaptive optics for ground-based telescopes, and interferometry is helping astronomers achieve unprecedented views of the heavens. Technology is rapidly advancing, and it seems reasonable to imagine that, within a few decades, new telescopes may offer views of distant objects that may be as detailed in comparison to Hubble Space Telescope images as Hubble's images are in comparison to those of the naked eye.

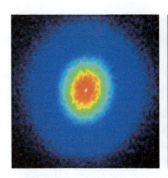 

a Atmospheric blurring makes this ground-based image of a double star look like a single star.

b When the same telescope is used with adaptive optics, the two stars are clearly distinguished. The angular separation between the two stars is 0.28 arcsecond.

▲ **FIGURE 5.30**
The technology of adaptive optics can enable a ground-based telescope to overcome most of the blurring caused by Earth's atmosphere. Both images were taken in near-infrared light with the Canada-France-Hawaii Telescope. The colors represent infrared brightness, with the brightest light shown in white (center of each star image) and the faintest light in blue to black.

---

the big picture Putting Chapter 5 into Perspective

This chapter's main purpose was to show how we learn about the universe by observing the light of distant objects. "Big picture" ideas that will help you keep your understanding in perspective include the following:

- Most of what we know about the universe comes from information that we receive from light, because matter leaves "fingerprints" that we can read by carefully analyzing the light of distant objects.

- The visible light that our eyes can see is only a small portion of the complete electromagnetic spectrum. Different portions of the spectrum contain different pieces

of the story of a distant object, so it is important to study all forms of light.

- Images are pretty, but we can often learn more by dispersing the light of a distant object into a spectrum, which can reveal the object's composition, surface temperature, motion toward or away from us, and more.

- Technology drives astronomical discovery. Every time we build a bigger telescope, develop a more sensitive detector, or open up a new wavelength region to study, we learn more about the universe.

my cosmic perspective The methods of learning from light that we've discussed for astronomy are also used to learn about many things on Earth. For example, we use light to learn about the atmosphere and global warming, to identify toxic chemicals in water, and for medical scans that help diagnose injuries and disease.

5.1 Basic Properties of Light and Matter

◆ What is light?

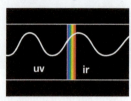

Light is an **electromagnetic wave,** but also comes in individual "pieces" called **photons.** Each photon has a precise wavelength, frequency, and energy: the shorter the wavelength, the higher the frequency and energy. In order of decreasing wavelength, the forms of light are **radio waves, microwaves, infrared, visible light, ultraviolet, X rays,** and **gamma rays.**

◆ What is matter?

Ordinary matter is made of **atoms,** which are made of **protons, neutrons,** and **electrons.** Atoms of different **chemical elements** have different numbers of protons. **Isotopes** of a particular chemical element all have the same number of protons but different numbers of neutrons. **Molecules** are made from two or more atoms.

◆ How do light and matter interact?

Matter can emit, absorb, transmit, or reflect (or scatter) light.

5.2 Learning from Light

◆ What are the three basic types of spectra?

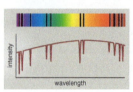

There are three basic types of spectra: a **continuous spectrum,** which looks like a rainbow; an **absorption line spectrum,** in which specific colors are missing from the rainbow; and an **emission line spectrum,** in which we see lines of specific colors against a black background.

◆ How does light tell us what things are made of?

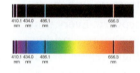

Emission or absorption lines occur only at specific wavelengths that correspond to particular energy level transitions in atoms or molecules. Every kind of atom, ion, and molecule produces a unique set of spectral lines, so we can determine an object's composition by identifying these lines.

◆ How does light tell us the temperatures of planets and stars?

Objects such as planets and stars produce **thermal radiation** spectra, the most common type of continuous spectra. We can determine temperature from these spectra because hotter objects emit more total radiation per unit area and emit photons with a higher average energy.

◆ How does light tell us the speed of a distant object?

The **Doppler effect** tells us how fast an object is moving toward or away from us. Spectral lines are shifted to shorter wavelengths (a **blueshift**) for objects moving toward us and to longer wavelengths (a **redshift**) for objects moving away from us.

5.3 Collecting Light with Telescopes

◆ How do telescopes help us learn about the universe?

Telescopes allow us to see fainter objects and to see more detail than we can see with our eyes, and to study light from all portions of the spectrum. **Light-collecting area** describes how much light a telescope can collect. **Angular resolution** determines the amount of detail in telescopic images; it is better (smaller) for larger telescopes and, for a given telescope size, better for shorter-wavelength light. Multiple telescopes can sometimes be used together through **interferometry** to achieve the angular resolution of a much larger telescope.

◆ Why do we put telescopes in space?

Telescopes in space are above Earth's atmosphere and not subject to problems caused by **light pollution,** atmospheric distortion and emission of light, or the fact that most forms of light do not penetrate through the atmosphere to the ground. However, for visible light, it is now possible to overcome some of the blurring effects of Earth's atmosphere through **adaptive optics.**

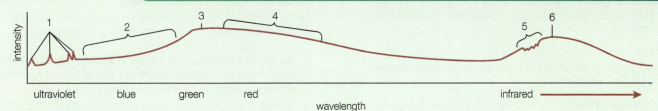

Refer to the numbered features on the above graph, which is a schematic spectrum of the planet Mars.

1. Which of the six numbered features represents emission lines?

2. Which of the six numbered features represents absorption lines?

3. Which portion(s) of the spectrum represent(s) reflected sunlight?
   a. 1 only
   b. 2, 3, and 4
   c. 3 and 6
   d. the entire spectrum

4. What does the wavelength of the peak labeled 6 tell us about Mars?
   a. its color
   b. its surface temperature
   c. its chemical composition
   d. its orbital speed

5. What feature(s) of this spectrum indicate(s) that Mars appears red in color?
   a. the wavelength of the peak labeled 3
   b. the wavelength of the peak labeled 6
   c. the fact that the intensity of region 4 is higher than that of region 2
   d. the fact that the peak labeled 3 is higher than the peak labeled 6

exercises and problems

MasteringAstronomy®   For instructor-assigned homework and other learning materials, go to MasteringAstronomy®.

Review Questions

1. Define *wavelength, frequency,* and *speed* for light waves. If light has a long wavelength, what can you say about its frequency? Explain.
2. What is a *photon*? In what way is a photon like a particle? In what way is it like a wave?
3. List the different forms of light in order from lowest to highest energy. Is the order the same from lowest to highest frequency? from shortest to longest wavelength?
4. Briefly describe the structure and size of an atom. How big is the *nucleus* compared to the entire atom?
5. Define *atomic number* and *atomic mass number*. Under what conditions are two atoms different *isotopes* of the same element?
6. What is *electrical charge*? Will an electron and a proton attract or repel one another? How about two electrons?
7. What are the four major ways in which light and matter can interact? Give an example of each from everyday life.
8. Describe the conditions that lead to each of the three basic types of spectra. Which type is the Sun's spectrum, and why?
9. Why do atoms emit or absorb light of specific wavelengths? How does this fact allow us to determine the chemical composition of a distant object?
10. Describe two ways in which the thermal radiation spectrum of an 8000 K star would differ from that of a 4000 K star.
11. Describe the *Doppler effect* for light and what we can learn from it. What does it mean to say that radio waves are *blueshifted*?
12. What are the two key properties of a telescope, and why is each important? Also distinguish between *refracting* and *reflecting* telescopes.
13. List at least three ways Earth's atmosphere can hinder astronomical observations.
14. Briefly describe how adaptive optics and interferometry can improve astronomical observations.

Test Your Understanding

⟳ Does It Make Sense?

Decide whether the statement makes sense (or is clearly true) or does not make sense (or is clearly false). Explain clearly; not all of these have definitive answers, so your explanation is more important than your chosen answer.

15. If you could view a spectrum of the light reflecting off a blue sweatshirt, you'd find the entire rainbow of color (looking the same as a spectrum of white light).
16. Because of their higher frequency, X rays must travel through space faster than radio waves.
17. Two isotopes of the element rubidium differ in their number of protons.
18. If the Sun's surface became much hotter (while the Sun's size remained the same), the Sun would emit more ultraviolet light but less visible light than it currently emits.
19. If you could see infrared light, you would see a glow from the backs of your eyelids when you closed your eyes.
20. If you had X-ray vision, then you could read this entire book without turning any pages.
21. If a distant galaxy has a substantial redshift (as viewed from our galaxy), then anyone living in that galaxy would see a substantial redshift in a spectrum of the Milky Way Galaxy.
22. Thanks to adaptive optics, telescopes on the ground can now make ultraviolet images of the cosmos.

23. Thanks to interferometry, a properly spaced set of 10-meter radio telescopes can achieve the angular resolution of a single 100-kilometer radio telescope.

24. If you lived on the Moon, you'd never see stars twinkle.

Quick Quiz

Choose the best answer to each of the following. Explain your reasoning with one or more complete sentences.

25. Why is a sunflower yellow? (a) It emits yellow light. (b) It absorbs yellow light. (c) It reflects yellow light.

26. Compared to red light, blue light has higher frequency and (a) higher energy and shorter wavelength. (b) higher energy and longer wavelength. (c) lower energy and shorter wavelength.

27. Radio waves are (a) a form of sound. (b) a form of light. (c) a type of spectrum.

28. Compared to an atom as a whole, an atomic nucleus is (a) very tiny but has most of the mass. (b) quite large and has most of the mass. (c) very tiny and has very little mass.

29. Some nitrogen atoms have seven neutrons and some have eight neutrons; these two forms of nitrogen are (a) ions of each other. (b) phases of each other. (c) isotopes of each other.

30. The set of spectral lines that we see in a star's spectrum depends on the star's (a) atomic structure. (b) chemical composition. (c) rotation rate.

31. A star whose spectrum peaks in the infrared is (a) cooler than our Sun. (b) hotter than our Sun. (c) larger than our Sun.

32. A spectral line that appears at a wavelength of 321 nm in the laboratory appears at a wavelength of 328 nm in the spectrum of a distant object. We say that the object's spectrum is (a) redshifted. (b) blueshifted. (c) whiteshifted.

33. How much greater is the light-collecting area of a 6-meter telescope than that of a 3-meter telescope? (a) two times (b) four times (c) six times

34. The Hubble Space Telescope obtains higher-resolution images than most ground-based telescopes because it is (a) larger. (b) closer to the stars. (c) above Earth's atmosphere.

Process of Science

35. *Elements in Space.* Astronomers claim that objects throughout the universe are made of the same chemical elements that exist here on Earth. Given that most of these objects are so far away that we can never hope to visit them, why are astronomers so confident that the objects are made from the same set of chemical elements, rather than completely different substances?

36. *Newton's Prisms.* Look back at the brief discussion in this chapter of how Newton demonstrated that the colors seen when sunlight passes through a prism come from the light itself rather than from the prism. Suppose you wanted to test Newton's findings. Assuming you have two prisms and a white screen, describe how you would arrange the prisms to duplicate Newton's discovery.

Group Work Exercise

37. *Which Telescope Would You Use?* **Roles:** *Scribe* (takes notes on the group's activities), *Proposer* (proposes explanations to the group), *Skeptic* (points out weaknesses in proposed explanations), *Moderator* (leads group discussion and makes sure everyone contributes). **Activity:** Your job is to choose a telescope for observing matter around a black hole; assume that the matter is emitting photons at

all wavelengths. Rank the following four telescopes from best to worst for this observing task. Explain your rankings.
   a. an X-ray telescope, 2 meters in diameter, located at the South Pole
   b. an infrared telescope, 2 meters in diameter, on a spacecraft in orbit around Earth and observing at a wavelength of 2 micrometers (2×10^{-6} m)
   c. an infrared telescope, 10 meters in diameter, equipped with adaptive optics, located on Mauna Kea in Hawaii and observing at a wavelength of 10 micrometers (10^{-5} m)
   d. a radio telescope, 300 meters in diameter, located in Puerto Rico

Investigate Further

Short-Answer/Essay Questions

38. *Atomic Terminology Practice I.*
   a. The most common form of iron has 26 protons and 30 neutrons. For a neutral atom of this form of iron, state the atomic number, atomic mass number, and number of electrons.
   b. Consider the following three atoms: Atom 1 has seven protons and eight neutrons; atom 2 has eight protons and seven neutrons; atom 3 has eight protons and eight neutrons. Which two are *isotopes* of the same element?
   c. Oxygen has atomic number 8. How many times must an oxygen atom be ionized to create an O^{+5} ion? How many electrons are in an O^{+5} ion?

39. *Atomic Terminology Practice II.*
   a. What are the atomic number and atomic mass number of fluorine atoms with 9 protons and 10 neutrons? If we could add a proton to a fluorine nucleus, would the result still be fluorine? What if we added a neutron to a fluorine nucleus? Explain.
   b. The most common isotope of gold has atomic number 79 and atomic mass number 197. How many protons and neutrons does the gold nucleus contain? If it is electrically neutral, how many electrons does it have? If it is triply ionized, how many electrons does it have?
   c. Uranium has atomic number 92. Its most common isotope is ^{238}U, but the form used in nuclear bombs and nuclear power plants is ^{235}U. How many neutrons are in each of these two isotopes of uranium?

40. *Energy Level Transitions.* The following labeled transitions represent an electron moving between energy levels in hydrogen. Answer each of the following questions and explain your answers.

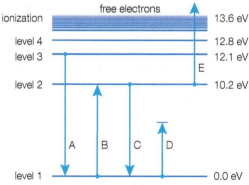

   a. Which transition could represent an atom that *absorbs* a photon with 10.2 eV of energy?
   b. Which transition could represent an atom that *emits* a photon with 10.2 eV of energy?
   c. Which transition represents an electron that is breaking free of the atom?

d. Which transition, as shown, is *not* possible?

e. Would transition A represent emission or absorption of light? How would the wavelength of the emitted or absorbed photon compare to that of the photon involved in transition C? Explain.

41. *Orion Nebula.* Viewed through a telescope, much of the Orion Nebula looks like a glowing cloud of gas. What type of spectrum would you expect to see from the glowing parts of the nebula? Why?

42. *The Doppler Effect.* In hydrogen, the transition from level 2 to level 1 has a rest wavelength of 121.6 nm. Suppose you see this line at a wavelength of 120.5 nm in Star A, at 121.2 nm in Star B, at 121.9 nm in Star C, and at 122.9 nm in Star D. Which stars are coming toward us? Which are moving away? Which star is moving fastest relative to us? Explain your answers without doing any calculations.

43. *Spectral Summary.* Clearly explain how studying an object's spectrum can allow us to determine each of the following properties of the object.
    a. The object's surface chemical composition
    b. The object's surface temperature
    c. Whether the object is a low-density cloud of gas or something more substantial
    d. The speed at which the object is moving toward or away from us

44. *Image Resolution.* What happens if you take a photograph from a newspaper, magazine, or book and blow it up to a larger size? Can you see more detail than you could before? Explain clearly, and relate your answer to the concepts of magnification and angular resolution in astronomical observations.

45. *Telescope Technology.* Suppose you were building a space-based observatory consisting of five individual telescopes. Which would be the best way to use these telescopes: as five individual telescopes with adaptive optics, or as five telescopes linked together for interferometry but without adaptive optics? Explain your reasoning clearly.

46. *Project: Twinkling Stars.* Using a star chart, identify 5–10 bright stars that should be visible in the early evening. On a clear night, observe each of these stars for a few minutes. Note the date and time, and for each star record the following information: approximate altitude and direction in your sky, brightness compared to other stars, color, and how much the star twinkles compared to other stars. Study your record. Can you draw any conclusions about how brightness and position in your sky affect twinkling?

Quantitative Problems

Be sure to show all calculations clearly and state your final answers in complete sentences.

47. *Thermal Radiation Laws.*
    a. Find the emitted power per square meter and wavelength of peak intensity for a 3000 K object that emits thermal radiation.
    b. Find the emitted power per square meter and wavelength of peak intensity for a 50,000 K object that emits thermal radiation.

48. *Hotter Sun.* Suppose the surface temperature of the Sun were about 12,000 K, rather than 6000 K.
    a. How much more thermal radiation would the Sun emit?
    b. What would happen to the Sun's wavelength of peak intensity?
    c. Do you think it would still be possible to have life on Earth? Explain.

49. *Doppler Calculations.* In hydrogen, the transition from level 2 to level 1 has a rest wavelength of 121.6 nm. Find the speed and direction (toward or away from us) for a star in which this line appears at wavelength
    a. 120.5 nm.
    b. 121.2 nm.
    c. 121.9 nm.
    d. 122.9 nm.

50. *Hubble's Field of View.* Large telescopes often have small fields of view. For example, the Hubble Space Telescope's (HST's) advanced camera has a field of view that is roughly square and about 0.06° on a side.
    a. Calculate the angular area of the HST's field of view in square degrees.
    b. The angular area of the entire sky is about 41,250 square degrees. How many pictures would the HST have to take with its camera to obtain a complete picture of the entire sky?

Discussion Questions

51. *The Changing Limitations of Science.* In 1835, French philosopher Auguste Comte stated that science would never allow us to learn the composition of stars. Although spectral lines had been seen in the Sun's spectrum at that time, it wasn't until the mid–19th century that scientists recognized (primarily through the work of Foucault and Kirchhoff) that spectral lines give clear information about chemical composition. Why might our present knowledge have seemed unattainable in 1835? Discuss how new discoveries can change the apparent limitations of science. Today, other questions seem beyond the reach of science, such as the question of how life began on Earth. Do you think such questions will ever be answerable through science? Defend your opinion.

52. *Science and Technology Funding.* Technological innovation clearly drives scientific discovery in astronomy, but the reverse is also true. For example, Newton made his discoveries in part because he wanted to explain the motions of the planets, but his discoveries have had far-reaching effects on our civilization. Congress often must decide between funding programs with purely scientific purposes ("basic research") and funding programs designed to develop new technologies. If you were a member of Congress, how would you try to allocate spending for basic research and technology? Why?

53. *Your Microwave Oven.* A *microwave oven* emits microwaves that have just the right wavelength needed to cause energy level changes in water molecules. Use this fact to explain how a microwave oven cooks your food. Why doesn't a microwave oven make a plastic dish get hot? Why do some clay dishes get hot in the microwave? Why do dishes that aren't themselves heated by the microwave oven sometimes still get hot when you heat food on them? (*Note:* It's not a good idea to put empty dishes in a microwave.)

Web Projects

54. *Kids and Light.* Visit one of the many websites designed to teach middle and high school students about light. Read the content, and try the activities. If you were a teacher, would you find the site useful for your students? Why or why not? Write a one-page summary of your conclusions.

55. *Major Ground-Based Observatories.* Take a virtual tour of one of the world's major astronomical observatories. Write a short report on why the observatory is useful to astronomy.

56. *Space Observatory.* Visit the website of a major space observatory, either existing or under development. Write a short report about the observatory, including its purpose, its orbit, and how it operates.

57. *Really Big Telescopes.* Learn about one of the projects to build a very large telescope (such as the Giant Magellan Telescope, the Thirty Meter Telescope, the European Extremely Large Telescope, or the High Definition Space Telescope). Write a short report about the telescope's current status and potential capabilities.

One of Isaac Newton's great insights was that physics is universal—the same physical laws govern both the motions of heavenly objects and the things we experience in everyday life. This illustration shows some of the key physical principles used in the study of astronomy, with examples of how they apply both on Earth and in space.

EXAMPLES ON EARTH

(1) Conservation of Energy: Energy can be transferred from one object to another or transformed from one type to another, but the total amount of energy is always conserved [Section 4.3].

kinetic energy

radiative energy

potential energy

Plants transform the energy of sunlight into food containing chemical potential energy, which our bodies can convert into energy of motion.

(2) Conservation of Angular Momentum: An object's angular momentum cannot change unless it transfers angular momentum to another object. Because angular momentum depends on the product of mass, velocity, and radius, a spinning object must spin faster as it shrinks in size and an orbiting object must move faster when its orbital distance is smaller [Section 4.3].

Conservation of angular momentum explains why a skater spins faster as she pulls in her arms.

(3) Gravity: Every mass in the universe attracts every other mass through the force called gravity. The strength of gravity between two objects depends on the product of the masses divided by the square of the distance between them [Section 4.4].

The force of gravity between a ball and Earth attracts both together, explaining why the ball accelerates as it falls.

(4) Thermal Radiation: Large objects emit a thermal radiation spectrum that depends on the object's temperature. Hotter objects emit photons with a higher average energy and emit radiation of greater intensity at all wavelengths [Section 5.2].

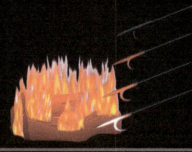

The glow you see from a hot fireplace poker is thermal radiation in the form of visible light.

(5) Electromagnetic Spectrum: Light is a wave that affects electrically charged particles and magnets. The wavelength and frequency of light waves range over a wide spectrum, consisting of gamma rays, X rays, ultraviolet light, visible light, infrared light, and radio waves. Visible light is only a small fraction of the entire spectrum [Section 5.1].

X-ray machines

light bulb

We encounter many different kinds of electromagnetic radiation in our everyday lives.

microwave oven

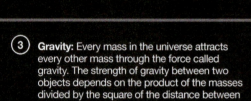

gamma rays | X rays | ultraviolet | infrared | radio

visible

microwaves

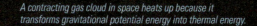

EXAMPLES IN SPACE

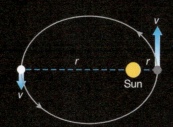

A contracting gas cloud in space heats up because it transforms gravitational potential energy into thermal energy.

Conservation of angular momentum also explains why a planet's orbital speed increases when it is closer to the Sun.

r

v

r

v

Sun

M_1

$$F_g = G \, \frac{M_1 M_2}{d^2}$$

M_2

d

Gravity also operates in space—its attractive force can act across great distances to pull objects closer together or to hold them in orbit.

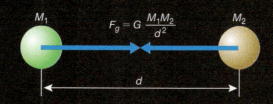

relative intensity per square meter of surface

10^8
10^6
10^4
10^2
10^0

10^1 10^2 10^3 10^4 10^5

visible light

15,000 K star

the Sun (5800 K)

3000 K star

wavelength (nm)

← ultraviolet   infrared →

Sunlight is also a visible form of thermal radiation. The Sun is much brighter and whiter than a fireplace poker because its surface is much hotter.

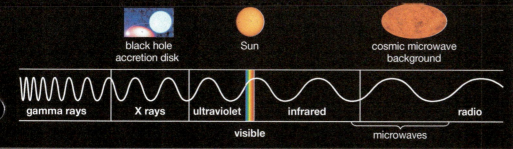

black hole accretion disk

Sun

cosmic microwave background

gamma rays | X rays | ultraviolet | infrared | radio

visible

microwaves

Many different forms of electromagnetic radiation are present in space. We therefore need to observe light of many different wavelengths to get a complete picture of the universe.

6 Formation of the Solar System

This is not an artist's conception! It is a real image of a disk in which planets are forming, taken by the Atacama Large Millimeter/submillimeter Array (ALMA); see Figure 6.17b for more details.

LEARNING GOALS

6.1 A Brief Tour of the Solar System
- What does the solar system look like?

6.2 The Nebular Theory of Solar System Formation
- What features of our solar system provide clues to how it formed?
- What is the nebular theory?

6.3 Explaining the Major Features of the Solar System
- What caused the orderly patterns of motion?
- Why are there two major types of planets?
- Where did asteroids and comets come from?
- How do we explain the "exceptions to the rules"?

6.4 The Age of the Solar System
- How do we know the age of the solar system?

How did Earth come to be? How old is it? Is it unique? Our ancestors could do little more than guess at the answers to these questions, but today we are able to address them scientifically. As we'll discuss in this chapter, careful study of the major features of our solar system has enabled scientists to put together a detailed theory of how Earth and our solar system were born. Understanding this theory will help us in later chapters as we explore in more detail both the planets of our solar system and other planetary systems.

ESSENTIAL PREPARATION

1. What is our place in the universe? [Section 1.1]
2. How big is the universe? [Section 1.1]
3. How did we come to be? [Section 1.2]
4. What keeps a planet rotating and orbiting the Sun? [Section 4.3]
5. What determines the strength of gravity? [Section 4.4]
6. How do gravity and energy allow us to understand orbits? [Section 4.4]

6.1 A Brief Tour of the Solar System

Our ancestors long ago recognized the motions of the planets through the sky, but it has been only a few hundred years since we learned that Earth is also a planet that orbits the Sun. Even then, we knew little about the other planets until the development of large telescopes. More recently, space exploration has brought us far greater understanding of other worlds. We've lived in this solar system all along, but only now are we getting to know it. Let's begin with a quick tour of our planetary system, which will provide context for the more detailed study that will follow.

◆ What does the solar system look like?

The first step in getting to know our solar system is to visualize what it looks like as a whole. Imagine viewing the solar system from beyond the orbits of the planets. What would we see?

Without a telescope, the answer would be "not much." Remember that the Sun and planets are all quite small compared to the distances between them [Section 1.1]—so small that if we viewed them from the outskirts of our solar system, the planets would be only pinpoints of light, and even the Sun would be just a small bright dot in the sky. But if we magnify the sizes of the planets about a thousand times compared to their distances from the Sun and show their orbital paths, we get the central picture in Figure 6.1 (pages 138–139).

The ten pages that follow Figure 6.1 offer a brief tour through our solar system, beginning at the Sun, continuing to each of the planets, and concluding with dwarf planets such as Pluto and Eris. The tour highlights a few of the most important features of each world we visit—just enough information so that you'll be ready for the comparative study we'll undertake in later chapters. The side of each page shows the planets to scale, using the 1-to-10-billion scale introduced in Chapter 1. The map along the bottom of each page shows the locations of the Sun and each of the planets in the Voyage scale model solar system, so that you can see their relative distances from the Sun. Table 6.1 follows the tour and summarizes key data.

As you study Figure 6.1, the tour pages, and Table 6.1, you'll quickly see that our solar system is *not* a random collection of worlds, but a system that exhibits many clear patterns. For example, Figure 6.1 shows that all **The planets exhibit clear patterns of composition and motion.** the planets orbit the Sun in the same direction and in nearly the same plane. In science, we always seek explanations for the existence of patterns like these. We will therefore devote most of this chapter to learning how our modern theory of solar system formation explains these and other features of the solar system.

The solar system's layout and composition offer four major clues to how it formed. The main illustration below shows the orbits of planets in the solar system from a perspective beyond Neptune, with the planets themselves magnified by about a thousand times relative to their orbits. (The Sun is not shown on the same scale as the planets; it would fill the page if it were.)

① **Large bodies in the solar system have orderly motions.** All planets have nearly circular orbits going in the same direction in nearly the same plane. Most large moons orbit their planets in this same direction, which is also the direction of the Sun's rotation.

Seen from above, planetary orbits are nearly circular.

Neptune

Mercury

Venus

Earth

Saturn

Jupiter

Mars

Uranus

White arrows indicate the rotation direction of the planets and Sun.

Red circles indicate the orbital direction of major moons around their planets.

Each planet's axis tilt is shown, with small circling arrows to indicate the direction of the planet's rotation.

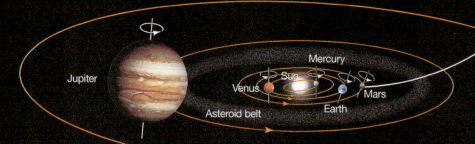

Orbits are shown to scale, but planet sizes are exaggerated about 1000 times and the Sun's size is exaggerated about 50 times relative to the orbits.

Jupiter

Mercury

Sun

Venus

Mars

Earth

Asteroid belt

Neptune

Orange arrows indicate the direction of orbital motion.

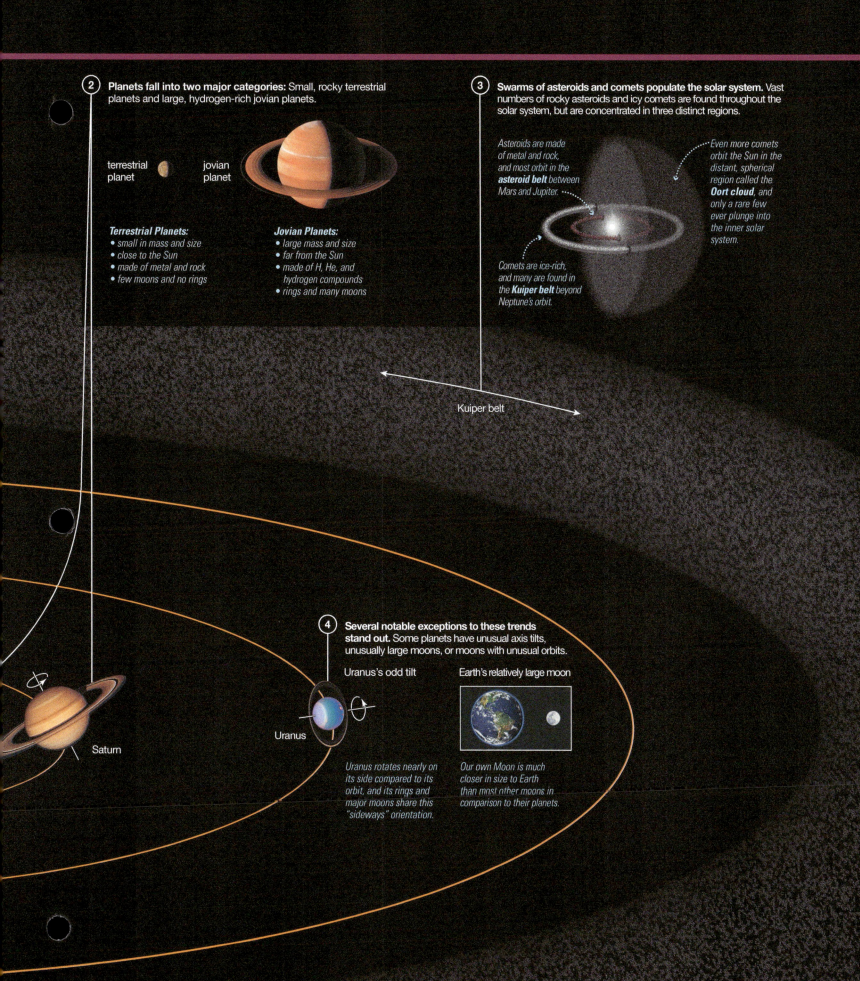

2 **Planets fall into two major categories:** Small, rocky terrestrial planets and large, hydrogen-rich jovian planets.

terrestrial planet   jovian planet

Terrestrial Planets:
- small in mass and size
- close to the Sun
- made of metal and rock
- few moons and no rings

Jovian Planets:
- large mass and size
- far from the Sun
- made of H, He, and hydrogen compounds
- rings and many moons

3 **Swarms of asteroids and comets populate the solar system.** Vast numbers of rocky asteroids and icy comets are found throughout the solar system, but are concentrated in three distinct regions.

Asteroids are made of metal and rock, and most orbit in the **asteroid belt** *between Mars and Jupiter.*

Even more comets orbit the Sun in the distant, spherical region called the **Oort cloud***, and only a rare few ever plunge into the inner solar system.*

Comets are ice-rich, and many are found in the **Kuiper belt** *beyond Neptune's orbit.*

Kuiper belt

4 **Several notable exceptions to these trends stand out.** Some planets have unusual axis tilts, unusually large moons, or moons with unusual orbits.

Uranus's odd tilt

Earth's relatively large moon

Uranus

Saturn

Uranus rotates nearly on its side compared to its orbit, and its rings and major moons share this "sideways" orientation.

Our own Moon is much closer in size to Earth than most other moons in comparison to their planets.

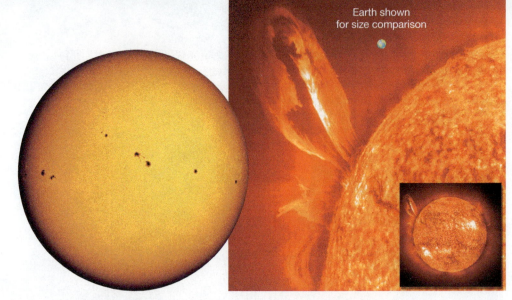

a A visible-light photograph of the Sun's surface. The dark splotches are sunspots—each large enough to swallow several Earths.

b This ultraviolet photograph, from the *SOHO* spacecraft, shows a huge streamer of hot gas on the Sun. The image of Earth was added for size comparison.

Earth shown for size comparison

◆ The Sun

- Radius: 696,000 km = $108R_{Earth}$
- Mass: $333,000M_{Earth}$
- Composition (by mass): 98% hydrogen and helium, 2% other elements

The Sun is by far the largest and brightest object in our solar system. It contains more than 99.8% of the solar system's total mass, making it nearly a thousand times as massive as everything else in the solar system combined.

The Sun's surface looks solid in photographs (Figure 6.2), but it is actually a roiling sea of hot (about 5800 K, or 5500°C or 10,000°F) hydrogen and helium gas. The surface is speckled with *sunspots* that appear dark in photographs only because they are slightly cooler than their surroundings. Solar storms sometimes send streamers of hot gas soaring far above the surface.

The Sun is gaseous throughout, and the temperature and pressure both increase with depth. The source of the Sun's energy lies deep in its core, where the temperatures and pressures are so high that the Sun is a nuclear fusion power plant.

Each second, fusion transforms about 600 million tons (600 billion kg) of the Sun's hydrogen into 596 million tons of helium. The "missing" 4 million tons becomes energy in accord with Einstein's famous formula, $E = mc^2$ [Section 4.3]. Despite losing 4 million tons of mass each second, the Sun contains so much hydrogen that it has already shone steadily for almost 5 billion years and will continue to shine for another 5 billion years.

The Sun is the most influential object in our solar system. Its gravity governs the orbits of the planets. Its heat is the primary influence on the temperatures of planetary surfaces and atmospheres. It is the source of virtually all the light in our solar system—planets and moons shine by virtue of the sunlight they reflect. In addition, charged particles flowing outward from the Sun make up the *solar wind* that interacts with planetary magnetic fields and influences planetary atmospheres. Nevertheless, we can understand almost all the present characteristics of the planets without knowing much more about the Sun than we have just discussed. We'll therefore save more detailed study of the Sun for Chapter 11, where we will study it as our prototype for understanding other stars.

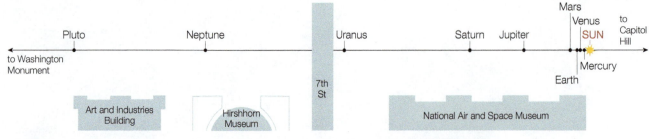

The Voyage scale model solar system represents sizes and distances in our solar system at one ten-billionth of their actual values (see Figure 1.5). The strip along the side of the page shows the sizes of the Sun and planets on this scale, and the map above shows their locations in the Voyage model on the National Mall in Washington, D.C. The Sun is about the size of a large grapefruit on this scale.

50 km

◆ Mercury

- Average distance from the Sun: 0.39 AU
- Radius: 2440 km $= 0.38R_{Earth}$
- Mass: $0.055M_{Earth}$
- Average density: 5.43 g/cm^3
- Composition: rocks, metals
- Average surface temperature: 700 K (day), 100 K (night)
- Moons: 0

Mercury is the innermost planet of our solar system, and the smallest of the eight official planets. It is a desolate, cratered world with no active volcanoes, no wind, no rain, and no life. Because there is virtually no air to scatter sunlight or color the sky, you could see stars even in the daytime if you stood on Mercury with your back toward the Sun.

You might expect Mercury to be very hot because of its closeness to the Sun, but in fact it is a world of both hot and cold extremes. Tidal forces from the Sun have forced Mercury into an unusual rotation pattern: Its 58.6-day rotation period means it rotates exactly three times for every two of its 87.9-day orbits of the Sun. This combination of rotation and orbit gives Mercury days and nights that last about 3 Earth months each. Daytime temperatures reach 425°C—nearly as hot as hot coals. At night or in shadow, the temperature falls below −150°C—far colder than Antarctica in winter.

Mercury's surface is heavily cratered (Figure 6.3), much like the surface of our Moon. But it also shows evidence of past geological activity, such as plains created by ancient lava flows and tall, steep cliffs that run hundreds of kilometers in length. These cliffs may be wrinkles from an episode of "planetary shrinking" early in Mercury's history. Mercury's high density (calculated from its mass and volume) indicates that it has a very large iron core, perhaps because it once suffered a huge impact that blasted its outer layers away.

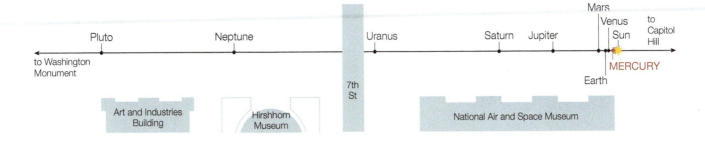

▲ **FIGURE 6.4**

The image above shows an artistic rendition of the surface of Venus as scientists think it would appear to our eyes. The surface topography is based on data from NASA's *Magellan* spacecraft. The inset (left) shows the full disk of Venus photographed by NASA's *Pioneer Venus Orbiter* with cameras sensitive to ultraviolet light. (Image above from the Voyage scale model solar system, developed by the Challenger Center for Space Science Education, the Smithsonian Institution, and NASA. Image by David P. Anderson, Southern Methodist University © 2001.)

◆ Venus

- Average distance from the Sun: 0.72 AU
- Radius: 6051 km $= 0.95 R_{Earth}$
- Mass: $0.82 M_{Earth}$
- Average density: 5.24 g/cm^3
- Composition: rocks, metals
- Average surface temperature: 740 K
- Moons: 0

Venus, the second planet from the Sun, is nearly identical in size to Earth. Before the era of spacecraft visits, Venus stood out largely for its strange rotation: It rotates on its axis very slowly and in the opposite direction of Earth, so days and nights are very long and the Sun rises in the west and sets in the east instead of rising in the east and setting in the west. Its surface is completely hidden from view by dense clouds, so we knew little about it until a few decades ago, when spacecraft began to map Venus with cloud-penetrating radar, discovering mountains, valleys, craters, and extensive evidence of past volcanic activity (Figure 6.4). Because we knew so little about it, some science fiction writers used its Earth-like size, thick atmosphere, and closer distance to the Sun to speculate that it might be a lush, tropical paradise—a "sister planet" to Earth.

The reality is far different. We now know that an extreme *greenhouse effect* bakes Venus's surface to an incredible 470°C (about 880°F), trapping heat so effectively that nighttime offers no relief. Day and night, Venus is hotter than a pizza oven, and the thick atmosphere bears down on the surface with a pressure equivalent to that nearly a kilometer (0.6 mile) beneath the ocean's surface on Earth. Far from being a beautiful sister planet to Earth, Venus resembles a traditional view of hell.

The fact that Venus and Earth are so similar in size and composition but so different in surface conditions suggests that Venus could teach us important lessons. In particular, Venus's greenhouse effect is caused by carbon dioxide, the same gas that is primarily responsible for global warming on Earth. Perhaps further study of Venus may help us better understand and solve some of the problems we face here at home.

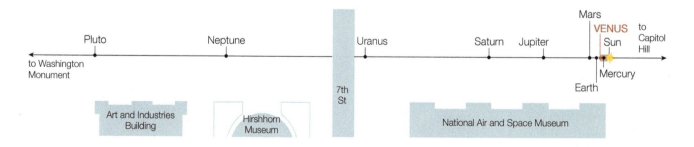

a This image (left), computer generated from satellite data, shows the striking contrast between the day and night hemispheres of Earth. The day side reveals little evidence of human presence, but at night our presence is revealed by the lights of human activity. (From the Voyage scale model solar system, developed by the Challenger Center for Space Science Education, the Smithsonian Institution, and NASA. Image created by ARC Science Simulations © 2001.)

b Earth and the Moon, shown to scale. The Moon is about 1/4 as large as Earth in diameter, while its mass is about 1/80 of Earth's mass. To show the distance between Earth and Moon on the same scale, you'd need to hold these two photographs about 1 meter (3 feet) apart.

▲ **FIGURE 6.5**
Earth, our home planet.

◆ Earth

- Average distance from the Sun: 1.00 AU
- Radius: 6378 km = $1R_{Earth}$
- Mass: $1.00M_{Earth}$
- Average density: 5.52 g/cm^3
- Composition: rocks, metals
- Average surface temperature: 290 K
- Moons: 1

Beyond Venus, we next encounter our home planet, Earth, the only known oasis of life in our solar system. Earth is also the only planet in our solar system with oxygen to breathe, ozone to shield the surface from deadly solar radiation, and abundant surface water. Temperatures are pleasant because Earth's atmosphere contains just enough carbon dioxide and water vapor to maintain a moderate greenhouse effect.

Despite Earth's small size, its beauty is striking (Figure 6.5a). Blue oceans cover nearly three-fourths of the surface, broken by the continental land masses and scattered islands. The polar caps are white with snow and ice, and white clouds are scattered above the surface. At night, the glow of artificial lights reveals the presence of an intelligent civilization.

Earth is the first planet on our tour with a moon. The Moon is surprisingly large compared with Earth (Figure 6.5b). Although it is not the largest moon in the solar system, almost all other moons are much smaller relative to the planets they orbit. As we'll discuss later in this chapter, the leading hypothesis holds that the Moon formed as a result of a giant impact early in Earth's history.

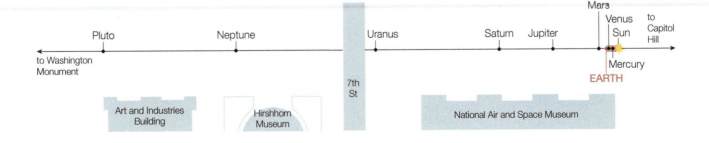

▲ **FIGURE 6.6**

The image on the right is a self-portrait of NASA's *Curiosity* rover on the floor of Gale Crater, assembled from dozens of separate images taken by the camera on the robot arm. Here, *Curiosity* is on the flank of its primary destination, Mount Sharp, with dark sand dunes visible to the left. The inset is a close-up of the disk of Mars photographed by the *Viking* orbiter; the horizontal "gash" across the center is the giant canyon Valles Marineris.

◆ Mars

- Average distance from the Sun: 1.52 AU
- Radius: 3397 km = $0.53R_{Earth}$
- Mass: $0.11M_{Earth}$
- Average density: 3.93 g/cm^3
- Composition: rocks, metals
- Average surface temperature: 220 K
- Moons: 2 (very small)

The next planet on our tour is Mars, the last of the four inner planets of our solar system (Figure 6.6). Mars is larger than Mercury and the Moon but only about half Earth's size in diameter; its mass is about 10% that of Earth. Mars has two tiny moons, Phobos and Deimos, that probably once were asteroids that were captured into Martian orbit early in the solar system's history.

Mars is a world of wonders, with ancient volcanoes that dwarf the largest mountains on Earth, a great canyon that runs nearly one-fifth of the way around the planet, and polar caps made of frozen water and carbon dioxide ("dry ice"). Although Mars is frozen today, the presence of dried-up riverbeds, rock-strewn floodplains, and minerals that form in water offers clear evidence that Mars had at least some warm and wet periods in the past. Major flows of liquid water probably ceased at least 3 billion years ago, but some liquid water could persist underground, perhaps flowing to the surface on occasion.

Mars's surface looks almost Earth-like, but you wouldn't want to visit without a spacesuit. The air pressure is far less than that on top of Mount Everest, the temperature is usually well below freezing, the trace amounts of oxygen would not be nearly enough to breathe, and the lack of atmospheric ozone would leave you exposed to deadly ultraviolet radiation from the Sun.

More than a dozen spacecraft have flown past, orbited, or landed on Mars, and plans are in the works for more. We may even send humans to Mars within the next few decades. By overturning rocks in ancient riverbeds or chipping away at ice in the polar caps, explorers will help us learn whether Mars has ever been home to life.

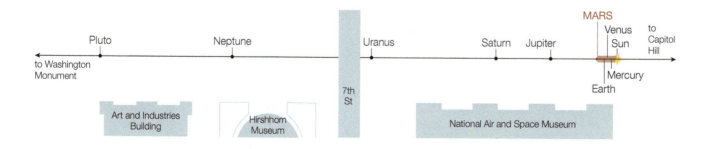

◀ **FIGURE 6.7**

This image shows what it would look like to be orbiting near Jupiter's moon Io as Jupiter comes into view. Notice the Great Red Spot to the left of Jupiter's center. The extraordinarily dark rings discovered during the *Voyager* missions are exaggerated to make them visible. This computer visualization was created using data from both NASA's *Voyager* and *Galileo* missions. (From the Voyage scale model solar system, developed by the Challenger Center for Space Science Education, the Smithsonian Institution, and NASA. Image created by ARC Science Simulations © 2001.)

◆ Jupiter

- Average distance from the Sun: 5.20 AU
- Radius: 71,492 km = $11.2R_{Earth}$
- Mass: $318M_{Earth}$
- Average density: 1.33 g/cm^3
- Composition: mostly hydrogen and helium
- Cloud-top temperature: 125 K
- Moons: at least 67

To reach the orbit of Jupiter from Mars, we must traverse a distance that is more than double the total distance from the Sun to Mars, passing through the *asteroid belt* along the way. Upon our arrival, we find a planet much larger than any we have seen so far (Figure 6.7).

Jupiter is so different from the planets of the inner solar system that we must adopt an entirely new mental image of the term *planet*. Its mass is more than 300 times that of Earth, and its volume is more than 1000 times that of Earth. Its most famous feature—a long-lived storm called the Great Red Spot—is itself large enough to swallow two or three Earths. Like the Sun, Jupiter is made primarily of hydrogen and helium and has no solid surface. If we plunged deep into Jupiter, the increasing gas pressure would crush us long before we ever reached its core.

Jupiter reigns over dozens of moons and a thin set of rings (too faint to be seen in most photographs). Most of the moons are very small, but four are large enough that we'd call them planets or dwarf planets if they orbited the Sun independently. These four moons—Io, Europa, Ganymede, and Callisto—are often called the *Galilean moons* (because Galileo discovered them), and they display varied and interesting geology. Io is the most volcanically active world in the solar system. Europa has an icy crust that may hide a subsurface ocean of liquid water, making it a promising place to search for life. Ganymede and Callisto may also have subsurface oceans, and their surfaces have many features that remain mysterious.

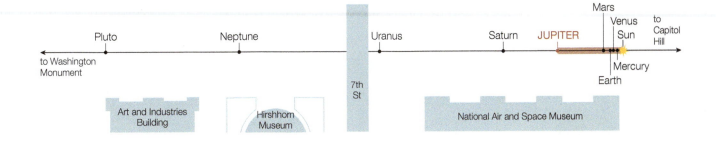

▲ **FIGURE 6.8**

Cassini's view of Saturn. We see the shadow of the rings on the upper right portion of Saturn's sunlit face, and the rings become lost in Saturn's shadow on the night side. The inset shows an infrared view of Titan, Saturn's large moon, shrouded in a thick, cloudy atmosphere.

◆ Saturn

- Average distance from the Sun: 9.54 AU
- Radius: 60,268 km = $9.4R_{Earth}$
- Mass: $95.2M_{Earth}$
- Average density: 0.70 g/cm^3
- Composition: mostly hydrogen and helium
- Cloud-top temperature: 95 K
- Moons: at least 62

The journey from Jupiter to Saturn is a long one: Saturn orbits nearly twice as far from the Sun as Jupiter. Saturn, the second-largest planet in our solar system, is only slightly smaller than Jupiter in diameter, but its lower density makes it considerably less massive (about one-third of Jupiter's mass). Like Jupiter, Saturn is made mostly of hydrogen and helium and has no solid surface.

Saturn is famous for its spectacular rings (Figure 6.8). All four of the giant outer planets have rings, but only Saturn's can be seen easily. Although the rings look solid from a distance, they are made of countless small particles, each of which orbits Saturn like a tiny moon. These particles of ice and rock range in size from dust grains to city blocks.

Saturn also has numerous moons, including at least two that are geologically active today: Enceladus, which has ice fountains spraying out from its southern hemisphere, and Titan, the only moon in the solar system with a thick atmosphere. Saturn and its moons are so far from the Sun that Titan's surface temperature is a frigid −180°C, making it far too cold for liquid water. However, studies by the *Cassini* spacecraft, which orbited Saturn from 2004 to 2017, and its *Huygens* probe, which landed on Titan in 2005, have revealed an erosion-carved surface with riverbeds and lakes—but the features are shaped by extremely cold liquid methane or ethane rather than liquid water.

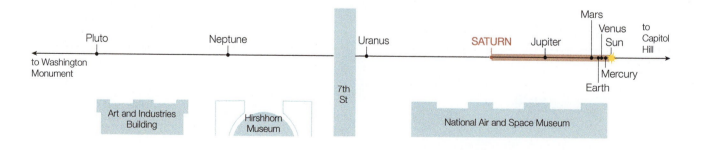

◀ **FIGURE 6.9**
This image shows a view of Uranus from high above its moon Ariel. The ring system is shown, although it would actually be too dark to see from this vantage point. This computer simulation is based on data from NASA's *Voyager 2* mission. (From the Voyage scale model solar system, developed by the Challenger Center for Space Science Education, the Smithsonian Institution, and NASA. Image created by ARC Science Simulations © 2001.)

◆ Uranus

- Average distance from the Sun: 19.2 AU
- Radius: 25,559 km = $4.0R_{Earth}$
- Mass: $14.5M_{Earth}$
- Average density: 1.32 g/cm^3
- Composition: hydrogen, helium, hydrogen compounds
- Cloud-top temperature: 60 K
- Moons: at least 27

It's another long journey to the next stop on our tour, as Uranus lies twice as far from the Sun as Saturn. Uranus (normally pronounced *YUR-uh-nus*) is much smaller than either Jupiter or Saturn but much larger than Earth. It is made largely of hydrogen, helium, and *hydrogen compounds* such as water (H_2O), ammonia (NH_3), and methane (CH_4). Methane gas gives Uranus its pale blue-green color (Figure 6.9). Like the other giants of the outer solar system, Uranus lacks a solid surface. More than two dozen moons orbit Uranus, along with a set of rings somewhat similar to those of Saturn but much darker and more difficult to see.

The entire Uranus system—planet, rings, and moon orbits—is tipped on its side compared to the rest of the planets. This extreme axis tilt may be the result of a cataclysmic collision that Uranus suffered as it was forming, and it gives Uranus the most extreme seasonal variations of any planet in our solar system. If you lived on a platform floating in Uranus's atmosphere near its north pole, you'd have continuous daylight for half of each orbit, or 42 years. Then, after a very gradual sunset, you'd enter into a 42-year-long night.

Only one spacecraft has visited Uranus: *Voyager 2*, which flew past all four of the giant outer planets before heading out of the solar system. Much of our current understanding of Uranus comes from that mission, though powerful new telescopes are also capable of studying it. Scientists hope it will not be too long before we can send another spacecraft to study Uranus and its rings and moons in greater detail.

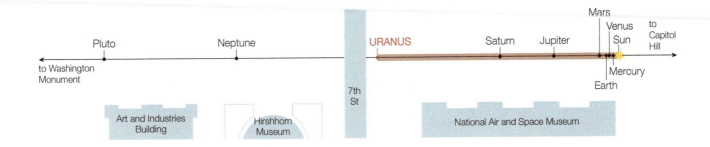

This image shows what it would look like to be orbiting Neptune's moon Triton as Neptune itself comes into view. The dark rings are exaggerated to make them visible in this computer simulation using data from NASA's *Voyager 2* mission. (From the Voyage scale model solar system, developed by the Challenger Center for Space Science Education, the Smithsonian Institution, and NASA. Image created by ARC Science Simulations © 2001.)

◆ Neptune

- Average distance from the Sun: 30.1 AU
- Radius: 24,764 km = $3.9R_{Earth}$
- Mass: $17.1M_{Earth}$
- Average density: 1.64 g/cm^3
- Composition: hydrogen, helium, hydrogen compounds
- Cloud-top temperature: 60 K
- Moons: at least 14

The journey from the orbit of Uranus to the orbit of Neptune is the longest yet in our tour, calling attention to the vast emptiness of the outer solar system. Nevertheless, Neptune looks nearly like a twin of Uranus, although it is more strikingly blue (Figure 6.10). It is slightly smaller than Uranus in size, but a higher density makes it slightly more massive even though the two planets share very similar compositions. Like Uranus, Neptune has been visited only by the *Voyager 2* spacecraft.

Neptune has rings and numerous moons. Its largest moon, Triton, is larger than Pluto and is one of the most fascinating moons in the solar system. Triton's icy surface has features that appear to be somewhat like geysers, although they spew nitrogen gas rather than water into the sky. Even more surprisingly, Triton is the only large moon in the solar system that orbits its planet "backward"—that is, in a direction opposite to the direction in which Neptune rotates. This backward orbit makes it a near certainty that Triton once orbited the Sun independently before somehow being captured into Neptune's orbit.

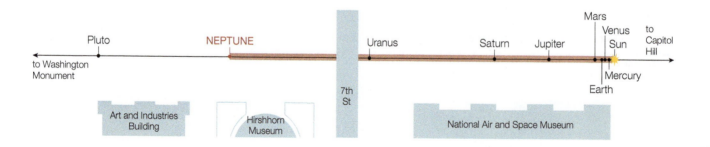

◀ **FIGURE 6.11**
Pluto and its largest moon, Charon,
photographed by the *New Horizons* spacecraft.
Although this image is a composite, it
approximates how the spacecraft saw the pair
while approaching them in July 2015.

• Dwarf planets: Pluto, Eris, and more

Pluto Data:

- Pluto's average distance from the Sun: 39.5 AU
- Radius: 1185 km $= 0.19R_{\text{Earth}}$
- Mass: $0.0022M_{\text{Earth}}$
- Average density: 1.9 g/cm^3
- Composition: ices, rock
- Average surface temperature: 40 K
- Moons: 5

We conclude our tour at Pluto (Figure 6.11), which reigned for some 75 years as the "ninth planet" in our solar system. However, the 2005 discovery of the slightly more massive Eris, and the fact that dozens of other objects are not much smaller than Pluto and Eris, led scientists to reconsider the definition of "planet." The result was that we now consider Pluto and Eris to be *dwarf planets,* too small to qualify as official planets but large enough to be round in shape. Several other solar system objects also qualify as dwarf planets, including Ceres, the largest asteroid of the asteroid belt.

Pluto and Eris belong to a vast collection of icy objects that orbit the Sun beyond Neptune, making up what we call the *Kuiper belt.* As you can see in Figure 6.1, the Kuiper belt is much like the asteroid belt, except it is farther from the Sun and composed of comet-like objects rather than rocky asteroids.

Pluto's characteristics help us to think about what it would be like to visit this distant realm. Pluto's average distance from the Sun lies as far beyond Neptune as Neptune lies beyond Uranus, making Pluto extremely cold and dimly lit even in daytime. From Pluto, the Sun is little more than a bright light among the stars. Pluto's largest moon, Charon, is locked together with it in synchronous rotation, so Charon dominates the sky on one side of Pluto but is never seen from the other side.

The great distances and small sizes of Pluto and other dwarf planets have made them difficult to study, but recent spacecraft missions are beginning to change that. The year 2015 was particularly exciting, featuring the *Dawn* spacecraft's arrival at Ceres and the *New Horizons* spacecraft's flyby of Pluto.

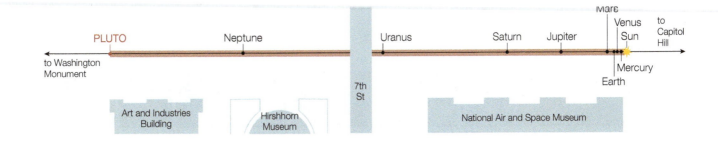

Photo	Planet	Relative Size	Average Distance from Sun (AU)	Average Equatorial Radius (km)	Mass (Earth = 1)	Average Density (g/cm³)	Orbital Period	Rotation Period	Axis Tilt	Average Surface (or Cloud-Top) Temperature[b]	Composition	Known Moons (2016)	Rings?
	Mercury	·	0.387	2440	0.055	5.43	87.9 days	58.6 days	0.0°	700 K (day) 100 K (night)	Rocks, metals	0	No
	Venus	●	0.723	6051	0.82	5.24	225 days	243 days	177.3°	740 K	Rocks, metals	0	No
	Earth	●	1.00	6378	1.00	5.52	1.00 year	23.93 hours	23.5°	290 K	Rocks, metals	1	No
	Mars	·	1.52	3397	0.11	3.93	1.88 years	24.6 hours	25.2°	220 K	Rocks, metals	2	No
	Jupiter	●	5.20	71,492	318	1.33	11.9 years	9.93 hours	3.1°	125 K	H, He, hydrogen compounds[c]	67	Yes
	Saturn	●	9.54	60,268	95.2	0.70	29.4 years	10.6 hours	26.7°	95 K	H, He, hydrogen compounds[c]	62	Yes
	Uranus	●	19.2	25,559	14.5	1.32	83.8 years	17.2 hours	97.9°	60 K	H, He, hydrogen compounds[c]	27	Yes
	Neptune	●	30.1	24,764	17.1	1.64	165 years	16.1 hours	29.6°	60 K	H, He, hydrogen compounds[c]	14	Yes
	Pluto	·	39.5	1185	0.0022	1.9	248 years	6.39 days	112.5°	44 K	Ices, rock	5	No
	Eris	·	67.7	1168	0.0028	2.3	557 years	1.08 days	78°	43 K	Ices, rock	1	No

[a]Including the dwarf planets Pluto and Eris; Appendix E gives a more complete list of planetary properties.

[b]Surface temperatures for all objects except Jupiter, Saturn, Uranus, and Neptune, for which cloud-top temperatures are listed.

[c]Includes water (H_2O), methane (CH_4), and ammonia (NH_3).

6.2 The Nebular Theory of Solar System Formation

Our major goal in this chapter is to understand the modern theory that we use to explain the origin of our solar system. To begin this process, let's investigate the criteria by which our theory can be judged.

◆ What features of our solar system provide clues to how it formed?

We have already seen that our solar system is not a random collection of worlds, but rather a family of worlds exhibiting many traits that would be difficult to attribute to coincidence. We could make a long list of such traits, but it is easier to develop a scientific theory by focusing on the more general structure of our solar system. For our purposes, four major features stand out, each corresponding to one of the numbered steps in Figure 6.1:

1. **Patterns of motion among large bodies.** The Sun, planets, and large moons generally orbit and rotate in a very organized way.

2. **Two major types of planets.** The eight planets* divide clearly into two groups: the small, rocky planets that are close together and close to the Sun, and the large, gas-rich planets that are farther apart and farther from the Sun.

3. **Asteroids and comets.** Between and beyond the planets, vast numbers of asteroids and comets orbit the Sun; some are large enough to qualify as dwarf planets. The locations, orbits, and compositions of these asteroids and comets follow distinct patterns.

4. **Exceptions to the rules.** The generally orderly solar system also has some notable exceptions. For example, among the inner planets only Earth has a large moon, and Uranus is tipped on its side. A successful theory must make allowances for exceptions even as it explains the general rules.

Because these four features are so important to our study of the solar system, let's investigate each of them in a little more detail.

Feature 1: Patterns of Motion Among Large Bodies   If you look back at Figure 6.1, you'll notice several clear patterns of motion among the large bodies of our solar system. (In this context, a "body" is simply an individual object such as the Sun, a planet, or a moon.) For example:

- All planetary orbits are nearly circular and lie nearly in the same plane.

- All planets orbit the Sun in the same direction: counterclockwise as viewed from high above Earth's North Pole.

- Most planets rotate in the same direction in which they orbit, with fairly small axis tilts. The Sun also rotates in this direction.

- Most of the solar system's large moons exhibit similar properties in their orbits around their planets, such as orbiting in their planet's equatorial plane in the same direction that the planet rotates.

---

*As this book goes to press, a search is on for a possible ninth planet orbiting far beyond Pluto. If this planet exists, it would have likely once orbited in the realm of the jovian planets, so most likely would have characteristics placing it in that group.

TABLE 6.2	Comparison of Terrestrial and Jovian Planets
Terrestrial Planets	**Jovian Planets**
Smaller size and mass	Larger size and mass
Higher density	Lower density
Made mostly of rock and metal	Made mostly of hydrogen, helium, and hydrogen compounds
Solid surface	No solid surface
Few (if any) moons and no rings	Rings and many moons
Closer to the Sun (and closer together), with warmer surfaces	Farther from the Sun (and farther apart), with cool temperatures at cloud tops

The Sun, planets, and large moons orbit and rotate in an organized way. We consider these orderly patterns together as the first major feature of our solar system. As we'll see shortly, our theory of solar system formation explains these patterns as consequences of processes that occurred during the early stages of the birth of our solar system.

Feature 2: Two Types of Planets   Our brief planetary tour showed that the four inner planets are quite different from the four outer planets. We say that these two groups represent two distinct planetary classes: *terrestrial* and *jovian*.

Terrestrial planets are small, rocky, and close to the Sun. *Jovian planets* are large, gas-rich, and far from the Sun. The **terrestrial planets** (*terrestrial* means "Earth-like") are the four planets of the inner solar system: Mercury, Venus, Earth, and Mars. These planets are relatively small and dense, with rocky surfaces and an abundance of metals in their cores. They have few moons, if any, and no rings. We count our Moon as a fifth terrestrial world, because its history has been shaped by the same processes that have shaped the terrestrial planets.

The **jovian planets** (*jovian* means "Jupiter-like") are the four large planets of the outer solar system: Jupiter, Saturn, Uranus, and Neptune. The jovian planets are much larger in size and lower in average density than the terrestrial planets, and they have rings and many moons. They lack solid surfaces and are made mostly of hydrogen, helium, and **hydrogen compounds**—compounds containing hydrogen, such as water (H_2O), ammonia (NH_3), and methane (CH_4). Because these substances are gases under earthly conditions, the jovian planets are sometimes called "gas giants." Table 6.2 contrasts the general traits of the terrestrial and jovian planets.

Feature 3: Asteroids and Comets   The third major feature of the solar system is the existence of vast numbers of small objects orbiting the Sun. These objects fall into two major groups: asteroids and comets.

Rocky asteroids and icy comets far outnumber the planets and their moons. **Asteroids** are rocky bodies that orbit the Sun much like planets, but they are much smaller (Figure 6.12). Even the largest asteroids are much smaller than our Moon. Most known asteroids are found within the **asteroid belt** between the orbits of Mars and Jupiter (see Figure 6.1).

Comets are also small objects that orbit the Sun, but they are made largely of ices (such as water ice, ammonia ice, and methane ice) mixed with rock. You are probably familiar with the occasional appearance of comets in the inner solar system, where they may become visible to the naked eye with long, beautiful tails (Figure 6.13). These visitors, which may delight sky watchers for a few weeks or months, are actually quite rare among comets. The vast majority of comets never visit the inner solar system. Instead, they orbit the Sun in one of the two distinct regions shown as feature 3 in Figure 6.1. The first is a donut-shaped region beyond the orbit of Neptune that we call the **Kuiper belt** (*Kuiper* rhymes with *piper*). The Kuiper belt contains at least 100,000 icy objects that are more than 100 kilometers in diameter, of which Pluto and Eris are the largest known. The second cometary region, called the **Oort cloud** (*Oort* rhymes with *court*), is much farther from the Sun and may contain a trillion comets. These comets have orbits randomly inclined to the ecliptic plane, giving the Oort cloud a roughly spherical shape.

Feature 4: Exceptions to the Rules   The fourth key feature of our solar system is that there are a few notable exceptions to the general rules.

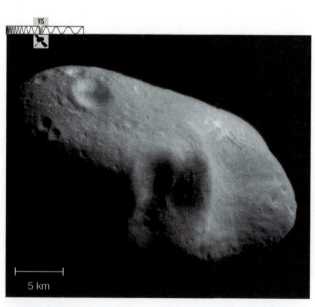

▲ **FIGURE 6.12**

The asteroid Eros (photographed from the *NEAR* spacecraft). Its appearance is probably typical of most asteroids. Eros is about 40 kilometers in length, and like other small objects in the solar system, it is not spherical.

For example, while most of the planets rotate in the same direction as they orbit, Uranus rotates nearly on its side, and Venus rotates "backward" (clockwise as viewed from high above Earth's North Pole). Similarly, while most large moons orbit their planets in the same direction as their planets rotate, many small moons have much more unusual orbits.

A successful theory of solar system formation must allow for exceptions to the general rules. One of the most interesting exceptions concerns our own Moon. While the other terrestrial planets have either no moons (Mercury and Venus) or very tiny moons (Mars), Earth has one of the largest moons in the solar system.

◆ What is the nebular theory?

Historically, scientists have proposed more than one hypothesis about the origin of the solar system. But only one hypothesis has survived the test of time.

From Hypothesis to Theory We generally trace the origins of our modern theory of solar system formation to around 1755, when German philosopher Immanuel Kant proposed that our solar system formed from the gravitational collapse of an interstellar cloud of gas. About 40 years later, French mathematician Pierre-Simon Laplace put forth the same idea independently. Because an interstellar cloud is usually called a *nebula* (Latin for "cloud"), this idea became known as the *nebular hypothesis*.

The nebular hypothesis remained popular throughout the 19th century. By the early 20th century, however, scientists had found a few aspects of our solar system that the hypothesis did not seem to explain well—at least in its original form as described by Kant and Laplace. While some scientists sought to modify the nebular hypothesis, others looked for alternative models of the solar system's birth.

During much of the first half of the 20th century, the nebular hypothesis faced stiff competition from a hypothesis proposing that the planets represent debris from a near-collision between the Sun and another star. According to this *close encounter hypothesis,* the planets formed from blobs of gas that had been gravitationally pulled out of the Sun during the near-collision.

Today, the close encounter hypothesis has been discarded. It began to lose favor when calculations showed that it could not account for either the observed orbital motions of the planets or the neat division of the planets into two major categories (terrestrial and jovian). Moreover, the close encounter hypothesis required a highly improbable event: a near-collision between our Sun and another star. Given the vast separation between star systems in our region of the galaxy, the chance of such an encounter is so small that it would be difficult to imagine it happening even once in order to form our solar system. It certainly could not account for the many other planetary systems that we have discovered in recent years.

The nebular theory holds that our solar system formed from the gravitational collapse of a great cloud of gas. While the close encounter hypothesis was losing favor, new discoveries about the physics of planet formation led to modifications of the nebular hypothesis. Using more sophisticated models of the processes that occur in a collapsing cloud of gas, scientists found that the nebular hypothesis offered natural explanations for all four general features of our solar system. Indeed, so much evidence has accumulated in favor of the nebular hypothesis that it has achieved the status of a scientific *theory* [Section 3.4]—the **nebular theory** of our solar system's birth.

▲ **FIGURE 6.13**
Comet McNaught over Patagonia, Argentina, in 2007. The fuzzy patches above the comet tail are the Magellanic Clouds, satellite galaxies of the Milky Way.

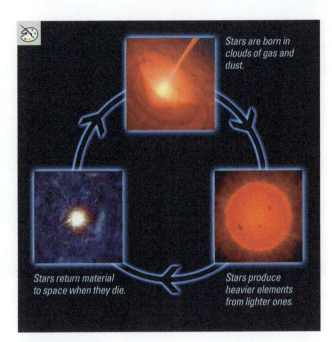

▲ **FIGURE 6.14**

This figure, which is a portion of Figure 1.9, summarizes the galactic recycling process.

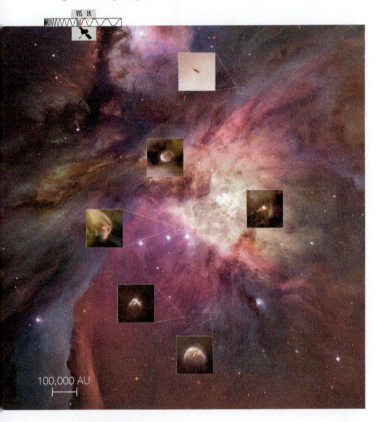

100,000 AU

▲ **FIGURE 6.15**

The Orion Nebula, an interstellar cloud in which new star systems are forming. The insets show close-ups of six protoplanetary disks (nicknamed "proplyds")—disks of material in which planets can form—each surrounding a young star that is being born. The nebula will ultimately give birth to thousands of stars and planetary systems over the next few million years. (Images from the Hubble Space Telescope.)

Origin of the Nebula In discussing the nebular theory, the particular cloud of gas from which our solar system was born is usually called the **solar nebula.** But where did this nebula come from?

The solar nebula contained hydrogen and helium from the Big Bang and heavier elements produced by stars.

Recall that the universe as a whole is thought to have been born in the Big Bang [Section 1.2], which essentially produced only two chemical elements: hydrogen and helium. Heavier elements were produced later by massive stars and released into space when the stars died. The heavy elements then mixed with other interstellar gas to form new generations of stars (Figure 6.14). Despite billions of years of heavy element creation by stars, the overall chemical composition of the universe remains predominantly hydrogen and helium. By studying the composition of the Sun, other stars of the same age, and interstellar gas clouds, we conclude that the gas that made up the solar nebula contained (by mass) about 98% hydrogen and helium and 2% all other elements combined.

think about it Could a solar system like ours have formed with the first generation of stars after the Big Bang? Explain.

Strong observational evidence supports this scenario. Spectroscopy shows that old stars have a smaller proportion of heavy elements than younger ones, just as we would expect if they were born at a time before many heavy elements had been manufactured. Moreover, visible and infrared telescopes allow us to study stars that are in the process of formation today. Figure 6.15 shows the Orion Nebula, in which many stars are in various stages of formation. Just as our scenario predicts, the forming stars are embedded within gas clouds like our solar nebula, and the characteristics of these clouds match what we expect if they are collapsing due to gravity.

6.3 Explaining the Major Features of the Solar System

We are now ready to look at the nebular theory in somewhat more detail. In the process, we'll see how it successfully accounts for all four major features of our solar system.

◆ What caused the orderly patterns of motion?

The solar nebula probably began as a large and roughly spherical cloud of very cold, low-density gas. Initially, this gas was so spread out—perhaps over a region a few light-years in diameter—that gravity alone may not have been strong enough to pull it together and start its collapse. Instead, the collapse may have been triggered by a cataclysmic event, such as the impact of a shock wave from the explosion of a nearby star (a supernova).

Once the collapse started, gravity enabled it to continue. Recall that the strength of gravity follows an inverse square law with distance [Section 4.4]. The mass of the cloud remained the same as it shrank, so the strength of gravity increased as the diameter of the cloud decreased.

Because gravity pulls inward in all directions, you might at first guess that the solar nebula would have remained spherical as it shrank. Indeed, the idea that gravity pulls in all directions explains why the Sun

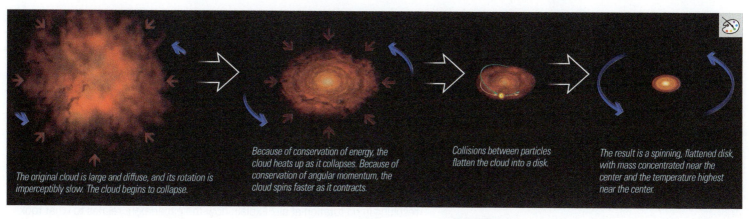

The original cloud is large and diffuse, and its rotation is imperceptibly slow. The cloud begins to collapse.

Because of conservation of energy, the cloud heats up as it collapses. Because of conservation of angular momentum, the cloud spins faster as it contracts.

Collisions between particles flatten the cloud into a disk.

The result is a spinning, flattened disk, with mass concentrated near the center and the temperature highest near the center.

▲ **FIGURE 6.16**
This sequence of illustrations shows how the gravitational collapse of a large cloud of gas causes it to become a spinning disk of matter. The hot, dense central bulge becomes a star, while planets can form in the surrounding disk.

and the planets are spherical. However, other physical laws also apply, and these explain how orderly motions arose in the solar nebula.

Heating, Spinning, and Flattening

As the solar nebula shrank in size, three important processes altered its density, temperature, and shape (Figure 6.16):

- **Heating.** The temperature of the solar nebula increased as it collapsed. Such heating represents energy conservation in action [Section 4.3]. As the cloud shrank, its gravitational potential energy was converted to the kinetic energy of individual gas particles falling inward. These particles crashed into one another, converting the kinetic energy of their inward fall to the random motions of thermal energy (see Figure 4.12b). The Sun formed in the center, where temperatures and densities were highest.

- **Spinning.** Like an ice skater pulling in her arms as she spins, the solar nebula rotated faster and faster as it shrank in radius. This increase in rotation rate represents conservation of angular momentum in action [Section 4.3]. The rotation of the cloud may have been imperceptibly slow before its collapse began, but the cloud's shrinkage made fast rotation inevitable. The rapid rotation helped ensure that not all the material in the solar nebula collapsed into the center: The greater the angular momentum of a rotating cloud, the more spread out it will be.

- **Flattening.** The solar nebula flattened into a disk. This flattening is a natural consequence of collisions between particles in a spinning cloud. A cloud may start with any size or shape, and different clumps of gas within the cloud may be moving in random directions at random speeds. These clumps collide and merge as the cloud collapses, and each new clump has the average velocity of the clumps that formed it. The random motions of the cloud therefore become more orderly as it collapses, transforming the cloud into a rotating, flattened disk. Similarly, collisions between clumps of material in highly elliptical orbits reduce their eccentricities, making their orbits more circular.

The orderly motions of our solar system are a direct result of its birth in a spinning, flattened cloud of gas. The formation of the spinning disk explains the orderly motions of our solar system. The planets all orbit the Sun in nearly the same plane because they formed in the flat disk. The direction in which the disk was spinning became the direction of the Sun's rotation and the orbits of the planets. Computer models show that planets would have tended to rotate in this same direction as they formed—which is why most planets rotate the same way—though the small sizes of planets compared to the entire disk allowed some exceptions to arise. The fact that collisions in the disk tended to make orbits more circular explains why the planets in our solar system have nearly circular orbits.

see it for yourself You can demonstrate the development of orderly motion by sprinkling pepper into a bowl of water and stirring it quickly in random directions. The water molecules constantly collide with one another, so the motion of the pepper grains will tend to settle into a slow rotation representing the average of the original, random velocities. Try the experiment several times, stirring the water differently each time. Do the random motions ever cancel out exactly, resulting in no rotation at all? Explain how the experiment relates to what took place in the solar nebula.

Testing the Model The same processes should affect other collapsing gas clouds, so we can test our model by searching for disks around other forming stars. Observational evidence does indeed support our model of spinning, heating, and flattening.

Observations of spinning disks of gas around other stars support the idea that our solar system formed from a similar disk.

The heating that occurs in a collapsing cloud of gas means the gas should emit thermal radiation [Section 5.2], primarily in the infrared. We've detected infrared radiation from many nebulae where star systems appear to be forming. More direct evidence comes from flattened, spinning disks around other stars (Figure 6.17), some of

▼ **FIGURE 6.17**

These images show flattened, spinning disks of material around other stars.

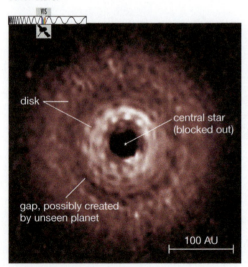

a This Hubble Space Telescope image shows a flattened, spinning disk around the star TW Hydrae. This particular disk also shows at least one circular "gap" in which material seems to have been cleared away, probably by a planet forming in the disk, which would have a gravitational attraction that would tend to sweep up material along its path.

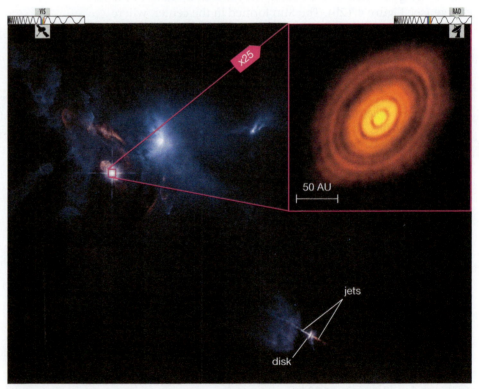

b The inset, from the Atacama Large Millimeter/submillimeter Array (ALMA), shows a disk around a star named HL Tauri; the concentric gaps in the disk are almost certainly regions being cleared as planets form. The disk diameter is about three times that of Neptune's orbit around the Sun. The background image, from the Hubble Space Telescope, shows the star-forming region in which this disk is located. Another disk, seen edge-on with jets extending outward, appears at lower right.

which appear to be ejecting jets of material perpendicular to their disks [Section 13.1]. These jets are thought to result from the flow of material from the disk onto the forming star, and they may influence the solar system formation processes.

Other support for the model comes from computer simulations of the formation process. A simulation begins with a set of data representing the conditions we observe in interstellar clouds. Then, with the aid of a computer, we apply the laws of physics to predict the changes that should occur over time. Computer simulations successfully reproduce most of the general characteristics of motion in our solar system, giving support to the nebular theory.

Additional evidence that our ideas about the formation of flattened disks are correct comes from many other structures in the universe. We expect flattening to occur anywhere orbiting particles can collide, which explains why we find so many cases of flat disks, including the disks of spiral galaxies like the Milky Way, the disks of planetary rings, and the *accretion disks* that surround neutron stars and black holes in close binary star systems [Section 14.2].

◆ Why are there two major types of planets?

The planets began to form after the solar nebula had collapsed into a flattened disk of perhaps 200 AU in diameter (about twice the present-day diameter of Pluto's orbit). The churning and mixing of gas in the solar nebula should have ensured that the nebula had the same composition throughout. How, then, did the terrestrial planets end up so different in composition from the jovian planets? The key clue comes from their locations: Terrestrial planets formed in the warm, inner regions of the swirling disk, while jovian planets formed in the colder, outer regions.

Condensation: Sowing the Seeds of Planets In the center of the collapsing solar nebula, gravity drew together enough material to form the Sun. In the surrounding disk, however, the gaseous material was too spread out for gravity alone to clump it together. Instead, material had to begin clumping in some other way and to grow in size until gravity could start pulling it together into planets. In essence, planet formation required the presence of "seeds"—solid bits of matter around which gravity could ultimately build planets.

Planet formation began around tiny "seeds" of solid metal, rock, or ice.

The basic process of seed formation was probably much like the formation of snowflakes in clouds on Earth: When the temperature is low enough, some atoms or molecules in a gas may bond and solidify. The general process in which solid (or liquid) particles form in a gas is called **condensation**—we say that the particles *condense* out of the gas. These particles start out microscopic in size, but they can grow larger with time.

Different materials condense at different temperatures. The ingredients of the solar nebula fell into four major categories (Table 6.3):

- **Hydrogen and helium gas (98% of the solar nebula's mass).** These gases never condense in interstellar space.

- **Hydrogen compounds (1.4% of the solar nebula).** Materials such as water (H_2O), methane (CH_4), and ammonia (NH_3) can solidify into **ices** at low temperatures (below about 150 K under the low pressure of the solar nebula).

- **Rock (0.4% of the solar nebula).** Rocky material is gaseous at high temperatures but condenses into solid form at temperatures below 500 K to 1300 K, depending on the type of rock.

TABLE 6.3 **Materials in the Solar Nebula**

A summary of the four types of materials present in the solar nebula. The squares represent the relative proportions of each type (by mass).

	Examples	Can condense at temperatures below	Relative abundance (by mass)
Hydrogen and Helium Gas	hydrogen, helium	do not condense in nebula	98%
Hydrogen Compounds	water (H_2O), methane (CH_4), ammonia (NH_3)	150 K	1.4%
Rock	various minerals	500–1300 K	0.4%
Metals	iron, nickel, aluminum	1000–1600 K	0.2%

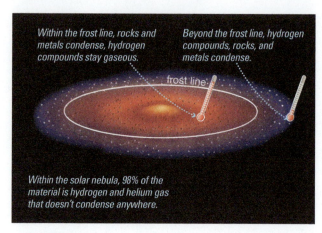

Within the frost line, rocks and metals condense, hydrogen compounds stay gaseous.

Beyond the frost line, hydrogen compounds, rocks, and metals condense.

frost line

Within the solar nebula, 98% of the material is hydrogen and helium gas that doesn't condense anywhere.

▲ **FIGURE 6.18**

Temperature differences in the solar nebula led to different kinds of condensed materials, sowing the seeds for two different kinds of planets.

- **Metals (0.2% of the solar nebula).** Metals such as iron, nickel, and aluminum are also gaseous at high temperatures but condense into solid form at temperatures below 1000 K to 1600 K (depending on the metal).

Because hydrogen and helium gas made up 98% of the solar nebula's mass and did not condense, the vast majority of the nebula remained gaseous at all times. However, other materials condensed wherever the temperature allowed (Figure 6.18). Close to the Sun, it was too hot for any material to condense. A little farther out (near Mercury's current orbit), it was cool enough for metals and some types of rock to condense into tiny solid particles, but other types of rock and all the hydrogen compounds remained gaseous. More types of rock could condense at the distances from the Sun where Venus, Earth, and Mars would form. In the region where the asteroid belt would eventually be located, temperatures were low enough to allow dark, carbon-rich minerals to condense along with minerals containing small amounts of water. It was cold enough for hydrogen compounds to condense into ices only beyond the **frost line,** which lay between the present-day orbits of Mars and Jupiter.

think about it Consider a region of the solar nebula in which the temperature was about 1300 K. What fraction of the material in this region was gaseous? What were the solid particles made of? Answer the same questions for a region with a temperature of 100 K. Was the 100 K region closer to or farther from the Sun? Explain.

The frost line marked the key transition between the warm inner regions of the solar system where terrestrial planets formed and the cool outer regions where jovian planets formed. Inside the frost line, only metal and rock could condense into solid "seeds." Beyond the frost line, the solid seeds were built of ice along with metal and rock. Moreover, because hydrogen compounds were nearly three times as abundant in the nebula as metal and rock combined (see Table 6.3), the total amount of solid material was far greater beyond the frost line than within it. The stage was set for the birth of two types of planets: planets born from seeds of metal and rock in the inner solar system and planets born from seeds of ice (as well as metal and rock) in the outer solar system.

> The solid seeds contained only metal and rock in the inner solar system, but also included ices in the outer solar system.

Building the Terrestrial Planets From this point, the story of the inner solar system seems fairly clear: The solid seeds of metal and rock in the inner solar system ultimately grew into the terrestrial planets we see today, but these planets ended up relatively small in size because rock and metal made up such a small amount of the material in the solar nebula.

The process by which small "seeds" grew into planets is called **accretion** (Figure 6.19). Accretion began with the microscopic solid particles that condensed from the gas of the solar nebula. These particles orbited the forming Sun with the same orderly, circular paths as the gas from which they condensed. Individual particles therefore moved at nearly the same speed as neighboring particles, so "collisions" were more like gentle touches. Although the particles were far too small to attract

> The terrestrial planets were made from metal and rock that condensed in the inner solar system.

common misconceptions

Solar Gravity and the Density of Planets

Some people guess that it was the Sun's gravity that pulled the dense rocky and metallic materials to the inner part of the solar nebula, or that gases escaped from the inner nebula because gravity couldn't hold them. But this is not the case—all the ingredients were orbiting the Sun together under the influence of the Sun's gravity. The orbit of a particle or a planet does not depend on its size or density, so the Sun's gravity cannot be the cause of the different kinds of planets. Rather, the different temperatures in the solar nebula are the cause.

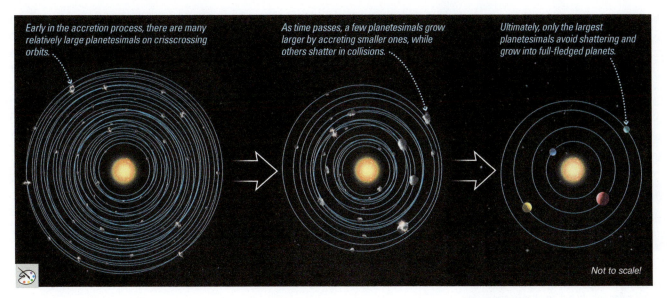

Early in the accretion process, there are many relatively large planetesimals on crisscrossing orbits.

As time passes, a few planetesimals grow larger by accreting smaller ones, while others shatter in collisions.

Ultimately, only the largest planetesimals avoid shattering and grow into full-fledged planets.

Not to scale!

▲ **FIGURE 6.19**
These diagrams show how planetesimals gradually accrete into terrestrial planets.

each other gravitationally at this point, they were able to move together through electrostatic forces—the same "static electricity" that makes hair stick to a comb. Small particles thereby began to combine into larger ones. As the particles grew in mass, they began to attract each other through gravity, accelerating their growth into boulders large enough to count as **planetesimals,** which means "pieces of planets."

The planetesimals grew rapidly at first. As they grew larger, they had both more surface area to make contact with other planetesimals and more gravity to attract them. Some planetesimals probably grew to hundreds of kilometers in size in only a few million years—a long time in human terms, but only about *one-thousandth* of the present age of the solar system. However, once the planetesimals reached these relatively large sizes, further growth became more difficult.

Gravitational encounters [Section 4.4] between planetesimals tended to alter their orbits, particularly those of the smaller ones. With different orbits crossing each other, collisions between planetesimals tended to occur at higher speeds and hence became more destructive. Such collisions tended to shatter planetesimals rather than help them grow. Only the largest planetesimals avoided being shattered and could grow into terrestrial planets.

Computer simulations support this model of the accretion process. Observational evidence comes from **meteorites,** rocks that have fallen to Earth from space. Meteorites that appear to be surviving fragments from the period of condensation contain metallic grains embedded in rocky minerals (Figure 6.20), just as we would expect if metal and rock condensed in the inner solar system. Meteorites thought to come from the outskirts of the asteroid belt contain abundant carbon-rich materials, and some contain water—again as we would expect for material that condensed in that region.

Making the Jovian Planets Accretion should have occurred similarly in the outer solar system, but condensation of ices meant both that there was more solid material and that this material contained ice in addition to metal and rock. The solid objects that reside in the outer solar system today, such as comets and the moons of the jovian planets, still show this ice-rich composition. However, the growth of icy planetesimals cannot be the whole story of jovian planet formation, because the jovian planets contain large amounts of hydrogen and helium gas.

▲ **FIGURE 6.20**
Shiny flakes of metal are clearly visible in this slice through a meteorite (a few centimeters across), mixed in among the rocky material. Such metallic flakes are just what we would expect to find if condensation really occurred in the solar nebula as described by the nebular theory.

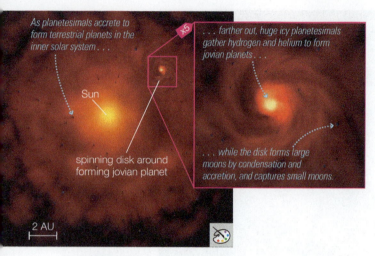

As planetesimals accrete to form terrestrial planets in the inner solar system . . .

Sun

spinning disk around forming jovian planet

. . . farther out, huge icy planetesimals gather hydrogen and helium to form jovian planets . . .

. . . while the disk forms large moons by condensation and accretion, and captures small moons.

2 AU

▲ **FIGURE 6.21**

The young jovian planets were surrounded by disks of gas, much like the disk of the entire solar nebula but smaller in size. According to the leading model, the planets grew as large, ice-rich planetesimals captured hydrogen and helium gas from the solar nebula. This painting shows the gas and planetesimals surrounding one jovian planet in the larger solar nebula.

The leading model for jovian planet formation holds that the largest ice-rich planetesimals became sufficiently massive for their gravity to capture some of the hydrogen and helium gas that made up the vast majority of the surrounding solar nebula. This added gas made their gravity even stronger, allowing them to capture even more gas. Ultimately, the jovian planets accreted so much gas that they bore little resemblance to the icy seeds from which they grew.

The jovian planets began as large, icy planetesimals, which then captured hydrogen and helium gas from the solar nebula.

This model also explains most of the large moons of the jovian planets. The same processes of heating, spinning, and flattening that made the disk of the solar nebula should also have affected the gas drawn by gravity to the young jovian planets. Each jovian planet came to be surrounded by its own disk of gas, spinning in the same direction as the planet rotated (Figure 6.21). Moons that accreted from ice-rich planetesimals within these disks ended up with nearly circular orbits going in the same direction as their planet's rotation and lying close to their planet's equatorial plane.

Clearing the Nebula The vast majority of the hydrogen and helium gas in the solar nebula never became part of any planet. So what happened to it? Apparently, it was cleared away by a combination of high-energy radiation (ultraviolet and X rays) from the young Sun and the *solar wind*— a stream of charged particles continually blown outward in all directions from the Sun. Observations show that stars tend to have much stronger winds and emit much more high-energy radiation when they are young, so the young Sun should have had a strong enough combination of radiation and wind to clear the solar system of its remaining gas.

The clearing of the gas sealed the compositional fate of the planets. If the gas had remained longer, it might have continued to cool until hydrogen compounds could have condensed into ices even in the inner solar system. In that case, the terrestrial planets might have accreted abundant ice, and perhaps hydrogen and helium gas as well, changing their basic nature. At the other extreme, if the gas had been blown out earlier, the raw materials of the planets might have been swept away

Remaining gas in the solar nebula was cleared away into space, ending the era of planet formation.

before the planets could fully form. Although these extreme scenarios did not occur in our solar system, they may sometimes occur around other stars. Planet formation may also sometimes be interrupted when radiation from hot, neighboring stars drives away material in a solar nebula.

◆ Where did asteroids and comets come from?

The process of planet formation also explains the origin of the many asteroids and comets in our solar system (including those large enough to qualify as dwarf planets): They are "leftovers" from the era of planet formation. Asteroids are the rocky leftover planetesimals of the inner solar system, while comets are the icy leftover planetesimals of the outer solar system. We'll see in Chapter 9 why most asteroids ended up in the asteroid belt while most comets ended up in either the Kuiper belt or the Oort cloud.

Rocky asteroids and icy comets are leftover planetesimals from the era of planet formation.

Evidence that asteroids and comets are leftover planetesimals comes from analysis of meteorites, spacecraft visits to comets and asteroids, and theoretical models of solar system formation. In fact, the nebular theory allowed scientists to predict the existence of comets in the Kuiper belt decades before any of them were discovered.

The asteroids and comets that exist today probably represent only a small fraction of the leftover planetesimals that roamed the young solar system. The rest are now gone. Some of these "lost" planetesimals may have been flung into deep space by gravitational encounters, but many others must have collided with the planets. When impacts occur on solid worlds, they leave behind **impact craters** as scars. Impacts have thereby transformed planetary landscapes and, in the case of Earth, have altered the course of evolution. For example, an impact is thought to have been responsible for the extinction of the dinosaurs [Section 9.5].

Although impacts occasionally still occur, the vast majority of these collisions occurred in the first few hundred million years of our solar system's history, during the period we call the **heavy bombardment.** Every world in our solar system must have been pelted by impacts during the heavy bombardment (Figure 6.22), and most of the craters on

Leftover planetesimals battered the planets during the solar system's first few hundred million years.

the Moon and other worlds date from this period.

These early impacts, including many that occurred before the planets finished forming, probably played a key role in making our existence possible. The metal and rock planetesimals that built the terrestrial planets should not have contained water or other hydrogen compounds, because it was too hot for these compounds to condense in our region of the solar nebula. How, then, did Earth come to have the water that makes up our oceans and the gases that first formed our atmosphere? The likely answer is that water, along with other hydrogen compounds, was brought to Earth and other terrestrial planets by the impacts of water-bearing planetesimals that formed farther from the Sun, probably in what is now the outer region of the asteroid belt. Remarkably, the water we drink and the air we breathe probably once were part of planetesimals that accreted beyond the orbit of Mars.

◆ How do we explain the "exceptions to the rules"?

We have now explained all the major features of our solar system except the exceptions to the rules, including our surprisingly large Moon. We now suspect that most of these exceptions arose from collisions or close gravitational encounters.

Captured Moons We have explained the orbits of most large jovian planet moons by their formation in a disk that swirled around the forming planet. But how do we explain moons with less orderly orbits, such as those that go in the "wrong" direction (opposite their planet's rotation) or that have large inclinations to their planet's equator? These moons are probably leftover planetesimals that originally orbited the Sun but were then captured into planetary orbit.

It's not easy for a planet to capture a moon. An object cannot switch from an unbound orbit (for example, an asteroid whizzing by Jupiter) to a bound orbit (for example, a moon orbiting Jupiter) unless it somehow loses orbital energy [Section 4.4]. For the jovian planets, captures probably occurred when passing planetesimals lost energy to friction in the extended and relatively dense gas that surrounded these planets as they formed. The planetesimals would have been slowed by friction with the gas, just as artificial satellites in low orbits are slowed by drag from Earth's atmosphere. If friction reduced a passing planetesimal's orbital

▲ **FIGURE 6.22**
Around 4 billion years ago, Earth, its Moon, and the other planets were heavily bombarded by leftover planetesimals. This painting shows the young Earth and Moon, with an impact in progress on Earth.

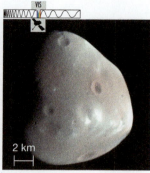

a Phobos   **b** Deimos

▲ **FIGURE 6.23**

The two moons of Mars are probably captured asteroids. Phobos is only about 13 kilometers across, and Deimos is only about 8 kilometers across—making each of these two moons small enough to fit within the boundaries of a typical large city. (Images from the *Mars Reconnaissance* orbiter.)

energy enough, it could have become an orbiting moon. Because of the random nature of the capture process, captured moons would not necessarily orbit in the same direction as their planet or in its equatorial plane. Most of the small moons of the jovian planets are thought to have been captured in this way. Mars may have similarly captured its two small moons, Phobos and Deimos, at a time when the planet had a much more extended atmosphere than it does today (Figure 6.23).

The Giant Impact Formation of Our Moon Capture processes cannot explain our own Moon, because it is much too large to have been captured by a small planet like Earth. We can also rule out the possibility that our Moon formed simultaneously with Earth, because if both had formed together, they would have accreted from planetesimals of the same type and should therefore have approximately the same composition and density. But this is not the case: The Moon's density is considerably lower than Earth's, indicating that it has a very different average composition. So how did we get our Moon? Today, the leading hypothesis suggests that it formed as the result of a **giant impact** between Earth and a huge planetesimal.

According to models, a few leftover planetesimals may have been as large as Mars. If one of these Mars-size objects struck a young planet, the blow might have tilted the planet's axis, changed the planet's rotation rate, or completely shattered the planet. The giant impact hypothe-

Our Moon is probably the result of a giant impact that blasted Earth's outer layers into orbit, where the material accreted to form the Moon.

sis holds that a Mars-size object hit Earth at a speed and angle that blasted Earth's outer layers into space. According to computer simulations, this material could have collected into orbit around our planet, and accretion within this ring of debris could have formed the Moon (Figure 6.24).

---

extraordinary claims   A Giant Impact Made Our Moon

In this chapter, you have seen that Earth's large moon is a notable "exception to the rules" of the nebular theory. But how did this exception arise? By the mid-20th century, four competing models had been suggested. The first held that the Moon formed along with Earth through the process of accretion. The second suggested that the Moon was born as an independent "planet" that was captured into Earth orbit. The third, or *fission* (splitting), model suggested that the young, molten Earth rotated so rapidly that it split into two pieces, with the smaller piece becoming the Moon. And a fourth suggested that the Moon had been formed by a giant impact between a wandering planetesimal and Earth. The first three models had weaknesses from the start: Accretion cannot account for the Moon's low density, capture is too difficult for a body as big as the Moon in comparison to Earth, and fission would require an unreasonable spin rate for Earth and an unknown mechanism to prevent the Moon from flying off into space after it split off. The fourth model seemed even weaker: Indeed, it was considered so outrageous at the time (it predated studies of how the planets grew through accretion) that it was completely forgotten by the scientific community.

The *Apollo* missions to the Moon (1969–1972) ruled out the first three models more conclusively based on study of Moon rocks

returned to Earth. Fortunately, the *Apollo* Moon rocks also helped scientists understand the history of cratering on the Moon, leading to a clearer picture of how many large objects had roamed the early solar system. With this in mind, in 1974, astronomers William Hartmann and Donald Davis proposed the giant impact hypothesis, unaware that a similar idea had been suggested previously. They recognized that this idea could explain why the Moon's composition closely matches that of Earth's mantle, and argued that enough large bodies had roamed the young solar system for such an impact to have been likely. This story is a clear example of how an extraordinary claim that may at first seem completely unreasonable will eventually be reconsidered if enough evidence accumulates for it.

We cannot go back in time to collect direct evidence of a giant impact, so we may never be able to establish with certainty that such an impact really occurred. Nevertheless, simulations of the event provide strong support for the idea. Many questions still remain to be answered, including the precise nature and size of the impactor, but most scientists now think it very likely that the Moon did indeed form in this way.

Verdict: Likely correct, though it may never be possible to support the hypothesis beyond reasonable doubt.

A Mars-sized planetesimal crashes into the young Earth, shattering both the planetesimal and our planet.

Hours later, our planet is completely molten and rotating very rapidly. Debris splashed out from Earth's outer layers is now in Earth orbit. Some debris rains back down on Earth, while some will gradually accrete to become the Moon.

Less than a thousand years later, the Moon's accretion is rapidly nearing its end, and relatively little debris still remains in Earth orbit.

▲ **FIGURE 6.24**

Artist's conception of the giant impact hypothesis for the formation of our Moon. The fact that ejected material came mostly from Earth's outer rocky layers explains why the Moon contains very little metal. The impact must have occurred more than 4.4 billion years ago, since that is the age of the oldest Moon rocks. As shown, the Moon formed quite close to a rapidly rotating Earth, but over billions of years, tidal forces have slowed Earth's rotation and moved the Moon's orbit outward (see Special Topic, page 99).

Strong support for the giant impact hypothesis comes from two features of the Moon's composition. First, the Moon's overall composition is quite similar to that of Earth's outer layers—just as we should expect if it were made from material blasted away from those layers. Second, the Moon has a much smaller proportion of easily vaporized ingredients (such as water) than Earth. This fact supports the hypothesis because the heat of the impact would have vaporized these ingredients. As gases, they would not have participated in the subsequent accretion of the Moon.

Other Exceptions   Giant impacts may also explain other exceptions to the general trends. For example, Pluto's moon Charon likely formed in a giant impact similar to the one thought to have formed our Moon, and Mercury's surprisingly high density may be the result of a giant impact that blasted away its lower-density outer layers. Giant impacts could also have been responsible for tilting the axes of many planets (including Earth) and perhaps for tipping Uranus on its side. Venus's slow and backward rotation could also be the result of a giant impact, though some scientists suspect it is a consequence of processes attributable to Venus's thick atmosphere.

Although we cannot definitively explain these exceptions to the general rules, the overall lesson is clear: The chaotic processes that accompanied planet formation, including the many collisions that surely occurred, are *expected* to have led to at least a few exceptions. We conclude that the nebular theory successfully accounts for all four major features of our solar system. Figure 6.25 summarizes the theory.

6.4 The Age of the Solar System

The nebular theory seems to explain *how* our solar system was born. But *when* was it born, and how do we know? The answer is that the planets began to form through accretion just over 4½ billion years ago, a fact we learned by determining the age of the oldest rocks in the solar system.

FIGURE 6.25

A summary of the process by which our solar system formed, according to the nebular theory.

A large, diffuse interstellar gas cloud (solar nebula) contracts due to gravity.

Contraction of Solar Nebula: As it contracts, the cloud heats, flattens, and spins faster, becoming a spinning disk of dust and gas.

The Sun will be born in the center.

Planets will form in the disk.

Warm temperatures allow only metal/rock "seeds" to condense in inner solar system.

Condensation of Solid Particles: Hydrogen and helium remain gaseous, but other materials can condense into solid "seeds" for building planets.

Cold temperatures allow "seeds" to contain abundant ice in the outer solar system.

Terrestrial planets are built from metal and rock.

Accretion of Planetesimals: Solid "seeds" collide and stick together. Larger ones attract others with their gravity, growing bigger still.

The seeds of jovian planets grow large enough to attract hydrogen and helium gas, making them into giant, mostly gaseous planets; moons form in disks of dust and gas that surround the planets.

Clearing the Nebula: The solar wind blows remaining gas into interstellar space.

Terrestrial planets remain in the inner solar system.

Jovian planets remain in the outer solar system.

"Leftovers" from the formation process become asteroids (metal/rock) and comets (mostly ice).

Not to scale

◆ How do we know the age of the solar system?

The first step in understanding how we've measured the age of the solar system is to understand how we determine the ages of individual rocks.

Dating Rocks The method by which we measure a rock's age is called **radiometric dating,** and it relies on careful measurement of the proportions of various atoms and isotopes in the rock. The method works because some atoms undergo changes with time that allow us to determine how long they have been held in place within the rock's solid structure. In other words, the age of a rock is the time since its atoms became locked together in their present arrangement, which in most cases means the time *since the rock last solidified*.

Recall that each chemical element is uniquely characterized by the number of protons in its nucleus, and that different *isotopes* of the same element differ in their number of neutrons [Section 5.1]. A **radioactive** isotope has a nucleus prone to spontaneous change, or *decay*, such as breaking apart or having one of its protons turn into a neutron. This decay always occurs at the same, measurable rate for any particular radioactive isotope. Decay rates are usually stated in terms of a **half-life**—the length of time it would take for half the nuclei in a collection to decay.

Consider the radioactive isotope potassium-40 (19 protons and 21 neutrons), which decays when one of its protons turns into a neutron, changing the potassium-40 into argon-40. The half-life for this decay process is 1.25 billion years. (Potassium-40 also decays by other paths that we will not discuss.) Imagine a small piece of rock that contained 1 microgram of potassium-40 and no argon-40 when it solidified long ago. The half-life of 1.25 billion years means that half the original potassium-40 had decayed into argon-40 by the time the rock was 1.25 billion years old, so at that time the rock contained ½ microgram of potassium-40 and ½ microgram of argon-40. Half of this remaining potassium-40 had then decayed by the end of the next 1.25 billion years, so after 2.5 billion years the rock contained ¼ microgram of potassium-40 and ¾ microgram of argon-40. After three half-lives, or 3.75 billion years, only ⅛ microgram of potassium-40 remained, while ⅞ microgram had become argon-40. Figure 6.26 summarizes the gradual decrease in the amount of potassium-40 and the corresponding rise in the amount of argon-40.

> We can determine the age of a rock through careful analysis of the proportions of various atoms and isotopes within it.

You can now see the essence of radiometric dating. Suppose you find a rock that contains equal numbers of atoms of potassium-40 and argon-40. If you assume that all the argon came from potassium decay (and if the rock shows no evidence of subsequent heating that could have allowed any argon to escape), then it must have taken precisely one half-life for the rock to end up with equal amounts of the two isotopes. You could therefore conclude that the rock is 1.25 billion years old. The only question is whether you are right in assuming that the rock lacked argon-40 when it formed. In this case, knowing a bit of "rock chemistry" helps. Potassium-40 is a natural ingredient of many minerals in rocks, but argon-40 is a gas that does not combine with other elements and did not condense in the solar nebula. If you find argon-40 gas trapped inside minerals, it must have come from radioactive decay of potassium-40.

Radiometric dating is possible with many other radioactive isotopes as well. In many cases, we can date a rock that contains more than one radioactive isotope, so agreement between the ages calculated from the different isotopes gives us confidence that we have dated the rock correctly.

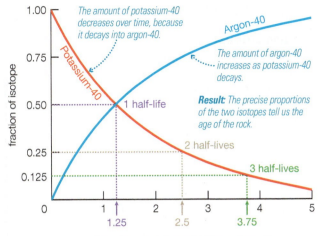

▲ **FIGURE 6.26**

Potassium-40 is radioactive, decaying into argon-40 with a half-life of 1.25 billion years. The red curve shows the decreasing amount of potassium-40, and the blue curve shows the increasing amount of argon-40. The remaining amount of potassium-40 drops in half with each successive half-life.

cosmic calculations 6.1

Radiometric Dating

From the definition of a half-life, it is possible to derive the following formula that allows you to calculate the age of a rock. To use the formula, you must measure the current amount of a radioactive substance in the rock and determine the original amount that was present long ago, which can often be done by measuring the abundance of the substance's decay products.

$$t = t_{half} \times \frac{\log_{10}\left(\dfrac{\text{current amount}}{\text{original amount}}\right)}{\log_{10}\left(\dfrac{1}{2}\right)}$$

where t is the time since the rock formed and t_{half} is the half-life of the radioactive substance; "$\log_{10}$" is the base-10 logarithm, which you can calculate with any scientific calculator.

Example: You chemically analyze a small sample of a meteorite. Potassium-40 and argon-40 are present in a ratio of 0.850 unit of potassium-40 atoms to 9.15 units of gaseous argon-40 atoms. (The units are unimportant, because only the relative amounts matter.) How old is the meteorite?

Solution: Because no argon gas could have been present in the meteorite when it formed, the 9.15 units of argon-40 must originally have been potassium-40. The sample must therefore have started with $0.850 + 9.15 = 10.0$ units of potassium-40 (the original amount), of which 0.85 unit remains (the current amount). The formula now reads

$$t = 1.25 \text{ billion yr} \times \frac{\log_{10}\left(\dfrac{0.850}{10.0}\right)}{\log_{10}\left(\dfrac{1}{2}\right)} = 4.45 \text{ billion yr}$$

This meteorite solidified about 4.45 billion years ago.

We can also check results from radiometric dating against those from other methods of measuring or estimating ages. For example, some fairly recent archaeological artifacts have original dates printed on them, and the dates agree with ages found by radiometric dating. We can validate the 4½-billion-year radiometric age for the solar system as a whole by comparing it to an age based on detailed study of the Sun. Theoretical models of the Sun, along with observations of other stars, show that stars slowly expand and brighten as they age. The model ages are not nearly as precise as radiometric ages, but they confirm that the Sun is between about 4 and 5 billion years old. Overall, the technique of radiometric dating has been checked in so many ways and relies on such basic scientific principles that there is no serious scientific debate about its validity.

Earth Rocks, Moon Rocks, and Meteorites   Radiometric dating tells us how long it has been since a rock solidified, which is not the same as the age of a planet as a whole. For example, we find rocks of many different ages on Earth. Some rocks are quite young because they formed recently from molten lava, while the oldest Earth rocks date to about 4 billion years ago (some small mineral grains are older). Earth itself must be even older.

Moon rocks brought back by the *Apollo* astronauts date to as far back as 4.4 billion years ago. Although they are older than Earth rocks, these Moon rocks must still be younger than the Moon itself. The ages of these rocks also tell us that the giant impact thought to have created the Moon must have occurred more than 4.4 billion years ago.

To go all the way back to the origin of the solar system, we must find rocks that have not melted or vaporized since they first condensed in the solar nebula. Meteorites that have fallen to Earth are our source of such rocks. Many meteorites appear to have remained unchanged since they condensed and accreted in the early solar system. Careful analysis of radioactive isotopes in these meteorites shows that a significant number of them share the oldest time of formation, which is about 4.56 billion years ago, so this time must mark the beginning of accretion in the solar nebula. Because the planets apparently accreted within about 50 million (0.05 billion) years, Earth and the other planets formed by about 4.5 billion years ago.

> Age dating of the oldest meteorites tells us that the solar system is about $4\frac{1}{2}$ billion years old.

---

the big picture ⟩ Putting Chapter 6 into Perspective

This chapter introduced the major features of our solar system and the current scientific theory of its formation. As you continue your study of the solar system, keep in mind the following "big picture" ideas:

- Our solar system is not a random collection of objects moving in random directions. Rather, it is highly organized, with clear patterns of motion and common traits among families of objects.

- We explain the origin of our solar system with the nebular theory, which holds that the solar system formed from the gravitational collapse of an interstellar gas cloud.

- Most of the general features of the solar system were determined by processes that occurred very early in the solar system's history.

- Our solar system is old—born about 4½ billion years ago—but it is still only about one-third as old as our 14-billion-year-old universe.

my cosmic perspective   Events that happened more than 4 billion years ago may seem distant from our lives, but we wouldn't be here without them; they explain not only how our planet came to exist, but where life-sustaining water comes from, and how we came to have the Moon in our sky.

6.1 A Brief Tour of the Solar System

◆ What does the solar system look like?

The planets are tiny compared to the distances between them. Our solar system consists of the Sun, the planets and their moons, and vast numbers of asteroids and comets. Each world has its own unique character, but there are many clear patterns among the worlds.

6.2 The Nebular Theory of Solar System Formation

◆ What features of our solar system provide clues to how it formed?

Four major features provide clues: (1) The Sun, planets, and large moons generally rotate and orbit in a very organized way. (2) The planets divide clearly into two groups: **terrestrial** and **jovian.** (3) The solar system contains vast numbers of asteroids and comets. (4) There are some notable exceptions to these general patterns.

◆ What is the nebular theory?

The **nebular theory** holds that the solar system formed from the gravitational collapse of a great cloud of gas and dust, which itself was the product of recycling of gas through many generations of stars within our galaxy.

6.3 Explaining the Major Features of the Solar System

◆ What caused the orderly patterns of motion?

As the solar nebula collapsed under gravity, natural processes caused it to heat up, spin faster, and flatten out as it shrank. The orderly motions we observe today all came from the orderly motion of this spinning disk.

◆ Why are there two major types of planets?

The inner regions of the solar nebula were relatively hot, so only metal and rock could condense into tiny, solid grains; these grains accreted into larger **planetesimals** that ultimately merged to make the terrestrial planets. Beyond the **frost line,** cooler temperatures also allowed more abundant **hydrogen compounds** to condense into ice, building ice-rich planetesimals; some of these grew large enough for their gravity to draw in hydrogen and helium gas, forming the jovian planets.

◆ Where did asteroids and comets come from?

Asteroids are the rocky leftover planetesimals of the inner solar system, and comets are the ice-rich leftover planetesimals of the outer solar system. These objects still occasionally collide with planets or moons, but the vast majority of impacts occurred during the **heavy bombardment** in the solar system's first few hundred million years.

◆ How do we explain the "exceptions to the rules"?

Most of the exceptions probably arose from collisions or close encounters with leftover planetesimals. Our Moon is most likely the result of a **giant impact** between a Mars-size planetesimal and the young Earth.

6.4 The Age of the Solar System

◆ How do we know the age of the solar system?

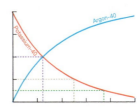

We can determine the age of a rock through **radiometric dating,** which is based on knowing the **half-life** of various radioactive isotopes and carefully measuring the proportions of these isotopes and their decay products within a rock. The oldest rocks are meteorites, and their age tells us that the accretion began in the solar nebula about 4.56 billion years ago.

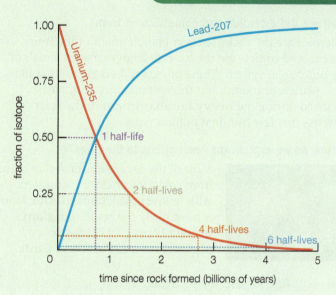

The graph shows the radioactive decay of uranium-235 to lead-207. (Uranium-235 also decays by another pathway, not shown.)

1. Compare the graph at the left to Figure 6.26, which shows the decay of potassium-40. Which element is more radioactive (undergoes radioactive decay more quickly)?
   a. uranium-235
   b. potassium-40
   c. Both are equally radioactive.

2. What fraction of the original uranium-235 should be left after 3.5 billion years?
   a. 1/2   b. 1/4
   c. 1/8   d. 1/32
   e. 1/64

3. You find a mysterious rock on the ground and determine that 60% of its uranium-235 has been converted into lead-207. What is the most likely origin of the rock, based on its radiometric age?
   a. It's older than our solar system, so it must have come from another solar system.
   b. It's a meteorite dating back to the formation of the solar system.
   c. It's a volcanic rock nearly a billion years old.
   d. It was just formed this year during the eruption of a nearby volcano.

exercises and problems

MasteringAstronomy®   For instructor-assigned homework and other learning materials, go to MasteringAstronomy®.

Review Questions

1. Briefly describe the layout of the solar system as it would appear from beyond the orbit of Neptune.
2. For each of the objects in the solar system tour (pages 140–149), describe at least two features that you find interesting.
3. Briefly describe the four major features of our solar system that provide clues to how it formed.
4. What are the basic differences between *terrestrial* and *jovian* planets? Which planets fall into each group?
5. What is the *nebular theory,* and why is it widely accepted by scientists today?
6. What do we mean by the *solar nebula*? What was it made of, and where did it come from?
7. Describe three key processes that led the solar nebula to take the form of a spinning disk. What evidence supports this scenario?
8. Describe the four categories of materials in the solar nebula by their condensation properties and abundance. Which ingredients condensed inside and outside the *frost line*?
9. Briefly describe the process by which terrestrial planets are thought to have formed.
10. How was the formation of the jovian planets similar to that of the terrestrial planets? How was it different? Why did the jovian planets end up with many moons?
11. What are asteroids and comets, and how did they come to exist? How and why are they different?

12. What was the *heavy bombardment,* and when did it occur?
13. What is the leading hypothesis for the Moon's formation? What evidence supports this hypothesis?
14. Briefly explain the technique of *radiometric dating*. What is a *half-life*? How do we use radiometric dating to determine the age of the solar system?

Test Your Understanding

○ Surprising Discoveries?

Suppose we found a solar system with the property described. (These are not real discoveries.) In light of what you've learned about the formation of our own solar system, decide whether the discovery should be considered reasonable or surprising. Explain your reasoning.

15. A solar system has five terrestrial planets in its inner solar system and three jovian planets in its outer solar system.
16. A solar system has four large jovian planets in its inner solar system and seven small terrestrial planets in its outer solar system.
17. A solar system has ten planets that all orbit the star in approximately the same plane. However, five planets orbit in one direction (e.g., counterclockwise), while the other five orbit in the opposite direction (e.g., clockwise).
18. A solar system has 12 planets that all orbit the star in the same direction and in nearly the same plane. The 15 largest moons in

this solar system orbit their planets in nearly the same direction and plane as well. However, several smaller moons have highly inclined orbits around their planets.

19. A solar system has six terrestrial planets and four jovian planets. Each of the six terrestrial planets has at least five moons, while the jovian planets have no moons at all.

20. A solar system has four Earth-size terrestrial planets. Each of the four planets has a single moon that is nearly identical in size to Earth's Moon.

21. A solar system has many rocky asteroids and many icy comets. However, most of the comets orbit in the inner solar system, while the asteroids orbit in far-flung regions much like the Kuiper belt and Oort cloud of our solar system.

22. A solar system has several planets similar in composition to the jovian planets of our solar system but similar in mass to the terrestrial planets of our solar system.

23. A solar system has several terrestrial planets and several larger planets made mostly of ice. (*Hint:* What would happen if the solar wind started earlier or later than in our solar system?)

24. Radiometric dating of the oldest meteorites from another solar system shows that they are a billion years younger than rocks from the terrestrial planets of the same system.

Quick Quiz

Choose the best answer to each of the following. Explain your reasoning with one or more complete sentences.

25. How many of the planets orbit the Sun in the same direction as Earth does? (a) a few (b) most (c) all

26. Planetary orbits in our solar system are (a) very eccentric (stretched-out) ellipses and in the same plane. (b) fairly circular and in the same plane. (c) fairly circular but oriented in every direction.

27. The solar nebula was 98% (a) rock and metal. (b) hydrogen compounds. (c) hydrogen and helium.

28. Which of the following did *not* occur during the collapse of the solar nebula? (a) spinning faster (b) heating up (c) concentrating denser materials nearer the Sun

29. What is Jupiter's main ingredient? (a) rock and metal (b) hydrogen compounds (c) hydrogen and helium

30. Which lists the major steps of solar system formation in the correct order? (a) collapse, accretion, condensation (b) collapse, condensation, accretion (c) accretion, condensation, collapse

31. Leftover ice-rich planetesimals are called (a) comets. (b) asteroids. (c) meteorites.

32. What's unusual about our Moon? (a) It's the only moon that orbits a terrestrial planet. (b) It's by far the largest moon in the solar system. (c) It's surprisingly large relative to the planet it orbits.

33. Are there any exceptions to the rule that planets rotate with small axis tilts and in the same direction as they orbit the Sun? (a) No (b) Venus is the only exception. (c) Venus and Uranus are exceptions.

34. About how old is the solar system? (a) 4.5 million years (b) 4.5 billion years (c) 4.5 trillion years

Process of Science

35. *Explaining the Past.* Is it really possible for science to inform us about things that may have happened billions of years ago? To address this question, test the nebular theory against each of the three hallmarks of science discussed in Chapter 3. Be as detailed as possible in explaining whether the theory does or does not satisfy these hallmarks. Use your explanations to decide whether the theory can really tell us about how our solar system formed. Defend your opinion.

36. *Unanswered Questions.* As discussed in this chapter, the nebular theory answers many but not all questions about the origin of our solar system. Choose one important but unanswered question about the origin of our solar system and write two or three paragraphs in which you discuss how we might answer this question in the future. Be as specific as possible, focusing on the type of evidence necessary to answer the question and how the evidence could be gathered. What are the benefits of finding answers to this question?

Group Work Exercise

37. *A Cold Solar Nebula.* **Roles:** *Scribe* (takes notes on the group's activities), *Proposer* (proposes explanations to the group), *Skeptic* (points out weaknesses in proposed explanations), *Moderator* (leads group discussion and makes sure everyone contributes). **Activity:** In our solar system, the frost line was located between Mars and Jupiter, but study of other solar systems suggests that our solar system could have turned out differently. Consider a hypothetical scenario in which the solar nebula was not cleared away by the solar wind until the entire disk of gas had cooled to 50 K.

    a. Make a list of ingredients that will condense at 50 K.

    b. Make a list of ways in which the terrestrial planets might have turned out differently under this alternative formation scenario.

    c. Repeat part (b) for the jovian planets.

    d. Discuss the likelihood that your predicted changes would match the actual characteristics of this alternative solar system.

    e. Come up with additional "what if" scenarios, discussing various other ways in which the planets might have turned out differently.

Investigate Further

Short-Answer/Essay Questions

38. *True or False.* Decide whether each statement is true or false, and explain why.

    a. On average, Venus has the hottest surface temperature of any planet in the solar system.

    b. Our Moon is about the same size as moons of the other terrestrial planets.

    c. The weather conditions on Mars today are much different than they were in the distant past.

    d. Moons cannot have atmospheres, active volcanoes, or liquid water.

    e. Saturn is the only planet in the solar system with rings.

    f. Neptune orbits the Sun in the opposite direction of all the other planets.

    g. If Pluto were as large as the planet Mercury, we would classify it as a terrestrial planet.

    h. Asteroids are made of essentially the same materials as the terrestrial planets.

    i. When scientists say that our solar system is about 4½ billion years old, they are making a rough estimate based on guesswork about how long it should have taken planets to form.

39. *Planetary Tour.* Based on the brief planetary tour in this chapter, which planet besides Earth do you think is the most interesting, and why? Defend your opinion clearly in two or three paragraphs.

40. *Patterns of Motion.* In one or two paragraphs, summarize the orderly patterns of motion in our solar system and explain why their existence should suggest that the Sun and the planets all formed at one time from one cloud of gas, rather than as individual objects at different times.

41. *Solar System Trends.* Study the planetary data in Table 6.1 to answer each of the following.
    a. Notice the relationship between distance from the Sun and surface temperature. Describe the trend, explain why it exists, and explain any notable exceptions to the trend.
    b. The text says that planets can be classified as either terrestrial or jovian. Describe in general how the columns for density, composition, and distance from the Sun support this classification.
    c. Describe the trend you see in orbital periods and explain the trend in terms of Kepler's third law.
    d. Which column tells you which planet has the shortest days? Are there notable differences in the length of a day for the different types of planets? Explain.
    e. Which planets would you expect not to have seasons? Why?

42. *Two Kinds of Planets.* The jovian planets differ from the terrestrial planets in a variety of ways. Using phrases or sentences that members of your family would understand, explain why the jovian planets differ from the terrestrial planets in each of the following: composition, size, density, distance from the Sun, and number of satellites.

43. *An Early Solar Wind.* Suppose the solar wind had cleared away the solar nebula before the seeds of the jovian planets could gravitationally draw in hydrogen and helium gas. How would the planets of the outer solar system be different? Would they still have many moons? Explain your answer in a few sentences.

44. *History of the Elements.* Our bodies (and most living things) are made mostly of water (H_2O). Summarize the "history" of a typical hydrogen atom from its creation to Earth's formation. Do the same for a typical oxygen atom. (*Hint:* Which elements were created in the Big Bang, and where were the others created?)

45. *Rocks from Other Solar Systems.* Many "leftovers" from planetary formation were likely ejected from our solar system, and the same has presumably happened in other star systems. Given that fact, should we expect to find meteorites that come from other star systems? How rare or common would you expect them to be? (Be sure to consider the distances between stars.) Suppose that we did find a meteorite identified as a leftover from another stellar system. What could we learn from it?

Quantitative Problems

Be sure to show all calculations clearly and state your final answers in complete sentences.

46. *Radiometric Dating.* You are dating rocks by their proportions of parent isotope potassium-40 (half-life 1.25 billion years) and daughter isotope argon-40. Find the age for each of the following.
    a. A rock that contains equal amounts of potassium-40 and argon-40
    b. A rock that contains three times as much argon-40 as potassium-40

47. *Lunar Rocks.* You are dating Moon rocks based on their proportions of uranium-238 (half-life of about 4.5 billion years) and its ultimate decay product, lead. Find the age for each of the following.
    a. A rock for which you determine that 55% of the original uranium-238 remains, while the other 45% has decayed into lead
    b. A rock for which you determine that 63% of the original uranium-238 remains, while the other 37% has decayed into lead

48. *Carbon-14 Dating.* The half-life of carbon-14 is about 5700 years.
    a. You find a piece of cloth painted with organic dye. By analyzing the dye, you find that only 77% of the carbon-14 originally in the dye remains. When was the cloth painted?
    b. A well-preserved piece of wood found at an archaeological site has 6.2% of the carbon-14 it must have had when it was living. Estimate when the wood was cut.
    c. Is carbon-14 useful for establishing Earth's age? Why or why not?

49. *What Are the Odds?* The fact that all the planets orbit the Sun in the same direction is cited as support for the nebular hypothesis. Imagine that there's a different hypothesis in which planets can be created orbiting the Sun in either direction. Under this hypothesis, what is the probability that eight planets would end up traveling in the same direction? (*Hint:* It's the same probability as that of flipping a coin eight times and getting all heads.)

Discussion Questions

50. *Planetary Priorities.* Suppose you were in charge of developing and prioritizing future planetary missions for NASA. What would you choose as your first priority for a new mission, and why?

51. *Lucky to Be Here?* Considering the overall process of solar system formation, do you think it was likely for a planet like Earth to have formed? Could random events in the early history of the solar system have prevented our being here today? What implications do your answers have for the possibility of Earth-like planets around other stars? Defend your opinions.

Web Projects

52. *Current Planetary Mission.* Find a list of planetary missions currently under way. Then choose one to learn about in detail. Write a one- to two-page summary of the mission's basic design, goals, and status.

53. *Dating the Past.* The method of radiometric dating that tells us the age of our solar system is also used to determine when many other past events occurred. For example, it is used to determine ages of fossils that tell us when humans first evolved and ages of relics that teach us about the rise of civilization. Research one key aspect of human history in which radiometric dating helps us piece the story together. Write two or three paragraphs on how radiometric dating is used in this case (such as what materials are dated and what radioactive elements are used) and what these studies have concluded. Does your understanding of the method lead you to accept the results? Why or why not?

7

Earth and the Terrestrial Worlds

Earth not only is our home planet but also serves as a model for understanding other terrestrial worlds.

LEARNING GOALS

7.1 Earth as a Planet
- ♦ Why is Earth geologically active?
- ♦ What processes shape Earth's surface?
- ♦ How does Earth's atmosphere affect the planet?

7.2 The Moon and Mercury: Geologically Dead
- ♦ Was there ever geological activity on the Moon or Mercury?

7.3 Mars: A Victim of Planetary Freeze-Drying
- ♦ What geological features tell us that water once flowed on Mars?
- ♦ Why did Mars change?

7.4 Venus: A Hothouse World
- ♦ Is Venus geologically active?
- ♦ Why is Venus so hot?

7.5 Earth as a Living Planet
- ♦ What unique features of Earth are important for life?
- ♦ How is human activity changing our planet?
- ♦ What makes a planet habitable?

t's easy to take for granted the qualities that make Earth so suitable for human life: a temperature neither boiling nor freezing, abundant water, a protective atmosphere, and a relatively stable environment. But we need look only as far as our neighboring terrestrial worlds to see how fortunate we are. The Moon is airless and barren, and Mercury is much the same. Venus is a searing hothouse, while Mars has an atmosphere so thin and cold that liquid water cannot last on its surface today.

How did the terrestrial worlds come to be so different, and why did Earth alone develop conditions that permit abundant life? We'll answer these questions by exploring key processes that have shaped Earth and the other terrestrial worlds over time. As we will see, the histories of the terrestrial worlds have been determined largely by properties endowed at their births.

7.1 Earth as a Planet

Earth's surface seems solid and steady, but every so often it offers us a reminder that nothing about it is permanent. If you live in Alaska or California, you've probably felt the ground shift beneath you in an earthquake. In Washington State, you may have witnessed the rumblings of Mount St. Helens. In Hawaii, a visit to the active Kilauea volcano will remind you that you are standing on mountains of volcanic rock protruding from the ocean floor.

Volcanoes and earthquakes are not the only processes acting to reshape Earth's surface. They are not even the most dramatic: Far greater change can occur on the rare occasions when an asteroid or a comet slams into Earth. More gradual processes can also have spectacular effects. The Colorado River causes only small changes in the landscape from year to year, but over millions of years its unrelenting flow carved the Grand Canyon. The Rocky Mountains were once twice as tall as they are today, but they have been cut down in size through tens of millions of years of erosion by wind, rain, and ice. Entire continents move slowly about, completely rearranging the map of Earth every few hundred million years.

Earth is not alone in having undergone tremendous change since its birth. The surfaces of all five terrestrial worlds—Mercury, Venus, Earth, the Moon, and Mars—must have looked quite similar when they were young, because all are made of rocky material and all were subjected to the impacts of the heavy bombardment [Section 6.3]. The great differences in their present-day appearances must therefore be the result of changes that have occurred through time.

Figure 7.1 shows global views of the terrestrial worlds to scale, along with sample surface views from orbit. Profound differences are immediately obvious. Mercury and the Moon still show the scars of their battering during the heavy bombardment. Venus is covered by a thick, cloudy atmosphere, but radar mapping reveals a surface dotted with volcanoes and other features indicating active geology. Mars has many features that appear to have been shaped by running water, but all are now dry. Earth has a diverse surface that reveals clear evidence of life.

Mercury

50 km

Heavily cratered Mercury has long steep cliffs (arrow).

Venus

100 km

Cloud-penetrating radar revealed this twin-peaked volcano on Venus.

Earth

100 km

A portion of Earth's surface as it appears without clouds.

Earth's Moon

100 km

The Moon's surface is heavily cratered in most places.

Mars

50 km

Mars has features that look like dry riverbeds; note the impact craters.

▲ **FIGURE 7.1**

Global views of the five terrestrial worlds to scale, along with sample close-ups viewed from orbit. All the images were taken with visible light except the Venus close-up, which is based on radar data.

Our primary goal in this chapter is to gain a deeper understanding of our own planet Earth by investigating how the terrestrial worlds came to be so different. We'll begin by examining the basic nature of our planet.

◆ Why is Earth geologically active?

All the terrestrial worlds have changed since their birth, but Earth is unique in the degree to which it continues to change today. We say that Earth is *geologically active,* meaning that its surface is continually being reshaped by volcanic eruptions, earthquakes, erosion, and other geological processes. Most geological activity is driven by processes that take place deep beneath the surface. Consequently, to understand why Earth is so much more geologically active than other worlds, we must examine what the terrestrial worlds are like inside.

Interior Structure We cannot see inside the terrestrial worlds, but a variety of clues tell us about their internal structures. For Earth, the most direct data come from *seismic waves,* vibrations that travel both through the interior and along the surface after an earthquake. We also have seismic data for the Moon, thanks to monitoring stations left behind by the *Apollo* astronauts, and the *Mars Insight* mission, scheduled for launch in 2018, includes a seismic station for Mars. We use less direct methods to study the interiors of other worlds. For example, comparing a world's overall average density to the density of its surface rock tells us how much more dense it must be inside, measurements of a world's gravity from spacecraft can tell us how mass is distributed inside it, studies of magnetic fields tell us about the interior layers in which these fields are generated, and volcanic rocks can tell us about interior composition.

Together, such studies have shown that all the terrestrial worlds have layered interiors. We divide these layers by density into three major categories:

- **Core:** The highest-density material, consisting primarily of metals such as nickel and iron, resides in a central core.

common misconceptions

Earth Is Not Full of Molten Lava

Many people guess that Earth is full of molten lava (or *magma*). This misconception may arise partially because we see molten lava emerging from inside Earth when a volcano erupts. However, Earth's mantle and crust are almost entirely solid. The lava that erupts from volcanoes comes from a narrow region of partially molten material beneath the lithosphere. The only part of Earth's interior that is fully molten is the outer core, which lies so deep within the planet that its material never erupts directly to the surface.

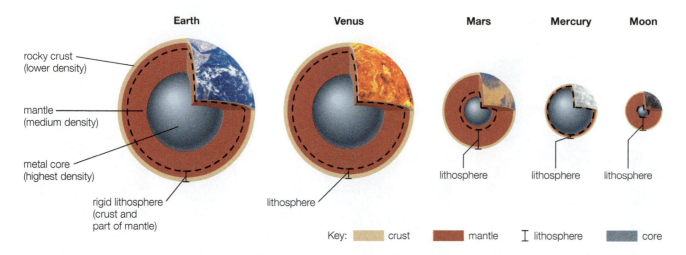

Earth   Venus   Mars   Mercury   Moon

rocky crust
(lower density)

mantle
(medium density)

metal core
(highest density)

rigid lithosphere
(crust and
part of mantle)

lithosphere

lithosphere   lithosphere   lithosphere

Key: crust   mantle   I lithosphere   core

▲ **FIGURE 7.2**

Interior structures of the terrestrial worlds, shown to scale and in order of decreasing size. Color coding shows the core-mantle-crust layering by density; a dashed circle represents the inner boundary of the lithosphere, defined by strength of rock rather than by density. The thicknesses of the crust and the lithosphere of Venus and Earth are exaggerated to make them visible in this figure.

▲ **FIGURE 7.3**

Silly Putty stretches when pulled slowly but breaks cleanly when pulled rapidly. Rock behaves just the same, but on a longer time scale.

- **Mantle:** Rocky material of moderate density—mostly minerals that contain silicon, oxygen, and other elements—forms a thick mantle that surrounds the core.

- **Crust:** The lowest-density rock (which includes the familiar rocks of Earth's surface) forms a thin crust, essentially representing the world's outer skin.

Figure 7.2 shows these layers for the five terrestrial worlds. Although not shown in the figure, Earth's metallic core actually consists of two distinct regions: a solid *inner core* and a molten (liquid) *outer core*.

It's often more useful to categorize interior layers by rock strength instead of density. The idea that rock can vary in strength may seem surprising, but like all matter built of atoms, rock is mostly empty space; its apparent solidity arises from electrical bonds between its atoms and molecules [Section 5.1]. Although these bonds are strong, they can still break and re-form when subjected to heat or sustained stress, which means that even solid rock can slowly deform and flow over millions and billions of years. The long-term behavior of rock is much like that of Silly Putty, which breaks like a brittle solid when you pull it sharply but deforms and stretches when you pull it slowly (Figure 7.3). Also like Silly Putty, rock becomes softer and easier to deform when it is warmer.

see it for yourself Roll some room-temperature Silly Putty into a ball and measure its diameter. Put the ball on a table and gently place a heavy book on top of it. After 5 seconds, measure the height of the squashed ball. Repeat the experiment, but warm the Silly Putty in hot water before you start. Repeat again, but cool the Silly Putty in ice water before you start. How does temperature affect the rate of "squashing"? How does the experiment relate to planetary geology?

A planet's *lithosphere* is its outer layer of cool, rigid rock. In terms of rock strength, Earth's outer layer consists of relatively cool and rigid rock, called the **lithosphere** (*lithos* is Greek for "stone"), that essentially "floats" on warmer, softer rock beneath. As shown by the dashed circles in Figure 7.2, the lithosphere encompasses the crust and part of the mantle of each world.

Differentiation and Internal Heat We can understand *why* the interiors are layered by thinking about what happens in a mixture of oil and water: Gravity pulls the denser water to the bottom, driving the less dense oil to the top. This process is called **differentiation,** because it

The terrestrial interiors were once hot enough to melt, allowing material to settle into layers of differing density.

results in layers made of *different* materials. The layered interiors of the terrestrial worlds tell us that they underwent differentiation at some time in the past, which means all these worlds must once have been hot enough inside for their interior rock and metal to melt. Dense metals like iron sank toward the center, driving less dense rocky material toward the surface.

Two major processes explain why all the terrestrial worlds once had hot interiors. First, as a planet formed through accretion [Section 6.3], the gravitational potential energy of incoming planetesimals was ultimately converted into thermal energy in the planet's interior. (The process of differentiation converted additional gravitational potential energy into thermal energy as denser materials sank to the core.) Second, the rock and metal that built the terrestrial worlds contained radioactive isotopes of elements such as uranium, potassium, and thorium. As these radioactive materials decay, they release heat directly into the planetary interiors, in essence converting some of the mass-energy ($E = mc^2$) of the radioactive

special topic   Seismic Waves

Seismic waves come in two basic types, analogous to the two ways you can generate waves in a Slinky (Figure 1). Pushing and pulling on one end of a Slinky (while someone holds the other end still) generates a wave in which the Slinky is bunched up in some places and stretched out in others. Waves like this in rock are called P waves. The *P* stands for *primary,* because these waves travel fastest and are the first to arrive after an earthquake, but it is easier to think of *P* as meaning *pressure* or *pushing.* P waves can travel through almost any material—whether solid, liquid, or gas—because molecules can always push on their neighbors no matter how weakly they are bound together. (Sound travels as a pressure wave quite similar to a P wave.)

Shaking a Slinky slightly up and down or side to side generates a different type of wave; in rock, such waves are called S waves. The *S* stands for *secondary* but is easier to remember as meaning *shear* or *side to side.* S waves travel only through solids, because the bonds between neighboring molecules in a liquid or gas are too weak to transmit up-and-down or sideways forces.

The speeds and directions of seismic waves depend on the composition, density, pressure, temperature, and phase (solid or liquid) of the material they pass through. For example, P waves reach the side of the world opposite an earthquake, but S waves do not. This tells us that a liquid layer has stopped the S waves, which is how we know that Earth has a liquid outer core (Figure 2). More careful analysis of seismic waves, combined with mathematical modeling, has allowed geologists to develop a detailed picture of Earth's interior structure.

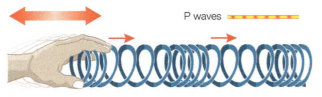

P waves

P waves result from compression and stretching in the direction of travel.

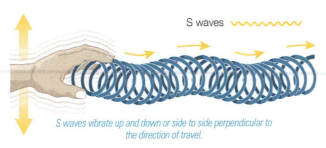

S waves

S waves vibrate up and down or side to side perpendicular to the direction of travel.

▲ **FIGURE 1**

Slinky examples demonstrating P and S waves.

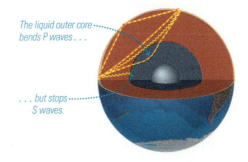

The liquid outer core bends P waves . . .

. . . but stops S waves.

▲ **FIGURE 2**

Because S waves do not reach the side of Earth opposite an earthquake, we infer that part of Earth's core is liquid.

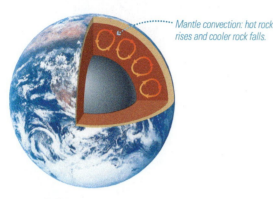

▲ FIGURE 7.4

Earth's hot interior allows the mantle to undergo convection, in which hot rock gradually rises upward while cool rock gradually falls. Arrows indicate the direction of flow in a portion of the mantle.

---

cosmic calculations 7.1

The Surface Area–to–Volume Ratio

The total amount of heat contained in a planet depends on its volume, but heat can escape into space only from the planet's *surface*. As heat escapes, more heat flows upward from the interior to replace it until the interior is no hotter than the surface. The time it takes a planet to lose its internal heat is related to the ratio of the *surface area* through which it loses heat to the *volume* that contains heat:

$$\text{surface area–to–volume ratio} = \frac{\text{surface area}}{\text{volume}}$$

A spherical planet (radius r) has surface area $4\pi r^2$ and volume $\frac{4}{3}\pi r^3$, so the ratio becomes

$$\underset{\text{(for a sphere)}}{\text{surface area–to–volume ratio}} = \frac{4\pi r^2}{\frac{4}{3}\pi r^3} = \frac{3}{r}$$

Because r appears in the denominator, we conclude that *larger objects have smaller surface area–to–volume ratios.*

Example: Compare the surface area–to–volume ratios of the Moon ($r_{\text{Moon}} = 1738$ km) and Earth ($r_{\text{Earth}} = 6378$ km).

Solution: We divide the two surface area–to–volume ratios:

$$\frac{\text{surface area–to–volume ratio (Moon)}}{\text{surface area–to–volume ratio (Earth)}} = \frac{3/r_{\text{Moon}}}{3/r_{\text{Earth}}} = \frac{r_{\text{Earth}}}{r_{\text{Moon}}}$$

$$= \frac{6378 \text{ km}}{1738 \text{ km}} = 3.7$$

The Moon's surface area–to–volume ratio is nearly four times as large as Earth's, which means the Moon would cool four times as fast if both worlds started with the same temperature and gained no additional heat.

---

nuclei to thermal energy. The amount of radioactive material declines with time as it decays, so this heat source supplied more heat when the planets were young than it does today.

The terrestrial interiors all retain at least some heat today, but the amount varies with size. Larger worlds tend to stay hot longer than smaller worlds for the same basic reason that a hot potato remains hot inside much longer than a hot pea.

see it for yourself It's easy to demonstrate that larger objects take longer to cool than smaller objects. The next time you eat something large and hot, cut off a small piece; notice how much more quickly the small piece cools than the rest of it. A similar experiment demonstrates the time it takes a cold object to warm up: Find two ice cubes of the same size. Crack one into small pieces, then compare the rates of melting. Explain your observations.

Internal Heat and Geological Activity Interior heat is the primary driver of geological activity. Temperature increases with depth inside a planet. If the interior is hot enough, the mantle can undergo **convection,** in which hot material gradually expands and cools as it rises upward, while cooler material from above gradually contracts and falls (Figure 7.4). Keep in mind that mantle convection primarily involves solid rock, which flows very slowly. At the typical rate of mantle convection on Earth—a few centimeters per year—it would take 100 million years for a piece of rock to be carried from the base of the mantle to the top.

Just as planetary size determines how long a planet stays hot, it is also the primary factor in the strength of mantle convection and lithospheric thickness. As a planet's interior cools, the rigid lithosphere grows thicker and convection occurs only deeper inside the planet. A thick lithosphere inhibits volcanic and tectonic activity, because any molten rock is too deeply buried to erupt to the surface and the strong lithosphere resists distortion by tectonic stresses. If the interior cools enough, convection may stop entirely, leaving the planet "geologically dead," with no eruptions or crustal movement.

Larger planets retain internal heat much longer than smaller ones, and this heat drives geological activity.

We can now understand the differences in lithospheric thickness shown in Figure 7.2, which go along with differences in geological activity. As the largest terrestrial planets, Earth and Venus remain quite hot inside and therefore have thin lithospheres and substantial geological activity. (Venus may have a thicker lithosphere than Earth for reasons we will discuss later.) With their small sizes, Mercury and the Moon have lost so much heat that they now have very thick lithospheres (Mercury's looks thin on Figure 7.2, but extends nearly down to its large core) and little if any geological activity. Mars, intermediate in size, retains enough internal heat for limited geological activity.

The Magnetic Field Interior heat is also responsible for Earth's global **magnetic field.** You are probably familiar with the general pattern of the magnetic field created by an iron bar (Figure 7.5a). Earth's magnetic field is generated by a process more similar to that of an *electromagnet,* in which the magnetic field arises as a battery forces charged particles to move along a coiled wire (Figure 7.5b). Earth does not contain a battery, but charged particles move with the molten metal in its liquid outer core (Figure 7.5c). Internal heat causes the liquid metal to rise and fall (convection), while Earth's rotation twists and distorts the convection pattern. The result is that electrons in the molten metal move within the outer core in much the same way they move in an electromagnet, generating Earth's magnetic field.

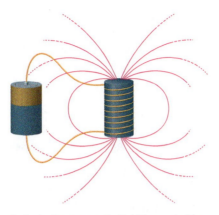

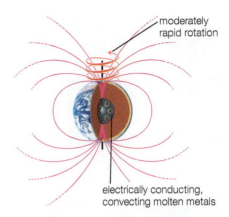

a This photo shows how a bar magnet influences iron filings (small black specks) around it. The *magnetic field lines* (red) represent this influence graphically.

b A similar magnetic field is created by an electromagnet, which is essentially a wire wrapped around a metal bar and attached to a battery. The field is created by the battery-forced motion of charged particles (electrons) along the wire.

c Earth's magnetic field also arises from the motion of charged particles. The charged particles move within Earth's liquid outer core, which is made of electrically conducting, convecting molten metals.

▲ **FIGURE 7.5**

Sources of magnetic fields.

The magnetic field is very important to life on Earth because it creates a **magnetosphere** that acts like a protective bubble surrounding our planet (Figure 7.6a), shielding Earth's surface from the energetic charged particles of the *solar wind* [Section 6.3]. These particles would otherwise strip away atmospheric gas and cause genetic damage to living organisms. The magnetosphere deflects most solar wind particles around our planet and channels the remaining particles toward the poles, where they can collide with atoms and molecules in our atmosphere and produce the beautiful lights of the **aurora** (Figure 7.6b).

> Earth's magnetic field, generated by convection of molten metal in its outer core, creates a *magnetosphere* that protects our planet from the solar wind.

▼ **FIGURE 7.6**

Earth's magnetosphere acts like a protective bubble, shielding our planet from charged particles coming from the solar wind.

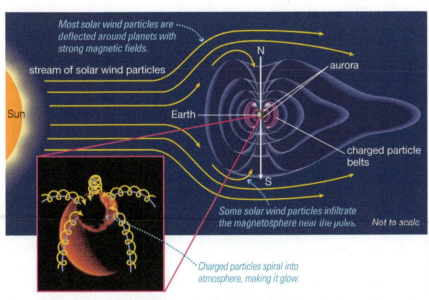

a This diagram shows how Earth's magnetosphere deflects solar wind particles. Some particles accumulate in charged particle belts encircling our planet. The inset is an ultraviolet image of a ring of auroras around the North Pole; the bright crescent at its left is part of the day side of Earth.

b This photograph shows the aurora near Yellowknife, Northwest Territories, Canada. In a video, you would see these lights dancing about in the sky.

▲ FIGURE 7.7

Artist's conception of the impact process.

None of the other terrestrial worlds have magnetic fields as strong as Earth's and they therefore lack protective magnetospheres. As we'll discuss later, this fact has profoundly affected the histories of Venus and Mars.

◆ What processes shape Earth's surface?

Although we find a huge variety of geological surface features on Earth and the other terrestrial worlds, nearly all of them can be explained by just four major geological processes:

- **Impact cratering:** the creation of bowl-shaped *impact craters* by asteroids or comets striking a planet's surface
- **Volcanism:** the eruption of molten rock, or *lava*, from a planet's interior onto its surface
- **Tectonics:** the disruption of a planet's surface by internal stresses
- **Erosion:** the wearing down or building up of geological features by wind, water, ice, and other phenomena of planetary weather

Nearly all geological features arise from impact cratering, volcanism, tectonics, or erosion. Note that impact cratering is the only one of the four processes with an external cause (impacts of objects from space). The other three are ultimately attributable to a planet's interior heat and represent what we usually define as *geological activity*.

Impact Cratering An impact crater forms when an asteroid or comet slams into a solid surface (Figure 7.7). An impacting object typically hits the surface at a speed between about 40,000 and 250,000 kilometers per hour. At such a tremendous speed, the impact releases enough energy to vaporize solid rock and blast out a *crater* (the Greek word for "cup"). Craters are usually circular, because an impact blasts out material in all directions, regardless of the incoming object's direction. Craters are typically about 10 times as wide as the objects that create them and about 10–20% as deep as they are wide. For example, an asteroid 1 kilometer in diameter will blast out a crater about 10 kilometers wide and 1–2 kilometers deep.

We have never witnessed a major impact on Earth, but geologists have identified more than 150 impact craters (Figure 7.8). While this may seem a substantial number, it is far smaller than the number of impact craters easily visible on the Moon, which leads to an important insight: Given their close proximity, we'd expect Earth and the Moon to have suffered similar numbers of impacts, so the relatively small number of craters on Earth suggests that Earth's craters have been erased with time by geological activity such as volcanic eruptions and erosion. This fact leads to a simple way to estimate the age of a world's surface, meaning the time since its current surface features formed.

We can estimate the geological age of any surface region from its number of impact craters, with more craters indicating an older surface. To understand the idea, recall that all the planets were battered by impacts during the heavy bombardment that ended almost 4 billion years ago [Section 6.3], but relatively few impacts have occurred since. In places where we see numerous craters, such as on much of the Moon's surface, we must be looking at a surface that has stayed virtually unchanged for at least 4 billion years. In contrast, when we see very few craters, as on Earth, we must be looking at a surface

▲ FIGURE 7.8

Meteor Crater in Arizona, made by an impact about 50,000 years ago, is one of the youngest of more than 150 known impact craters on Earth. It is more than a kilometer across and nearly 200 meters deep, implying the impact of an asteroid about 50 meters across.

that has undergone recent change. More detailed analysis (particularly of craters on the Moon) has allowed scientists to estimate the cratering rate throughout solar system history. As a result, scientists can estimate the age of a planetary surface just by photographing it from orbit and counting its craters.

Volcanism Volcanism occurs when underground molten rock finds a path to the surface (Figure 7.9). The "thickness" (or *viscosity*) of molten lava can vary, depending on the composition of the molten rock. Thick lava tends to make tall, steep volcanoes, while runny lava can make broad, flat lava plains. Earth is the most volcanically active of the terrestrial worlds, a fact traceable to its relatively large size, which has allowed it to retain internal heat.

Volcanic mountains are the most obvious result of volcanism, but volcanism has had a much more profound effect on our planet: It explains the existence of our atmosphere and oceans. Recall that Earth accreted from rocky and metallic planetesimals, while water and other ices were brought in by planetesimals from more distant reaches of the solar system [Section 6.3]. Water and gases became trapped beneath the surface in much the same way the gas in a carbonated beverage is trapped in a pressurized bottle. Volcanic eruptions later released some of this gas into the atmosphere in a process known as **outgassing** (Figure 7.10).

Earth's atmosphere and oceans were made from gases released from the interior by volcanic outgassing.

On Earth, the major gases released by outgassing have been water vapor (H_2O), carbon dioxide (CO_2), and nitrogen (N_2). Most of the outgassed water vapor condensed to form our oceans, and much of the nitrogen remains as the dominant ingredient (77%) of our atmosphere. We'll discuss how the atmosphere lost its carbon dioxide and gained its oxygen in Section 7.5.

Tectonics *Tectonics* refers to any surface reshaping that results from stretching, compression, or other forces acting on the lithosphere. Figure 7.11 shows two examples of tectonic features on Earth, one created as the surface is pushed upward by compression (the Himalayas) and the other created by surface stretching (the Red Sea).

Tectonic activity usually goes hand in hand with volcanism, because both require internal heat and therefore depend on planetary size. While tectonic activity has occurred on every terrestrial world, it has been particularly important on Earth, where the underlying mantle convection fractured the lithosphere into more than a dozen pieces, or *plates*. These plates move over, under, and around each other, leading to a special type of tectonics—**plate tectonics**—that appears to be unique to Earth in our solar system. We'll discuss plate tectonics further in Section 7.5.

Tectonics and volcanism both require internal heat and therefore depend on a planet's size.

Erosion Erosion begins with the breakdown and transport of surface rock through the action of ice, liquid, or gas. Familiar features of erosion include valleys carved by glacial ice, canyons carved by rivers, sand dunes shaped by wind, and river deltas made of sediments carried downstream (Figure 7.12). Note that while we usually associate erosion with breakdown, sand dunes and river deltas are examples of features built up by erosion. Indeed, erosion has built much of Earth's surface rock: Over long periods of time, erosion piled sediments into layers on the

molten rock in the crust

▲ **FIGURE 7.9**

Volcanism. The photo shows the eruption of an active volcano on the flanks of Kilauea on the Big Island of Hawaii. The inset shows the underlying process: Molten rock collects in a chamber from which it can erupt upward.

▲ **FIGURE 7.10**

This photo shows the eruption of Mount St. Helens (Washington State) on May 18, 1980. Note the tremendous outgassing that accompanied the eruption.

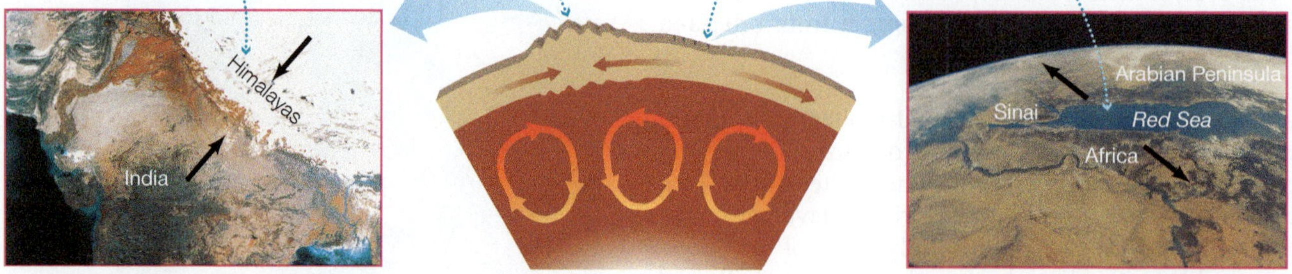

Internal stresses can cause compression in the crust . . .

. . . creating mountains like the Himalayas.

Internal stresses can also pull the crust apart . . .

. . . creating cracks and seas.

Earth's tallest mountain range, the Himalayas, created as India pushes into the rest of Asia.

The Red Sea, created as the Arabian Peninsula was torn away from Africa.

▲ **FIGURE 7.11**

Tectonic forces can produce a wide variety of features. Mountains created by tectonic compression and valleys or seas created by tectonic stretching are among the most common. Both images are satellite photos.

a The Colorado River has been carving the Grand Canyon for millions of years.

b Glaciers created Yosemite Valley during ice ages.

c Wind erosion wears away rocks and builds up sand dunes.

d This river delta is built from sediments worn away by wind and rain and then carried downstream.

▲ **FIGURE 7.12**

A few examples of erosion on Earth.

floors of oceans and seas, forming *sedimentary rock*. The layered rock of the Grand Canyon is an example, built up by erosion long before the Colorado River carved the canyon.

Erosion can both break down and build up geological features.

Erosion plays a far more important role on Earth than on any other terrestrial world, primarily because our planet has both strong winds and liquid water. Note that both winds and water are ultimately traceable to volcanic outgassing, which means that like volcanism and tectonics, erosion requires interior heat. However, interior heat alone is not enough. Venus lacks substantial erosion because it has weak winds and is too hot for liquid water, while water erosion no longer occurs on Mars because it is too cold.

◆ How does Earth's atmosphere affect the planet?

Most people are aware that the atmosphere supplies the oxygen we breathe, but Earth's atmosphere also affects our planet in several deeper ways. We've already discussed its role in weather and erosion. In addition, the atmosphere protects the surface from dangerous solar radiation and makes our planet warm enough for liquid water to exist.

As you study these effects, keep two other important ideas in mind. First, the atmosphere is a mix of gases; its overall composition is approximately 77% nitrogen (N_2) and 21% oxygen (O_2), with small amounts of argon, water vapor, carbon dioxide, and other gases. Second, the atmosphere is remarkably thin: Most of the air lies within 10 kilometers of the surface, which means it makes a layer around Earth equivalent to the thickness of a dollar bill laid over a standard globe.

Surface Protection The Sun emits the visible light that allows us to see, but it also emits dangerous ultraviolet and X-ray radiation. Our atmosphere protects us against this radiation by absorbing it high above the ground (Figure 7.13). We can understand how by considering interactions between air and different forms of light.

X rays can be absorbed by almost any atom or molecule. As a result, X rays from the Sun are absorbed where they first strike air, which is high in the atmosphere. No solar X rays reach the ground, which is why X-ray telescopes must be placed in space.

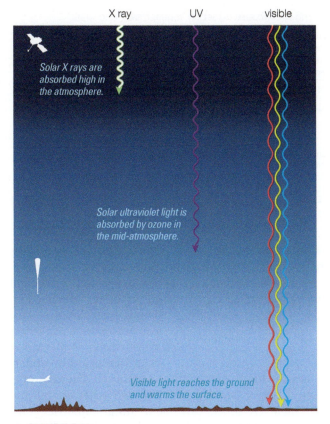

X ray UV visible

Solar X rays are absorbed high in the atmosphere.

Solar ultraviolet light is absorbed by ozone in the mid-atmosphere.

Visible light reaches the ground and warms the surface.

▲ **FIGURE 7.13**

This diagram summarizes how different forms of light from the Sun are affected by Earth's atmosphere.

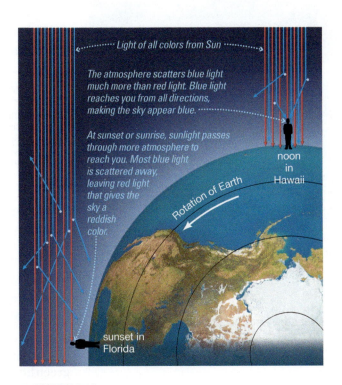

Light of all colors from Sun

The atmosphere scatters blue light much more than red light. Blue light reaches you from all directions, making the sky appear blue.

At sunset or sunrise, sunlight passes through more atmosphere to reach you. Most blue light is scattered away, leaving red light that gives the sky a reddish color.

noon in Hawaii

Rotation of Earth

sunset in Florida

▲ **FIGURE 7.14**

This diagram summarizes why the sky is blue and sunsets (and sunrises) are red.

Ultraviolet light is not so easily absorbed, because most gases are transparent to ultraviolet light. We owe our protection from ultraviolet light to a relatively rare gas, called **ozone** (O_3), that resides primarily in the middle layer of Earth's atmosphere (the *stratosphere*). Without this ultraviolet-absorbing ozone, life on land could not survive.

Solar X rays are absorbed by atoms and molecules high in the atmosphere, and ultraviolet radiation is absorbed by ozone in the middle atmosphere.

Visible light can pass through the atmosphere to the ground, making it the primary source of heat for Earth's surface, the energy source for photosynthesis, and the light that allows us to see. However, not all visible light photons take a straight path to the ground. Some are scattered randomly around the sky, which is why our sky is bright in the daytime. Without this scattering, our sky would look like the lunar sky does to an astronaut, with the Sun a very bright circle set against a black background. Scattering also explains the color of our sky (Figure 7.14). Visible light consists of all the colors of the rainbow, but the colors are not all scattered equally. Gas molecules scatter blue light (shorter wavelength and higher energy) so much more effectively than red light (longer wavelength and lower energy) that, for practical purposes, we can imagine that only the blue light gets scattered. When the Sun is overhead, this scattered blue light reaches our eyes from all directions, so the sky appears blue. At sunset or sunrise, when sunlight passes through more air on its way to our eyes, so much of the blue light is scattered away that we are left primarily with red light to color the sky.

common misconceptions

Why Is the Sky Blue?

If you ask around, you'll find a wide variety of misconceptions about why the sky is blue. Some people guess that the sky is blue because of light reflecting from the oceans, but that could not explain blue skies over inland areas. Others claim that "air is blue," a vague statement that is clearly wrong: If air molecules emitted blue light, then air would glow blue even in the dark; if they were blue because they reflected blue light and absorbed red light, then no red light could reach us at sunset. The real explanation for the blue sky is light scattering, as shown in Figure 7.14, which also explains our red sunsets.

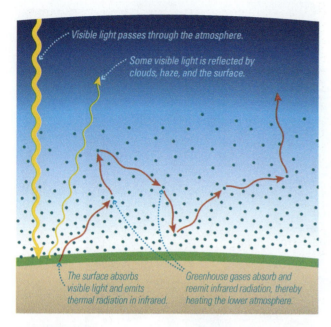

Visible light passes through the atmosphere.

Some visible light is reflected by clouds, haze, and the surface.

The surface absorbs visible light and emits thermal radiation in infrared.

Greenhouse gases absorb and reemit infrared radiation, thereby heating the lower atmosphere.

▲ FIGURE 7.15

The greenhouse effect. The lower atmosphere becomes warmer than it would be if it had no greenhouse gases such as water vapor, carbon dioxide, and methane.

TABLE 7.1	The Greenhouse Effect on Venus, Earth, and Mars		
World	Calculated "No Greenhouse" Average Surface Temperature	Actual Average Surface Temperature*	Greenhouse Warming
Venus	−40°C	470°C	510°C
Earth	−16°C	15°C	31°C
Mars	−56°C	−50°C	6°C

*The "no greenhouse" temperature is calculated by assuming no change to the atmosphere other than lack of greenhouse warming. For example, Venus has a lower "no greenhouse" temperature than Earth even though it is closer to the Sun, because the high reflectivity of its bright clouds means that it absorbs less sunlight than Earth.

The Greenhouse Effect   Visible light warms Earth's surface, but not as much as you might guess. It's easy for scientists to calculate the temperature we'd expect for any world based on its distance from the Sun and the percentage of incoming sunlight that it absorbs (the rest is reflected).

The greenhouse effect keeps Earth's surface much warmer than it would be otherwise, allowing water to stay liquid over most of the surface.

These calculations give correct answers for airless worlds such as the Moon and Mercury, but underestimate the temperatures for worlds with atmospheres. The reason is that certain atmospheric gases can trap heat through what we call the **greenhouse effect.** Table 7.1 shows calculated "no greenhouse" temperatures for Venus, Earth, and Mars, along with actual temperatures. Note, for example, that Earth's temperature would be well below freezing without the greenhouse effect, which means this effect is crucial to the existence of life on Earth.

Figure 7.15 shows the basic idea behind the greenhouse effect. Some of the visible light that reaches the ground is reflected and some is absorbed. The absorbed energy must be returned to space—otherwise the ground would rapidly heat up—and planetary surfaces return energy by emitting infrared light (because planetary surface temperatures are in the range in which their thermal radiation peaks in the infrared [Section 5.2]). The greenhouse effect occurs when the atmosphere temporarily "traps" some of this infrared light, slowing its return to space.

Atmospheric gases that are particularly good at absorbing infrared light are called **greenhouse gases,** and they include water vapor (H_2O), carbon dioxide (CO_2), and methane (CH_4). A gas molecule that has absorbed a photon of infrared light then re-emits it in some random direction. This photon can then be absorbed by another gas molecule, which does the same thing. The net result is that greenhouse gases tend to slow the escape of infrared radiation from the lower atmosphere, while their molecular motions heat the surrounding air. In this way, the greenhouse effect makes the surface and the lower atmosphere warmer than they would be from sunlight alone. The more greenhouse gases present, the greater the degree of surface warming.

think about it   Molecules made from two atoms of the same type—such as the N_2 and O_2 molecules that together make up 98% of Earth's atmosphere—are poor infrared absorbers. But imagine this were not the case. How would Earth be different if nitrogen and oxygen absorbed infrared light?

7.2 The Moon and Mercury: Geologically Dead

In the rest of this chapter, we will investigate the histories of the terrestrial worlds, with the goal of learning how and why Earth became unique. We'll start with the two worlds that have the simplest histories: the Moon and Mercury (Figure 7.16).

◆ Was there ever geological activity on the Moon or Mercury?

The simple histories of the Moon and Mercury are a direct consequence of their small sizes. Both lost most of their internal heat long ago, leaving them without any significant geological activity and without outgassing to supply atmospheric gas. Their small sizes also mean that their gravity is too weak to hold gas they may have gained in the past. This fact combined with the lack of outgassing explains why neither the Moon nor Mercury has a significant atmosphere.

The fact that these worlds have been "geologically dead" for billions of years explains why the major features of their surfaces are impact craters formed during the heavy bombardment. Nevertheless, we expect that both worlds should have had hot interiors early in their histories, and observations confirm that both have at least some volcanic and tectonic features.

Geological Features of the Moon The familiar face of the full moon shows that not all regions of the surface look the same (Figure 7.17). Some regions are heavily cratered while others, known as the **lunar maria,** look smooth and dark. Indeed, the maria (Latin for "seas") got their name because they look much like oceans when seen from afar. Today, we know that the maria are actually the sites of large impacts that were later flooded by lava.

Figure 7.18 shows how the maria probably formed. During the heavy bombardment, craters covered the Moon's entire surface. The largest impacts were violent enough to fracture the Moon's lithosphere beneath the huge craters they created. Although there was no molten rock to flood the craters immediately, the ongoing decay of radioactive elements in the Moon's interior ultimately built up enough heat to cause mantle melting about 3 to 4 billion years ago. Molten rock then welled up through the cracks in the lithosphere, flooding the largest impact craters with lava. The maria are generally circular because they are flooded craters (and craters are almost always round), and they are dark because the lava consisted of dark, iron-rich rock. The relative lack of craters within the maria is a consequence of the fact that the lava floods occurred after the heavy bombardment subsided, and relatively few impacts have occurred since that time.

The Moon's dark, smooth maria were made by floods of lava billions of years ago, when the Moon's interior was heated by radioactive decay.

The Moon's era of geological activity is long gone. Today, the Moon is a desolate and nearly unchanging place (Figure 7.19a). The only ongoing geological change on the Moon is a very slow "sandblasting" of the surface by *micrometeorites,* sand-size particles from space—the same types of particles that burn up as meteors in Earth's atmosphere. The micrometeorites gradually pulverize the surface rock, which explains why the lunar surface is covered

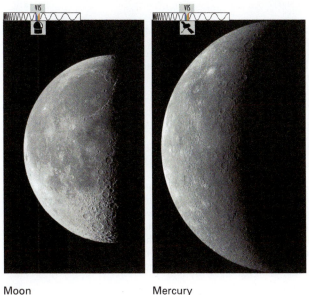

Moon　　　　　　　　Mercury

▲ **FIGURE 7.16**
Views of the Moon and Mercury, shown to scale. The Mercury photo is from *MESSENGER*.

▲ **FIGURE 7.17**
The familiar face of the full Moon shows numerous dark, smooth maria.

   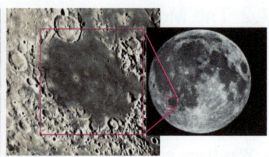

a This illustration shows the Mare Humorum region as it probably looked about 4 billion years ago, when it would have been completely covered in craters.

b Around that time, a huge impact excavated the crater that would later become Mare Humorum. The impact fractured the Moon's lithosphere and erased the many craters that had existed earlier.

c A few hundred million years later, heat from radioactive decay built up enough to melt the Moon's upper mantle. Molten lava welled up through the lithospheric cracks, flooding the impact crater.

d This photo shows Mare Humorum as it appears today, and the inset shows its location on the Moon.

▲ **FIGURE 7.18**

The lunar maria formed between 3 and 4 billion years ago, when molten lava flooded large craters that had formed hundreds of millions of years earlier. This sequence of diagrams represents the formation of Mare Humorum.

a An astronaut takes the Lunar Roving Vehicle for a spin during the final *Apollo* mission to the Moon (*Apollo 17*, December 1972).

b The *Apollo* astronauts left footprints, like this one, in the Moon's powdery "soil." Micrometeorites will eventually erase the footprints, but not for millions of years.

▲ **FIGURE 7.19**

The Moon today is geologically dead, but it can still tell us a lot about the history of our solar system.

by a layer of powdery "soil." You can see this powdery surface in photos from the *Apollo* missions to the Moon (Figure 7.19b). Pulverization by micrometeorites is a very slow process, so the rover tracks and astronaut footprints could remain for millions of years.

Humans have not returned to the Moon in the more than four decades since *Apollo*, but numerous robotic spacecraft have studied the Moon from orbit. These spacecraft have identified numerous small-scale features attesting to volcanic and tectonic activity when the Moon was young. They've also confirmed an important prediction: Water ice, presumably deposited by comet impacts over millions of years, resides in permanently shadowed craters near the lunar poles. Ice was first confirmed in 2009, when scientists sent the rocket from the *LCROSS* spacecraft crashing into a crater near the Moon's south pole so that they could study the debris splashed upward. Not long after, a radar sensor aboard India's *Chandrayaan-1* spacecraft detected ice deposits in similar craters near the Moon's north pole. There is apparently enough water ice to be useful to future colonists, at least if the colonies are built near the lunar poles.

think about it Do you think we should send humans back to the Moon? Why or why not?

Geological Features of Mercury Mercury's crater-covered surface looks so much like the Moon's that it's often difficult to tell them apart in surface photos. Nevertheless, these two worlds have a few important differences.

In many places, Mercury's craters are less crowded together than the craters in the most ancient regions of the Moon, suggesting that molten lava later covered up some of the craters that formed on Mercury during the heavy bombardment (Figure 7.20a). As on the Moon, these lava flows probably occurred when heat from radioactive decay accumulated enough to melt part of the mantle. The largest impact craters, called *basins* (Figure 7.20b), were made by impacts that released enough heat to fill the basins with molten rock. Many of these basins have few craters within them, indicating that they must have formed at a time when the heavy bombardment was already subsiding. However, Mercury does not appear to have any features formed by much later lava flows, like those of the lunar maria.

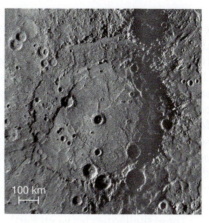

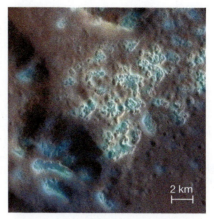

a A close-up view of Mercury's surface, showing impact craters and smooth regions where lava apparently covered up craters.

b The Rembrandt Basin, a large impact crater on Mercury.

c The light-colored "hollows" on this crater floor (shown with colors enhanced) are thought to have formed as easy-to-vaporize materials escaped over millions of years, causing the rock to crumble and make hollowed-out pits.

▲ **FIGURE 7.20**

MESSENGER images of geological features on Mercury.

Although there has been no more volcanism on Mercury for billions of years, its surface still displays one odd form of geological activity: Some crater floors appear to be releasing easily vaporized materials from the rock, causing the rock to crumble and make pits nicknamed "hollows" (see Figure 7.20c). The release of the vaporized gases leaves behind a light-colored coating, whose composition remains unknown. Spacecraft observations (by *MESSENGER*) also show evidence of water ice in permanently shadowed craters near Mercury's poles. As in polar craters on the moon, the ice probably came from impacts of comets.

Mercury's most surprising feature is a set of tremendous cliffs that appear to be distributed all over the planet. These cliffs have vertical faces up to 3 or more kilometers high and typically run for hundreds of kilometers across the surface. They probably formed when tectonic forces compressed the crust, causing the surface to crumple. Because crumpling would have shrunk the portions of the surface it affected, Mercury as a whole could not have stayed the same size unless other parts of the surface expanded. However, we find no evidence of large-scale "stretch marks" on Mercury. Can it be that the whole planet simply shrank?

Apparently so. Recall that in addition to being larger than the Moon, Mercury also has a surprisingly large iron core. Mercury therefore gained and retained more internal heat than the Moon, and this heat caused Mercury's core to swell in size. Later, as the core cooled, it contracted by perhaps

The planet Mercury appears to have shrunk long ago, leaving behind long, steep cliffs.

as much as 20 kilometers in radius (Figure 7.21). The mantle and lithosphere must have contracted along with the core, generating the tectonic stresses that created the great cliffs. The contraction probably also closed off any remaining volcanic vents, ending Mercury's period of volcanism.

Mercury's core and mantle shrank . . .

Some portions of the crust were forced to slide under others.

Today we see long, steep cliffs created by this crustal movement.

. . . causing Mercury's crust to contract.

Not to scale!

a This diagram shows how Mercury's cliffs probably formed as the core shrank and the surface crumpled.

b This cliff extends about 100 kilometers in length, and its vertical face is as much as 2 kilometers tall. (Photo from *Mariner 10*.)

▲ **FIGURE 7.21**

Long cliffs on Mercury offer evidence that the entire planet shrank early in its history, perhaps by as much as 20 kilometers in radius.

7.3 Mars: A Victim of Planetary Freeze-Drying

Based on its size (see Figure 7.1), we expect Mars to be more geologically active than the Moon or Mercury but less active than Earth or Venus. Observations confirm this basic picture, though Mars's greater distance from the Sun—about 50% farther than Earth—has also played a role in its geological history.

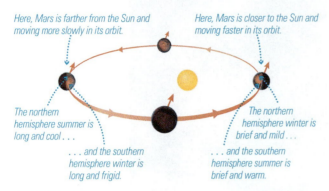

Seasons on Mars

Here, Mars is farther from the Sun and moving more slowly in its orbit.

Here, Mars is closer to the Sun and moving faster in its orbit.

The northern hemisphere summer is long and cool . . .

The northern hemisphere winter is brief and mild . . .

. . . and the southern hemisphere winter is long and frigid.

. . . and the southern hemisphere summer is brief and warm.

▲ **FIGURE 7.22**

The ellipticity of Mars's orbit makes seasons more extreme (warmer summers and colder winters) in the southern hemisphere than in the northern hemisphere.

Mars bears many superficial similarities to Earth. Surface views might easily be mistaken for deserts or volcanic plains on Earth (see Figure 6.6), the Martian day is less than an hour longer than an Earth day, and Mars has polar caps that resemble Earth's, although they contain frozen carbon dioxide in addition to water ice. Mars even has Earth-like seasons as a result of having a similar axis tilt, though its more elliptical orbit puts it significantly closer to the Sun during southern hemisphere summer (and farther from the Sun during southern hemisphere winter), giving its southern hemisphere more extreme seasons (Figure 7.22). In fact, winds associated with seasonal changes sometimes lead to global dust storms that can change the appearance of Martian surface features in telescopic views from Earth. Such changes fooled astronomers of the past into thinking they were seeing seasonal changes in vegetation.

These similarities between Earth and Mars have long made the idea of life on Mars a staple of science fiction, but there are also very important differences. The Martian atmosphere is so thin that the atmospheric pressure is less than 1% of that on the surface of Earth; as a result, you could not survive without a spacesuit. The thin atmosphere also means a weak greenhouse effect, even though carbon dioxide is the main component of the atmosphere. This fact, combined with Mars's distance from the Sun, gives it a global average temperature of about −50°C (−58°F). The lack of oxygen means that Mars lacks an ozone layer, so much of the Sun's damaging ultraviolet radiation passes unhindered to the surface.

All in all, surface conditions on Mars today make it seem utterly inhospitable to life. However, careful study of Martian geology offers evidence that Mars had a warmer and wetter past. If so, it might have had conditions under which life could have arisen, and it's conceivable that such life could still survive under the surface. The search for past or present life is a major reason why we have sent more spacecraft to Mars than to any other planet.

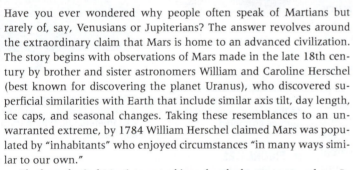

extraordinary claims Martians!

Have you ever wondered why people often speak of Martians but rarely of, say, Venusians or Jupiterians? The answer revolves around the extraordinary claim that Mars is home to an advanced civilization. The story begins with observations of Mars made in the late 18th century by brother and sister astronomers William and Caroline Herschel (best known for discovering the planet Uranus), who discovered superficial similarities with Earth that include similar axis tilt, day length, ice caps, and seasonal changes. Taking these resemblances to an unwarranted extreme, by 1784 William Herschel claimed Mars was populated by "inhabitants" who enjoyed circumstances "in many ways similar to our own."

The hypothetical Martians got a bigger break about a century later. In 1879, Italian astronomer Giovanni Schiaparelli reported seeing a network of linear features on Mars through his telescope. He named these features *canali*, by which he meant the Italian word for "channels," but it was frequently translated as "canals." Motivated by what sounded like evidence of intelligent life, wealthy American astronomer Percival Lowell built an observatory (still operating in Flagstaff, Arizona) for the study of Mars. He published his first maps of the canals barely a year later, and eventually mapped almost 200 canals. Besides claiming them to be the work of an advanced civilization, he also claimed they were needed to transport water from the polar caps to drying cities.

Lowell's claims generated great public interest, but his fellow scientists were far more skeptical. They did not see the canals through their own telescopes, and pointed out that Lowell's straight-line canals made no sense, because real canals would follow natural contours of topography (for example, to go around mountains). Nevertheless, much of the public didn't let go of the Martians until 1965, when NASA's *Mariner 4* spacecraft flew past Mars and sent back photos of a barren, cratered surface.

Even then, not everyone gave up. In the late 1970s, NASA's *Viking* orbiters snapped tens of thousands of photographs of Mars, and among them were a few that looked like artifacts of a civilization, including one that resembled a human face. The "face on Mars" became a cottage industry for those who still wished to believe in Martians, but higher-resolution images from later missions confirmed what every scientist already knew: The face and other supposed artifacts of civilization were nothing more than plays of light and shadow, along with the human tendency to see patterns where none exist. In the end, visions of Martian civilization proved to be no more real than the shapes that children see in clouds, or the patterns among stars that led our ancestors to see the mythical gods in the constellations.

Verdict: Rejected.

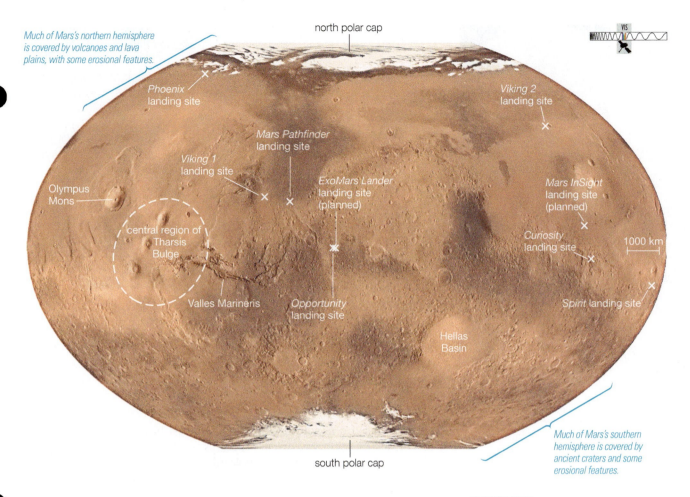

Much of Mars's northern hemisphere is covered by volcanoes and lava plains, with some erosional features.

north polar cap

Phoenix landing site

Viking 2 landing site

Mars Pathfinder landing site

Viking 1 landing site

ExoMars Lander landing site (planned)

Mars InSight landing site (planned)

Olympus Mons

central region of Tharsis Bulge

Curiosity landing site

1000 km

Valles Marineris

Opportunity landing site

Spirit landing site

Hellas Basin

Much of Mars's southern hemisphere is covered by ancient craters and some erosional features.

south polar cap

▲ **FIGURE 7.23**

This image showing the full surface of Mars is a composite made by combining more than 1000 images with more than 200 million altitude measurements from the *Mars Global Surveyor* mission. Several key geological features are labeled, and landing sites of Mars missions are marked. On the same scale, a map of Earth would be about twice as tall and twice as wide, giving it four times the area; however, because Earth's surface is about three-quarters ocean, both planets have nearly the same land area.

◆ What geological features tell us that water once flowed on Mars?

There are no lakes, rivers, or even puddles of liquid water on the surface of Mars today. We know this not only because we've photographed most of the surface in fairly high resolution but also because the surface conditions do not allow it. In most places and at most times, Mars is so cold that any liquid water would immediately freeze into ice. Even when the temperature rises above freezing, as it often does at midday near the equator, the air pressure is so low that liquid water would quickly evaporate. In other words, liquid water is *unstable* on Mars today: If you put on a spacesuit and took a cup of water outside your pressurized spaceship, the water would rapidly either freeze or boil away (or some combination of both). Nevertheless, Mars offers ample geological evidence of past water flows.

The Geology of Mars Figure 7.23 shows a map of the full surface of Mars. Aside from the polar caps, the most striking feature is the dramatic difference in terrain around different parts of the planet. Much of the southern hemisphere has relatively high elevation and is scarred by numerous large impact craters, including the large crater known as the Hellas Basin. In contrast, the northern plains show few impact craters and tend to be below the average Martian surface level. The differences in crater crowding tell us that the southern highlands are a much older

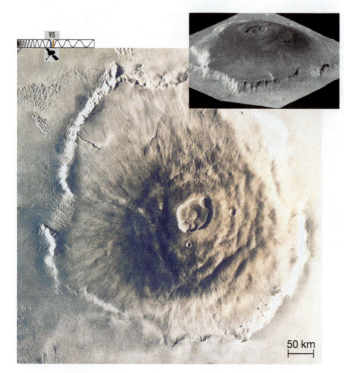

▲ **FIGURE 7.24**

Olympus Mons, photographed from orbit; note the tall cliff around its rim and the central volcanic crater from which lava erupted. The inset shows a 3-D perspective on this immense volcano.

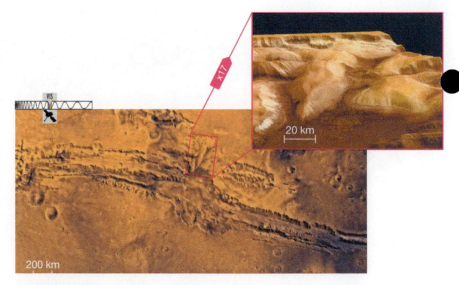

▲ **FIGURE 7.25**

Valles Marineris is a huge valley on Mars created in part by tectonic stresses. The inset shows a perspective view looking north across the center of the canyon, obtained by *Mars Express*.

surface than the northern plains, which must have had their early craters erased by other geological processes.

Mars has clearly had active volcanism in the past, as it is dotted with towering volcanoes. One, called *Olympus Mons* (Figure 7.24), is the tallest known mountain in the solar system. Its peak stands about 26 kilometers above the average Martian surface level—about three times as high as Mount Everest stands above sea level on Earth—and its base is large enough to cover an area the size of Arizona. Much of the base is rimmed by a cliff that in places is 6 kilometers high. Several other large volcanoes are also found on or near the continent-size *Tharsis Bulge*, which was probably created by a long-lived plume of rising mantle material that pushed the surface upward and provided the molten rock for the eruptions that built the giant volcanoes.

Mars has had active volcanism in the past, and its surface is dotted with numerous large volcanoes.

Mars also has tectonic features, of which the most prominent is a long, deep system of valleys called *Valles Marineris* (Figure 7.25). Valles Marineris extends almost a fifth of the way along the planet's equator. It is as long as the United States is wide and almost four times as deep as Earth's Grand Canyon. No one knows exactly how Valles Marineris formed, but its location suggests a link to the Tharsis Bulge. Perhaps it formed because of tectonic stresses accompanying the uplift of material that created Tharsis, cracking the surface and leaving the tall cliff walls of the valleys.

We have not witnessed any recent volcanic or tectonic activity on Mars, but Mars is not yet "geologically dead" like the Moon or Mercury. It probably still retains underground volcanic heat, and it is possible that some of its volcanoes may erupt again someday. Nevertheless, because of Mars's relatively small size, its interior is presumably cooling and its lithosphere thickening, and its era of geological activity will likely come to an end within a few billion years.

Ancient Water Flows Strong evidence that Mars had flowing water in the distant past comes from both orbital and surface studies. For example, Figure 7.26 shows a dry riverbed that was almost certainly carved by running water, although we cannot yet say whether the water came from runoff after rainfall, from erosion by water-rich debris flows, or from an underground source. Regardless of the specific mechanism, the water apparently stopped flowing long ago. Notice that a few impact craters lie on top of the channels. Based on counts of impact craters in and near the river channels, it must have been at least 2–3 billion years since water last flowed through them.

Dried-up riverbeds and other geological features show that water flowed on Mars in the distant past.

Other orbital evidence also argues for rain and surface water in the Martian past. Figure 7.27a shows a broad region of the ancient, heavily cratered southern highlands. Notice the indistinct rims of many large craters and the relative lack of small craters. Both facts argue for ancient rainfall, which would have eroded crater rims and erased small craters altogether. Figure 7.27b shows a three-dimensional perspective of the surface that suggests water once flowed between two ancient crater lakes. Figure 7.27c shows what looks like a river delta where water flowed into an ancient crater; spectra indicate clay minerals on the crater floor, which were probably deposited by sediments flowing down the river. Thousands of other orbital images show additional evidence of water erosion.

Surface studies further strengthen the case for past water. In 2004, the robotic rovers *Spirit* and *Opportunity* landed on nearly opposite sides of Mars (see Figure 7.23). The twin rovers carried cameras, instruments to identify rock composition, and a grinder to expose fresh rock for analysis. The rovers long outlasted their design lifetime of 3 months, with *Opportunity* still going as this book was being written, more than 12 years

▲ **FIGURE 7.26**

This photo, taken by the *Mars Express* orbiter, shows what appears to be a dried-up meandering riverbed, now filled with windblown dust. The small impact craters scattered about the site allow scientists to estimate how long ago the water flowed.

▼ **FIGURE 7.27**
More evidence of past water on Mars.

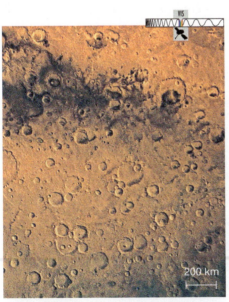

a This photo shows a broad region of the southern highlands on Mars. The eroded rims of large craters and the relative lack of small craters suggest erosion by rainfall.

landing site of *Spirit* rover
100 km

b This computer-generated perspective view shows how a Martian valley forms a natural passage between two possible ancient lakes (shaded blue). Vertical relief is exaggerated 14 times to reveal the topography.

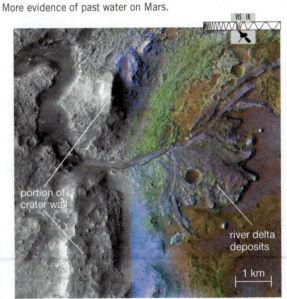

portion of crater wall
river delta deposits
1 km

c Combined visible/infrared image of an ancient river delta that formed where water flowing down a valley emptied into a lake filling a large crater (portions of the crater wall are identified). Clay minerals are identified in green.

Mars (Endurance Crater)   **Earth (northern Arizona)**

▲ **FIGURE 7.28**

"Blueberries" on two planets. In both cases, the foreground shows hematite "blueberries," which formed within sedimentary rock layers like those in the background, then eroded out and rolled downhill; the varying tilts of the rock layers hint at changing winds or waves during formation. The background rocks are about twice as far away from the camera in the Earth photo as in the Mars photo (taken by the *Opportunity* rover).

▼ **FIGURE 7.29**

Evidence of past water in Gale Crater on Mars, from the *Curiosity* rover.

after arrival. The latest rover, *Curiosity* (see Figure 6.6), landed in 2012 at a site called Gale Crater. *Curiosity* carries the most powerful set of scientific instruments ever landed on another world, including cameras, drills, microscopes, rock and soil analyzers, a laser to vaporize rock (see Figure 19.14), and a spectrograph to analyze the vaporized material.

All three rovers have found abundant mineral evidence of past liquid water. For example, rocks at the *Opportunity* landing site contain tiny spheres—nicknamed "blueberries"—composed of minerals (such

Mineral evidence found by robotic rovers provides further indication of liquid water in Mars's past.

as hematite and jarosite) that suggest formation in a salty environment such as a pond or a lake (Figure 7.28). *Curiosity* drove through an ancient streambed, where clumps of pebbles with rounded surfaces and sedimentary layers clearly indicate formation in flowing water (Figure 7.29). Moreover, while the mineral evidence found by *Opportunity* suggests that the region was covered by highly acidic water, chemical analysis of the region studied by *Curiosity* suggests that it was once covered by a lake of relatively pure ("drinkable") water, meaning water like that found in lakes on Earth.

Curiosity is continuing its journey through Gale Crater, and has begun climbing the 5-kilometer-tall Mount Sharp (Figure 7.30). Mount Sharp was chosen as a destination because images from orbit indicate that it contains sedimentary rock layers dating to many different times over the past several billion years. Scientists hope that careful study of these layers by *Curiosity* will help us learn much more about Mars's geological and climatological history, and perhaps even shed light on whether Mars has ever been home to life.

think about it Look for the latest results from *Curiosity*. How far has it traveled? Has it made any new discoveries concerning water or life on Mars?

Martian Water Today If water once flowed over the surface of Mars, where did it all go? As we'll discuss shortly, much of the water was probably lost to space forever. However, significant amounts of water ice still

Mars   **Earth**

a Clumps of rounded pebbles found by *Curiosity* in an ancient streambed on Mars show a structure nearly identical to that found in a typical streambed on Earth, providing strong evidence that the pebbles were rounded by flowing water.

b The even layers of the foreground rocks in this image from *Curiosity* are characteristic of sediment deposited over time in the delta of a river that emptied into a lake.

▲ **FIGURE 7.30**

The view of Mount Sharp in Gale Crater, taken from the Bradbury landing site by NASA's *Curiosity* rover. In the main image, parts of the rover and its shadow are visible in the foreground. The zoom-in on Mount Sharp, more than 20 kilometers away, shows the tilted sedimentary rock layers that scientists hope will reveal whether Mars was once habitable.

remain. Some is found in the polar caps, which are made mostly of water ice overlaid with a thin layer (perhaps a few meters thick) of frozen carbon dioxide. The full extent of Martian water ice is only beginning to become clear, but it now seems that if all the ice melted, it could make an ocean at least about 10 meters deep over the entire planet. It is even possible that some liquid water exists underground near sources of volcanic heat, providing a potential home to microscopic life.

Although we have found no geological evidence to suggest that any large-scale water flows have occurred on Mars in the past half-billion years, orbital photographs offer tantalizing hints of smaller-scale water flows in much more recent times. The strongest evidence comes from photos show-

Streaks on crater walls suggest that water might still occasionally flow on Mars today.

ing dark streaks on crater walls (Figure 7.31). The streaks appear to grow during the warmest times of year—when ice would tend to melt—suggesting that they are created by flowing water. However, no one has yet explained how water or ice might have gotten trapped in the rock layers from which the streaks appear to flow. Other changing features are less clearly connected to liquid water but still demonstrate effects of warming temperature changes on Martian ice. For example, Figure 7.32 shows a massive landslide that occurred as polar ice warmed during a recent spring.

◆ Why did Mars change?

There seems little doubt that Mars had wetter and possibly warmer periods, probably with rainfall, before about 3 billion years ago. The full extent of these periods is a topic of considerable scientific debate. Some scientists suspect that Mars was continuously warm and wet for much of its

▲ **FIGURE 7.31**

This image shows a crater wall with dark streaks that may be created by flowing water. The streaks, which originate below the rock outcrops visible across the center of the image, change from season to season, becoming more prominent in spring and summer. The image, constructed from data from the *Mars Reconnaissance Orbiter*, is shown in a perspective view with enhanced color.

▲ **FIGURE 7.32**

The *Mars Reconnaissance Orbiter* captured this landslide in Mars's north polar region. In a cliff over 700 meters high, layers of dusty ice thawed during the northern spring, causing the landslide.

▲ **FIGURE 7.33**

Mars's northern hemisphere may once have held a vast ocean; this artist's conception shows what it might have looked like some 4 billion years ago.

first billion years of existence, and some evidence even suggests an ocean may have covered much of the northern hemisphere (Figure 7.33). Others think that Mars had only intermittent periods of rainfall, perhaps triggered by the heat of large impacts, and that ancient lakes, ponds, or oceans may have been completely ice-covered. Either way, Mars apparently underwent major and permanent climate change, turning a world that could sustain liquid water into a frozen wasteland.

The idea that Mars once had a thicker atmosphere makes sense, because we would expect that its many volcanoes outgassed plenty of atmospheric gas. Much of this gas should have been water vapor and carbon dioxide, and these greenhouse gases would have warmed the planet. If Martian volcanoes outgassed greenhouse gases in the same proportions as Earth's volcanoes, Mars would have had enough water to fill oceans tens or even hundreds of meters deep.

Early in its history, Mars probably had a dense atmosphere from volcanic outgassing, with a stronger greenhouse effect than it has today.

The bigger question is not whether Mars once had a denser atmosphere, but what happened to it. Mars must have somehow lost most of its carbon dioxide gas. This loss would have weakened the greenhouse effect until the planet essentially froze over. Some of the carbon dioxide condensed and became part of the polar caps, and some may be chemically bound to surface rock. However, the bulk of the gas was probably lost to space.

The precise way in which Mars lost its carbon dioxide gas is not fully settled, but the leading hypothesis suggests a close link to a change in Mars's magnetic field (Figure 7.34). Early in its history, Mars probably had molten convecting metals in its core, much like Earth today. The combination of this convecting metal with Mars's rotation should have produced a magnetic field and a protective magnetosphere. However,

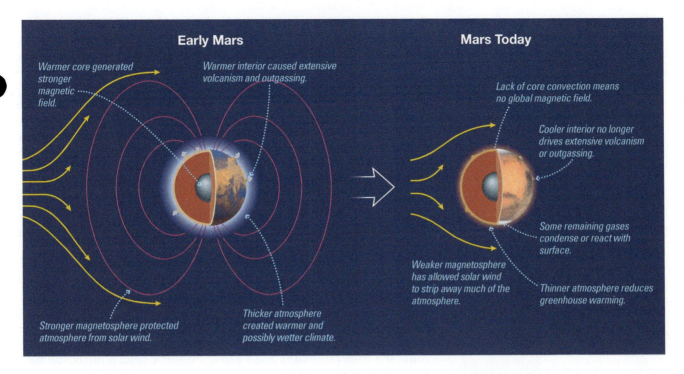

Early Mars

Warmer core generated stronger magnetic field.

Warmer interior caused extensive volcanism and outgassing.

Stronger magnetosphere protected atmosphere from solar wind.

Thicker atmosphere created warmer and possibly wetter climate.

Mars Today

Lack of core convection means no global magnetic field.

Cooler interior no longer drives extensive volcanism or outgassing.

Some remaining gases condense or react with surface.

Weaker magnetosphere has allowed solar wind to strip away much of the atmosphere.

Thinner atmosphere reduces greenhouse warming.

▲ **FIGURE 7.34**
Some 3 billion years ago, Mars underwent dramatic climate change, ensuring that rain could never fall again.

the magnetic field would have weakened as the small planet cooled and core convection ceased, leaving atmospheric gases vulnerable to solar wind particles. More specifically, the hypothesis suggests that carbon dioxide molecules were dissociated into carbon and oxygen atoms by sunlight or chemical processes, and the resulting atoms were then stripped away by the solar wind.

This hypothesis is being put to the test by the *MAVEN* mission, which has been orbiting Mars since 2014 and is measuring the escape of gases from Mars's atmosphere today. The first two panels of Figure 7.35 show results demonstrating that Mars is indeed losing the carbon and oxygen atoms that once made up carbon dioxide molecules, supporting the idea that Mars has lost substantial amounts of carbon dioxide gas through this process over the past few billion years.

▼ **FIGURE 7.35**
These ultraviolet images from NASA's *MAVEN* spacecraft show carbon, oxygen, and hydrogen atoms (which came from dissociated carbon dioxide and water molecules) in the Martian atmosphere. The red circle shows the size of Mars; black regions indicate a lack of data. Notice that the atomic gases extend high above the surface, where some of the carbon and oxygen is escaping through solar wind stripping, and hydrogen is being lost to thermal escape. Hydrogen extends the highest because light gases move fastest.

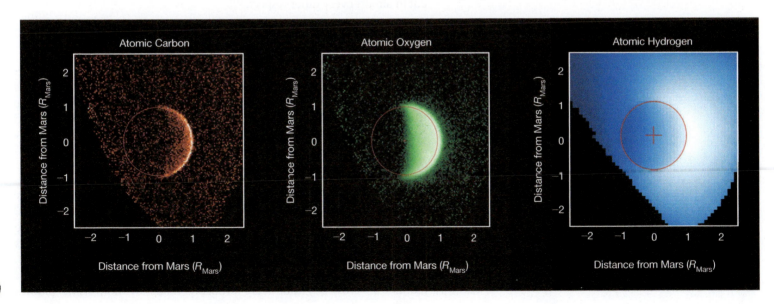

Much of the water once present on Mars is also probably gone for good. Like the carbon dioxide, some water vapor may have been stripped away by the solar wind. However, Mars also lost water in another way. Because the Martian atmosphere lacks ultraviolet-absorbing gases, atmospheric water molecules would have been easily broken apart by ultraviolet photons. The hydrogen atoms that broke away from the water molecules would have been lost rapidly to space. With these hydrogen atoms gone, the water molecules could not be made whole again. Initially, oxygen from the water molecules would have remained in the atmosphere, but over time this oxygen was lost too. Some was probably stripped away by the solar wind, and the rest was drawn out of the atmosphere through chemical reactions with surface rock. This process literally rusted the Martian rocks, giving the "red planet" its distinctive tint.

Mars underwent permanent climate change about 3 billion years ago, when it lost much of its atmospheric carbon dioxide and water to space.

In summary, Mars's fate was probably sealed by its relatively small size. It was big enough for volcanism and outgassing to release water and atmospheric gas early in its history, but too small to maintain the internal heat needed to keep this water and gas. As its interior cooled, its volcanoes quieted and it lost its magnetic field, allowing gas to be stripped away to space. If Mars had been as large as Earth, so that it could still have outgassing and a global magnetic field, it might still have a moderate climate today.

think about it Could Mars still have flowing water today if it had formed closer to the Sun? Use your answer to explain how Mars's fate has been shaped not just by its size, but also by its distance from the Sun.

7.4 Venus: A Hothouse World

On the basis of size alone, we would expect Venus and Earth to be quite similar: Venus is only about 5% smaller than Earth in radius (see Figure 7.1), and its overall composition is about the same as that of Earth. However, as we saw in our planetary tour [Section 6.1], the surface of Venus is a searing hothouse, quite unlike the surface of Earth. In this section, we'll investigate how a planet so similar in size and composition to Earth ended up so different in almost every other respect.

◆ Is Venus geologically active?

Venus's thick cloud cover prevents us from seeing through to its surface, but we can study its geological features with *radar mapping*, which bounces radio waves off the surface and uses the reflections to create three-dimensional images. Figure 7.36 shows a global map and selected close-ups made with radar mapping from the *Magellan* spacecraft. Careful study suggests that Venus is geologically active, just as we would expect for a planet almost as large as Earth.

Geological Features of Venus The images in Figure 7.36 show many geological features similar to Earth's, including occasional impact craters, volcanoes, and evidence of surface contortions caused by tectonic forces.

Venus shows features of volcanism and tectonics, just as we expect for a planet of similar size to Earth.

Venus also has some unique features, such as the large, circular *coronae* (Latin for "crowns") that were probably made by hot, rising plumes

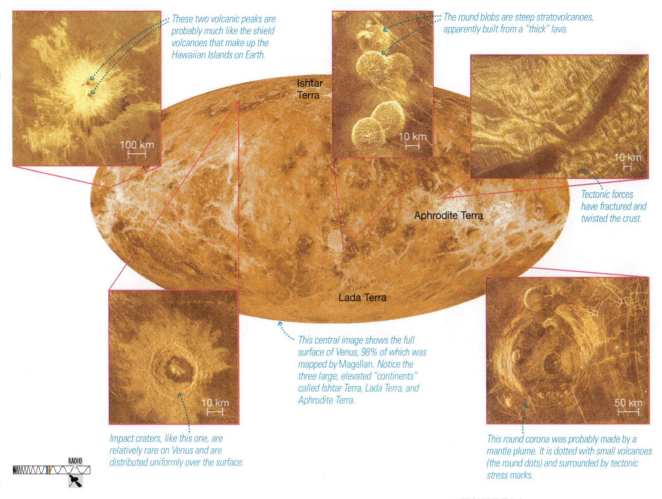

These two volcanic peaks are probably much like the shield volcanoes that make up the Hawaiian Islands on Earth.

100 km

The round blobs are steep stratovolcanoes, apparently built from a "thick" lava.

Ishtar Terra

10 km

Tectonic forces have fractured and twisted the crust.

10 km

Aphrodite Terra

Lada Terra

This central image shows the full surface of Venus, 98% of which was mapped by Magellan. Notice the three large, elevated "continents" called Ishtar Terra, Lada Terra, and Aphrodite Terra.

10 km

Impact craters, like this one, are relatively rare on Venus and are distributed uniformly over the surface.

RADIO

This round corona was probably made by a mantle plume. It is dotted with small volcanoes (the round dots) and surrounded by tectonic stress marks.

50 km

▲ **FIGURE 7.36**

The global map and close-ups show Venus's surface as revealed through radar mapping with the *Magellan* spacecraft, which operated from 1990 to 1993 and had resolution good enough to "see" features as small as 100 meters across. Bright regions represent rough areas or higher altitudes.

of mantle rock. These plumes probably also forced lava to the surface, explaining the volcanoes found near coronae.

Venus almost certainly remains geologically active today, since it should still retain nearly as much internal heat as Earth. Its relatively few impact craters confirm that its surface is geologically young. In addition, the composition of Venus's clouds suggests that volcanoes must still be active on geological time scales (erupting within the past 100 million years). The clouds contain sulfuric acid, which is made from sulfur dioxide (SO_2) and water released into the atmosphere by volcanic outgassing. Because sulfur dioxide is gradually removed from the atmosphere by chemical reactions with surface rocks, the existence of sulfuric acid clouds means that volcanic outgassing must have occurred within about the past 100 million years. Observations from the European Space Agency's *Venus Express* spacecraft (which orbited Venus from 2006 until early 2015) narrow the timeline much more. *Venus Express* detected an infrared spectral feature from rocks on three volcanoes that suggests they erupted within about the past 250,000 years (Figure 7.37). The Japanese *Akatsuki* spacecraft, which entered Venus orbit in 2015, is conducting further studies of Venus's atmosphere and searching for signs of volcanic activity.

Venus lacks significant erosion despite its thick atmosphere. The former Soviet Union sent several landers to Venus in the 1970s and early 1980s. Before the intense surface heat destroyed them, the probes returned images of a bleak, volcanic landscape with little evidence of

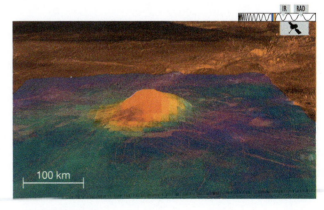

IR RAD

100 km

▲ **FIGURE 7.37**

This composite image shows a volcano called Idunn Mons on Venus. Surface topography details are from NASA's *Magellan* radar mapper (enlarged about 30 times to make the volcano easier to see), and colors represent infrared data from the *Venus Express* spacecraft. Red colors indicate relatively new rock that has not been chemically altered by Venus's harsh atmosphere, suggesting that lava flows occurred within the past 250,000 years.

▲ FIGURE 7.38

This photo from one of the former Soviet Union's *Venera* landers shows Venus's surface; part of the lander is in the foreground, with sky in the background. Many volcanic rocks are visible, hardly affected by erosion despite their presumed age of about 750 million years (the age of the entire surface).

erosion (Figure 7.38). We can trace the lack of erosion on Venus to two facts. First, Venus is far too hot for any type of rain or snow on its surface. Second, Venus has virtually no wind or weather because its slow rotation—once every 243 days—means that its atmosphere barely stirs the surface.

The Absence of Plate Tectonics   We can easily explain the lack of erosion on Venus, but another "missing feature" of its geology is more surprising: Venus shows no evidence of Earth-like plate tectonics. As we'll discuss in the next section, plate tectonics shapes nearly all of Earth's major geological features, including mid-ocean ridges, deep ocean trenches, and long mountain ranges. Venus lacks any similar features.

Instead, Venus shows evidence of a very different type of global geological change. On Earth, plate tectonics resculpts the surface gradually, so different regions have different ages. In contrast, Venus's relatively few impact craters are distributed fairly uniformly over the entire planet, suggesting that the surface is about the same age everywhere. Crater counts suggest a surface age of about 750 million years, leading us to conclude that the entire surface was somehow "repaved" at that time.

The absence of present-day plate tectonics on Venus is a major mystery, though scientists have a few ideas that might explain it. Earth's lithosphere was broken into plates by forces due to the underlying mantle convection. The lack of plate tectonics on Venus therefore suggests either that it has weaker mantle convection or that its lithosphere resists fracturing. The first possibility seems unlikely, because Venus's similarity to Earth in size and density lead us to expect it to have a similar level of mantle convection. Most scientists therefore suspect that Venus's lithosphere resists fracturing into plates because it is thicker and stronger than Earth's.

Venus's lack of Earth-like plate tectonics poses a scientific mystery, but may arise because Venus has a thicker and stronger lithosphere than Earth.

Even if a thicker and stronger lithosphere explains the lack of plate tectonics on Venus, we are still left with the question of why the lithospheres of Venus and Earth should differ. One possible answer invokes Venus's high surface temperature. Venus is so hot that any water in its crust and mantle has probably been baked out over time. Water tends to soften and lubricate rock, so its loss would have thickened and strengthened Venus's lithosphere. If this hypothesis is correct, then Venus might have had plate tectonics if it had not become so hot in the first place.

◆ Why is Venus so hot?

It's tempting to attribute Venus's high surface temperature solely to the fact that it is closer than Earth to the Sun, but Venus's clouds reflect so much sunlight back to space that its surface actually absorbs less sunlight than Earth. As a result, Venus would be quite cold without its strong greenhouse effect: Calculations show that its average surface temperature would be a frigid −40°C (−40°F) in that case, while its actual temperature is about 470°C (880°F). The real question is why Venus has such a strong greenhouse effect.

The simple answer is that Venus has a huge amount of carbon dioxide in its atmosphere—nearly 200,000 times as much as in Earth's atmosphere. However, a deeper question still remains. Given their similar sizes and compositions, we expect Venus and

Venus's thick carbon dioxide atmosphere creates the extremely strong greenhouse effect that makes Venus so hot.

Earth to have had similar levels of volcanic outgassing, and the released gas ought to have had about the same composition on both worlds. Why, then, is Venus's atmosphere so different from Earth's?

The Fate of Outgassed Water and Carbon Dioxide We expect that huge amounts of water and carbon dioxide should have been outgassed into the atmospheres of both Venus and Earth. Venus's atmosphere does indeed have huge amounts of carbon dioxide, but it has virtually no water. Earth's atmosphere has very little of either gas. We conclude that Venus must have somehow lost its outgassed water, while Earth has lost both water vapor and carbon dioxide. But where did these gases go?

We can easily account for both missing gases on Earth. The huge amount of water vapor released into our atmosphere condensed into rain, forming our oceans. In other words, the water is still here, but mostly in liquid rather than gaseous form. The huge amount of carbon dioxide released into our atmosphere is also still here, but in solid form: Carbon dioxide dissolves in water, where it can undergo chemical reactions to make **carbonate rocks** (rocks rich in carbon and oxygen) such as limestone. Earth has about 170,000 times as much carbon dioxide locked up in rocks as in its atmosphere—which means that Earth does indeed have almost as much total carbon dioxide as Venus. Of course, the fact that Earth's carbon dioxide is mostly in rocks rather than in the atmosphere makes all the difference in the world: If this carbon dioxide were in our atmosphere, our planet would be nearly as hot as Venus and certainly uninhabitable.

Earth has as much carbon dioxide as Venus, but it is mostly locked away in rocks rather than in our atmosphere.

We are left with the question of what happened to Venus's water. Venus has no oceans and little atmospheric water, and as noted earlier, any water in its crust and mantle was probably baked out long ago. This absence of water explains why Venus retains so much carbon dioxide in its atmosphere: Without oceans, carbon dioxide cannot dissolve or become locked away in carbonate rocks. If it is true that a huge amount of water was outgassed on Venus, the water molecules have somehow disappeared.

Venus retains carbon dioxide in its atmosphere because it lacks oceans to dissolve the carbon dioxide and lock it away in rock.

The leading hypothesis for the disappearance of Venus's water invokes one of the same processes thought to have removed water from Mars. Ultraviolet light from the Sun breaks apart water molecules in Venus's atmosphere. The hydrogen atoms then escape to space, ensuring that the water molecules can never re-form. The oxygen from the water molecules is lost to a combination of chemical reactions with surface rocks and stripping by the solar wind; Venus's lack of magnetic field leaves its atmosphere vulnerable to the solar wind.

Acting over billions of years, the breakdown of water molecules and the escape of hydrogen can easily explain the loss of an ocean's worth of water from Venus—as long as the water was in the atmosphere rather than in liquid oceans. Our quest to understand Venus's high temperature thereby leads to one more question: Why didn't Venus end up with oceans like Earth, which would have prevented its water from being lost to space?

The Runaway Greenhouse Effect To understand why Venus does not have oceans, let's consider what would happen if we could magically move Earth to the orbit of Venus (Figure 7.39).

If Earth moved to Venus's orbit

More intense sunlight . . .

. . . would raise surface temperature by about 30°C.

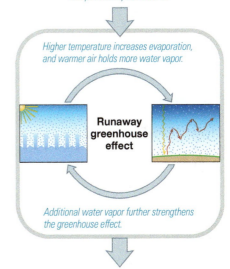

Higher temperature increases evaporation, and warmer air holds more water vapor.

Runaway greenhouse effect

Additional water vapor further strengthens the greenhouse effect.

Result: Oceans evaporate and carbonate rocks decompose, releasing CO_2 . . .

. . . making Earth hotter than Venus.

▲ **FIGURE 7.39**
This diagram shows how, if Earth were placed at Venus's distance from the Sun, the runaway greenhouse effect would cause the oceans to evaporate completely.

The greater intensity of sunlight would almost immediately raise Earth's global average temperature by about 30°C, from its current 15°C to about 45°C (113°F). Although this is still well below the boiling point of water, the higher temperature would lead to increased evaporation of water from the oceans. The higher temperature would also allow the atmosphere to hold more water vapor before the vapor condensed to make rain. The combination of more evaporation and greater atmospheric capacity for water vapor would substantially increase the total amount of water vapor in Earth's atmosphere. Now, remember that water vapor, like carbon dioxide, is a greenhouse gas. The added water vapor would therefore strengthen the greenhouse effect, driving temperatures a little higher. The higher temperatures, in turn, would lead to even more ocean evaporation and more water vapor in the atmosphere—strengthening the greenhouse effect even further. In other words, we'd have a positive feedback loop in which each little bit of additional water vapor in the atmosphere would mean higher temperature and even more water vapor. The process would career rapidly out of control as a **runaway greenhouse effect.**

The runaway process would cause Earth to heat up until the oceans completely evaporated and the carbonate rocks released all their carbon dioxide back into the atmosphere. By the time the process was complete, temperatures on our "moved Earth" would be even higher than they are on Venus today, thanks to the combined greenhouse effects of carbon dioxide and water vapor in the atmosphere. The water vapor would then gradually disappear, as ultraviolet light broke water molecules apart and the hydrogen escaped to space. In short, moving Earth to Venus's orbit would essentially turn our planet into another Venus.

We have arrived at a simple explanation of why Venus is so much hotter than Earth. Even though Venus is only about 30% closer to the Sun than Earth, this difference was enough to be critical. On Earth, it was cool enough for water to rain down to make oceans. The oceans then dissolved carbon dioxide and chemical reactions locked it away in carbonate rocks, leaving our atmosphere with only enough greenhouse gases to make our planet pleasantly warm. On Venus, the greater intensity of sunlight made it just enough warmer that oceans either never formed or soon evaporated, leaving Venus with a thick atmosphere full of greenhouse gases.

Venus is too close to the Sun to have liquid water oceans. Without water to dissolve carbon dioxide gas, Venus was doomed to its runaway greenhouse effect.

The next time you see Venus shining brightly as the morning or evening "star," consider the radically different path it has taken from that taken by Earth—and thank your lucky star. If Earth had formed a bit closer to the Sun or if the Sun had been slightly hotter, our planet might have suffered the same greenhouse-baked fate.

think about it We've seen that moving Earth to Venus's orbit would cause our planet to become Venus-like. If we could somehow move Venus to Earth's orbit, would it become Earth-like? Why or why not?

7.5 Earth as a Living Planet

Now that we have discussed general features of terrestrial planet geology and atmospheres, along with a brief overview of the histories of the other terrestrial worlds, we are ready to explore our own planet Earth in more detail. In this section, we'll discuss the features of Earth that make it so hospitable to life, the ways in which human activity may be affecting this hospitability, and general lessons about what may make some worlds habitable and others not.

◆ What unique features of Earth are important for life?

If you think about all we've learned about the terrestrial worlds, you can probably identify a number of features that are unique to Earth. Four unique features turn out to be particularly important to life on Earth:

- **Surface liquid water:** Earth has a surface temperature and atmospheric pressure that allow liquid water to be stable on its surface.
- **Atmospheric oxygen:** Earth has atmospheric oxygen and an ozone layer.
- **Plate tectonics:** Earth's surface is shaped largely by this distinctive type of tectonics.
- **Climate stability:** Earth's climate has remained stable enough for liquid water to persist for billions of years.

Our Unique Oceans and Atmosphere The first and second items in our list—abundant liquid water and atmospheric oxygen—are clearly important to our existence. Life as we know it requires water [Section 19.1], and animal life requires oxygen. We have already explained the origin of Earth's water: Water vapor outgassed from volcanoes rained down on the surface to make the oceans and neither froze nor evaporated thanks to our moderate greenhouse effect and distance from the Sun. But where did the oxygen in our atmosphere come from?

Oxygen (O_2) is not a product of volcanic outgassing. In fact, no geological process can explain how oxygen came to make up such a large fraction (21%) of Earth's atmosphere. Moreover, oxygen is a highly reactive gas that would disappear from the atmosphere in just a few million years if it were not continuously resupplied. Fire, rust, and the discoloration of freshly cut fruits and vegetables are everyday examples of chemical reactions that remove oxygen from the atmosphere. Similar reactions between oxygen and surface materials give rise to the reddish appearance of much of Earth's rock and clay. So we must explain not only how oxygen got into Earth's atmosphere in the first place, but also how the amount of oxygen remains relatively steady even though chemical reactions remove it rapidly from the atmosphere.

The answer to the oxygen mystery is life itself (Figure 7.40). Plants and many microorganisms release oxygen through photosynthesis. Photosynthesis takes in CO_2 and releases O_2 (while the carbon becomes incorporated into living tissues).
Virtually all Earth's oxygen was originally released into the atmosphere by photosynthetic life. Today, photosynthetic organisms return oxygen to the atmosphere in approximate balance with the rate at which animals and chemical reactions consume it, which is why the oxygen concentration remains relatively steady. This oxygen is also what makes possible Earth's protective ozone layer, since ozone (O_3) is produced from ordinary oxygen (O_2).

Without life, Earth would lack atmospheric oxygen and an ozone layer.

think about it Suppose that, somehow, all photosynthetic life (such as plants) died out. What would happen to the oxygen in our atmosphere? Could animals, including us, still survive?

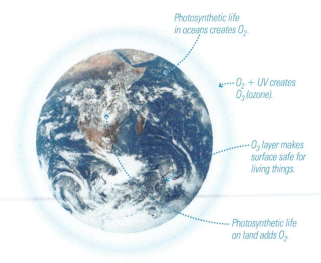

Photosynthetic life in oceans creates O_2.

O_2 + UV creates O_3 (ozone).

O_3 layer makes surface safe for living things.

Photosynthetic life on land adds O_2.

▲ **FIGURE 7.40**
The origin of oxygen and ozone in Earth's atmosphere can be traced to life.

▶ **FIGURE 7.41**

This relief map shows plate boundaries (solid yellow lines), with arrows to represent directions of plate motion. Color represents elevation, progressing from blue (lowest) to red (highest).

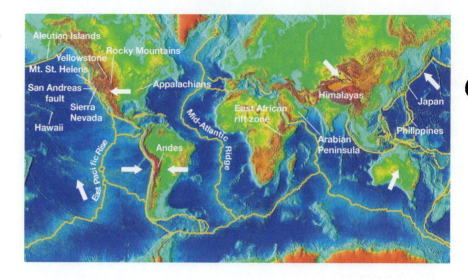

Plate Tectonics

The third and fourth items on our list—plate tectonics and climate stability—turn out to be closely linked. Recall that Earth's lithosphere is broken into more than a dozen plates that slowly move about through the action we call *plate tectonics* (Figure 7.41). The plates

Plate tectonics acts like a giant conveyor belt for Earth's lithosphere, continually recycling the seafloor and building up the continents.

move at average speeds of only a few centimeters per year (about the speed at which your fingernails grow), but over millions of years their motions act like a giant conveyor belt for Earth's lithosphere, creating new crust and recycling old crust back into the mantle (Figure 7.42). Mantle material rises upward and erupts to the surface along *mid-ocean ridges*, becoming new crust for the seafloor (more formally called oceanic crust). This newly emerging material causes the seafloor to spread away from the ridge, which is why the ridges are found in the middle of the ocean. Over tens of millions of years, any piece of seafloor gradually makes its way across the ocean bottom, then finally gets recycled into the mantle.

The seafloor recycling occurs at places where the seafloor descends under continental plates in a process called *subduction*. The descending seafloor crust heats up and may partially melt, with the molten rock then

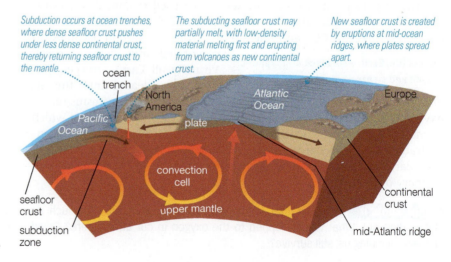

▶ **FIGURE 7.42**

Plate tectonics acts like a giant conveyor belt for Earth's lithosphere.

erupting to the surface. That is why active volcanoes tend to be found along the edges of continents. Moreover, the lowest-density material tends to melt first, so the material emerging from these landlocked volcanoes is much lower in density than seafloor crust. This lower density explains why continents rise above the seafloor. Radiometric dating confirms this picture of plate tectonics. Seafloor crust is never more than a couple of hundred million years old (because it is recycled at the subduction zones) and is younger near the mid-ocean ridges (where it first emerges and solidifies).

In fact, almost all of Earth's active geology is tied to plate tectonics. For example, mountain building occurs where continental plates are pushed together, and valleys or seas can form where continental plates are pulling apart (see Figure 7.11). Earthquakes tend to occur when two plates get "stuck" against one another and then lurch violently when the pressure builds to the breaking point. Even the present arrangement of the continents is due to plate tectonics, since the continents are slowly pushed around as seafloors spread and subduct (Figure 7.43).

think about it Explain why the west coast states of California, Oregon, and Washington are prone to more earthquakes and volcanoes than other parts of the United States. Find the locations of recent earthquakes and volcanic eruptions worldwide. Do the locations fit the pattern you expect?

Climate Stability

Earth's long-term climate stability has clearly been important to the ongoing evolution of life—and hence to our own relatively recent arrival as a species (see Figure 1.10). Had our planet undergone a runaway greenhouse effect like Venus, life would certainly have been extinguished. If Earth had suffered loss of atmosphere and a global freezing like Mars, any surviving life would have been driven to hide in underground pockets of liquid water.

Earth's climate is not perfectly stable—our planet has endured numerous ice ages and warm periods in the past. Nevertheless, even in the deepest ice ages and warmest warm periods, Earth's temperature has remained in a range in which some liquid water could still exist and harbor life. This long-term climate stability is even more remarkable, because models suggest that the Sun has brightened substantially (about 30%) over the past 4 billion years, yet Earth's temperature has managed to stay in nearly the same range throughout this time. Apparently, the strength of the greenhouse effect self-adjusts to keep the climate stable.

The mechanism by which Earth self-regulates its temperature is called the **carbon dioxide cycle,** or the **CO_2 cycle** for short. As shown in Figure 7.44, carbon dioxide enters the atmosphere through outgassing, then returns to Earth's interior by dissolving in water and forming carbonate rock on the seafloor that ultimately subducts back into the mantle. The CO_2 cycle acts as a long-term thermostat for Earth, because the overall rate at which carbon dioxide is pulled from the atmosphere is very sensitive to temperature: The higher the temperature, the higher the rate at which carbon dioxide is removed.

Earth has remained habitable for billions of years because its climate is kept stable by the natural action of the carbon dioxide cycle.

To see how this self-regulation occurs, consider what happens if Earth warms up a bit. The warmer temperature means more evaporation and rainfall, pulling more CO_2 out of the atmosphere. The reduced atmospheric CO_2 concentration leads to a weakened greenhouse effect that counteracts the initial warming and cools the planet back down. Similarly, if Earth cools a bit, rainfall decreases and less CO_2 is dissolved, allowing the CO_2 released by volcanism to build back up in the atmosphere. The

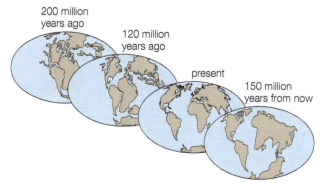

▲ **FIGURE 7.43**

Past, present, and future arrangements of Earth's continents. Notice how the present shapes of South America and Africa reflect the way they fit together in the past. The continents are always in motion, so the arrangement looked different at earlier times and will look different in the future.

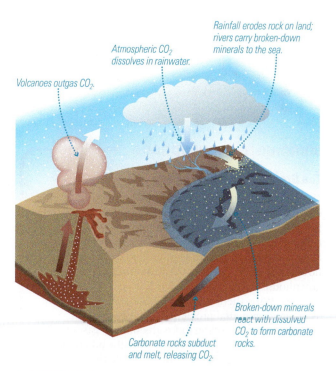

▲ **FIGURE 7.44**

This diagram shows how the CO_2 cycle continually moves carbon dioxide from the atmosphere to the ocean to rock and back to the atmosphere. Notice that plate tectonics (subduction in particular) plays a crucial role in the cycle.

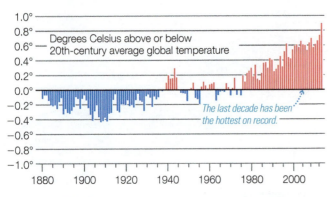

▲ FIGURE 7.45

Average global temperatures from 1880 through 2015. Notice the clear global warming trend of the past few decades. (Data from the National Climate Data Center.)

increased CO_2 concentration strengthens the greenhouse effect and warms the planet back up.

We can now see why plate tectonics is so intimately connected to our existence. Plate tectonics is a crucial part of the CO_2 cycle; without plate tectonics, Earth would lack a mechanism for self-regulating its climate, and Earth might then have undergone changes as dramatic as those that occurred on Venus and Mars. Note, however, that the role of plate tectonics also means that this climate regulation occurs only on long time scales of hundreds of thousands to millions of years. It therefore has no effect on short-term changes, such as those we will discuss next.

◆ How is human activity changing our planet?

Our planet may regulate its own climate quite effectively over long time scales, but fossil and geological evidence tells us that substantial and rapid swings in global climate can occur on shorter ones. For example, Earth cycles in and out of ice ages on geologically short time scales of tens of thousands of years, and in some cases the geological evidence indicates that the climate warmed several degrees Celsius in just decades. Past climate changes have obviously been due to natural causes, but Earth is now undergoing climate change for a new reason: Human activity is rapidly increasing the atmospheric concentration of greenhouse gases, particularly of carbon dioxide. Effects of this increase in greenhouse gas concentration are apparent already: Global average temperatures have risen by about 0.85°C (1.5°F) in the past century (Figure 7.45). This **global warming** is one of the most important issues of our time.

Global Warming Global warming has been a hot political issue, both because some people debate its cause and because efforts to slow or stop the warming would require finding new energy sources and making other changes that could dramatically affect the world's economy. However, a major research effort has gradually added to our understanding of the potential threat, particularly in the past two decades. The case linking global warming with human activity rests on three basic facts:

Human activity is rapidly increasing the atmospheric concentrations of greenhouse gases, causing the global average temperature to rise.

1. The greenhouse effect is a simple and well-understood scientific model. We can be confident in our understanding of it because it so successfully explains the observed surface temperatures of other planets. Given this basic model, there is no doubt that a rising concentration of greenhouse gases would make our planet warm up more than it would otherwise; the only debate is about how soon and how much.

2. Human activity such as the burning of fossil fuels is clearly increasing the amounts of greenhouse gases in the atmosphere. Observations show that the atmospheric concentration of carbon dioxide is now about 40% higher than it was before the industrial revolution began or at any other time during the past million years, and it is continuing to rise rapidly (Figure 7.46). We can be confident that the rise is a result of human activity, because the atmosphere is becoming enriched in molecules of CO_2 carrying the distinct ratio of isotopes present in fossil fuels.

3. Climate models that ignore human activity fail to match the observed rise in global temperatures. In contrast, climate models that

common misconceptions

The Greenhouse Effect Is Bad

The greenhouse effect is often in the news, usually in discussions about environmental problems, but in itself the greenhouse effect is not a bad thing. In fact, we could not exist without it, since it is responsible for keeping our planet warm enough for liquid water to flow in the oceans and on the surface. The "no greenhouse" temperature of Earth is well below freezing. Why, then, is the greenhouse effect discussed as an environmental problem? The reason is that human activity is adding more greenhouse gases to the atmosphere—and scientists agree that the additional gases are changing Earth's climate. While the greenhouse effect makes Earth livable, it is also responsible for the searing 470°C temperature of Venus—proving that it's possible to have too much of a good thing.

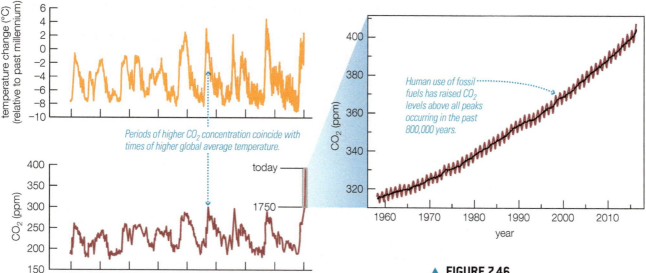

Periods of higher CO₂ concentration coincide with times of higher global average temperature.

Human use of fossil fuels has raised CO₂ levels above all peaks occurring in the past 800,000 years.

▲ **FIGURE 7.46**

This diagram shows the atmospheric concentration of carbon dioxide and global average temperature over the past 800,000 years. Data for the past few decades come from direct measurements; earlier data come from studies of air bubbles trapped in Antarctic ice (ice core samples). The CO₂ concentration is measured in parts per million (ppm), which is the number of CO₂ molecules among every 1 million air molecules.

include the enhanced greenhouse effect from human production of greenhouse gases match the observed temperature trend quite well (Figure 7.47), providing additional support for the idea that global warming results from human activity.

These facts offer convincing evidence that we humans are now tinkering with the climate in a way that may cause major changes not just in the distant future, but in our own lifetimes. The same models that convince scientists of the reality of human-induced global warming tell us that if current trends in the greenhouse gas concentration continue—that is, if we do nothing to slow our emissions of carbon dioxide and other greenhouse gases—the warming trend will continue to accelerate. By the end of this century, the global average temperature would be 2°C–5°C (4°F–10°F) higher than it is now, giving our children and grandchildren the warmest climate that any generation of *Homo sapiens* has ever experienced.

think about it Based on the rate of rise in the CO₂ concentration shown in Figure 7.46, how long will it be until we reach a doubling of the pre-industrial-age concentration of 280 ppm? Discuss the implications of your answer.

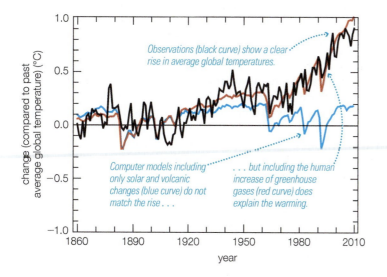

Observations (black curve) show a clear rise in average global temperatures.

Computer models including only solar and volcanic changes (blue curve) do not match the rise . . .

. . . but including the human increase of greenhouse gases (red curve) does explain the warming.

◀ **FIGURE 7.47**

This graph compares observed temperature changes (black curve) with the predictions of climate models that include only natural factors, such as changes in the brightness of the Sun and effects of volcanoes (blue curve) and models that also include the human contribution to increasing greenhouse gas concentration (red curve). Only the red curve matches the observations well. (The red and blue model curves are averages of many scientists' independent models of global warming, which generally agree with each other within 0.1°C–0.2°C.)

Regional Temperature Change: Average for 2011–2015 Compared to Average 1951–1980

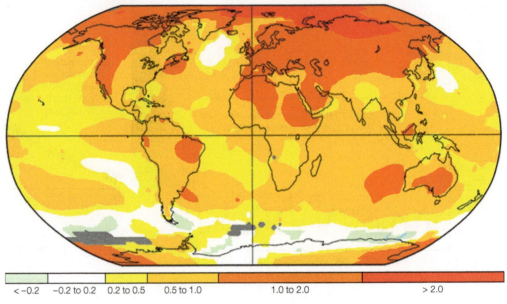

| < −0.2 | −0.2 to 0.2 | 0.2 to 0.5 | 0.5 to 1.0 | 1.0 to 2.0 | > 2.0 |

temperature difference (°C)

Consequences of Global Warming A temperature increase of a few degrees might not sound so bad, but small changes in *average* temperature can lead to much more dramatic changes in climate patterns. Figure 7.48 compares recent regional temperatures to the averages from 1951–1980. Notice that almost all regions of the world are now significantly warmer than they were a few decades ago, but some regions warmed much more than others. This is what we mean by "climate change"—the idea that regional effects can be very different from the global average. Models suggest that regional climate changes will be further amplified as the global average temperature continues to increase.

The consequences of global warming are not simply hotter weather. Changes in rainfall patterns are also expected; for example, many scientists suspect that recent drought and wildfires in the United States and Canada are a result of the changing climate, in which case we can expect these regions to experience more periods of prolonged drought and more wildfires in the future. The warmer weather is also causing spring to arrive earlier, which can further contribute to drier summers and wildfire risk. Moreover, the general warming of the atmosphere means more total energy in the atmosphere and increased evaporation from the oceans, leading to more intense storms—perhaps including more frequent or more severe hurricanes. The same phenomena can also increase the severity of winter storms, leading to the somewhat surprising fact that global warming can cause more severe winter blizzards, a pattern that we may already be seeing in recent winters in the eastern United States.

Global warming is also expected to cause a rise in sea level. The oceans expand very slightly as they warm, an effect that has already caused sea level to rise about 20 centimeters in the past century and is expected to cause a rise of another 30 centimeters by about 2100. Melting ice may have a much greater effect. While the melting of ice in the

Arctic Ocean does not alter sea level—it is already floating—melting of landlocked ice does. Recent data suggest a surprisingly rapid change in Greenland's ice sheet, leading some scientists to worry that ice melt could cause sea level to rise as much as several *meters*—enough to flood much of Florida—by the end of this century. The oceans are also suffering in another way. Some of the carbon dioxide released into the atmosphere by human activity (roughly 30% of it) ends up dissolving in the oceans, where it undergoes chemical reactions that make the oceans more acidic. This "ocean acidification" has been tied to the demise of many coral reefs around the world and to less productive fisheries, and even reduces the ability of the oceans to absorb more carbon dioxide.

Figure 7.49 summarizes the evidence for and consequences of global warming. Fortunately, most scientists believe that we still have time to avert the most serious consequences of global warming, provided we dramatically and rapidly curtail our greenhouse gas emissions. The most obvious way to cut back on these emissions is to improve energy efficiency. For example, doubling the average gas mileage of cars would immediately cut automobile-related carbon dioxide emissions in half. Other tactics could include replacing fossil fuels with alternative energy sources—such as biofuel, solar, wind, or nuclear energy—or finding ways to capture and bury the carbon dioxide from the fossil fuels that we still use. The key idea to keep in mind is that global warming is a problem of our own making, and it is therefore up to us to find a way to solve it.

think about it If you were a political leader, how would *you* deal with the threat of global warming?

---

extraordinary claims | Human Activity Can Change the Climate

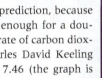

The topic of global warming and associated climate changes is so commonplace in today's news that it's tempting to think of it as a relatively new idea. But the first clear claim that human emissions of carbon dioxide might change the climate dates back to 1896, when Swedish scientist Svante Arrhenius was contemplating possible causes of ice ages, which had recently been identified in the geological record.

The greenhouse effect of gases like carbon dioxide and water vapor had already been discovered, and Arrhenius decided to check whether changes in the carbon dioxide concentration might trigger ice ages. Calculating how a change in carbon dioxide concentration will change the temperature requires complex calculations of the many feedbacks involved. Today scientists do such calculations with detailed computer models, but Arrhenius had only pencil and paper. Nevertheless, after months of tedious calculations, he concluded that a 50% drop in the carbon dioxide concentration would cause Earth's temperature to drop about 5°C—which was indeed enough to bring about an ice age—while doubling it would cause a rise of about 5°C. Moreover, with the help of a colleague, he made an astonishing discovery: Human activities such as the burning of coal were already adding about as much carbon dioxide to the atmosphere each year as volcanoes. This insight led Arrhenius to make an extraordinary claim: Human activity might eventually raise the carbon dioxide concentration enough to cause significant global warming.

Arrhenius was not particularly concerned by his prediction, because he did not think we could add carbon dioxide fast enough for a doubling to occur in less than several centuries. But the rate of carbon dioxide emissions grew rapidly. In 1958, scientist Charles David Keeling began making the measurements shown in Figure 7.46 (the graph is often called the *Keeling curve* in his honor). These measurements, along with measurements of isotope ratios that show the added carbon dioxide comes from human activity, leave no doubt that we are causing a substantial rise in the carbon dioxide concentration.

The only question that remained was whether this increase would really cause global warming, as Arrhenius had predicted. As discussed in the chapter, modern computer models do an excellent job of reproducing past climate, and while there are uncertainties in the precise values, these models generally agree with Arrhenius's prediction that a doubling of the carbon dioxide concentration should cause the global average temperature to rise by a few degrees Celsius. This fact, along with data showing that global warming is already well under way, explain why Arrhenius's extraordinary claim is now well accepted by the vast majority of scientists.

Verdict: Strongly supported.

Scientific studies of global warming apply the same basic approach used in all areas of science: We create models of nature, compare the predictions of those models with observations, and use our comparisons to improve the models. We have found that climate models agree more closely with observations if they include human production of greenhouse gases like carbon dioxide, making scientists confident that human activity is indeed causing global warming.

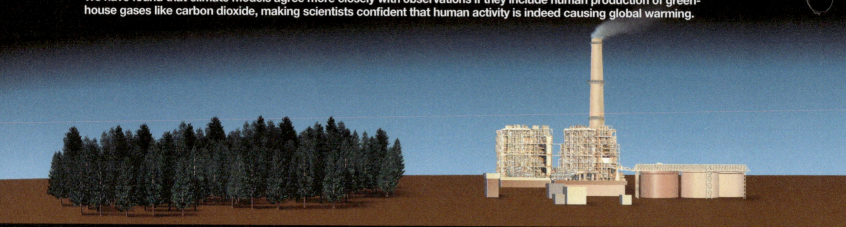

1 The greenhouse effect makes a planetary surface warmer than it would be otherwise because greenhouse gases such as carbon dioxide, methane, and water vapor slow the escape of infrared light radiated by the planet. Scientists have great confidence in models of the greenhouse effect because they successfully predict the surface temperatures of Venus, Earth, and Mars.

2 Human activity is adding carbon dioxide and other greenhouse gases to the atmosphere. While the carbon dioxide concentration also varies naturally, its concentration is now much higher than it has been at any time in the previous million years, and it is continuing to rise rapidly.

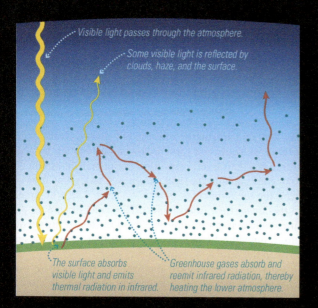

Visible light passes through the atmosphere.

Some visible light is reflected by clouds, haze, and the surface.

The surface absorbs visible light and emits thermal radiation in infrared.

Greenhouse gases absorb and reemit infrared radiation, thereby heating the lower atmosphere.

The graph shows that today's CO$_2$ levels are higher than at any point in the past 800,000 years.

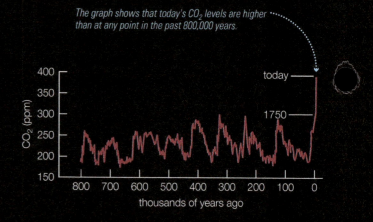

Global Average Surface Temperature

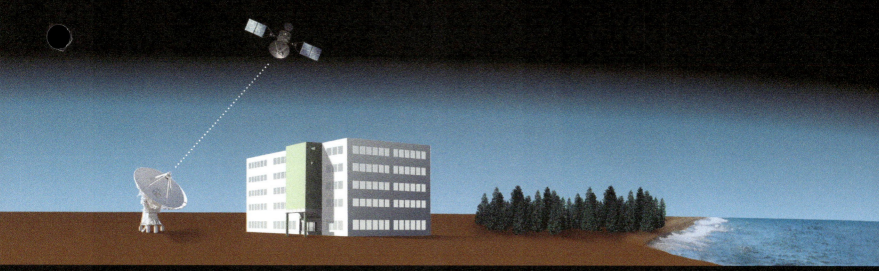

3 Observations show that Earth's average surface temperature has risen during the last several decades. Computer models of Earth's climate show that an increased greenhouse effect triggered by CO_2 from human activities can explain the observed temperature increase.

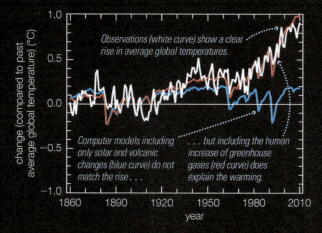

Observations (white curve) show a clear rise in average global temperatures.

Computer models including only solar and volcanic changes (blue curve) do not match the rise . . .

. . . but including the human increase of greenhouse gases (red curve) does explain the warming.

HALLMARK OF SCIENCE **Science progresses through creation and testing of models of nature that explain the observations as simply as possible.** Observations showing a rise in Earth's temperature demand a scientific explanation. Models that include an increased greenhouse effect due to human activity explain those observations better than models without human activity.

4 Models can also be used to predict the consequences of a continued rise in greenhouse gas concentrations. These models show that, without significant reductions in greenhouse gas emissions, we should expect further increases in global average temperature, rising sea levels, and more intense and destructive weather patterns.

This diagram shows the change in Florida's coastline that would occur if sea levels rose by 1 meter. Some models predict that this rise could occur within a century. The light blue regions show portions of the existing coastline that would be flooded.

The Role of Planetary Size

Small Terrestrial Planets

Interior cools rapidly . . .

. . . so that tectonic and volcanic activity cease after a billion years or so. Many ancient craters therefore remain.

Lack of volcanism means little outgassing, and low gravity allows gas to escape more easily; no atmosphere means no erosion.

Large Terrestrial Planets

Warm interior causes mantle convection . . .

. . . leading to ongoing tectonic and volcanic activity; most ancient craters have been erased.

Outgassing produces an atmosphere and strong gravity holds it, so that erosion is possible.

Core may be molten, producing a magnetic field if rotation is fast enough, and a magnetosphere that can shield an atmosphere from the solar wind.

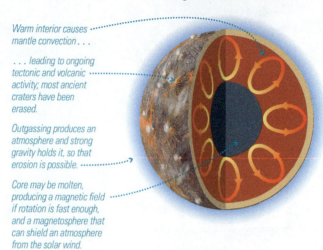

The Role of Distance from the Sun

Planets Close to the Sun

Surface is too hot for rain, snow, or ice, so little erosion occurs.

High atmospheric temperature allows gas to escape more easily.

Planets at Intermediate Distances from the Sun

Moderate surface temperatures can allow for oceans, rain, snow, and ice, leading to substantial erosion.

Gravity can more easily hold atmospheric gases.

Planets Far from the Sun

Low surface temperatures can allow for ice and snow, but no rain or oceans, limiting erosion.

Atmosphere may exist, but gases can more easily condense to make surface ice.

Sun

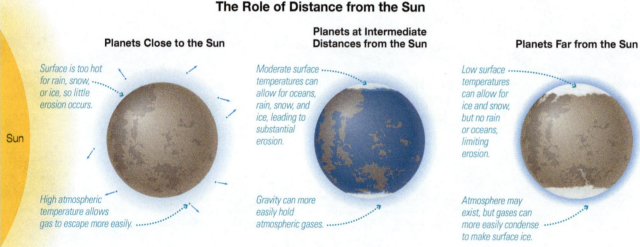

▲ **FIGURE 7.50**

This illustration shows how a terrestrial world's size and distance from the Sun help determine its geological history and whether it has conditions suitable for life. Earth is habitable because it is large enough and at a suitably moderate distance from the Sun.

◆ What makes a planet habitable?

We have discussed the features of our planet that have made it habitable for a great variety of life, including us. But why is Earth the only terrestrial world that has these features?

Our comparative study of the terrestrial worlds tells us there are two primary answers, both summarized in Figure 7.50. First, Earth is habitable because it is large enough to have remained geologically active since its birth, so that outgassing could release the water and gases that formed our atmosphere and oceans. In addition, the core has remained hot enough that Earth has retained a global magnetic field, which generates a magnetosphere that protects our atmosphere from the solar wind. Second, we are located at a distance from the Sun at which outgassed water vapor was able to

Earth is habitable because it is large enough to remain geologically active and located at a distance from the Sun where oceans were able to form.

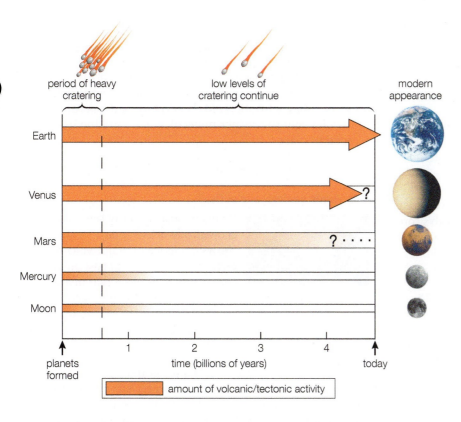

Earth

Venus

Mars

Mercury

Moon

period of heavy cratering

low levels of cratering continue

modern appearance

?

? · · · ·

planets formed

time (billions of years)

today

1   2   3   4

amount of volcanic/tectonic activity

◄ **FIGURE 7.51**

This diagram summarizes the geological histories of the terrestrial worlds. The brackets along the top indicate that impact cratering has affected all worlds similarly. The arrows represent volcanic and tectonic activity. A thicker and darker arrow means more volcanic/tectonic activity, and the arrow length tells us how long this activity persisted. Notice the trend's relationship to planetary size: Earth remains active to this day; Venus has also been active, though we are uncertain whether it remains so; Mars has had an intermediate level of activity and might still have low-level volcanism; Mercury and the Moon have had very little volcanic/tectonic activity. Erosion is not shown, because it has played a significant role only on Earth (ongoing) and on Mars (in the past and at low levels today).

condense and rain down to form oceans, making possible the carbon dioxide cycle that regulates our climate.

Figure 7.51 shows the trends we've seen for the terrestrial planets in our solar system, which should make sense as you examine the role of planetary size. In principle, these lessons mean we can now predict the geological and atmospheric properties of terrestrial worlds around other stars. Only a suitably large terrestrial planet located at an intermediate distance from its star is likely to have conditions under which life could thrive. We will discuss these conditions—along with possibilities for life—further in Chapter 19.

the big picture   Putting Chapter 7 into Perspective

In this chapter, we have explored the histories of the terrestrial worlds. As you think about the details you have learned, keep the following "big picture" ideas in mind:

• The terrestrial worlds all looked much the same when they were born, so their present-day differences are a result of geological processes that occurred in the ensuing 4½ billion years.

• The primary factor in a terrestrial world's geological history is its size: Only a relatively large world can retain internal heat long enough for ongoing geological activity. However, distance from the Sun also plays an important role in determining whether temperatures will allow for liquid water.

• Although distance from the Sun helps determine surface temperature, the strength of the greenhouse effect can play an even bigger role. Humans are currently altering the balance of greenhouse gases in Earth's atmosphere, with potentially dire consequences.

• The histories of Venus and Mars show that a stable climate like Earth's is more the exception than the rule. The stable climate that makes our existence possible is a direct consequence of our planet's unique geology, including its plate tectonics and carbon dioxide cycle.

my cosmic perspective Our presence on Earth has been made possible both by the obvious "hazards" of a geologically active planet, and the less visible mechanisms that render Earth's climate stable.

7.1 Earth as a Planet

◆ Why is Earth geologically active?

Internal heat drives geological activity, and Earth retains internal heat because of its relatively large size for a terrestrial world. This heat causes mantle **convection** and keeps Earth's **lithosphere** thin, ensuring active surface geology. It also keeps part of Earth's core melted, and circulation of this molten metal creates Earth's magnetic field.

◆ What processes shape Earth's surface?

The four major geological processes are **impact cratering, volcanism, tectonics,** and **erosion.** Earth has experienced many impacts, but most craters have been erased by other processes. We owe the existence of our atmosphere and oceans to volcanic **outgassing.** A special type of tectonics—**plate tectonics**—shapes much of Earth's surface. Ice, water, and wind drive rampant erosion on our planet.

◆ How does Earth's atmosphere affect the planet?

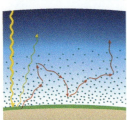

Two crucial effects are: (1) protecting the surface from dangerous solar radiation—ultraviolet is absorbed by ozone and X rays are absorbed high in the atmosphere; and (2) the **greenhouse effect,** without which Earth's surface temperature would be below freezing.

7.2 The Moon and Mercury: Geologically Dead

◆ Was there ever geological activity on the Moon or Mercury?

Both the Moon and Mercury had limited volcanism and tectonics when they were young. However, because of their small sizes, their interiors long ago cooled too much for ongoing geological activity.

7.3 Mars: A Victim of Planetary Freeze-Drying

◆ What geological features tell us that water once flowed on Mars?

Dry riverbeds, eroded craters, and studies of Martian minerals all show that water once flowed on Mars, though any periods of rainfall seem to have ended at least 3 billion years ago. Mars today still has water ice underground and in its polar caps and could possibly have pockets of underground liquid water.

◆ Why did Mars change?

Mars's atmosphere must once have been thicker with a stronger greenhouse effect, so Mars must have lost a great deal of atmospheric gas. Much of the lost gas probably was stripped away by the solar wind, after Mars lost its magnetic field and protective magnetosphere. Mars also lost water, because solar ultraviolet light split water molecules apart and the hydrogen escaped to space.

7.4 Venus: A Hothouse World

◆ Is Venus geologically active?

Venus almost certainly remains geologically active today. Its surface shows evidence of major volcanic or tectonic activity in the past billion years, and it should retain nearly as much internal heat as Earth. However, geological activity on Venus differs from that on Earth in at least two key ways: lack of erosion and lack of plate tectonics.

◆ Why is Venus so hot?

Venus's extreme surface heat is a result of its thick, carbon dioxide atmosphere, which creates a very strong greenhouse effect. The reason Venus has such a thick atmosphere is its distance from the Sun: It was too close to develop liquid oceans like those on Earth, where most of the outgassed carbon dioxide dissolved in water and became locked away in **carbonate rock.** Carbon dioxide remained in Venus's atmosphere, creating a **runaway greenhouse effect.**

7.5 Earth as a Living Planet

◆ What unique features of Earth are important for life?

Unique features of Earth on which we depend for survival are (1) surface liquid water, made possible by Earth's moderate temperature; (2) atmospheric oxygen, a product of photosynthetic life; (3) plate tectonics, driven by internal heat; and (4) climate stability, a result of the **carbon dioxide cycle,** which in turn requires plate tectonics.

◆ How is human activity changing our planet?

The global average temperature has risen about 0.8°C over the past hundred years, accompanied by an even larger rise in the atmospheric CO_2 concentration—a result of fossil fuel burning and other human activity. The current CO_2 concentration is higher than at any time in the past million years, and strong evidence indicates that this higher concentration is indeed the cause of **global warming.**

◆ What makes a planet habitable?

We can trace Earth's habitability to its relatively large size and its distance from the Sun. Its size maintains the internal heat that allowed volcanic outgassing to lead to our oceans and atmosphere, and also drives the plate tectonics that helps regulate our climate through the carbon dioxide cycle. Earth's distance from the Sun is neither too close nor too far, thereby allowing liquid water to exist on its surface.

Refer to this image of Mercury from *MESSENGER* to answer the following questions.

1. Label 1a lies on the rim of a large crater, and 1b lies on the rim of a smaller one. Which crater must have formed first?
   a. crater 1a
   b. crater 1b
   c. cannot be determined

2. The region around 2b has far fewer craters than the region around 2c. The crater floor at 2a is also flat and smooth, without many smaller craters on it. Why are regions 2a and 2b so smooth?
   a. Few small craters ever formed in these regions.
   b. Erosion erased craters that once existed in these regions.
   c. Lava flows covered craters that once existed in these regions.

3. A tectonic ridge appears to connect points 3a and 3b, crossing several craters. From its appearance, we can conclude that it must have formed
   a. before the area was cratered.
   b. after the area was cratered.
   c. at the same time the area was cratered.

4. Using your answers from questions 1–3, list the following features in order from oldest to youngest:
   a. the tectonic ridge from 3a to 3b
   b. crater 1a
   c. the smooth floor of crater 1b

exercises and problems

MasteringAstronomy®   For instructor-assigned homework and other learning materials, go to MasteringAstronomy®.

Review Questions

1. Describe the core-mantle-crust structures of the terrestrial worlds. What is a *lithosphere*? What is *differentiation*?
2. Why did the terrestrial worlds undergo *differentiation*? Why have larger worlds retained more internal heat than smaller ones?
3. Why does Earth have a global *magnetic field*? What is the *magnetosphere*?
4. Define the four major geological processes, giving examples of features on Earth shaped by each process.
5. How do crater counts tell us the age of a surface? Explain why the Moon has so many more craters than Earth.
6. What is *outgassing*, and how did it lead to the existence of Earth's atmosphere and oceans?
7. Describe the key ways in which the atmosphere affects Earth. What is the *greenhouse effect*, and how does it work?
8. Briefly summarize the geological histories of the Moon and Mercury. How did the lunar maria form? How are Mercury's great cliffs thought to have formed?
9. Choose at least three features on the global map of Mars (Figure 7.23), and explain the nature and likely origin of each.
10. Explain why liquid water is not stable on Mars today, but why we nonetheless think it flowed there in the distant past.
11. Describe the leading hypothesis for how Mars lost much of its atmosphere some 3 billion years ago, and identify the role played by Mars's size.

12. Describe the basic geology of Venus. Why is it surprising that Venus lacks plate tectonics? What might explain this lack?
13. What do we mean by a *runaway greenhouse effect*? Explain why this process occurred on Venus but not on Earth.
14. Describe four ways in which Earth is unique among the terrestrial worlds, and how each is important to life.
15. Describe the conveyor-like action of *plate tectonics*, and how it changes the arrangement of the continents with time.
16. What is the *carbon dioxide cycle*, and why is it so crucial to life on Earth?
17. Briefly summarize the evidence linking human activity to *global warming*. What are its potential consequences?
18. Based on Figure 7.50, summarize the roles of planetary size and distance from the Sun in explaining the histories of the terrestrial worlds.

Test Your Understanding

Surprising Discoveries?

Suppose we were to make the following discoveries. (These are not real discoveries.) In light of your understanding of planetary geology, decide whether the discovery should be considered reasonable or surprising. Explain your reasoning clearly, if possible tracing your logic back to basic planetary properties of size or distance from the Sun; because not all of these have definitive answers, your explanation is more important than your chosen answer.

19. New photographs reveal sand dunes on Mercury.
20. A new orbiter observes a volcanic eruption on Venus.
21. Radiometric dating of rocks brought back from one lunar crater shows it formed only 10 million years ago.
22. A new orbital photograph of Mars shows a crater bottom filled with a lake of liquid water.
23. Clear-cutting in the Amazon rain forest on Earth exposes terrain that is as heavily cratered as the Moon.
24. Drilling into the Martian surface, a robotic spacecraft discovers liquid water a few meters beneath the slopes of a Martian volcano.
25. Seismic studies on Earth reveal a "lost continent" that held great human cities just a few thousand years ago but is now buried beneath the Atlantic seafloor.
26. We find a planet in another solar system that has an Earth-like atmosphere with plentiful oxygen but no life of any kind.
27. We find a planet in another solar system that has Earth-like plate tectonics; the planet is the size of the Moon and orbits 1 AU from its star.
28. We find evidence that during dinosaur times, when Earth had no polar ice caps, the atmosphere had more carbon dioxide than it does today.

Quick Quiz

Choose the best answer to each of the following. Explain your reasoning with one or more complete sentences.

29. Which heat source continues to contribute to Earth's *internal* heat? (a) accretion (b) radioactive decay (c) sunlight
30. In general, what kind of terrestrial planet would you expect to have the thickest lithosphere? (a) a large planet (b) a small planet (c) a planet located far from the Sun
31. Which of a planet's fundamental properties has the greatest effect on its level of volcanic and tectonic activity? (a) size (b) distance from the Sun (c) rotation rate

32. Which describes our understanding of flowing water on Mars? (a) It was never important. (b) It was important once, but no longer. (c) It is a major process on the Martian surface today.
33. What do we conclude if a planet has few impact craters of any size? (a) The planet was never bombarded by asteroids or comets. (b) Its atmosphere stopped impactors of all sizes. (c) Geological processes have erased craters.
34. How many of the five terrestrial worlds are considered "geologically dead"? (a) none (b) two (c) four
35. Which terrestrial world has the most atmospheric gas? (a) Venus (b) Earth (c) Mars
36. Which of the following is a strong greenhouse gas? (a) nitrogen (b) carbon dioxide (c) oxygen
37. The oxygen in Earth's atmosphere was released by (a) volcanic outgassing. (b) the CO_2 cycle. (c) life.
38. Where is most of the CO_2 that has outgassed from Earth's volcanoes? (a) in the atmosphere (b) escaped to space (c) locked up in rocks

Process of Science

39. *What Is Predictable?* Briefly explain why much of a planet's geological history is destined from its birth, and discuss the level of detail that is predictable. For example, was Mars's general level of volcanism predictable? Could we have predicted a mountain as tall as Olympus Mons or a canyon as long as Valles Marineris? Explain.
40. *Science with Consequences.* A small, vocal group of people still dispute that humans are causing global warming. Do some research to find the basis of their claims. Then defend or refute their findings based on your own studies and your understanding of the hallmarks of science discussed in Chapter 3.
41. *Unanswered Questions.* Choose one important but unanswered question about Mars's past, and write two or three paragraphs discussing how we might answer this question in the future. Be as specific as possible, focusing on the type of evidence necessary to answer the question and on how the evidence could be gathered. What are the benefits of finding answers to this question?

Group Work Exercise

42. *Are We Causing Global Warming?* **Roles:** *Scribe* (takes notes on the group's activities), *Advocate* (argues in favor of the claim that human activity is causing global warming), *Skeptic* (points out weaknesses in the arguments made by the *Advocate*), *Moderator* (leads group discussion and makes sure everyone contributes). **Activity:**
    a. Work together to make a list of scientific observations that have been proposed as evidence that humans are causing global warming. Your list should include, but not be limited to, the evidence in Figures 7.45–7.49.
    b. *Advocate* presents the case that humans are causing global warming, drawing on the evidence from part (a).
    c. *Skeptic* attempts to refute *Advocate*'s case using scientific arguments.
    d. After hearing these arguments, *Moderator* and *Scribe* decide whose arguments were more persuasive and explain their reasoning.
    e. Each person in the group writes up a summary of the discussion.

Investigate Further

Short-Answer/Essay Questions

43. *Miniature Mars.* Suppose Mars had turned out to be significantly smaller than its current size—say, the size of our Moon. How would this have affected the number of geological features due to each of the four major geological processes? Do you think Mars would still be a good candidate for harboring extraterrestrial life? Summarize your answers in two or three paragraphs.

44. *Two Paths Diverged.* Briefly explain how the different atmospheric properties of Earth and Venus can be explained by the fundamental properties of size and distance from the Sun.

45. *Change in Formation Properties.* Consider either Earth's size or its distance from the Sun, and suppose that it had been different (for example, smaller size or greater distance). Describe how this change might have affected Earth's subsequent history and the possibility of life on Earth.

46. *Experiment: Planetary Cooling in a Freezer.* Fill two small plastic containers of similar shape but different size with cold water and put both into the freezer at the same time. Every hour or so, record the time and your estimate of the thickness of the "lithosphere" (the frozen layer) in each container. How long does it take the water in each container to freeze completely? Describe the relevance of your experiment to planetary geology. Extra credit: Plot your results on a graph with time on the *x*-axis and lithospheric thickness on the *y*-axis. What is the ratio of the two freezing times?

47. *Amateur Astronomy: Observing the Moon.* Any amateur telescope has a resolution adequate to identify geological features on the Moon. The light highlands and dark maria should be evident, and shadowing is visible near the line between night and day. Try to observe the Moon near the first- or third-quarter phase. Sketch or photograph the Moon at low magnification, and then zoom in on a region of interest. Again sketch or photograph your field of view, label its features, and identify the geological process that created them. Look for craters, volcanic plains, and tectonic features. Estimate the size of each feature by comparing it to the size of the whole Moon (radius = 1738 kilometers).

48. *Global Warming.* What, if anything, do you think we should be doing to alleviate the threat of global warming? Write a one-page editorial defending your opinion.

Quantitative Problems

Be sure to show all calculations clearly and state your final answers in complete sentences.

49. *Surface Area–to–Volume Ratio.* Compare the surface area–to–volume ratios of
    a. the Moon and Mars.
    b. Earth and Venus.
    In each case, use your answer to discuss differences in internal heat on the two worlds.

50. *Doubling Your Size.* Just as the surface area–to–volume ratio depends on size, so can other properties. To see how, suppose that your size suddenly doubled; that is, your height, width, and depth all doubled. (For example, if you are 5 feet tall, you would become 10 feet tall.)
    a. By what factor has your waist size increased?
    b. How much more material will be required for your clothes? (*Hint:* Clothes cover the *surface area* of your body.)
    c. By what factor has your weight increased? (*Hint:* Weight depends on the *volume* of your body.)
    d. The pressure on your weight-bearing joints depends on how much *weight* is supported by the *surface area* of each joint. How has this pressure changed?

51. *Plate Tectonics.* Typical motions of one plate relative to another are 1 centimeter per year. At this rate, how long would it take for two continents 3000 kilometers apart to collide? What are the global consequences of motions like this?

52. *Planet Berth.* Imagine a planet, which we'll call *Berth*, orbiting a star identical to the Sun at a distance of 1 AU. Assume that Berth has eight times as much mass as Earth and is twice as large as Earth in diameter.
    a. How does Berth's density compare to Earth's?
    b. How does Berth's surface area compare to Earth's?
    c. Based on your answers to (a) and (b), discuss how Berth's geological history is likely to have differed from Earth's.

Discussion Questions

53. *Worth the Effort?* Politicians often debate whether planetary missions are worth the expense involved. If you were in Congress, would you support more or fewer missions? Why?

54. *Lucky Earth.* The climate histories of Venus and Mars make it clear that it's not "easy" to get a pleasant climate like that of Earth. How does this affect your opinion about whether Earth-like planets might exist around other stars? Explain.

55. *Terraforming Mars.* Some people have suggested that we might be able to carry out planetwide engineering of Mars that would cause its climate to warm and its atmosphere to thicken. This type of planet engineering is called *terraforming,* because its objective is to make a planet more Earth-like and therefore easier for humans to live on. Discuss possible ways to terraform Mars. Do any of these ideas seem practical? Do they seem like good ideas? Defend your opinions.

Web Projects

56. *"Coolest" Surface Photo.* Visit the Astronomy Picture of the Day website, and search for past images of the terrestrial worlds. After looking at many of the images, choose the one you think is the "coolest." Make a printout, write a short description of what it shows, and explain what you like about it.

57. *Planetary Mission.* Choose a current mission that is studying one of the terrestrial worlds, visit its website, and write a brief summary of its goals, current status, and most important discoveries to date.

58. *Mars Colonization.* Visit the website of a group that advocates human colonization of Mars, such as the Mars Society. Learn about the challenges of human survival on Mars and about prospects for terraforming Mars. Do you think colonization of Mars is a good idea? Write a short essay describing what you've learned and defend your opinions.

8 Jovian Planet Systems

Saturn, photographed by the *Cassini* spacecraft while it was in Saturn's shadow. The small blue dot of light just inside Saturn's rings at the left (about the 10 o'clock position) is Earth, far in the distance.

n Roman mythology, the namesakes of the jovian planets are rulers among gods: Jupiter is the king of the gods, Saturn is Jupiter's father, Uranus is the lord of the sky, and Neptune rules the sea. However, our ancestors could not have foreseen the true majesty of the four jovian planets. The smallest, Neptune, is large enough to contain the volume of more than 50 Earths. The largest, Jupiter, has a volume some 1400 times that of Earth. These worlds lack solid surfaces and are totally unlike the terrestrial planets. Their many moons and rings only add to their intrigue.

Why should we care about a set of worlds so different from our own? Apart from satisfying natural curiosity, studies of the jovian planets and their moons help us understand the birth and evolution of planetary systems—which in turn helps us understand our own planet Earth. In this chapter, we'll explore the jovian planet systems, first focusing on the planets themselves, then on their many moons, and finally on their beautifully complex rings.

ESSENTIAL PREPARATION

1. How does gravity cause tides? [Section 4.4]
2. What does the solar system look like? [Section 6.1]
3. Why are there two major types of planets? [Section 6.3]
4. Why is Earth geologically active? [Section 7.1]

8.1 A Different Kind of Planet

Figure 8.1 shows a montage of the jovian planets compiled by the *Voyager* spacecraft. Their immense sizes are apparent in the comparison to Earth. But while all four are enormous, the given data also reveal important differences between them. We toured the jovian planets briefly in Section 6.1. Now we are ready to explore them in more depth.

◆ What are jovian planets made of?

As we discussed in Chapter 6, the jovian planets are made mostly of hydrogen, helium, and hydrogen compounds, making them very different from the rocky terrestrial worlds. However, the jovian planets do not all have exactly the same composition, and they also differ in their interior structures.

General Composition Jupiter and Saturn are made mostly of hydrogen and helium, giving them compositions much more similar to that of the Sun than to that of the terrestrial planets. Some people even call Jupiter a "failed star" because it has a starlike composition but lacks the nuclear fusion needed to make it shine. This is a consequence of its size: Although Jupiter is large for a planet, it is only about 1/80 as massive as the lowest-mass stars. As a result, its gravity is too weak to compress its interior to the extreme temperatures and densities needed for nuclear fusion. Uranus and Neptune differ from Jupiter and Saturn both in being much smaller and in having much smaller proportions of hydrogen and helium, leaving them with compositions consisting primarily of hydrogen compounds—especially water (H_2O), methane (CH_4), and ammonia (NH_3)—along with smaller amounts of metal and rock.

The differences in composition among the jovian planets can probably be traced to their origins. Recall that the jovian planets formed in the outer solar system, where it was cold enough for hydrogen compounds to condense into ices [Section 6.3]. Because hydrogen

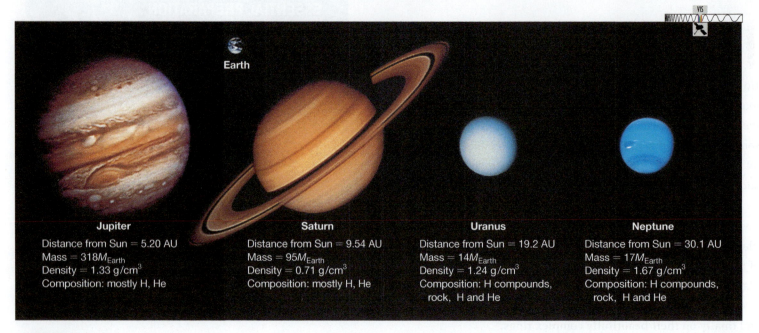

▲ FIGURE 8.1

Jupiter, Saturn, Uranus, and Neptune, shown to scale with Earth for comparison.

Jupiter	
Distance from Sun = 5.20 AU	
Mass = $318M_{Earth}$	
Density = 1.33 g/cm^3	
Composition: mostly H, He	

Earth

Jupiter
Distance from Sun = 5.20 AU
Mass = $318M_{Earth}$
Density = 1.33 g/cm^3
Composition: mostly H, He

Saturn
Distance from Sun = 9.54 AU
Mass = $95M_{Earth}$
Density = 0.71 g/cm^3
Composition: mostly H, He

Uranus
Distance from Sun = 19.2 AU
Mass = $14M_{Earth}$
Density = 1.24 g/cm^3
Composition: H compounds, rock, H and He

Neptune
Distance from Sun = 30.1 AU
Mass = $17M_{Earth}$
Density = 1.67 g/cm^3
Composition: H compounds, rock, H and He

compounds were so much more abundant than metal and rock, some of the ice-rich planetesimals of the outer solar system grew to great size. Once these planetesimals became sufficiently massive, their gravity allowed them to draw in the hydrogen and helium gas that surrounded them. Models suggest that all four jovian planets grew from ice-rich planetesimals of about the same mass—roughly 10 times the mass of Earth—but they captured different amounts of hydrogen and helium gas from the surrounding solar nebula.

The jovian planets are made mostly of hydrogen, helium, and hydrogen compounds, and differ primarily in their relative proportions of hydrogen compounds.

Jupiter and Saturn captured so much hydrogen and helium gas that these gases now make up the vast majority of their masses. The ice-rich planetesimals from which they grew now represent only about 3% of Jupiter's mass and about 10% of Saturn's mass. Uranus and Neptune pulled in much less gas from the solar nebula, so hydrogen and helium make up less than half of their total masses.

Why did the different planets capture different amounts of gas? The answer probably lies in their distances from the Sun as they formed. The solid particles that condensed farther from the Sun should have been more widely spread out than those that condensed nearer to the Sun, which means it would have taken longer for them to accrete into large, icy planetesimals. As the nearest jovian planet to the Sun, Jupiter would have been the first to get a planetesimal large enough for its gravity to start drawing in gas, followed by Saturn, Uranus, and Neptune. Because all the planets stopped accreting gas at the same time—when the solar wind blew all the remaining gas into interstellar space—the more distant planets had less time to capture gas and ended up smaller in size. Similar processes, played out to different extents in different planetary systems, may explain much of the wider variety of planetary types that scientists have discovered among planets around other stars [Section 10.2].

The jovian planets farther from the Sun took longer to form and captured less hydrogen and helium gas, explaining why Uranus and Neptune have larger proportions of hydrogen compounds, rock, and metal.

Density Differences Figure 8.1 shows that Saturn is considerably less dense than Uranus or Neptune. This should make sense, because the hydrogen compounds, rock, and metal that make up Uranus and Neptune are normally much more dense than hydrogen or helium gas. However, by the same logic we'd expect Jupiter to be even less dense than Saturn—but it's not.

think about it Saturn's average density of 0.71 g/cm^3 is less than that of water. As a result, it is sometimes said that Saturn could float on a giant ocean. Suppose there really were a gigantic planet with a gigantic ocean and we put Saturn on the ocean's surface. Would it float? If not, what would happen?

We can understand Jupiter's surprisingly high density by thinking about how massive planets are affected by their own gravity. Building a planet of hydrogen and helium is a bit like making one out of fluffy pillows. Imagine assembling a planet pillow by pillow. As each new pillow is added, those on the bottom are compressed more by those above. As the lower pillows are forced closer together, their mutual gravitational attraction increases, compressing them even further. At first the stack grows substantially with each pillow, but eventually the growth slows until adding pillows barely increases the height of the stack (Figure 8.2a).

This analogy explains why Jupiter is only slightly larger than Saturn in radius even though it is more than three times as massive. The extra mass of Jupiter compresses its interior to a much higher density. More precise calculations show that Jupiter's radius is almost the maximum possible radius for a jovian planet. If much more gas were added to Jupiter, its weight would actually compress the interior enough to make the planet *smaller* rather than larger (Figure 8.2b). Some extrasolar planets that are larger in mass than Jupiter are therefore smaller in size.

see it for yourself Measure the thickness of your pillow; then put it at the bottom of a stack of other pillows, folded blankets, or clothing. How much has the pillow been compressed by the stack above it? Insert your hand between the different layers to feel the pressure differences—and imagine the kind of pressures and compression you'd find in a stack tens of thousands of kilometers tall!

Jovian Planet Interiors The jovian planets are often called "gas giants," making it sound as if they were entirely gaseous like air on Earth. However, while it is true that they are made mostly of materials that would be gases on Earth, the high pressures in jovian interiors compress these materials to different forms. Let's first consider Jupiter's interior.

Jupiter's lack of a solid surface makes it tempting to think of the planet as "all atmosphere," but you could not fly through Jupiter's interior in the way airplanes fly through air. A spacecraft plunging into Jupiter would find increasingly higher temperatures and pressures as it descended. The *Galileo* spacecraft dropped a scientific probe into Jupiter in 1995 that collected measurements for about an hour before the ever-increasing pressures and temperatures destroyed it. The probe provided valuable data about Jupiter's atmosphere but didn't last long enough to sample the interior: It survived to a depth of only about 200 kilometers, or about 0.3% of Jupiter's radius.

Computer models tell us that Jupiter has fairly distinct interior layers (Figure 8.3). The layers do not differ much in composition—all except the core are mostly hydrogen and helium. Instead they differ in the phase

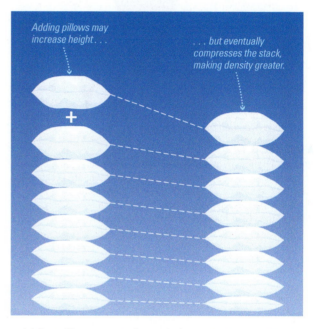

a Adding pillows to a stack may increase its height at first but eventually compresses the stack, making its density greater.

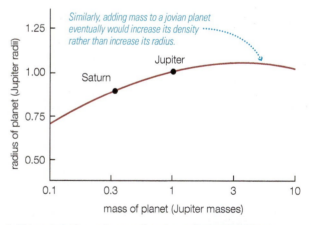

b This graph shows how radius depends on mass for a hydrogen/helium planet. Notice that Jupiter is only slightly larger in radius than Saturn, despite being three times as massive. Gravitational compression of a planet much more massive than Jupiter would actually make its radius smaller.

▲ **FIGURE 8.2**
The relationship between mass and radius for a planet made of hydrogen and helium.

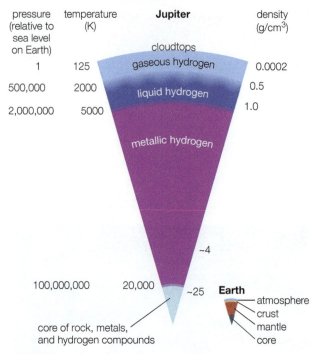

pressure (relative to sea level on Earth)	temperature (K)		density (g/cm^3)
		Jupiter	
		cloudtops	
1	125	gaseous hydrogen	0.0002
500,000	2000	liquid hydrogen	0.5
2,000,000	5000		1.0
		metallic hydrogen	
			~4
100,000,000	20,000		~25

core of rock, metals, and hydrogen compounds

Earth
atmosphere
crust
mantle
core

▲ **FIGURE 8.3**

Jupiter's interior structure, labeled with the pressure, temperature, and density at various depths. Earth's interior structure is shown to scale for comparison. (The thicknesses of Earth's crust and atmosphere are exaggerated.) Note that Jupiter's core is only slightly larger than Earth but is about 10 times as massive.

▼ **FIGURE 8.4**

These diagrams compare the interior structures of the jovian planets, shown approximately to scale. All four planets have cores of rock, metal, and hydrogen compounds, with masses about 10 times the mass of Earth. They differ primarily in the depth of the hydrogen/helium layers that surround their cores. The cores of Uranus and Neptune are differentiated into separate layers of rock/metal and hydrogen compounds.

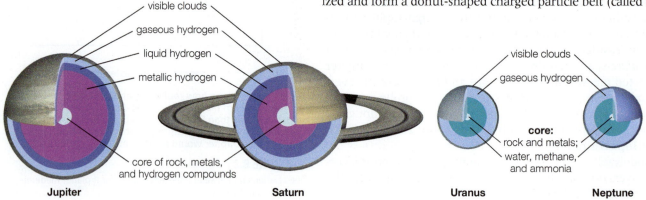

visible clouds
gaseous hydrogen
liquid hydrogen
metallic hydrogen
core of rock, metals, and hydrogen compounds

Jupiter **Saturn**

visible clouds
gaseous hydrogen
core:
rock and metals;
water, methane, and ammonia

Uranus **Neptune**

If you plunged below Jupiter's clouds, you'd never encounter a solid surface—just ever denser and hotter hydrogen/helium compressed into bizarre liquid and metallic phases.

(such as liquid or gas) of their hydrogen. The outer layer is the only region in which conditions are moderate enough for hydrogen to remain in its familiar, gaseous form. This layer makes up about the outer 10% of Jupiter and represents what we usually think of as Jupiter's atmosphere. As we look deeper, the high pressures compress the hydrogen into liquid form in the next 10% of Jupiter's depth, and below that the temperatures and pressures become so extreme that hydrogen is forced into a compact, liquid metallic (electrically conducting) form. Finally, the models tell us that Jupiter's core is a mix of hydrogen compounds, rock, and metal, but compressed to such extremes of temperature and density that this mix bears little resemblance to familiar solids or liquids.

We can extend the ideas from Jupiter to the other jovian planets. Because all four jovian planets have cores of about the same mass, their interiors differ mainly in the hydrogen/helium layers that surround their cores (Figure 8.4). Remember that while the outer layers are named for the phase of their hydrogen, they also contain helium and hydrogen compounds.

Saturn has the same basic layering as Jupiter, but its lower mass makes the weight of the overlying layers less than on Jupiter. As a result, we must look deeper into Saturn to find each level where pressure changes hydrogen from one phase to another. Uranus and Neptune have somewhat different layering because their internal pressures never become high enough to form liquid or metallic hydrogen, so they have only a thick layer of gaseous hydrogen surrounding their cores. Models suggest that their core mix of hydrogen compounds, rock, and metal may be liquid, making for very odd "oceans" buried deep inside Uranus and Neptune; the cores may also be differentiated so that hydrogen compounds reside in an outer core above an inner core of rock and metal.

Magnetic Fields Recall that Earth has a global magnetic field generated by the movements of charged particles in our planet's outer core of molten metal [Section 7.1]. The jovian planets also have global magnetic fields generated by motions of charged particles deep in their interiors.

Jupiter's magnetic field is by far the strongest—some 20,000 times as strong as Earth's. This strong field, generated in Jupiter's thick layer of metallic hydrogen, creates an enormous magnetosphere that begins to deflect the solar wind some 3 million kilometers in front of Jupiter (Figure 8.5). If we could see Jupiter's magnetosphere, it would be larger than the full moon in our sky. Gases escaping from volcanoes on Jupiter's moon Io feed vast numbers of particles into the magnetosphere, where they become ionized and form a donut-shaped charged particle belt (called the *Io torus*) that

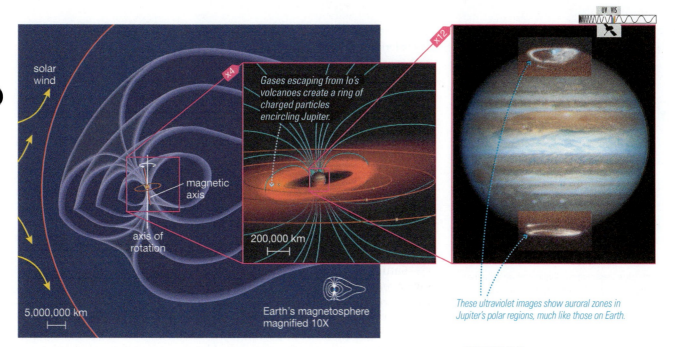

These ultraviolet images show auroral zones in Jupiter's polar regions, much like those on Earth.

▲ **FIGURE 8.5**

Jupiter's strong magnetic field gives it an enormous magnetosphere. Gases escaping from Io feed the donut-shaped Io torus, and particles entering Jupiter's atmosphere near its magnetic poles contribute to auroras on Jupiter. The image at the right is a composite of ultraviolet images of the polar regions overlaid on a visible image of the whole planet, all taken by the Hubble Space Telescope.

approximately traces Io's orbit. The particles also create belts of intense radiation and contribute to auroras on Jupiter.

think about it The *Juno* mission entered Jupiter orbit in July 2016 to study Jupiter's interior, atmosphere, and magnetosphere. Find and briefly describe its major discoveries to date.

The other jovian planets also have magnetic fields and magnetospheres, but theirs are much weaker than Jupiter's (although still much stronger than Earth's). Saturn's magnetic field is weaker than Jupiter's because it has a thinner layer of electrically conducting metallic hydrogen. Uranus and Neptune, smaller still, have no metallic hydrogen at all. Their relatively weak magnetic fields must be generated in their core "oceans" of hydrogen compounds, rock, and metal.

◆ What is the weather like on jovian planets?

Jovian atmospheres have dynamic winds and weather, with colorful clouds and enormous storms. Weather on these planets is driven not only by energy from the Sun (as on the terrestrial planets), but also by heat generated within the planets themselves. All but Uranus generate a great deal of internal heat. No one knows the precise source of the internal heat on jovian planets, but it probably comes from the conversion of gravitational potential energy to thermal energy inside them. The best guesses are that this conversion comes from a slow but imperceptible contraction in overall size, or from ongoing differentiation as heavier materials continue to sink toward the core.

Clouds and Colors Many mysteries remain about the colors of the jovian planets, but clouds play a major role. Earth's clouds look white from space because they are made of water that reflects the white light of the Sun. The jovian planets have clouds of several different types, and some of these reflect light of other colors.

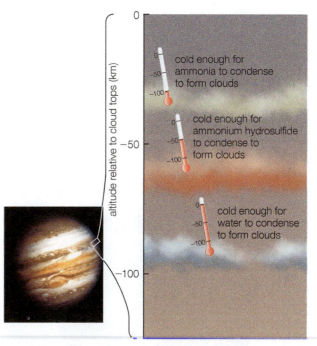

▲ **FIGURE 8.6**

This diagram shows the temperature structure of Jupiter's atmosphere. Jupiter has at least three distinct cloud layers because different atmospheric gases condense at different temperatures and hence at different altitudes. The tops of the ammonia clouds are usually considered the zero altitude for Jupiter, which is why lower altitudes are negative.

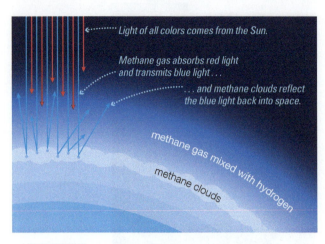

▲ FIGURE 8.7

Neptune and Uranus look blue because methane gas absorbs red light but transmits blue light. Clouds of methane snowflakes reflect the transmitted blue light back to space.

Clouds of different compositions and colors form at different altitudes in each jovian planet's atmosphere.

Clouds form when a gas condenses to make tiny liquid droplets or solid flakes. Water vapor is the only gas that can condense to form clouds in Earth's atmosphere, but Jupiter's atmosphere has three major types of clouds, each of which forms at different altitudes. Looking from the bottom up in Figure 8.6, Jupiter's lowest cloud layer occurs about 100 kilometers below the highest cloudtops. At this depth, temperatures are nearly Earth-like and water can condense to form clouds. The temperature drops as you go higher, and about 50 kilometers above the water clouds it is cold enough for a gas called ammonium hydrosulfide (NH_4SH) to condense into clouds. These ammonium hydrosulfide clouds reflect brown and red light, producing many of the dark colors of Jupiter. Higher still, the temperature is so cold that ammonia (NH_3) condenses to make an upper layer of white clouds.

Saturn has the same set of three cloud layers as Jupiter, but Saturn's lower temperatures (due to both its greater distance from the Sun and its weaker gravity) mean these layers occur deeper in Saturn's atmosphere. This fact probably explains Saturn's more subdued colors: Less light penetrates to the depths at which Saturn's clouds are found, and the light they reflect is more obscured by the atmosphere above them.

Uranus and Neptune are so cold that any cloud layers similar to those of Jupiter or Saturn would be buried too deep in their atmospheres for us to see. However, the cold temperatures allow some of their abundant methane gas to condense into clouds. Methane gas also absorbs red light, allowing only blue light to penetrate to the level at which the methane clouds form. The methane clouds reflect this blue light upward, giving these two planets their blue colors (Figure 8.7).

Global Winds and Storms One of the most striking visual features of the jovian planets is the striped appearance of Jupiter. The stripes represent alternating bands of rising and falling air, and their colors arise from the clouds that we see. Bands of rising air are white because the rising air condenses and forms white ammonia clouds at high altitudes (see Figure 8.6). Bands of falling air do not contain ammonia clouds, allowing us to see the reddish-brown ammonium hydrosulfide clouds that lie at lower altitudes. The fact that these bands wrap around Jupiter is a consequence of the same effect

▼ FIGURE 8.8

Wind patterns on both Earth and Jupiter arise from the way planetary rotation affects moving air.

a This photograph shows how storms circulate around low-pressure regions (L) on Earth. Earth's rotation causes this circulation, which is in opposite directions in the two hemispheres.

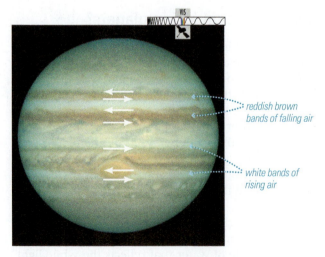

reddish brown bands of falling air

white bands of rising air

b Jupiter's faster rotation and larger size essentially stretch out its circulation patterns into planet-wide bands of fast-moving air.

(called the *Coriolis effect*) that makes storms around low-pressure regions on Earth circulate counterclockwise in the Northern Hemisphere and clockwise in the Southern Hemisphere (Figure 8.8a). Jupiter's faster rotation and larger size make this effect much stronger, in essence stretching the circulation east-west to the point that it wraps all the way around the planet (Figure 8.8b). Jupiter therefore has very high east-west wind speeds—sometimes more than 400 kilometers per hour (250 miles per hour).

Jovian planets also show numerous storms, the most famous being Jupiter's **Great Red Spot,** which is more than twice as wide as Earth. The

The rapid rotation of the jovian planets helps drive strong winds, creating their banded appearances and sometimes giving rise to huge storms.

Great Red Spot is somewhat like a hurricane on Earth, except that its winds circulate around a high- rather than low-pressure region (Figure 8.9). It is also extremely long-lived compared to storms on Earth: Astronomers have seen it throughout the three centuries during which telescopes have been powerful enough to detect it. No one knows why the Great Red Spot has lasted so long. However, storms on Earth tend to lose their strength when they pass over land. Perhaps Jupiter's biggest storms last for centuries simply because there's no solid surface effect to sap their energy. Two other long-lived storms have recently been observed to undergo a mysterious change, turning from white to red. The more recent one was torn apart by the Great Red Spot as it passed nearby (Figure 8.10a).

The other jovian planets also have dramatic weather patterns (Figure 8.10b–d). As on Jupiter, Saturn's rapid rotation creates alternating bands of rising and falling air, along with rapid east-west winds. In fact, Saturn's winds are even faster than Jupiter's—a surprise that scientists have yet to explain. Neptune's atmosphere is also banded, and we have seen a high-pressure storm, called the Great Dark Spot, similar to Jupiter's Great Red Spot. However, the Great Dark Spot did not last as long; it disappeared from view just 6 years after its discovery. Uranus had a more subdued appearance when *Voyager 2* flew past it in 1986, but now seems to have more frequent storms, probably due to the slow but dramatic change of seasons caused by its extreme axis tilt and 84-year orbit around the Sun.

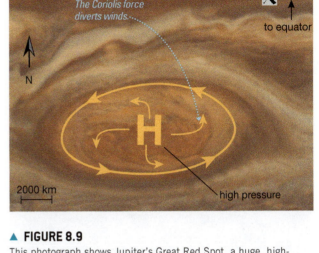

▲ **FIGURE 8.9**

This photograph shows Jupiter's Great Red Spot, a huge, high-pressure storm that is large enough to swallow two or three Earths. The overlaid diagram shows a weather map of the region.

▼ **FIGURE 8.10**

Selected views of weather patterns on the four jovian planets.

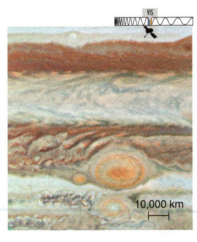

a This Hubble Space Telescope image shows Jupiter's southern hemisphere with the Great Red Spot, "Baby Red" (to its left), and "Red Jr." (below). Baby Red was torn apart by the Great Red Spot a few days later.

b Saturn's atmosphere, photographed by *Voyager 1*. Its banded appearance is very similar to that of Jupiter.

c This infrared image of Uranus from the Keck Telescope shows several storms (the bright blotches) and Uranus's thin rings (red).

d Neptune's atmosphere, viewed from *Voyager 2*, shows bands and occasional strong storms. The large storm (white arrow) was called the Great Dark Spot.

Medium-Size and Large Moons of the Jovian Planets

Jupiter

Io   Europa   Ganymede   Callisto

Saturn

Mimas Enceladus Tethys Dione Rhea Titan Iapetus

Uranus

Miranda   Ariel   Umbriel   Titania   Oberon

Neptune

Triton   Nereid

Other objects for comparison

Mercury   Moon   Pluto   3000 km

▲ **FIGURE 8.11**

The medium-size and large moons of the jovian planets, with sizes (but not distances) shown to scale. Mercury, the Moon, and Pluto are included for comparison.

8.2 A Wealth of Worlds: Satellites of Ice and Rock

The jovian planets are majestic and fascinating, but they are only the beginning of our exploration of jovian planet *systems*. Each of the four jovian systems includes numerous moons and a set of rings. The total mass of all the moons and rings put together is minuscule compared to that of any one of the jovian planets, but the remarkable diversity of these satellites makes up for their lack of size.

◆ What kinds of moons orbit the jovian planets?

We now know of at least 170 moons orbiting the jovian planets. Jupiter and Saturn have the most, each with more than 60 known moons. It's helpful to organize these moons into three groups by size: small moons less than about 300 kilometers in diameter, medium-size moons ranging from about 300 to 1500 kilometers in diameter, and large moons more than 1500 kilometers in diameter. These categories are useful because size relates to geological activity. In general, larger moons are more likely to show evidence of past or present geological activity.

Figure 8.11 shows a montage of all the medium-size and large moons. These moons resemble the terrestrial planets in many ways. Each is spherical with a solid surface and its own unique geology. Some possess atmospheres, hot interiors, and even magnetic fields. The two largest—Jupiter's moon Ganymede and Saturn's moon Titan—are larger than the planet Mercury, while four others (Jupiter's moons Io, Europa, and Callisto and Neptune's moon Triton) are larger than the largest known dwarf planets, Pluto and Eris. However, these moons differ from terrestrial worlds in their compositions: Because they formed in the cold outer solar system, most of them contain substantial amounts of ice in addition to metal and rock.

Most of the medium-size and large moons probably formed by accretion within the disks of gas surrounding individual jovian planets [Section 6.3]. That explains why their orbits are almost circular and lie close to the equatorial plane of their parent planet, and also why these moons orbit in the same direction in which their planet rotates. In contrast, many small moons are probably captured asteroids or comets, and therefore do not follow any particular orbital patterns; many even orbit backward relative to their planet's rotation.

A few jovian moons rival the smallest planets in size and geological interest, while vast numbers of smaller moons are captured asteroids and comets.

The small moons have irregular shapes, much like potatoes (Figure 8.12), because their gravities are too weak to force their rigid material into spheres. We do not expect these moons to have any significant geological activity, so they are essentially just chunks of ice and rock held captive by the gravity of their parent planet.

◆ Why are Jupiter's Galilean moons geologically active?

We now embark on a brief tour of the most interesting moons of the jovian planets. Our first stop is Jupiter, where the four *Galilean*

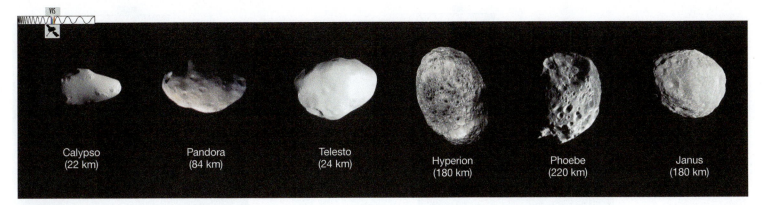

Calypso
(22 km)

Pandora
(84 km)

Telesto
(24 km)

Hyperion
(180 km)

Phoebe
(220 km)

Janus
(180 km)

▲ **FIGURE 8.12**

These photos from the *Cassini* spacecraft show six of Saturn's smaller moons. All are much smaller than the smallest moons shown in Figure 8.11. Their irregular shapes are due to their small size, which makes their gravities too weak to force them into spheres. The sizes in parentheses represent approximate lengths along their longest axes.

moons (discovered by Galileo [Section 3.3]) are each large enough that they would count as planets or dwarf planets if they orbited the Sun (Figure 8.13).

Io: The Volcano World For anyone who thinks of moons as barren, geologically dead places like our own Moon, Io shatters the stereotype. Io is by far the most volcanically active world in our solar system. Large volcanoes pockmark its entire surface (Figure 8.14), and eruptions are so frequent that they have buried virtually every impact crater. Io probably also has tectonic activity, because tectonics and volcanism generally go hand in hand. However, debris from volcanic eruptions has probably buried most tectonic features.

Io's active volcanoes tell us that it must be quite hot inside. However, Io is only about the size of our geologically dead Moon, so it should have

▼ **FIGURE 8.13**

This set of photos, taken by the *Galileo* spacecraft, shows global views of the four Galilean moons as we know them today. Sizes are shown to scale. (Io is about the size of Earth's Moon.)

1000 km

Io

Europa

Ganymede

Callisto

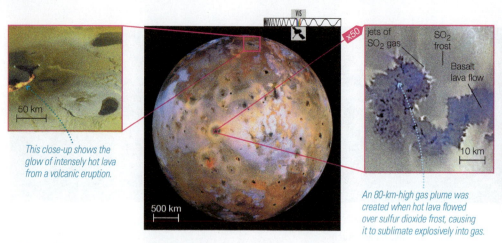

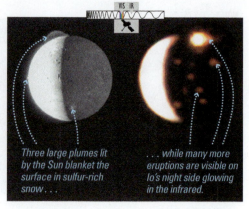

This close-up shows the glow of intensely hot lava from a volcanic eruption.

50 km

500 km

x50

jets of SO₂ gas

SO₂ frost

Basalt lava flow

10 km

An 80-km-high gas plume was created when hot lava flowed over sulfur dioxide frost, causing it to sublimate explosively into gas.

Three large plumes lit by the Sun blanket the surface in sulfur-rich snow . . .

. . . while many more eruptions are visible on Io's night side glowing in the infrared.

a Most of the black, brown, and red spots on Io's surface are recently active volcanic features. White and yellow areas are sulfur dioxide (SO_2) and sulfur deposits, respectively, from volcanic gases. (Photographs from the *Galileo* spacecraft; some colors slightly enhanced or altered.)

b Two views of Io's volcanoes taken by *New Horizons* on its way to Pluto.

▲ **FIGURE 8.14**

Io is the most volcanically active body in the solar system.

long ago lost any heat from its birth and is too small for radioactivity to provide much ongoing heat. How, then, can Io be so hot inside? The only possible answer is that some other ongoing process must be heating Io's interior. Scientists have identified this process and call it **tidal heating,** because it arises from effects of tidal forces exerted by Jupiter.

Just as Earth exerts a tidal force that causes the Moon to keep the same face toward us at all times [Section 4.4], a tidal force from Jupiter makes Io keep the same face toward Jupiter as it orbits. But Jupiter's mass makes this tidal force far larger than the tidal force that Earth exerts on the Moon. Moreover, Io's orbit is slightly elliptical, so its orbital speed and distance from Jupiter vary. This variation means that the strength and direction of the tidal force change slightly as Io moves through each orbit, which in turn changes the size and orientation of Io's tidal bulges (Figure 8.15a). The result is that Io is continuously being

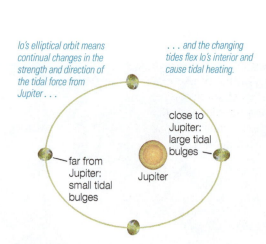

Io's elliptical orbit means continual changes in the strength and direction of the tidal force from Jupiter . . .

. . . and the changing tides flex Io's interior and cause tidal heating.

far from Jupiter: small tidal bulges

close to Jupiter: large tidal bulges

Jupiter

a Tidal heating arises because Io's elliptical orbit (exaggerated in this diagram) causes varying tides.

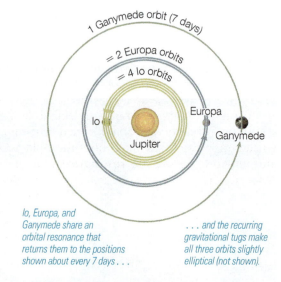

1 Ganymede orbit (7 days)

= 2 Europa orbits

= 4 Io orbits

Io

Europa

Ganymede

Jupiter

Io, Europa, and Ganymede share an orbital resonance that returns them to the positions shown about every 7 days . . .

. . . and the recurring gravitational tugs make all three orbits slightly elliptical (not shown).

b Io's orbit is elliptical because of the orbital resonance Io shares with Europa and Ganymede.

▲ **FIGURE 8.15**

These diagrams explain the cause of Io's tidal heating. Tidal heating has a weaker effect on Europa and Ganymede, because they are farther from Jupiter and tidal forces weaken with distance.

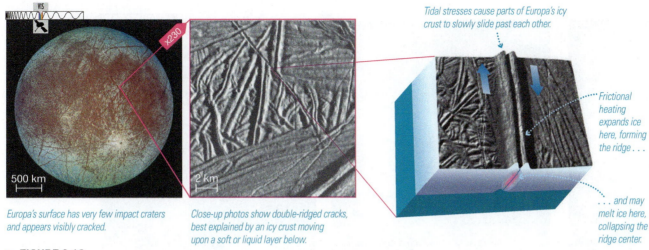

Tidal stresses cause parts of Europa's icy crust to slowly slide past each other.

Frictional heating expands ice here, forming the ridge . . .

. . . and may melt ice here, collapsing the ridge center.

Europa's surface has very few impact craters and appears visibly cracked.

Close-up photos show double-ridged cracks, best explained by an icy crust moving upon a soft or liquid layer below.

▲ **FIGURE 8.16**

Europa's icy crust probably hides a deep, liquid water ocean beneath its surface. These photos are from the *Galileo* spacecraft; colors are enhanced in the global view.

flexed in different directions, which generates friction inside it. The flexing heats the interior in the same way that flexing warms Silly Putty. Tidal heating generates tremendous heat on Io, and calculations show that this heat can explain Io's incredible volcanic activity.

see it for yourself   Pry apart the overlapping ends of a paper clip, and hold one end in each hand. Flex the ends apart and together until the paper clip breaks. Lightly touch the broken end to your finger or lips—can you feel the warmth produced by flexing? How is this heating similar to the tidal heating of Io?

However, we are still left with a deeper question: Why is Io's orbit slightly elliptical, when almost all other large satellites have nearly circular orbits? The answer lies in an interesting dance executed by Io and its neighboring moons (Figure 8.15b). During the time Ganymede takes to complete one orbit of Jupiter, Europa completes exactly two orbits and Io completes exactly four orbits. The three moons therefore line up periodically, and the gravitational tugs they exert on one another add up over time. Because the tugs are always in the same direction with each alignment, they tend to stretch out the orbits, making them slightly elliptical. The effect is much like that of pushing a child on a swing. If timed properly, a series of small pushes can add up to a *resonance* that causes the child to swing quite high. For the three moons, the **orbital resonance** that makes their orbits elliptical comes from the small gravitational tugs that repeat at each alignment.

Orbital resonances among the Galilean moons make Io's orbit slightly elliptical, leading to tidal heating that explains Io's volcanic activity.

Europa: The Water World?   Europa's surface is made almost entirely of water ice, with very few impact craters and extensive cracks (Figure 8.16). These features represent clear signs of ongoing geological activity, presumably involving flows of either liquid water or ice that is soft enough to undergo convection. The driver for this activity is the same type of tidal heating that drives Io's volcanoes, except it is weaker because of Europa's greater distance from Jupiter. Still, calculations suggest that Europa should have enough tidal heating to melt a subsurface layer of ice into liquid water.

These facts led scientists to suspect that Europa may hide a deep ocean of liquid water between its rocky mantle and its icy crust (Figure 8.17). Data

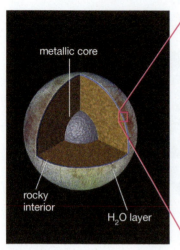

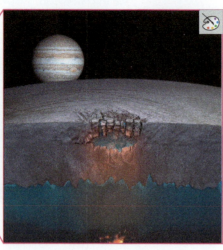

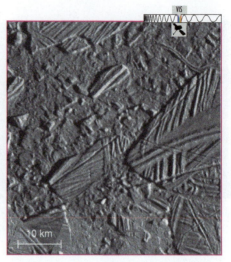

Europa may have a 100-km-thick ocean under an icy crust.

Tides or upwelling warm ice may sometimes create lakes within the ice, causing the crust above to crack . . .

. . . explaining surface terrain that looks like a jumble of icebergs suspended in a place where liquid or slushy water froze.

▲ **FIGURE 8.17**

This diagram shows the leading model of Europa's interior structure. There is little doubt that the H_2O layer is real, but some questions remain about whether the material beneath the icy crust is liquid water, relatively warm convecting ice, or some of each.

collected by the *Galileo* spacecraft support this hypothesis in two primary ways. First, many surface features are difficult to explain unless there is a subsurface ocean. Second, Europa has a magnetic field that changes in a way that suggests it is generated in a liquid layer of electrically conducting material, and a salty ocean would fit the bill.

Tidal heating may create a deep ocean of liquid water beneath Europa's icy crust.

If it really exists, Europa's liquid ocean may be more than 100 kilometers deep and contain more than twice as much liquid water as all of Earth's oceans combined. It is this possibility—along with the likelihood that the seafloor is dotted with undersea volcanoes—that makes scientists wonder whether Europa might harbor life, a possibility we will discuss in Chapter 19.

Ganymede and Callisto Jupiter's two other large moons, Ganymede and Callisto, also show intriguing geology. Like Europa, both have surfaces of water ice.

Ganymede, the largest moon in the solar system, appears to have a dual personality (Figure 8.18). Some regions are dark and densely cratered, suggesting that they look much the same today as they did billions of years ago. Other regions are light-colored with very few craters, suggesting that liquid water has recently erupted and refrozen. Moreover, magnetic field data indicate that Ganymede, like Europa, could have a subsurface ocean of liquid water. If so, we'd need to explain the source of the heat that melts Ganymede's subsurface ice. Ganymede has some tidal heating, but calculations suggest that it is not strong enough to account for an ocean. Perhaps ongoing radioactive decay supplies enough additional heat to make an ocean. Or perhaps not—no one yet knows what secrets Ganymede hides.

Callisto, the outermost Galilean moon, looks most like what scientists originally expected for an outer solar system satellite: a heavily cratered iceball (Figure 8.19). The bright patches on its surface are impact craters. However, the surface also holds some surprises. Close-up images show a dark, powdery substance concentrated in low-lying areas, leaving ridges and crests bright white. The nature of this material and how it got there are unknown. Even more surprising, magnetic field data suggest that Callisto, too,

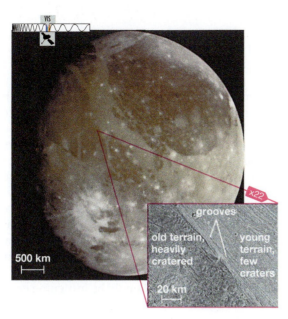

▲ FIGURE 8.18

Ganymede, the largest moon in the solar system, has both old and young regions on its surface of water ice. The dark regions are heavily cratered and must be billions of years old, while the light regions are younger landscapes where eruptions of water have presumably erased ancient craters; the long grooves in the light regions were probably formed by water erupting along surface cracks. Notice that the boundary between the two types of terrain can be quite sharp.

could hide a subsurface ocean. No one knows what might heat the interior of Callisto, since it does not participate in the orbital resonances of the other

> **Tidal heating is weak on Ganymede and absent on Callisto, yet both moons show some evidence of subsurface oceans.**

Galilean moons and therefore has no tidal heating at all. Nevertheless, the potential for an ocean raises the intriguing possibility that there could be oceans on three worlds orbiting Jupiter—with far more total water than we have here on Earth.

◆ What geological activity do we see on Titan and other moons?

Aside from the four Galilean moons, the rest of Jupiter's moons fall in the small category, for which we expect no geological activity. However, if you look back at Figure 8.11, you'll see that the remaining jovian planets have 14 more medium-size or large moons among them: seven for Saturn, five for Uranus, and two for Neptune. Thanks especially to the *Cassini* spacecraft, which began orbiting Saturn in 2004, we have learned that some of these moons have surprisingly active geology.

Titan Saturn's moon Titan is the second-largest moon in the solar system (after Ganymede). It is also unique among the moons of our solar system in having a thick atmosphere—so thick that we cannot see through it with visible light (Figure 8.20). Titan's reddish color comes from chemicals in its atmosphere much like those that make smog over cities on Earth. The atmosphere is more than 95% nitrogen, not that different from the 77% nitrogen content of Earth's atmosphere. However, the rest of Earth's atmosphere is mostly oxygen, while the rest of Titan's consists of argon and methane (CH_4), ethane (C_2H_6), and other hydrogen compounds.

Callisto is heavily cratered, indicating an old surface that nonetheless may hide a deeply buried ocean.

Close-up photo shows a dark powder overlaying the low areas of the surface.

▲ FIGURE 8.19

Callisto, the outermost of the four Galilean moons, has a heavily cratered icy surface.

▲ FIGURE 8.20

Titan, as photographed by the *Cassini* spacecraft, is enshrouded by a thick atmosphere with clouds and haze. *Cassini* was outfitted with filters designed to peer through the atmosphere at the near-infrared wavelengths of light that are least affected by the atmosphere. The inset shows sunlight reflecting off a lake of liquid methane in Titan's north polar region.

Titan's atmosphere is thought to be a product of methane and ammonia gas released from Titan's interior and surface. Some is released by vaporization of methane and ammonia ices, and some may be released volcanically. Solar ultraviolet light can split these gas molecules, releasing hydrogen atoms that escape to space and leaving highly reactive compounds containing carbon and nitrogen, which react to make the other ingredients of Titan's atmosphere. For example, the abundant molecular nitrogen is made after ultraviolet light breaks down ammonia (NH_3) molecules, and ethane is made from methane.

Methane and ethane are both greenhouse gases and therefore give Titan a greenhouse effect [Section 7.1] that makes it warmer than it would be otherwise. Still, because of its great distance from the Sun, its surface temperature is a frigid 93 K (−180°C). The surface pressure on Titan is about 1.5 times the sea level pressure on Earth, which would be fairly comfortable if not for the lack of oxygen and the cold temperatures.

A moon with a thick atmosphere would be intriguing enough, but we have at least two other reasons for special interest in Titan. First, its complex atmospheric chemistry produces numerous carbon compounds—the chemicals that are the basis of life. Second, although it is far too cold for liquid water to exist on Titan, conditions are right for methane or ethane rain, which creates rivers flowing into lakes and seas. These facts led NASA and the European Space Agency (ESA) to combine forces to explore Titan. In 2005, NASA's *Cassini* "mother ship" released the ESA-built probe called *Huygens* (pronounced "Hoy-guns"), which parachuted to the surface (Figure 8.21). During its descent, the probe photographed river valleys merging together, flowing down to what looks like an ancient shoreline. On landing, instruments on the probe discovered that the surface has a hard crust but is a bit squishy below, perhaps like sand with liquid mixed in, and photos showed "ice boulders" rounded by erosion. All these results support the idea of a wet climate—but wet with liquid methane rather than liquid water.

Titan has a thick atmosphere, leading to methane rain and surprising erosional geology.

Subsequent *Cassini* observations taught us more about Titan. The brighter regions in the central image of Figure 8.21 are icy hills that may have been made by ice volcanoes. The dark valleys were probably created

▼ **FIGURE 8.21**

This sequence zooms in on the *Huygens* landing site on Titan. Left: a global view taken by the orbiting *Cassini* spacecraft. Center: an aerial view from the descending probe. Right: a surface view taken by the probe after landing; the "rocks," which are 10–20 centimeters across, are presumably made of water ice.

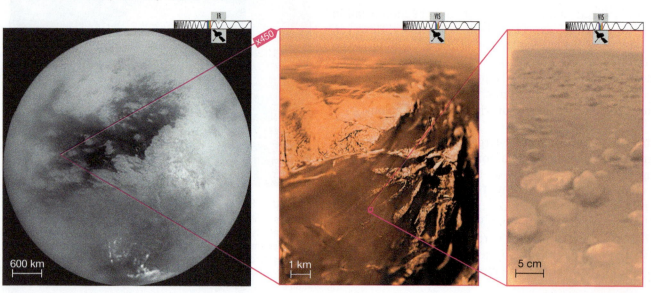

600 km

1 km

5 cm

when methane rain carried down "smog particles" that concentrated on river bottoms. The vast plains into which the valleys appear to empty are low-lying regions, but they do not appear to be liquid. Instead, they are probably covered by smog particles carried down by the rivers and then sculpted into vast dune fields by Titan's global winds. All in all, conditions in Titan's equatorial regions appear to be analogous to those in the desert southwest of the United States, where infrequent rainfall carves valleys and creates vast dry lakes called *playas* where the water evaporates or soaks into the ground. The polar regions of Titan, revealed by *Cassini* radar, contain numerous lakes of liquid methane or ethane (Figure 8.22). Images also reveal polar storm clouds and riverbeds leading into the lakes, suggesting that Titan has a methane/ethane cycle resembling the water cycle on Earth.

Perhaps the most astonishing result from the *Cassini/Huygens* mission is how familiar the landscape looks in this alien environment with unfamiliar materials. Instead of liquid water, Titan has liquid methane and ethane. Instead of rock, Titan has ice. Instead of molten lava, Titan has a slush of water ice mixed with ammonia. Instead of surface dirt, Titan's surface has smog-like particles that rain out of the sky and accumulate on the ground. The similarities between the physical processes that occur on Titan and Earth appear to be far more important in shaping the landscapes than the fact that the two worlds have very different compositions and temperatures.

think about it What other geological features might you expect on Titan, given its similarities to Earth? How might those features be different, given the differences in temperature and composition?

Saturn's Medium-Size Moons

The *Cassini* mission also flew past other moons of Saturn. The six medium-size moons reveal complex histories (Figure 8.23).

Only Mimas, the smallest of these six moons, shows little evidence of past volcanism or tectonics. It is essentially a heavily cratered iceball, with one huge crater nicknamed "Darth Crater" because of Mimas's resemblance to the Death Star in the *Star Wars* movies. Most of Saturn's other medium-size moons also have heavily cratered surfaces, confirming that they lack global geological activity today. However, we find abundant evidence of past volcanism and/or tectonics. Smooth regions appear to be places where icy lava once flowed, and close-up views of the bright streaks (such as the long streaks visible on Dione) show them to be vast sets of tectonic cliffs running parallel to one another.

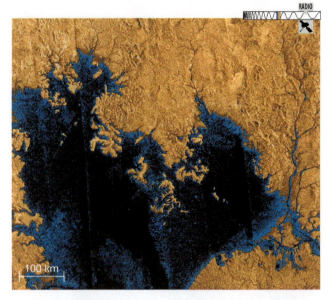

▲ **FIGURE 8.22**

Radar image of Ligeia Mare near Titan's north pole, showing lakes of liquid methane or ethane at a temperature of −180°C. Most solid surfaces reflect radar well, and these regions are artificially shaded tan to suggest land. The liquid surfaces reflect radar poorly, and these regions are shaded blue and black to suggest lakes.

▼ **FIGURE 8.23**

Portraits taken by the *Cassini* spacecraft of Saturn's medium-size moons (not to scale). All but Mimas show evidence of past volcanism and/or tectonics.

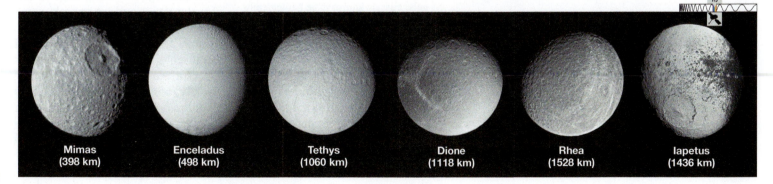

Mimas (398 km)   Enceladus (498 km)   Tethys (1060 km)   Dione (1118 km)   Rhea (1528 km)   Iapetus (1436 km)

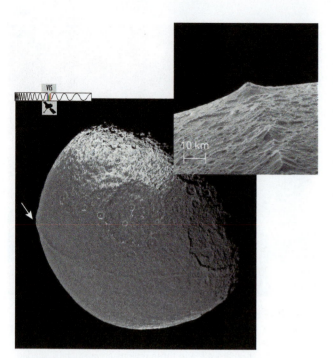

▲ FIGURE 8.24

Saturn's moon Iapetus has a 10-kilometer-tall equatorial ridge (white arrow) that spans nearly half its circumference. The inset shows a portion of the ridge in perspective.

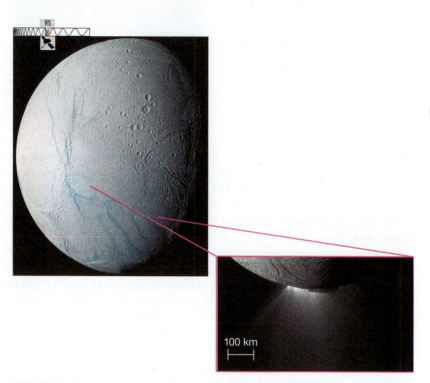

▲ FIGURE 8.25

Cassini photo of Saturn's moon Enceladus. The blue "tiger stripes" near the bottom of the main photo are regions of fresh ice that must have recently emerged from below. The colors are exaggerated; the image is a composite made at near-ultraviolet, visible, and near-infrared wavelengths. The inset shows Enceladus backlit by the Sun, with fountains of ice particles (and water vapor) clearly visible as they spray out of the south polar region.

Iapetus is particularly bizarre (Figure 8.24). It has an astonishing ridge more than 10 kilometers high that spans nearly half its circumference, curiously aligned along the equator. No one knows its origin, but it is likely the result of tectonic activity. Moreover, much of Iapetus appears coated in dark dust that apparently comes from Phoebe (see Figure 8.12), a small moon on which impacts can easily eject dust that then spirals inward toward Iapetus.

Enceladus provided an even bigger surprise: This moon is barely 500 kilometers across—small enough to fit inside the borders of Colorado—and yet it shows clear evidence of *ongoing* geological activity (Figure 8.25). Its surface has very few impact craters—and some regions have none at all—telling us that recent geological activity has erased older craters. The strange grooves near its south pole are measurably warmer than the surrounding terrain, and photographs show this region venting huge clouds of water vapor and ice crystals, some containing salt and rock dust. These fountains must have some subsurface source, and careful measurements of the way Enceladus wobbles on its axis as it orbits Saturn have led scientists to suspect that Enceladus has a global, subsurface ocean of liquid water or of a colder water/ammonia mixture. The ocean is estimated to lie 30 to 40 kilometers beneath the moon's surface and may be up to 30 kilometers in depth. This, in turn, makes us wonder about possible life on Enceladus. Internal heat on Enceladus comes from tidal heating through an orbital resonance (with Dione), though scientists were surprised to learn that the heating is enough to make Enceladus active today.

Enceladus is the smallest moon in the solar system known to be geologically active today.

The *Cassini* mission was scheduled to end shortly after this book was published in 2017, with scientists taking some risks with the spacecraft during its final year in hopes of gathering spectacular new results. What did the spacecraft discover during its final year? What happened to it at the end, and why?

Moons of Uranus and Neptune   We know far less about the moons of Uranus and Neptune, because they have been photographed close-up only once each, during the *Voyager 2* flybys of the 1980s. Nevertheless, we again see evidence of surprising geological activity.

Uranus has five medium-size moons (and no large moons), and at least three of them show evidence of past volcanism or tectonics. Miranda, the smallest of the five, is the most surprising (Figure 8.26). Despite its small size, it shows tremendous tectonic features and relatively few craters. Apparently, it underwent geological activity well after the heavy bombardment ended [Section 6.3], erasing its early craters.

The surprises continue with Neptune's moon Triton (Figure 8.27). Triton is a strange moon to begin with: It is a large moon, but it orbits Neptune "backward" (opposite to Neptune's rotation) and at a high inclination to Neptune's equator. These are telltale signs of a moon that was captured rather than formed in the disk of gas around its planet. No one knows how a moon as large as Triton could have been captured, but models suggest one possible mechanism: Triton may have once been a member of a binary Kuiper belt object that passed so close to Neptune that Triton lost energy and was captured while its companion gained energy and was flung away at high speed.

Triton's geology is just as surprising as its origin. It is smaller than our own Moon, yet its surface shows evidence of relatively recent geo-

Triton orbits Neptune "backward" and shows evidence of relatively recent geological activity.

logical activity. Some regions show signs of past volcanism, while others show wrinkly ridges (nicknamed "cantaloupe terrain") that appear tectonic in nature. Triton even has a very thin atmosphere that has left some wind streaks on its surface. It's likely that Triton was originally captured into an elliptical orbit, which may have led to enough tidal heating to explain its geological activity.

◆ Why are jovian moons more geologically active than small rocky planets?

Based on what we learned when studying the geology of the terrestrial worlds, the active geology of the jovian moons seems out of character with their sizes. Numerous jovian moons remained geologically active

Tidal heating plus the easy melting and deformation of ices means that even small icy moons can sustain geological activity.

far longer than Mercury or our Moon, yet they are no bigger and in many cases much smaller in size. However, there are two crucial differences between the jovian moons and the terrestrial worlds: composition and tidal heating.

Because they formed far from the Sun, most of the jovian moons contain ices that can melt or deform at far lower temperatures than rock. As a result, they can experience geological activity even when their interiors have cooled to temperatures far below those of rocky worlds. Indeed, except on Io, most of the volcanism that has occurred in the outer solar system is probably "ice volcanism" that produces a

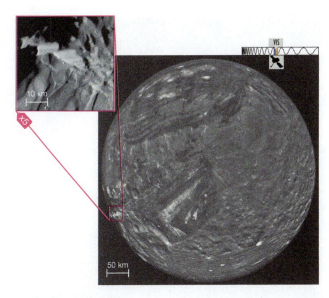

▲ **FIGURE 8.26**

The surface of Uranus's moon Miranda shows astonishing tectonic activity despite its small size. The cliff walls seen in the inset are higher than those of the Grand Canyon on Earth.

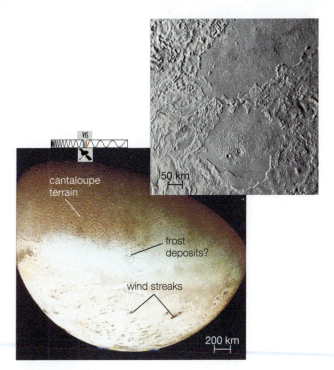

cantaloupe terrain

frost deposits?

wind streaks

50 km

200 km

▲ **FIGURE 8.27**

Neptune's moon Triton shows evidence of a surprising level of past geological activity. Main image: Triton's southern hemisphere as seen by *Voyager 2*. Inset: The close-up shows lava-filled impact basins similar to the lunar maria, but the lava was water or slush rather than molten rock.

Terrestrial Planet Geology

- Internal heat, primarily from radioactive decay, can cause volcanic and tectonic activity.

- Only large planets retain enough internal heat to stay geologically active today.

- Example: Mars (photo above) probably retains some internal heat. If it had been smaller, like Mercury, it would be geologically "dead" today. If it had been larger, like Earth, it would probably have much more active and ongoing tectonics and volcanism.

Jovian Moon Geology

- Tidal heating can cause tremendous geological activity on moons with elliptical orbits around massive planets.

- Even without tidal heating, icy materials can melt and deform at lower temperatures than rock, increasing the likelihood of geological activity.

- Together, these effects explain why icy moons are much more likely to have ongoing geological activity than rocky terrestrial worlds of the same size.

- Example: Ganymede (photo above) shows evidence of recent geological activity, even though it is similar in size to the geologically dead terrestrial planet Mercury.

▲ **FIGURE 8.28**

Jovian moons can be much more geologically active than terrestrial worlds of similar size because of their icy compositions and tidal heating, which is not an important factor on the terrestrial worlds.

lava composed mainly of water, perhaps mixed with methane and ammonia.

The major lesson, then, is that "ice geology" is possible at far lower temperatures than "rock geology." This fact, combined in many cases with tidal heating, explains how the jovian moons have had such interesting geological histories despite their small sizes. Figure 8.28 summarizes the differences between the geology of jovian moons and that of the terrestrial worlds. We will see the role of ice geology again when we discuss Pluto in Chapter 9.

8.3 Jovian Planet Rings

The jovian planet systems have three major components: the planets themselves, the moons, and their rings. We have already studied the planets and their moons, so we now turn our attention to their amazing rings.

◆ What are Saturn's rings like?

You can see Saturn's rings through a backyard telescope, but learning their true nature requires higher resolution (Figure 8.29). From Earth, the rings appear to be continuous, concentric sheets of material separated by a large gap (called the *Cassini division*). Spacecraft images reveal

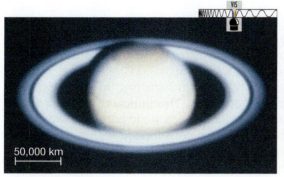

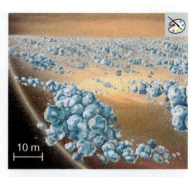

a This Earth-based telescopic view of Saturn makes the rings look like large, concentric sheets. The dark gap within the rings is called the *Cassini division*.

b This image of Saturn's rings from the *Cassini* spacecraft reveals many individual rings separated by narrow gaps.

c Artist's conception of particles in a ring system. Particles clump together because of gravity, but small random velocities cause collisions that break them up.

▲ **FIGURE 8.29**
Zooming in on Saturn's rings.

these "sheets" to be made of many individual rings, each separated from the next by a narrow gap. But even these appearances are somewhat deceiving. If we could wander into Saturn's rings, we'd find that they are made of countless icy particles ranging in size from dust grains to large boulders, sometimes clumped together by their mutual gravity. All are far too small to be photographed even from spacecraft passing nearby.

Ring Particle Characteristics   Spectroscopy reveals that Saturn's ring particles are mostly made of relatively reflective water ice. The rings look bright where they contain enough particles to intercept sunlight and scatter it back toward us. We see gaps in places where there are few particles to reflect sunlight.

Each individual ring particle orbits Saturn independently in accord with Kepler's laws, so the rings are much like myriad tiny moons. The individual ring particles are so close together that they collide frequently. In the densest parts of the rings, each particle collides with another every few hours. However, the collisions are fairly gentle: Despite the high orbital speeds of the ring particles, nearby ring particles orbit at nearly the same speed and in the same direction, and therefore touch only gently when they collide.

> **Saturn's rings are made of vast numbers of icy particles ranging in size from dust grains to boulders, each circling Saturn according to Kepler's laws.**

think about it   Which ring particles travel faster: those closer to Saturn or those farther away? Explain why. (*Hint:* Review Kepler's third law.)

The frequent collisions explain why Saturn's rings are one of the thinnest known astronomical structures. They span more than 270,000 kilometers in diameter but are only a few tens of *meters* thick. To understand how collisions keep the rings thin, imagine what would happen to a ring particle on an orbit slightly inclined to the central ring plane. The particle would collide with other particles every time its orbit intersected the ring plane, and its orbital tilt would be reduced with every collision. Before long, these collisions would force the particle to conform to the orbital pattern of the other particles, and any particle that moved away from the narrow ring plane would soon be brought back within it.

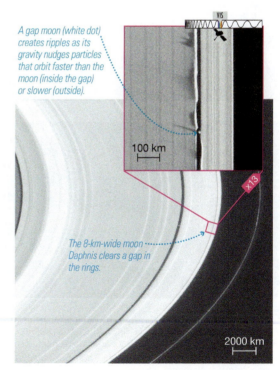

A gap moon (white dot) creates ripples as its gravity nudges particles that orbit faster than the moon (inside the gap) or slower (outside).

100 km

The 8-km-wide moon Daphnis clears a gap in the rings.

2000 km

▲ **FIGURE 8.30**
Small moons within the rings have important effects on ring structure (*Cassini* photos). The inset image was taken near Saturn's equinox, so the moon and ripples cast long shadows to the left.

Jupiter

Saturn

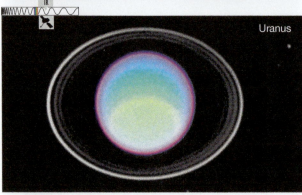

Uranus

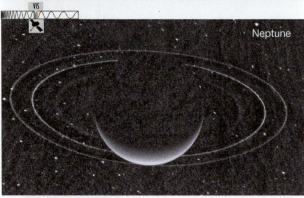

Neptune

▲ **FIGURE 8.31**

Four ring systems (not to scale). The rings differ in appearance and in the composition and sizes of the ring particles. (Jupiter: Keck telescope, infrared; Saturn: *Cassini*, visible; Uranus: Hubble Space Telescope, infrared; Neptune: *Voyager 2*, visible.)

Rings and Gaps   Close-up photographs show an astonishing number of rings, gaps, ripples, and other features—as many as 100,000 altogether (Figure 8.30). Scientists are still struggling to explain all the features, but some general ideas are now clear.

Rings and gaps are caused by particles bunching up at some orbital distances and being forced out at others. This bunching happens when gravity nudges the orbits of ring particles in some particular way. One source of nudging comes from small moons located within the gaps in the rings themselves, sometimes called *gap moons*. The gravity of a gap moon can effectively keep the gap clear of smaller ring particles. In some cases, two nearby gap moons can force particles between them into a very narrow ring. (The gap moons are often called *shepherd moons* in those cases, because they shepherd particles into line.)

Ring particles also may be nudged by the gravity from larger, more distant moons. For example, a ring particle orbiting about 120,000 kilometers from Saturn's center will circle the planet in exactly half the time it takes the moon Mimas to orbit. Every time Mimas returns to a certain location, the ring particle will also be at its original location and therefore will experience the same gravitational nudge from Mimas. The periodic nudges reinforce one another and clear a gap in the rings—in this case, the Cassini division. This type of reinforcement due to repeated gravitational tugs is another example of an *orbital resonance,* much like the orbital resonance that makes Io's orbit elliptical (see Figure 8.15b). Other orbital resonances, caused by moons both within the rings and farther out from Saturn, probably explain most of the intricate structures we see.

◆ Why do the jovian planets have rings?

Saturn's rings were once thought to be unique in the solar system, leading scientists to assume they were formed by some kind of rare event, such as a moon wandering too close to Saturn and being torn apart by tidal forces. However, we now know that all four jovian planets have rings (Figure 8.31), though Saturn's rings have more numerous and more reflective particles than the other ring systems. We therefore need an explanation for rings that doesn't require rare events to have happened for all four planets.

Some scientists once guessed that the ring particles might be leftover chunks of rock and ice that condensed in the disks of gas that orbited each jovian planet when it was young. This would explain why all four jovian planets have rings, because tidal forces near each planet would have prevented these chunks from accreting into a full-fledged moon. However, we now know that the ring particles cannot be leftovers from the birth of the planets, because they could not have survived for billions of years. Ring particles are continually being ground down in size, primarily by the impacts of the countless sand-size particles that orbit the Sun—the same types of particles that become meteors in Earth's atmosphere and cause micrometeorite impacts on the Moon [Section 7.2]. Millions of years of such tiny impacts would have ground the existing ring particles to dust long ago.

We are left with only one reasonable possibility: New particles must be continually supplied to the rings to replace those that are destroyed. These new particles must come from a source that lies in each planet's equatorial plane. The most likely source is numerous small "moonlets"—moons the size of gap moons (see Figure 8.30)—that

Ring particles cannot last for billions of years, so the rings we see today must be made of particles created recently.

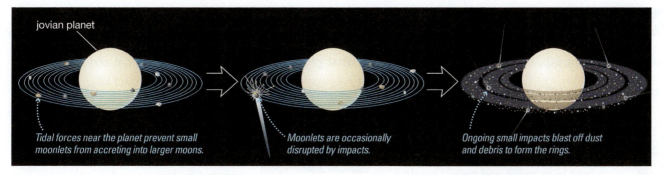

jovian planet

Tidal forces near the planet prevent small moonlets from accreting into larger moons.

Moonlets are occasionally disrupted by impacts.

Ongoing small impacts blast off dust and debris to form the rings.

▲ **FIGURE 8.32**
This illustration summarizes the current model for the origin of rings around the jovian planets.

formed in the disks of material orbiting the young jovian planets. As with the ring particles themselves, tiny impacts are gradually grinding away these small moons, but they are large enough to still exist despite 4½ billion years of such sandblasting.

The small moons contribute ring particles in two ways. First, each tiny impact releases particles from a small moon's surface, and these released particles become new, dust-size ring particles. Ongoing impacts ensure that some ring particles are present at all times. Second, occasional larger impacts can shatter a small moon completely, creating a supply of boulder-size ring particles. The frequent tiny impacts then slowly grind these boulders into smaller ring particles. Some of these particles are "recycled" by forming into small clumps, only to come apart again later on; others are ground down to dust and slowly spiral onto their planet. In summary, all ring particles ultimately come from the gradual dismantling of small moons that formed during the birth of the solar system (Figure 8.32).

New ring particles are released by impacts on small moons within the rings.

the big picture — Putting Chapter 8 into Perspective

In this chapter, we saw that the jovian planets really are a different kind of planet and, indeed, a different kind of planetary system. As you continue your study of the solar system, keep in mind the following "big picture" ideas:

- The jovian planets dwarf the terrestrial planets. Even some of their moons are as large as terrestrial worlds.

- The jovian planets may lack solid surfaces on which geology can occur, but they are interesting and dynamic worlds with rapid winds, huge storms, strong magnetic fields, and interiors in which common materials behave in unfamiliar ways.

- Despite their relatively small sizes and frigid temperatures, many jovian moons are geologically active by virtue of their icy compositions—a result of their formation in the outer regions of the solar nebula—and tidal heating.

- Ring systems probably owe their existence to small moons formed in the disks of gas that produced the jovian planets billions of years ago. The rings we see today are composed of particles liberated from those moons quite recently.

- Understanding jovian planet systems forced us to modify many of our earlier ideas about the solar system by adding the concepts of ice geology, tidal heating, and orbital resonances. Each new set of circumstances we discover offers new opportunities to learn how our universe works.

my cosmic perspective Jovian moons like Europa, once considered geologically dead and uninteresting because of their small size and great distance, could turn out to be our nearest neighbors in a biological sense.

8.1 A Different Kind of Planet

◆ What are jovian planets made of?

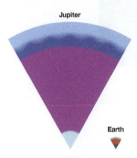

Jupiter and Saturn are made almost entirely of hydrogen and helium, while Uranus and Neptune are made mostly of hydrogen compounds mixed with metal and rock. They lack solid surfaces but have very high internal pressures and densities. Each jovian planet has a core about 10 times as massive as Earth, consisting of hydrogen compounds, metals, and rock, so they differ mainly in their surrounding layers of hydrogen and helium.

◆ What is the weather like on jovian planets?

The jovian planets all have multiple cloud layers that give them distinctive colors, fast winds, and large storms. Some storms, such as the **Great Red Spot,** can apparently rage for centuries or longer.

8.2 A Wealth of Worlds: Satellites of Ice and Rock

◆ What kinds of moons orbit the jovian planets?

We can categorize the sizes of the many known moons as small, medium, or large. Most of the medium and large moons probably formed in the disks of gas that surrounded the jovian planets when they were young. Smaller moons are often captured asteroids or comets.

◆ Why are Jupiter's Galilean moons geologically active?

Io is the most volcanically active object in the solar system, thanks to an interior kept hot by **tidal heating**—which occurs because Io's close orbit is made elliptical by **orbital resonance** with other moons of Jupiter. Europa (and possibly Ganymede) may have a deep, liquid water ocean under its icy crust, also due to tidal heating. Callisto is the least geologically active, since it has no orbital resonance or tidal heating, but it may also have a subsurface ocean.

◆ What geological activity do we see on Titan and other moons?

Many medium-size and large moons show a surprisingly high level of past or present volcanism or tectonics. Titan has a thick atmosphere and ongoing erosion, and Enceladus is also geologically active today. Triton, which apparently was captured by Neptune, also shows signs of recent geological activity.

◆ Why are jovian moons more geologically active than small rocky planets?

Ices deform and melt at much lower temperatures than rock, allowing icy volcanism and tectonics at surprisingly low temperatures. In addition, some jovian moons have a heat source—tidal heating—that is not important for the terrestrial worlds.

8.3 Jovian Planet Rings

◆ What are Saturn's rings like?

Saturn's rings are made up of countless individual particles, each orbiting Saturn independently like a tiny moon. The rings lie in Saturn's equatorial plane, and they are extremely thin.

◆ Why do the jovian planets have rings?

Ring particles probably come from the dismantling of small moons formed in the disks of gas that surrounded the jovian planets billions of years ago. Small ring particles come from countless tiny impacts on the surfaces of these moons, while larger ones come from impacts that shatter the moons.

visual skills check

Check your understanding of some of the many types of visual information used in astronomy. For additional practice, try the Chapter 8 Visual Quiz at MasteringAstronomy®.

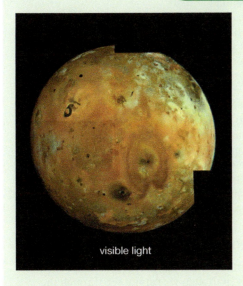

visible light

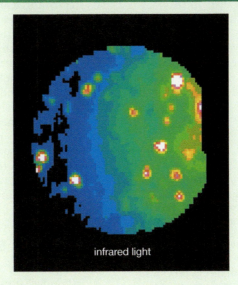

infrared light

Left: approximate colors of Io in visible light; black spots are volcanoes that are still active or have recently gone inactive. Right: infrared thermal emission from Io; the bright spots are active volcanoes. (Both images are composed from Galileo photos, but taken at different times.)

1. What do the colors represent in the right image?
   a. the actual colors of Io's surface
   b. the colors we would see if our eyes were sensitive to infrared
   c. the intensity of the infrared light
   d. regions of different chemical composition on the surface

2. Which color in the right image indicates regions with the highest temperature?
   a. blue   b. green   c. orange
   d. red   e. white

3. The right image was obtained when only part of Io was in sunlight. Based on the colors, which part of the surface was in sunlight?
   a. the left side
   b. the right side
   c. only the peaks of the volcanoes

4. By comparing the two images, what can you conclude about Io's volcanoes?
   a. Every black spot in the visible image has a bright spot in the infrared image, so all of Io's volcanoes were active when the photos were taken.
   b. There are more black spots in the visible image than bright spots in the infrared image, so many of Io's volcanoes were inactive when the photos were taken.
   c. There are more bright spots in the infrared image than black spots in the visible image, so new eruptions must have started after the visible photo was taken.

exercises and problems

MasteringAstronomy® For instructor-assigned homework and other learning materials, go to MasteringAstronomy®.

Review Questions

1. Briefly describe how differences in composition among the jovian planets can be traced to their formation.
2. Why is Jupiter so much more dense than Saturn? Could a planet be smaller in size than Jupiter but greater in mass? Explain.
3. Briefly describe the interior structure of Jupiter and explain why it is layered in this way. How do the interiors of the other jovian planets compare to that of Jupiter?
4. Why does Jupiter have such a strong magnetic field? Describe a few features of Jupiter's magnetosphere.
5. Briefly describe Jupiter's cloud layers. How do the cloud layers help explain Jupiter's colors? Why are Saturn's colors more subdued? Why are Uranus and Neptune blue?
6. Briefly describe Jupiter's weather patterns and contrast them with those on the other jovian planets. What is the *Great Red Spot*?
7. Briefly describe how we categorize jovian moons by size. What is the origin of most of the medium and large moons? What is the origin of many of the small moons?
8. Describe key features of Jupiter's four Galilean moons and Enceladus. Explain the roles of tidal heating and orbital resonances in explaining these features.

9. Describe the atmosphere and surface features of Titan.
10. Why do we think Triton is a captured moon? How might its capture be relevant to its geological activity?
11. Briefly explain why icy moons can have active geology at much smaller sizes than rocky worlds.
12. What are planetary rings made of, and how do they differ among the four jovian planets? Briefly describe the effects of gap moons and orbital resonance on ring systems.
13. Explain why ring particles must be replenished over time, and where we think ring particles come from.

Test Your Understanding

Surprising Discoveries?

Suppose someone claimed to make the discoveries described below. (These are not real discoveries.) Decide whether each discovery should be considered reasonable or surprising. Explain clearly; not all of these have definitive answers, so your explanation is more important than your chosen answer.

14. Saturn's core is pockmarked with impact craters and dotted with volcanoes erupting lava.
15. Neptune's deep blue color is not due to methane, as previously thought, but instead is due to its surface being covered with an ocean of liquid water.
16. A jovian planet in another star system has a moon as big as Mars.
17. A planet orbiting another star is made primarily of hydrogen and helium and has approximately the same mass as Jupiter but the same size as Neptune.
18. A previously unknown moon orbits Jupiter outside the orbits of other known moons. It is the smallest of Jupiter's moons but has several large, active volcanoes.
19. A previously unknown moon orbits Neptune in the planet's equatorial plane and in the same direction that Neptune rotates, but it is made almost entirely of metals such as iron and nickel.
20. An icy, medium-size moon orbits a jovian planet in a star system that is only a few hundred million years old. The moon shows evidence of active tectonics.
21. A jovian planet is discovered in a star system that is much older than our solar system. The planet has no moons at all, but it has a system of rings as spectacular as the rings of Saturn.
22. Future observations discover rainfall of liquid water on Titan.
23. During a future mission to Uranus, scientists discover it is orbited by another 20 previously unknown moons.

Quick Quiz

Choose the best answer to each of the following. Explain your reasoning with one or more complete sentences.

24. Which lists the jovian planets in order of increasing distance from the Sun? (a) Jupiter, Saturn, Uranus, Pluto (b) Saturn, Jupiter, Uranus, Neptune (c) Jupiter, Saturn, Uranus, Neptune
25. Why does Neptune appear blue and Jupiter red? (a) Neptune is hotter, which means bluer thermal emission. (b) Methane in Neptune's atmosphere absorbs red light. (c) Neptune's air molecules scatter blue light, much as Earth's atmosphere does.
26. Why is Jupiter denser than Saturn? (a) It has a larger proportion of rock and metal. (b) It has a larger proportion of hydrogen. (c) Its higher mass and gravity compress its interior.
27. Some jovian planets give off more energy than they receive because of (a) fusion in their cores. (b) tidal heating. (c) ongoing contraction or differentiation.

28. The main ingredients of most moons of the jovian planets are (a) rock and metal. (b) frozen hydrogen compounds. (c) hydrogen and helium.
29. Why is Io more volcanically active than our moon? (a) Io is much larger. (b) Io has a higher concentration of radioactive elements. (c) Io has a different internal heat source.
30. What is unusual about Triton? (a) It orbits its planet backward. (b) It does not keep the same face toward its planet. (c) It is the only moon with its own rings.
31. Which moon shows evidence of rainfall and erosion by some liquid substance? (a) Europa (b) Titan (c) Ganymede
32. Saturn's many moons affect its rings through (a) tidal forces. (b) orbital resonances. (c) magnetic field interactions.
33. Saturn's rings (a) have looked basically the same since they formed along with Saturn. (b) were created long ago when tidal forces tore apart a large moon. (c) are continually supplied with new particles by impacts with small moons.

Process of Science

34. *Europan Ocean.* Scientists strongly suspect that Europa has a subsurface ocean, even though we cannot see through the surface ice. Briefly explain why scientists think this ocean exists. Is the "belief" in a Europan ocean scientific? Explain.
35. *Breaking the Rules.* As discussed in Chapter 7, the geological "rules" for the terrestrial worlds tell us that a world as small as Io should not have any geological activity. However, the *Voyager* images of Io's volcanoes proved that the old "rules" had been wrong. Based on your understanding of the nature of science [Section 3.4], should this be seen as a failure in the process of science? Defend your opinion.
36. *Unanswered Question.* Choose one unanswered question about a jovian planet or moon. Write a few paragraphs discussing the question and the specific types of evidence needed to answer it.

Group Work Exercise

37. *Comparing Jovian Moons.* **Roles:** *Scribe* (collects data and takes notes on the group's activities), *Proposer* (proposes hypotheses and explanations of the data), *Skeptic* (points out weaknesses in the hypotheses and explanations), *Moderator* (leads group discussion and makes sure everyone contributes). **Activity:** Compare the moons of Jupiter, drawing on the data in Appendix E.
    a. Collect data on Jupiter's four largest moons from Table E.3 in Appendix E and determine which moon has the greatest density.
    b. Use Table E.3 to determine what other solar system moon most resembles the moon from part (a) in mass, radius, and density.
    c. Propose a hypothesis about the composition of the moon from part (a), based on its resemblance to the moon from part (b), and examine potential concerns about the viability of the hypothesis.
    d. Use Table E.3 to determine whether there is a trend in density with orbital distance among the major moons of Jupiter; briefly describe any trends.
    e. Suggest a hypothesis that accounts for any trend found in part (d), and discuss potential concerns with the hypothesis.
    f. Develop and describe an experiment that could test the hypotheses in parts (c) and (e).

Investigate Further

Short-Answer/Essay Questions

38. *The Importance of Rotation.* Suppose the material that formed Jupiter came together without any rotation so that no "jovian nebula" formed and the planet today wasn't spinning. How else would the jovian system be different? Think of as many effects as you can, and explain each in a sentence.

39. *Comparing Jovian Planets.* You can do comparative planetology armed only with telescopes and an understanding of gravity.
    a. The small moon Amalthea orbits Jupiter at about the same distance in kilometers as Mimas orbits Saturn, yet Mimas takes almost twice as long to orbit. From this observation, what can you conclude about how Jupiter and Saturn differ? Explain.
    b. Jupiter and Saturn are not very different in radius. When you combine this information with your answer to part (a), what can you conclude? Explain.

40. *Minor Ingredients Matter.* Suppose the jovian planets' atmospheres were composed only of hydrogen and helium, with no hydrogen compounds at all. How would the atmospheres be different in terms of clouds, color, and weather? Explain.

41. *Observing Project: Jupiter's Moons.* Using binoculars or a small telescope, view the moons of Jupiter. Make a sketch of what you see, or take a photograph. Repeat your observations several times (nightly, if possible) over a period of a couple of weeks. Can you determine which moon is which? Can you measure the moons' orbital periods? Can you determine their approximate distances from Jupiter? Explain.

42. *Observing Project: Saturn's Rings.* Using binoculars or a small telescope, view the rings of Saturn. Make a sketch of what you see, or take a photograph. What season is it in Saturn's northern hemisphere? How far do the rings extend above Saturn's atmosphere? Can you identify any gaps in the rings? Describe any other features you notice.

Quantitative Problems

Be sure to show all calculations clearly and state your final answers in complete sentences.

43. *Disappearing Moon.* Io loses about a ton (1000 kilograms) of sulfur dioxide per second to Jupiter's magnetosphere.
    a. At this rate, what fraction of its mass would Io lose in 4½ billion years?
    b. Suppose sulfur dioxide currently makes up 1% of Io's mass. When will Io run out of this gas at the current loss rate?

44. *Ring Particle Collisions.* Each ring particle in the densest part of Saturn's rings collides with another about every 5 hours. If a ring particle survived for the age of the solar system, how many collisions would it undergo?

45. *Prometheus and Pandora.* These two moons orbit Saturn at average distances of 139,350 and 141,700 kilometers, respectively.

a. Using Newton's version of Kepler's third law, find their two orbital periods. Find the percent difference in their distances and in their orbital periods.
b. Consider the two in a race around Saturn: In one Prometheus orbit, how far behind is Pandora (in units of time)? In how many Prometheus orbits will Pandora have fallen behind by one of its own orbital periods? Convert this number of periods back into units of time. This is how often the satellites pass by each other.

46. *Orbital Resonances.* Using the data in Appendix E, identify the orbital resonance relationship between Titan and Hyperion. (*Hint:* If the orbital period of one were 1.5 times that of the other, we would say that they were in a 3:2 resonance.) Which medium-size moon is in a 2:1 resonance with Enceladus?

47. *Titanic Titan.* What is the ratio of Titan's mass to that of all the other satellites of Saturn whose masses are listed in Appendix E? Calculate the strength of gravity on Titan compared to that on Mimas. Comment on how this affects the possibility of atmospheres on each.

48. *Saturn's Thin Rings.* Saturn's ring system is over 270,000 kilometers wide and approximately 50 meters thick. Assuming the rings could be shrunk down so that their diameter was the width of a dollar bill (6.6 centimeters), how thick would the rings be? Compare your answer to the actual thickness of a dollar bill (0.01 centimeter).

Discussion Questions

49. *Jovian Planet Mission.* We can study terrestrial planets up close by landing on them, but jovian planets have no surfaces to land on. Suppose that you were in charge of planning a long-term mission to "float" in the atmosphere of a jovian planet. Describe the technology you would use and how you would ensure survival for any people assigned to this mission.

50. *Pick a Moon.* Suppose you could choose any one moon to visit in the solar system. Which one would you pick, and why? What dangers would you face in your visit to this moon? What kinds of scientific instruments would you want to bring along for studies?

Web Projects

51. *News from Cassini.* Find the latest news about the *Cassini* mission to Saturn. What is the current mission status? Write a short report about the mission's status and results too recent to be in the textbooks.

52. *Oceans of Europa.* The possibility of a subsurface ocean on Europa holds great scientific interest. Investigate plans for future study of Europa, either from Earth or with spacecraft. Write a short summary of the plans and how they might help us learn whether Europa really has an ocean and, if so, what it might contain.

9 Asteroids, Comets, and Dwarf Planets
Their Nature, Orbits, and Impacts

1930
Lowell
Observatory

2010
Hubble Space Telescope

2015
New Horizons

Pluto as we've seen it through time. Left: Pluto is circled in this discovery image from 1930. Center: Pluto as revealed in 2010, based on computer image processing of Hubble Space Telescope images. Right: Pluto revealed by the *New Horizons* spacecraft in 2015, showing the same hemisphere as in the Hubble image.

LEARNING GOALS

9.1 Classifying Small Bodies
- ◆ How do we classify small bodies?

9.2 Asteroids
- ◆ What are asteroids like?
- ◆ What do meteorites tell us about asteroids and the early solar system?
- ◆ Why is there an asteroid belt?

9.3 Comets
- ◆ Why do comets grow tails?
- ◆ Where do comets come from?

9.4 Pluto and the Kuiper Belt
- ◆ What is Pluto like?
- ◆ What do we know about other Kuiper belt comets?

9.5 Cosmic Collisions: Small Bodies Versus the Planets
- ◆ Did an impact kill the dinosaurs?
- ◆ How great is the impact risk today?
- ◆ How do the jovian planets affect impact rates and life on Earth?

A steroids and comets might at first seem insignificant compared to the planets and moons we've discussed so far, but there is strength in their large numbers. The appearance of a comet has more than once altered the course of human history when our ancestors acted on superstitions related to the sighting. More profoundly, asteroids or comets brought water and other chemical ingredients that helped make life on Earth possible, while their occasional impacts have scarred our planet with impact craters and sometimes altered the course of biological evolution.

In this chapter, we will explore the small bodies of our solar system. We'll consider asteroids and the pieces of them that fall to Earth as meteorites, along with comets and dwarf planets like Pluto, Eris, and Ceres. We'll also explore the dramatic effects of the occasional collisions between small bodies and large planets.

ESSENTIAL PREPARATION

1. What does the solar system look like? [Section 6.1]
2. Where did asteroids and comets come from? [Section 6.3]
3. What processes shape Earth's surface? [Section 7.1]

9.1 Classifying Small Bodies

Vast numbers of small bodies orbit our Sun, ranging in size from specks of dust to dwarf planets like Pluto. These objects vary greatly in many properties, making it a challenge to classify them neatly.

◆ How do we classify small bodies?

Recall that, today, we use relatively simple definitions of asteroids and comets: Both orbit the Sun and are too small to be considered planets, but asteroids are rocky while comets are ice-rich [Section 6.2]. But these definitions have changed over time (and are still subject to some debate), with the "dwarf planet" category being used only since 2006. Moreover, pieces of asteroids or comets sometimes fall to Earth, and we then give them different names. We therefore begin with a brief overview of the most commonly used classifications.

Asteroids The word *asteroid*, which means "starlike," is an artifact from the time when all we knew about asteroids was that they appeared as star-like points of light in telescopes. Even at that time, however, astronomers recognized that their motion relative to stars showed them to be in our solar system (Figure 9.1). The first asteroid discovered was Ceres, in 1801, and it was initially called a "planet." Three more asteroids (Pallas, Juno, and Vesta) were discovered over the next seven years, and as astronomers realized how small these objects were compared to the other planets, they came to be called "minor planets." The term *asteroid* did not come into wide use until many decades later, after asteroid discoveries began to come at a more rapid pace and it became clear that most of them lie in the *asteroid belt* between Mars and Jupiter (see Figure 6.1).

Comets For most of human history, comets were familiar only from their occasional presence in the night sky. Every few years, a comet becomes visible to the naked eye, appearing as a fuzzy ball with a long tail. Indeed, the word *comet* comes from the Greek word for "hair," a reference to the appearance of their tails in our sky (Figure 9.2). In photographs, the tails seem to suggest that comets are racing across the sky, but they are

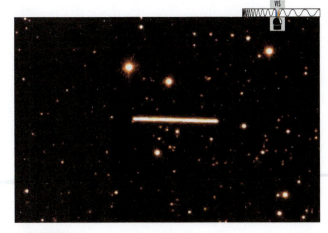

▲ **FIGURE 9.1**

Because asteroids orbit the Sun, they move through our sky relative to the stars. In this long-exposure photograph, stars show up as distinct dots, while the motion of an asteroid relative to the stars makes it show up as a short streak.

a Comet Hyakutake.

b Comet Hale-Bopp, photographed over Boulder, Colorado.

▲ **FIGURE 9.2**
Brilliant comets can appear at almost any time, as demonstrated by the back-to-back appearances of Comet Hyakutake in 1996 and Comet Hale-Bopp in 1997.

not. If you watch a comet for minutes or hours, it will remain nearly stationary relative to the stars around it in the sky. You'll notice its gradual motion relative to the constellations only over a period of many days, and a comet may remain visible for weeks before it fades from view.

We now know that the vast majority of comets do not have tails and never venture anywhere close to Earth. Instead, they remain in the outer reaches of our solar system, orbiting the Sun far beyond the orbit of Neptune in the two vast reservoirs we call the *Kuiper belt* and the *Oort cloud* (see Figure 6.1, Step 3). The comets that appear with tails in the night sky are rare ones that have had their orbits changed by the gravitational influences of planets, other comets, or stars passing by in the distance, causing them to venture into the inner solar system. Most of these comets will not return to the inner solar system for thousands of years, if ever. A few happen to pass near enough to a planet to have their orbits changed further, and some end up on elliptical orbits that periodically bring them close to the Sun. The most famous example is Halley's Comet, which orbits the Sun every 76 years and will next be seen in 2061.

> Most comets remain perpetually frozen in the outer solar system. Only a few enter the inner solar system, where they can grow tails.

see it for yourself Bright comets can be quite photogenic. Search the Web for comet images taken from Earth. Which is your favorite? Which has the best combination of beauty and scientifically interesting detail?

Dwarf Planets Just as the first asteroids were called planets, Pluto too was called a planet after its discovery in 1930. However, as scientists learned more about it, Pluto was recognized as a misfit among the planets because of its small size (a mass only about 0.2% that of Earth), ice-rich composition, and an orbit much more eccentric and more inclined to the ecliptic plane than that of any of the other planets.

Further questions about Pluto arose as astronomers better understood the origin of comets. By the 1950s, astronomers realized that many of the comets that visit the inner solar system must be coming from the region of the *Kuiper belt*, and that Pluto orbits the Sun near the middle of this region. In other words, Pluto began to seem more and more like an unusually large comet. In the 1990s, astronomers began to discover other Pluto-like objects in this region, with the only major difference being that these other objects were smaller than Pluto. Apparently, Pluto was not so unusual in this part of the solar system.

As time passed, larger and larger objects were found in the Kuiper belt, culminating with the 2005 discovery of Eris (Figure 9.3), which is about the same size as Pluto but about 27% larger in mass. Eris is named for a Greek goddess who caused strife and arguments among humans, a commentary on the arguments its discovery caused about the definition of "planet." After all, if Pluto was a planet, then surely the more massive Eris must be one as well. This also led scientists to question the status of the numerous objects only slightly smaller than Pluto. Would they too become "planets"?

In 2006, the International Astronomical Union created the *dwarf planet* category to accommodate Pluto, Eris, and other "small bodies" that are large enough to be round (see Special Topic, page 7). Two other members of the Kuiper belt—Makemake and Haumea*—have also been designated dwarf planets, as has the asteroid Ceres. However, because the definition depends on roundness and we may not always know the precise shape of a distant object, dozens of other objects may yet join the list. Figure 9.4 compares the sizes of several dwarf planets and potential dwarf planets to that of Earth.

For most practical purposes, we now have straightforward definitions: Asteroids are rocky leftover planetesimals that orbit the Sun, comets are icy leftover planetesimals, and dwarf planets can be either asteroids or comets that are large enough to be round. However, even these seemingly simple boundaries can be fuzzy. One difficult case concerns the Kuiper belt, where all the objects from the smallest boulders to the largest dwarf planets probably share the same basic composition of ice and rock. In other words, they are all essentially comets of different sizes. That is why we often refer to all of them as *comets* of the Kuiper belt. However, some astronomers object to calling objects "comets" if they never venture into the inner solar system and show tails. As a result, you may also hear these objects referred to as *Kuiper belt objects* (*KBOs*) or *trans-Neptunian objects* (*TNOs*).

> In terms of composition, Pluto, Eris, and other large objects of the Kuiper belt are essentially large comets.

think about it Suppose that we someday discover an object in the Kuiper belt (or in the similar zone of another star system) that is Pluto-like in composition but as large in size as Earth. Under current definitions, the fact that it orbits in the same region as many similar objects would qualify it as a dwarf planet. Does that seem reasonable for an object the size of Earth? How would *you* classify it? Defend your opinion.

Meteors and Meteorites In everyday language, we often use the terms *meteors* and *meteorites* interchangeably. Technically, however, a **meteor** (which means "a thing in the air") is only a flash of light caused by a particle of dust or rock entering our atmosphere at high speed, not the particle itself. Meteors are sometimes called *shooting stars* or *falling stars*, because some people once thought they really were stars falling from the sky.

The vast majority of the particles that make meteors are no larger than peas and burn up completely before reaching the ground. Only in rare cases is a meteor caused by a chunk of rock large enough to survive

---

*Haumea is actually oblong, but it counts as a dwarf planet because it would be round if not for its high rotation rate.

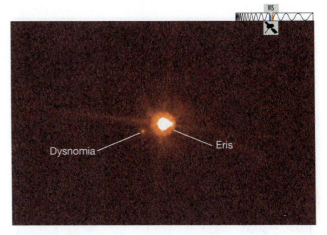

▲ **FIGURE 9.3**
Eris and its moon, photographed by the Hubble Space Telescope.

▲ **FIGURE 9.4**
The largest known objects of the Kuiper belt (as of 2016) and asteroid belt, along with their known moons, shown to scale with Earth for comparison. All except Pluto, Ceres, and Vesta are paintings based on guesses about their appearances. Pluto, Eris, Makemake, Haumea, and Ceres are officially considered dwarf planets, but some of the others may yet join the list.

▲ **FIGURE 9.5**

This large meteorite, called the Ahnighito Meteorite, is located at the American Museum of Natural History in New York. Its dark, pitted surface is a result of its fiery passage through Earth's atmosphere.

the plunge through our atmosphere and leave a **meteorite** (which means "associated with meteors") on the ground. Meteorites are usually covered with a dark, pitted crust resulting from their fiery passage through the atmosphere (Figure 9.5).

The origin of meteorites was long a mystery, but we now know that the vast majority of meteorites come from the asteroid belt, which means they are essentially either very small asteroids or pieces of asteroids. However, in a few cases, scientists have identified meteorites with compositions that appear to match either the Moon or Mars, and careful analysis makes us very confident that these meteorites were indeed chipped off these worlds. This makes sense: Moderately large impacts can blast surface material from terrestrial worlds into interplanetary space, where the rocks can orbit the Sun until they come crashing down on another world. Calculations show that it is not surprising that we should have found a few meteorites from the Moon and Mars in this way. These *lunar meteorites* and *Martian meteorites* therefore represent direct samples from these worlds, which makes them very valuable for scientific study.

9.2 Asteroids

We are now ready to turn our attention to studying asteroids. Scientifically, these small bodies are very important, because many of them remain much as they were when they first formed, some 4½ billion years ago. They therefore have clues to the story of our solar system's birth encoded in their compositions, locations, and numbers.

◆ What are asteroids like?

Asteroids come in a wide variety of sizes and shapes. The largest—dwarf planet Ceres—contains nearly as much mass as all other asteroids put together and is just under 1000 kilometers in diameter, or a little less than one-third the diameter of our Moon. Its low density and location in the outer part of the asteroid belt suggest that water ice makes up a significant fraction of its composition. The *Dawn* spacecraft entered orbit of Ceres in March 2015, giving us our first close-up look at this remarkable world (Figure 9.6). As expected, Ceres is covered by craters left by many past impacts. But it also held some surprises, including bright spots on its surface, which might be made of salt or other mineral deposits left behind as water vaporized and disappeared.

Prior to reaching Ceres, the *Dawn* spacecraft spent about 14 months (during 2011 and 2012) orbiting Vesta, the second-largest asteroid by mass. *Dawn* images revealed a battered world that is nearly 25% wider than it is tall (Figure 9.7). This irregular shape is probably a result of an impact that gouged out a huge crater near the south pole, which means that Vesta was probably once much more spherical in shape. Vesta may also have been volcanically active early in its history.

Besides Ceres and Vesta, about a dozen other asteroids are large enough that we would call them medium-size moons if they orbited a planet, but most asteroids are far smaller. Scientists estimate that there are more than a million asteroids that measure at least 1 kilometer across, and many more millions that are smaller in size. Despite their large numbers, asteroids don't add up to much in total mass. If we could put all the asteroids together (including Ceres and Vesta) and allow gravity to compress them into a

The total mass of all asteroids is much less than the mass of any terrestrial planet.

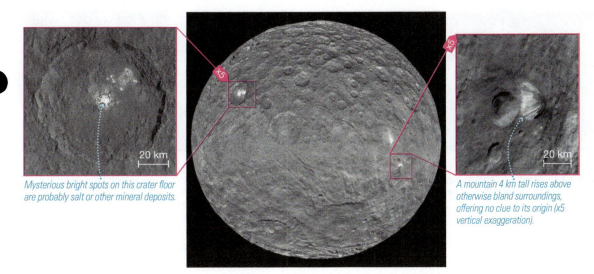

Mysterious bright spots on this crater floor are probably salt or other mineral deposits.

A mountain 4 km tall rises above otherwise bland surroundings, offering no clue to its origin (x5 vertical exaggeration).

20 km

20 km

▲ **FIGURE 9.6**
The dwarf planet Ceres, imaged by the *Dawn* spacecraft. The global view shows abundant craters and unexplained cracks and faults, while the insets focus on some of the most intriguing mysteries.

sphere, they'd make an object less than 2000 kilometers in diameter—just over half the diameter of our Moon.

Spacecraft visits have begun to reveal the character of small asteroids as well as large ones. Small asteroids come in diverse shapes (Figure 9.8), because their gravity is too weak to reshape them into spheres. Many may also be fragments of larger asteroids that were shattered in collisions. Some spacecraft have even been designed to return asteroid samples to Earth. The first was Japan's *Hayabusa* mission, which in 2010 returned a small dust sample from the asteroid Itokawa (see Figure 9.8d). The follow-up *Hayabusa 2*, launched in 2014, is on its way to collecting an asteroid sample in 2018 that is scheduled to be returned to Earth in 2020. NASA's ambitious *Osiris-REX* mission, launched in 2016, aims to return a sample in 2023 from an asteroid called 101955 Bennu, thought to be a carbon-rich asteroid that may be similar to the types of objects that brought many of the ingredients for oceans, atmosphere, and life to Earth.

◆ What do meteorites tell us about asteroids and the early solar system?

We can learn even more about asteroids by studying samples of them, and while spacecraft are only beginning to collect samples directly, we

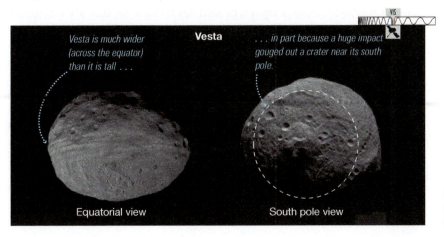

VIS

Vesta

Vesta is much wider (across the equator) than it is tall . . .

. . . in part because a huge impact gouged out a crater near its south pole.

Equatorial view

South pole view

◀ **FIGURE 9.7**
Global images of Vesta taken by the *Dawn* spacecraft. The dashed circle is the outline of the huge polar crater.

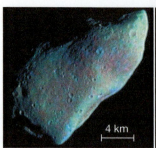

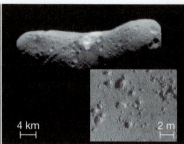
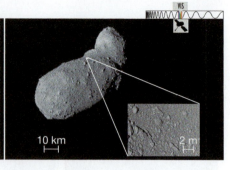

a Gaspra, photographed by the *Galileo* spacecraft. Colors are exaggerated to show detail.

b Mathilde, photographed by the *Near-Earth Asteroid Rendezvous* (*NEAR*) spacecraft on its way to Eros.

c Eros, photographed by the *NEAR* spacecraft, which orbited Eros for a year before ending its mission with a soft landing on the asteroid's surface.

d Itokawa, photographed by the Japanese *Hayabusa* mission, which landed on the surface and captured a sample that it returned to Earth.

▲ **FIGURE 9.8**

Close-up views of selected asteroids studied by spacecraft.

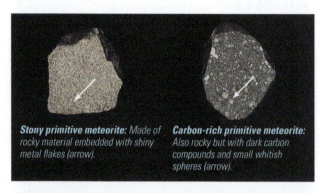

Stony primitive meteorite: Made of rocky material embedded with shiny metal flakes (arrow).

Carbon-rich primitive meteorite: Also rocky but with dark carbon compounds and small whitish spheres (arrow).

a Primitive meteorites.

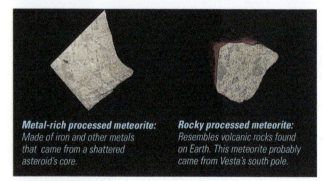

Metal-rich processed meteorite: Made of iron and other metals that came from a shattered asteroid's core.

Rocky processed meteorite: Resembles volcanic rocks found on Earth. This meteorite probably came from Vesta's south pole.

b Processed meteorites.

▲ **FIGURE 9.9**

There are two basic types of meteorites: primitive and processed. Each also has two subtypes. They are shown slightly smaller than actual size. (The meteorites have flat faces because they have been sliced with rock saws.)

already have tens of thousands of other asteroid samples—the rocks called *meteorites* that fall from the sky. Detailed analysis of thousands of meteorites shows that they come in two basic types:

- **Primitive meteorites** (Figure 9.9a) are "primitive" in the sense of being remnants from the time when solid material first condensed from the solar nebula. Radiometric dating confirms them to be the oldest rocks in the solar system. Primitive meteorites come in two major subtypes, which have differences in composition that probably reflect where they condensed: Those containing carbon compounds must have condensed farther out in the asteroid belt, where it was cool enough for such compounds and even some water to condense.

- **Processed meteorites** (Figure 9.9b) are "processed" in the sense that, unlike the primitive meteorites, they have been remade over time. Radiometric dating confirms that processed meteorites are slightly younger than primitive meteorites, just as we would expect. More specifically, processed meteorites appear to have come from asteroids that, like the terrestrial worlds, were large enough to have undergone differentiation into a core-mantle-crust structure [Section 7.1]. Processed meteorites also come in two major subtypes. Some are metal-rich, suggesting that they are fragments of the cores of shattered asteroids. Others have rocky compositions suggesting that they are fragments from mantles or crusts; some even appear volcanic in origin. Many of these appear to have come from Vesta, presumably blasted out by the impact that formed Vesta's large south polar crater.

Both types of meteorites teach us important lessons about our solar system. Primitive meteorites represent samples of material that are essentially unchanged since they first accreted in the solar nebula. They therefore provide information about the composition of the solidified material from which the planets formed, and their ages tell us the age of the solar system [Section 6.4]. Processed meteorites provide detailed information about the larger asteroids from which they came. Those that appear volcanic in origin tell us that some asteroids were geologically active when they were young. Those that come from the cores or mantles of shattered asteroids essentially present us with an opportunity to study a "dissected planet," and they represent a form of direct proof that large worlds really do undergo differentiation, confirming what we infer from seismic studies of Earth.

Most meteorites are pieces of asteroids, and they teach us much about the early history of our solar system.

◆ Why is there an asteroid belt?

The vast majority of asteroids orbit the Sun in the asteroid belt between the orbits of Mars and Jupiter (Figure 9.10). But why are asteroids concentrated in this region, and why didn't a full-fledged planet form instead?

think about it Why don't we find asteroids beyond the orbit of Jupiter? (*Hint:* Recall the effects of the frost line in the young solar system.)

The answer lies with gravitational effects of Jupiter. Virtually all planetesimals that formed inside the orbit of Mars eventually accreted onto one of the inner planets. But those that formed between Mars and Jupiter were strongly influenced by *orbital resonances* with Jupiter, and only a small fraction of these planetesimals ended up with orbits that have allowed them to survive as asteroids to this day. Recall that an orbital resonance occurs whenever two objects periodically line up with each other [Section 8.2]. In the asteroid belt, an orbital resonance occurs whenever an asteroid has an orbital period that is a simple fraction of Jupiter's orbital period, such as $\frac{1}{2}$, $\frac{1}{4}$, or $\frac{2}{5}$. In those cases, the asteroid experiences repeated tugs from Jupiter that tend to nudge it out of that orbit. For example, an asteroid with an orbital period of 6 years—half of Jupiter's 12-year period—would receive the same gravitational nudge from Jupiter every 12 years and therefore would soon be pushed out of this orbit. We can see the effect of orbital resonances on a graph showing the numbers of asteroids with various orbital periods (Figure 9.11). Notice the gaps that indicate a lack of asteroids with periods in resonance with Jupiter, confirming that these orbits have been cleared by the resonances. (The gaps are often called *Kirkwood gaps,* after their discoverer.) Asteroids that don't have an orbital resonance with Jupiter can last much longer—in some cases for billions of years—but in the long run are likely to suffer a collision with other asteroids that have been nudged by Jupiter.

Orbital resonances probably also explain why no planet formed between Mars and Jupiter. Early in the solar system's history, this region probably contained more than enough rocky material to form another terrestrial planet. However, resonances with the young Jupiter disrupted the orbits of planetesimals in this region, sometimes sending them crashing into each other and sometimes kicking them out of the region. Once kicked out, the planetesimals ultimately either crashed into a planet or moon or were flung out of the solar system or into the Sun. Over the next 4½ billion years, these ongoing orbital disruptions caused the asteroid belt to lose most of its original mass.

Jupiter's gravity, through the influence of orbital resonances, prevented asteroids from accreting into a planet and still shapes their orbits today.

The asteroid belt is still undergoing slow change. Jupiter's gravity continues to nudge asteroid orbits, sending asteroids on collision courses with each other and occasionally the planets. A major collision occurs somewhere in the asteroid belt every 100,000 years or so. Over long periods of time, larger asteroids continue to be broken into smaller ones, with each collision also creating numerous dust-size particles. The asteroid belt has been grinding itself down for more than 4 billion years and will continue to do so for as long as the solar system exists.

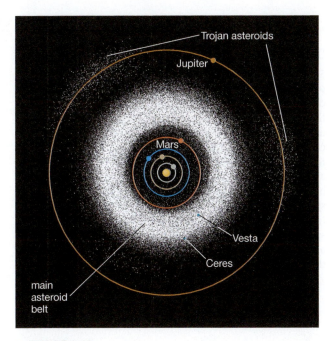

▲ **FIGURE 9.10**
This figure shows the positions of more than 150,000 asteroids on a single night, and the locations of Vesta and Ceres when the *Dawn* mission reached each of them. To scale, the asteroids themselves would be much smaller than shown. The asteroids that share Jupiter's orbit, found 60° ahead of and behind Jupiter, are called *Trojan asteroids.*

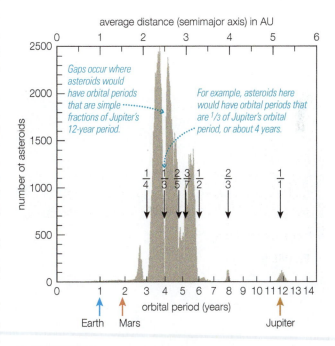

▲ **FIGURE 9.11**
This graph shows the numbers of asteroids with various orbital periods, which correspond to different average distances from the Sun (labeled along the top). Notice the gaps created by orbital resonances with Jupiter. (Some resonances, such as $\frac{1}{1}$ or $\frac{2}{3}$, are stable and tend to collect asteroids; $\frac{1}{1}$ represents the Trojan asteroids, which have the same orbital period as Jupiter.)

common misconceptions

Dodge Those Asteroids!

Science fiction movies often show brave spacecraft pilots navigating through crowded fields of asteroids, dodging this way and that as they heroically pass through with only a few bumps and bruises. It's great drama, but not very realistic. The asteroid belt looks crowded in Figure 9.10, but in reality it is an enormous region of space. Despite their large numbers, asteroids are so far apart on average that it would take incredibly bad luck to crash into one by accident. Indeed, spacecraft must be carefully guided to fly close enough to an asteroid to take a decent photograph. Future space travelers will have plenty of dangers to worry about, but dodging asteroids is not likely to be one of them.

9.3 Comets

We now turn our attention to comets. Although we now know that Pluto and other large objects of the Kuiper belt are essentially large comets, for most of human history the only known comets were the relatively small ones that sometimes enter the inner solar system and grow long tails. In this section, we'll explore modern understanding of small comets and how we learned that they come from the Kuiper belt and Oort cloud.

◆ Why do comets grow tails?

Far from the Sun, small comets must look much like small asteroids, with assorted shapes made possible by the fact that gravity is too weak to compress them into spheres. But they differ from asteroids in composition, as we expect for objects that formed in the cold outer solar system: Comets are basically chunks of ice mixed with rocky dust and some more complex chemicals, and hence they are often described as "dirty snowballs."

The Flashy Lives of Comets   The "dirty snowball" idea explains how comets grow tails when they are heated by the Sun. To see what happens, let's follow the comet path shown in Figure 9.12.

Far from the Sun, the comet is completely frozen—in essence, the "dirty snowball" in solid form. For a comet plunging inward, we call this frozen center the **nucleus** of the comet. As a comet accelerates toward the Sun, its surface temperature increases, and ices begin to vaporize into gas that easily escapes the comet's weak gravity. Some of the escaping gas drags away dust particles from the nucleus, and the gas and dust create a huge, dusty atmosphere called a **coma.** The coma grows as the comet continues into the inner solar system, and some of the gas

When a comet nears the Sun, its ices can vaporize into gas and carry off dust, creating a coma and long tails.

▼ **FIGURE 9.12**

A comet grows a coma and tail around its nucleus only if it happens to come close to the Sun. Most comets never do this, instead remaining perpetually frozen in the far outer solar system.

Far from the Sun, the nucleus is frozen.

Nucleus warms and ice begins to vaporize into gas.

Gas coma begins to form around nucleus when comet is about 5 AU from Sun.

Tails form by about 1 AU from Sun; tails point away from Sun.

Dust tail is pushed by sunlight.

Earth's orbit

solar radiation

solar wind

Solar heating diminishes; coma and tail disappear between 3 and 5 AU from Sun.

Larger particles are unaffected by sunlight or solar wind.

Plasma tail is swept back by solar wind.

a This diagram (not to scale) shows the changes that occur when a comet's orbit takes it on a passage into the inner solar system.

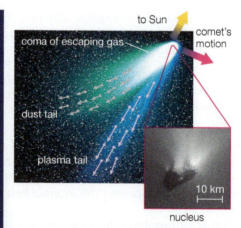

to Sun

comet's motion

coma of escaping gas

dust tail

plasma tail

10 km

nucleus

b Anatomy of a comet. The larger image is a ground-based photo of Comet Hale-Bopp. The inset shows the nucleus of Halley's Comet photographed by the *Giotto* spacecraft.

and dust is pushed away from the Sun, forming the comet's tails. Comet tails can be hundreds of millions of kilometers in length.

Comets have two visible tails. The **plasma tail** consists of gas that is ionized by ultraviolet light from the Sun and pushed outward by the solar wind. That is why the plasma tail extends almost directly away from the Sun at all times. The **dust tail** consists of dust-size particles that are unaffected by the solar wind and instead are pushed outward by the much weaker pressure of sunlight itself (*radiation pressure*). The dust tail therefore points generally away from the Sun, but has a slight curve back in the direction the comet came from.

After the comet loops around the Sun and begins to head back outward, vaporization declines, the coma dissipates, and the tails disappear. Nothing happens until the comet again comes sunward—in a century, a millennium, a million years, or perhaps never. Comets that do return eventually use up their ices, some crumbling into pieces and others becoming inactive nuclei.

Spacecraft missions have taught us more about comets. Our first clear view of a comet nucleus came from the European Space Agency's *Giotto* spacecraft, which flew past Halley's Comet during its 1986 visit to the inner solar system (see Figure 9.12b). In 2004, NASA's *Stardust* spacecraft captured dust particles from Comet Wild 2 (pronounced "vilt two"), returning them to Earth for scientific study. In 2005, NASA's *Deep Impact* mission sent a 370-kilogram projectile crashing into Comet Tempel 1 at 37,000 kilometers per hour. The impact created a plume of dust from the surface—so much dust that the comet must have been dust-covered to a depth of tens of meters—and hot gas composed of material vaporized from deep within the comet. Spectroscopy of this gas showed that the comet contains many complex organic molecules.

More recently, the European Space Agency's *Rosetta* mission began orbiting Comet 67P/Churyumov-Gerasimenko (Comet C-G for short) when it was still almost fully frozen in 2014, then stayed with the comet for more than a year as it approached and passed perihelion (Figure 9.13). *Rosetta* also sent a small lander, named *Philae*, to the surface of the comet, though it did not anchor itself as planned and therefore returned less data than had been hoped. Nevertheless, *Rosetta* and *Philae* already provided a wealth of information on the nature and behavior of comets. One of the most

a September 2014, before substantial comet activity had begun.

b November 2014.

c Close-up of a jet shortly before perihelion in August 2015.

▲ **FIGURE 9.13**
Comet "C-G" (67P/Churyumov-Gerasimenko) as imaged by the *Rosetta* spacecraft.

interesting results shows that the water ice that drives cometary activity near the Sun is hidden under a substantial crust of dusty material that is composed of rock and carbon-bearing molecules. Virtually no ice is exposed at the surface. This explains the striking features in Figure 9.13c, in which we see jets of water vapor emanating from pits, or "sinkholes," where the crust has collapsed, exposing the ice at depth.

Comet Tails and Meteor Showers Comets also eject sand- to pebble-size particles that are too big to be affected by either the solar wind or sunlight. These particles essentially form a third, invisible tail that follows the comet around its orbit. They are also the particles responsible for most meteors and meteor showers.

The sand- to pebble-size particles are much too small to be seen themselves, but they enter the atmosphere at such high speeds that they make the surrounding air glow with heat. It is this glow that we see as the brief but brilliant flash of a *meteor*, lasting only until the particle is vaporized by the heat.

Comet dust is sprinkled throughout the inner solar system, but the "third tails" of ejected particles are concentrated along the orbits of comets. As a result, while you can typically see a few meteors on any clear night, many more are visible on those nights when our planet is crossing a comet's orbit. You may see dozens of meteors per hour during one of these **meteor showers,** which recur at about the same time each year because the orbiting Earth passes through a particular comet's orbit at the same time each year. The meteors of a meteor shower generally appear to radiate from a particular direction in the sky, for essentially the same reason that snow or heavy rain seems to come from a particular direction in front of a moving car (Figure 9.14). Because more meteors hit Earth from

> A comet ejects small particles that cause meteor showers when Earth crosses the comet's orbit.

▼ **FIGURE 9.14**
The geometry of meteor showers.

Snowflakes and meteors appear to radiate from a single direction based on our motion relative to them.

a Meteors appear to radiate from a particular point in the sky for the same reason that we see snow or heavy rain come from a single point in front of a moving car.

b This digital composite photo, taken in Australia during the 2001 Leonid meteor shower, shows meteors as streaks of light radiating from the same point in the sky. The large rock is Uluru, also known as Ayers Rock. Note that, by eye, you would not see so many meteors all at once; each meteor would flash across your sky for a few seconds, and even in the best meteor showers you'd likely see one only every few minutes.

the front than from behind (just as more snow hits the front windshield of a moving car), meteor showers are best observed in the predawn sky, when part of the sky faces in the direction of Earth's motion. Table 9.1 lists major annual meteor showers and their parent comet, if known.

see it for yourself Try to observe the next meteor shower (see Table 9.1); be prepared with a star chart, a marker pen, and a dim (preferably red) flashlight. Each time you see a meteor, record its path on your star chart. Record at least a dozen meteors, and try to determine the *radiant* of the shower—that is, the constellation from which the meteors appear to radiate. Does the meteor shower live up to its name?

◆ Where do comets come from?

We've stated that comets we see in the inner solar system come from two vast reservoirs of comets in the distant outer solar system. But we've never actually seen any small comets at such great distances from the Sun, so you may wonder how we know they are out there. The answer is relatively simple: The comets that we see in the inner solar system must come from *somewhere*, and we can figure out where that somewhere must be by tracing their orbits back. We can then estimate numbers by figuring out how many must reside at great distances to explain the average number that enter the inner solar system each year.

Most comets that visit the inner solar system do not orbit the Sun in the same direction as the planets, and their elliptical orbits have random orientations. Tracing their orbits back shows that they come from far beyond the planets—sometimes nearly a quarter of the distance to the nearest star. These comets must come plunging sunward from the vast, spherical region of space that we call the *Oort cloud*. Be sure to note that the Oort cloud is *not* a cloud of gas, but rather a collection of many individual comets. Based on the number of Oort cloud comets that visit the inner solar system, the Oort cloud must contain about a trillion (10^{12}) comets.

A smaller number of the comets that visit the inner solar system travel around the Sun in the same direction and in nearly the same plane as the planets, and their elliptical orbits carry them no more than about twice as far from the Sun as Neptune. These comets must come from the donut-shaped *Kuiper belt* that lies beyond the orbit of Neptune. Figure 9.15 contrasts the general features of the Kuiper belt and the Oort cloud.

How did comets end up in these far-flung regions of the solar system? The only answer that makes scientific sense comes from thinking about what happened to the leftover icy planetesimals that roamed the region in which the jovian planets formed.

The leftover planetesimals that cruised the spaces between Jupiter, Saturn, Uranus, and Neptune were doomed to suffer either a collision or a close gravitational encounter with one of the young jovian planets. Recall that when a small object passes near a large planet, the planet is hardly affected but the small object may be flung off at high speed [Section 4.4]. The planetesimals that escaped being swallowed therefore tended to be flung off in all directions. Some may have been cast away at such high speeds that they completely escaped the solar system. The rest ended up on orbits with very large average distances from the Sun, becoming the comets of the Oort cloud. The random directions in which these comets were flung explain why the Oort cloud is roughly spherical in shape. Oort cloud comets are so far from the Sun that they can be nudged by the gravity of nearby stars (and even by the mass of the

TABLE 9.1	Major Annual Meteor Showers	
Shower Name	**Approximate Date**	**Associated Comet**
Quadrantids	January 3	?
Lyrids	April 22	Thatcher
Eta Aquarids	May 5	Halley
Delta Aquarids	July 28	?
Perseids	August 12	Swift-Tuttle
Orionids	October 22	Halley
Taurids	November 3	Encke
Leonids	November 17	Tempel-Tuttle
Geminids	December 14	Phaeton
Ursids	December 23	Tuttle

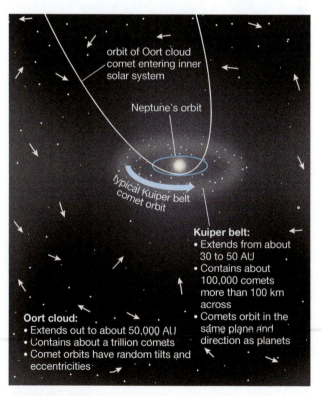

orbit of Oort cloud comet entering inner solar system

Neptune's orbit

typical Kuiper belt comet orbit

Kuiper belt:
• Extends from about 30 to 50 AU
• Contains about 100,000 comets more than 100 km across
• Comets orbit in the same plane and direction as planets

Oort cloud:
• Extends out to about 50,000 AU
• Contains about a trillion comets
• Comet orbits have random tilts and eccentricities

▲ **FIGURE 9.15**
The comets we occasionally see in the inner solar system come from two major reservoirs in the outer solar system: the Kuiper belt and the Oort cloud.

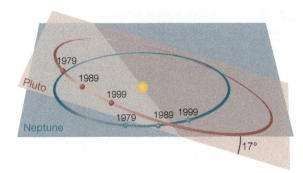

▲ FIGURE 9.16

Pluto's orbit is significantly elliptical and tilted relative to the ecliptic. Pluto comes closer to the Sun than Neptune for 20 years in each 248-year orbit, as was the case between 1979 and 1999. There's no danger of a collision, however, thanks to an orbital resonance in which Neptune completes three orbits for every two of Pluto's.

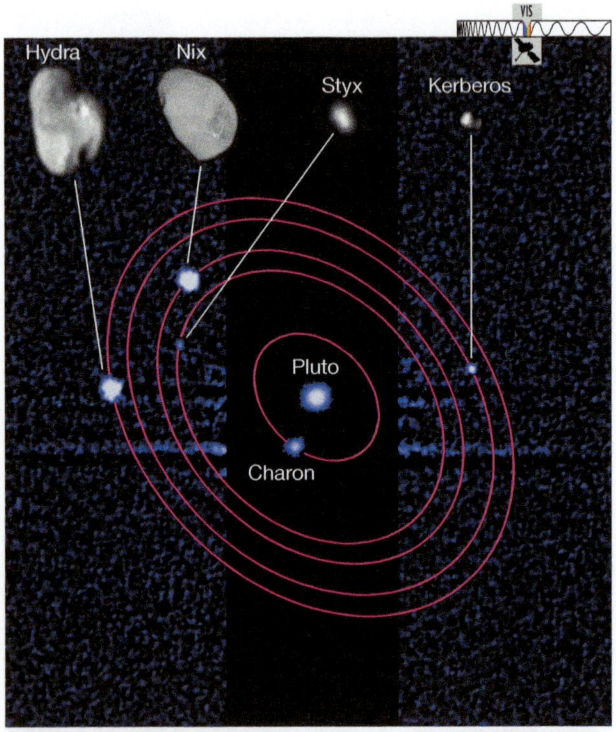

▲ FIGURE 9.17

The main image shows a "family portrait" of Pluto and its five moons as viewed by the Hubble Space Telescope, along with orbital paths for the moons. The insets show images of the four smaller moons taken by *New Horizons* during its flyby of Pluto. The images are shown to scale, and are magnified about 1000 times compared to the scale on which the orbits are shown. (The bluish dots and stripes in the main image come from scattered light within the camera.)

galaxy as a whole), preventing some of them from ever returning to the region of the planets and sending others plummeting toward the Sun.

Beyond the orbit of Neptune, the icy planetesimals were much less likely to be cast off by gravitational encounters. Instead, they remained in orbits going in the same directions as planetary orbits and concentrated relatively near the ecliptic plane. These are the comets of the Kuiper belt. Kuiper belt comets can be nudged by the gravity of the jovian planets through orbital resonances, sending some on orbits that pass through the inner solar system.

Kuiper belt comets orbit in the region in which they formed, just beyond Neptune's orbit. The more distant Oort cloud contains comets that once orbited among the jovian planets.

To summarize, the comets of the Kuiper belt seem to have originated farther from the Sun than the comets of the Oort cloud, even though the Oort cloud comets are now much more distant. The Oort cloud consists of leftover planetesimals that were flung outward after forming between the jovian planets, while the Kuiper belt consists of leftover planetesimals that formed and still remain in the outskirts of the planetary realm.

9.4 Pluto and the Kuiper Belt

The small comets we have studied in the inner solar system are presumably representative of their frozen cousins in the Kuiper belt and Oort cloud, though our current telescopes are not up to the challenge of detecting such small objects at such great distances. In fact, we've never directly detected an object of any size in the Oort cloud. We have, however, detected many moderately large objects in the Kuiper belt. As of 2016, more than 1200 icy objects had been directly observed in the Kuiper belt, allowing scientists to infer that the region contains at least 100,000 objects more than 100 kilometers across. Still, there's only one object in the Kuiper belt that we know much about: Pluto. We'll therefore use Pluto as our case study into the nature of large Kuiper belt objects, and see what we may infer about the Kuiper belt more generally.

◆ What is Pluto like?

Pluto orbits the Sun once every 248 years, and its orbit is much more elliptical and inclined to the ecliptic plane than that of any of the eight planets (Figure 9.16). In fact, Pluto sometimes comes closer to the Sun than Neptune, although there is no danger of collision: Neptune orbits the Sun precisely three times for every two Pluto orbits, and this stable orbital resonance means that Neptune is always a safe distance away whenever Pluto approaches its orbit.

Pluto has five known moons (Figure 9.17). The largest, Charon, has more than half the diameter and about $\frac{1}{8}$ the mass of Pluto, and orbits only 20,000 kilometers away. (For comparison, our Moon has a mass $\frac{1}{80}$ of Earth's and orbits 400,000 kilometers away.) This suggests that Charon—as well as the smaller moons—formed as the result of a *giant impact* similar to the one thought to have formed our Moon [Section 6.3]. Such an impact may also explain why Pluto rotates almost on its side.

Pluto Before *New Horizons* Pluto is so far from the Sun that it took astronomers a long time to learn much about it. You can understand why with a simple analogy: Trying to see Pluto from Earth is equivalent to looking for a snowball the size of your fist that is about 600 kilometers away—and in very dim light. Nevertheless, astronomers managed to learn a lot even before the 2015 flyby of the *New Horizons* spacecraft.

Much of what we learned came after the 1978 discovery of Charon (other moons were discovered later). Observations of Charon's orbit allowed scientists to calculate Pluto's precise mass by applying Newton's version of Kepler's third law [Section 4.4]. Charon soon provided another learning opportunity through some good luck in timing: From 1985 to 1990, Pluto and Charon happened to be aligned in a way that made them eclipse each other every few days as seen from Earth—something that happens only about every 120 years. Detailed analysis of brightness variations during these eclipses allowed the calculation of sizes, masses, and densities for both Pluto and Charon, confirming that both have comet-like compositions of ice and rock. The eclipse data even allowed astronomers to construct rough maps of Pluto's surface markings, and improved telescopic observations, including many with the Hubble Space Telescope, showed a varied surface suggestive of unknown forms of activity.

Pluto is very cold, with an average temperature of only 40 K, as we would expect at its great distance from the Sun. Nevertheless, Earth-based observations showed that Pluto has a thin atmosphere of nitrogen, methane, and carbon monoxide formed by vaporization of surface ices. The amount and composition of atmospheric gas can change as Pluto's distance from the Sun varies along its elliptical orbit, and because of seasonal effects on condensation and vaporization that arise from Pluto's large axis tilt.

Despite the cold, the view from Pluto would be stunning. Charon would dominate the sky, appearing almost 10 times as large in angular size as our Moon appears from Earth. The mutual tidal forces acting between Pluto and Charon long ago made them rotate synchronously with each other [Section 4.4], so Charon is visible from only one side of Pluto and always shows the same face to Pluto. This synchronous rotation also means that Pluto's "day" is the same length as Charon's "month" (orbital period) of 6.4 Earth days. Viewed from Pluto's surface, Charon neither rises nor sets but instead hangs motionless as it cycles through its phases every 6.4 days. The Sun would appear more than a thousand times fainter than it appears here on Earth and would be no larger in angular size than Jupiter appears in our skies.

New Horizons at Pluto   The *New Horizons* spacecraft was launched into space in January 2006 at a higher speed than any previous spacecraft, and it used Jupiter's gravity to gain an additional speed boost when it flew past the giant planet just 13 months later. Still, Pluto is so far from the Sun that *New Horizons* had been traveling through space for 9½ years when it flew past Pluto in July 2015 at a speed of about 50,000 kilometers per hour. The results were spectacular, dramatically improving our view and understanding of this distant world (see the chapter-opening photos, page 240).

New Horizons images show clear evidence of geological activity on Pluto (Figure 9.18), some of which must have occurred within the past 100 million years—so recent compared to the age of the solar system that it is likely the activity continues to this day. Vast regions of Pluto lack craters, implying that ancient craters have been erased. The heart-shaped "Tombaugh Regio" shows wide, smooth ice plains that may be flowing

The *New Horizons* mission revealed Pluto and Charon to have surprisingly high levels of geological activity.

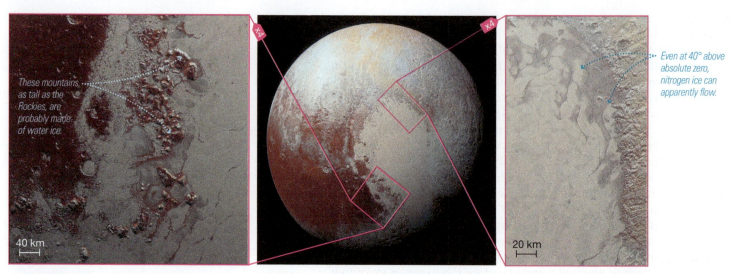

These mountains, as tall as the Rockies, are probably made of water ice.

40 km

Even at 40° above absolute zero, nitrogen ice can apparently flow.

20 km

▲ **FIGURE 9.18**

New Horizons images of Pluto, with enhanced color to identify different regions. The bright, heart-shaped region is called Tombaugh Regio after Pluto's discoverer.

Smooth regions resemble the lunar maria, suggesting the eruption of "lava," which in this case might be water.

▲ **FIGURE 9.19**

New Horizons images of Pluto's largest moon, Charon.

glaciers of nitrogen ice. Pluto also has mountains several kilometers tall, comparable in height to the Rocky Mountains on Earth. These cannot be composed of nitrogen ice, which isn't rigid enough to maintain the steep slopes seen. Instead, the mountains must be made of water ice, which is rigid and strong at Pluto's low temperature.

Charon also has vast, smooth plains that are the hallmark of recent geological activity, as well as canyons comparable in length and depth to Earth's Grand Canyon (Figure 9.19). Scientists are still trying to understand the heat source that drives the geological activity of Pluto and Charon. Neither world is expected to have much heat from radioactive decay, and the tidal lock between Pluto and Charon means that no tidal heating occurs. Perhaps the activity occurs because even the frigid 40 K temperature isn't cold enough to make nitrogen ice very rigid, so a small amount of heat from radioactive decay and/or the changing ice thickness over Pluto's elliptical orbit is enough to cause the flow.

At the end of its brief visit, *New Horizons* looked back to capture a stunning view of Pluto's thin atmosphere above jagged mountains and smooth plains (Figure 9.20). The atmosphere is visible when looking back toward the Sun because it contains enough haze to scatter sunlight forward. The haze probably forms as solar ultraviolet light breaks down methane and nitrogen molecules in the atmosphere, which can then react to make longer molecular chains and haze particles. These particles probably explain Pluto's pale brown color: They drift down through the atmosphere and coat the surface.

Overall, while scientists were surprised by the level of geological activity at Pluto, it reinforced what we had already learned elsewhere. In particular, Pluto seems to be a more extreme example of the "ice geology" that allows jovian moons like Europa, Titan, and Enceladus to be active at lower temperatures than the terrestrial worlds with their "rock geology" (see Figure 8.28). The general lesson appears to be that worlds can have astonishing similarities in geological activity despite tremendous differences in composition and temperature. It will be interesting to see if this lesson applies to worlds not yet explored.

think about it Following its Pluto encounter, scientists aimed the *New Horizons* spacecraft toward a much smaller Kuiper belt comet, with the flyby to occur in January 2019. Find the current status of the mission; if the flyby has already occurred, what did we learn?

▲ **FIGURE 9.20**
Pluto's jagged mountains, smooth plains, and hazy atmosphere, seen as *New Horizons* looked back toward the Sun after the encounter.

◆ What do we know about other Kuiper belt comets?

We don't know nearly as much about any of the other large comets of the Kuiper belt as we do about Pluto, but because they all are thought to have formed in the same distant region of the solar system, we expect them to be similar in nature and composition to Pluto. One key piece of evidence of similarities comes from careful study of orbits. Like Pluto, many Kuiper belt comets have stable orbital resonances with Neptune. In fact, hundreds of Kuiper belt comets have the *same* orbital period and average distance from the Sun as Pluto itself. Other evidence comes from the fact that several other Kuiper belt comets (including Eris) have known moons, so in those cases we can calculate masses and densities. These results, along with spectral information, confirm the idea that these objects have comet-like compositions of ice and rock.

In addition, there's almost certainly one other large object from the Kuiper belt (besides Pluto) that we've photographed up close: Neptune's moon Triton. Recall that Triton's "backward" orbit indicates that it must be a captured object [Section 8.2], and it was almost certainly captured from the Kuiper belt. Therefore, *Voyager* images of Triton (see Figure 8.27) probably represent images of a *former* member of the Kuiper belt. Scientifically, Triton suggests at least two important ideas about the Kuiper belt. First, Triton is larger than both Pluto and Eris (about 15% larger in diameter), suggesting that the Kuiper belt may once have contained more and larger objects than it does today. Second, Triton shows signs of significant past or present geological activity, reinforcing the idea that distant, ice-rich worlds can be much more geologically active than we might have guessed from their sizes alone.

Pluto and other larger Kuiper belt objects are smaller, icier, and more distant than any of the planets. They can have moons, atmospheres, and possibly geological activity.

a Jupiter's tidal forces ripped apart the single comet nucleus of SL9 into a chain of smaller nuclei.

100,000 km

b This painting shows how the SL9 impacts might have looked from the surface of Io. The impacts occurred on Jupiter's night side.

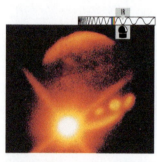

c This infrared photo shows the brilliant glow of a rising fireball from the impact of one SL9 nucleus in 1994. Jupiter is the round disk, with the impact occurring near the lower left.

d The black spot in this Hubble Space Telescope photo is a scar from the impact of an unknown object that struck Jupiter in July 2009.

▲ **FIGURE 9.21**
The impacts of Comet Shoemaker-Levy 9 on Jupiter allowed astronomers their first direct view of a cosmic collision.

9.5 Cosmic Collisions: Small Bodies Versus the Planets

Small meteorites probably crash down somewhere on Earth nearly every day, and the presence of impact craters [Section 7.1] attests to the fact that much larger impacts have happened in the past. Fortunately, the vast majority of impacts happened during the heavy bombardment, which ended nearly 4 billion years ago. Nevertheless, plenty of small bodies still roam the solar system, and cosmic collisions still occur on occasion.

Direct proof that large impacts still occur comes from the fact that we have witnessed a few. The Sun is sometimes hit by comets, though only the comets suffer in the process. More dramatically, in 1994 we witnessed a major impact by a comet on Jupiter. The comet, named Shoemaker-Levy 9, or SL9 for short, had already been ripped apart by tidal forces during a previous pass near Jupiter, so it consisted of a string of nuclei rather than a single nucleus (Figure 9.21a). Comet SL9 was discovered more than a year before it collided with Jupiter, and orbital calculations told astronomers precisely when the collision would occur. When the impacts began, they were observed with nearly every major telescope in existence, as well as by spacecraft that were in position to get a view. Each of the individual nuclei crashed into Jupiter with an energy equivalent to that of a million hydrogen bombs (Figure 9.21b, c). Comet nuclei barely a kilometer across left scars—some large enough to swallow Earth—that lasted for months before dissipating with Jupiter's strong winds. We've since observed the aftermaths of at least five more impacts on Jupiter, one of which is shown in Figure 9.21d.

◆ Did an impact kill the dinosaurs?

There's no doubt that major impacts have occurred on Earth in the past: Geologists have identified more than 150 impact craters on our planet. So before we consider whether an impact might occur in our lifetimes, it's worth examining the potential consequences if it did. Clearly, an impact could cause widespread physical damage. But a growing body of evidence, accumulated over the past three decades, suggests that an impact can do much more—in some cases, large impacts may have altered the entire course of evolution.

In 1978, while analyzing geological samples collected in Italy, a scientific team led by father and son Luis and Walter Alvarez made a startling discovery. They found that a thin layer of dark sediments deposited about 65 million years ago—about the time the dinosaurs went extinct—was unusually rich in the element iridium. Iridium is a metal that is rare on Earth's surface (because it sank to Earth's core when our planet underwent differentiation) but common in meteorites. Subsequent studies found the same iridium-rich layer in 65-million-year-old sediments around the world (Figure 9.22). The Alvarez team suggested a stunning hypothesis: The extinction of the dinosaurs was caused by the impact of an asteroid or comet.

In fact, the death of the dinosaurs was only a small part of the biological devastation that seems to have occurred 65 million years ago. The fossil record suggests that up to 99% of all living organisms died around that time and that up to 75% of all existing *species* were driven to extinction. This makes the event a clear example of a **mass extinction**—the rapid extinction of a large fraction of all living species.

There's still debate about whether a period of active volcanism also contributed to the mass extinction, but there's little doubt that a major

An iridium-rich sediment layer and a 65-million-year-old crater show that a large impact occurred at the time the dinosaurs died out.

impact coincided with the death of the dinosaurs. In addition to extensive evidence from within the sediments, scientists have identified a 65-million-year-old crater (Figure 9.23), apparently created by the impact of an asteroid or a comet measuring about 10 kilometers across.

If the impact was indeed the cause of the mass extinction, here's how it probably happened. On that fateful day some 65 million years ago, the asteroid or comet slammed into Mexico with the force of a hundred million hydrogen bombs (Figure 9.24). North America may have been devastated immediately. Not long after, the hot debris rained around the rest of the world, igniting fires that killed many more living organisms.

The longer-term effects were even more severe. Dust and smoke remained in the atmosphere for weeks or months, blocking sunlight and causing temperatures to fall as if Earth were experiencing a harsh global winter. The reduced sunlight would have stopped photosynthesis for up to a year, killing large numbers of species throughout the food chain. Acid rain may have been another by-product, killing vegetation and acidifying lakes around the world. Chemical reactions in the atmosphere probably produced nitrous oxides and other compounds that dissolved in the oceans and killed marine organisms. Recent evidence also suggests that a period of intense volcanic activity followed the impact—and perhaps was caused by it—which may have further exacerbated the immediate effects of the impact.

Perhaps the most astonishing fact is not that so many plant and animal species died but that some survived. Among the survivors were a few small mammals. These mammals may have survived in part because they lived in underground burrows and managed to store enough food to outlast the global winter that immediately followed the impact.

The evolutionary impact of the extinctions was profound. For 180 million years, dinosaurs had diversified into a great many species large and small, while mammals (which had arisen at almost the same time as the dinosaurs) had generally remained small and rodent-like. With the dinosaurs gone, mammals became the new kings of the planet. Over the next 65 million years, the mammals rapidly evolved into an assortment of much larger mammals—ultimately including us.

▲ **FIGURE 9.22**

Around the world, sedimentary rock layers dating to 65 million years ago share the evidence of the impact of a comet or asteroid. Fossils of dinosaurs and many other species appear only in rocks below the iridium-rich layer.

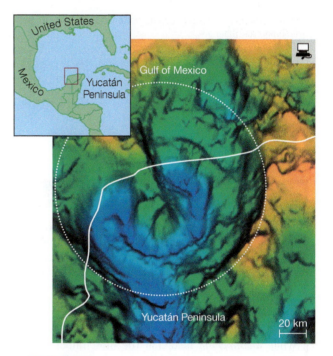

▲ **FIGURE 9.23**

This computer-generated image, based on measurements of small local variations in the strength of gravity, shows an impact crater (dashed circle) in the northwest corner of Mexico's Yucatán Peninsula; the inset shows the location.

◄ **FIGURE 9.24**

This painting shows an asteroid or comet moments before its impact on Earth, some 65 million years ago. The impact probably caused the extinction of the dinosaurs, and if it hadn't occurred the dinosaurs might still rule Earth today.

How great is the impact risk today?

Impacts as large as the one implicated in the demise of the dinosaurs are very rare events. But smaller impacts can still have devastating effects. For example, in 1908, a small asteroid apparently exploded in midair over Tunguska, Siberia, releasing energy equivalent to that of several atomic bombs (Figure 9.25). If the asteroid had exploded over a major city instead of Siberia, it would have been the worst natural disaster in human history.

For anyone who still doubted the reality of an ongoing threat, February 15, 2013 was a date to remember. Astronomers were already aware that a 40-meter-long asteroid would pass just 28,000 kilometers above Earth's surface on that day. But as observers around the world prepared to watch the event, the people of Chelyabinsk, Russia suddenly saw a brilliant flash of light as a smaller, previously undetected asteroid entered the atmosphere above them at a speed of more than 60,000 kilometers per hour (Figure 9.26). Later estimated to have been about 20 meters long and to have had a mass of 10,000 tons, the asteroid streaked across the sky as a giant meteor until friction with the atmosphere made it detonate with the power of a 500-kiloton nuclear bomb. More than a thousand people were injured, mostly by glass shattered by the shock wave. This was in some sense lucky: Had the asteroid's trajectory been directed more vertically toward the ground, the detonation would have occurred at lower altitude, causing much greater damage.

> Damaging impacts have occurred at least twice in recent history: in 1908 over Tunguska and in 2013 over Chelyabinsk, Russia.

More impacts are virtually guaranteed to occur in the future. Figure 9.27 shows how often, on average, we expect Earth to be hit by objects of different sizes, based on geological data from past impacts. The good news is that we are highly unlikely to be hit by an asteroid as large as the one that killed the

▲ **FIGURE 9.25**

This photo shows forests burned and flattened by the 1908 impact over Tunguska, Siberia.

---

extraordinary claims · The Death of the Dinosaurs Was Catastrophic, Not Gradual

By the mid-19th century, geologists recognized that Earth is shaped by both gradual and sudden changes. For example, sediments build up gradually as they are deposited, while volcanic eruptions can spew out vast amounts of rock in a very short time. But which type of process has been more important over the history of Earth? This question became a topic of such great debate among scientists that the two views were given names: *Uniformitarianism* was the view that today's Earth was shaped primarily by slow, gradual changes, while *catastrophism* held that the most important processes were dramatic events like global floods and gigantic volcanic eruptions.

A key issue in the debate was Earth's age. Geologists were already aware that major changes had occurred in the past. For example, the commonplace finding of seashells on mountaintops shows that seas can be transformed over time into mountains. If Earth was very old, then there had been plenty of time for such changes to occur gradually, but if Earth was young, catastrophic changes would have been necessary. During the latter half of the 19th century, evidence began to accumulate for a very old Earth, a case that was sealed in the mid-20th century when radiometric dating showed our planet to be more than 4 billion years old. There had clearly been enough time for uniformitarianism to explain most of Earth's features. But could it explain all of them?

The extinction of the dinosaurs posed at least a potential challenge. Dinosaur fossils are abundant in sediment layers dating to more than about 65 million years ago but absent from younger layers, indicating

that the dinosaur extinction occurred in a "geological blink of the eye." But that still could have meant a period of a few million years, and most scientists assumed that the extinction occurred through relatively gradual change, such as changes in climate that were detrimental to dinosaur survival. So when Luis and Walter Alvarez proposed that the extinction had been a sudden event due to an asteroid impact, it was an extraordinary claim not only in terms of the dinosaur extinction but also in that it reintroduced the idea of catastrophism as an important part of Earth's history.

The initial evidence, which was from the iridium-rich sediment layer (see Figure 9.22), was strong enough to generate widespread scientific interest in the impact hypothesis, but many scientists (including the Alvarezes themselves) pointed out a major flaw: An impact large enough to deposit so much iridium should have left a noticeable crater, and no such crater was known at the time. That changed with the discovery of the *Chicxulub crater* in 1991 (see Figure 9.23). As discussed in this chapter, there is still scientific debate about whether or not the impact was the sole cause of the mass extinction, but at least part of the Alvarezes' extraordinary claim is not subject to any dispute at all: While most of Earth's history can be explained through the gradual processes of uniformitarianism, occasional catastrophic events can and do occur, sometimes with crucial consequences.

Verdict: Very likely.

dinosaurs. Impacts of that size occur tens of millions of years apart on average, which means we face a far greater danger of doing ourselves in than of being done in by a large asteroid or comet. The bad news is that smaller impacts occur much more frequently. Objects a few meters across probably enter Earth's atmosphere every week or so, and objects of the size that caused the Tunguska event probably strike our planet every couple hundred years or so. Until we have more closely monitored the skies for potential threats, we cannot discount the possibility of a very damaging impact.

Several efforts to search for potential impact threats are currently under way, though it's not clear we could do anything about an impending impact even if we saw it coming. Some people have proposed schemes to save Earth by using nuclear weapons or other means to demolish or divert an incoming asteroid, but no one knows whether current technology is really up to the task. We can only hope that the threat doesn't become a reality before we're ready.

think about it After the Chelyabinsk impact, the following statement circulated widely: "Meteors are nature's way of asking 'How's that space program coming?'" Comment on the meaning of this statement. How much time and money do *you* think we should be spending to counter the impact threat? Defend your opinion.

▲ **FIGURE 9.26**
This photo shows the meteor trail of the previously undetected 10,000-ton asteroid that detonated in the sky above Chelyabinsk, Russia, on February 15, 2013. The explosion had the power of a 500-kiloton nuclear bomb and caused injuries to more than a thousand people.

◆ How do the jovian planets affect impact rates and life on Earth?

Ancient people imagined that the mere movement of planets relative to the visible stars in our sky could somehow have an astrological influence on our lives. Scientists no longer give credence to this ancient superstition, but we now know that planets can have a real effect on life on Earth. By shaping the orbits of asteroids and comets, the planets have contributed to cosmic collisions that helped shape our destiny.

The jovian planets, especially Jupiter, have had the greatest effects (Figure 9.28). As we've discussed, Jupiter disturbed the orbits of rocky planetesimals outside Mars's orbit, preventing a planet from forming and creating the asteroid belt. The jovian planets also ejected icy planetesimals to create the distant Oort cloud of comets, and orbital resonances with Neptune shape the orbits of the comets in the Kuiper belt. Ultimately, every asteroid or comet that has impacted Earth since the end of the heavy bombardment was in some sense sent our way by the influence of Jupiter or one of the other jovian planets.

This connection between the jovian planets and impacts leads us to wonder whether life as we know it would still exist if the solar system were laid out differently. Impacts have clearly played a major role in evolution (just ask the dinosaurs), and the impact rate might have been quite different with a different set of jovian planets. For example, if Jupiter did not exist, the threat from asteroids might be much smaller, since the objects that make up the asteroid belt might instead have become part of a planet. Of course, the threat from comets might then be correspondingly greater: Jupiter probably ejected more comets to the Oort cloud than any other jovian planet, and without Jupiter, those comets might have remained dangerously close to Earth. We may never know precisely how life might have differed with a different set of jovian planets, but there is no doubt that these planets have had profound effects on life on Earth.

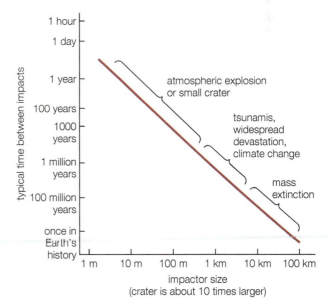

▲ **FIGURE 9.27**
This graph shows that larger objects (asteroids or comets) hit Earth less frequently than smaller ones. The labels describe the effects of impacts of different sizes.

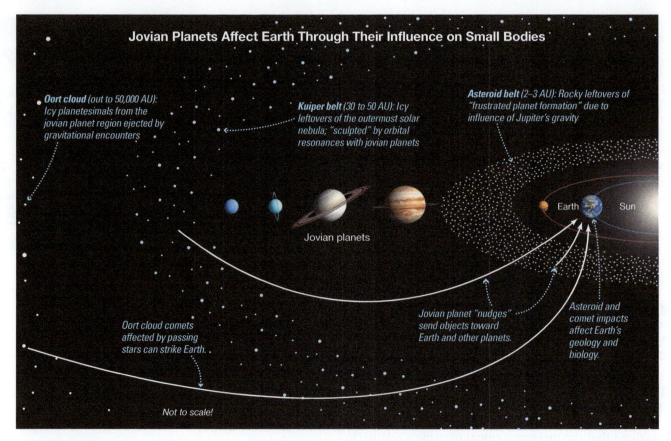

Jovian Planets Affect Earth Through Their Influence on Small Bodies

Oort cloud (out to 50,000 AU): Icy planetesimals from the jovian planet region ejected by gravitational encounters

Kuiper belt (30 to 50 AU): Icy leftovers of the outermost solar nebula; "sculpted" by orbital resonances with jovian planets

Asteroid belt (2–3 AU): Rocky leftovers of "frustrated planet formation" due to influence of Jupiter's gravity

Earth   Sun

Jovian planets

Oort cloud comets affected by passing stars can strike Earth.

Jovian planet "nudges" send objects toward Earth and other planets.

Asteroid and comet impacts affect Earth's geology and biology.

Not to scale!

▲ **FIGURE 9.28**

The connections between the jovian planets, small bodies, and Earth. The gravity of the jovian planets helped shape both the asteroid belt and the Kuiper belt, and the Oort cloud consists of comets ejected from the jovian planet region by gravitational encounters with these large planets. Ongoing gravitational influences sometimes send asteroids or comets toward Earth.

---

the big picture   **Putting Chapter 9 into Context**

In this chapter we concluded the study of our own solar system by focusing on its smallest objects, finding that these objects can have big consequences. Keep in mind the following "big picture" ideas:

- Asteroids and comets may be small compared to planets, but they are very important scientifically, because they provide evidence that has helped us understand how the solar system formed.

- Pluto, once considered a "misfit" among the planets, is now recognized as just one of many moderately sized objects in the Kuiper belt. In terms of composition, the dwarf planets Pluto and Eris—along with other similar objects—are essentially comets of unusually large size.

- The small bodies are subject to the gravitational whims of the largest. The jovian planets shaped the asteroid belt, the Kuiper belt, and the Oort cloud, and they continue to nudge objects onto collision courses with the planets.

- Collisions not only bring meteorites and leave impact craters but also can profoundly affect life on Earth. An impact probably wiped out the dinosaurs, and future impacts pose a threat that we cannot ignore.

my cosmic perspective   Perhaps more so than any other type of astronomical object, the small bodies of the solar system have the potential to touch our lives directly. We can hold pieces of them as meteorites, their impacts pose at least some threat to our survival, their compositions make them potentially useful, and, some 65 million years ago, one of them hit the Earth in an event that may have paved the way for our existence.

9.1 Classifying Small Bodies

◆ How do we classify small bodies?

An **asteroid** is a rocky leftover planetesimal that orbits the Sun. A **comet** is similar but ice-rich in composition. A **dwarf planet** can be either an asteroid or a comet that is large enough to be round. When a small piece of dust or rock enters Earth's atmosphere, the air around it will glow with the flash of what we call a **meteor**; if the rock is large enough to survive and reach the ground, we call it a **meteorite**.

9.2 Asteroids

◆ What are asteroids like?

Asteroids come in a wide range of sizes and shapes, though small ones are far more common. All are rocky in composition, but those at greater distances from the Sun contain more carbon compounds and water. Despite their enormous numbers, the total mass of all asteroids combined is less than that of our Moon.

◆ What do meteorites tell us about asteroids and the early solar system?

Meteorites represent samples of asteroids. **Primitive meteorites** are essentially unchanged since the birth of the solar system and tell us about the material that accreted to make asteroids and planets. **Processed meteorites** are fragments of larger asteroids that underwent differentiation, telling us that many asteroids had substantial interior heat and volcanism.

◆ Why is there an asteroid belt?

The asteroid belt is all that remains of the swarm of planetesimals that once lay between Mars and Jupiter. Orbital resonances nudged orbits in this region, leading to collisions and gradually ejecting material, so that the region lost most of its original mass. Resonances continue to shape asteroid orbits and lead to occasional collisions today.

9.3 Comets

◆ Why do comets grow tails?

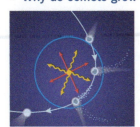

Comet composition makes them much like "dirty snowballs." Far from the Sun, the snowball is frozen as the comet's **nucleus**. As a comet comes inward, ice vaporizes into gas, which along with escaping dust forms a **coma** and two tails: a **plasma tail** of ionized gas and a **dust tail**. Larger particles can also escape, becoming the particles that cause **meteor showers** on Earth.

◆ Where do comets come from?

Comets come from two reservoirs: the **Kuiper belt** and the **Oort cloud**. The Kuiper belt comets still reside in the region beyond Neptune in which they formed. The Oort cloud comets formed between the jovian planets, and were flung out to great distances from the Sun by gravitational encounters with the planets.

9.4 Pluto and the Kuiper Belt

◆ What is Pluto like?

Pluto is an ice-rich world that orbits the Sun in resonance with Neptune, so that it sometimes comes closer to the Sun than Neptune, but there is no danger of a collision. It has a large moon, Charon, possibly formed in a giant impact along with several smaller moons. Pluto has a thin atmosphere, and the *New Horizons* mission revealed a more varied and more active surface than had been expected.

◆ What do we know about other Kuiper belt comets?

Most other Kuiper belt comets likely have an ice-rich composition similar to that of Pluto. Neptune's moon Triton, which probably formed in the Kuiper belt, offers additional evidence that some of the large objects may be geologically active like Pluto.

9.5 Cosmic Collisions: Small Bodies Versus the Planets

◆ Did an impact kill the dinosaurs?

It may not have been the sole cause, but a major impact clearly coincided with the **mass extinction** in which the dinosaurs died out, about 65 million years ago. Sediments from the time contain iridium and other clear evidence of an impact, and an impact crater of the right age lies near the coast of Mexico.

◆ How great is the impact risk today?

Impacts certainly pose a threat, though the probability of a major impact in our lifetimes is fairly low. Impacts like the Tunguska event may occur every couple of hundred years, and would be catastrophic if they occurred in populated areas.

◆ How do the jovian planets affect impact rates and life on Earth?

Impacts are always linked in at least some way to the gravitational influences of Jupiter and the other jovian planets. These influences have shaped the asteroid belt, the Kuiper belt, and the Oort cloud, and continue to determine when an object is flung our way.

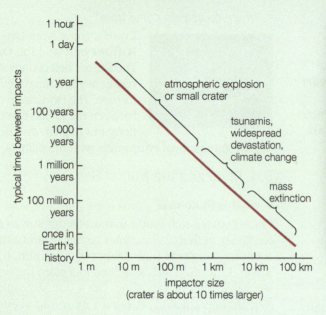

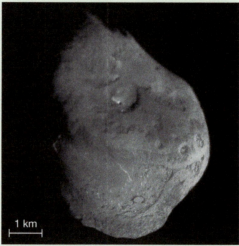

The graph above (Figure 9.27) shows how often impacts occur for objects of different sizes. The photo above shows Comet Tempel 1 moments before the *Deep Impact* spacecraft crashed into it.

1. Estimate Comet Tempel 1's diameter, using the scale bar in the photo above.

2. According to the graph above, how frequently do objects the size of Comet Tempel 1 strike Earth?
   a. once in Earth's history
   b. about once every hundred million years
   c. about once every million years
   d. about once every thousand years

3. Consider an object twice the size of Comet Tempel 1. What kind of damage would this object cause if it hit our planet?
   a. mass extinction
   b. widespread devastation and climate change
   c. atmospheric explosion or a small crater

4. Meteor Crater in Arizona is about 1.2 kilometers across. According to the graph, about how big was the object that made this crater? (*Note*: Be sure to read the axis labels carefully.)
   a. 1 meter   b. 10 meters
   c. 100 meters   d. 1 kilometer

5. How often do objects big enough to create craters like Meteor Crater impact Earth?
   a. once in Earth's history
   b. about once every ten thousand years
   c. about once every few million years
   d. about once every day, but most burn up in the atmosphere or land in the ocean

exercises and problems

MasteringAstronomy®   For instructor-assigned homework and other learning materials, go to MasteringAstronomy®.

Review Questions

Short-Answer Questions Based on the Reading

1. Briefly define *asteroid*, *comet*, *dwarf planet*, *meteor*, and *meteorite*. How did the discovery of Eris force astronomers to reconsider the definition of *planet*?

2. How large are asteroids? How does the total mass of all asteroids compare to the mass of a terrestrial world?

3. Distinguish between *primitive meteorites* and *processed meteorites* in terms of both composition and origin.

4. What do meteorites and spacecraft observations tell us about the geology of asteroids?

5. Where is the *asteroid belt* located, and why? Explain how orbital resonances with Jupiter affect the asteroid belt.

6. What produces the *coma* and *tails* of a comet? What is the *nucleus*? Why do the tails point away from the Sun?

7. How are *meteor showers* linked to comets, and why do they recur at about the same time each year?

8. Describe the *Kuiper belt* and *Oort cloud* in terms of their locations and the orbits of comets within them. How did comets come to exist in these two regions?

9. Briefly describe Pluto and Charon. Why won't Pluto collide with Neptune? How do we think Charon formed?

10. Briefly describe the evidence suggesting that an impact caused the *mass extinction* that killed off the dinosaurs. How might the impact have led to the mass extinction?

11. How often should we expect impacts of various sizes on Earth? How serious a threat do we face from these impacts?

12. Briefly summarize the role of the jovian planets in shaping the orbits of small bodies in the solar system and in influencing life on Earth.

Test Your Understanding

Surprising Discoveries?

Suppose someone claimed to make the discoveries described below. (These are not real discoveries.) Decide whether each discovery should be considered reasonable or surprising. More than one right answer may be possible, so explain your answer clearly.

13. A small asteroid that orbits within the asteroid belt has an active volcano.

14. Scientists discover a meteorite that, based on radiometric dating, is 7.9 billion years old.

15. An object that resembles a comet in size and composition is discovered orbiting in the inner solar system.

16. Studies of a large object in the Kuiper belt reveal that it is made almost entirely of rocky (as opposed to icy) material.

17. Astronomers discover a previously unknown comet that will be brightly visible in our night sky about 2 years from now.

18. A mission to Eris finds that it has lakes of liquid water on its surface.

19. Geologists discover a crater from a 5-kilometer object that impacted Earth more than 100 million years ago.

20. Archaeologists learn that the fall of ancient Rome was caused in large part by an asteroid impact in Asia.

21. In another solar system, astronomers discover an object the size of Earth orbiting its star at the distance of the Kuiper belt.

22. Astronomers discover an asteroid with an orbit suggesting that it will impact Earth in the year 2064.

Quick Quiz

Choose the best answer to each of the following. Explain your reasoning with one or more complete sentences.

23. The asteroid belt lies between the orbits of (a) Earth and Mars. (b) Mars and Jupiter. (c) Jupiter and Saturn.

24. Jupiter nudges the asteroids through the influence of (a) tidal forces. (b) orbital resonances. (c) magnetic fields.

25. Can an asteroid be pure metal? (a) No; all asteroids contain rock. (b) Yes; it must have formed where only metal could condense in the solar nebula. (c) Yes; it must be from the core of a shattered asteroid.

26. Did a large terrestrial planet ever form in the region of the asteroid belt? (a) No, because there was never enough mass there. (b) No, because Jupiter prevented one from accreting. (c) Yes, but it was shattered by a giant impact.

27. What does Pluto most resemble? (a) a terrestrial planet (b) a jovian planet (c) a comet

28. How big an object causes a typical "shooting star"? (a) a grain of sand or a small pebble (b) a boulder (c) an object the size of a car

29. Which have the most elliptical and tilted orbits? (a) asteroids (b) Kuiper belt comets (c) Oort cloud comets

30. Which are thought to have formed farthest from the Sun? (a) asteroids (b) Kuiper belt comets (c) Oort cloud comets

31. About how often does a 1-kilometer object strike Earth? (a) every year (b) every million years (c) every billion years

32. What would happen if a 1-kilometer object struck Earth? (a) It would break up in the atmosphere without causing widespread damage. (b) It would cause widespread devastation and climate change. (c) It would cause a mass extinction.

Process of Science

33. *The Pluto Debate.* Research the decision to demote Pluto to dwarf planet. In your opinion, is this a good example of the scientific process? Does it exhibit the hallmarks of science described in Chapter 3? Compare your conclusions to opinions you find about the debate, and describe how you think astronomers should handle this or similar debates in the future.

34. *Life or Death Astronomy.* In most cases, the study of the solar system has little direct effect on our lives. But the discovery of an asteroid or comet on a collision course with Earth is another matter. How should the standards for verifiable observations described in Chapter 3 apply in this case? Is the potential danger so great that any astronomer with any evidence of an impending impact should spread the word as soon as possible? Or is the potential for panic so great that even higher standards of verification ought to be applied? What kind of review process, if any, would you set in place? Who should be informed of an impact threat, and when?

Group Work Exercise

35. *Assessing Impact Danger.* **Roles:** *Scribe* (collects data and takes notes on the group's activities), *Proposer* (proposes hypotheses and explanations of the data), *Skeptic* (points out weaknesses in the hypotheses and explanations), *Moderator* (leads group discussion and makes sure everyone contributes). **Activity:** Assess the risks we face on Earth from meteorite and comet impacts as follows.

   a. Consider the question of the odds that human civilization will be destroyed by an impact during your lifetime. Determine the kinds of information you would need, develop a method for making the estimate, and write down your method.

   b. Analyze Figure 9.27 and determine whether it contains any of the necessary information.

   c. Apply your group's method to estimate the probability that civilization will be destroyed by an impact during your lifetime, which you can assume to be 100 years for the purpose of this exercise.

   d. Estimate the probability that an impact will cause "widespread devastation" somewhere on Earth during your lifetime.

   e. Finding near-Earth asteroids early greatly increases our chances of deflecting them. Given the probabilities from parts c and d and considering the damage these events would cause, decide as a group how much money per year should be spent on finding near-Earth asteroids and explain your reasoning.

Investigate Further

Short-Answer/Essay Questions

36. *The Role of Jupiter.* Suppose that Jupiter had never existed. Describe at least three ways in which our solar system would be different, and clearly explain why.

37. *Life Story of an Iron Atom.* Imagine that you are an iron atom in a processed meteorite made mostly of iron. Tell the story of how you got to Earth, beginning from the time you were part of the gas in the solar nebula 4.6 billion years ago. Include as much detail as possible. Your story should be scientifically accurate but also creative and interesting.

38. *Asteroids vs. Comets.* Contrast the compositions and locations of comets and asteroids, and explain in your own words why they have turned out differently.

39. *Comet Tails.* Describe in your own words why comets have tails. Why do most comets have two distinct visible tails, and why do the tails go in different directions? Why is the third, invisible tail of small pebbles of interest to us on Earth?

40. *Oort Cloud vs. Kuiper Belt.* Explain in your own words how and why there are two different reservoirs of comets. Be sure to discuss where the two groups of comets formed and what kinds of orbits they travel on.

41. *Project: Dirty Snowballs.* If there is snow where you live or study, make a dirty snowball. (The ice chunks that form behind tires work well.) How much dirt does it take to darken snow? Find out by allowing your dirty snowball to melt in a container and measuring the approximate proportions of water and dirt afterward.

Quantitative Problems

Be sure to show all calculations clearly and state your final answers in complete sentences.

42. *Adding Up Asteroids.* It's estimated that there are a million asteroids 1 kilometer across or larger. If a million asteroids 1 kilometer across were all combined into one object, how big would it be? How many 1-kilometer asteroids would it take to make an object as large as Earth? (*Hint:* You can assume they're spherical. The expression for the volume of a sphere is $\frac{4}{3}\pi r^3$, where r is the radius.)

43. *Impact Energies.* A relatively small impact crater 20 kilometers in diameter could be made by a comet 2 kilometers in diameter traveling at 30 kilometers per second (30,000 m/s).
    a. Assume that the comet has a total mass of 4.2×10^{12} kilograms. What is its total kinetic energy? (*Hint:* The kinetic energy is equal to $\frac{1}{2}mv^2$, where m is the comet's mass and v is its speed. If you use mass in kilograms and velocity in m/s, the answer for kinetic energy will have units of joules.)
    b. Convert your answer from part a to an equivalent in megatons of TNT, the unit used for nuclear bombs. Comment on the degree of devastation such a comet could cause if it struck a populated region on Earth. (*Hint:* One megaton of TNT releases 4.2×10^{15} joules of energy.)

44. *The "Near Miss" of Toutatis.* The 5-kilometer asteroid Toutatis passed a mere 1.5 million kilometers from Earth in 2004. Suppose Toutatis were destined to pass *somewhere* within 1.5 million kilometers of Earth. Calculate the probability that this "somewhere" would have meant that it slammed into Earth. Based on your result, do you think it is fair to call the 2004 passage a "near miss"? Explain. (*Hint:* You can calculate the probability by considering an imaginary dartboard of radius 1.5 million kilometers in which the bull's-eye has Earth's radius, 6378 kilometers.)

45. *Room to Roam.* It's estimated that there are a trillion comets in the Oort cloud, which extends out to about 50,000 AU. What is the total volume of the Oort cloud, in cubic AU? How much space does each comet have in cubic AU, on average? Take the cube root of the average volume per comet to find the comets' typical spacing in AU. (*Hints:* For the purpose of this calculation, you can assume the Oort cloud fills the whole sphere out to 50,000 AU. The volume of a sphere is given by $\frac{4}{3}\pi r^3$, where r is the radius.)

46. *Comet Dust Accumulation.* A few hundred tons of comet dust are added to Earth daily from the millions of meteors that enter our atmosphere. Estimate the time it would take for Earth to get 0.1% heavier at this rate. Is this mass accumulation significant for Earth as a planet? Explain.

Discussion Questions

47. *Rise of the Mammals.* Suppose the impact 65 million years ago had not occurred. How do you think our planet would be different? For example, do you think that mammals still would eventually have come to dominate Earth? Would we be here? Defend your opinions.

48. *How Should Kids Count Planets?* The demotion of Pluto from planet to dwarf planet affected many teachers because they had to change from talking about nine planets to eight, and even a decade after the demotion many teachers still haven't fully accepted it. Should they? How would you recommend that school teachers deal with the new definitions? Be sure to consider both alternative definitions and the fact that it remains possible that the International Astronomical Union may change the definitions again in the future.

Web Projects

49. *Asteroid and Comet Missions.* Learn about a current or planned space mission to study asteroids or comets. What are its scientific goals? Write a one- to two-page summary of your findings.

50. *Impact Hazards.* Many groups are searching for near-Earth asteroids that might impact our planet. They use something called the Torino Scale to evaluate the possible danger posed by an asteroid based on how well we know its orbit. What is this scale? What object has reached the highest level on this scale? What were the estimated chances of impact, and when?

51. *Beneficial Asteroids.* Learn about one of several efforts under way to mine asteroids for human benefit. How far along are the efforts? Discuss the issues that will determine if they succeed.

10 Other Planetary Systems
The New Science of Distant Worlds

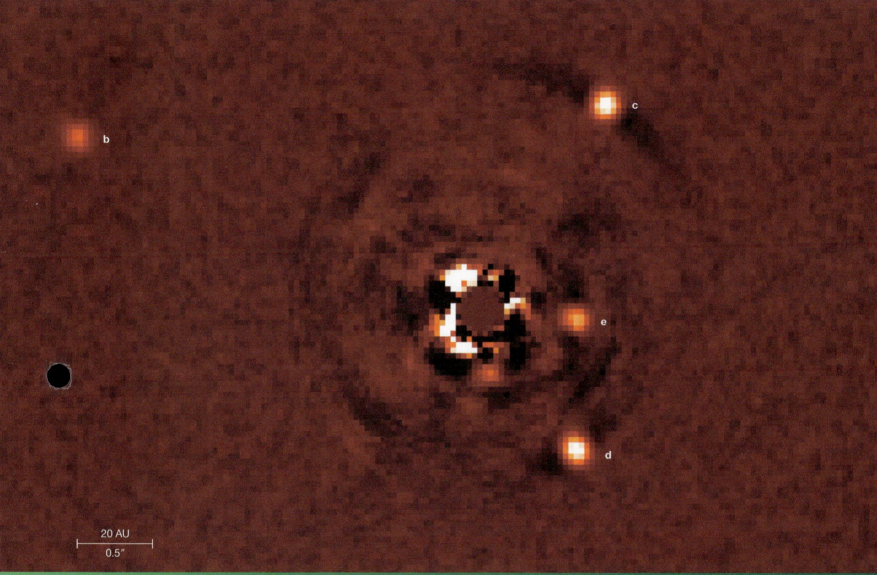

20 AU
0.5″

This infrared image from the Large Binocular Telescope shows direct detection of four planets (marked b, c, d, e) orbiting the star HR 8799. Light from the star itself (center) was mostly blocked out during the exposure, as indicated by the solid red circle.

LEARNING GOALS

10.1 Detecting Planets Around Other Stars
- ◆ How do we detect planets around other stars?

10.2 The Nature of Planets Around Other Stars
- ◆ What properties of extrasolar planets can we measure?
- ◆ How do extrasolar planets compare with planets in our solar system?

10.3 The Formation of Other Planetary Systems
- ◆ Do we need to modify our theory of solar system formation?
- ◆ Are planetary systems like ours common?

The Copernican revolution, which taught us that Earth is a planet orbiting the Sun, opened up the possibility that planets might also orbit other stars. Still, until the 1990s, no such planets were known. But then another scientific revolution began, and the list of known extrasolar planets now is in the thousands and growing rapidly.

The advancing science of extrasolar planets has dramatic implications for our understanding of our place in the universe. The fact that planets are common in the universe makes it seem more likely that we might someday find life elsewhere, perhaps even intelligent life. Moreover, having many more worlds to compare to our own vastly enhances our ability to learn how planets work, which may help us better understand our home planet, Earth. It also allows us to test the nebular theory of solar system formation in new settings. In this chapter, we'll focus our attention on the exciting new science of other planetary systems.

10.1 Detecting Planets Around Other Stars

We've known for centuries that other stars are distant suns, making it natural to suspect that they would have their own planetary systems. The nebular theory of solar system formation, well established decades ago, made such systems seem even more likely, because it predicts that planetary systems should form as a natural outcome of the processes that accompany the birth of any star. However, it took many decades for technology to reach the point at which this prediction could be put to the test through a search for planets around other stars, or **extrasolar planets** for short. The result has been an astounding success for the theory.

◆ How do we detect planets around other stars?

Detecting extrasolar planets poses a huge technological challenge for two basic reasons. First, as we discussed in Chapter 1, planets are extremely tiny compared to the vast distances between stars. Second, stars are typically *a billion times* brighter than the light reflected by any orbiting planets, so starlight tends to overwhelm any planetary light in photographs. This problem is somewhat lessened—but not eliminated—if we observe in infrared light, because planets emit their own infrared light and stars are usually dimmer in the infrared.

The first step in understanding how scientists are overcoming these challenges is to recognize that there are two general ways of learning about a distant object: *directly*, which means by obtaining images or spectra of the object, and *indirectly*, which means by inferring the object's existence or properties without actually seeing it. While scientists have achieved some success in direct detection (such as the image that opens this chapter), we will need significant advances in technology before we will be able to obtain higher-resolution images or spectra. As a result, nearly all current understanding of extrasolar planets comes from indirect study. There are two major indirect approaches to finding and studying extrasolar planets:

1. Observing the motion of a star to detect the subtle gravitational tugs of orbiting planets

2. Observing changes to a star's brightness that occur when one of its planets passes in front of the star as viewed from Earth

Almost all extrasolar planets detected to date have been found indirectly rather than through direct observation.

The earliest discoveries of extrasolar planets came primarily through the first approach, but the current champion of extrasolar planet discoveries—NASA's *Kepler* mission—used the second. We can learn even more when we can combine both indirect approaches, because each can provide different information about a distant planet.

Gravitational Tugs Although we usually think of a star as remaining still while planets orbit around it, that is only approximately correct. In reality, all the objects in a star system, including the star itself, orbit the system's "balance point," or *center of mass*. To understand how this fact allows us to discover extrasolar planets, imagine the viewpoint of extraterrestrial astronomers observing our solar system from afar.

Let's start by considering only the influence of Jupiter (Figure 10.1). Jupiter is the most massive planet, but the Sun still outweighs it by about a thousand times. As a result, the center of mass between the Sun and Jupiter lies about one-thousandth of the way from the center of the Sun to the center of Jupiter, putting it just outside the Sun's visible surface. In other words, what we usually think of as Jupiter's 12-year orbit around the Sun is really a 12-year orbit around their center of mass. Because the Sun and Jupiter are always on opposite sides of the center of mass (otherwise it wouldn't be a "center"), the Sun must orbit this point with the same 12-year period. The Sun's orbit traces out only a small ellipse with each 12-year period, because the Sun's average orbital distance is barely larger than its own radius. Nevertheless, with sufficiently precise measurements, extraterrestrial astronomers could detect this orbital movement of the Sun and thereby deduce the existence of Jupiter, even without having observed Jupiter itself. They could even determine Jupiter's mass from the orbital characteristics of the Sun as it goes around the center of mass. A more massive planet at the same distance would pull the center of mass farther from the Sun's center, giving the Sun a larger orbit. Because the Sun's period around the center would still be 12 years, the larger orbit would mean a faster orbital speed around the center of mass.

see it for yourself To see how a small planet can make a big star wobble, find a pencil and tape a heavy object (such as a set of keys) and a lighter object (perhaps a few coins) to opposite ends. Tie a string (or piece of floss) at the balance point—the center of mass—so that the pencil is horizontal. Then tap the lighter object into "orbit" around the heavier object. What does the heavier object do, and why? How does your setup correspond to a planet orbiting a star? Experiment further with objects of different weights or shorter pencils; try to explain the differences you see.

Planets exert gravitational tugs on their star, causing the star to orbit around the system's center of mass.

The other planets also exert gravitational tugs on the Sun, each adding a small additional effect to the effects of Jupiter (Figure 10.2). In principle, with

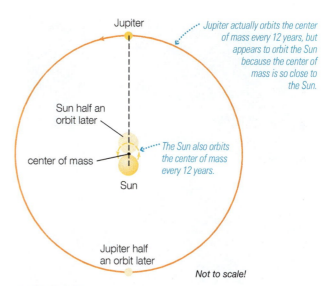

Jupiter actually orbits the center of mass every 12 years, but appears to orbit the Sun because the center of mass is so close to the Sun.

The Sun also orbits the center of mass every 12 years.

Not to scale!

▲ **FIGURE 10.1**

This diagram shows how both the Sun and Jupiter orbit around their mutual center of mass, which lies very close to the Sun. The diagram is not to scale; the sizes of the Sun and its orbit are exaggerated about 100 times compared to the size shown for Jupiter's orbit, and Jupiter's size is exaggerated even more.

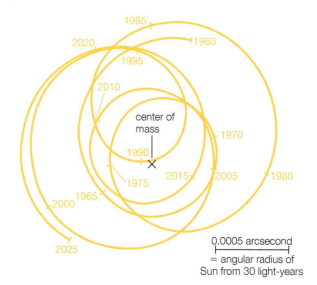

0.0005 arcsecond = angular radius of Sun from 30 light-years

▲ **FIGURE 10.2**

This diagram shows the orbital path of the Sun around the center of mass of our solar system as it would appear from a distance of 30 light-years away, for the period 1960–2025. Notice that the entire range of motion during this period is only about 0.0015 arcsecond, which is almost 100 times smaller than the angular resolution of the Hubble Space Telescope. Nevertheless, if alien astronomers could measure this motion, they could learn of the existence of planets in our solar system.

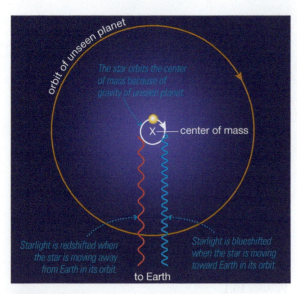

▲ **FIGURE 10.3**

The Doppler method for discovering extrasolar planets: The star's Doppler shift alternates toward the blue and toward the red, allowing us to detect its slight motion—caused by an orbiting planet—around the center of mass.

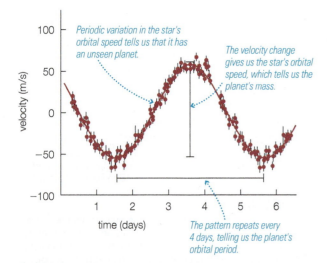

▲ **FIGURE 10.4**

A periodic Doppler shift in the spectrum of the star 51 Pegasi reveals the presence of a large planet with an orbital period of about 4 days. Dots are actual data points; bars through dots represent measurement uncertainty.

sufficiently precise measurements of the Sun's orbital motion made over many decades, an extraterrestrial astronomer could deduce the existence of all the planets of our solar system. If we turn this idea around, it means we can search for planets in other star systems by carefully watching for the tiny orbital motion of a star around the center of mass of its star system.

The Astrometric Method Astronomers use two distinct observation methods to search for the gravitational tugs of planets on stars. The first, called the **astrometric method** (*astrometry* means "measurement of the stars"), uses very precise measurements of stellar positions in the sky to look for the slight motion caused by orbiting planets. If a star "wobbles" gradually around its average position (the center of mass), we must be observing the influence of unseen planets.

The primary difficulty with the astrometric method is that we are looking for changes to position that are very small even for nearby stars, and these changes become smaller for more distant stars. In addition,

> The *astrometric method* searches for small "wobbles" in a star's position.

the stellar motions are largest for massive planets orbiting far from their star, but the long orbital periods of such planets mean that it can take decades to notice the motion. As a result, the astrometric method has been of only limited use to date. However, the European Space Agency's *GAIA* mission, launched in 2013, should soon change this. *GAIA* has a goal of obtaining astrometric observations of a billion stars in our galaxy with an accuracy of 10 microarcseconds, which should enable it to detect thousands of extrasolar planets.

The Doppler Method The other method of searching for gravitational tugs due to orbiting planets is the **Doppler method**, which searches for a star's orbital movement around the center of mass by looking for changing Doppler shifts in its spectrum [Section 5.2]. Recall that the Doppler effect causes a blueshift when a star is moving toward us and a redshift when it is moving away from us, so alternating blueshifts and redshifts (relative to a star's average Doppler shift) indicate orbital motion around a center of mass (Figure 10.3).

As an example, Figure 10.4 shows data from the first discovery of an extrasolar planet around a Sun-like star: the 1995 discovery of a planet orbiting the star called 51 Pegasi. The 4-day period of the star's motion

> The *Doppler method* searches for alternating blueshifts and redshifts in a star's spectrum.

must be the orbital period of its planet. We therefore know that the planet lies so close to the star that its "year" lasts only 4 Earth days, which means its surface temperature is probably over 1000 K. The Doppler data also allow us to determine the planet's approximate mass, because a more massive planet has a greater gravitational effect on the star (for a given orbital distance) and therefore causes the star to move at higher speed around the system's center of mass. In the case of 51 Pegasi, the data show that the planet has about half the mass of Jupiter. Scientists therefore refer to this planet as a **hot Jupiter**, because it has a Jupiter-like mass but a much higher surface temperature.

The Doppler method has proven very successful. As of 2016, it has been used to detect about 700 planets, including more than 100 in multiplanet systems. Keep in mind that because the Doppler method searches for gravitational tugs from orbiting planets, it is much better for finding massive planets like Jupiter than small planets like Earth. It's also best suited to identifying planets that orbit relatively close to their

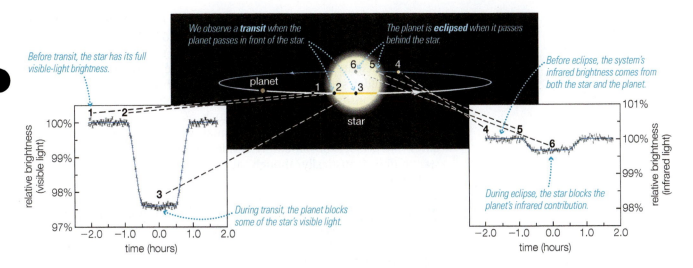

We observe a **transit** when the planet passes in front of the star.

The planet is **eclipsed** when it passes behind the star.

Before transit, the star has its full visible-light brightness.

Before eclipse, the system's infrared brightness comes from both the star and the planet.

During transit, the planet blocks some of the star's visible light.

During eclipse, the star blocks the planet's infrared contribution.

star

planet

▲ FIGURE 10.5

This diagram represents transits and eclipses of the planet orbiting the star HD 189733. The graph shows that each transit lasts about 2 hours, during which the star's visible-light brightness dips by about 2.5%. The eclipses are observable in the infrared, because the star blocks the infrared contribution of the planet. The transits and eclipses each occur once during every 2.2-day orbit of the planet.

star, because being closer means a stronger gravitational tug and hence a greater velocity for the star as it orbits the system's center of mass. Shorter orbits also allow easier confirmation of a planet's existence, because it is easier to record multiple repeats of the Doppler curve.

Transits The second general approach to detecting planets indirectly relies on searching for slight changes in a star's brightness caused by orbiting planets. If we were to examine a large sample of stars with planets, a small number of these systems—probably about 1%—would by chance be aligned in such a way that one or more of the star's planets would pass directly between us and the star once each orbit. The result is a **transit**, in which the planet appears to move across the face of the star, causing a small, temporary dip in the system's brightness (Figure 10.5). The larger the planet, the more dimming it will cause. Some transiting planets also undergo a measurable **eclipse** as the planet goes behind the star. Eclipse observations are more easily accomplished in the infrared, because planets contribute a greater proportion of a system's infrared brightness than its visible-light brightness.

The **transit method** searches for transits and eclipses by carefully monitoring a star system's brightness over an extended period of time. Because most stars exhibit intrinsic variations in brightness, we can be confident that we have detected a planet only if we observe the dips in brightness to repeat with a regular period, indicating that the same planet is passing in front of the star at the same time in each orbit.

If a planet happens to orbit edge-on as seen from Earth, its star will dim slightly as the planet passes in front of its star.

The period of the repeated transits is the orbital period of the planet. Scientists generally require at least three repeated transits before concluding that a planet is probably responsible, meaning at least about 3 years of monitoring to detect a planet in an Earth-like orbit around a Sun-like star.

think about it Suppose you were an alien astronomer searching for planets going around our Sun. Where would your star system have to be located in order for you to observe transits across the Sun? Which planets in our solar system would be the easiest to discover through transits? Explain.

Transits have been observed with many telescopes (including some as small as backyard telescopes), but the vast majority of transit detections to date came from NASA's *Kepler* mission, which searched for planetary transits from 2009 to 2013. *Kepler* monitored about 150,000 stars for transits, measuring their brightnesses about every 30 minutes. Based on data that had been analyzed by mid-2016, scientists had already identified about 2000 stars being orbited by a total of more than 4000 planet "candidates," including dozens that are as small as Earth. The term "candidates" is used because even though *Kepler* data show three (or more) transits, the existence of these planets has not yet been confirmed by follow-up observations (such as detection by the Doppler method); as a result, there's still a small possibility that some "candidates" are artifacts of something besides an orbiting planet that causes a star to dim. Statistically, though, at least about 90% of the candidates will turn out to be real planets.

think about it As this book went to press, two new transit missions—NASA's *TESS* and the European Space Agency's *CHEOPS*—were slated for launch in late 2017. Find the goals and current status of these missions. How will they build on *Kepler*'s discoveries?

Other Strategies Cosmic Context Figure 10.6 (pages 272–273) summarizes the major planet detection methods, both indirect and direct. However, several other methods have also had some success. For example, more than a dozen planets have been detected by what is sometimes called *microlensing*, which is a type of gravitational lensing [Section 18.2] in which the light of a star is temporarily magnified as another star passes in front of it and bends its light. Careful study of the microlensing event can reveal whether the foreground star has planets. The major drawback to this method is that the special alignment necessary for microlensing is a one-time event, which generally means that there is no opportunity for confirmation or follow-up observations. A different strategy takes advantage of the fact that many stars are surrounded by disks of dust. A planet within such a disk can exert small gravitational tugs on dust particles to produce gaps, waves, or ripples that may be detectable. As we learn more about extrasolar planets, new search methods are sure to arise.

10.2 The Nature of Planets Around Other Stars

The mere existence of planets around other stars has changed our perception of our place in the universe, because it shows that our planetary system is not unique. Scientifically, however, we want to know much more than just that these planets exist. In this section, we'll explore what we have learned to date about planets around other stars.

◆ What properties of extrasolar planets can we measure?

Despite the challenge of detecting extrasolar planets, we can learn a surprising amount about them. Depending on the method or methods, we can determine such planetary characteristics as orbital period and distance, orbital eccentricity, mass, size, density, and even a little bit about a planet's atmospheric composition and temperature.

Orbital Period and Distance All three of the major indirect detection methods tell us a planet's orbital period. The astrometric method allows us to observe the star's orbital motion around the system's center of mass, which means we know the star's orbital period; the planet's orbital period must be the same. For the Doppler method, a detected planet's orbital period is simply the time between peaks in the star's velocity curve (see Figure 10.4), and for the transit method, the orbital period is the time between repeated transits.

Once we know the orbital period, we can determine average orbital distance (semimajor axis) with Newton's version of Kepler's third law [Section 4.4]. Recall that for a small object like a planet orbiting a much more massive object like a star, this law expresses a relationship between the star's mass, the planet's orbital period, and the planet's average distance. We generally know the masses of the stars with extrasolar planets (through methods we'll discuss in Chapter 12). Therefore, using the star's mass and the planet's orbital period, we can calculate the planet's average orbital distance.

The major detection methods all tell us orbital period, from which we can calculate orbital distance.

Orbital Eccentricity   All planetary orbits are ellipses, but recall that ellipses vary in eccentricity, which is a measure of how "stretched out" they are (see Figure 3.13). The planets in our solar system all have nearly circular orbits (low eccentricity), which means that their actual distances from the Sun are always relatively close to their average distances. Planets with higher eccentricity swing in close to their star on one side of their orbit and go much farther from their star on the other side.

We can determine eccentricity from both the astrometric method and the Doppler method, though most measurements to date come from Doppler data. A planet with a perfectly circular orbit travels at a constant speed around its star, so its velocity curve is perfectly symmetric. Any asymmetry in the velocity curve tells us that the planet is moving with varying speed along the orbit and therefore must have a more eccentric elliptical orbit.

Planetary Mass   Both the astrometric method and the Doppler method measure motions caused by the gravitational tug of a planet, so both can in principle allow us to estimate planetary masses. These methods tell us about planetary mass because, for a given orbital distance, a more massive planet will cause its star to move at higher velocity around the center of mass.

There is an important caveat for the Doppler method. Recall that Doppler shifts reveal only the part of a star's motion directed toward or away from us (see Figure 5.15). As a result, the Doppler shift tells us a star's full orbital velocity only when we view its orbit precisely edge-on. In all other cases, the velocity inferred from Doppler shifts will be less than the full orbital velocity, which means that the mass we calculate will be the planet's *minimum* possible mass (or a "lower limit" mass). Statistically, however, we expect that the actual planetary mass should be no more than double the minimum mass in at least about 85% of all cases, so masses from the Doppler method provide relatively good estimates for most planets.

Planetary Size   Transit observations are presently the only means by which we can measure a planet's size or radius. The basic idea is easy to understand: The more of a star's light that a planet blocks during a transit, the larger the planet must be.

The search for planets around other stars is one of the fastest growing and most exciting areas of astronomy, with known extrasolar planets already numbering well into the thousands. This figure summarizes major techniques that astronomers use to search for and study extrasolar planets.

① **Gravitational Tugs:** We can detect a planet by observing the small orbital motion of its star as both the star and its planet orbit their mutual center of mass. The star's orbital period is the same as that of its planet, and the star's orbital speed depends on the planet's distance and mass. Any additional planets around the star will produce additional features in the star's orbital motion.

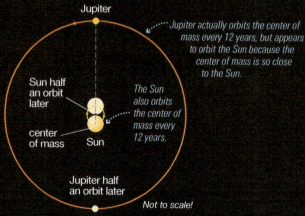

Jupiter

Jupiter actually orbits the center of mass every 12 years, but appears to orbit the Sun because the center of mass is so close to the Sun.

Sun half an orbit later

The Sun also orbits the center of mass every 12 years.

center of mass   Sun

Jupiter half an orbit later

Not to scale!

①a **The Doppler Method:** As a star moves alternately toward and away from us around the center of mass, we can detect its motion by observing alternating Doppler shifts in the star's spectrum: a blueshift as the star approaches and a redshift as it recedes.

①b **The Astrometric Method:** A star's orbit around the center of mass leads to tiny changes in the star's position in the sky. The *GAIA* mission is expected to discover many new planets with this method.

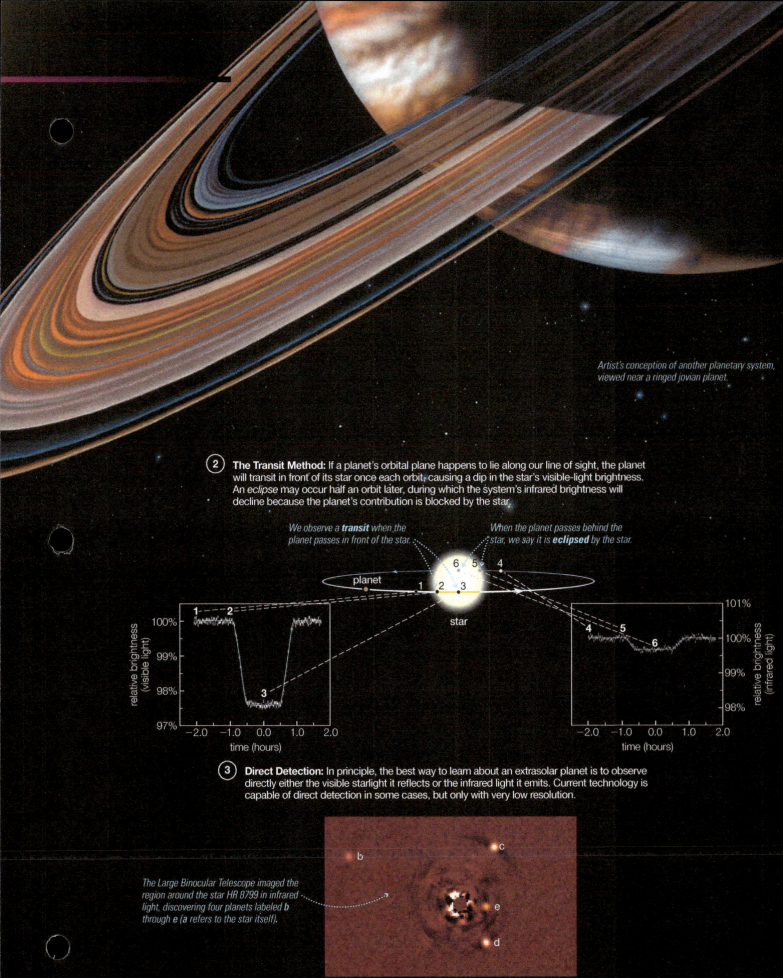

Artist's conception of another planetary system, viewed near a ringed jovian planet.

2 **The Transit Method:** If a planet's orbital plane happens to lie along our line of sight, the planet will transit in front of its star once each orbit, causing a dip in the star's visible-light brightness. An *eclipse* may occur half an orbit later, during which the system's infrared brightness will decline because the planet's contribution is blocked by the star.

We observe a **transit** when the planet passes in front of the star.

When the planet passes behind the star, we say it is **eclipsed** by the star.

6 5 4

planet   1 2 3

star

3 **Direct Detection:** In principle, the best way to learn about an extrasolar planet is to observe directly either the visible starlight it reflects or the infrared light it emits. Current technology is capable of direct detection in some cases, but only with very low resolution.

The Large Binocular Telescope imaged the region around the star HR 8799 in infrared light, discovering four planets labeled b through e (a refers to the star itself).

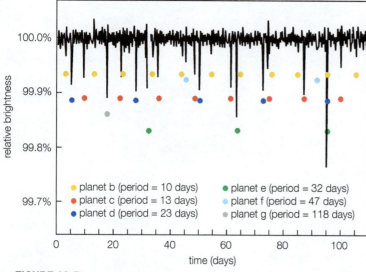

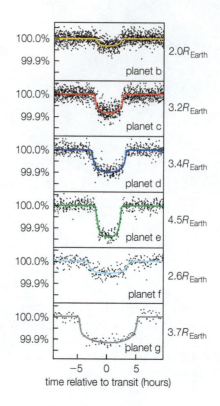

▲ FIGURE 10.7

Transits from the six planets of the Kepler 11 system, known as Kepler 11b through Kepler 11g. (Kepler 11a is the star.) The black line shows the brightness of the star Kepler 11 over a 110-day period. Colored dots indicate transits by six different planets; in some cases, more than one planet transits at the same time. The panels at right show the dips due to each planet separately, and indicate the planet's size in Earth radii.

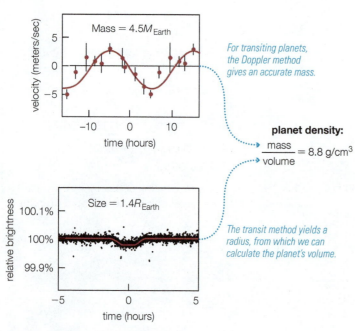

▲ FIGURE 10.8

This diagram summarizes how we can combine data from the Doppler and transit methods to calculate a planet's average density. Data are for the planet Kepler 10b. (The final density calculation requires converting the mass to grams and the volume to units of cubic centimeters.)

The Kepler 11 planetary system offers a great example. Figure 10.7 shows the brightness of the star Kepler 11 over a 110-day period. Each downward dip represents a transit in which a planet blocks a small fraction of the starlight. Careful study shows that there are six different planets (represented by the colored dots) that transit the star. The panels show a transit of each planet in more detail, along with the planet's size calculated from the depth of the dip. When we combine these sizes with the orbital periods and distances, we find that this system's planets have sizes ranging from 2 to 5 times Earth's size and all orbit between 0.1 and 0.5 AU from their star. Transported to our solar system, the six planets would lie inside Venus's orbit.

The astrometric and Doppler methods tell us about planetary mass, while the transit method tells us planetary size.

think about it New data are coming in so quickly that the records for lowest-mass planet and smallest-size planet change rapidly. Do a Web search to find the extrasolar planet that currently holds the record for lowest confirmed *mass* and for smallest measured *size*.

Planetary Density No single method measures a planet's average density (mass divided by volume), but we can calculate it if we know a planet's size from the transit method and its mass from the Doppler method. Note that we can get a precise density value in this case, because transiting planets must have edge-on orbits, so the Doppler method gives an exact (rather than minimum) mass in these cases. Figure 10.8 shows the process applied to a planet known as Kepler 10b.

Atmospheric Composition and Temperature Although detailed understanding of the atmospheres and temperatures of extrasolar planets will have to await technology capable of direct observations

TABLE 10.1 Major Ways of Measuring Properties of Extrasolar Planets

	Planetary Property	Method(s) Used	Explanation
Orbital Properties	period	Doppler, astrometric, or transit	We directly measure orbital period.
	distance	Doppler, astrometric, or transit	We calculate orbital distance from orbital period using Newton's version of Kepler's third law.
	eccentricity	Doppler or astrometric	Velocity curves and astrometric star positions reveal eccentricity.
Physical Properties	mass	Doppler or astrometric	We calculate mass based on the amount of stellar motion caused by a planet's gravitational tug.
	size (radius)	transit	We calculate size based on the amount of dip in a star's brightness during a transit.
	density	transit plus Doppler	We calculate density by dividing the mass by the volume (using the size from the transit method).
	atmospheric composition, temperature	transit or direct detection	Transits and eclipses provide data on atmospheric composition and temperature.

and spectra, careful analysis of data from transits and eclipses can provide some information.

Infrared observations of eclipses provide the key data for temperature. Recall that planets generally emit infrared light, and the amount of infrared emission (per unit area) depends on a planet's temperature. As a planet goes behind its star, the system's infrared brightness will drop because we are no longer seeing the planet's infrared emission. The extent of the drop tells us how much infrared the planet emits, and we can combine this amount of infrared emission with the planet's radius (measured by the transits) to calculate an approximate temperature.

Transits and eclipses can also provide limited information about atmospheric composition. For a planet with an atmosphere, astronomers compare spectra of the system taken with the planet in front of (for transits) or behind (for eclipses) its star to spectra taken at other times. Careful analysis of differences between these spectra can reveal spectral lines caused by the planet's atmosphere. Because we generally know what gases are responsible for particular spectral lines, we can infer the existence of these gases in the planet's atmosphere.

Table 10.1 summarizes how we measure planetary properties.

◆ How do extrasolar planets compare with planets in our solar system?

The number of known extrasolar planets for which we have measured key properties is now large enough that we are beginning to gain insight into how these planets compare to the planets of our own solar system. Let's explore the general features of extrasolar planets as we know them today.

Orbital Properties A few extrasolar planets have been found with sizes and orbits similar to those of Earth (Figure 10.9). However, many others exhibit one or both of two orbital properties that seem surprising in light of what we know about our own solar system. First, many of them orbit quite close to their stars—in many cases much closer than Mercury orbits the Sun—even though they seem to be Jupiter-like in mass or size. Second, many have relatively large eccentricities, a clear contrast with the nearly circular orbits of planets in our solar system.

Part of the reason for these surprising properties is the nature of our detection methods. Recall that all the methods require much more time to discover planets with long orbital periods—those orbiting far from their stars—than planets with short orbital periods. Since we haven't been

cosmic calculations 10.1

Finding Sizes of Extrasolar Planets

We determine a planet's radius from the fraction of a star's light blocked during a transit. Viewed against the sky, both the star and the planet appear as tiny circular disks. These disks are far too small for our telescopes to resolve, but the fraction of the star's light that is blocked must be equal to the area of the planet's disk (r_{planet}^2) divided by the area of the star's disk (r_{star}^2):

$$\frac{\text{fraction of}}{\text{light blocked}} = \frac{\text{area of planet's disk}}{\text{area of star's disk}} = \frac{\pi r_{planet}^2}{\pi r_{star}^2} = \frac{r_{planet}^2}{r_{star}^2}$$

Solving for the planet's radius, we find

$$r_{planet} = r_{star} \times \sqrt{\text{fraction of light blocked}}$$

Example: Figure 10.5 shows a transit of the star HD 189733. The star's radius is about 800,000 kilometers ($1.15R_{Sun}$), and the planet blocks 1.7% of the star's light during a transit. What is the planet's radius?

Solution: We simply plug the given data into the above formula:

$$r_{planet} = r_{star} \times \sqrt{\text{fraction of light blocked}}$$
$$= 800,000 \text{ km} \times \sqrt{0.017}$$
$$\approx 100,000 \text{ km}$$

The planet's radius is about 100,000 kilometers, which makes it about 40% larger in radius than Jupiter (radius 71,500 kilometers).

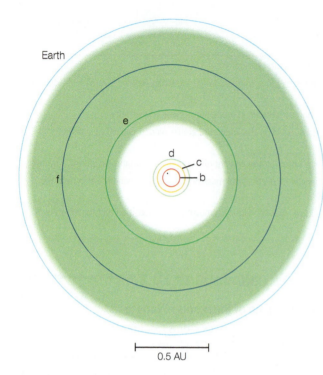

Earth

0.5 AU

▲ **FIGURE 10.9**

A top-down view of planetary orbits in the Kepler 11 system, compared to Earth's orbit. The green region shows the *habitable zone* [Section 19.3] of this planetary system; the habitable zone is interior to that around our Sun because the Kepler 11 star is one-fifth as bright as the Sun. The two planets in the habitable zone (shaded green) are "super-Earths," having radii 40% and 60% greater than Earth's.

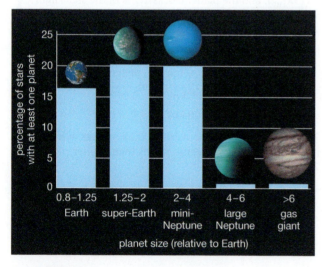

▲ **FIGURE 10.10**

This bar chart shows the estimated proportions of all stars that have planets of different size categories, based on *Kepler* results. Because *Kepler* data have been analyzed only for planets with relatively short orbital periods to date, these estimates are nearly certain to increase as additional data are studied.

observing extrasolar planets for very long, it's only natural that most of our discoveries would be of planets that are close-in with short orbital periods.

It's more challenging to understand why Jupiter-size planets are found close-in, since the nebular theory predicts that large planets should form **Many extrasolar planets orbit surprisingly close to their stars or with surprisingly high orbital eccentricity.** far from their star. High eccentricities also pose a challenge, since the nebular theory provides a natural explanation for the nearly circular orbits in our solar system. Indeed, as we'll discuss in the next section, these observations forced scientists to reconsider the nebular theory, and while it has survived this scrutiny and even been strengthened, we have added features to it that we had not recognized when studying only the planets in our own solar system.

Another interesting surprise came from studying binary star systems. Scientists had not been sure whether planets could form in binary star systems, but data from the *Kepler* mission have shown that they can, at least in some cases. So some worlds have two (or more) "suns" in the sky, much like the fictional planet Tatooine depicted in *Star Wars*.

Sizes, Masses, and Densities The *Kepler* mission detected so many planets that scientists can use statistics to estimate the proportions of all stars that have planets of various sizes. Figure 10.10 shows results from these statistics. Two remarkable conclusions are already apparent. First, planets are common: By looking across all size categories, astronomers conclude that at least 70% of all stars harbor at least one planet. Second, small planets appear to outnumber large planets by a significant margin, suggesting that Earth-size planets are also very common. Keep in mind that the statistics remain incomplete: Because planets with larger orbits take longer to make the repeated transits necessary for confidence in a *Kepler* detection, and because small planets are more difficult to detect, the current statistics reflect only planets with relatively short orbital periods (in most cases, periods less than about 50 days for small planets and up to 250 days for large planets). As more data from the mission are analyzed, the estimated fractions of stars with planets can only increase.

The fact that the current statistics are dominated by planets that orbit relatively close to their stars means that most of the planets represented in Figure 10.10 are probably too hot to harbor life. However, Figure 10.11 **In general, smaller planets seem to be more common than larger ones.** shows *Kepler* data in a different way: as a graph of planet size versus orbital period. At first glance, the empty space at the lower right might seem to suggest that there are few Earth-size planets in Earth-size orbits. But the lack of data points in this region of the graph stems from the extreme difficulty in detecting the infrequent, shallow transits such planets make. Accounting statistically for these effects, scientists now estimate that about 20% of stars are likely to have a planet less than twice Earth's size in an orbit within the region around the star in which liquid water could potentially exist on the planet's surface (the *habitable zone* [Section 19.3]). These planets could potentially be homes to life.

In order to determine the general nature of an extrasolar planet, such as whether it is terrestrial or jovian, we need to determine its mass as well as its size, so that we can calculate its average density. Although we currently have both size and mass data for only a limited set of planets, the results already reveal a surprise: The density range for extrasolar planets is significantly wider than that for the planets in our own solar system. At the extremes, we've identified planets with average densities as great as that of iron and as low as that of Styrofoam, and we've found planets with almost every average density in between.

The Nature of Extrasolar Planets

We now come to the key question about extrasolar planets: Do they fall into the same terrestrial and jovian categories as the planets in our solar system, or do we find additional types of planets?

We cannot yet know for certain, because we do not yet have direct ways of obtaining spectra with sufficient detail to determine the compositions of extrasolar planets. Nevertheless, because we have measured the abundances of different chemical elements among other stars, we know that all planetary systems must start out from gas clouds generally similar in composition to the solar nebula. That is, all star systems are born from gas clouds containing at least about 98% hydrogen and helium, sprinkled with much smaller amounts of hydrogen compounds, rock, and metal (see Table 6.3). Given this fact, knowing a planet's average density gives us great insight into its likely composition, even if we cannot observe it directly.

More specifically, we can use our understanding of the behavior of different materials to create *models* that will tell us the expected composition of a planet based on its mass and radius, from which we can also calculate its average density. The results are shown in Figure 10.12 for a sample of planets for which both mass (usually from the Doppler method) and radius (from transits) are known. Be sure you understand the following key features of the figure:

- The horizontal axis shows planetary mass, in units of Earth masses. (The top of the graph shows the equivalent values in Jupiter masses.) Notice that this axis uses a scale that rises by powers of 10 because the masses vary over such a wide range.

- The vertical axis shows planetary radius in units of Earth radii. (The right side of the graph shows the equivalent values in Jupiter radii.)

- Each dot represents one planet for which both mass and radius have been measured. Planets discussed in this chapter are called out by name. Planets of our solar system are marked in green.

- The paintings around the graph show artist conceptions of what representative worlds might look like.

- Average density is easy to calculate from mass and radius, but the different scales used on the two axes make it difficult to read average density directly from the graph. To help with that, the three curves extending from the lower left to the top show three representative average densities.

- The colored regions indicate models representing the expected compositions of planets with the indicated combinations of mass and radius.

As you study Figure 10.12, you'll notice that extrasolar planets show much more variety than the planets of our own solar system. For example, HAT-P-32b has more than twice Jupiter's radius despite having the same mass, giving it an average density of about 0.14 g/cm^3—similar to that of Styrofoam. This low average density is probably a result of the fact that this planet orbits only 0.035 AU from its star, putting it more than 10 times closer to its star than Mercury is to the Sun. The close-in orbit gives the planet a very high temperature, which should puff up the planet's atmosphere and may explain why it has such a large size relative to its mass. Near the other extreme, the planet COROT 14b is only slightly larger than Jupiter but several times as massive, giving it an average density near 8 g/cm^3, about the same as the density of iron. Although such a

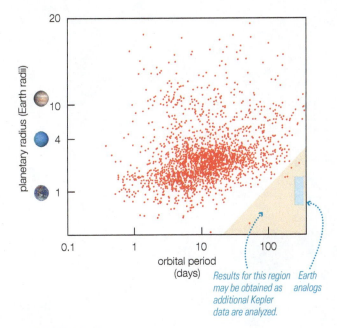

Results for this region may be obtained as additional Kepler data are analyzed. Earth analogs

▲ **FIGURE 10.11**

This figure shows the orbital periods and sizes of candidate planets identified from *Kepler* data. Although detecting the "Earth analogs" at the lower right is very difficult, the overall statistics suggest that about 20% of stars are likely to have such a planet.

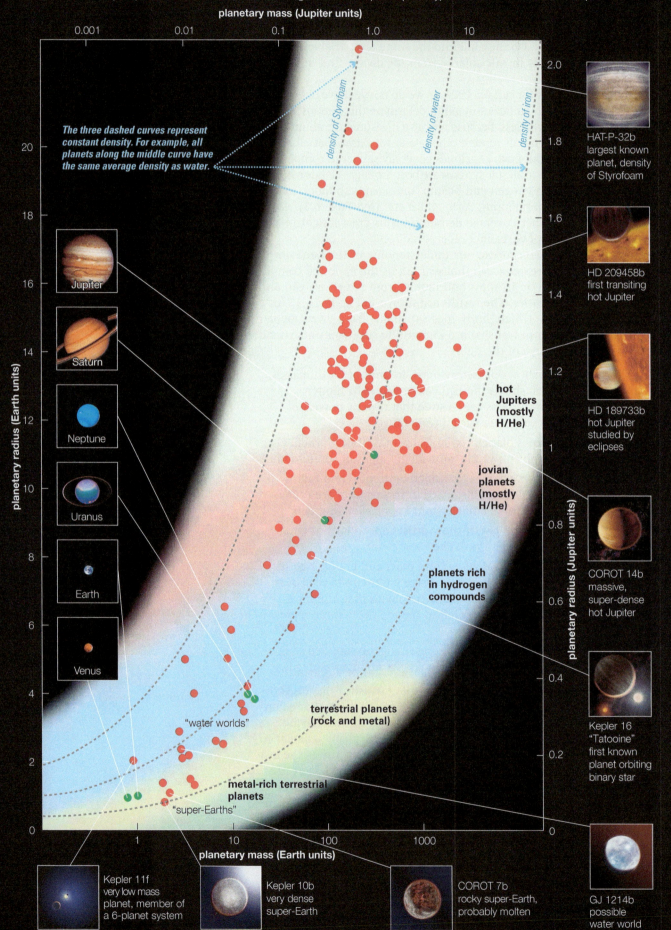

FIGURE 10.12

Masses and sizes of a sample of extrasolar planets for which both have been measured, compared to those of planets in our solar system. Each dot represents one planet. Dashed lines are lines of constant density for planets of different masses. Colored regions indicate expected planet types based on models of their compositions.

planetary mass (Jupiter units)

The three dashed curves represent constant density. For example, all planets along the middle curve have the same average density as water.

density of Styrofoam

density of water

density of iron

planetary radius (Earth units)

Jupiter

Saturn

Neptune

Uranus

Earth

Venus

hot Jupiters (mostly H/He)

jovian planets (mostly H/He)

planets rich in hydrogen compounds

terrestrial planets (rock and metal)

"water worlds"

metal-rich terrestrial planets

"super-Earths"

planetary radius (Jupiter units)

planetary mass (Earth units)

HAT-P-32b largest known planet, density of Styrofoam

HD 209458b first transiting hot Jupiter

HD 189733b hot Jupiter studied by eclipses

COROT 14b massive, super-dense hot Jupiter

Kepler 16 "Tatooine" first known planet orbiting binary star

GJ 1214b possible water world

Kepler 11f very low mass planet, member of a 6-planet system

Kepler 10b very dense super-Earth

COROT 7b rocky super-Earth, probably molten

high average density might seem surprising, it is not totally unexpected. Recall that jovian planets more massive than Jupiter are expected to have such strong gravity that they will be compressed to smaller sizes and much higher densities (see Figure 8.2b). Despite the wide spread in their densities, both HAT-P-32b and COROT 14b seem clearly to fall into the jovian planet category, being made largely of hydrogen and helium.

We also see many planets that appear to be terrestrial in nature, with compositions of rock and metal. For example, COROT 7b has an average density near 5 g/cm^3, comparable to Earth's density. Because it has a mass about 5 times that of Earth, COROT 7b is an example of what is sometimes called a "super-Earth." It orbits very close to its star, so its surface is probably molten. This and the many other known super-Earths are likely to have a rock/metal composition similar to that of the terrestrial worlds in our solar system.

Perhaps the most surprising planets shown in Figure 10.12 are the ones that are not clearly in either the terrestrial or the jovian category. Several planets cluster in the region of Uranus and Neptune, and perhaps share their composition of hydrogen compounds shrouded in an envelope of hydrogen and helium gas. Others (perhaps GJ 1214b) appear to fit the model for "water worlds" and may be made predominantly of water, either in liquid form or as a high-pressure solid, or perhaps of other hydrogen compounds. Alternatively, some of these worlds might be composed of a dense rocky/metallic core and a thick layer of low-density hydrogen and helium gas.

The bottom line is that extrasolar planets are abundant and diverse. Unlike our solar system, with only two clear types of "typical planets," other planetary systems have additional types of planets that defy easy categorization.

10.3 The Formation of Other Planetary Systems

The discovery of extrasolar planets presents us with an opportunity to test our theory of solar system formation. Can our existing theory explain other planetary systems, or do we have to go back to the drawing board?

◆ Do we need to modify our theory of solar system formation?

As we discussed in Chapter 6, the nebular theory holds that our solar system's planets formed as a natural consequence of processes that accompanied the formation of our Sun. If the theory is correct, then the same processes should accompany the births of other stars, so the nebular theory clearly predicts the existence of other planetary systems. In that sense, the discovery of extrasolar planets means the theory has passed a major test, because its most basic prediction has been verified. Other key details of the theory also seem supported. For example, the nebular theory says that jovian planet formation begins with condensation of solid particles of rock and ice (see Figure 6.20), which then accrete to larger sizes and capture nebular gas. We therefore expect that such planets should form more easily in a nebula with a higher proportion of rock and ice, and in fact more large planets have been found around stars richer in the elements that make these ingredients.

Nevertheless, extrasolar planets have already presented at least two significant challenges to our theory. One concerns the categories of planets; as we saw in Figure 10.12, many extrasolar planets do not fall neatly into either the terrestrial or the jovian category. An even more significant challenge is posed by the orbits of extrasolar planets. According to the nebular theory, jovian planets form as gravity pulls in gas around large, icy planetesimals that accrete in a spinning disk of material around a young star. The theory therefore predicts that jovian planets should form only in the cold outer regions of star systems (because it must be cold for ice to condense). The many known hot Jupiters that appear jovian in nature but have close-in orbits present a direct challenge to these ideas.

Explaining Planetary Orbits The nature of science demands that we question the validity of a theory whenever it is challenged by any observation or experiment [Section 3.4]. If the theory cannot explain the new observations, then we must revise or discard it. The surprising orbits of many known extrasolar planets have indeed caused scientists to reexamine the nebular theory of solar system formation.

Questioning began almost immediately once the first extrasolar planets were discovered. These planets were massive and had close-in orbits, making scientists wonder whether something might be fundamentally wrong with the nebular theory. For example, is it possible for jovian planets to form very close to a star? Astronomers addressed this question by studying many possible models of planet formation and reexamining the entire basis of the nebular theory. Several years of such reexamination did not turn up any good reasons to discard the basic theory, or any alternative means of forming jovian planets close to their stars. While we can't completely rule out the possibility that a major flaw has gone undetected, it now seems much more likely that the basic outline of the nebular theory is correct. Scientists therefore suspect that extrasolar jovian planets were indeed born with circular orbits far from their stars, and that those that now have close-in orbits underwent some sort of "planetary migration."

The idea of migration is not as strange as it may sound, and computer models of solar system formation suggest at least one likely mechanism: Migration may be caused by waves passing through a gaseous disk (Figure 10.13). The gravity of a planet moving through a disk can create waves that propagate through the disk, causing material to bunch up as the waves pass by. This bunched-up matter (in the wave peaks) then exerts a gravitational pull on the planet that tends to reduce its orbital energy, causing the planet to migrate inward toward its star.

This type of migration is not thought to have played a significant role in our own solar system, because the nebular gas was cleared out before it could have much effect. However, planets may form earlier in some other planetary systems, or the nebular gas may be cleared out later, allowing time for jovian planets to migrate substantially inward. In a few cases, the planets may form so early that they end up spiraling all the way into their stars. Indeed, astronomers have noted that some stars have an unusual assortment of elements in their outer layers, suggesting that they may have swallowed entire planets (including migrating jovian planets and possibly terrestrial planets shepherded inward along with the jovian planets). These ideas are not just hypothetical: One recently discovered planet appears to be on a million-year death spiral into its star.

Related mechanisms may explain the surprisingly high orbital eccentricities of many extrasolar planets. For example, planetary migration increases the chances that planets will influence each other gravitationally. In some cases, planets may pass close enough for a gravitational encounter [Section 4.4] in which one planet gains enough energy to escape from the star system entirely while the other is flung inward into a highly

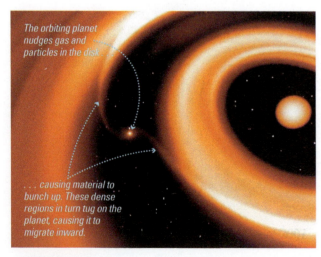

The orbiting planet nudges gas and particles in the disk...

... causing material to bunch up. These dense regions in turn tug on the planet, causing it to migrate inward.

▲ **FIGURE 10.13**
This figure shows a simulation of waves created by a planet embedded in a disk of material surrounding its star.

elliptical orbit. In other cases, continuing gravitational tugs may lead to orbital resonances much like those that cause the orbits of Jupiter's moons Io, Europa, and Ganymede to be more elliptical than they would be otherwise (see Figure 8.15b).

Explaining Planetary Types   Assuming we are correct about the role of planetary migration in explaining the surprising orbits of hot Jupiters, the basic tenets of the nebular theory seem to hold. That is, we expect rocky terrestrial worlds to form in the inner regions of solar systems and hydrogen-rich jovian planets to form in the outer regions, although migration may later cause some jovian planets to spiral inward (and likely alter the orbits of smaller planets as well). The remaining mystery, then, is why other systems seem to have planetary types that don't fall neatly into the terrestrial and jovian categories that we identify in our solar system.

Scientists still cannot fully explain the wide range of extrasolar planet properties, but we can envision possible explanations that seem to make sense and that would not fundamentally alter the nebular theory. For example, hydrogen-rich extrasolar planets vary in density by a factor of 100 (see Figure 10.12)—a range far greater than the density range we observe in our solar system—but it seems reasonable to think that much of this range is attributable to increases in temperature caused by some jovian planets being very close to their stars. As we've discussed, heating may puff up their atmospheres to large sizes and low densities, though models cannot yet account for the full density range. Higher densities can be explained by planets capturing even more hydrogen and helium gas than Jupiter, and compressing to smaller sizes due to higher gravity (see Figure 8.2b).

Similarly, the lack of "water world" planets in our solar system may not be as mysterious as it seems. Water worlds may be similar to Uranus and Neptune, though in some cases much smaller. These worlds may be much like the ice-rich planetesimals that seeded the formation of jovian planets in our solar system. In that case, perhaps whether water world planets exist depends on when a star clears its nebular gas, halting the epoch of planet formation. In our solar system, this did not occur until the ice-rich planetesimals pulled in vast quantities of hydrogen and helium gas from the solar nebula. Perhaps in other systems, an early solar wind blasted out the hydrogen and helium gas before it could be captured.

Super-Earths pose a different mystery: How did these planets gather so much rocky material? The answer is not yet known, though perhaps we should not be too surprised. After all, even though rocky material comprised less than 2% of the solar nebula, this still in principle was enough to build planets much larger than Earth. We may also need a better understanding of the conditions under which super-Earths or water worlds may capture hydrogen and helium gas, since such capture could lead to planets whose true nature is hiding under enormous envelopes of gas. With hundreds of new planets being discovered every year, other surprising planetary types are likely to be found in the future.

An Improved Nebular Theory   The bottom line is that discoveries of extrasolar planets have shown us that the nebular theory was incomplete. It explained the formation of planets and the simple layout of a solar system such as ours. However, it needs new features—such as planetary migration and variations in the basic planetary types—to explain the differing layouts of other planetary systems. A much wider variety of system arrangements seems possible than we had guessed before the discovery of extrasolar planets.

◆ Are planetary systems like ours common?

The validation of the nebular theory leads us to perhaps the most profound question still to be addressed in our study of extrasolar planets: Given that we have not yet discovered other planetary systems that are quite like ours, does this mean that our solar system is of a rare type, or does it simply mean that we have not yet acquired enough data to see how common systems like ours really are?

This is a profound question because of its implications for the way we view our place in the universe. If planetary systems like ours are common, then it seems reasonable to imagine that Earth-like planets—and perhaps life and civilizations—might also be common. But if our solar system is a rarity or even unique, then Earth might be the lone inhabited planet in our galaxy or even the universe.

As we've discussed, current evidence from both transit detections (see Figure 10.7) and Doppler discoveries already strongly supports the idea that planetary systems are common. We've also discussed statistical evidence indicating that as many as 20% of stars may be orbited by an approximately Earth-size planet in the habitable zone. Even the nearest star besides the Sun, Proxima Centauri, has a planet not too much larger than Earth within its habitable zone. However, we do not yet know the actual nature of these worlds, and despite the large number of known planetary systems, it is still too soon to say whether any of these planetary systems are genuinely "like ours." The bottom line is that we do not yet know whether planetary systems like ours are common, but more definitive answers should be coming over the next decade or two.

think about it Not long before most of today's college students were born, the only known planets were those of our own solar system. Today, the evidence suggests that many or most stars have planets. For the galaxy as a whole, that's a change in estimated number of planets from less than 10 to more than 100 billion. Do you think this change should alter our perspective on our place in the universe? Defend your opinion.

the big picture ❯ Putting Chapter 10 into Perspective

In this chapter, we have explored one of the newest areas of astronomy—the study of planetary systems beyond our own. As you continue your studies, please keep in mind the following important ideas:

- In a period of barely two decades, we have gone from knowing of no other planets around other stars to knowing that many or most stars have one or more planets. As a result, there is no longer any question that planets are common in the universe.

- The discovery of other planetary systems represents a striking confirmation of a key prediction of the nebular theory of solar system formation. Nevertheless, the precise characteristics of other planets and planetary systems pose challenges to details of the theory that scientists are still investigating.

- While we have already identified thousands of extrasolar planets or candidate planets, nearly all of these have been found with indirect methods. These methods allow us to determine many properties of the planets, but we will need direct images or spectra to learn about them in much more detail.

- It is too soon to know if planetary systems with layouts like ours are rare or common, but technology already exists that could allow us to answer this question and to learn if Earth-like planets are also common. It is only a matter of time until we know the answer to these fundamental questions.

my cosmic perspective The discovery that habitable worlds may be common even among relatively nearby stars means that, for the first time, we have evidence that could support the reality of many science fiction dreams of our future. Does this fact change the way you think about whether and how we should begin to reach for the stars?

10.1 Detecting Planets Around Other Stars

◆ How do we detect planets around other stars?

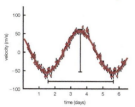

We can look for a planet's gravitational effect on its star through the **astrometric method**, which looks for small shifts in stellar position, or the **Doppler method**, which looks for the Doppler shifts that reveal the back-and-forth motion of stars. For the small fraction of planetary systems with orbits aligned edge-on to Earth, we can search for **transits**, in which a planet blocks a little of its star's light as it passes in front of the star.

10.2 The Nature of Planets Around Other Stars

◆ What properties of extrasolar planets can we measure?

All detection methods allow us to determine a planet's orbital period and distance from its star. The astrometric and Doppler methods can provide masses (or minimum masses), while the transit method can provide sizes. In cases where transit and Doppler methods are used together, we can determine average density. In some cases, transits (and eclipses) can provide other data, including limited data about atmospheric composition and temperature.

◆ How do extrasolar planets compare with planets in our solar system?

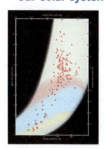

The known extrasolar planets have a much wider range of properties than the planets in our solar system. Many orbit much closer to their stars and with more eccentric orbital paths; some extrasolar jovian planets, called *hot Jupiters*, are also found close to their stars. We have also observed properties indicating planetary types, such as water worlds, that do not fall neatly into the traditional terrestrial and jovian categories.

10.3 The Formation of Other Planetary Systems

◆ Do we need to modify our theory of solar system formation?

Our basic theory seems sound, but we have had to modify it to allow for planetary migration and a wider range of planetary types than we find in our solar system. Many mysteries remain, but they are unlikely to require major change to the nebular theory of solar system formation.

◆ Are planetary systems like ours common?

Current evidence indicates that planetary systems are very common, though we do not yet have enough data to know for sure whether systems with layouts like ours—and possibly Earth-like planets—are also common.

---

visual skills check

Check your understanding of some of the many types of visual information used in astronomy. For additional practice, try the Chapter 10 Visual Quiz at MasteringAstronomy .

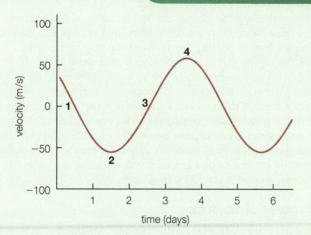

This graph shows the periodic variations in the Doppler shift of a star caused by a planet orbiting around it. Positive velocities mean the star is moving away from Earth, and negative velocities mean the star is moving toward Earth. (You can assume that the orbit is viewed edge-on from Earth.)

1. How long does it take the star and planet to complete one orbit around their center of mass?

2. What maximum velocity does the star attain?

3. Match the star's positions at points 1, 2, 3, and 4 in the plot with the descriptions below.
   a. headed straight toward Earth
   b. headed straight away from Earth
   c. closest to Earth
   d. farthest from Earth

4. Match the planet's positions at points 1, 2, 3, and 4 in the plot with the descriptions in question 3.

5. How would the plot change if the planet were more massive (but remained at the same orbital distance)?
   a. It would not change, because it describes the motion of the star, not the planet.
   b. The peaks and valleys would get significantly larger (greater positive and negative velocities) because of larger gravitational tugs.
   c. The peaks and valleys would get significantly closer together (shorter period) because of larger gravitational tugs.

MasteringAstronomy® For instructor-assigned homework and other learning materials, go to MasteringAstronomy®.

Review Questions

1. Why are *extrasolar planets* hard to detect directly?
2. What are the two current major approaches to detecting extrasolar planets indirectly?
3. How can gravitational tugs from orbiting planets affect the motion of a star? Explain how alien astronomers could deduce the existence of planets in our solar system by observing the Sun's motion.
4. Briefly describe the *astrometric method*. What is the *GAIA* mission?
5. Briefly describe the *Doppler method*. Summarize the evidence that the planet orbiting 51 Pegasi is a *hot Jupiter*.
6. How does the *transit method* work? What was the *Kepler* mission?
7. Briefly summarize the planetary properties we can measure with current detection methods.
8. Why does the Doppler method generally allow us to determine only *minimum* planetary masses? In what cases can we be confident that we know precise masses? Explain.
9. How does the transit method tell us planetary size, and in what cases can we also learn mass and density?
10. How do the orbits of known extrasolar planets differ from those of planets in our solar system? Why are these orbits surprising?
11. Summarize the current state of knowledge about extrasolar planet masses and sizes. Based on the evidence, is it likely that smaller planets or larger planets are more common?
12. Summarize the key features shown in Figure 10.12, and briefly describe the nature of planets that would fit each of the model regions shown on the graph.
13. What is planetary migration, and how may it account for the surprising orbits of many extrasolar planets?
14. How can scientists account for the fact that extrasolar planets seem to come in a wider range of types than the planets of our solar system?
15. Overall, does the nebular theory seem adequate for describing the origins of other planetary systems? Explain.
16. Based on current evidence, how common are planetary systems?

Test Your Understanding

🌀 Does It Make Sense?

Decide whether the statement makes sense (or is clearly true) or does not make sense (or is clearly false). Explain clearly; not all these have definitive answers, so your explanation is more important than your chosen answer.

17. An extraterrestrial astronomer surveying our solar system with the Doppler method could discover the existence of Jupiter with just a few days of observation.
18. The fact that we have not yet discovered an Earth-size extrasolar planet in an Earth-like orbit tells us that such planets must be very rare.
19. Within the next few years, astronomers expect to confirm all the planet detections made with the astrometric and Doppler methods by observing transits of these same planets.
20. The infrared brightness of a star system decreases when a planet goes into eclipse.
21. Some extrasolar planets are likely to be made mostly of water.
22. Some extrasolar planets are likely to be made mostly of gold.
23. Current evidence suggests that there could be 100 billion or more planets in the Milky Way Galaxy.

24. It's the year 2018: The *Kepler 2* mission has announced the discovery of numerous planets with Neptune-like orbits around their stars.
25. It's the year 2020: Astronomers have successfully photographed an Earth-size planet, showing that it has oceans and continents.
26. It's the year 2040: Scientists announce that our first spacecraft to reach an extrasolar planet is now orbiting a planet in a planetary system located near the center of the Milky Way Galaxy.

Quick Quiz

Choose the best answer to each of the following. Explain your reasoning with one or more complete sentences.

27. Which method could detect a planet in an orbit that is face-on to Earth? (a) Doppler method (b) transit method (c) astrometric method
28. Which detection method(s) measure gravitational tug of a planet on its star, allowing us to estimate planetary mass? (a) the transit method only (b) the Doppler method only (c) the astrometric and Doppler methods
29. Which one of the following can the transit method tell us about a planet? (a) its mass (b) its size (c) the eccentricity of its orbit
30. To determine a planet's average density, we can use (a) the transit method alone. (b) the astrometric and Doppler methods together. (c) the transit and Doppler methods together.
31. Based on the model types shown in Figure 10.12, a planet made almost entirely of hydrogen compounds would be considered a (a) terrestrial planet. (b) jovian planet. (c) "water world."
32. Look at the dot for Jupiter in Figure 10.12, then at the red dot directly to Jupiter's left. Compared to the density of Jupiter, the density of the planet represented by that dot is (a) higher. (b) lower. (c) the same.
33. The term "super-Earth" refers to a planet that is (a) the size of Earth but with more water. (b) larger than Earth but on a close-in orbit that makes it much hotter than Earth. (c) similar in composition to Earth but larger in size.
34. What's the best explanation for the location of hot Jupiters? (a) They formed closer to their stars than Jupiter did. (b) They formed farther out like Jupiter but then migrated inward. (c) The strong gravity of their stars pulled them in close.
35. Based on computer models, when is planetary migration most likely to occur in a planetary system? (a) early in its history, when there is still a gaseous disk around the star (b) shortly after a stellar wind clears the gaseous disk away (c) late in its history, when asteroids and comets occasionally collide with planets
36. Based on current data, planetary systems appear to be (a) extremely rare. (b) present around at least about one-third of all stars. (c) present around at least 99% of all stars.

🌀 Process of Science

37. *When Is a Theory Wrong?* As discussed in this chapter, in its original form the nebular theory of solar system formation does not explain the orbits of many known extrasolar planets, but it can explain them with modifications such as allowing for planetary migration. Does this mean the theory was "wrong" or only "incomplete" before the modifications were made? Explain. Be sure to look back at the discussion in Chapter 3 of the nature of science and scientific theories.

38. *Unanswered Questions.* As discussed in this chapter, we are only just beginning to learn about extrasolar planets. Briefly describe one important but unanswered question related to the study of planets around other stars. Then write 2–3 paragraphs in which you discuss how we might answer this question in the future. Be as specific as possible, focusing on the type of evidence necessary to answer the question and how the evidence could be gathered. What are the benefits of finding answers to this question?

Group Work Exercise

39. *Time to Move On.* **Roles:** *Scribe* (takes notes on the group's activities), *Proposer* (proposes explanations to the group), *Skeptic* (points out weaknesses in proposed explanations), *Moderator* (leads group discussion and makes sure everyone contributes). **Activity:** A common theme in science fiction is "leaving home" to find a new planet for humans to live on. Now that we know about thousands of planets, we can start imagining how to choose.
    a. Make a list of characteristics that you would look for in a planet that might make a good home.
    b. Examine the planets in Figure 10.12. Does this graph give enough information to determine which planets might make good homes, or poor ones? If not, what's missing?
    c. Suppose you also knew the orbital distance for each of the planets in Figure 10.12. Would that make it easier to find potential good homes? Why or why not?

Investigate Further

Short-Answer/Essay Questions

40. *Explaining the Doppler Method.* Explain how the Doppler method works in terms an elementary school child would understand. It may help to use an analogy to explain the difficulty of direct detection and the general phenomenon of the Doppler shift.
41. *Comparing Methods.* What are the strengths and limitations of the Doppler and transit methods? What kinds of planets are easiest to detect with each method? Are there certain planets that each method cannot detect, even if the planets are very large? Explain. What advantages are gained if a planet can be detected by both methods?
42. *No Hot Jupiters Here.* How do we think hot Jupiters formed? Why didn't one form in our solar system?
43. *Low-Density Planets.* Only one planet in our solar system has a density less than 1 g/cm³, but many extrasolar planets do. Explain why in a few sentences. (*Hint:* Consider the densities of the jovian planets in our solar system, given in Figure 8.1.)
44. *Detect an Extrasolar Planet for Yourself.* Most colleges and many amateur astronomers have the equipment necessary to detect known extrasolar planets using the transit method. All that's required is a telescope 10 or more inches in diameter, a CCD camera system, and a computer system for data analysis. The basic method is to take exposures of a few minutes' duration over a period of several hours around the times of predicted transit, and to compare the brightness of the star being transited to that of other stars in the same CCD frame. For complete instructions, see the study area of MasteringAstronomy.

Quantitative Problems

Be sure to show all calculations clearly and state your final answers in complete sentences.

45. *Lost in the Glare.* How hard would it be for an alien astronomer to detect the light from planets in our solar system compared to the light from the Sun itself?

a. Calculate the fraction of the total emitted sunlight that reaches Earth. (*Hint:* Find the area of a sphere around the Sun with a radius equal to Earth's average orbital distance area $= 4\pi a^2$, then calculate the fraction of that area taken up by the disk of Earth [area $= \pi r_{Earth}^2$].)
b. Earth reflects 29% of the Sun's light. Based on this fact and your answer from part a, what fraction of total sunlight is *reflected* by Earth? (*Hint:* Your answer will simply be the overall fraction of all the Sun's light that is reflected by Earth.)
c. Would detecting Jupiter be easier or harder than detecting Earth? Comment on whether you think Jupiter's larger size or greater distance has a stronger effect on its detectability. You may neglect any difference in reflectivity between Earth and Jupiter.
46. *Transit of TrES-1.* The planet TrES-1, orbiting a distant star, has been detected by both the transit and the Doppler technique, so we can calculate its density and get an idea of what kind of planet it is.
    a. Using the method of Cosmic Calculations 10.1, calculate the radius of the transiting planet. The planetary transits block 2% of the star's light. The star that TrES-1 orbits has a radius of about 85% of our Sun's radius.
    b. The mass of the planet is approximately 0.75 times the mass of Jupiter, and Jupiter's mass is about 1.9×10^{27} kilograms. Calculate the average density of the planet. Give your answer in grams per cubic centimeter. Compare this density to the average densities of Saturn (0.7 g/cm³) and Earth (5.5 g/cm³). Is the planet terrestrial or jovian in nature? (*Hint:* To find the volume of the planet, use the formula for the volume of a sphere: $V = \frac{4}{3}\pi r^3$. Be careful with unit conversions.)
47. *Planet Around 51 Pegasi.* The star 51 Pegasi has about the same mass as our Sun. A planet discovered orbiting it has an orbital period of 4.23 days. The mass of the planet is estimated to be 0.6 times the mass of Jupiter. Use Kepler's third law to find the planet's average distance (semimajor axis) from its star. (*Hint:* Because the mass of 51 Pegasi is about the same as the mass of our Sun, you can use Kepler's third law in its original form, $p^2 = a^3$ [Section 3.3]. Be sure to convert the period into years before using this equation.)

Discussion Questions

48. *So What?* What is the significance of the discovery of extrasolar planets, if any? Justify your answer in the context of this book's discussion of the history of astronomy.
49. *Is It Worth It?* Thanks to rapidly advancing technology, we could probably now build space observatories capable of obtaining images and spectra of Earth-size planets around other stars, with enough resolution to be able to determine whether they are Earth-like, and perhaps even to detect spectral signatures that would indicate the presence of life. However, such observatories would likely cost several billion dollars. Suppose you were a member of the U.S. Congress. How much would you be willing to spend on such observatories? Defend your opinion.

Web Projects

50. *New Planets.* Research the latest extrasolar planet discoveries. Create a "planet journal," complete with illustrations as needed, with a page for each of at least three recently discovered planets. On each page, note the method that was used to find the planet, list any information we have about the nature of the planet, and discuss how the planet does or does not fit in with our current understanding of planetary systems.
51. *Extrasolar Planet Mission.* Learn about a proposed future mission to study extrasolar planets, including its proposed design, capabilities, and goals. Write a short report on your findings.

Comparing the worlds in the solar system has taught us important lessons about Earth and why it is so suitable for life. This illustration summarizes some of the major lessons we've learned by studying other worlds both in our own solar system and beyond it.

(1) Comparing the terrestrial worlds shows that a planet's size and distance from the Sun are the primary factors that determine how it evolves through time [Chapter 7].

Venus demonstrates the importance of distance from the Sun: If Earth were moved to the orbit of Venus, it would suffer a runaway greenhouse effect and become too hot for life.

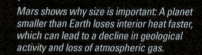

Mars shows why size is important: A planet smaller than Earth loses interior heat faster, which can lead to a decline in geological activity and loss of atmospheric gas.

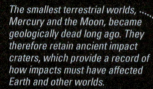

The smallest terrestrial worlds, Mercury and the Moon, became geologically dead long ago. They therefore retain ancient impact craters, which provide a record of how impacts must have affected Earth and other worlds.

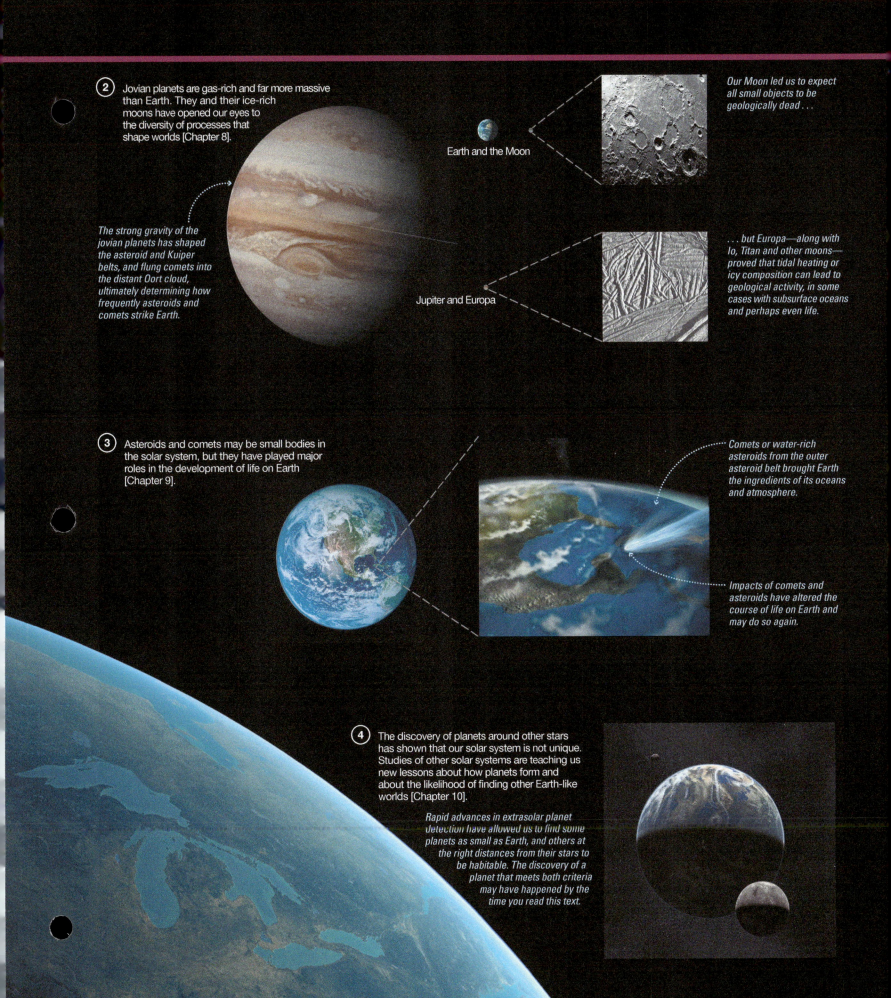

2 Jovian planets are gas-rich and far more massive than Earth. They and their ice-rich moons have opened our eyes to the diversity of processes that shape worlds [Chapter 8].

The strong gravity of the jovian planets has shaped the asteroid and Kuiper belts, and flung comets into the distant Oort cloud, ultimately determining how frequently asteroids and comets strike Earth.

Earth and the Moon

Our Moon led us to expect all small objects to be geologically dead . . .

Jupiter and Europa

. . . but Europa—along with Io, Titan and other moons—proved that tidal heating or icy composition can lead to geological activity, in some cases with subsurface oceans and perhaps even life.

3 Asteroids and comets may be small bodies in the solar system, but they have played major roles in the development of life on Earth [Chapter 9].

Comets or water-rich asteroids from the outer asteroid belt brought Earth the ingredients of its oceans and atmosphere.

Impacts of comets and asteroids have altered the course of life on Earth and may do so again.

4 The discovery of planets around other stars has shown that our solar system is not unique. Studies of other solar systems are teaching us new lessons about how planets form and about the likelihood of finding other Earth-like worlds [Chapter 10].

Rapid advances in extrasolar planet detection have allowed us to find some planets as small as Earth, and others at the right distances from their stars to be habitable. The discovery of a planet that meets both criteria may have happened by the time you read this text.

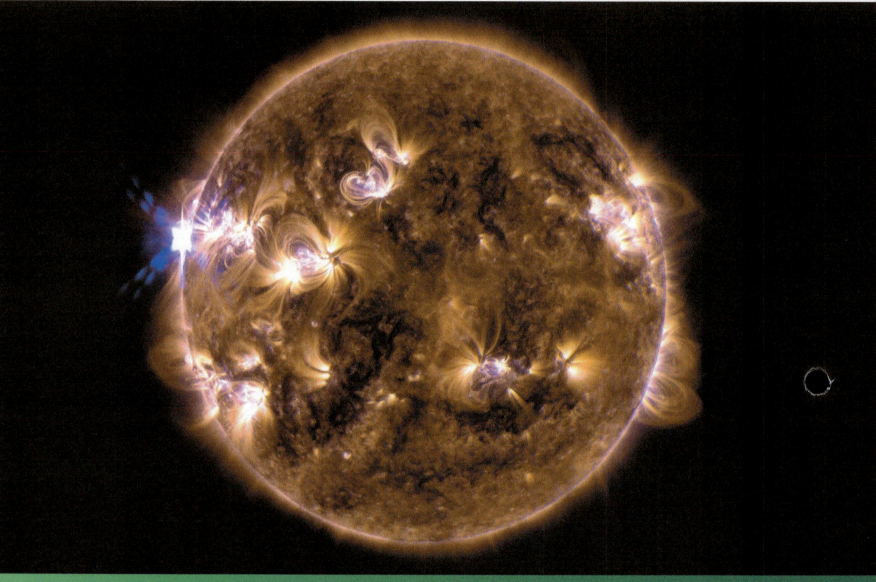

11 Our Star

NASA's *Solar Dynamics Observatory* captured this X-ray image of our Sun, with a solar flare in progress (bright spot at left just above center). Colors correspond to the strength of the X-ray emission, with the brightest emission in blue and purple.

A stronomy today encompasses the study of the entire universe, but the root of the word *astronomy* comes from the Greek word for "star." Although we have learned a lot about the universe up to this point in the book, only now do we turn our attention to the study of the stars, the namesakes of astronomy.

When we think of stars, we usually think of the beautiful points of light visible on a clear night. But the nearest and most easily studied star is visible only in the daytime—our Sun. In this chapter, we will study the Sun in some detail. We will see how the Sun generates the energy that supports life on Earth. Equally important, we will study our Sun as a star so that it can serve as an introduction to subsequent chapters in which we will study stars throughout the universe.

ESSENTIAL PREPARATION

1. Where do objects get their energy? [Section 4.3]
2. What is matter? [Section 5.1]
3. How does light tell us the temperatures of planets and stars? [Section 5.2]

11.1 A Closer Look at the Sun

We discussed the general features of the Sun in our tour of the solar system in Chapter 6 (see page 140). Now it's time to get better acquainted with our nearest star.

◆ Why does the Sun shine?

Ancient thinkers often imagined the Sun to be some type of fire, perhaps a lump of burning coal or wood. It was a reasonable suggestion for the times, since the idea could not yet be tested. This situation changed in the mid-19th century, by which time the Sun's size and distance had been measured with reasonable accuracy. Scientists could then calculate the true energy output of the Sun, and this quickly ruled out coal, wood, or any other type of chemical burning: There is simply no way that chemical processes can account for the Sun's huge energy output.

In the late 19th century, astronomers came up with an idea that seemed more plausible, at least at first. They suggested that the Sun generates energy by slowly contracting in size, a process called **gravitational contraction** (or *Kelvin-Helmholtz contraction,* after the scientists who suggested it). Recall that a shrinking gas cloud heats up because the gravitational potential energy of gas particles far from the cloud center is converted into thermal energy as the gas moves inward (see Figure 4.12b). A gradually shrinking Sun would always have some gas moving inward, converting gravitational potential energy into thermal energy. This thermal energy would keep the inside of the Sun hot.

Because of its large mass, the Sun would need to contract only very slightly each year to maintain its temperature—so slightly that the contraction would have been unnoticeable to 19th-century astronomers. Calculations showed that gravitational contraction could have kept the Sun shining steadily for up to about 25 million years. For a while, some astronomers thought that this idea had solved the ancient mystery of how the Sun shines. However, geologists pointed out a fatal flaw: Studies of rocks and fossils had already suggested that Earth was far older than 25 million years, which meant that gravitational contraction could not account for the Sun's long-term energy generation.

With both chemical processes and gravitational contraction ruled out as possible explanations for why the Sun shines, scientists were at a loss. There was no known way that an object the size of the Sun could generate so much energy for billions of years. A completely new type of explanation was needed, and it came with Einstein's publication of his special theory of relativity in 1905.

Einstein's theory included his famous equation $E = mc^2$, which tells us that mass itself contains an enormous amount of potential energy [Section 4.3]. Calculations showed that the Sun's mass contained more than enough energy to account for billions of years of sunshine, if the Sun could somehow convert some of its mass into thermal energy. It took a few decades for scientists to work out the details, but by the end of the 1930s we had learned that the Sun converts mass into energy through the process of *nuclear fusion* [Section 1.2].

The Sun shines by converting mass into energy through nuclear fusion.

How Fusion Started Nuclear fusion requires extremely high temperatures and densities (for reasons we will discuss in the next section). In the Sun, these conditions are found deep in the core. But how did the Sun become hot enough for fusion to begin in the first place?

The answer invokes the mechanism of gravitational contraction. Recall that our Sun was born about 4½ billion years ago from a collapsing cloud of interstellar gas [Section 6.2]. The contraction of the cloud released gravitational potential energy, raising the interior temperature and pressure. This process continued until the core finally became hot enough to sustain nuclear fusion, because only then did the Sun produce enough energy to give it the stability that it has today.

Gravitational contraction released the energy that made the Sun's core hot enough for fusion.

The Stable Sun The Sun continues to shine steadily today because it has achieved two kinds of balance that keep its size and energy output stable. The first kind of balance, called **gravitational equilibrium** (or *hydrostatic equilibrium*), is between the outward push of internal gas pressure and the inward pull of gravity. A stack of acrobats provides a simple example of gravitational equilibrium (Figure 11.1). The bottom person supports the weight of everybody above him, so his arms must push upward with enough pressure to support all this weight. At each higher level, the overlying weight is less, so it's a little easier for each additional person to hold up the rest of the stack.

Gravitational equilibrium works much the same in the Sun, except the outward push against gravity comes from internal gas pressure rather than an acrobat's arms. The Sun's internal pressure precisely balances gravity at every point within it, thereby keeping the Sun stable in size (Figure 11.2). Because the weight of overlying layers is greater as we look deeper into the Sun, the pressure must increase with depth. Deep in the Sun's core, the pressure makes the gas hot and dense enough to sustain nuclear fusion. The energy released by fusion, in turn, heats the gas and maintains the pressure that keeps the Sun in balance against the inward pull of gravity.

think about it Earth's atmosphere is also in gravitational equilibrium, with the weight of upper layers supported by the pressure in lower layers. Use this idea to explain why the air gets thinner at higher altitudes.

▲ **FIGURE 11.1**

An acrobat stack is in gravitational equilibrium: The lowest person supports the most weight and feels the greatest pressure, and the overlying weight and underlying pressure decrease for those higher up.

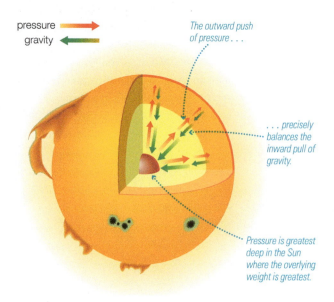

pressure ⟶
gravity ⟵

The outward push of pressure . . .

. . . precisely balances the inward pull of gravity.

Pressure is greatest deep in the Sun where the overlying weight is greatest.

▲ **FIGURE 11.2**

Gravitational equilibrium in the Sun: At each point inside, the pressure pushing outward balances the weight of the overlying layers.

The second kind of balance is **energy balance** between the rate at which fusion releases energy in the Sun's core and the rate at which the Sun's surface radiates this energy into space (Figure 11.3). Energy balance is important, because without it the balance between pressure and gravity would not remain steady. If fusion in the core did not replace the energy radiated from the surface, thereby keeping the total thermal energy content constant, then gravitational contraction would cause the Sun to shrink and force its core temperature to rise.

Two types of balance keep the Sun stable: Pressure balances gravity and fusion energy balances the energy radiated into space.

In summary, the answer to the question "Why does the Sun shine?" is that about 4½ billion years ago *gravitational contraction* made the Sun hot enough to sustain nuclear fusion in its core. Ever since, energy liberated by fusion has maintained *gravitational equilibrium* and *energy balance* within the Sun, keeping it shining steadily. The Sun was born with enough nuclear fuel to last about 10 billion years, which means it is now only about halfway through this 10-billion-year lifetime.

◆ What is the Sun's structure?

The Sun is essentially a giant ball of hot gas or, more technically, *plasma*—a gas in which atoms are ionized because of the high temperature. The differing temperatures and densities of the plasma at different depths give the Sun the layered structure shown in Figure 11.4. To help you make sense of the layering, let's imagine that you have a spaceship that can somehow withstand the immense heat and pressure as you take an imaginary journey from Earth to the center of the Sun.

Basic Properties of the Sun As you begin your journey from Earth, the Sun appears as a whitish ball of glowing gas. Just as astronomers have done in real life, you can use simple observations to determine basic properties of the Sun. Spectroscopy [Section 5.2] tells you that the Sun is made almost entirely of hydrogen and helium. From the Sun's angular size and distance, you can determine that its radius is just under 700,000 kilometers, or more than 100 times the radius of Earth. Even **sunspots,** which appear as dark splotches on the Sun's surface, can be larger in size than Earth.

You can measure the Sun's mass using Newton's version of Kepler's third law. It is about 2×10^{30} kilograms, which is some 300,000 times the mass of Earth and nearly 1000 times the mass of all the planets in our solar system put together. You can observe the Sun's rotation by tracking the motion of sunspots or by measuring Doppler shifts [Section 5.2] on opposite sides of the Sun. Unlike a spinning ball, the entire Sun does *not* rotate at the same rate: The solar equator completes one rotation in about 25 days, and the rotation period increases with latitude to about 30 days near the solar poles.

think about it As a brief review, describe how astronomers use Newton's version of Kepler's third law to determine the mass of the Sun. What two properties of Earth's orbit do we need to know?

The Sun releases an enormous amount of radiative energy into space. In science, we measure energy in units of *joules* [Section 4.3]. We define **power** as the *rate* at which energy is used or released. The standard unit of power is the **watt,** defined as 1 joule of energy per second; that is, 1 watt = 1 joule/s. For example, a 100-watt light bulb requires 100 joules of energy for every second it is left turned on. The

Energy released by fusion in the Sun's core . . .

. . . balances the radiative energy emitted from the Sun's surface.

▲ **FIGURE 11.3**
Energy balance in the Sun: Fusion supplies energy in the core at the same rate as the Sun radiates energy from its surface.

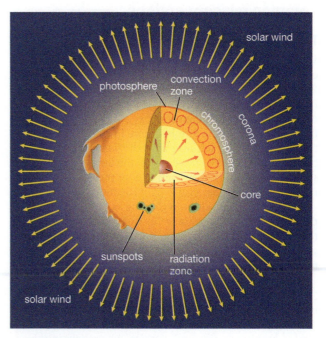

▲ **FIGURE 11.4**
The basic structure of the Sun.

TABLE 11.1 Basic Properties of the Sun

Radius (R_{Sun})	696,000 km (about 109 times the radius of Earth)
Mass (M_{Sun})	2×10^{30} kg (about 300,000 times the mass of Earth)
Luminosity (L_{Sun})	3.8×10^{26} watts
Composition (by percentage of mass)	70% hydrogen, 28% helium, 2% heavier elements
Rotation rate	25 days (equator) to 30 days (poles)
Surface temperature	5800 K (average); 4000 K (sunspots)
Core temperature	15 million K

Sun's total power output, or **luminosity,** is an incredible 3.8×10^{26} watts. Table 11.1 summarizes the basic properties of the Sun.

The Sun's Atmosphere   Even at a great distance from the Sun, you and your spacecraft can feel slight effects from the **solar wind**—the stream of charged particles continually blown outward in all directions from the Sun. The solar wind helps shape the magnetospheres of planets (see Figure 7.6) and blows back the material that forms the plasma tails of comets [Section 9.3].

As you approach the Sun more closely, you begin to encounter the low-density gas that represents what we usually think of as the Sun's atmosphere. The outermost layer of this atmosphere, called the **corona,** extends several million kilometers above the visible surface of the Sun. The temperature of the corona is astonishingly high—about 1 million K—explaining why this region emits most of the Sun's X rays. However, the corona's density is so low that your spaceship feels relatively little heat, despite the million-degree temperature [Section 4.3].

Nearer the surface, the temperature suddenly drops to about 10,000 K in the **chromosphere,** the middle layer of the solar atmosphere and the region that radiates most of the Sun's ultraviolet light. Then you plunge through the lowest layer of the atmosphere, the **photosphere,** which is the visible surface of the Sun. Although the photosphere looks like a well-defined surface from Earth, it consists of gas far less dense than Earth's atmosphere. The temperature of the photosphere averages just under 6000 K, and its surface seethes and churns like a pot of boiling water. The photosphere is also where you'll find sunspots, regions of intense magnetic fields that would cause your compass needle to swing wildly about.

The Sun's upper atmosphere is much hotter than the visible surface, or photosphere, but its density is much lower.

The Sun's Interior   Up to this point in your journey, you may have seen Earth and the stars when you looked back. But blazing light engulfs you as you slip beneath the photosphere. You are inside the Sun, and incredible turbulence tosses your spacecraft about. If you can hold steady long enough to see what is going on around you, you'll notice spouts of hot gas rising upward, surrounded by cooler gas cascading down from above. You are in the **convection zone,** where energy generated in the solar core travels outward, transported by the rising of hot gas and falling of cool gas called *convection* [Section 7.1]. The photosphere above you is the top of the convection zone, and convection is the cause of the Sun's seething, churning surface.

About a third of the way down to the Sun's center, the turbulence of the convection zone gives way to the calmer plasma of the **radiation zone,** where energy moves outward primarily in the form of photons of light. The temperature rises to almost 10 million K, and your spacecraft is bathed in X rays trillions of times more intense than the visible light at the solar surface.

Inside the Sun, temperature rises with depth, reaching 15 million K in the core.

No real spacecraft could survive, but your imaginary one keeps plunging straight down to the solar **core.** There you finally find the source of the Sun's energy: nuclear fusion transforming hydrogen into helium. At the Sun's center, the temperature is about 15 million K, the density is more than 100 times that of water, and the pressure is 200 billion times that on Earth's surface.

common misconceptions

The Sun Is Not on Fire

We often say that the Sun is "burning," a term that conjures up images of a giant bonfire in the sky. However, the Sun does not burn in the same sense as a fire burns on Earth. Fires generate light through chemical changes that consume oxygen and produce a flame. The glow of the Sun has more in common with the glowing embers left over after the flames have burned out. Much like hot embers, the Sun's surface shines because it is hot enough to emit thermal radiation that includes visible light [Section 5.2].

Hot embers quickly stop glowing as they cool, but the Sun keeps shining because its surface is kept hot by the energy rising from its core. Because this energy is generated by nuclear fusion, we sometimes say that it is the result of "nuclear burning"—a term intended to suggest nuclear changes in much the same way that "chemical burning" suggests chemical changes. While it is reasonable to say that the Sun undergoes nuclear burning in its core, it is not accurate to speak of any kind of burning on the Sun's surface, where light is produced primarily by thermal radiation.

The energy produced in the core today will take a few hundred thousand years to reach the surface.

With your journey complete, it's time to turn around and head back home. We'll continue this chapter by studying fusion in the solar core and then tracing the flow of the energy generated by fusion as it moves outward through the Sun.

11.2 Nuclear Fusion in the Sun

We've seen that the Sun shines because of energy generated by nuclear fusion, and that this fusion occurs under the extreme temperatures and densities found deep in the Sun's core. But exactly how does fusion occur and release energy? And how can we claim to know about something taking place out of sight in the Sun's interior?

Before we begin to answer these questions, it's important to realize that the nuclear reactions that generate energy in the Sun are very different from those used to generate energy in human-built nuclear reactors on Earth. Our nuclear power plants generate energy by splitting large nuclei—such as those of uranium or plutonium—into smaller ones. The process of splitting an atomic nucleus is called *nuclear fission*. In contrast, the Sun makes energy by combining, or fusing, two or more small nuclei into a larger one. That is why we call the process *nuclear fusion*. Figure 11.5 summarizes the difference between fission and fusion.

◆ How does nuclear fusion occur in the Sun?

Fusion occurs within the Sun because the 15 million K plasma in the solar core is like a "soup" of hot gas full of bare, positively charged atomic nuclei (and negatively charged electrons) whizzing about at extremely high speeds. At any time, some of these nuclei are on high-speed collision courses with each other. In most cases, electromagnetic forces deflect the nuclei, preventing collisions, because positive charges repel one another. However, if nuclei collide with sufficient energy, they can stick together (fuse) to form a heavier nucleus.

Sticking positively charged nuclei together is not easy. The **strong force,** which binds protons and neutrons together in atomic nuclei, is the only force in nature that can overcome the electromagnetic repulsion between two positively charged nuclei. In contrast to gravitational and electromagnetic forces, which drop off gradually as the distances between particles increase (by an inverse square law [Section 4.4]), the strong force is more like glue or Velcro: It overpowers the electromagnetic force over very small distances, but is insignificant when the distances between particles exceed the typical sizes of atomic nuclei. The key to nuclear fusion is pushing the positively charged nuclei close enough together for the strong force to outmuscle electromagnetic repulsion (Figure 11.6).

The high pressures and temperatures in the solar core are just right for fusion of hydrogen nuclei into helium nuclei. The high temperature is important because the nuclei must collide at very high speeds if they are to come close enough together to fuse. The higher the temperature, the more energetic the collisions, making fusion reactions more likely at higher temperatures. The high pressure of the overlying

Positively charged nuclei fuse together if they pass close enough for the strong force to overpower electromagnetic repulsion.

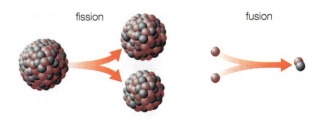

▲ FIGURE 11.5
Nuclear fission splits a nucleus into smaller nuclei (not usually of equal size), while nuclear fusion combines smaller nuclei into a larger nucleus.

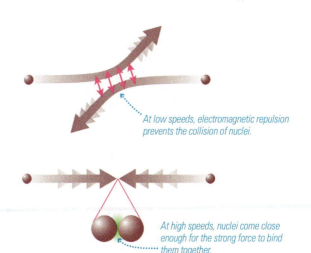

At low speeds, electromagnetic repulsion prevents the collision of nuclei.

At high speeds, nuclei come close enough for the strong force to bind them together.

▲ FIGURE 11.6
Positively charged nuclei can fuse only if a high-speed collision brings them close enough for the strong force to come into play.

Hydrogen Fusion by the Proton–Proton Chain

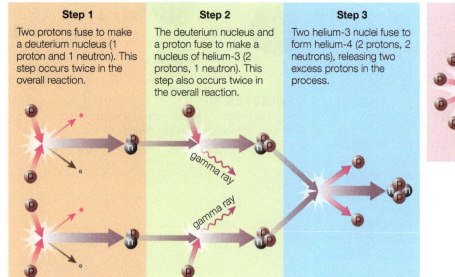

Step 1
Two protons fuse to make a deuterium nucleus (1 proton and 1 neutron). This step occurs twice in the overall reaction.

Step 2
The deuterium nucleus and a proton fuse to make a nucleus of helium-3 (2 protons, 1 neutron). This step also occurs twice in the overall reaction.

Step 3
Two helium-3 nuclei fuse to form helium-4 (2 protons, 2 neutrons), releasing two excess protons in the process.

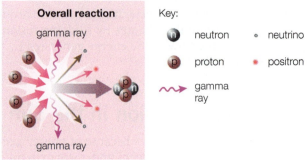

Overall reaction

Key:
- neutron
- proton
- gamma ray
- neutrino
- positron

▲ **FIGURE 11.7**
In the Sun, four hydrogen nuclei (protons) fuse into one helium-4 nucleus by way of the proton–proton chain. Gamma rays and subatomic particles known as neutrinos and positrons carry off the energy released in the reaction.

---

cosmic calculations 11.1

The Ideal Gas Law

The pressure (P) of a gas depends on both its temperature (T) and its *number density* (n)—the number of particles contained in each cubic centimeter of gas. The *ideal gas law* expresses the relationship:

$$P = nkT$$

where $k = 1.38 \times 10^{-23}$ joule/K is *Boltzmann's constant*.

Example: The Sun's core density is about 10^{26} particles per cubic centimeter (which we write as 10^{26} cm^{-3}) and its temperature is about 15 million K (1.5×10^7 K). Compare the Sun's core pressure to that of Earth's atmosphere at sea level, where the density is about 2.4×10^{19} particles per cubic centimeter and the temperature is about 300 K.

Solution: Dividing the Sun's core pressure by Earth's atmospheric pressure, we find that their ratio is

$$\frac{P_{\text{Sun(core)}}}{P_{\text{Earth(atmos)}}} = \frac{n_{\text{Sun}} k\, T_{\text{Sun}}}{n_{\text{Earth}} k\, T_{\text{Earth}}}$$

$$= \frac{10^{26}\ \text{cm}^{-3} \times (1.5 \times 10^7\ \text{K})}{(2.4 \times 10^{19}\ \text{cm}^{-3}) \times 300\ \text{K}}$$

$$= 2 \times 10^{11}$$

The Sun's core pressure is about 200 billion (2×10^{11}) times Earth's atmospheric pressure.

---

layers is necessary because, without it, the hot plasma of the solar core would simply explode into space, shutting off the nuclear reactions.

think about it The Sun generates energy by fusing hydrogen into helium, but as we'll see in later chapters, some stars fuse helium or even heavier elements. Do temperatures need to be higher or lower for the fusion of heavier elements? Why? (*Hint:* How does the positive charge of a nucleus affect its ability to fuse with another nucleus?)

The Proton–Proton Chain Let's investigate the Sun's fusion process in a little more detail. Recall that hydrogen nuclei are simply individual protons, while the most common form of helium consists of two protons and two neutrons(see Figure 5.6). The overall hydrogen fusion reaction therefore transforms four individual protons into a helium nucleus containing two protons and two neutrons:

$4\ ^1\text{H}$ $1\ ^4\text{He}$

This overall reaction actually proceeds through several steps involving just two nuclei at a time. The sequence of steps that occurs in the Sun is called the **proton–proton chain,** because it begins with collisions between individual protons (hydrogen nuclei). Figure 11.7 illustrates the steps in the proton–proton chain. Notice that the overall reaction is just as described above, with four protons combining to make one helium nucleus. Energy is carried off by the gamma rays and subatomic particles (neutrinos and positrons) released in the process.

Nuclear fusion in the Sun combines four hydrogen nuclei into one helium nucleus, releasing energy in the process.

Fusion of hydrogen into helium generates energy because a helium nucleus has a mass slightly less (by about 0.7%) than the combined

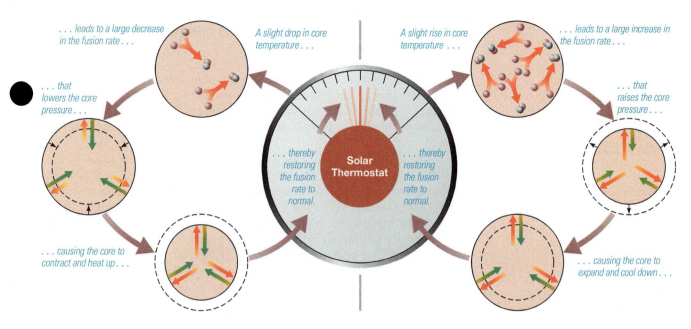

The following labels appear around the diagram:

. . . leads to a large decrease in the fusion rate . . .

A slight drop in core temperature . . .

A slight rise in core temperature . . .

. . . leads to a large increase in the fusion rate . . .

. . . that lowers the core pressure . . .

. . . that raises the core pressure . . .

Solar Thermostat

. . . thereby restoring the fusion rate to normal.

. . . thereby restoring the fusion rate to normal.

. . . causing the core to contract and heat up . . .

. . . causing the core to expand and cool down . . .

mass of four hydrogen nuclei. That is, when four hydrogen nuclei fuse into a helium nucleus, a little bit of mass disappears. This disappearing mass becomes energy in accord with Einstein's formula $E = mc^2$. Overall, fusion in the Sun converts about 600 million tons of hydrogen into 596 million tons of helium every second, which means that 4 million tons of matter is turned into energy each second. Although this sounds like a lot, it is a minuscule fraction of the Sun's total mass and does not affect the overall mass of the Sun in any measurable way.

The Solar Thermostat Nuclear fusion is the source of all the energy the Sun releases into space. If the fusion rate varied, so would the Sun's energy output, and large variations in the Sun's luminosity would almost surely be lethal to life on Earth. Fortunately, the Sun fuses hydrogen at a steady rate, thanks to a natural feedback process that acts as a thermostat for the Sun's interior. To see how it works, let's examine what would happen if a small change were to occur in the core temperature (Figure 11.8).

Suppose the Sun's core temperature rose very slightly. The rate of nuclear fusion is very sensitive to temperature; a slight temperature increase would cause the fusion rate to soar as protons in the core collided more frequently and with more energy. Because energy moves slowly through the Sun's interior, this extra energy would be bottled up in the core, temporarily forcing the Sun out of energy balance and raising the core pressure. The push of this pressure would temporarily exceed the pull of gravity, causing the core to expand and cool. This cooling, in turn, would cause the fusion rate to drop back down until the core returned to its original size and temperature, restoring both gravitational equilibrium and energy balance.

Gravitational equilibrium and energy balance act together as a thermostat to keep the Sun's core temperature and fusion rate steady.

A slight drop in the Sun's core temperature would trigger an opposite chain of events. The reduced core temperature would lead to a decrease in the rate of nuclear fusion, causing a drop in pressure and contraction of the core. As the core shrank, its temperature would rise until the fusion rate returned to normal and restored the core to its original size and temperature.

▲ **FIGURE 11.8**

The solar thermostat. Gravitational equilibrium regulates the Sun's core temperature. Everything is in balance if the amount of energy leaving the core equals the amount of energy produced by fusion. A rise in core temperature triggers a chain of events that causes the core to expand, lowering its temperature to the original value. A decrease in core temperature triggers an opposite chain of events, also restoring the original core temperature.

▲ **FIGURE 11.9**

A photon in the solar interior bounces randomly among electrons, slowly working its way outward.

◆ How does the energy from fusion get out of the Sun?

The solar thermostat balances the Sun's fusion rate so that the amount of nuclear energy generated in the core equals the amount of energy radiated from the surface as sunlight. However, the journey of solar energy from the core to the photosphere takes hundreds of thousands of years.

Most of the energy released by fusion starts its journey out of the solar core in the form of photons. Although photons travel at the speed of light, the paths they take through the Sun's interior zigzag so much that it takes them a very long time to make any outward progress. Deep in the solar interior, the plasma is so dense that a photon can travel only a fraction of a millimeter in any direction before it interacts with an electron. Each time a photon collides with an electron, the photon gets deflected into a new random direction. The photon therefore bounces around the dense interior in a haphazard way (sometimes called a *random walk*) and only very gradually works its way outward from the Sun's center (Figure 11.9).

Energy released by fusion moves outward through the Sun's radiation zone (see Figure 11.4) primarily by way of these randomly bouncing photons. At the top of the radiation zone, where the temperature

Randomly bouncing photons carry energy through the deepest layers of the Sun, and convection carries energy through the upper layers to the surface.

drops to about 2 million K, the solar plasma absorbs photons more readily, rather than just bouncing them around. This absorption creates the conditions needed for convection, so above this level we find the Sun's convection zone. The rising of hot plasma and sinking of cool plasma form a cycle that transports energy outward from the base of the convection zone to the photosphere (Figure 11.10).

In the photosphere, the density of the gas is low enough that photons can escape to space. The energy produced hundreds of thousands of years earlier in the solar core then finally emerges from the Sun as thermal radiation [Section 5.2] coming from the almost 6000 K gas of the photosphere. Once in space, the photons travel away at the speed of light, bathing the planets in sunlight.

▼ **FIGURE 11.10**

The Sun's photosphere churns with rising hot gas and falling cool gas as a result of underlying convection.

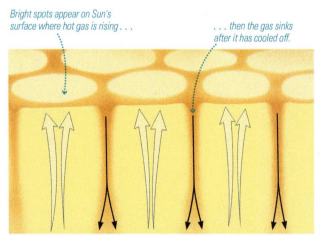

Bright spots appear on Sun's surface where hot gas is rising . . .

. . . then the gas sinks after it has cooled off.

a This diagram shows convection beneath the Sun's surface: hot gas (yellow arrows) rises while cooler gas (black arrows) descends around it.

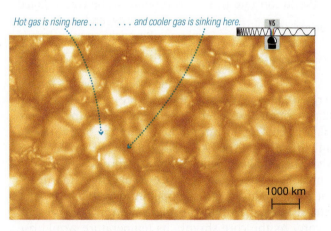

Hot gas is rising here . . .

. . . and cooler gas is sinking here.

1000 km

b This photograph shows the mottled appearance of the Sun's photosphere. The bright spots correspond to the rising plumes of hot gas in the diagram in part a.

◆ How do we know what is happening inside the Sun?

We cannot see inside the Sun, so you may wonder how we can claim to know so much about what goes on underneath its surface. We can study the Sun's interior in three basic ways: through mathematical models, observations of solar vibrations, and observations of solar neutrinos.

Mathematical Models The primary way we learn about the interior of the Sun (and other stars) is by creating *mathematical models* that use the laws of physics to predict internal conditions. A basic model starts with the Sun's observed composition and mass and then solves equations that describe gravitational equilibrium and energy balance. With the aid of a computer, we can use the model to calculate the Sun's temperature, pressure, and density at any depth. We can then predict the rate of nuclear fusion in the solar core by combining these calculations with knowledge about nuclear fusion gathered in laboratories here on Earth.

If a model is a good description of the Sun's interior, it should correctly predict the radius, surface temperature, luminosity, age, and other observable properties of the Sun. Current models predict these properties quite accurately, giving us confidence that we really do understand what is going on inside the Sun.

Solar Vibrations A second way to learn about the inside of the Sun is to observe vibrations of the Sun's surface similar to the vibrations that earthquakes cause on Earth. Gas moving around in the solar interior generates vibrations that travel through the Sun like sound waves moving through air. We can observe these vibrations on the Sun's surface by looking for Doppler shifts [Section 5.2]. Light from portions of the surface that are rising toward us is slightly blueshifted, while light from portions that are falling away from us is slightly redshifted. The vibrations are relatively small but measurable (Figure 11.11).

We can deduce a great deal about the solar interior by carefully analyzing these vibrations. (By analogy to seismology on Earth, this type of study of the Sun is called *helioseismology—helios* means "sun.") Results to date confirm that our mathematical models of the solar interior are on the right track, while also providing data that help us to improve the models further.

The characteristics of solar vibrations support our mathematical models of the Sun's interior.

Solar Neutrinos A third way to study the solar interior is to observe **neutrinos**, a type of subatomic particle produced by fusion in the core (see Figure 11.7). Don't panic, but about a *thousand trillion* of these solar neutrinos will zip through your body as you read this sentence—but they will do no damage at all. The reason is that neutrinos interact with other matter only through the *weak force* [Section 17.1] and gravity, not through the electromagnetic force. As a result, they can pass through almost anything. For example, an inch of lead will stop an X ray, but stopping an average neutrino would require a slab of lead more than a light-year thick!

In principle, neutrinos give us a direct way to study fusion in the Sun's core, because nearly all of them pass straight through the solar interior into space. Traveling at nearly the speed of light, they reach us just minutes after they were produced. In practice, their elusiveness makes neutrinos dauntingly difficult to detect. Nevertheless, neutrinos *do* occasionally interact with matter, and it is possible to capture a few solar neutrinos using a large enough detector (Figure 11.12). So that neutrino captures can be

-2500 -2000 -1500 -1000 -500   0   500 1000 1500 2000
Velocity (m/s)

▲ **FIGURE 11.11**

This image shows vibrations on the Sun's surface that have been measured from Doppler shifts. Shades of orange show how quickly each spot on the Sun's surface is moving toward or away from us at a particular moment. Dark shades (negative velocities) represent motion toward us; light shades (positive velocities) represent motion away from us. The large-scale change in color from left to right reflects the Sun's rotation, and the small-scale ripples reflect the surface vibrations.

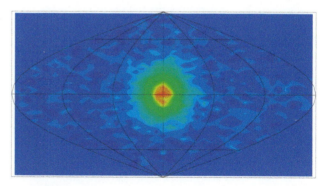

▲ **FIGURE 11.12**

This image shows neutrinos detected by the Super-Kamiokande neutrino observatory in Japan. Levels of color indicate the brightness of the neutrino signal, with red representing the strongest signal and blue representing the weakest. The grid shows the full sky, centered on the Sun, making it clear that the Sun is the source of these neutrinos. The neutrinos are produced by fusion and therefore represent direct proof that fusion really is responsible for the Sun's energy output.

▲ FIGURE 11.13

This photo shows the inside of the main vessel of the Borexino neutrino detector, located in a mine 1400 meters underground at Italy's Gran Sasso National Lab. During operations, the vessel is filled with 300 tons of a fluid that emits tiny flashes of light when a neutrino is captured; these flashes are recorded by the detectors visible on the inside of the sphere. Borexino has helped confirm that the number of neutrinos coming from the Sun agrees with the predictions of solar fusion models.

distinguished from reactions caused by other particles, neutrino detectors are usually placed deep underground or under the Antarctic ice. The overlying rock or ice blocks most other particles, but neutrinos pass right through.

The fact that we detect solar neutrinos provides direct proof that nuclear fusion is responsible for the Sun's energy. However, early attempts to detect solar neutrinos found only about one-third of the number expected based on the Sun's energy output. This disagreement between model predictions and observations came to be called the *solar neutrino problem*. It was solved in the early 2000s. The solution is this: Neutrinos come in three types, called *electron neutrinos*, *muon neutrinos*, and *tau neutrinos*. The early solar neutrino detectors could detect only electron neutrinos, which didn't seem like a problem at first, because fusion should produce only this type. However, we now know that neutrinos can change among the three types while passing through matter, so by the time solar neutrinos reach our detectors, only about one-third of them are still electron neutrinos. We can be confident in this solution, because more recent detectors that can detect all three neutrino types have confirmed that the total number of solar neutrinos matches predictions (Figure 11.13).

> Neutrinos provide a direct way to measure nuclear fusion in the Sun, and results indicate that fusion occurs as our models predict.

11.3 The Sun–Earth Connection

Energy liberated by nuclear fusion in the Sun's core eventually reaches the solar surface, where it helps create a wide variety of phenomena that we can observe from Earth. Sunspots are the most obvious of these phenomena. Because sunspots and other features of the Sun's surface change with time, they constitute what we call *solar weather*, or **solar activity**. The "storms" associated with solar weather are not just of academic interest. Sometimes they affect our day-to-day life on Earth. In this section, we'll explore solar activity and its far-reaching effects.

◆ What causes solar activity?

Most of the Sun's surface churns constantly with rising and falling gas, so it looks like the close-up photo shown in Figure 11.10b. However, larger features sometimes appear, including sunspots, huge explosions known as *solar flares*, and gigantic loops of hot gas extending high into the Sun's corona. All these features are created by magnetic fields, which form and change easily in the convecting plasma in the outer layers of the Sun.

Sunspots and Magnetic Fields Sunspots are the most striking features of the solar surface (Figure 11.14a). If you could look directly at a sunspot without damaging your eyes, it would be blindingly bright. Sunspots appear dark in photographs only because they are *less* bright than the surrounding photosphere. They are less bright because they are cooler: The temperature of the plasma in sunspots is about 4000 K, significantly cooler than the 5800 K plasma that surrounds them.

You may wonder how sunspots can be so much cooler than their surroundings. Gas usually flows easily, so you might expect the hotter gas around a sunspot to mix with the cooler gas within it, quickly warming the sunspot. The fact that sunspots stay relatively cool means that something must prevent hot plasma from entering them, and that something turns out to be magnetic fields.

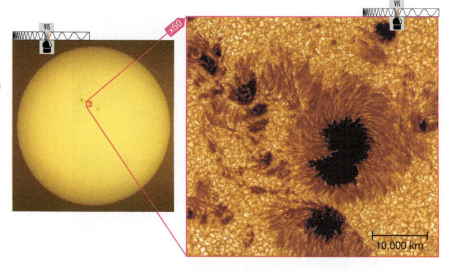

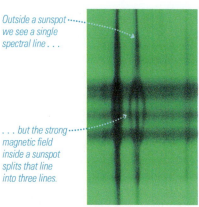

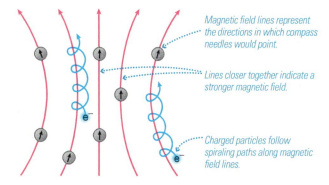

b Very strong magnetic fields split the absorption lines in spectra of sunspot regions. The dark vertical bands are absorption lines in a spectrum of the Sun. Notice that these lines split where they cross the dark horizontal bands corresponding to sunspots.

10,000 km

a This close-up view of the Sun's surface shows two large sunspots and several smaller ones. Each of the big sunspots is roughly as large as Earth.

▲ **FIGURE 11.14**
Sunspots are regions of strong magnetic fields.

Detailed observations of the Sun's spectral lines reveal sunspots to be regions with strong magnetic fields. These magnetic fields can alter the energy levels in atoms and ions, causing some spectral lines to split into two or more closely spaced lines (Figure 11.14b). Wherever we see this effect (called the *Zeeman effect*), magnetic fields must be present. Scientists can therefore use this effect to map magnetic fields on the Sun.

To understand how sunspots stay cooler than their surroundings, we must investigate the nature of magnetic fields in a little more depth. Magnetic fields are invisible, but we can represent them by drawing **magnetic field lines** (Figure 11.15). These lines represent the directions in which compass needles would point if we placed them within the magnetic field. The lines are closer together where the field is stronger and farther apart where the field is weaker. Because these imaginary field lines are easier to visualize than the magnetic field itself, we usually discuss magnetic fields by talking about how the field lines would appear. Charged particles, such as the ions and electrons in the solar plasma, cannot easily move perpendicular to the field lines; instead, they follow spiraling paths along them.

Sunspots are regions kept cooler than the surrounding photosphere by strong magnetic fields.

Solar magnetic field lines act somewhat like elastic bands, twisted into contortions and knots by turbulent motions in the solar atmosphere. Sunspots occur where tightly wound magnetic fields poke nearly straight out from the solar interior (Figure 11.16a). These tight magnetic field lines suppress convection within the sunspot and prevent surrounding plasma from entering the sunspot. With hot plasma unable to enter the region, the sunspot plasma becomes cooler than that of the rest of the photosphere. Individual sunspots typically last up to a few weeks, dissolving when their magnetic fields weaken and allow hotter plasma to flow in.

Sunspots tend to occur in pairs, connected by a loop of magnetic field lines that can arc high above the Sun's surface (Figure 11.16b). Gas in the Sun's chromosphere and corona becomes trapped in these giant loops, called **solar prominences.** Some prominences rise to heights of more than 100,000 km above the Sun's surface. Individual prominences can last for days or even weeks.

Magnetic field lines represent the directions in which compass needles would point.

Lines closer together indicate a stronger magnetic field.

Charged particles follow spiraling paths along magnetic field lines.

▲ **FIGURE 11.15**
We draw magnetic field lines (red) to represent invisible magnetic fields.

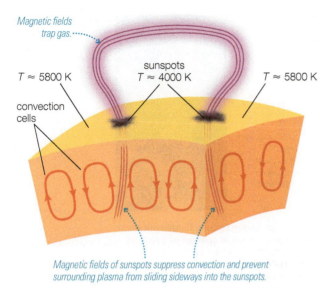

a Pairs of sunspots are connected by tightly wound magnetic field lines.

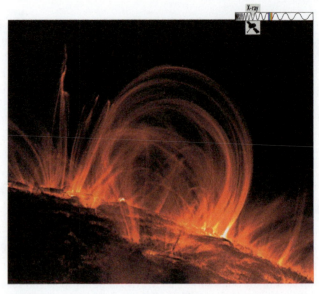

b This X-ray photo (from NASA's *TRACE* mission) shows hot gas trapped within looped magnetic field lines.

▲ **FIGURE 11.16**
Strong magnetic fields keep sunspots cooler than the surrounding photosphere, while magnetic loops can arch from the sunspots to great heights above the Sun's surface.

Magnetic fields trap gas.

$T \approx 5800$ K   sunspots $T \approx 4000$ K $T \approx 5800$ K

convection cells

Magnetic fields of sunspots suppress convection and prevent surrounding plasma from sliding sideways into the sunspots.

Solar Storms The magnetic fields winding through sunspots and prominences sometimes undergo dramatic and sudden changes, producing short-lived but intense storms on the Sun. The most dramatic of these storms are **solar flares,** which emit bursts of ultraviolet and X-ray light along with charged particles moving at nearly the speed of light (Figure 11.17).

Flares generally occur in the vicinity of sunspots, which is why we think they are created by changes in magnetic fields. The leading model suggests that solar flares occur when magnetic field lines become so twisted and knotted that they can no longer bear the tension, causing them to snap suddenly and reorganize themselves into a less twisted configuration. The energy released in the process heats the nearby plasma to 100 million K over the next few minutes to hours, generating the intense radiation and high-speed particles that we see from solar flares.

Energy released when magnetic field lines snap can lead to dramatic solar storms, which sometimes eject bursts of energetic particles into space.

Heating of the Chromosphere and Corona As we've seen, many of the most dramatic weather patterns and storms on the Sun involve the very hot gas of the Sun's chromosphere and corona. But why is this gas so hot in the first place?

Recall that temperatures gradually decline as we move outward from the Sun's core to its photosphere. We might expect the decline to continue above the photosphere, but instead it reverses, making the chromosphere and corona much hotter than the Sun's surface. Some aspects of this heating remain a mystery today, but we have at least a general explanation: Strong magnetic fields carry energy upward from the churning solar surface to the chromosphere and corona. More specifically, the rising and falling of gas in the convection zone probably shakes tightly wound magnetic field lines beneath the solar surface, transmitting energy upward until it is ultimately deposited as heat. The same magnetic fields that keep sunspots cool therefore make the overlying plasma of the chromosphere and corona hot.

Observations support this model. The gas density in the chromosphere and corona is so low that we cannot see visible light from these

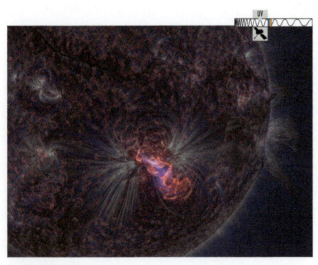

▲ **FIGURE 11.17**
This ultraviolet image (from *Solar Dynamics Observatory*) shows a solar flare erupting from the Sun's surface.

layers (except during a total eclipse when the bright photosphere is blocked from view; see Figure 2.26). However, the roughly 10,000 K plasma of the chromosphere emits strongly in the ultraviolet and the 1 million K plasma of the corona is the source of virtually all X rays from the Sun. X-ray images show that bright spots in the corona tend to be directly above sunspots in the photosphere (Figure 11.18), indicating that both are created by the same magnetic fields.

Magnetic fields deposit energy above the Sun's surface, heating the chromosphere and corona.

Notice that some regions of the corona barely show up in X-ray images; these regions, called **coronal holes,** are nearly devoid of hot coronal gas. More detailed analyses show that the magnetic field lines in coronal holes project out into space like broken rubber bands, allowing particles spiraling along them to escape the Sun altogether. These particles streaming outward from the corona are the source of the solar wind.

Flares and other solar storms sometimes eject large numbers of highly energetic charged particles from the Sun's corona. These particles travel outward from the Sun in huge bubbles that we call **coronal mass ejections** (Figure 11.19). These bubbles have strong magnetic fields and can reach Earth in a couple of days if they happen to be aimed in our direction.

Particles ejected from the Sun during periods of high activity can hamper radio communications, disrupt power delivery, and damage orbiting satellites.

Once a coronal mass ejection reaches Earth, it can create a *geomagnetic storm* in Earth's magnetosphere. On the positive side, these storms can lead to unusually strong auroras (see Figure 7.6) that can be visible throughout much of the United States. On the negative side, they can hamper radio communications, disrupt electrical power delivery, and damage the electronic components in orbiting satellites. They can also heat Earth's upper atmosphere, causing it to expand in a way that can increase drag on low-orbiting satellites, sometimes enough that they eventually plummet back to Earth.

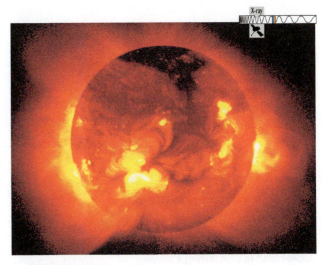

▲ **FIGURE 11.18**
An X-ray image of the Sun reveals the 1 million K gas of the corona. Brighter regions (yellow) correspond to regions of stronger X-ray emission. The darker regions (such as near the north pole at the top of this photo) are *coronal holes* from which the solar wind escapes. (From the *Yohkoh* Space Observatory.)

◆ How does solar activity vary with time?

Solar weather is just as unpredictable as weather on Earth. Individual sunspots can appear or disappear at almost any time, and we have no way to know that a solar storm is coming until we observe it through our telescopes. However, long-term observations have revealed patterns in solar activity indicating that sunspots and solar storms are more common at some times than at others. They've also helped us understand the origin of the Sun's strong magnetic fields, which are responsible for all the solar activity we've discussed.

The Sunspot Cycle The most notable pattern in solar activity is the **sunspot cycle**—a cycle in which the average number of sunspots on the Sun gradually rises and falls (Figure 11.20). At the time of *solar maximum,* when sunspots are most numerous, we may see dozens of sunspots on the Sun at one time. In contrast, we may see few, if any, sunspots at the time of *solar minimum.* The frequency of prominences, flares, and coronal mass ejections also follows the sunspot cycle, with these events being most common at solar maximum and least common at solar minimum.

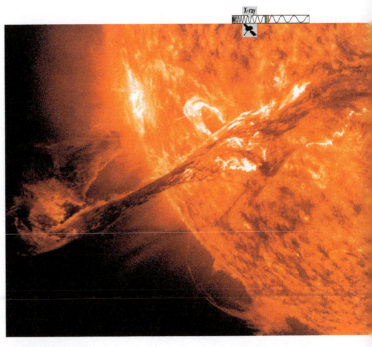

▲ **FIGURE 11.19**
This X-ray image from the *Solar Dynamics Observatory* spacecraft shows a solar eruption that led to a coronal mass ejection on August 31, 2012.

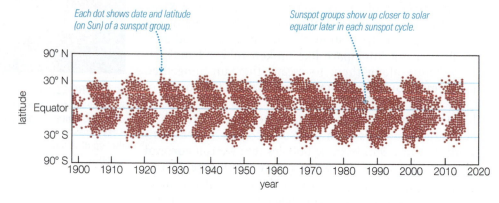

a This graph shows how the number of sunspots on the Sun changes with time. The vertical axis shows the percentage of the Sun's surface covered by sunspots. The cycle has a period of approximately 11 years.

Sunspot activity peaks at solar maximum.

Sunspots are rare during solar minimum.

Each dot shows date and latitude (on Sun) of a sunspot group.

Sunspot groups show up closer to solar equator later in each sunspot cycle.

b This graph shows how the latitudes at which sunspot groups appear tend to shift during a single sunspot cycle.

▲ **FIGURE 11.20**

Sunspot cycle since about 1900.

Notice that the sunspot cycle varies from one period to the next (Figure 11.20a). The average length of time between maximums is 11 years, but we have observed it to be as short as 7 years and as long as 15 years. Observations going further back in time suggest that sunspot activity can sometimes cease almost entirely. For example, astronomers observed virtually no sunspots between the years 1645 and 1715, a period sometimes called the *Maunder minimum* (after E. W. Maunder, who identified it in historical sunspot records).

The average number of sunspots on the Sun rises and falls in an approximately 11-year cycle.

The locations of sunspots on the Sun also vary with the sunspot cycle (Figure 11.20b). As a cycle begins at solar minimum, sunspots form primarily at mid-latitudes (30° to 40°) on the Sun. The sunspots tend to form at lower latitudes as the cycle progresses, appearing very close to the solar equator as the next solar minimum approaches. Then the sunspots of the next cycle begin to form near mid-latitudes again.

A less obvious feature of the sunspot cycle is that something peculiar happens to the Sun's magnetic field at each solar maximum: The Sun's entire magnetic field starts to flip, turning magnetic north into magnetic south and vice versa. We know this because the magnetic field lines connecting pairs of sunspots (see Figure 11.16a) on the same side of the solar equator all tend to point in the same direction throughout an 11-year cycle. For example, all compass needles might point from the easternmost sunspot to the westernmost sunspot in each sunspot pair north of the solar equator. However, by the time the cycle ends at solar minimum, the magnetic field has reversed: In the subsequent solar cycle, the field lines connecting pairs of sunspots point in the opposite direction. The Sun's complete magnetic cycle (sometimes called the *solar cycle*) therefore averages 22 years, since it takes two 11-year sunspot cycles for the magnetic field to return to the way it started.

The Cause of the Sunspot Cycle   The precise reasons for the sunspot cycle are not fully understood, but the leading model ties it to a

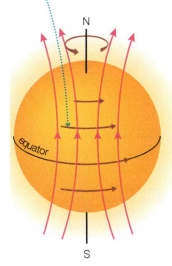

Charged particles tend to push the field lines around with the Sun's rotation.

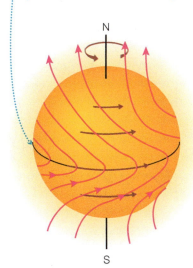

Because the Sun rotates faster near its equator than at its poles, the field lines bend ahead at the equator.

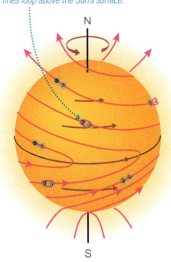

The field lines become more and more twisted with time, and sunspots form when the twisted lines loop above the Sun's surface.

▲ **FIGURE 11.21**

The Sun rotates more quickly at its equator than it does near its poles. Because gas circles the Sun faster at the equator, it drags the Sun's north–south magnetic field lines into a more twisted configuration. The magnetic field lines linking pairs of sunspots, depicted here as green and black blobs, trace out the directions of these stretched and distorted field lines.

combination of convection and the Sun's rotation. Convection is thought to dredge up weak magnetic fields generated in the solar interior, amplifying them as they rise. The Sun's rotation—faster at its equator than near its poles—then stretches and shapes these fields.

think about it Suppose you observe two sunspots: one near the Sun's equator and one directly north of it at 20° latitude. Where would you expect to see the two spots if you observed the Sun again a few days later? Would one still be directly north of the other? Explain.

Imagine what happens to magnetic field lines that start out running along the Sun's surface from south to north (Figure 11.21). At the equator, the lines circle the Sun every 25 days, but at higher latitudes, they lag behind. As a result, the lines gradually get wound more and more tightly around the Sun. This process, operating at all times over the entire Sun, produces the contorted field lines that generate sunspots and other solar activity.

The detailed behavior of these magnetic fields is quite complex, so scientists attempt to study it with sophisticated computer models. Using these models, scientists have successfully replicated many features of the sunspot cycle, including changes in the number and latitude of sunspots and the magnetic field reversals that occur about every 11 years. However, much remains mysterious, including why the period of the sunspot cycle varies and why solar activity is different from one cycle to the next.

The Sunspot Cycle and Earth's Climate Despite the changes that occur during the sunspot cycle, the Sun's total output of energy barely changes at all—the largest measured changes have been less than 0.1% of the Sun's average luminosity. However, the ultraviolet and X-ray output of the Sun, which comes from the magnetically heated gas of the chromosphere and corona, can vary much more significantly. Could any of these changes affect the weather or climate on Earth?

Some data suggest connections between solar activity and Earth's climate. For example, the Maunder minimum from 1645 to 1715, when solar activity seems to have virtually ceased, was a time of exceptionally low temperatures in Europe and North America known as the *Little Ice*

This graph compares the global average temperature and the amount of sunlight reaching Earth (the solar *irradiance*, measured in watts per square meter) since 1880. The light red and blue curves show annual changes while the dark curves show 11-year averages, which smooth out the effects of the 11-year solar cycle. Notice that, for recent decades, the amount of sunlight has remained approximately constant while Earth has warmed, ruling out the Sun as the cause of recent global warming.

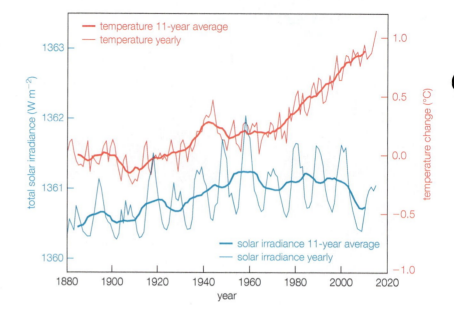

Age. However, no one knows whether the low solar activity caused these low temperatures or whether it was just a coincidence. Similarly, some researchers have claimed that certain weather phenomena, such as drought cycles or frequencies of storms, are correlated with the 11- or 22-year cycle of solar activity. A few scientists have even claimed that changes in the Sun may be responsible for Earth's recent global warming, but the Sun's energy output has actually remained nearly steady while the temperature has continued to rise (Figure 11.22), leaving little doubt that the warming is a result of human activity [Section 7.5]. Nevertheless, the study of solar activity's possible effects on climate remains an active field of research.

the big picture   Putting Chapter 11 into Perspective

In this chapter, we have examined our Sun, the nearest star. When you look back at this chapter, make sure you understand the following "big picture" ideas:

- The ancient mystery of why the Sun shines is now solved. The Sun shines with energy generated by fusion of hydrogen into helium in the Sun's core. After a journey through the solar interior lasting several hundred thousand years and an 8-minute journey through space, a small fraction of this energy reaches Earth and supplies sunlight and heat.

- The Sun shines steadily thanks to the balance between pressure and gravity (gravitational equilibrium) and the balance between energy production in the core and energy release at the surface (energy balance). These two

kinds of balance create a natural thermostat that regulates the Sun's fusion rate, keeping the Sun shining steadily and allowing life to flourish on Earth.

- The Sun's atmosphere displays its own version of weather and climate, governed by solar magnetic fields. Some solar weather, such as coronal mass ejections, clearly affects Earth's magnetosphere. Other claimed connections between solar activity and Earth's climate may or may not be real.

- The Sun is important not only as our source of light and heat, but also because it is the only star near enough for us to study in great detail. In the coming chapters, we will use what we've learned about the Sun to help us understand other stars.

my cosmic perspective   The Sun affects us more directly than does any other astronomical object. It is the source of all our heat and light, and its "solar weather" can even affect our use of technology.

11.1 A Closer Look at the Sun

◆ Why does the Sun shine?

The Sun began to shine about 4½ billion years ago when **gravitational contraction** made its core hot enough to sustain nuclear fusion. It has shined steadily ever since because of two types of balance: (1) **gravitational equilibrium,** a balance between the outward push of pressure and the inward pull of gravity, and (2) **energy balance** between the energy released by fusion in the core and the energy radiated into space from the Sun's surface.

◆ What is the Sun's structure?

The Sun's interior layers, from the inside out, are the **core,** the **radiation zone,** and the **convection zone.** Atop the convection zone lies the **photosphere,** the surface layer from which photons can freely escape into space. Above the photosphere are the warmer **chromosphere** and the very hot **corona.**

11.2 Nuclear Fusion in the Sun

◆ How does nuclear fusion occur in the Sun?

The core's extreme temperature and density are just right for fusion of hydrogen into helium, which occurs via the **proton–**

proton chain. Because the fusion rate is so sensitive to temperature, gravitational equilibrium and energy balance act together as a thermostat to keep that rate steady.

◆ How does the energy from fusion get out of the Sun?

Energy moves through the deepest layers of the Sun—the core and the radiation zone—in the form of randomly bouncing photons. After energy emerges from the radiation zone, convection carries it the rest of the way to the photosphere, where it is radiated into space as sunlight. Energy produced in the core takes hundreds of thousands of years to reach the photosphere.

◆ How do we know what is happening inside the Sun?

We can construct theoretical models of the solar interior using known laws of physics and then check the models against observations of the Sun's size, surface temperature, and energy output. We also use studies of solar vibrations and solar **neutrinos.**

11.3 The Sun–Earth Connection

◆ What causes solar activity?

Sunspots and other changing features of the Sun constitute **solar activity,** which is caused by strong magnetic fields that contort and sometimes snap, creating phenomena that include **flares, prominences,** and **coronal mass ejections.** The magnetic fields also carry energy upward, depositing the heat that explains the high temperatures of the chromosphere and corona.

◆ How does solar activity vary with time?

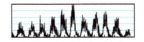

The **sunspot cycle,** or the variation in the number of sunspots on the Sun's surface, has an average period of 11 years. The magnetic field flip-flops every 11 years or so, resulting in a 22-year magnetic cycle. The number of sunspots can vary dramatically from one cycle to the next, and sometimes sunspots seem to be absent altogether. The sunspot cycle and other solar activity are tied to the Sun's ever-changing magnetic field, which is created by the combination of convection and the Sun's rotation pattern (faster at the equator than at the poles).

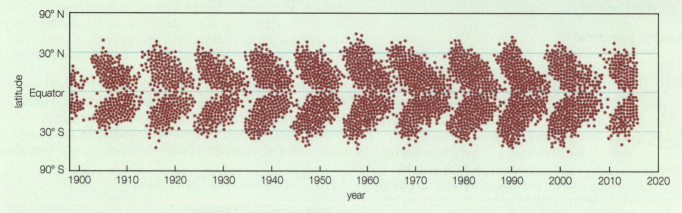

Figure 11.20b, repeated above, shows the latitudes at which sunspots appeared on the surface of the Sun during the 20th century. Answer the following questions, using the information provided in the figure.

1. Which of the following years had the least sunspot activity?
   a. 1930
   b. 1949
   c. 1961
   d. 1987

2. What is the approximate range in latitude over which sunspots appear?

3. According to the figure, how do the positions of sunspots appear to change during one sunspot cycle? Do they get closer to or farther from the equator with time?

exercises and problems

MasteringAstronomy® For instructor-assigned homework and other learning materials, go to MasteringAstronomy®.

Review Questions

1. Briefly describe how *gravitational contraction* generates energy. When was it important in the Sun's history? Explain.
2. What two forces are balanced in *gravitational equilibrium*? What does it mean for the Sun to be in *energy balance*?
3. State the Sun's luminosity, mass, radius, and average surface temperature, and put the numbers in perspective.
4. Briefly describe the distinguishing features of each of the layers of the Sun shown in Figure 11.4.
5. Distinguish between nuclear *fission* and *fusion*. Which one is used in nuclear power plants? Which one is used by the Sun?
6. Why does nuclear fusion require high temperatures and pressures?
7. What is the overall nuclear fusion reaction in the Sun? Briefly describe the proton–proton chain.
8. Describe how a natural "solar thermostat" keeps the core fusion rate steady in the Sun.
9. Describe how energy generated by fusion makes its way to the Sun's surface. How long does it take?
10. How do mathematical models help us learn about conditions inside the Sun, and what gives us confidence in the models?
11. What are *neutrinos*? What was the *solar neutrino problem,* and how was it solved?

12. What is *solar activity*? Describe key features including *sunspots, solar prominences, solar flares,* and *coronal mass ejections.*
13. How do magnetic fields keep sunspots cooler than the surrounding plasma? Explain.
14. Why are the chromosphere and corona best viewed with ultraviolet and X-ray telescopes, respectively? Briefly explain how we think the chromosphere and corona are heated.
15. What is the *sunspot cycle*? Describe the leading model for explaining it. Does the sunspot cycle influence Earth's climate?

Test Your Understanding

 Does It Make Sense?

Decide whether the statement makes sense (or is clearly true) or does not make sense (or is clearly false). Explain clearly; not all of these have definitive answers, so your explanation is more important than your chosen answer.

16. Before Einstein, gravitational contraction appeared to be a perfectly plausible mechanism for solar energy generation.
17. The solar wind usually flows outward from the Sun, but sometimes it turns around and flows backward.
18. If fusion in the solar core ceased today, worldwide panic would break out tomorrow as the Sun began to grow dimmer.

19. Astronomers have recently photographed magnetic fields churning deep beneath the solar photosphere.
20. I wear a lead vest to protect myself from solar neutrinos.
21. There haven't been many sunspots this year, but there ought to be many more in about 5 years.
22. News of a solar flare caused concern among businesses involved in communication and electrical power generation.
23. By observing solar neutrinos, we can learn about nuclear fusion deep in the Sun's core.
24. If the Sun's magnetic field somehow disappeared, there would be no more sunspots on the Sun.
25. Scientists are currently building an infrared telescope designed to observe fusion reactions in the Sun's core.

Quick Quiz

Choose the best answer to each of the following. Explain your reasoning with one or more complete sentences.

26. Which of these groups of particles has the greatest mass? (a) a helium nucleus with two protons and two neutrons (b) four electrons (c) four individual protons
27. Which of these layers of the Sun is coolest? (a) core (b) radiation zone (c) photosphere
28. X-ray images of the Sun generally show the (a) photosphere. (b) chromosphere. (c) corona.
29. Scientists estimate the central temperature of the Sun using (a) probes that measure changes in Earth's atmosphere. (b) mathematical models of the Sun. (c) laboratories that create miniature versions of the Sun.
30. Sunspots appear darker than their surroundings because they (a) are cooler than their surroundings. (b) block some of the sunlight from the photosphere. (c) do not emit any light.
31. At the center of the Sun, fusion converts hydrogen into (a) plasma. (b) radiation and elements like carbon and nitrogen. (c) helium, energy, and neutrinos.
32. Solar energy leaves the core of the Sun in the form of (a) photons. (b) rising hot gas. (c) sound waves.
33. The fact that we observe neutrinos from the Sun provides direct evidence of (a) fusion in the Sun's core. (b) convection in the Sun's interior. (c) the existence of the solar wind.
34. What causes the cycle of solar activity? (a) changes in the Sun's fusion rate (b) changes in the organization of the Sun's magnetic field (c) changes in the speed of the solar wind
35. Which of these things poses the greatest hazard to communication satellites? (a) photons from the Sun (b) solar magnetic fields (c) particles from the Sun

🌀 Process of Science

36. *Inside the Sun.* Scientists claim to know what is going on inside the Sun, even though we cannot directly observe the Sun's interior. What is the basis for these claims, and how are they aligned with the hallmarks of science outlined in Section 3.4?
37. *The Solar Neutrino Problem.* Early solar neutrino experiments detected only about a third of the number of neutrinos predicted by the theory of fusion in the Sun. Why didn't scientists simply abandon their model at this point? What features of the Sun did the model get right? What alternatives were there for explaining the mismatch between the predictions and the observations?

Group Work Exercise

38. *The Sun's Future.* **Roles:** *Scribe* (takes notes on the group's activities), *Proposer* (proposes explanations to the group), *Skeptic* (points out weaknesses in proposed explanations), *Moderator* (leads group discussion and makes sure everyone contributes). **Activity:** Consider what you have learned about how the Sun came to shine steadily with energy from fusion in its core, then discuss the following questions:
    a. What will happen to the core temperature of the Sun *after* its core runs out of hydrogen for fusion? Will the temperature go up or down?
    b. If you think the temperature will go up, will it rise forever? What could eventually stop the temperature from rising? If you think the temperature will go down, will it decrease forever? What could eventually stop it from falling?
    c. Propose and describe an Earth-based experiment or a set of stellar observations that could test your hypothesis from part (b).

Investigate Further

Short-Answer/Essay Questions

39. *The End of Fusion I.* Describe what would happen in the Sun if fusion reactions abruptly ceased.
40. *The End of Fusion II.* If fusion reactions in the Sun were to suddenly cease, would we be able to tell? If so, how?
41. *A Really Strong Force.* How would the interior temperature of the Sun be different if the strong force that binds nuclei together were 10 times as strong?
42. *Covered with Sunspots.* Describe what the Sun would look like from Earth if the entire photosphere were the same temperature as a sunspot.
43. *Inside the Sun.* Describe how scientists determine what the interior of the Sun is like. Could we send a probe into the Sun to measure what is happening there?
44. *Solar Energy Output.* Observations over the past century show that the Sun's visible-light output varies by less than 1%, but its X-ray output can vary by a factor of 10 or more. Explain why the changes in X-ray output can be so much more pronounced than the changes in visible-light output.
45. *An Angry Sun.* A *Time* magazine cover once suggested that an "angry Sun" was becoming more active as human activity changed Earth's climate. It's certainly possible for the Sun to become more active at the same time that humans are affecting Earth, but is it possible that the Sun could be responding to human activity? Can humans affect the Sun in any significant way? Explain.

Quantitative Problems

Be sure to show all calculations clearly and state your final answers in complete sentences.

46. *The Color of the Sun.* The Sun's average surface temperature is about 5800 K. Use Wien's law (see Cosmic Calculations 5.1) to calculate the wavelength of peak thermal emission from the Sun. What color does this wavelength correspond to in the visible-light spectrum? Why do you think the Sun appears white or yellow to our eyes?
47. *The Color of a Sunspot.* The typical temperature of a sunspot is about 4000 K. Use Wien's law (see Cosmic Calculations 5.1) to

calculate the wavelength of peak thermal emission from a sunspot. What color does this wavelength correspond to in the visible-light spectrum? How does this color compare with that of the Sun?

48. *Solar Mass Loss.* Estimate how much mass the Sun will lose through fusion reactions during its 10-billion-year life. You can simplify the problem by assuming the Sun's energy output remains constant. Compare the amount of mass lost with Earth's mass.

49. *Pressure of the Photosphere.* The gas pressure of the photosphere changes substantially from its upper levels to its lower levels. Near the top of the photosphere, the temperature is about 4500 K and there are about 1.6×10^{16} gas particles per cubic centimeter. In the middle, the temperature is about 5800 K and there are about 1.0×10^{17} gas particles per cubic centimeter. At the bottom of the photosphere, the temperature is about 7000 K and there are about 1.5×10^{17} gas particles per cubic centimeter. Compare the pressures of each of these layers and explain the reason for the trend in pressure that you find. How do these gas pressures compare with Earth's atmospheric pressure at sea level? (*Hint:* See Cosmic Calculations 11.1.)

50. *The Lifetime of the Sun.* The total mass of the Sun is about 2×10^{30} kg, of which about 75% was hydrogen when the Sun formed. However, only about 13% of this hydrogen ever becomes available for fusion in the core. The rest remains in layers of the Sun where the temperature is too low for fusion.
    a. Based on the given information, calculate the total mass of hydrogen available for fusion over the lifetime of the Sun.
    b. Combine your results from part (a) and the fact that the Sun fuses about 600 billion kg of hydrogen each second to calculate how long the Sun's initial supply of hydrogen can last. Give your answer in both seconds and years.
    c. Given that our solar system is now about 4.6 billion years old, when will we need to start worrying about the Sun running out of hydrogen for fusion?

51. *Solar Power Collectors.* This problem leads you through the calculation and discussion of how much solar power can be collected by solar cells on Earth.
    a. Imagine a giant sphere with a radius of 1 AU surrounding the Sun. What is the surface area of this sphere in square meters? (*Hint:* The formula for the surface area of a sphere is $4\pi r^2$.)
    b. Because this imaginary giant sphere surrounds the Sun, the Sun's entire luminosity of 3.8×10^{26} watts must pass through

it. Calculate the power passing through each square meter of this imaginary sphere in *watts per square meter*. Explain why this number represents the maximum power per square meter that a solar collector in Earth orbit can collect.
    c. List several reasons why the average power per square meter collected by a solar collector on the ground will always be less than what you found in part (b).
    d. Suppose you want to put a solar collector on your roof. If you want to optimize the amount of power you can collect, how should you orient the collector? (*Hint:* The optimum orientation depends on both your latitude and the time of year and day.)

Discussion Questions

52. *The Role of the Sun.* Briefly discuss how the Sun affects us here on Earth. Be sure to consider not only factors such as its light and warmth, but also how the study of the Sun has led us to new understandings in science and to technological developments. Overall, how important has solar research been to our lives?

53. *The Sun and Global Warming.* One of the most pressing environmental issues on Earth is the extent to which human emissions of greenhouse gases are warming our planet. Some people claim that part or all of the observed warming over the past century may be due to changes in the Sun. However, the data in Figure 11.22 show that global temperatures have risen since the 1970s without much change in the overall amount of light Earth receives from the Sun. Discuss how people make decisions about the causes of global warming and whether they are rooted in scientific data or depend primarily on other factors.

Web Projects

54. *Current Solar Weather.* Daily information about solar activity is available at numerous websites. Where are we in the sunspot cycle right now? When is the next solar maximum or minimum expected? Have there been any major solar storms in the past few months? If so, did they have any significant effects on Earth? Summarize your findings in a one- to two-page report.

55. *Solar Observatories in Space.* Visit the website for a space mission designed to observe the Sun, and compile a short album of images from the mission. Briefly describe what each image shows.

.12 Surveying the Stars

This single image captures many types of stars, as well as some of the gas between them. The bright foreground star at the bottom, surrounded by glowing orange gas, is Antares, a supergiant star that marks the "heart" of the scorpion in the constellation Scorpius. To the right of Antares is a much more distant cluster of stars—the globular cluster M4. Figure 12.18 shows the properties of this cluster's stars.

LEARNING GOALS

12.1 Properties of Stars

- ◆ How do we measure stellar luminosities?
- ◆ How do we measure stellar temperatures?
- ◆ How do we measure stellar masses?

12.2 Patterns Among Stars

- ◆ What is a Hertzsprung-Russell diagram?
- ◆ What is the significance of the main sequence?
- ◆ What are giants, supergiants, and white dwarfs?

12.3 Star Clusters

- ◆ What are the two types of star clusters?
- ◆ How do we measure the age of a star cluster?

ESSENTIAL PREPARATION

1. How does Newton's law of gravity extend Kepler's laws? [Section 4.4]

2. How does light tell us what things are made of? [Section 5.2]

3. How does light tell us the temperatures of planets and stars? [Section 5.2]

4. How does light tell us the speed of a distant object? [Section 5.2]

On a clear, dark night, a few thousand stars are visible to the naked eye. Many more become visible through binoculars, and a powerful telescope reveals so many stars that we could never hope to count them. Like each individual person, each individual star is unique. Like the human family, all stars have much in common.

Today, we know that stars are born from clouds of interstellar gas, shine brilliantly by nuclear fusion for millions to billions of years, and then die, sometimes in dramatic ways. In this chapter, we'll discuss how we study and categorize stars and how we have come to realize that stars, like people, change over their lifetimes.

12.1 Properties of Stars

Imagine that an alien spaceship flies by Earth on a simple but short mission: The visitors have just 1 minute to learn everything they can about the human race. In 60 seconds, they will see next to nothing of any individual person's life. Instead, they will obtain a collective "snapshot" of humanity showing people from all stages of life engaged in their daily activities. From this snapshot alone, they must piece together their entire understanding of human beings and their lives, from birth to death.

We face a similar problem when we look at the stars. Compared with stellar lifetimes of millions or billions of years, the few hundred years humans have spent studying stars with telescopes is rather like the aliens' 1-minute glimpse of humanity. We see only a brief moment in any star's life, and our collective snapshot of the heavens consists of such frozen moments for billions of stars. From this snapshot, we have learned to reconstruct the life cycles of stars.

We now know that all stars have a lot in common with the Sun. They all form in great clouds of gas and dust and begin their lives with roughly the same chemical composition as the Sun: about three-quarters hydrogen and one-quarter helium (by mass), with no more than about 2% consisting of elements heavier than helium. Nevertheless, stars are not all the same; they differ in size, age, brightness, and temperature. We'll devote most of this and the next chapter to understanding how and why stars differ. First, however, let's explore how we measure three of the most fundamental properties of stars: luminosity, surface temperature, and mass.

◆ How do we measure stellar luminosities?

If you go outside on any clear night, you'll immediately see that stars differ in brightness. Some stars are so bright that we can use them to identify constellations [Section 2.1]. Others are so dim that our naked eyes cannot see them at all. However, these differences in brightness do not by themselves tell us anything about how much light these stars are generating, because the brightness of a star depends on its distance as well as on how much light it actually emits. For example, the stars Procyon and Betelgeuse, which make up two of the three corners of the Winter Triangle (see Figure 2.2), appear about equally bright in our sky. However, Betelgeuse actually emits about 15,000 times as much light as Procyon. It has about the same brightness in our sky because it is much farther away.

Until the 20th century, people classified stars primarily by their brightness and location in our sky. On the next clear night, find your favorite constellation and visually rank its stars by brightness. Then look to see how that constellation is represented on the star charts in Appendix I. Why do the star charts use different size dots for different stars? Do the brightness rankings on the chart differ from what you see?

Because two similar-looking stars can be generating very different amounts of light, we need to distinguish clearly between a star's brightness in our sky and the actual amount of light that it emits into space (Figure 12.1):

- When we talk about how bright stars look in our sky, we are talking about **apparent brightness.** More specifically, we define the apparent brightness of any star in our sky as the amount of power (energy per second) reaching us *per unit area*.

- When we talk about how bright stars are in an absolute sense, regardless of their distance, we are talking about **luminosity**—the total amount of power that a star emits into space.

A star's apparent brightness in the sky depends on both its true light output, or luminosity, and its distance from us.

We can understand the difference between apparent brightness and luminosity by thinking about a standard light bulb. The bulb always puts out the same amount of light, so its luminosity doesn't vary. However, the bulb's apparent brightness depends on your distance from it: It will look quite bright if you stand very close to it, but quite dim if you are far away.

The Inverse Square Law for Light The apparent brightness of a star or any other light source obeys an *inverse square law* with distance, much like the inverse square law that describes the force of gravity [Section 4.4]. For example, if we viewed the Sun from twice Earth's distance, it would appear dimmer by a factor of $2^2 = 4$. If we viewed it from 10 times Earth's distance, it would appear $10^2 = 100$ times dimmer.

Figure 12.2 shows why apparent brightness follows an inverse square law. The same total amount of light must pass through each imaginary sphere surrounding the star. If we focus on the light passing through the small square on the sphere located at 1 AU, we see that the same amount of light must pass through *four* squares of the same size on the sphere located at 2 AU. Each square on the sphere at 2 AU therefore receives only $1/2^2 = 1/4$ as much light as the square on the sphere at 1 AU. Similarly, the same amount of light passes through *nine* squares of the same size on the sphere located at 3 AU, so each of these squares receives only $1/3^2 = 1/9$ as much light as the square on the sphere at 1 AU. Generalizing, the amount of light received per unit area decreases with increasing distance by the square of the distance, thereby obeying an inverse square law.

Doubling the distance to a star would decrease its apparent brightness by a factor of 2^2, or 4.

This inverse square law leads to a very simple and important formula relating the apparent brightness, luminosity, and distance of any light source. We will call it the **inverse square law for light:**

$$\text{apparent brightness} = \frac{\text{luminosity}}{4\pi \times \text{distance}^2}$$

Because the standard units of luminosity are watts [Section 11.1], the units of apparent brightness are *watts per square meter*. (The 4π in the

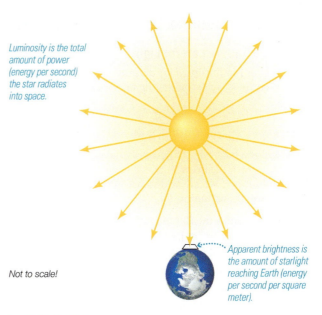

Luminosity is the total amount of power (energy per second) the star radiates into space.

Not to scale!

Apparent brightness is the amount of starlight reaching Earth (energy per second per square meter).

▲ **FIGURE 12.1**
Luminosity is a measure of power, and apparent brightness is a measure of power per unit area.

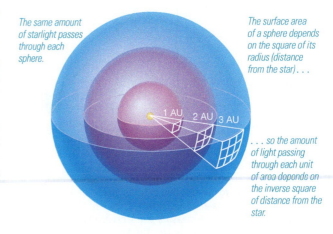

The same amount of starlight passes through each sphere.

The surface area of a sphere depends on the square of its radius (distance from the star) . . .

. . . so the amount of light passing through each unit of area depends on the inverse square of distance from the star.

▲ **FIGURE 12.2**
The inverse square law for light: The apparent brightness of a star declines with the square of its distance.

The Inverse Square Law for Light

If we use L for luminosity, d for distance, and b for apparent brightness, we can write the *inverse square law for light* as

$$b = \frac{L}{4\pi \times d^2}$$

As noted in the text, we can generally measure apparent brightness, which means we can use this formula to calculate luminosity if we know distance and to calculate distance if we know luminosity.

Example: The Sun's measured apparent brightness is 1.36×10^3 watts/m² at Earth's distance from the Sun, 1.5×10^{11} m. What is the Sun's luminosity?

Solution: In order to solve the inverse square law formula for the luminosity L, we first multiply both sides by $4\pi \times d^2$. Then we plug in the given values:

$$L = 4\pi \times d^2 \times b$$
$$= 4\pi \times (1.5 \times 10^{11}\,\text{m})^2 \times \left(1.36 \times 10^3\,\frac{\text{watts}}{\text{m}^2}\right)$$
$$= 3.8 \times 10^{26}\,\text{watts}$$

Just by measuring the Sun's apparent brightness and distance, we can calculate the total power that it radiates into all directions in space.

formula comes from the fact that the surface area of a sphere is given by $4\pi \times \text{radius}^2$.)

In principle, we can always determine a star's apparent brightness by carefully measuring the amount of light per square meter we receive from the star. We can then use the inverse square law to calculate a star's luminosity if we can first measure its distance, or to calculate a star's distance if we know its luminosity.

think about it Suppose Star A is four times as luminous as Star B. How will their apparent brightnesses compare if they are both the same distance from Earth? How will their apparent brightnesses compare if Star A is twice as far from Earth as Star B? Explain.

Measuring Distance Through Stellar Parallax

The most direct way to measure a star's distance is with *stellar parallax,* the small annual shifts in a star's apparent position caused by Earth's motion around the Sun [Section 2.4]. Astronomers measure stellar parallax by comparing observations of a nearby star made 6 months apart (Figure 12.3). The nearby star appears to shift against the background of more distant stars because we are observing it from two opposite points of Earth's orbit.

We can measure the distance to a nearby star by observing how its apparent location shifts as Earth orbits the Sun.

We can calculate a star's distance if we know the precise amount of the star's annual shift due to parallax. This means measuring the angle p in Figure 12.3, which we call the star's *parallax angle.* Notice that this angle would be smaller if the star were farther away, meaning that more distant stars have *smaller* parallax angles. Real stellar parallax angles are very small: Even the nearest stars have parallax angles of less than 1 arcsecond, well below the approximately 1 arcminute angular resolution of the naked eye.

By definition, the distance to an object with a parallax angle of 1 arcsecond is 1 parsec (pc), which is equivalent to 3.26 light-years. (The word *parsec* comes from combining the words *parallax* and *arcsecond.*) This leads to a simple formula for a star's distance: If we measure the parallax angle p in arcseconds, the star's distance d in parsecs is $d = 1/p$; we just multiply by 3.26 to convert from parsecs to light-years. For example, a star with a parallax angle $p = \frac{1}{10}$ arcsecond is 10 parsecs away, or $10 \times 3.26 = 32.6$ light-years. You may hear astronomers state distances in parsecs, kiloparsecs (1000 parsecs), or megaparsecs (1 million parsecs), but in this book we'll stick to distances in light-years.

think about it Suppose Star A has a parallax angle of 0.2 arcsecond and Star B has a parallax angle of 0.4 arcsecond. How does the distance of Star A from Earth compare to that of Star B?

Parallax was the first reliable technique astronomers developed for measuring distances to stars, and it remains the only technique that tells us stellar distances without any assumptions about the nature of stars. If we know a star's distance from its parallax angle, we can calculate its luminosity with the inverse square law for light. In fact, parallax measurements are the key to all other distance measurements in the universe, because astronomers use parallax measurements of the distances of nearby stars as the base of a chain of techniques that allows measurement of much greater distances [Section 16.2]. As of

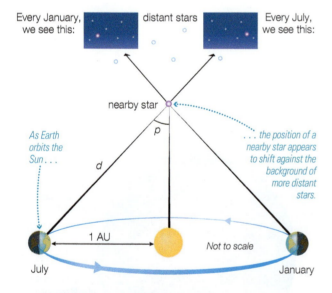

Every January, we see this: distant stars Every July, we see this:

nearby star

As Earth orbits the Sun...

p

d

... the position of a nearby star appears to shift against the background of more distant stars.

1 AU

Not to scale

July January

▲ **FIGURE 12.3**

Parallax makes the apparent position of a nearby star shift back and forth with respect to distant stars over the course of each year. The angle p, called the *parallax angle*, represents half the total parallax shift each year. If we measure p in arcseconds, the distance d to the star in parsecs is $1/p$. The angle in this figure is greatly exaggerated: All stars have parallax angles of less than 1 arcsecond.

2016, astronomers have made parallax measurements for more than 100,000 stars out to distances of more than 1500 light-years, and the European *GAIA* spacecraft is in the process of making parallax measurements for up to a billion stars, out to distances of tens of thousands of light-years.

The Luminosity Range of Stars   Now that we have discussed how we determine stellar luminosities, it's time to take a quick look at the results. We usually state stellar luminosities in comparison to the Sun's luminosity, L_{Sun}. For example, Proxima Centauri, the nearest of the three stars in the Alpha Centauri system and hence the nearest star besides our Sun, has only about 0.0006 times the luminosity of the Sun, or $0.0006L_{Sun}$. Betelgeuse, the bright left-shoulder star of Orion, has a luminosity of $120,000L_{Sun}$, meaning that it is 120,000 times as luminous as the Sun. Overall, studies of the luminosities of many stars have taught us two particularly important lessons:

- Stars come in a wide range of luminosities, with our Sun somewhere in the middle. The dimmest stars have luminosities $\frac{1}{10,000}$ times that of the Sun ($10^{-4}L_{Sun}$) while the brightest stars are about 1 million times as luminous as the Sun (10^6L_{Sun}).
- Dim stars are far more common than bright stars. For example, even though our Sun is roughly in the middle of the overall range of stellar luminosities, it is brighter than the vast majority of stars in our galaxy.

The Magnitude System   Many astronomy resources (including star charts) describe the apparent brightnesses and luminosities of stars in an alternative way: They use the ancient *magnitude system* devised by the Greek astronomer Hipparchus (c. 190–120 B.C.). Although we won't make much use of this system in this book, it's useful to be familiar with it in case you encounter it elsewhere.

Hipparchus designated the brightest stars in the sky as "first magnitude," the next brightest "second magnitude," and so on. The faintest visible stars were magnitude 6. Today we call this type of designation an **apparent magnitude** because it describes how bright a star *appears* in the sky. Notice that the magnitude scale runs backward: A larger apparent magnitude means a dimmer apparent brightness. For example, a star of magnitude 4 is dimmer than a star of magnitude 1.

In modern times, the magnitude system has been extended and more precisely defined: Each difference of five magnitudes is defined to represent a factor of exactly 100 in brightness. For example, a magnitude 1 star is 100 times as bright as a magnitude 6 star, and a magnitude 3 star is 100 times as bright as a magnitude 8 star. As a result of this precise definition, stars can have fractional apparent magnitudes and a few bright stars have apparent magnitudes *less than* 1—which means *brighter* than magnitude 1. For example, the brightest star in the night sky, Sirius, has an apparent magnitude of –1.46.

The modern magnitude system also defines **absolute magnitude** as a way of describing stellar luminosity. A star's absolute magnitude is the apparent magnitude it would have *if* it were at a distance of 10 parsecs (32.6 light-years) from Earth. For example, the Sun's absolute magnitude is about 4.8, meaning that the Sun would have an apparent magnitude of 4.8 *if* it were 10 parsecs away from us—bright enough to be visible but not conspicuous on a dark night.

► **FIGURE 12.4**

This Hubble Space Telescope photo shows a wide variety of stars that differ in color and brightness. Most of the stars in this photo lie about 2000 light-years from the center of our galaxy. We can see these stars despite the fact that they are near the galactic center because of a gap (known as "Baade's window") in the dusty clouds that obscure our view of most other such stars.

common misconceptions

Photos of Stars

Astronomical photographs convey a great deal of information, but they also contain artifacts that are not real. For example, the spikes visible around bright stars are an artifact created by the way starlight interacts with the supports holding the secondary mirror in a telescope [Section 5.3], and stars appear to have different sizes in photographs because overexposure of bright stars causes their light to spill over a larger region of an image than does the light of dimmer stars. Overexposure also explains why the centers of globular clusters and galaxies usually look like big blobs in photographs: The central regions of these objects contain many more stars than the outskirts, and the combined light of so many stars tends to get overexposed, making a blended patch of light. The artifacts can sometimes be useful: The size issue makes it easy to identify the brighter stars, and the spikes generally occur only with point sources of light like stars—which means you can use them to distinguish stars (which have the spikes) from distant galaxies (which don't) in photos like the Hubble Extreme Deep Field on page 1 of this book.

◆ How do we measure stellar temperatures?

A second fundamental property of a star is its surface temperature. You might wonder why we emphasize *surface* temperature rather than interior temperature. The answer is that only surface temperature is directly measurable; interior temperatures are inferred from mathematical models of stellar interiors [Section 11.2]. Whenever you hear astronomers speak of the "temperature" of a star, you can be pretty sure they mean surface temperature unless they state otherwise.

Measuring a star's surface temperature is somewhat easier than measuring its luminosity, because the star's distance doesn't affect the measurement. Instead, we determine surface temperature from either the star's color or its spectrum.

Color and Temperature   Take a careful look at Figure 12.4. Notice that stars come in almost every color of the rainbow. Simply looking at the colors tells us something about the surface temperatures of the stars. For example, a red star is cooler than a blue star.

Stars come in different colors because they emit thermal radiation [Section 5.2]. Recall that a thermal radiation spectrum depends only on the (surface) temperature of the object that emits it (see Figure 5.11). For example, the Sun's 5800 K surface temperature causes it to emit most strongly in the middle of the visible portion of the spectrum, which is why the Sun looks yellow or white in color. A cooler star, such as Betelgeuse (surface temperature 3650 K), looks red because it emits much more red light than blue light. A hotter star, such as Sirius (surface temperature 9400 K), emits a little more blue light than red light and therefore has a slightly blue color to it.

Astronomers can measure surface temperature fairly precisely by comparing a star's apparent brightness in two different colors of light.

TABLE 12.1 **The Spectral Sequence**

Spectral Type	Example(s)	Temperature Range	Key Absorption Line Features	Brightest Wavelength (color)	Typical Spectrum (selected lines labeled)
O	Stars of Orion's Belt	>33,000 K	Lines of ionized helium, weak hydrogen lines	< 89 nm (ultraviolet)*	
B	Rigel	33,000 K– 10,000 K	Lines of neutral helium, moderate hydrogen lines	89–290 nm (ultraviolet)*	
A	Sirius	10,000 K– 7500 K	Very strong hydrogen lines	290–390 nm (violet)*	
F	Polaris	7500 K–6000 K	Moderate hydrogen lines, moderate lines of ionized calcium	390–480 nm (blue)*	
G	Sun, Alpha Centauri A	6000 K–5200 K	Weak hydrogen lines, strong lines of ionized calcium	480–560 nm (yellow)	
K	Arcturus	5200 K–3700 K	Lines of neutral and singly ionized metals, some molecules	560–780 nm (red)	
M	Betelgeuse, Proxima Centauri	<3700 K	Strong molecular lines	>780 nm (infrared)	

*All stars above 6000 K look more or less white to the human eye because they emit plenty of radiation at all visible wavelengths.

For example, by comparing the amount of blue light and red light coming from Sirius, astronomers can measure how much more blue light it emits than red light. Because thermal radiation spectra have a distinctive shape (again, see Figure 5.11), the difference between blue and red light output allows astronomers to calculate a surface temperature.

Spectral Type and Temperature A star's spectral lines provide a second way to measure its surface temperature. In fact, because interstellar dust can affect the apparent colors of stars, temperatures determined from spectral lines are generally more accurate than temperatures determined from colors alone. Stars displaying spectral lines of highly ionized elements must be fairly hot, because it takes a high temperature to ionize atoms. Stars displaying spectral lines of molecules must be relatively cool, because molecules break apart into individual atoms unless they are at relatively cool temperatures. The types of spectral lines present in a star's spectrum therefore provide a direct measure of the star's surface temperature.

Astronomers classify stars according to surface temperature by assigning a **spectral type** determined from the spectral lines present in a star's spectrum. The hottest stars, with the bluest colors, are called spectral type O, followed in order of declining surface temperature by spectral types B, A, F, G, K, and M. The traditional mnemonic for remembering this sequence, OBAFGKM, is "Oh Be A Fine Girl/Guy, Kiss Me!" Table 12.1 summarizes the characteristics of each spectral type. (The sequence of spectral types is sometimes extended beyond type M, with spectral types L, T, and Y representing starlike objects— usually brown dwarfs [Section 13.1]—that are cooler than M stars.)

Each spectral type is subdivided into numbered subcategories (such as B0, B1, ..., B9). The larger the number, the cooler the star. For example, the Sun is designated spectral type G2, which means it is slightly hotter than a G3 star but cooler than a G1 star.

Spectra of stars show that their surface temperatures range from more than 40,000 K to less than 3000 K, corresponding to the sequence of spectral types OBAFGKM.

The range of surface temperatures for stars is much narrower than the range of luminosities. The coolest stars, of spectral type M, have surface temperatures below 3700 K. The hottest stars, of spectral type O, have surface temperatures that can exceed 40,000 K. Cool, red stars are much more common than hot, blue stars.

think about it Invent your own mnemonic for the OBAFGKM sequence. To help get you thinking, here are two examples: (1) Only Bungling Astronomers Forget Generally Known Mnemonics and (2) Only Business Acts For Good, Karl Marx.

History of the Spectral Sequence

You may wonder why the spectral types follow the peculiar order of OBAFGKM. The answer lies in the history of stellar spectroscopy.

In the late 19th century, Harvard College Observatory Director Edward Pickering (1846–1919) began a project of studying and classifying stellar spectra. There was a lot of work to be done, so Pickering hired assistants whom he called "computers." In part because of the lack of other opportunities for women at the time, most of the computers were women who had studied physics or astronomy at women's colleges such as Wellesley and Radcliffe.

One of the first computers was Williamina Fleming (1857–1911), who classified stellar spectra according to the strength of their hydrogen lines: type A for the strongest hydrogen lines, type B for slightly weaker hydrogen lines, and so on, to type O, for stars with the weakest hydrogen lines. Pickering published Fleming's classifications of more than 10,000 stars in 1890.

As more stellar spectra were obtained and the spectra were studied in greater detail, it became clear that a classification scheme based solely on hydrogen lines was inadequate. Ultimately, the task of finding a better classification scheme fell to Annie Jump Cannon (1863–1941), who joined Pickering's team in 1896 (Figure 12.5). Building on the work of Fleming and another computer, Antonia Maury (1866–1952), Cannon soon realized that the spectral classes fell into a natural order—but not the alphabetical order determined by hydrogen lines alone. Moreover, she found that some of the original classes overlapped others and could be eliminated. Cannon discovered that the natural sequence consisted of just a few of Pickering and Fleming's original classes in the order OBAFGKM; she also added the subdivisions by number.

The astronomical community adopted Cannon's system of stellar classification in 1910. However, no one at that time knew *why* spectra followed the OBAFGKM sequence. Many astronomers guessed, incorrectly, that the different sets of spectral lines indicated different compositions for the stars. The correct answer—that all stars are made primarily of hydrogen and helium and that a star's surface temperature determines the strength of its spectral lines—was discovered at Harvard Observatory in 1925 by Cecilia Payne-Gaposchkin (1900–1979). Relying on insights from what was then the newly developing science of quantum mechanics, Payne-Gaposchkin showed that the differences in spectral lines from star to star merely reflected changes in the ionization level of the emitting

▲ **FIGURE 12.5**
Edward Pickering and his "computers" pose at Harvard College Observatory in 1913. Annie Jump Cannon is fifth from the left in the back row.

atoms. For example, O stars have weak hydrogen lines because, at their high surface temperatures, nearly all their hydrogen is ionized. Without an electron to "jump" between energy levels, ionized hydrogen can neither emit nor absorb its usual specific wavelengths of light. At the other end of the spectral sequence, M stars are cool enough for some particularly stable molecules to form, explaining their strong molecular absorption lines.

◆ How do we measure stellar masses?

Mass is generally more difficult to measure than surface temperature or luminosity. The most dependable method for "weighing" a star relies on Newton's version of Kepler's third law [Section 4.4]. Recall that this law can be applied only when we can observe one object orbiting another, and it requires that we measure both the orbital period and the average orbital distance of the orbiting object. For stars, these requirements generally mean that we can apply the law to measure masses only in **binary star systems**—systems in which two stars continually orbit each other. Before we consider how we determine the orbital periods and distances needed to use Newton's version of Kepler's third law, let's look briefly at the different types of binary star systems that we can observe.

Types of Binary Star Systems Surveys show that about half of all stars orbit a companion star of some kind and are therefore members of binary star systems. These star systems fall into three classes:

- A *visual binary* is a pair of stars that we can see distinctly (with a telescope) as the stars orbit each other. Sometimes we observe a star slowly shifting position in the sky as if it were a member of a visual binary, but its companion is too dim to be seen. For example, slow shifts in the position of Sirius, the brightest star in the night sky, revealed it to be a binary star long before its companion was discovered (Figure 12.6).

- A *spectroscopic binary* is identified through observations of Doppler shifts in its spectral lines [Section 5.2]. If one star is orbiting another, it periodically moves toward us and away from us in its orbit, which means its spectral lines will show alternating blueshifts and redshifts (Figure 12.7). Sometimes we see two sets of lines shifting back and forth—one set from each of the two stars in the system (a *double-lined* spectroscopic binary). Other times we see a set of shifting lines from only one star because its companion is too dim to be detected (a *single-lined* spectroscopic binary).

- An *eclipsing binary* is a pair of stars that orbit in the plane of our line of sight (Figure 12.8). When neither star is eclipsed, we see the combined light of both stars. When one star eclipses the other, the apparent brightness of the system drops because some of the light is

▼ **FIGURE 12.6**

Each frame represents the relative positions of Sirius A and Sirius B at 10-year intervals from 1900 to 1970. The back-and-forth "wobble" of Sirius A allowed astronomers to infer the existence of Sirius B even before the two stars could be resolved in telescopic photos. The average orbital separation of the binary system is about 20 AU.

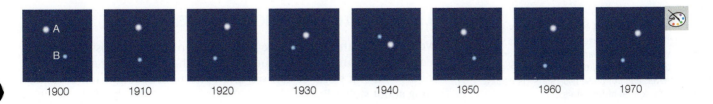

| 1900 | 1910 | 1920 | 1930 | 1940 | 1950 | 1960 | 1970 |

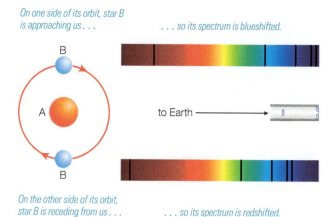

On one side of its orbit, star B is approaching us . . .

. . . so its spectrum is blueshifted.

to Earth ⟶

On the other side of its orbit, star B is receding from us . . .

. . . so its spectrum is redshifted.

▲ **FIGURE 12.7**

The spectral lines of a star in a binary system are periodically blueshifted as it moves toward us in its orbit and redshifted as it moves away from us.

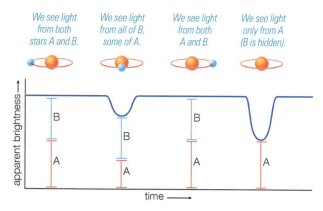

We see light from both stars A and B.

We see light from all of B, some of A.

We see light from both A and B.

We see light only from A (B is hidden).

apparent brightness ⟶

time ⟶

▲ **FIGURE 12.8**

The apparent brightness of an eclipsing binary system drops when one star eclipses the other.

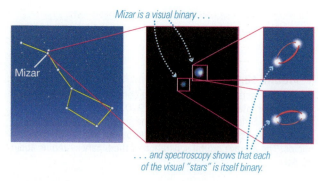

Mizar is a visual binary . . .

Mizar

. . . and spectroscopy shows that each of the visual "stars" is itself binary.

▲ **FIGURE 12.9**

Mizar looks like one star to the naked eye but is actually a system of four stars. Through a telescope, Mizar appears to be a visual binary made up of two stars, Mizar A and Mizar B, that gradually change positions, indicating that they orbit each other every few thousand years. Moreover, each of these two "stars" is itself a spectroscopic binary, for a total of four stars.

blocked from our view. A *light curve,* or graph of apparent brightness against time, reveals the pattern of the eclipses. The most famous example of an eclipsing binary is Algol (Arabic for "the ghoul"), the "demon star" in the constellation Perseus. Algol's brightness drops to only a third of its usual level for a few hours about every 3 days as the brighter of its two stars is eclipsed by its dimmer companion.

Note that these three ways of identifying binaries are essentially the same as the three major methods used to detect extrasolar planets [Section 10.1]: Observing a visual binary means watching changes in position and hence is equivalent to the *astrometric method*; looking at spectral changes in spectroscopic binaries is equivalent to the *Doppler method*; and eclipsing binaries are essentially undergoing both *transits* and *eclipses*.

Some star systems combine two or more of these binary types. For example, telescopic observations reveal Mizar (the second star in the handle of the Big Dipper) to be a visual binary. Spectroscopy then shows that each of the two stars in the visual binary is itself a spectroscopic binary (Figure 12.9).

Masses in Binary Systems Even for a binary system, we can apply Newton's version of Kepler's third law to calculate mass only if we can measure both the orbital period and the separation of the two stars. We can directly measure orbital period for all three binary types, but determining the average separation of binary stars requires that we know precisely how the stellar orbits are oriented to our line of sight. While we can sometimes determine this orientation for visual and spectroscopic binaries, we always know what it is for eclipsing binaries.

> We can determine the masses of stars in binary systems if we can measure both their orbital period and the separation between them.

Much like transits of extrasolar planets, eclipses in a binary system tell us that the two stars orbit edge-on to our line of sight. Doppler shift measurements therefore tell us the true orbital velocities of the stars (see Figure 5.15), and we can use these velocities with the orbital periods to calculate the orbital separation. We can then apply Newton's version of Kepler's third law to determine the masses of the stars. As an added bonus, eclipsing binaries allow us to measure stellar radii directly. Because we know how fast the stars are moving across our line of sight as one eclipses the other, we can determine their radii by timing how long each eclipse lasts.

Through careful observations of eclipsing binaries and other binary star systems, astronomers have established the masses of many different kinds of stars. The overall range extends from as little as 0.08 times the mass of the Sun ($0.08 M_{Sun}$) to at least 150 times the mass of the Sun ($150 M_{Sun}$). We'll discuss the reasons for this mass range in Chapter 13.

12.2 Patterns Among Stars

We have seen that stars come in a wide range of luminosities, surface temperatures, and masses. But are these characteristics randomly distributed among stars, or can we find patterns that might tell us something about stellar lives?

Before reading any further, take another look at Figure 12.4 and think about how you would classify these stars. Almost all of them are at nearly the same distance from Earth, so we can compare their true luminosities by looking at their apparent brightnesses in the photograph. If you look closely, you might notice a couple of important patterns:

- Most of the very brightest stars are reddish in color.
- If you ignore those relatively few bright red stars, there's a general trend to the luminosities and colors among all the rest of the stars: The brighter ones are white with a little bit of blue tint, the more modest ones are similar to our Sun in color with a yellowish white tint, and the dimmest ones are barely visible specks of red.

Keeping in mind that colors tell us about surface temperature—blue is hotter and red is cooler—you can see how these patterns tell us about relationships between surface temperature and luminosity.

Danish astronomer Ejnar Hertzsprung and American astronomer Henry Norris Russell recognized these relationships in the first decade of the 20th century. Building upon the work of Annie Jump Cannon and others, Hertzsprung and Russell independently decided to make graphs of stellar properties by plotting stellar luminosities on one axis and spectral types on the other. These graphs revealed previously unsuspected patterns among the properties of stars and ultimately unlocked the secrets of stellar life cycles.

◆ What is a Hertzsprung-Russell diagram?

Graphs of the type made by Hertzsprung and Russell are now called **Hertzsprung-Russell (H-R) diagrams.** These diagrams quickly became one of the most important tools in astronomical research, and they remain central to the study of stars.

Basics of the H-R Diagram   Figure 12.10 (pages 320–321) shows how we construct an H-R diagram, with a complete diagram on the right-hand page. Note that on an H-R diagram:

- The horizontal axis represents stellar surface temperature, which, as we've discussed, corresponds to spectral type. Temperature *decreases* from left to right because Hertzsprung and Russell based their diagrams on the spectral sequence OBAFGKM.
- The vertical axis represents stellar luminosity, in units of the Sun's luminosity (L_{Sun}). Stellar luminosities span a wide range, so we keep the graph compact by making each tick mark represent a luminosity 10 times as large as that of the prior tick mark.

Each location on the H-R diagram represents a unique combination of spectral type and luminosity. For example, the dot representing the Sun in Figure 12.10 corresponds to the Sun's spectral type, G2, and its luminosity, $1L_{Sun}$. Because luminosity in-

An H-R diagram plots the surface temperatures of stars against their luminosities.

creases upward on the diagram and surface temperature in-creases leftward, stars near the upper left are hot and luminous. Similarly, stars near the upper right are cool and luminous, stars near the lower right are cool and dim, and stars near the lower left are hot and dim.

> **think about it**   Explain how the colors of the stars in Figure 12.10 help indicate stellar surface temperature. Do these colors tell us anything about *interior* temperatures? Why or why not?

cosmic calculations 12.2

Radius of a Star

Almost all stars are too distant for us to measure their radii directly. So how do we know that some are small and some are supergiants? We can calculate a star's radius from its luminosity and surface temperature. Recall from Cosmic Calculations 5.1 that the amount of thermal radiation emitted *per unit area* by a star of temperature T is σT^4, where the constant $\sigma = 5.7 \times 10^{-8}$ watt/($m^2 \times$ Kelvin^4). The star's total luminosity L is equal to this power per unit area times the star's surface area, which is $4\pi r^2$ for a star of radius r:

$$L = 4\pi r^2 \times \sigma T^4$$

With a bit of algebra, we can solve this formula for the star's radius r:

$$r = \sqrt{\frac{L}{4\pi\sigma T^4}}$$

Example: The star Betelgeuse has a surface temperature of about 3650 K and a luminosity of $120{,}000 L_{Sun}$, which is about 4.6×10^{31} watts. What is its radius?

Solution: We use the formula given above with $L = 4.6 \times 10^{31}$ watts and $T = 3650$ K:

$$r = \sqrt{\frac{L}{4\pi\sigma T^4}}$$

$$= \sqrt{\frac{4.6 \times 10^{31} \text{ watts}}{4\pi \times \left(5.7 \times 10^{-8} \dfrac{\text{watt}}{m^2 \times K^4}\right) \times (3650 \text{ K})^4}}$$

$$= \sqrt{\frac{4.6 \times 10^{31} \text{ watts}}{1.3 \times 10^{8} \dfrac{\text{watt}}{m^2}}} = 5.9 \times 10^{11} \text{ m}$$

The radius of Betelgeuse is about 590 billion meters, or 590 million kilometers. This is almost four times the Earth–Sun distance (1 AU ≈ 150 million km). That is why we call Betelgeuse a supergiant.

Hertzsprung-Russell (H-R) diagrams are very important tools in astronomy because they reveal key relationships among the properties of stars. An H-R diagram is made by plotting stars according to their surface temperatures and luminosities. This figure shows a step-by-step approach to building an H-R diagram.

① An H-R Diagram Is a Graph: A star's position along the horizontal axis indicates its surface temperature, which is closely related to its color and spectral type. Its position along the vertical axis indicates its luminosity.

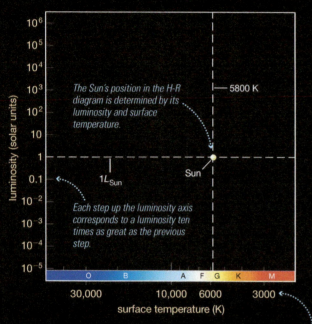

The Sun's position in the H-R diagram is determined by its luminosity and surface temperature.

5800 K

$1L_{Sun}$

Sun

Each step up the luminosity axis corresponds to a luminosity ten times as great as the previous step.

Temperature runs backward on the horizontal axis, with hot blue stars on the left and cool red stars on the right.

② Main Sequence: Our Sun falls along the main sequence, a line of stars extending from the upper left of the diagram to the lower right. Most stars are main-sequence stars, which shine by fusing hydrogen into helium in their cores.

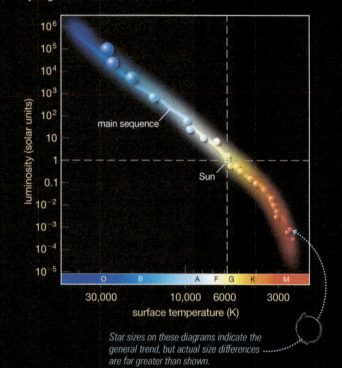

main sequence

Sun

Star sizes on these diagrams indicate the general trend, but actual size differences are far greater than shown.

③ Giants and Supergiants: Stars in the upper right of an H-R diagram are more luminous than main-sequence stars of the same surface temperature. They must therefore be very large in radius, which is why they are known as *giants* and *supergiants*.

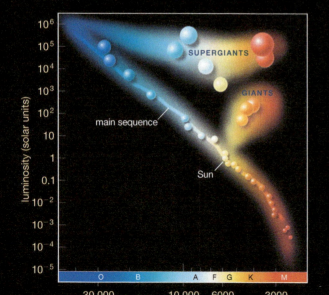

SUPERGIANTS

GIANTS

main sequence

Sun

④ White Dwarfs: Stars in the lower left have high surface temperatures, dim luminosities, and small radii. These stars are known as *white dwarfs*.

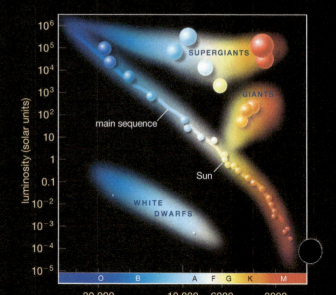

SUPERGIANTS

GIANTS

main sequence

Sun

WHITE DWARFS

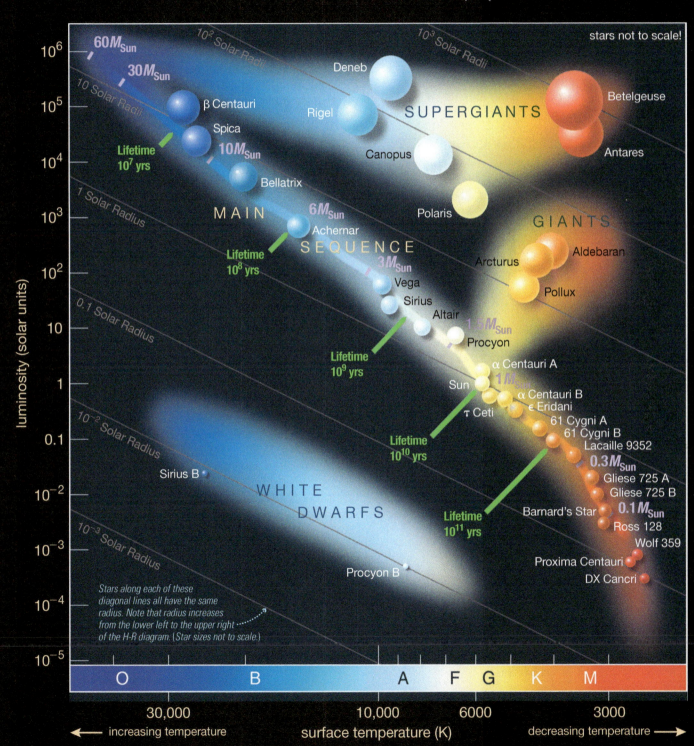

⑤ **Masses on the Main Sequence:** Stellar masses (purple labels) decrease from the upper left to the lower right on the main sequence.

⑥ **Lifetimes on the Main Sequence:** Stellar lifetimes (green labels) increase from the upper left to lower right on the main sequence: High-mass stars live shorter lives because their high luminosities mean they consume their nuclear fuel more quickly.

10^2 Solar Radii

10^3 Solar Radii

stars not to scale!

10 Solar Radii

$60 M_{Sun}$

$30 M_{Sun}$

Deneb

Betelgeuse

β Centauri

Rigel

SUPERGIANTS

Spica

$10 M_{Sun}$

Canopus

Antares

Lifetime 10^7 yrs

Bellatrix

1 Solar Radius

MAIN

$6 M_{Sun}$

Polaris

GIANTS

Achernar

SEQUENCE

$3 M_{Sun}$

Arcturus

Aldebaran

Lifetime 10^8 yrs

0.1 Solar Radius

Vega

Pollux

Sirius

Altair

$1.5 M_{Sun}$

Procyon

Lifetime 10^9 yrs

α Centauri A

$1 M_{Sun}$

Sun

α Centauri B

ε Eridani

τ Ceti

61 Cygni A

10^{-2} Solar Radius

61 Cygni B

Lacaille 9352

Lifetime 10^{10} yrs

$0.3 M_{Sun}$

Gliese 725 A

Sirius B

Gliese 725 B

WHITE

DWARFS

Barnard's Star

$0.1 M_{Sun}$

Ross 128

Lifetime 10^{11} yrs

Wolf 359

10^{-3} Solar Radius

Proxima Centauri

DX Cancri

Procyon B

Stars along each of these diagonal lines all have the same radius. Note that radius increases from the lower left to the upper right of the H-R diagram. (Star sizes not to scale.)

luminosity (solar units)

10^6

10^5

10^4

10^3

10^2

10

1

0.1

10^{-2}

10^{-3}

10^{-4}

10^{-5}

O B A F G K M

30,000 10,000 6000 3000

← increasing temperature

surface temperature (K)

decreasing temperature →

The H-R diagram also provides information about stellar radii, because a star's luminosity depends on both its surface temperature and its surface area or radius. If two stars have the same surface temperature, one can be more luminous than the other only if it is larger in size. Stellar radii therefore must increase as we go from the high-temperature, low-luminosity corner on the lower left of the H-R diagram to the low-temperature, high-luminosity corner on the upper right. Notice the diagonal lines that represent different stellar radii in Figure 12.10.

Patterns in the H-R Diagram Stars do not fall randomly throughout an H-R diagram like Figure 12.10 but instead cluster into four major groups:

- Most stars fall somewhere along the **main sequence,** the prominent streak running from the upper left to the lower right on the H-R diagram. Notice that our Sun is one of these *main-sequence stars.*

- The stars in the upper right are called **supergiants** because they are very large in addition to being very bright.

- Just below the supergiants are the **giants,** which are somewhat smaller in radius and lower in luminosity (but still much larger and brighter than main-sequence stars of the same spectral type).

- The stars near the lower left are small in radius and appear white in color because of their high temperatures. We call these stars **white dwarfs.**

Luminosity Classes In addition to the four major groups we've just listed, stars sometimes fall into "in-between" categories. For more precise work, astronomers therefore assign each star to a **luminosity class,** designated with a roman numeral from I to V. The luminosity class describes the region of the H-R diagram in which the star falls; so despite the name, a star's luminosity class is more closely related to its size than to its luminosity. The basic luminosity classes are I for supergiants, III for giants, and V for main-sequence stars. Luminosity classes II and IV are intermediate to the others. For example, luminosity class IV represents stars with radii larger than those of main-sequence stars but not quite large enough to qualify them as giants. Table 12.2 summarizes the luminosity classes. White dwarfs fall outside this classification system and instead are often assigned the luminosity class "wd."

Complete Stellar Classification We have now described two different ways of categorizing stars:

- A star's *spectral type,* designated by one of the letters OBAFGKM, tells us its surface temperature and color. O stars are the hottest and bluest, while M stars are the coolest and reddest.

- A star's *luminosity class,* designated by a roman numeral, is based on its luminosity but also tells us about the star's radius. Luminosity class I stars have the largest radii, with radii decreasing to luminosity class V.

The full classification of a star includes both a spectral type (OBAFGKM) and a luminosity class.

We use both spectral type and luminosity class to fully classify a star. For example, the complete classification of our Sun is G2 V. The G2 spectral type means it is yellow-white in color, and the luminosity class V means it is a hydrogen-fusing, main-sequence star.

TABLE 12.2 Stellar Luminosity Classes

Class	Description
I	Supergiants
II	Bright giants
III	Giants
IV	Subgiants
V	Main-sequence stars

Betelgeuse is M2 I, making it a red supergiant. Proxima Centauri is M5 V—similar in color and surface temperature to Betelgeuse, but far dimmer because of its much smaller size.

think about it By studying Figure 12.10, determine the approximate spectral type, luminosity class, and radius of the following stars: Bellatrix, Vega, Antares, Pollux, and Proxima Centauri.

◆ What is the significance of the main sequence?

Most stars, including our Sun, have properties that place them on the main sequence of the H-R diagram. You can see in Figure 12.10 that high-luminosity main-sequence stars have hot surfaces and low-luminosity main-sequence stars have cooler surfaces. This relationship between luminosity and surface temperature comes about because a star's position along the main sequence is closely related to its mass. All stars along the main sequence are fusing hydrogen into helium in their cores, just like the Sun, and mass determines both surface temperature and luminosity because it is the key factor in the star's rate of hydrogen fusion.

Masses Along the Main Sequence If you look along the main sequence in Figure 12.10, you'll notice purple labels indicating stellar masses and green labels indicating stellar lifetimes. To make them easier to see, Figure 12.11 repeats the same data but shows only the main sequence rather than the entire H-R diagram. Let's focus first on mass.

Notice that *stellar masses decrease downward along the main sequence.* At the upper end of the main sequence, the hot, luminous O stars can have masses as high as 150 or more times that of the Sun. On the lower end, cool, dim M stars may have as little as 0.08 times the mass of the Sun ($0.08 M_{Sun}$). Many more stars fall on the lower end of the main sequence than on the upper end, which tells us that low-mass stars are much more common than high-mass stars.

The orderly arrangement of stellar masses along the main sequence tells us that *mass* is the most important attribute of a hydrogen-fusing star. The reason is that mass determines the balancing point at which the energy released by hydrogen fusion in the core equals the energy lost from the star's surface. The great range of stellar luminosities on the H-R diagram shows that the point of energy balance is very sensitive to mass. For example, a $10 M_{Sun}$ main-sequence star is about 10,000 times as luminous as the Sun.

The relationship between mass and surface temperature is a little more subtle. In general, a very luminous star must be very large or have a very high surface temperature, or some combination of both. The most massive main-sequence stars are many thousands of times as luminous as the Sun but only about

A main-sequence star's mass determines both its luminosity and its surface temperature.

10 times the size of the Sun in radius. Their surfaces must be significantly hotter than the Sun's surface to account for their high luminosities. Main-sequence stars more massive than the Sun therefore have higher surface temperatures than the Sun, and those less massive than the Sun have lower surface temperatures. That is why the main sequence slices diagonally from the upper left to the lower right on the H-R diagram.

The fact that mass, surface temperature, and luminosity are all related means that we can estimate a main-sequence star's mass just by knowing its spectral type. For example, any hydrogen-fusing, main-sequence star that has the same spectral type as the Sun (G2) must have

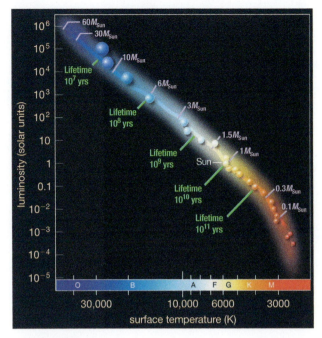

▲ **FIGURE 12.11**

The main sequence from Figure 12.10 is isolated here so that you can more easily see how masses and lifetimes vary along it. Notice that more massive hydrogen-fusing stars are brighter and hotter than less massive ones, but have shorter lifetimes. (Stellar masses are given in units of solar masses: $1 M_{Sun} = 2 \times 10^{30}$ kg.)

about the same mass and luminosity as the Sun. Similarly, any main-sequence star of spectral type B1 must have about the same mass and luminosity as Spica (see Figure 12.10). Note that only main-sequence stars follow this simple relationship between mass, temperature, and luminosity; it does *not* hold for giants, supergiants, or white dwarfs.

Lifetimes Along the Main Sequence A star is born with a limited supply of core hydrogen and therefore can remain as a hydrogen-fusing, main-sequence star for only a limited time—the star's **main-sequence lifetime.** Because stars spend the vast majority of their lives as main-sequence stars, we sometimes refer to the main-sequence lifetime as simply the "lifetime." Like masses, stellar lifetimes vary in an orderly way as we move up the main sequence: Massive stars near the upper end of the main sequence have *shorter* lives than less massive stars near the lower end (see Figure 12.11).

Why do more massive stars have shorter lives? A star's lifetime depends on both its mass and its luminosity. Its mass determines how much hydrogen fuel the star initially contains in its core. Its luminosity determines how rapidly the star uses up its fuel. Massive stars start their lives with a larger supply of hydrogen, but they fuse this hydrogen into helium so rapidly that they end up with shorter lives. For example, a 10-solar-mass star ($10M_{Sun}$) is born with 10 times as much hydrogen as the Sun. However, its luminosity of $10,000L_{Sun}$ means that it uses up this hydrogen at a rate 10,000 times as fast as the rate in the Sun. With 10 times as much hydrogen being consumed at 10,000 times the rate, the lifetime of a $10M_{Sun}$ star would be only about $\frac{10}{10,000} = \frac{1}{1000}$ as long as the Sun's lifetime, or about 10 billion $\div$ 1000 = 10 million years. (Its actual lifetime is a little longer than this, because it can use more of its core hydrogen for fusion than the Sun can.)

> More massive stars live much shorter lives because they fuse hydrogen at a much greater rate.

On the other end of the scale, a 0.3-solar-mass main-sequence star emits a luminosity just 0.01 times that of the Sun and consequently lives about $\frac{0.3}{0.01} = 30$ times as long as the Sun, or some 300 billion years. In a universe that is now about 14 billion years old, even the most ancient of these small, dim, red stars of spectral type M still survive, and they will continue to shine faintly for hundreds of billions of years to come.

Mass: A Star's Most Fundamental Property Astronomers began classifying stars by spectral type and luminosity class before they understood why stars vary in these properties. Today, we know that the most fundamental property of any star is its *mass* (Figure 12.12). A star's mass determines both its surface temperature and its luminosity throughout the main-sequence portion of its life, and these properties in turn explain why higher-mass stars have shorter lifetimes.

think about it Which of the stars labeled in Figure 12.10 has the longest lifetime? Explain.

◆ What are giants, supergiants, and white dwarfs?

Main-sequence stars fuse hydrogen into helium in their cores, but what about the other classes of stars on the H-R diagram? These other classes all represent stars that have exhausted the supply of hydrogen in their central cores, so they can no longer generate energy in the same way as our Sun.

B1 V
Spica $11M_{Sun}$
Lifetime 10^7 yrs

A1 V
Sirius $2M_{Sun}$
Lifetime 10^9 yrs

G2 V
Sun $1M_{Sun}$
Lifetime 10^{10} yrs

M5.5 V
Proxima
Centauri $0.12M_{Sun}$
Lifetime 10^{12} yrs

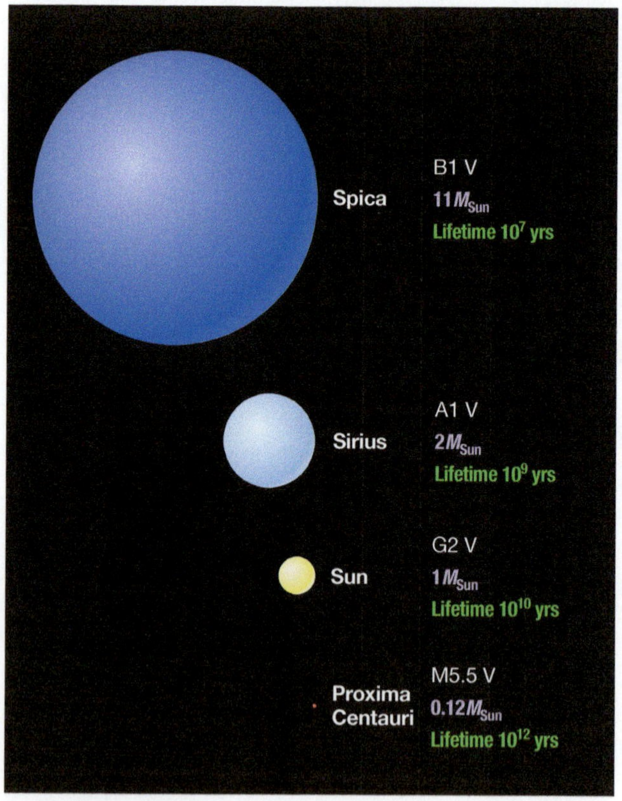

▲ **FIGURE 12.12**
Four main-sequence stars shown to scale. The mass of a main-sequence star determines its fundamental properties of luminosity, surface temperature, radius, and lifetime. More massive main-sequence stars are hotter and brighter than less massive ones, but have shorter lifetimes.

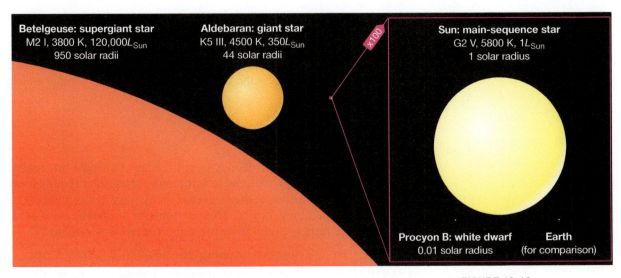

Betelgeuse: supergiant star
M2 I, 3800 K, 120,000L_{Sun}
950 solar radii

Aldebaran: giant star
K5 III, 4500 K, 350L_{Sun}
44 solar radii

×100

Sun: main-sequence star
G2 V, 5800 K, 1L_{Sun}
1 solar radius

Procyon B: white dwarf
0.01 solar radius

Earth
(for comparison)

▲ **FIGURE 12.13**
The relative sizes of stars. A supergiant like Betelgeuse would fill the inner solar system and extend more than 80% of the way to Jupiter's orbit. A giant like Aldebaran would fill the inner half of Mercury's orbit. The Sun is a hundred times larger in radius than a white dwarf, which is roughly the same size as Earth.

Giants and Supergiants The bright red stars in Figure 12.4 are giants and supergiants whose properties place them to the upper right of the main sequence in an H-R diagram. The fact that these stars are cooler but much more luminous than the Sun tells us that they must be much larger in radius than the Sun. Remember that a star's surface temperature determines the amount of light it emits per unit surface area [Section 5.2]: Hotter stars emit much more light per unit surface area than cooler stars. For example, a blue star emits far more total light than a red star of the same size. A star that is red and cool can be bright only if it has a very large surface area, which means it must be enormous in size.

As we'll discuss in Chapter 13, we now know that giants and supergiants are stars nearing the ends of their lives. They have already exhausted the supply of hydrogen fuel in their central cores and are in essence facing an energy crisis as they try to stave off the inevitable crushing force of gravity. This crisis causes these stars to release fusion energy at a furious rate, which accounts for their high luminosities and causes their surfaces to expand to enormous size (Figure 12.13). For example, Arcturus and Aldebaran (the eye of the bull in the constellation Taurus) are giant stars more than 10 times as large in radius as our Sun. Betelgeuse, the left shoulder in the constellation Orion, is an enormous supergiant with a radius roughly 1000 times that of the Sun, equivalent to almost four times the Earth–Sun distance.

> Giants and supergiants are stars that are nearing the ends of their lives.

Because giants and supergiants are so bright, we can see them even if they are not especially close to us. Many of the brightest stars in our sky are giants or supergiants, often identifiable by their reddish colors. Overall, however, giants and supergiants are considerably rarer than main-sequence stars. In our snapshot of the heavens, we catch most stars in the act of hydrogen fusion and relatively few in a later stage of life.

White Dwarfs Giants and supergiants eventually run out of fuel entirely. A giant with a mass similar to that of our Sun ultimately ejects its outer layers, leaving behind a "dead" core in which all nuclear fusion has ceased. White dwarfs are these remaining embers of former giants.

White dwarfs are hot because they are essentially exposed stellar cores, but they are dim because they lack an energy source and radiate only their leftover heat into space. A typical white dwarf is no larger in size than Earth, but has a mass similar to that of our Sun. Clearly, white dwarfs must be made of matter compressed to an extremely high density, unlike anything found on Earth. We'll discuss the nature of white dwarfs and other stellar corpses in Chapter 14.

12.3 Star Clusters

All stars are born from giant clouds of gas. Because a single interstellar cloud can contain enough material to form many stars, stars usually form in groups. In our snapshot of the heavens, many stars still congregate in the groups in which they formed. These groups are known as *star clusters*, and they are extremely useful to astronomers for two key reasons:

1. All the stars in a cluster lie at about the same distance from Earth.

2. All the stars in a cluster formed at about the same time (within a few million years of one another).

Astronomers can therefore use star clusters to compare the properties of stars that all have similar ages.

◆ What are the two types of star clusters?

Star clusters come in two basic types: modest-size **open clusters** and densely packed **globular clusters.** The two types differ not only in how densely they are packed with stars, but also in their locations and ages.

Recall that most of the stars, gas, and dust in the Milky Way Galaxy, including our Sun, lie in the relatively flat *galactic disk;* the region above and below the disk is called the *halo* of the galaxy (see Figure 1.14). Open clusters are always found in the disk of the galaxy and tend to be young in age. They can contain up to several thousand stars and typically are about 30 light-years across. The most famous open cluster, the *Pleiades* (Figure 12.14), is often called the *Seven Sisters,* although only six of the cluster's several thousand stars are easily visible to the naked eye. Other cultures have other names for this beautiful group of stars. In Japanese, it is called *Subaru,* which is why the logo for Subaru automobiles is a diagram of the Pleiades.

In contrast, most globular clusters are found in the halo, and their stars are among the oldest in the universe. A globular cluster can contain more than a million stars concentrated in a ball typically 60 to 150 light-years across. Its central region can have 10,000 stars packed into a space just a few light-years across (Figure 12.15). The view from a planet in a globular cluster would be marvelous, with thousands of stars lying closer to that planet than Alpha Centauri is to the Sun.

◆ How do we measure the age of a star cluster?

We can determine the age of a star cluster by plotting its stars on an H-R diagram. Figure 12.16 shows the process for the Pleiades. Note that most of the stars in this cluster lie along the main sequence, with one important exception: At the upper end of the main sequence, stars trail away to the right. That is, the hot, short-lived stars of spectral type O are missing from the main sequence. Apparently, the Pleiades cluster is old enough for hydrogen fusion to have ended in the cores of its main-sequence O stars, but young enough that hydrogen fusion is still happening in the cores of some stars of spectral type B that still reside on the main sequence.

▲ FIGURE 12.14

A photo of the Pleiades, a nearby open cluster of stars in the constellation Taurus. The most prominent stars in this open cluster are of spectral type B, indicating that the Pleiades are no more than 100 million years old, relatively young for a star cluster. The region shown is about 11 light-years across.

▲ FIGURE 12.15

The globular cluster M80 is more than 12 billion years old. The prominent reddish stars in this Hubble Space Telescope photo are red giants nearing the ends of their lives. The central region pictured here is about 15 light-years across.

The precise point on the H-R diagram at which a cluster's stars diverge from the main sequence is called the **main-sequence turnoff point.** For the Pleiades, it occurs around spectral type B6. The main-sequence lifetime of a B6 star is roughly 100 million years, so this must be the age of the Pleiades. Any star in the Pleiades that was born with a main-sequence spectral type hotter than B6 had a lifetime shorter than 100 million years and is no longer found on the main sequence. Over the next few billion years, the B stars in the Pleiades will die out, followed by the A stars and the F stars. If we could make an H-R diagram for the Pleiades every few million years, we would find its main sequence gradually growing shorter.

The age of a star cluster approximately equals the hydrogen core-fusion lifetime of the most massive main-sequence stars remaining within it.

Figure 12.17 shows this idea by comparing the main sequences of several open clusters. In each case, *the age of the cluster is equal to the lifetime of stars at its main-sequence turnoff point.* Stars in a particular cluster that once resided above the turnoff point on the main sequence have already exhausted their core supply of hydrogen, while stars below the turnoff point remain on the main sequence.

think about it Suppose a star cluster is precisely 10 billion years old. Where would you expect to find its main-sequence turnoff point? Would you expect this cluster to have any main-sequence stars of spectral type A? Would you expect it to have main-sequence stars of spectral type K? Explain. (*Hint*: What is the lifetime of our Sun?)

The technique of identifying main-sequence turnoff points is our most powerful tool for evaluating the ages of star clusters. We've learned that most open clusters are relatively young, with very few older than about 5 billion years. In contrast, the stars at the main-sequence turnoff points in globular clusters are usually less massive than our Sun (Figure 12.18). Because stars like our Sun have a lifetime of about 10 billion years and these stars have already died in globular clusters, we conclude that globular

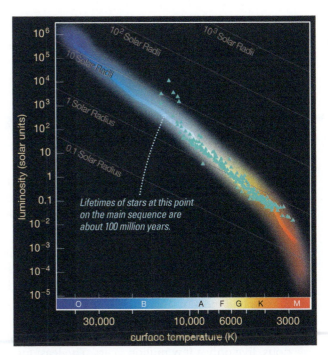

▲ FIGURE 12.16

An H-R diagram for the Pleiades. Triangles represent individual stars. The Pleiades cluster is missing its upper main-sequence stars, indicating that hydrogen core fusion in these stars has already ended. The main-sequence turnoff point at about spectral type B6 tells us that the Pleiades are about 100 million years old.

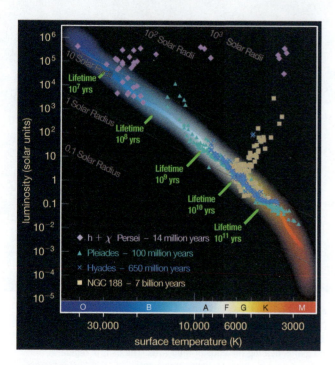

▲ **FIGURE 12.17**

This H-R diagram shows stars from the Pleiades and three other clusters. Their differing main-sequence turnoff points indicate very different ages.

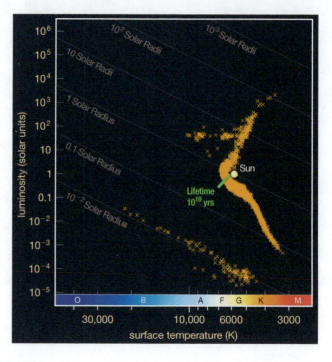

▲ **FIGURE 12.18**

This H-R diagram shows stars from the globular cluster M4. The main-sequence turnoff point is in the vicinity of stars like our Sun, indicating an age for this cluster of around 10 billion years. A more technical analysis of this cluster (which accounts for the lower abundance of heavy elements in these old stars) places its age at around 13 billion years.

cluster stars are older than 10 billion years. More precise studies of the turnoff points in globular clusters, coupled with theoretical calculations of stellar lifetimes, place the ages of these clusters at about 13 billion years, making them the oldest known objects in the galaxy and implying that they began to form within the first billion years of the universe's roughly 14-billion-year history.

the big picture   Putting Chapter 12 into Perspective

We have classified the diverse families of stars visible in the night sky. Much of what we know about stars, galaxies, and the universe itself is based on the fundamental properties of stars introduced in this chapter. Make sure you understand the following "big picture" ideas:

- All stars are made primarily of hydrogen and helium at the time they form. The differences between stars are primarily due to differences in mass and stage of life.

- Stars spend most of their lives as main-sequence stars that fuse hydrogen into helium in their cores. The most massive stars, which are also the hottest and most luminous, live only a few million years. The least massive stars, which are coolest and dimmest, will survive until the universe is many times its present age.

- The key to recognizing the patterns among stars was the H-R diagram, which shows stellar surface temperatures on the horizontal axis and luminosities on the vertical axis. The H-R diagram is one of the most important tools of modern astronomy.

- Much of what we know about the universe comes from studies of star clusters. We can measure a star cluster's age by plotting its stars in an H-R diagram and determining the core hydrogen-fusion lifetime of the brightest and most massive stars still on the main sequence.

my cosmic perspective   You may be used to seeing stars on clear, dark nights, but you probably never realized how diverse they are in size, color, and age. Next time you're looking up at the stars on a clear, dark night, try to find one of the brightest red supergiant stars, such as Betelgeuse or Antares, and notice its redder color.

12.1 Properties of Stars

◆ How do we measure stellar luminosities?

 The **apparent brightness** of a star in our sky depends on both its **luminosity**—the total amount of light it emits into space—and its distance, as expressed by the **inverse square law for light.** We can therefore calculate luminosity from apparent brightness and distance; we can measure the latter through stellar parallax.

◆ How do we measure stellar temperatures?

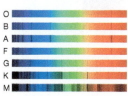 We measure a star's surface temperature from its color or spectrum, and we classify stars according to the sequence of **spectral types** OBAFGKM, which runs from hottest to coolest. Cool, red stars of spectral type M are much more common than hot, blue stars of spectral type O.

◆ How do we measure stellar masses?

We can calculate the masses of stars in **binary star systems** using Newton's version of Kepler's third law if we can measure the orbital period and separation of the two stars.

12.2 Patterns Among Stars

◆ What is a Hertzsprung-Russell diagram?

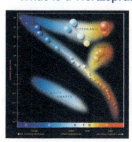 An **H-R diagram** plots stars according to their surface temperatures (or spectral types) and luminosities. Most stars appear on the H-R diagram in the narrow strip known as the **main sequence. Giants** and **supergiants** are to the upper right of the main sequence and **white dwarfs** are to the lower left.

◆ What is the significance of the main sequence?

Stars on the main sequence are all fusing hydrogen into helium in their cores. A star's position along the main sequence depends on its mass: High-mass stars are at the upper left and masses of stars become progressively smaller as we move toward the lower right. Lifetimes vary in the opposite way, because higher-mass stars live shorter lives.

◆ What are giants, supergiants, and white dwarfs?

Giants and supergiants are stars that have exhausted their central core supplies of hydrogen for fusion and are undergoing other forms of fusion at a prodigious rate as they near the ends of their lives. White dwarfs are the exposed cores of stars that have already died, meaning they have no further means of generating energy through fusion.

12.3 Star Clusters

◆ What are the two types of star clusters?

  Open clusters contain up to several thousand stars and are found in the disk of our galaxy. **Globular clusters** contain hundreds of thousands of stars, all closely packed together, and are found mainly in the halo of the galaxy.

◆ How do we measure the age of a star cluster?

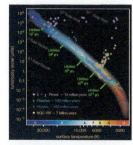 Because all of a cluster's stars were born at about the same time, we can measure a cluster's age by finding the **main-sequence turnoff point** on an H-R diagram of its stars. The cluster's age is equal to the hydrogen core-fusion lifetime of the hottest, most luminous stars that remain on the main sequence. Open clusters are much younger than globular clusters, which can be as old as about 13 billion years.

visual skills check

Check your understanding of some of the many types of visual information used in astronomy. For additional practice, try the Chapter 12 Visual Quiz at MasteringAstronomy®.

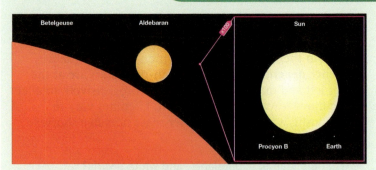

The figure above, similar to Figure 12.13, uses zoom-ins to compare the sizes of giant and supergiant stars to the sizes of Earth and the Sun.

1. Suppose we wanted to represent all of these objects on the 1-to-10-billion scale from Chapter 1, on which the Sun is about the size of a grapefruit. Approximately how large in diameter would the star Aldebaran be on this scale?
   a. 40 centimeters (the size of a typical beach ball)
   b. 6 meters (roughly the size of a dorm room)
   c. 15 meters (roughly the size of a typical house)
   d. 70 meters (slightly smaller than a football field)

2. Approximately how large in diameter would the star Betelgeuse be on this same scale?
   a. 40 centimeters (the size of a typical beach ball)
   b. 6 meters (roughly the size of a dorm room)
   c. 130 meters (slightly larger than a football field)
   d. 3 kilometers (the size of a small town)

3. Approximately how large in diameter would the star Procyon B be on this same scale?
   a. 10 centimeters (the size of a large grapefruit)
   b. 1 centimeter (the size of a grape)
   c. 1 millimeter (the size of a grape seed)
   d. 0.1 millimeter (roughly the width of a human hair)

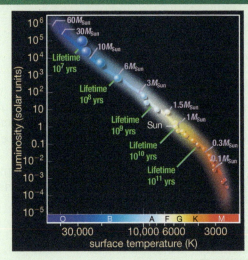

The H-R diagram above is identical to Figure 12.11. Answer the following questions based on the information given in the figure.

4. What are the approximate luminosity and lifetime of a star whose mass is 10 times that of the Sun?

5. What are the approximate luminosity and lifetime of a star whose mass is 3 times that of the Sun?

6. What are the approximate luminosity and lifetime of a star whose mass is twice that of the Sun?

exercises and problems

MasteringAstronomy® For instructor-assigned homework and other learning materials, go to MasteringAstronomy®.

Review Questions

1. Briefly explain how we can learn about the lives of stars even though their lives are far longer than human lives.

2. In what ways are all stars similar? In what ways can stars differ?

3. How is a star's *apparent brightness* related to its *luminosity*? Explain by describing the *inverse square law for light*.

4. How do we use *stellar parallax* to determine a star's distance, and how can we then determine its luminosity?

5. What do we mean by a star's *spectral type*, and how is it related to the star's surface temperature and color? Which types of stars are hottest and coolest in the spectral sequence OBAFGKM?

6. What are the three basic types of *binary star systems*? Why are *eclipsing binaries* so important for measuring masses of stars?

7. Draw a sketch of an Hertzsprung-Russell (H-R) diagram. Label the *main sequence, giants, supergiants,* and *white dwarfs*. Where on this diagram do we find stars that are cool and dim? Cool and luminous? Hot and dim? Hot and luminous?

8. What do we mean by a star's *luminosity class*? Briefly explain how we classify stars by spectral type and luminosity class.

9. What is the defining characteristic of a *main-sequence* star? Briefly explain why massive main-sequence stars are more luminous and have hotter surfaces than less massive main-sequence stars.

10. Which stars have longer lifetimes: massive stars or less massive stars? Explain why.

11. How do giants and supergiants differ from main-sequence stars? What are white dwarfs?

12. Why is a star's birth mass its most fundamental property?
13. Describe in general terms how *open clusters* and *globular clusters* differ in their numbers of stars, ages, and locations in a galaxy.
14. Explain why H-R diagrams look different for star clusters of different ages. How does the location of the *main-sequence turnoff point* tell us the age of the star cluster?

Test Your Understanding

Does It Make Sense?

Decide whether the statement makes sense (or is clearly true) or does not make sense (or is clearly false). Explain clearly; not all of these have definitive answers, so your explanation is more important than your chosen answer.

15. Two stars that look very different must be made of different kinds of elements.
16. Two stars that have the same apparent brightness in the sky must also have the same luminosity.
17. Sirius looks brighter than Alpha Centauri, but we know that Alpha Centauri is closer because its apparent position in the sky shifts by a larger amount as Earth orbits the Sun.
18. Stars that look red-hot have hotter surfaces than stars that look blue.
19. Some of the stars on the main sequence of the H-R diagram are not converting hydrogen into helium.
20. The smallest, hottest stars are plotted in the lower left portion of the H-R diagram.
21. Stars that begin their lives with the most mass live longer than less massive stars because they have so much more hydrogen fuel.
22. Star clusters with lots of bright, blue stars of spectral types O and B are generally younger than clusters that don't have any such stars.
23. All giants, supergiants, and white dwarfs were once main-sequence stars.
24. Most of the stars in the sky are more massive than the Sun.

Quick Quiz

Choose the best answer to each of the following. Explain your reasoning with one or more complete sentences.

25. If the star Alpha Centauri were moved to a distance 10 times farther from Earth than it is now, its parallax angle would (a) get larger. (b) get smaller. (c) stay the same.
26. What do we need to measure in order to determine a star's luminosity? (a) apparent brightness and mass (b) apparent brightness and temperature (c) apparent brightness and distance
27. What two pieces of information would you need in order to measure the masses of stars in an eclipsing binary system? (a) the time between eclipses and the average distance between the stars (b) the period of the binary system and its distance from the Sun (c) the velocities of the stars and the Doppler shifts of their absorption lines
28. Which of these stars has the coolest surface temperature? (a) an A star (b) an F star (c) a K star
29. Which of these stars is the most massive? (a) a main-sequence A star (b) a main-sequence G star (c) a main-sequence M star
30. Which of these stars has the longest lifetime? (a) a main-sequence A star (b) a main-sequence G star (c) a main-sequence M star
31. Which of these stars has the largest radius? (a) a supergiant A star (b) a giant K star (c) a supergiant M star
32. Which of these stars has the greatest surface temperature? (a) a main-sequence B star (b) a supergiant A star (c) a giant K star

33. Which of these star clusters is youngest? (a) a cluster whose brightest main-sequence stars are white (b) a cluster whose brightest stars are red (c) a cluster containing stars of all colors
34. Which of these star clusters is oldest? (a) a cluster whose brightest main-sequence stars are white (b) a cluster whose brightest main-sequence stars are yellow (c) a cluster containing stars of all colors

Process of Science

35. *Classification.* As discussed in the text, Annie Jump Cannon and her colleagues developed our modern system of stellar classification. Why do you think rapid advances in our understanding of stars followed so quickly on the heels of this effort? What other areas of science have had huge advances in understanding following an improved system of classification?
36. *Life Spans of Stars.* Scientists estimate the life span of a star by dividing the total amount of energy available for fusion by the rate at which the star radiates energy into space. Such calculations predict that the life spans of high-mass stars are shorter than those of low-mass stars. Describe one type of observation that can test this prediction and verify that it is correct.

Group Work Exercise

37. *Comparing Stellar Properties.* **Roles:** *Analyst 1* (analyzes data in Table F.1), *Scribe 1* (records results and conclusions about Table F.1), *Analyst 2* (analyzes data in Table F.2), *Scribe 2* (records results and conclusions about Table F.2). **Activity:** Work in pairs to answer the following questions, with Pair 1 using the data in Appendix Table F.1 and Pair 2 using the data in Appendix Table F.2.
    a. Count and record the number of stars of each spectral type (OBAFGKM) for both tables.
    b. Count and record the number of stars of each luminosity class (I, II, III, IV, V) for both tables.
    c. Compare the counts of spectral types in the two tables. Discuss any differences you find and develop a hypothesis that could explain them.
    d. Compare the counts of luminosity classes in the two tables. Discuss any differences you find and develop a hypothesis that could explain them.
    e. Write down the nearest star of each spectral type; if it cannot be determined, explain why.
    f. Determine whether there could be an O star within 200 light-years of the Sun based on the data in the tables, and explain your reasoning.

Investigate Further

Short-Answer/Essay Questions

38. *Stellar Data.* Consider the following data table for several bright stars. M_v is absolute magnitude, and m_v is apparent magnitude. Use these data to answer the following questions, and include a brief explanation with each answer. (*Hint:* Remember that the magnitude scale runs backward, so brighter stars have smaller or more negative magnitudes.)
    a. Which star appears brightest in our sky?
    b. Which star appears faintest in our sky?
    c. Which star has the greatest luminosity?
    d. Which star has the least luminosity?
    e. Which star has the highest surface temperature?
    f. Which star has the lowest surface temperature?

g. Which star is most similar to the Sun?

h. Which star is a red supergiant?

i. Which star has the largest radius?

j. Which stars have finished fusing hydrogen in their cores?

k. Among the main-sequence stars listed, which one is the most massive?

l. Among the main-sequence stars listed, which one has the longest lifetime?

Star	M_v	m_v	Spectral Type	Luminosity Class
Aldebaran	−0.2	+0.9	K5	III
Alpha Centauri A	+4.4	0.0	G2	V
Antares	−4.5	+0.9	M1	I
Canopus	−3.1	−0.7	F0	II
Fomalhaut	+2.0	+1.2	A3	V
Regulus	−0.6	+1.4	B7	V
Sirius	+1.4	−1.4	A1	V
Spica	−3.6	+0.9	B1	V

39. *Data Tables.* Study the spectral types listed in Appendix F for the 20 brightest stars and for the stars within 12 light-years of Earth. Why do you think the two lists are so different? Explain.

40. *Interpreting the H-R Diagram.* Using the information in Figure 12.10, describe how Proxima Centauri differs from Sirius.

41. *Parallax from Jupiter.* Suppose you could travel to Jupiter and observe changes in positions of nearby stars during one orbit of Jupiter around the Sun. Describe how those changes would differ from what we would measure from Earth. How would your ability to measure the distances to stars from the vantage point of Jupiter be different?

42. *An Expanding Star.* Describe what would happen to the surface temperature of a star if its radius doubled in size with no change in luminosity.

43. *Colors of Eclipsing Binaries.* Figure 12.8 shows an eclipsing binary system consisting of a small blue star and a larger red star. Explain why the decrease in apparent brightness of the combined system is greater when the blue star is eclipsed than when the red star is eclipsed.

44. *Visual and Spectroscopic Binaries.* Suppose you are observing two binary star systems at the same distance from Earth. Both are spectroscopic binaries consisting of similar types of stars, but only one of the binary systems is a visual binary. Which of these star systems would you expect to have the greater Doppler shifts in its spectra? Explain your reasoning.

45. *Life of a Star Cluster.* Imagine you could watch a star cluster from the time of its birth to an age of 13 billion years. Describe in one or two paragraphs what you would see happening during that time.

Quantitative Problems

Be sure to show all calculations clearly and state your final answers in complete sentences.

46. *The Inverse Square Law for Light.* Earth is about 150 million km from the Sun, and the apparent brightness of the Sun in our sky is about 1300 watts/m². Using these two facts and the inverse square law for light, determine the apparent brightness that we would measure for the Sun *if* we were located at the following positions.

a. Half Earth's distance from the Sun

b. Twice Earth's distance from the Sun

c. Five times Earth's distance from the Sun

47. *The Luminosity of Alpha Centauri A.* Alpha Centauri A lies at a distance of 4.4 light-years from Earth and has an apparent brightness in our night sky of 2.7×10^{-8} watt/m². Recall that 1 light-year = 9.5×10^{15} m.

a. Use the inverse square law for light to calculate the luminosity of Alpha Centauri A.

b. Suppose you have a light bulb that emits 100 watts of visible light. How far away would you have to put the light bulb for it to have the same apparent brightness as Alpha Centauri A in our sky? (*Hint:* Use 100 watts as *L* in the inverse square law for light, and use the apparent brightness given above for Alpha Centauri A. Then solve for the distance.)

48. *More Practice with the Inverse Square Law for Light.* Use the inverse square law for light to answer each of the following questions.

a. Suppose a star has the same luminosity as our Sun (3.8×10^{26} watts) but is located at a distance of 10 light-years from Earth. What is its apparent brightness?

b. Suppose a star has the same apparent brightness as Alpha Centauri A (2.7×10^{-8} watt/m²) but is located at a distance of 200 light-years from Earth. What is its luminosity?

c. Suppose a star has a luminosity of 8×10^{26} watts and an apparent brightness of 3.5×10^{-12} watt/m². How far away is it from Earth? Give your answer in both kilometers and light-years.

d. Suppose a star has a luminosity of 5×10^{29} watts and an apparent brightness of 9×10^{-15} watt/m². How far away is it from Earth? Give your answer in both kilometers and light-years.

49. *Parallax and Distance.* Use the parallax formula to calculate the distance to each of the following stars. Give your answers in both parsecs and light-years.

a. Alpha Centauri: parallax angle of 0.7420 arcsecond

b. Procyon: parallax angle of 0.2860 arcsecond

50. *Radius of a Star.* Sirius A has a luminosity of $26L_{Sun}$ and a surface temperature of about 9400 K. What is its radius? (*Hint:* See Cosmic Calculations 12.2.)

🌐 Discussion Question

51. *Snapshot of the Heavens.* The beginning of the chapter likened the problem of studying the lives of stars to learning about human beings from a 1-minute glance at human life. What could you learn about human life by looking at a single snapshot of a large family, including babies, parents, and grandparents? How is the study of such a snapshot similar to what scientists do when they study the lives of stars? How is it different?

Web Projects

52. *Women in Astronomy.* Until fairly recently, men greatly outnumbered women in professional astronomy. Nevertheless, many women made crucial discoveries in astronomy throughout history—including the discovery of the spectral sequence for stars. Do some research about the life and discoveries of a female astronomer from any time period and write a two- to three-page scientific biography.

53. *Parallax Missions.* The European Space Agency's *Hipparcos* spacecraft made precise parallax measurements for more than 40,000 stars from 1989 to 1993, and the agency's *GAIA* mission is designed to measure parallax for many more stars. Learn about how these satellites can measure much smaller parallax angles than is possible from the ground and how those measurements affect our knowledge of the universe. Write a one- to two-page report on your findings.

13 Star Stuff

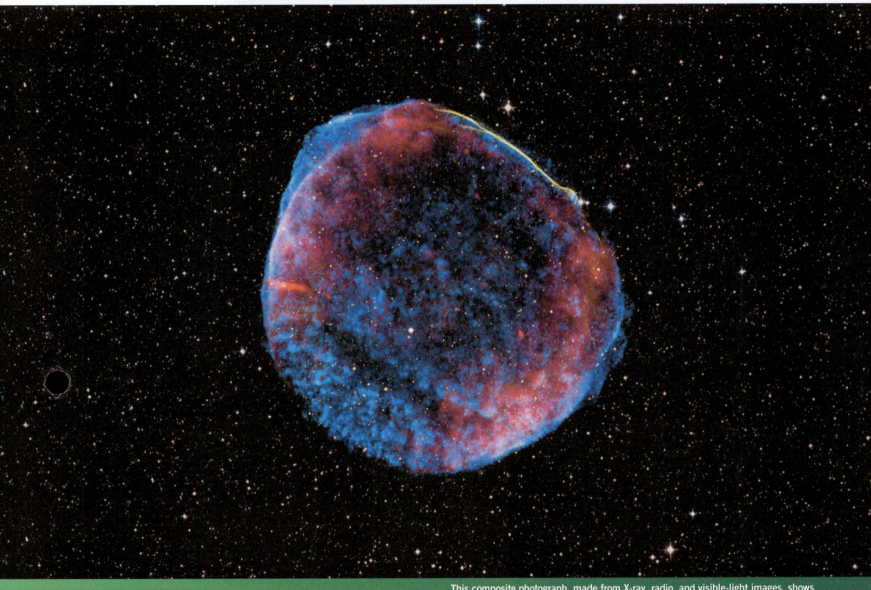

This composite photograph, made from X-ray, radio, and visible-light images, shows an exploding star that is ejecting newly made elements into interstellar space.

LEARNING GOALS

13.1 Star Birth
- ◆ How do stars form?
- ◆ How massive are newborn stars?

13.2 Life as a Low-Mass Star
- ◆ What are the life stages of a low-mass star?
- ◆ How does a low-mass star die?

13.3 Life as a High-Mass Star
- ◆ What are the life stages of a high-mass star?
- ◆ How do high-mass stars make the elements necessary for life?
- ◆ How does a high-mass star die?

13.4 Stars in Close Binaries
- ◆ How are the lives of stars with close companions different?

We inhale oxygen with every breath. Iron-bearing hemoglobin in our blood carries this oxygen through our bodies. Chains of carbon and nitrogen form the backbone of the proteins, fats, and carbohydrates in our cells. Calcium strengthens our bones, while sodium and potassium ions communicate signals through our nerves. What does all this biology have to do with astronomy? The profound answer is that life is based on elements created by stars.

In this chapter, we will discuss the origins of the elements in some detail by delving into the lives of stars. As you read, keep in mind that no matter how far removed the stars may seem from our everyday lives, they are connected to us in the most intimate way possible: Without the births, lives, and deaths of stars, none of us would be here. We are truly made from "star stuff."

13.1 Star Birth

The story of a star's life begins in the murky depths of interstellar space. Using main-sequence turnoff points [Section 12.3] to determine the ages of star clusters, we have found that the youngest clusters are always associated with dark clouds of gas and dust, indicating that these interstellar clouds are the birthplaces of stars. Theoretical models of star formation confirm this idea, telling us how the physical processes that occur in gas clouds ultimately give birth to stars. In this section, we'll explore the processes that give birth to stars, setting the stage for us to explore the rest of stellar life cycles in later sections of this chapter.

◆ How do stars form?

A star is born when gravity causes a cloud of interstellar gas to contract to the point at which the central object becomes hot enough to sustain nuclear fusion in its core. However, gravity does not always succeed in making an interstellar cloud contract, because the cloud's internal gas pressure can resist gravity.

Recall that our Sun remains stable in size because the inward pull of gravity is balanced by the outward push of gas pressure, a balance that we call *gravitational equilibrium* [Section 11.1]. The pressure is far weaker in the low-density gas of interstellar space, but so is the pull of gravity. In most places within our galaxy, gravity is not strong enough to overcome the internal pressure of interstellar gas, which is why star formation does not occur everywhere.

If you think about it, you'll realize that two things can help gravity win out over pressure and start the collapse of a cloud of gas: (1) higher density, because packing the gas particles closer together makes the gravitational forces between them stronger, and (2) lower temperature, because lowering a cloud's temperature reduces the gas pressure. We therefore expect star-forming clouds to be colder and denser than most other interstellar gas.

Stars are born in cold, dense clouds of gas whose pressure cannot resist gravitational contraction.

Observations confirm the idea that stars are born within the coldest and densest clouds, usually called **molecular clouds** because they are cold enough and dense enough to allow atoms to combine and form

A star-forming molecular cloud in the constellation Scorpius. The region pictured here is about 50 light-years across.

Newborn stars produce white patches in the cloud where starlight illuminates surrounding gas.

The cloud looks dark where dust particles block the light from more distant stars.

molecules (Figure 13.1). Typical molecular cloud temperatures are only 10–30 K. (Recall that 0 K is absolute zero; see Figure 4.10.) And while their densities are low enough that they would qualify as superb vacuums by earthly standards, molecular clouds are hundreds to thousands of times as dense as other regions of interstellar space. The molecular clouds that give birth to stars tend to be quite large, because more total mass also helps gravity overcome gas pressure. A typical star-forming cloud is thousands of times as massive as a typical star, and can give birth to many stars at the same time. That is why stars are usually born in clusters.

From Cloud to Protostar Once a large molecular cloud begins to collapse, gravity pulls the gas toward the cloud's densest regions, causing it to fragment into smaller pieces that each form one or more new stars (Figure 13.2). Each shrinking cloud fragment heats up as it contracts.

▼ **FIGURE 13.2**

Computer simulation of a fragmenting molecular cloud. Gravity attracts matter to the densest regions of a molecular cloud. If gravity can overcome pressure in these dense regions, they collapse to form even denser knots of gas. Each of these knots can form one or more new stars.

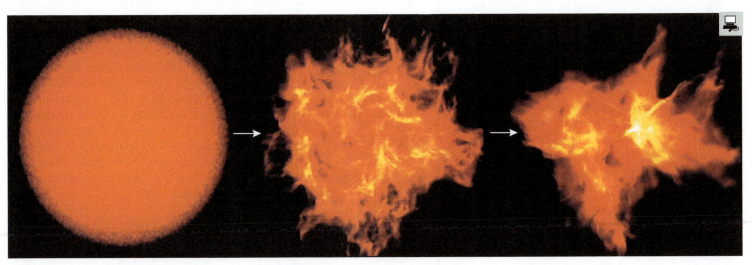

a The simulation begins with a turbulent gas cloud 1.2 light-years across, containing $50 M_{Sun}$ of gas.

b Random motions in the cloud cause it to become lumpy, with some regions denser than others. If gravity can overcome pressure in these dense regions, they can collapse to form even denser lumps of matter.

c The large cloud therefore fragments into many smaller lumps of matter, and each lump can go on to form one or more new stars.

a This visible-light image from the Hubble Space Telescope shows part of the Eagle Nebula, a gas cloud in which stars are currently forming.

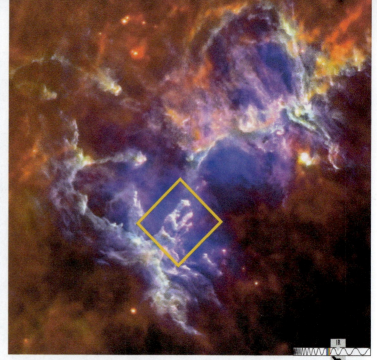

b This image from the Herschel Space Telescope shows infrared light from the Eagle Nebula, with the portion shown in part a outlined by the yellow square. Notice that the dark clouds in the visible-light image are glowing in the infrared image.

▲ **FIGURE 13.3**

Star-forming molecular clouds are cooler than stars and therefore emit infrared rather than visible light.

cosmic calculations 13.1

Conditions for Star Birth

As discussed in the text, gravity can win out over pressure to start the collapse of an interstellar cloud only when the gas temperature (T) is relatively low and the gas density (n) is relatively high. Exactly how cold and dense the cloud must be depends on the total amount of matter it contains: The more massive the cloud, the more easily gravity can win out. Astronomers have found a simple formula that tells us how massive a cloud must be for gravity to start its collapse:

$$M_{minimum} = 18 M_{Sun} \sqrt{\frac{T^3}{n}}$$

where T is the gas temperature in Kelvin and n is the gas density in particles per cubic centimeter. $M_{minimum}$ is the minimum mass for making stars: Gravity can make a cloud collapse and give birth to stars only if the cloud's total mass is *greater* than $M_{minimum}$.

Example: A cloud has a temperature of 30 K and an average density of 300 particles per cubic centimeter. What mass must this cloud have in order to form stars?

Solution: We insert the values $T = 30$ K and $n = 300$ particles per cubic centimeter into the $M_{minimum}$ formula:

$$M_{minimum} = 18 M_{Sun} \sqrt{\frac{30^3}{300}} = 18 M_{Sun} \sqrt{90} \approx 171 M_{Sun}$$

Gravity can cause this cloud to collapse and form stars only if the cloud's mass is greater than about $171 M_{Sun}$.

The source of this heat is the gravitational potential energy [Section 4.3] released as gravity pulls each part of the cloud fragment closer to the center of the fragment. Early in the process of star formation, the contracting gas quickly radiates away much of this energy, preventing the temperature and pressure from building enough to resist gravity. The temperature of the cloud remains below 100 K, so it glows in long-wavelength infrared light (Figure 13.3).

As the cloud continues to contract, the growing density makes it increasingly difficult for radiation to escape, especially from the center. Once the central region of the cloud fragment becomes dense enough to trap infrared radiation so that it can no longer radiate away its heat, the central temperature and pressure begin to rise dramatically. The rising pressure pushes back against the crush of gravity, slowing the contraction. The dense center of the cloud fragment is now a **protostar**—the clump of gas that will become a new star. Gas from the surrounding cloud continues to rain down on the protostar, increasing its mass. Protostars are not yet true stars because their cores are not hot enough for nuclear fusion. Nevertheless, they can be quite luminous because of all the energy released by the infalling matter.

A molecular cloud fragment heats up as gravity makes it contract, producing a protostar at its center.

Disks and Jets   Protostars rotate rapidly. Random motions of gas particles inevitably give a gas cloud some overall rotation, although it may be imperceptibly slow. Like an ice skater pulling in her arms, a shrinking cloud rotates increasingly fast in order to keep its total angular momentum the same [Section 4.3].

The rotation of the shrinking cloud fragment also produces a spinning disk of gas around the protostar (Figure 13.4). A disk forms because collisions between gas particles cause the cloud fragment to flatten along its rotation axis [Section 6.3], and the disk keeps spinning because angular momentum must be conserved. Planets may later form in the disk.

Observations show that many young protostars fire high-speed streams of gas, or **jets,** into interstellar space (Figure 13.5). We generally see two jets, shooting in opposite directions along the protostar's rotation axis. Sometimes the jets are lined with glowing blobs of gas, which are presumably clumps of matter swept up as the jets plow into the surrounding interstellar material.

Conservation of angular momentum ensures that protostars rotate rapidly and are surrounded by spinning disks of gas.

No one knows exactly how protostars generate these jets, but magnetic fields probably play an important role. A protostar's rapid rotation helps it generate a strong magnetic field, and this field may channel the jets along the rotation axis. In addition, the strong magnetic field helps the protostar generate a strong *protostellar wind*—an outward flow of particles similar to the *solar wind* [Section 11.1] but much stronger. Together, winds and jets probably help clear away the cocoon of gas that surrounds a forming star, revealing the protostar within. They also help the protostar shed some of its angular momentum by carrying material off into interstellar space, causing the protostar's rotation to slow down.

Single Star or Binary?   Angular momentum is also part of the reason so many stars belong to binary star systems. As a molecular cloud contracts, breaks up into fragments, and forms protostars, some of those protostars end up quite close to one another. Gravity can then pull two neighboring protostars closer together, but they don't crash into each other. Instead, they go into orbit around each other, because each pair of protostars has a certain amount of angular momentum. Pairs with large amounts of angular momentum have large orbits, and those with smaller amounts orbit closer together. If the two stars end up quite close to each other, they form what we call a **close binary** system. Stars in a close binary are typically separated by less than 0.1 AU and orbit each other every few days.

Neighboring protostars sometimes end up orbiting each other in binary star systems.

From Protostar to the Main Sequence   A protostar becomes a true star when its core temperature reaches 10 million K, hot enough for hydrogen fusion to operate efficiently. The core temperature continues to rise until fusion in the core generates enough energy to balance the energy lost from the surface in the form of radiation. Gravitational contraction then stops, because the star has achieved energy balance. It is now a main-sequence star [Section 12.2].

The length of time from the formation of a protostar to the birth of a main-sequence star depends on the star's mass. Massive stars do everything faster. The contraction of a high-mass protostar into a main-sequence star of spectral type O or B may take only a million years or less. A star like our Sun takes about 30 million years from the beginning of the protostellar stage to become a main-sequence star. A very-low-mass star of spectral type M may spend more than 100 million years as a protostar. The most massive stars in a young star cluster therefore may live and die before the smallest stars even begin to fuse hydrogen in their cores.

A protostar becomes a main-sequence star when it achieves energy balance between hydrogen fusion in its core and radiation from its surface.

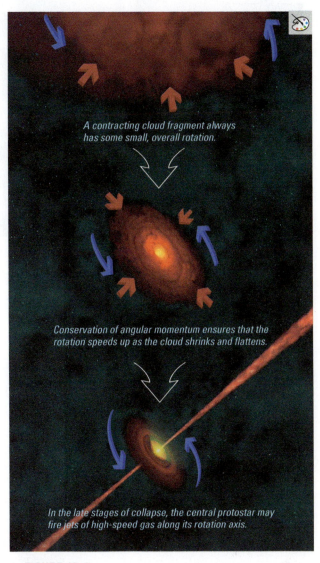

A contracting cloud fragment always has some small, overall rotation.

Conservation of angular momentum ensures that the rotation speeds up as the cloud shrinks and flattens.

In the late stages of collapse, the central protostar may fire jets of high-speed gas along its rotation axis.

▲ **FIGURE 13.4**
Artist's conception of star birth.

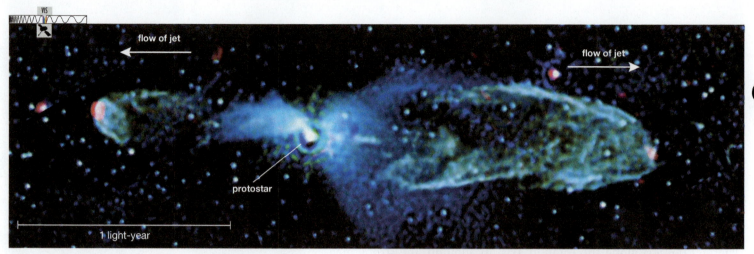

a This photograph shows two jets of material being shot in opposite directions by a protostar. The structures to the left and right of the protostar are formed as the jet material rams into surrounding interstellar gas.

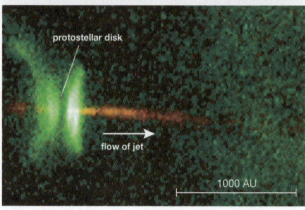

b This photograph shows a close-up view of a jet (red) and a disk of gas (green) around a protostar. We are seeing the disk nearly edge-on. The top and bottom surfaces of the disk are glowing, but we cannot see the darker middle layers of the disk.

▲ **FIGURE 13.5**
These photos show jets of gas shot from protostars into interstellar space.

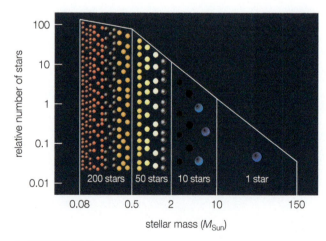

▲ **FIGURE 13.6**
Demographics of newborn stars. This schematic diagram shows how many stars in each mass range are produced for every star greater than $10M_{Sun}$ in an episode of star formation. Very massive stars are relatively rare; lower-mass stars are much more common.

◆ How massive are newborn stars?

The masses of newborn stars seem to depend on the processes that govern clumping and fragmentation in star-forming clouds. We do not yet fully understand those processes, but we can observe the results.

Observations of young star clusters show that stars with low masses greatly outnumber stars with high masses. For every star with a mass above $10M_{Sun}$, there are typically 10 stars with masses between $2M_{Sun}$ and $10M_{Sun}$, 50 stars with masses between $0.5M_{Sun}$ and $2M_{Sun}$, and a few hundred stars with masses below $0.5M_{Sun}$ (Figure 13.6). Note that although the Sun lies toward the middle of the overall range of stellar masses, most stars in a new star cluster are less massive than the Sun. With the passage of time, the balance tilts even more in favor of the low-mass stars, because high-mass stars live short lives and die off early.

Limits on Stellar Masses There are limits on both the minimum and the maximum possible masses of stars. On the high-mass end, observations indicate that the maximum mass is around $150M_{Sun}$, though it might be higher. In any event, theoretical models suggest that stars with masses above about $100M_{Sun}$ should have such enormous energy output from their cores that they will ultimately drive their outer layers into space.

a Artist's conception of a brown dwarf, orbited by a planet (to its left) in a system with multiple stars. The reddish color approximates how a brown dwarf would appear to human eyes. The bands are shown because we expect brown dwarfs to look more like giant jovian planets than stars.

b An infrared image showing brown dwarfs (circled) in the constellation Orion. They are easier to spot in star-forming regions like this one than elsewhere in our galaxy, because young brown dwarfs still have much of the thermal energy left by the process of gravitational contraction. They therefore emit measurable amounts of infrared light.

▲ **FIGURE 13.7**
Brown dwarfs are "failed stars" that have masses lower than the $0.08M_{Sun}$ required to sustain core fusion.

On the low-mass end, calculations show that a protostar cannot become a true star unless its mass is at least $0.08M_{Sun}$, which is equivalent to about 80 times the mass of Jupiter. Below this mass, the protostar's gravity will be too weak to contract the core enough to reach the 10 million K threshold for efficient hydrogen fusion. In that case, the protostar will stabilize as a sort of "failed star" known as a **brown dwarf.** Brown dwarfs radiate primarily in the infrared and actually look deep red or magenta in color rather than brown (Figure 13.7). Because brown dwarfs do not sustain steady fusion in their cores, they cool with time as they radiate away their internal thermal energy. In essence, brown dwarfs occupy a fuzzy gap between what we call a planet and what we call a star.

> Stars more massive than about $100M_{Sun}$ blow off their outer layers, while protostars smaller than $0.08M_{Sun}$ become *brown dwarfs* that never get hot enough for efficient hydrogen fusion.

Pressure in Brown Dwarfs The source of the pressure that stops gravity from squeezing a brown dwarf's core to the point at which it could sustain fusion is quite different from the type of pressure we encounter in most other situations. Ordinary gas pressure is often called **thermal pressure** because it is closely linked to temperature: Raising the temperature increases the particle speeds and thereby raises the thermal pressure. However, calculations show that thermal pressure is insufficient to hold off gravity in a brown dwarf. Instead, the crush of gravity in a brown dwarf is halted by **degeneracy pressure,** a type of pressure that does not depend on temperature at all. It depends instead on the laws of *quantum mechanics* that also give rise to distinct energy levels in atoms [Section 5.2].

In much the same way that electrons in atoms are restricted to occupying only particular energy levels, quantum mechanics places restrictions

a When there are many more available places (chairs) than particles (people), a particle is unlikely to try to occupy the same place as another particle. The only pressure comes from the temperature-related motion of the particles.

b When the number of particles (people) approaches the number of available places (chairs), finding an available place requires that the particles move faster than they would otherwise. The extra motion creates degeneracy pressure.

▲ **FIGURE 13.8**

The auditorium analogy for degeneracy pressure. Chairs represent available places (quantum states) for electrons, and people who must keep moving from chair to chair represent electrons.

on how closely together electrons (and some other subatomic particles) can be packed in a gas. Under most circumstances, these restrictions have little effect on the motions or locations of the electrons, and hence little effect on the pressure. However, in a protostar with a mass below $0.08M_{Sun}$, the electrons become packed closely enough for these restrictions to matter. We can see how this fact leads to degeneracy pressure with a simple analogy.

Imagine an auditorium in which the laws of quantum mechanics dictate the spacing between chairs and people represent electrons (Figure 13.8). As though playing a game of musical chairs, the people are always moving from seat to seat, just as electrons must remain constantly in motion. Most objects are like auditoriums with many more available chairs than people, so the people (electrons) can easily find chairs as they move about. However, the cores of protostars with masses below $0.08M_{Sun}$ are like much smaller auditoriums with so few chairs that the people (electrons) fill nearly all of them. Because there are virtually no open seats, the people (electrons) can't all squeeze into a smaller section of the auditorium. This resistance to squeezing is the origin of degeneracy pressure. If the people were really like electrons, the laws of quantum mechanics would also require them to move faster and faster to find open seats as they were squeezed into a smaller section. However, their speeds would have nothing to do with temperature. Degeneracy pressure and the particle motion that goes with it arise *only* because of the restrictions on where the particles can go, which is why temperature does not affect them.

> Brown dwarfs are supported against gravity by *degeneracy pressure*, which does not weaken with decreasing temperature.

The fact that degeneracy pressure does not rise and fall with temperature means that a brown dwarf's interior pressure remains stable even as it cools with time. As a result, gravity can never gain the upper hand and cause a brown dwarf to contract further, no matter how much the brown dwarf cools. In the constant battle of any "star" to resist the crush of gravity, brown dwarfs are winners, albeit dim ones. As we'll see in Chapter 14, degeneracy is also very important in the stellar corpses known as white dwarfs and neutron stars.

13.2 Life as a Low-Mass Star

A star's mass is its most important property. It determines not only a star's luminosity and surface temperature during its main-sequence (hydrogen core fusion) life [Section 12.2], but also its main-sequence lifetime and its ultimate fate. Stars of similar birth mass therefore lead similar lives and die in similar ways. However, the life stories of stars on the low end of the overall range of stellar masses are quite different from those of stars on the high end. As a result, we can simplify our discussion of stellar lives by dividing stars into three basic groups by mass:

- **Low-mass stars** are stars born with less than about 2 solar masses ($2M_{Sun}$) of material.

- **Intermediate-mass stars** have birth masses between about 2 and 8 solar masses.

- **High-mass stars** are those stars born with masses greater than about 8 solar masses.

We will focus primarily on the dramatic differences between the lives of low- and high-mass stars, beginning in this section with low-mass stars. The life stages of intermediate-mass stars are quite similar to the corresponding stages of high-mass stars until the very ends of their lives, so we include them in our discussion of high-mass stars.

You may wonder how we can be confident that we know the life stories of stars. Our confidence comes from a combination of theoretical modeling and observations. Astronomers use the known laws of physics to create mathematical models of stellar interiors, just as we use models to determine what is going on inside the Sun [Section 11.2]. These models predict both what is happening in a star at any stage of its life and how the star must change as its life continues. The models also predict what a star should look like from the outside at various stages of its life. By observing stars of different ages, we can test whether our models are successful in their predictions. The close correspondence between these models and observations gives us great confidence that we really do understand the lives of stars.

◆ What are the life stages of a low-mass star?

All low-mass stars go through life stages similar to those of the low-mass star that we know as our Sun. Let's therefore take our Sun as an example of a low-mass star, investigating how it will change over the course of its life.

Main-Sequence Stage In the grand hierarchy of stars, our Sun ranks as rather mediocre. We should be thankful for this mediocrity: If the Sun had been a high-mass star, it would have lasted only a few million years, dying before life could have arisen on Earth. Instead, the Sun has shone steadily for nearly 5 billion years as a main-sequence star, providing the light and heat that have allowed life to thrive on our planet. Its main-sequence life will continue for another 5 billion years or so.

As we discussed in Chapter 11, our Sun fuses hydrogen into helium in its core via the *proton–proton chain*. The Sun shines steadily because *gravita-tional equilibrium* (see Figure 11.2) and *energy balance* (see Figure 11.3) work together as a self-regulating *solar thermostat* that keeps the Sun's fusion rate and overall luminosity quite steady. Steady hydrogen fusion in the core is the basic characteristic of a star's main-sequence stage, which occupies about 90% of any star's total lifetime.

> Stars spend about 90% of their lives shining steadily as main-sequence stars.

Other low-mass stars have similarly long lives, following the usual rule: Stars with mass greater than the Sun have main-sequence lifetimes shorter than the Sun's 10 billion years, and stars with mass lower than the Sun have longer lifetimes.

Red Giant Stage Hydrogen fusion supplies the thermal energy that keeps a main-sequence star in balance. But when the Sun's core hydrogen is finally depleted, nuclear fusion will cease. With no fusion to replace the energy the star radiates from its surface, the core will no longer be able to resist the inward pull of gravity, and it will begin to shrink. After 10 billion years of shining steadily, the Sun will enter an entirely new phase of life.

Somewhat surprisingly, the Sun's outer layers will expand outward during this new life stage, even though its core will be shrinking under the crush of gravity. Over a period of about a billion years—or about

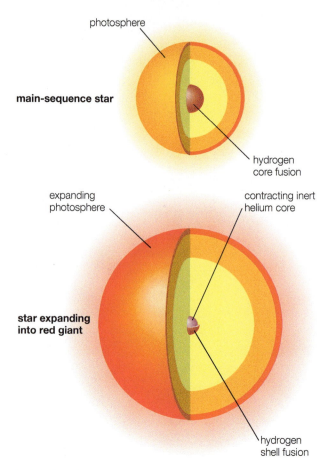

photosphere

main-sequence star

hydrogen
core fusion

expanding
photosphere

contracting inert
helium core

**star expanding
into red giant**

hydrogen
shell fusion

▲ **FIGURE 13.9**

After a star ends its main-sequence life, its inert helium core contracts while hydrogen shell fusion begins. The high rate of fusion in the hydrogen shell forces the star's upper layers to expand outward. (Not to scale.)

10% as long as its main-sequence lifetime—the Sun will grow in size and luminosity to become a **red giant.** At the peak of its red giant stage, the Sun will be more than 100 times as large in radius and more than 1000 times as luminous as it is today.

To understand why the Sun's outer layers will expand even while its core is shrinking, we need to think about the composition of the core at the end of the Sun's main-sequence life. After the core exhausts its hydrogen, it will be made almost entirely of helium, because helium is the product of hydrogen fusion. However, the gas surrounding the core will still contain hydrogen that has never previously undergone fusion. Because gravity will shrink both the *inert* (non-fusing) helium core and the surrounding *shell* of hydrogen, the hydrogen shell will soon become hot enough for **hydrogen shell fusion**—hydrogen fusion in a shell around the core (Figure 13.9). In fact, the shell will become so hot that hydrogen shell fusion will proceed at a much higher rate than core hydrogen fusion does today. The higher fusion rate will generate enough energy to dramatically increase the Sun's luminosity and enough pressure to push the surrounding layers of gas outward. This expansion will also weaken the pull of gravity at the surface, allowing large amounts of mass to escape in a *stellar wind.* Observations of winds from red giants show that they carry away much more matter than the solar wind carries away from the Sun today.

After exhausting its core hydrogen, the Sun will expand to become a red giant, powered by rapid hydrogen fusion in a shell surrounding the core.

The situation will grow more extreme as long as the helium core remains inert. Today, the self-correcting feedback process of the solar thermostat regulates the Sun's fusion rate: A rise in the fusion rate causes the core to inflate and cool until the fusion rate drops back down (see Figure 11.8). In contrast, thermal energy generated in the hydrogen-fusing shell of a red giant cannot do anything to inflate the inert core that lies underneath. Instead, newly produced helium keeps adding to the mass of the helium core, amplifying its gravitational pull and shrinking it further. The hydrogen-fusing shell shrinks along with the core, growing hotter and denser. The fusion rate in the shell consequently rises, feeding even more helium to the core. The star is caught in a vicious circle with a broken thermostat.

The core and shell will continue to contract and heat up—with the Sun as a whole continuing to grow larger and more luminous—until the temperature in the inert helium core reaches about 100 million K. At that point, it will be hot enough for helium nuclei to begin to fuse together, and the Sun will enter the next stage of its life. Other low-mass stars expand into red giants in the same way, though the time required depends on the star's mass: Like all other phases of stellar lives, the process occurs faster for more massive stars and more slowly for less massive stars. In some very-low-mass stars, the inert helium core may never become hot enough to fuse helium. In that case, the core collapse will be halted by degeneracy pressure, leaving the star's corpse as a *helium white dwarf.*

The Sun's core will continue to shrink and hydrogen shell fusion will continue to intensify as the Sun grows into a red giant.

think about it Before you read on, briefly summarize why a star grows larger and brighter after it exhausts its core hydrogen. When does the growth of a red giant finally halt, and why? How would a star's red giant stage be different if the temperature required for helium fusion were around 200 million K, rather than 100 million K? Why?

Helium Fusion Recall that fusion occurs only when two nuclei come close enough together for the attractive *strong force* to overcome electromagnetic repulsion [Section 11.2]. Helium nuclei have two protons (and two neutrons) and hence a greater positive charge than a hydrogen nucleus with its single proton. The greater charge means that helium nuclei repel one another more strongly than hydrogen nuclei. **Helium fusion** therefore occurs only when nuclei slam into one another at much higher speeds than those needed for hydrogen fusion, which means that helium fusion requires much higher temperatures.

The helium fusion process (often called the *triple-alpha reaction* because helium nuclei are sometimes called *alpha particles*) converts three helium nuclei into one carbon nucleus:

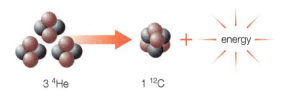

3 ⁴He   1 ¹²C

Energy is released because the carbon-12 nucleus has a slightly lower mass than the three helium-4 nuclei, and the lost mass becomes energy in accord with $E = mc^2$.

The ignition of helium fusion in a low-mass star like the Sun has one subtlety. Theoretical models show that the thermal pressure in the inert helium core is too low to counteract gravity. Instead, the pressure fighting against gravity is *degeneracy pressure*—the same type of pressure that supports brown dwarfs. Because degeneracy pressure does *not* increase with temperature, the onset of helium fusion heats the core rapidly without causing it to expand in size, at least at first. Instead, the rising temperature causes the helium fusion rate to spike drastically in what is called a **helium flash.**

The helium flash releases an enormous amount of energy into the core. In a matter of seconds, the temperature rises so much that thermal pressure again becomes dominant and degeneracy pressure is no longer important. In fact, the thermal pressure becomes strong enough to push back against gravity. The core begins to expand and pushes the hydrogen-fusing shell outward, lowering its temperature and its fusion rate. The result is that, even though helium core fusion and hydrogen shell fusion are taking place simultaneously in the star (Figure 13.10), energy production falls from the peak it reached during the red giant stage, reducing the star's luminosity and allowing its outer layers to contract somewhat. As the outer layers contract, the star's surface temperature increases, so its color turns back toward yellow from red. After having spent about a billion years expanding into a luminous red giant, the Sun will decline in size and luminosity as it becomes a *helium core-fusion star.*

The sudden onset of helium fusion in the Sun's core will stop core shrinkage, and the Sun will become smaller and less luminous than it was as a red giant.

We can see examples of low-mass stars in all the life stages we have discussed so far in the H-R diagram of a globular cluster (Figure 13.11). Stars along the lower right, below the main-sequence turnoff point, are still in their hydrogen-fusing, main-sequence stage. Just above and to the right of the main-sequence turnoff point we see *subgiants*—stars that have just begun their expansion into red giants as their cores have shut down and hydrogen shell fusion has begun. The longer a star undergoes

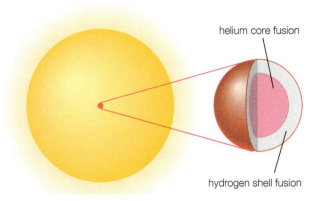

▲ **FIGURE 13.10**
Core structure of a helium core-fusion star. Helium fusion causes the helium core and hydrogen-fusing shell to expand and slightly cool, thereby reducing the overall energy generation rate in comparison to the rate during the red giant stage. The outer layers shrink, so a helium core-fusion star is smaller than a red giant of the same mass.

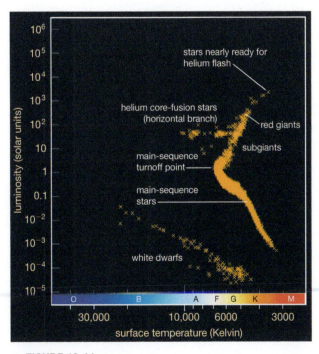

▲ **FIGURE 13.11**
An H-R diagram of a globular cluster showing low-mass stars in several different life stages.

hydrogen shell fusion, the larger and more luminous it becomes, which is why we see a continuous line of stars right up to the most luminous red giants. These are the red giants on the verge of a helium flash. The stars that have already undergone the helium flash and become helium core-fusion stars appear below and to the left of the red giants, because they are somewhat smaller, hotter, and less luminous than they were at the moment of helium flash. Because these helium core-fusion stars all have about the same luminosity but can differ in surface temperature, they trace out a horizontal line on the H-R diagram known as the *horizontal branch*.

◆ How does a low-mass star die?

It is only a matter of time until a helium core-fusion star converts all its core helium into carbon. In the Sun, the core helium will run out after about 100 million years—only about 1% as long as the Sun's 10-billion-year main-sequence lifetime. When the core helium is exhausted, fusion will again cease. The core, now made of the carbon produced by helium fusion, will begin to shrink once more under the crush of gravity.

Last Gasps The exhaustion of core helium will cause the Sun to expand once again, just as it did when it became a red giant. This time, the trigger for the expansion will be helium fusion in a shell around the inert carbon core. Meanwhile, hydrogen fusion will continue in a shell atop the helium layer. The Sun will have become a *double shell–fusion star*. Both shells will contract along with the inert core, driving their temperatures and fusion rates so high that the Sun will expand to an even greater size and luminosity than in its first red giant stage.

The furious fusion rates in the helium and hydrogen shells cannot last for more than a few million years. The Sun's only hope of extending its life at this stage lies with its carbon core, but this is a false hope for a low-mass star like the Sun. Carbon fusion is possible only at temperatures above about 600 million K, and degeneracy pressure will halt the collapse of the Sun's core before it ever gets that hot. With the carbon core unable to undergo fusion and provide a new source of energy, the Sun will finally have reached the end of its life.

Core shrinkage will resume after helium core fusion ends, while both helium-fusing and hydrogen-fusing shells make the Sun bigger and more luminous than ever.

think about it Suppose the universe contained only low-mass stars. Would elements heavier than carbon exist? Why or why not?

The huge size of the dying Sun will give it only a very weak grip on its outer layers. As the Sun's luminosity and radius keep rising, its wind will grow stronger. Observations of other stars in this late stage of life show that their winds are an important source of the *interstellar dust grains* found in star-forming clouds. These dust grains form because the wind cools as it flows away from the star. At the point where the gas temperature has dropped to between about 1000 and 2000 K, some of the heavier elements in the wind begin to condense into microscopic clusters, forming small, solid particles of dust. This process of dust formation is much like the condensation that occurred in the solar nebula before the planets formed [Section 6.3]. The dust particles drift with the stellar wind into interstellar space, where they mix with other gas and dust in the galaxy.

Planetary Nebula   The Sun's final end will be beautiful to those who witness it, as long as they stay far away. Through winds and other processes, the Sun will eject its outer layers into space, creating a huge shell of gas expanding away from the inert carbon core. The exposed core will still be very hot and will therefore emit intense ultraviolet radiation. This radiation will ionize the gas in the expanding shell, making it glow brightly as a **planetary nebula.** Figure 13.12 shows two of many known examples of planetary nebulae around other low-mass stars that have recently died in this way. Note that, despite their name, planetary nebulae have nothing to do with planets. The name comes from the fact that nearby planetary nebulae look much like planets through small telescopes, appearing as simple disks.

When it dies, the Sun will eject its outer layers into space as a planetary nebula, leaving its exposed core behind as a white dwarf.

The glow of the Sun's planetary nebula will fade as the exposed core cools and the ejected gas disperses into space. The nebula will disappear within about 100,000 years, leaving the carbon core behind as a *white dwarf.* Recall from Chapter 12 that white dwarfs are small in radius and often quite hot. We can now understand why: They are small in radius because they are the exposed cores of dead stars, supported against the crush of gravity by degeneracy pressure. They are often hot because some of them were only recently the centers of stars, though over time they will cool and fade from view.

The Fate of Earth   The death of the Sun will obviously have consequences for Earth, and some of these consequences will begin even before the Sun enters the final stages of its life. Although the Sun will shine steadily for the next 5 billion years of its hydrogen core-fusion life, theoretical models show that it will become slightly more luminous with time. This gradual rise in luminosity will be small compared to what will happen during the red giant stage, but it will probably be enough to cause a *runaway greenhouse effect* [Section 7.4] on Earth somewhere between about 1 and 4 billion years from now, making Earth's oceans boil away.

Temperatures on Earth will rise even more dramatically when the Sun finally exhausts its core supply of hydrogen, somewhere around the year A.D. 5,000,000,000, and conditions will become even worse as the Sun grows into a red giant over the next several hundred million years. Just before the helium flash, the Sun will be more than 1000 times as luminous as it is today, and this huge luminosity will heat Earth's surface to more than 1000 K. Clearly, any surviving humans will need to have found a new home.

The Sun will shrink and cool somewhat after the helium flash turns it into a helium core-fusion star, providing a temporary lull in Earth's incineration. However, this respite will last only 100 million years or so, and then Earth will suffer one final disaster. After exhausting its core helium, the Sun will expand again during its last million years. Its luminosity will soar to thousands of times what it is today, and its radius will grow to nearly the present radius of Earth's orbit—so large that solar prominences might lap at Earth's surface. Finally, the Sun will eject its outer layers as a planetary nebula that will drift away into interstellar space. If Earth is not destroyed, its charred surface will be

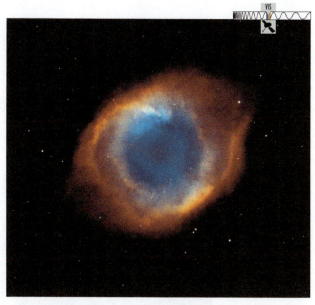

a Helix Nebula. The central white dot is the hot white dwarf.

b Butterfly Nebula. The hot white dwarf is hidden in the dark ring of dust at the center.

▲ **FIGURE 13.12**
Hubble Space Telescope photos of planetary nebulae, which form when low-mass stars in their final death throes cast off their outer layers of gas, leaving behind the hot core that ejected the gas. The hot core ionizes and energizes the shells of gas that surround it. As the gas disperses into space, the hot core remains as a white dwarf.

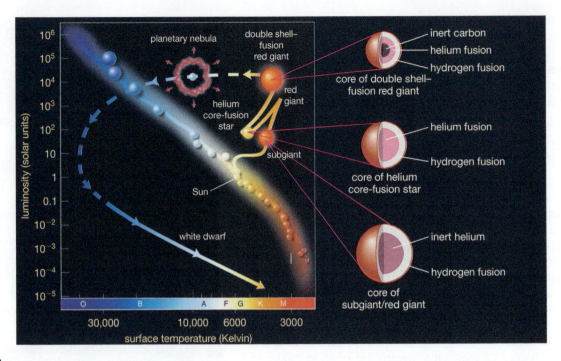

▲ **FIGURE 13.13**

The life track of our Sun from its main-sequence stage to the white dwarf stage. Core structure is shown at key stages.

cold and dark in the faint, fading light of the white dwarf that the Sun will have become.

The Sun's Life on an H-R Diagram

We have discussed the Sun's life stages in general terms, but astronomers can determine with computer models how the Sun's surface temperature and luminosity will change during each stage of its life. We can plot the results on an H-R diagram, producing a **life track** (also called an *evolutionary track*) that shows the Sun's luminosity and surface temperature at each point in its life (Figure 13.13).

A star's life track on an H-R diagram shows how its properties change with time.

Remember that, although the life track cannot show it, the Sun spends most of its life as a main-sequence star fusing hydrogen in its core. The life track shows what happens once the Sun exhausts its core hydrogen supply and is no longer a main-sequence star. The life track goes generally upward as the Sun becomes a subgiant and then a red giant, because the Sun expands in size and luminosity while it undergoes hydrogen shell fusion. The track also goes slightly to the right during this stage, because the Sun's surface temperature falls a bit. The tip of the red giant stage represents the moment of helium flash, after which the Sun contracts in size, shown by the track going down and left, as the Sun becomes a helium core-fusion star. The life track again turns upward as the Sun enters its second red giant phase, this time with energy generated by fusion in shells of both helium and hydrogen. As the Sun ejects its planetary nebula, the dashed curve indicates that we are shifting from plotting the surface temperature of a red giant to plotting the surface temperature of the exposed stellar core left behind. The curve becomes solid again near the lower left, indicating that this core is a hot white dwarf. From that point, the curve continues downward and to the right as the white dwarf cools and fades.

13.3 Life as a High-Mass Star

Human life would be impossible without both low-mass stars and high-mass stars. The long lives of low-mass stars allow evolution to proceed for billions of years, but only high-mass stars produce the full array of elements on which life depends.

The early stages of a high-mass star's life are similar to the early stages of the Sun's life, except they proceed much more rapidly. But the late stages of life are quite different for high-mass stars. The cores of low-mass stars never become hot enough to fuse elements heavier than helium. Heavier nuclei contain more positively charged protons and therefore repel each other more strongly than lighter nuclei. As a result, these nuclei can fuse only at extremely high temperatures—temperatures that occur only in the cores of high-mass stars nearing the end of their lives, when the immense weight of their overlying layers bears down on a core that has already exhausted its hydrogen fuel.

The highest-mass stars proceed to fuse increasingly heavy elements until they have exhausted all possible fusion sources. When fusion finally stops for good, gravity causes the core to implode suddenly. As we will soon see, the implosion of the core causes the star to self-destruct in the titanic explosion we call a *supernova*. The fast-paced life and cataclysmic death of a high-mass star are surely among the great dramas of the universe.

◆ What are the life stages of a high-mass star?

Like all other stars, a high-mass star forms out of a cloud fragment that gravity forces to contract into a protostar. Just as in a low-mass star, hydrogen fusion begins when the gravitational potential energy released by the contracting protostar makes the core hot enough. However, hydrogen fusion proceeds through a different set of steps inside high-mass stars, which is part of the reason these stars live such brief but brilliant lives. As a specific example, let's investigate the life stages of a star born with 25 times the mass of the Sun ($25M_{Sun}$).

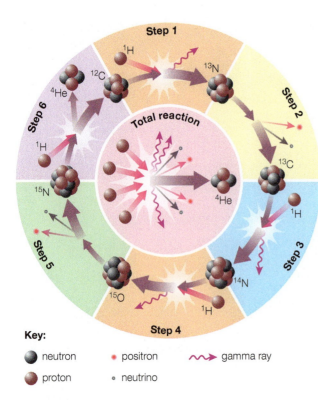

Step 1

Step 2

Step 3

Step 4

Step 5

Step 6

Total reaction

^{1}H

^{12}C

^{4}He

^{1}H

^{15}N

^{15}O

^{1}H

^{14}N

^{1}H

^{13}C

^{4}He

^{13}N

Key:

● neutron ● positron 〰 gamma ray

● proton ∘ neutrino

▲ **FIGURE 13.14**

This diagram illustrates the six steps of the CNO cycle by which massive stars fuse hydrogen into helium. Note that the overall result is the same as that of the proton–proton chain: Four hydrogen nuclei fuse to make one helium nucleus. The carbon, nitrogen, and oxygen nuclei help the overall cycle proceed, but these nuclei are neither consumed nor created.

Hydrogen Fusion in a High-Mass Star Recall that a low-mass star like our Sun fuses hydrogen into helium through the *proton–proton chain* (see Figure 11.7). In a high-mass star, gravitational compression makes the hydrogen core hotter than in lower-mass stars. The greater core temperature makes it possible for protons to slam into carbon, oxygen, or nitrogen nuclei as well as into other protons. As a result, hydrogen fusion in high-mass stars occurs through a chain of reactions called the **CNO cycle** (the letters *CNO* stand for carbon, nitrogen, and oxygen). Figure 13.14 shows the six steps of the CNO cycle.

Notice that the overall reaction of the CNO cycle is the same as that of the proton–proton chain: Four hydrogen nuclei fuse into one helium-4 nucleus. The amount of energy generated in each reaction cycle therefore is

A high-mass star lives a short life, rapidly fusing its core hydrogen into helium via the CNO cycle.

also the same—it is equal to the difference in mass between the four hydrogen nuclei and the one helium nucleus multiplied by c^2. However, hydrogen fusion proceeds much more rapidly through the CNO cycle than through the proton-proton chain in the cores of high-mass stars.

Becoming a Supergiant Our $25 M_{Sun}$ star will begin to run low on hydrogen fuel after only a few million years. As its core hydrogen runs out, the star responds much like a low-mass star, but much faster. It develops a hydrogen-fusing shell, and its outer layers begin to expand outward, ultimately turning it into a *supergiant*. At the same time, the core contracts, and this gravitational contraction releases energy that raises the core temperature until it becomes hot enough to fuse helium into carbon. However, there is no helium flash in high-mass stars. Their core temperatures are so high that thermal pressure remains strong, preventing degeneracy pressure from becoming a factor. Helium core fusion therefore ignites gradually, just as hydrogen core fusion did at the beginning of the star's life.

Our high-mass star fuses helium into carbon so rapidly that it is left with an inert carbon core after just a few hundred thousand years. Once again, the absence of fusion leaves the core without an energy source to fight off the crush of gravity. The inert carbon core shrinks, the crush of

Near the end of its life, a high-mass star expands to become a supergiant as fusion proceeds furiously in its core and surrounding shells.

gravity intensifies, and the core pressure, temperature, and density all rise. Meanwhile, a helium-fusing shell forms between the inert core and the hydrogen-fusing shell, and the star's outer layers swell further. Eventually, the core will become hot enough to fuse carbon into heavier elements, and the star will undergo another cycle of core fusion followed by shell fusion.

The high-mass star's outward appearance responds relatively slowly to these interior changes. As each stage of core fusion ceases, fusion in the surrounding shells intensifies and further inflates the star's outer layers. Each time the core flares up, the outer layers contract somewhat, but the star's overall luminosity remains about the same. The result is that the star's life track zigzags across the top of the H-R diagram (Figure 13.15). In the most massive stars, the core changes so quickly that the outer layers have virtually no time to respond, and the star progresses steadily toward becoming a red supergiant.

One of these massive, red supergiant stars happens to be the star Betelgeuse, the upper left shoulder of Orion. Its radius is about 900 solar radii, about four times the Sun–Earth distance, and it is a relatively nearby 600 light-years away. We have no way of knowing what stage of nuclear fusion is taking place in Betelgeuse's core as we currently see it in the sky. It may have thousands of years of nuclear fusion still ahead, or it may be on the verge of death. If the latter is the case, then sometime soon we may witness one of the most dramatic events that ever occurs in the universe—a supernova explosion. To understand why a supernova happens, we need to look more closely at how high-mass stars make elements heavier than carbon.

◆ How do high-mass stars make the elements necessary for life?

A low-mass star can't make elements heavier than carbon because degeneracy pressure halts the contraction of its inert carbon core before it can get hot enough for fusion. The life of an intermediate-mass star ($2M_{Sun}$–$8M_{Sun}$) ends the same way. A high-mass star has no such problem, because the crush of gravity in a high-mass star keeps its carbon core so hot that degeneracy pressure never comes into play. After helium fusion stops, the gravitational contraction of the carbon core continues until it reaches the 600 million K required to fuse carbon into heavier elements.

Carbon fusion provides the core with a new source of energy that restores gravitational equilibrium, but only temporarily. In a $25M_{Sun}$ star, carbon fusion may last only a few hundred years. When the core carbon has been depleted, the core again begins to collapse, shrinking and heating once more until it can fuse a still heavier element. The star is engaged in the final phases of a desperate battle against the ever-strengthening crush of gravity. The star will ultimately lose the battle, but it will be a victory for life in the universe: In the process of its struggle against gravity, the star will produce the heavy elements of which Earth-like planets and living things are made.

Fusion of Heavier Nuclei The nuclear reactions in a high-mass star's final stages of life become quite complex, and many different reactions may take place simultaneously. The simplest sequence of fusion stages occurs through successive **helium-capture reactions**—reactions in which a helium nucleus fuses into some other nucleus (Figure 13.16a). Helium capture can fuse carbon into oxygen, oxygen into neon, neon into magnesium, and so on. At high enough temperatures, a star's core can fuse heavy nuclei to one another. For example, fusing carbon to oxygen creates silicon, fusing two oxygen nuclei creates sulfur, and fusing two silicon nuclei generates iron (Figure 13.16b). Some of these heavy-element reactions release free neutrons, which may fuse with heavy nuclei to make still rarer elements. The star is forging the variety of elements that, on Earth at least, became the stuff of life.

Each time the core depletes the elements it is fusing, it shrinks and heats until it becomes hot enough for other fusion reactions. Meanwhile, a new type of shell fusion ignites between the core and the overlying shells. Near the end, the star's central region resembles the inside of an onion, with layer upon layer of shells, each fusing different elements (Figure 13.17). During the star's final few days, iron begins to pile up in the silicon-fusing core.

The core of a high-mass star eventually becomes hot enough for fusion to produce the elements of which we and Earth are made.

Iron: Bad News for the Stellar Core The core continues shrinking and heating while iron accumulates within it. If iron were like the elements in prior stages of nuclear fusion, this core contraction would stop when iron fusion ignited. However, iron is unique among the elements

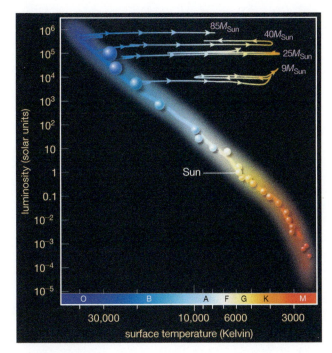

▲ **FIGURE 13.15**

Life tracks on the H-R diagram from main-sequence star to red supergiant for selected high-mass stars. Labels on the tracks give the star's mass at the beginning of its main-sequence life. (Based on models from A. Maeder and G. Meynet.)

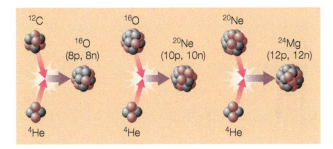

a Helium-capture reactions.

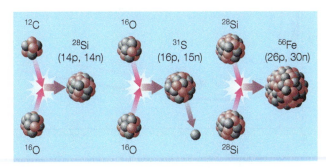

b Other reactions. (Note: Fusion of two silicon nuclei first produces nickel-56, which decays rapidly to cobalt-56 and then to iron-56.)

▲ **FIGURE 13.16**

A few of the many nuclear reactions that occur in the final stages of a high-mass star's life.

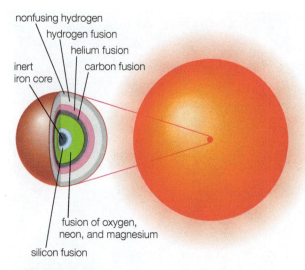

▲ FIGURE 13.17

The multiple layers of nuclear shell fusion in the core of a high-mass star during the final days of its life.

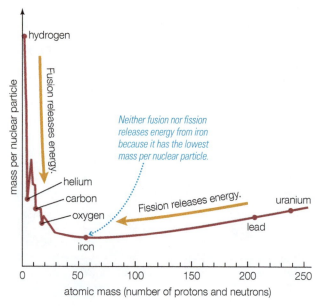

▲ FIGURE 13.18

Overall, the average mass per nuclear particle declines from hydrogen to iron and then increases. Selected nuclei are labeled to provide reference points. (This graph shows the most general trends only. A more detailed graph would show numerous up-and-down bumps superimposed on the general trends. The vertical scale is arbitrary, but shows the general idea.)

in a very important way: It is the one element from which it is *not* possible to generate any kind of nuclear energy.

To understand why iron is unique, remember that only two basic processes can release nuclear energy: *fusion* of light elements into heavier ones and *fission* of very heavy elements into not-so-heavy ones (see Figure 11.5). Recall that hydrogen fusion converts four protons (hydrogen nuclei) into a helium nucleus that consists of two protons and two neutrons. This fusion reaction generates energy (in accord with $E = mc^2$) because the *mass* of the helium nucleus is less than the combined mass of the four hydrogen nuclei that fused to create it, even though the total number of *nuclear particles* (protons and neutrons combined) does not change.

In other words, fusing hydrogen into helium generates energy because helium has a lower *mass per nuclear particle* than hydrogen. Similarly, fusing three helium-4 nuclei into one carbon-12 nucleus generates energy because carbon has a lower mass per nuclear particle than helium, which means that some mass disappears and becomes energy in this fusion reaction. The decrease in mass per nuclear particle from hydrogen to helium to carbon is part of a general trend shown in Figure 13.18.

The mass per nuclear particle tends to decrease as we go from light elements to iron, which means that fusion of light nuclei into heavier nuclei generates energy. This trend reverses beyond iron: The mass per nuclear particle tends to *increase* as we look to still heavier elements. As a result, elements heavier than iron can generate nuclear energy only through fission into lighter elements. For example, uranium has a greater mass per nuclear particle than lead, so uranium fission (which ultimately leaves lead as a by-product) must convert some mass into energy.

Iron has the lowest mass per nuclear particle of all nuclei and therefore cannot release energy by either fusion or fission. Once the matter in a stellar core turns to iron, it can generate no further energy. The core's only hope of resisting the crush of gravity lies with degeneracy pressure, but iron keeps piling up until even degeneracy pressure cannot support the core. What ensues is the ultimate nuclear waste catastrophe: The star explodes as a supernova, scattering all the newly made elements into interstellar space.

> A high-mass star's death is imminent when iron piles up in its core, because fusion of iron releases no energy.

think about it How would the universe be different if hydrogen, rather than iron, had the lowest mass per nuclear particle? Why?

Evidence for the Origin of Elements Before we look at how a supernova happens, let's consider the evidence that indicates we actually understand the origin of the elements. We cannot see inside stars, so we cannot directly observe elements being created. However, the signature of nuclear reactions in massive stars is written in the patterns of elemental abundances across the universe.

For example, if massive stars really produce heavy elements (that is, elements heavier than hydrogen and helium) and scatter these elements into space when they die, the total amount of these heavy elements in interstellar gas should gradually increase with time (because additional massive stars have died). We expect stars born recently to contain a greater proportion of heavy elements than stars born in the distant past, because the younger stars formed from interstellar gas that contained more heavy elements. Stellar spectra confirm this prediction: Elements besides hydrogen and helium can make up as little as 0.1% of the total mass of very old stars in globular clusters, while these elements can make up as much as 2–3% of the mass of young stars in open clusters.

We gain even more confidence in our model of elemental creation when we compare the abundances of various elements in the cosmos. Because helium-capture reactions add two protons (and two neutrons) at a time, we expect nuclei with even numbers of protons to outnumber those with odd numbers of protons that fall between them.

Measurements of element abundances in the cosmos confirm our models of how high-mass stars produce heavy elements.

Observations confirm that even-numbered nuclei such as carbon, oxygen, and neon have higher abundances than the elements in between them (Figure 13.19). Similarly, we expect elements heavier than iron to be extremely rare, because they are made primarily by rare fusion reactions (most of which occur shortly before or during a supernova). Again, observations verify this prediction.

◆ How does a high-mass star die?

Let's return now to our high-mass star, with iron piling up in its core. As we've discussed, it has no hope of generating any energy by fusing this iron. After shining brilliantly for a few million years, the star will not live to see another day.

The Supernova Explosion   The degeneracy pressure that briefly supports the inert iron core arises because the laws of quantum mechanics prohibit electrons from getting too close together. Once gravity pushes the electrons past the quantum mechanical limit, however, they can no longer exist freely. In an instant, the electrons disappear by combining with protons to form neutrons, releasing neutrinos in the process (Figure 13.20). The degeneracy pressure provided by the electrons instantly vanishes, and gravity has free rein.

In a fraction of a second, an iron core with a mass comparable to that of our Sun and a size larger than that of Earth collapses into a ball of neutrons just a few kilometers across. The collapse halts only because the neutrons have a degeneracy pressure of their own. The entire core then resembles a giant atomic nucleus. If you recall that ordinary atoms are made almost entirely of empty space [Section 5.1] and that almost all their mass is in their nuclei, you'll realize that a giant atomic nucleus must have an astoundingly high density.

The gravitational collapse of the core releases an enormous amount of energy—more than 100 times what the Sun will radiate over its entire 10-billion-year lifetime. Where does this energy go? It drives the star's outer layers into space in a titanic explosion called a **supernova.** The ball of neutrons left behind is called a **neutron star.** In some cases, the remaining mass may be so large that gravity also overcomes the degeneracy pressure associated with neutrons, and the core continues to collapse until it becomes a *black hole* [Section 14.3].

Theoretical models of supernovae successfully reproduce the observed energy outputs of real supernovae, but the precise mechanism of the explosion is not yet clear. Two general processes could contribute to the explosion. In the first process, degeneracy pressure halts the gravitational collapse by preventing neutrons from getting too close together and causes the core to rebound slightly and ram

When gravity overcomes degeneracy pressure in the iron core, the core collapses into a ball of neutrons and the star explodes in a supernova.

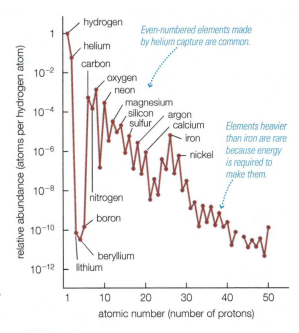

▲ **FIGURE 13.19**

This graph shows the observed relative abundances of elements in the galaxy compared to the abundance of hydrogen (given as 1). For example, the abundance of nitrogen is about 10^{-4}, or 1/10,000, which means that there are about 10,000 times as many hydrogen atoms in the galaxy as nitrogen atoms.

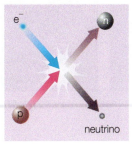

▲ **FIGURE 13.20**

During the final, catastrophic collapse of a high-mass stellar core, electrons and protons combine to form neutrons, accompanied by the release of neutrinos.

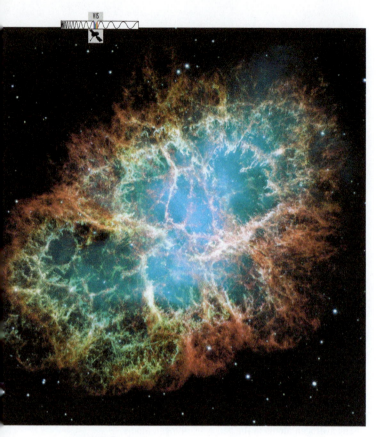

▲ **FIGURE 13.21**

The Crab Nebula is the remnant of the supernova observed in A.D. 1054.

Before	After

▲ **FIGURE 13.22**

Before and after photos of the location of Supernova 1987A. The arrow in the left-hand photograph points to the star observed to explode in 1987. The supernova actually appeared as a bright point of light; it appears larger than a point in the right-hand photograph only because of overexposure.

into overlying material that is still falling inward. However, current models of supernovae suggest that rebound also needs to be energized by the neutrinos formed when electrons and protons combine to make neutrons. Although neutrinos rarely interact with anything [Section 11.2], so many are produced when the core implodes that they propel the star's upper layers into space at a speed of 10,000 kilometers per second—fast enough to travel the distance from the Sun to Earth in only about 4 hours.

The heat of the explosion makes the gas shine with dazzling brilliance. For about a week, a supernova blazes as powerfully as 10 billion Suns, rivaling the luminosity of a moderate-size galaxy. The ejected gases slowly cool and fade in brightness over the next several months, continuing to expand outward until they eventually mix with other gases in interstellar space. The scattered debris from the supernova carries with it the variety of elements produced by nuclear fusion in the star, as well as additional elements created when some of the neutrons produced during the core collapse slammed into other nuclei. Millions or billions of years later, this debris may be incorporated into a new generation of stars. We are truly "star stuff," because we and our planet were built from the debris of stars that exploded long ago.

The supernova scatters the elements produced by the star into space and leaves behind either a neutron star or a black hole.

see it for yourself To see an effect similar to the core-bounce process in a supernova, find a tennis ball and a basketball. Then place the tennis ball directly on top of the basketball and drop them together on a hard floor. How does the speed at which the tennis ball bounces up compare to the speed at which it fell? How is the response of the tennis ball like the response of the supernova's outer layers to the rebound of the core?

Supernova Observations The study of supernovae owes a great debt to astronomers of many different epochs and cultures. Careful scrutiny of the night skies allowed ancient peoples to identify several supernovae whose remains can still be seen. The most famous example is the Crab Nebula in the constellation Taurus. The Crab Nebula is a **supernova remnant**—an expanding cloud of debris from a supernova explosion (Figure 13.21). A spinning neutron star lies at the center of the Crab Nebula, providing evidence that supernovae really do create neutron stars. Photographs taken years apart show that the nebula is growing larger at a rate of several thousand kilometers per second. Calculating backward from its present size, we can trace the nebula's birth to sometime near A.D. 1100. Chinese observers recorded a "guest star" near this location on July 4, 1054, undoubtedly the supernova that created this nebula.

No supernova has been seen in our own galaxy since 1604, but modern astronomers routinely discover supernovae in other galaxies. The nearest of these extragalactic supernovae burst into view in 1987. Because it was the first supernova detected that year, it was given the name **Supernova 1987A** (Figure 13.22). This stellar explosion was located in the *Large Magellanic Cloud,* a small galaxy that orbits the Milky Way, and was visible to the naked eye from southern latitudes. As the nearest supernova witnessed in four centuries, Supernova 1987A provided a unique opportunity to study a supernova and its debris in detail.

Observations of supernova remnants confirm many aspects of our models of how high-mass stars live and die.

When Betelgeuse explodes as a supernova, it will be more than 10 times as bright as the full moon in our sky. If our ancestors had seen Betelgeuse explode a few hundred or a few thousand years ago, do you think it could have had any effect on human history? How do you think our modern society would react if we saw Betelgeuse explode tomorrow?

Summary of Stellar Lives Throughout our discussion of stellar life stories, we've found that a star's birth mass is the major factor that determines its entire life. Figure 13.23 summarizes the life cycles of the $1M_{Sun}$ and $25M_{Sun}$ stars we have focused on in this chapter.

The key point is that fusion proceeds relatively slowly in low-mass stars during their main-sequence lifetimes, giving them long lives, and even at the ends of their lives the cores of these stars never become hot enough to fuse carbon. These stars die in planetary nebulae, leaving their dead cores behind as white dwarfs. Fusion reactions proceed much faster in the cores of high-mass stars, giving them shorter lives. Moreover, their high masses mean gravity is strong enough to keep compressing the core as each stage of fusion ends, until they ultimately reach the point at which iron piles up in the core. Because iron fusion does not produce energy, the core catastrophically collapses and releases energy that causes the rest of the star to explode as a supernova. The dead core remains behind as either a neutron star or a black hole [Section 14.3].

> All low- and intermediate-mass stars follow life stages similar to those of our Sun, while high-mass stars live short but brilliant lives and die in supernova explosions.

13.4 Stars in Close Binaries

We have so far treated stars as if they all lived in isolation, but nearly half the stars we see in the sky are actually binary star systems. In this final section, we will examine how a star's life story can change if it happens to orbit another star closely enough that mass can sometimes flow from one star to the other.

◆ How are the lives of stars with close companions different?

For the most part, stars in binary systems proceed from birth to death as if they were isolated. However, exceptions can occur in close binary star systems. Algol, the "demon star" in the constellation Perseus, is a good example. It appears as a single star to our eyes and telescopes, but it is actually an eclipsing binary star system [Section 12.1] consisting of two stars that orbit each other closely: a $3.7M_{Sun}$ main-sequence star and a $0.8M_{Sun}$ subgiant.

A moment's thought reveals that something quite strange is going on. The stars of a binary system are born at the same time and therefore must both be the same age. We know that more massive stars live shorter lives, and therefore the more massive star must exhaust its core hydrogen and become a subgiant before the less massive star does. How, then, can Algol's less massive star be a subgiant while the more massive star is still fusing hydrogen in its core as a main-sequence star?

This *Algol paradox* reveals some of the complications in stellar life cycles that can arise in close binary systems. The two stars in a close

All stars spend most of their time as main-sequence stars and then change dramatically near the ends of their lives. This figure shows the life stages of a high-mass star and a low-mass star, using the cosmic calendar from Chapter 1 to illustrate the relative lengths of these life stages. On this calendar, the 14-billion-year lifetime of the universe corresponds to a single year.

LIFE OF A HIGH-MASS STAR (25M_{Sun})

This high-mass star goes from protostar to supernova in about 6 million years, corresponding to less than 4 hours on the cosmic calendar.

① Protostar: A star system forms when a cloud of interstellar gas collapses under gravity.

② Blue main-sequence star: In the core of a high-mass star, four hydrogen nuclei fuse into a single helium nucleus by the series of reactions known as the CNO cycle.

③ Red supergiant: After core hydrogen is exhausted, the core shrinks and heats. Hydrogen fusion begins around the inert helium core, causing the star to expand into a red supergiant.

Actual Length of Stage	40,000 years	5 million years	100,000 years
Time on Cosmic Calendar	12:00:00 AM → 12:01:30 AM	12:01:30 AM → 3:10:00 AM	3:10:00 AM → 3:14:00 AM

These times correspond to the life stages of a 25M_{Sun} star born around midnight on a typical day of the cosmic calendar.

LIFE OF A LOW-MASS STAR (1M_{Sun})

This low-mass star goes from protostar to planetary nebula in about 11.5 billion years, corresponding to 10 months on the cosmic calendar.

① Protostar: A star system forms when a cloud of interstellar gas collapses under gravity.

② Yellow main-sequence star: In the core of a low-mass star, four hydrogen nuclei fuse into a single helium nucleus by the series of reactions known as the proton–proton chain.

③ Red giant star: After core hydrogen is exhausted, the core shrinks and heats. Hydrogen fusion begins around the inert helium core, causing the star to expand into a red giant.

Actual Length of Stage	30 million years	10 billion years	1 billion years
Time on Cosmic Calendar	March 1 → March 2	March 2 → November 30	November 30 → December 27

These dates correspond to the life stages of a 1M_{Sun} star born in early March on the cosmic calendar.

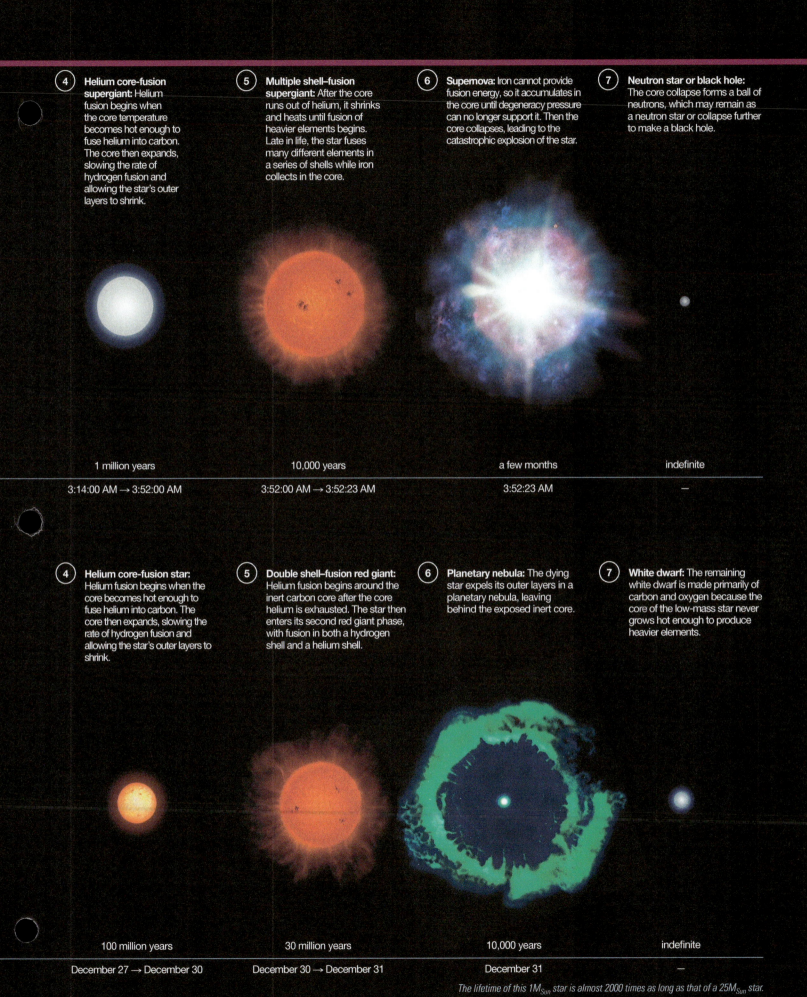

④ Helium core-fusion supergiant: Helium fusion begins when the core temperature becomes hot enough to fuse helium into carbon. The core then expands, slowing the rate of hydrogen fusion and allowing the star's outer layers to shrink.

⑤ Multiple shell–fusion supergiant: After the core runs out of helium, it shrinks and heats until fusion of heavier elements begins. Late in life, the star fuses many different elements in a series of shells while iron collects in the core.

⑥ Supernova: Iron cannot provide fusion energy, so it accumulates in the core until degeneracy pressure can no longer support it. Then the core collapses, leading to the catastrophic explosion of the star.

⑦ Neutron star or black hole: The core collapse forms a ball of neutrons, which may remain as a neutron star or collapse further to make a black hole.

| 1 million years | 10,000 years | a few months | indefinite |
| 3:14:00 AM → 3:52:00 AM | 3:52:00 AM → 3:52:23 AM | 3:52:23 AM | — |

④ Helium core-fusion star: Helium fusion begins when the core becomes hot enough to fuse helium into carbon. The core then expands, slowing the rate of hydrogen fusion and allowing the star's outer layers to shrink.

⑤ Double shell–fusion red giant: Helium fusion begins around the inert carbon core after the core helium is exhausted. The star then enters its second red giant phase, with fusion in both a hydrogen shell and a helium shell.

⑥ Planetary nebula: The dying star expels its outer layers in a planetary nebula, leaving behind the exposed inert core.

⑦ White dwarf: The remaining white dwarf is made primarily of carbon and oxygen because the core of the low-mass star never grows hot enough to produce heavier elements.

| 100 million years | 30 million years | 10,000 years | indefinite |
| December 27 → December 30 | December 30 → December 31 | December 31 | — |

The lifetime of this $1M_{Sun}$ star is almost 2000 times as long as that of a $25M_{Sun}$ star.

Algol shortly after its birth. The higher-mass star (left) evolved more quickly than its lower-mass companion (right).

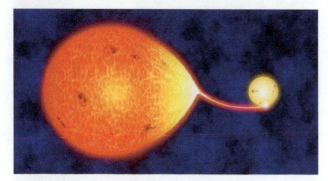

Algol at onset of mass transfer. When the more massive star expanded into a red giant, it began losing some of its mass to its normal, hydrogen core-fusion companion.

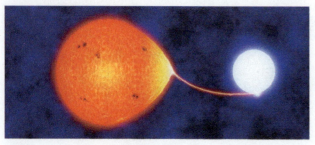

Algol today. As a result of the mass transfer, the red giant has shrunk to a subgiant, and the normal star on the right is now the more massive of the two stars.

▲ **FIGURE 13.24**
Artist's conception of the development of the Algol close binary system.

binary system are near enough to exert significant tidal forces on each other [Section 4.4]. The gravity of each star attracts the near side of the other star more strongly than it attracts the far side. The stars therefore stretch into football-like shapes rather than remaining spherical. In addition, the stars become *tidally locked* so that they always show the same face to each other, much as the Moon always shows the same face to Earth.

During the time that both stars are main-sequence stars, the tidal forces have little effect on their lives. However, when the more massive star (which exhausts its core hydrogen sooner) begins to expand into a red giant, gas from its outer layers can spill over onto its companion. This **mass exchange** occurs when the giant grows so large that its tidally distorted outer layers succumb to the gravitational attraction of the smaller companion star. The companion then begins to gain mass at the expense of the giant.

Stars in close binary systems can exchange mass with each other, altering their life histories.

The solution to the Algol paradox should now be clear (Figure 13.24). The $0.8M_{Sun}$ subgiant *used to be* much more massive. As the more massive star, it was the first to begin expanding into a red giant. As it expanded, however, so much of its matter spilled over onto its companion that it is now the less massive star.

The future may hold even more interesting events for Algol. The $3.7M_{Sun}$ star is still gaining mass from its subgiant companion. Its life cycle is therefore actually accelerating as its increasing gravity raises its hydrogen core-fusion rate. Millions of years from now, it will exhaust its hydrogen and begin to expand into a red giant itself. At that point, it can begin to transfer mass *back* to its companion. Even more amazing things can happen in other mass-exchange systems, particularly when one of the stars is a white dwarf or a neutron star. But that is a topic for the next chapter.

---

the big picture | Putting Chapter 13 into Perspective

In this chapter, we have seen how the origin of the elements, first discussed in Chapter 1, is linked to the lives and deaths of stars. As you look back, keep in mind these "big picture" ideas:

- Virtually all elements in the universe besides hydrogen and helium were forged in stars. We and our planet are therefore made of stuff produced by stars that lived and died long ago.

- Low-mass stars like our Sun live long lives and die with the ejection of planetary nebulae, leaving behind white dwarfs.

- High-mass stars live fast and die young, exploding dramatically as supernovae and leaving behind neutron stars or black holes.

- Close binary stars can exchange mass, altering the usual course of stellar evolution.

my cosmic perspective One of the deepest connections we have with the cosmos is through the many elements that make up our bodies, which were formed by nuclear fusion in stars.

13.1 Star Birth

◆ How do stars form?

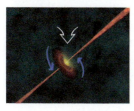

Stars are born in cold, relatively dense **molecular clouds.** As a cloud fragment collapses under gravity, it becomes a rapidly rotating **protostar** surrounded by a spinning disk of gas in which planets may form. The protostar may also fire **jets** of matter outward along its poles.

◆ How massive are newborn stars?

On the upper end, the most massive newborn stars are about $150M_{Sun}$. On the lower end, stars cannot be less massive than $0.08M_{Sun}$; below this mass, **degeneracy pressure** prevents gravity from making the core hot enough for efficient hydrogen fusion, and the object becomes a **brown dwarf.** Low-mass stars far outnumber high-mass stars.

13.2 Life as a Low-Mass Star

◆ What are the life stages of a low-mass star?

A low-mass star spends most of its life generating energy by fusing hydrogen in its core. When the core's hydrogen is exhausted, the core begins to shrink while the star as a whole expands to become a **red giant,** with **hydrogen shell fusion** occurring around an inert helium core. When the core becomes hot enough, a **helium flash** initiates **helium fusion** in the core, which fuses helium into carbon; the star shrinks somewhat in size and luminosity during this time. The core shrinks again when helium core fusion ceases, while both helium and hydrogen fusion occur in shells around the inert carbon core and cause the outer layers to expand once more.

◆ How does a low-mass star die?

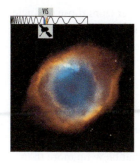

A low-mass star like the Sun never gets hot enough to fuse carbon in its core, because degeneracy pressure stops the gravitational collapse of the core. The star expels its outer layers into space as a **planetary nebula,** leaving its exposed core behind as a white dwarf.

13.3 Life as a High-Mass Star

◆ What are the life stages of a high-mass star?

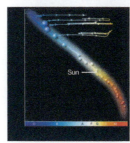

A high-mass star lives a much shorter life than a low-mass star, fusing hydrogen into helium via the **CNO cycle.** After exhausting its core hydrogen, a high-mass star begins hydrogen shell fusion and then goes through a series of stages, fusing successively heavier elements. As a result, the star swells in size to become a supergiant.

◆ How do high-mass stars make the elements necessary for life?

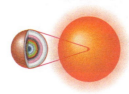

In its final stages of life, a high-mass star's core becomes hot enough to fuse carbon and other heavy elements. The variety of different fusion reactions produces a wide range of elements—including all the elements necessary for life—that are then released into space when the star dies.

◆ How does a high-mass star die?

A high-mass star dies in a cataclysmic explosion called a **supernova,** scattering newly produced elements into space and leaving behind a neutron star or black hole. The supernova occurs after fusion begins to pile up iron in the high-mass star's core. Because iron fusion cannot release energy, the core cannot hold off the crush of gravity for long. In the instant that gravity overcomes degeneracy pressure, the core collapses and the star explodes. The expelled gas may be visible for a few thousand years as a **supernova remnant.**

13.4 Stars in Close Binaries

◆ How are the lives of stars with close companions different?

When one star in a close binary system begins to swell in size at the end of its main-sequence stage, it can begin to transfer mass to its companion. This **mass exchange** can then change the remaining life histories of both stars.

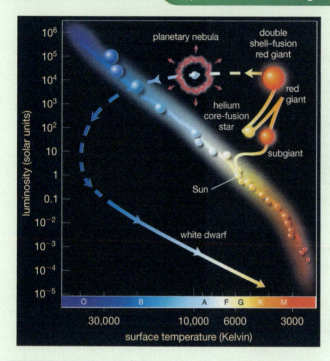

This figure, similar to the left side of Figure 13.13, shows the future life stages of the Sun on an H-R diagram. Use the figure to answer the following questions.

1. What will the Sun's approximate luminosity be during the subgiant stage?

2. When the Sun is a red giant, what will its approximate surface temperature be?

3. Just before the Sun produces a planetary nebula, what will its approximate luminosity be?

4. When the Sun becomes a white dwarf with a surface temperature similar to its current surface temperature, what will its luminosity be?

exercises and problems

 **Mastering Astronomy®**   For instructor-assigned homework and other learning materials, go to MasteringAstronomy®.

Review Questions

1. What is a *molecular cloud*? Briefly describe the process by which a protostar forms from gas in a molecular cloud.

2. Why do protostars rotate rapidly? How can a close binary star system form?

3. Why does a spinning disk of gas surround a protostar? Describe key phenomena seen among protostars, such as strong *stellar winds* and *jets*.

4. What are the minimum and maximum masses for a star, and why do these limits occur? What is a *brown dwarf*?

5. What is *degeneracy pressure,* and how does it differ from *thermal pressure*? Explain why degeneracy pressure can support a stellar core against gravity even when the core becomes cold.

6. Briefly describe the Sun's life stages after it exhausts its core hydrogen. Discuss both the changes occurring in the Sun's core and the changes visible from outside the Sun.

7. Why does helium fusion require much higher temperatures than hydrogen fusion? Briefly explain why helium fusion in the Sun will begin with a *helium flash*.

8. What is a *planetary nebula*? What happens to the core of a star after a planetary nebula occurs?

9. What will happen to Earth as the Sun changes in the future?

10. What do we mean by a star's *life track* on an H-R diagram? Summarize the stages of the Sun's life track in Figure 13.13.

11. In broad terms, explain how the life of a high-mass star differs from that of a low-mass star.

12. Describe some of the nuclear reactions that can occur in high-mass stars after they exhaust their core helium. Why do these reactions occur in high-mass but not in low-mass stars?

13. Why can't iron be fused to release energy?

14. Summarize some of the observational evidence supporting our ideas about how the elements form in massive stars.

15. What event initiates a *supernova,* and why is a neutron star or black hole left behind? What observational evidence supports our understanding of supernovae?

16. Why can the lives of close binary stars differ from those of single stars? Describe the *Algol paradox* and its resolution.

Test Your Understanding

○ Does It Make Sense?

Decide whether the statement makes sense (or is clearly true) or does not make sense (or is clearly false). Explain clearly; not all of these have definitive answers, so your explanation is more important than your chosen answer.

17. The iron in my blood came from a star that blew up more than 4 billion years ago.

18. I discovered stars being born within a patch of extremely low-density, hot interstellar gas.

19. Humanity will eventually have to find another planet to live on, because one day the Sun will blow up as a supernova.
20. I sure am glad hydrogen has a higher mass per nuclear particle than many other elements. If it had the lowest mass per nuclear particle, none of us would be here.
21. If the Sun had been born 4½ billion years ago as a high-mass star rather than as a low-mass star, Jupiter would have Earth-like conditions today, while Earth would be hot like Venus.
22. If you could look inside the Sun today, you'd find that its core contains a much higher proportion of helium and a lower proportion of hydrogen than it did when the Sun was born.
23. I just discovered a $3.5M_{Sun}$ main-sequence star orbiting a $2.5M_{Sun}$ red giant. I'll bet that red giant was more massive than the $3.5M_{Sun}$ star when it was a main-sequence star.
24. Globular clusters generally contain lots of white dwarfs.
25. After hydrogen fusion stops in a low-mass star, its core cools off until the star becomes a red giant.
26. The uranium in nuclear reactors comes from supernova explosions.

Quick Quiz

Choose the best answer to each of the following. Explain your reasoning with one or more complete sentences.

27. Stars can form most easily in clouds that are (a) cold and dense. (b) warm and dense. (c) hot and low-density.
28. A brown dwarf is (a) an object not quite massive enough to be a star. (b) a white dwarf that has cooled off. (c) a starlike object that is less massive than Jupiter.
29. Which of these stars has the hottest *core*? (a) a blue main-sequence star (b) a red supergiant (c) a red main-sequence star
30. Which of these stars does *not* have fusion occurring in its core? (a) a red giant (b) a red main-sequence star (c) a blue main-sequence star
31. After the helium flash in a low-mass star, the star's luminosity (a) goes up. (b) goes down. (c) stays the same.
32. What would stars be like if hydrogen had the smallest mass per nuclear particle? (a) Stars would be brighter. (b) All stars would be red giants. (c) Nuclear fusion would not occur in stars of any mass.
33. What would stars be like if carbon had the smallest mass per nuclear particle? (a) Supernovae would be more common. (b) Supernovae would never occur. (c) High-mass stars would be hotter.
34. What would you be most likely to find if you returned to the solar system in 10 billion years? (a) a neutron star (b) a white dwarf (c) a black hole
35. Which of these stars has the shortest life expectancy? (a) an isolated $1M_{Sun}$ star (b) a $1M_{Sun}$ star in a close binary system with a $0.8M_{Sun}$ star (c) a $1M_{Sun}$ star in a close binary system with a $2M_{Sun}$ star
36. What happens to the core of a high-mass star after it runs out of hydrogen? (a) It shrinks and heats up. (b) It shrinks and cools down. (c) Helium fusion begins right away.

🌀 Process of Science

37. *Predicting the Sun's Future.* Models of stellar evolution make detailed predictions about the fate of the Sun. Describe one piece of evidence that supports each of the following model predictions:
    a. The Sun cannot continue supplying Earth with light and heat forever.
    b. The Sun will become a red giant before the end of its life.
    c. The Sun will leave behind a white dwarf after it dies.
38. *Predicting the Properties of Brown Dwarfs.* Models of star formation predict that objects less massive than $0.08M_{Sun}$ become brown dwarfs instead of true hydrogen-fusing stars. Once they have formed, these objects cool because no fusion is occurring to replace the thermal energy lost from their surfaces. How would you expect the properties of brown dwarfs in an older star cluster to compare with those of brown dwarfs in a younger one? Propose an observing program that could test your hypothesis.

Group Work Exercise

39. *Comparing Models of Stars with Data.* **Roles:** *Proposer* (starts the discussion of each prediction by trying to explain how the data in these particular figures support it), *Skeptic* (tries to rebut that explanation), *Moderator* (decides whether the data provide strong support for the prediction, weak support, or no support), *Scribe* (records the discussion and the decision of the Moderator). **Activity:** Evaluate each of the following predictions using the information provided in Figures 12.10, 12.17, and 12.18.
    a. Type O stars have shorter lives than type G stars.
    b. Type K supergiant stars produce iron before they explode as supernovae.
    c. Type F stars become much more luminous near the ends of their lives than they were as main-sequence stars.
    d. Type O stars do not become more luminous near the ends of their lives but do become redder.
    e. Type M stars have longer lives than type K stars.
    f. Stars similar to the Sun reach a maximum size of about 100 solar radii during the red giant stage.
    g. Type K main-sequence stars will become red giants when their cores run out of hydrogen.
    h. Some stars become white dwarfs at the ends of their lives.
    i. White dwarfs cool with time but do not change much in radius.

Investigate Further

Short-Answer/Essay Questions

40. *Brown Dwarfs.* How are brown dwarfs like jovian planets? In what ways are brown dwarfs like stars?
41. *Homes to Civilization?* We do not yet know how many stars have Earth-like planets, nor do we know the likelihood that such planets might harbor advanced civilizations like our own. However, some stars can probably be ruled out as candidates for advanced civilizations. For example, given that it took a few billion years for humans to evolve on Earth, it seems unlikely that advanced life would have had time to evolve around a star that is only a few million years old. For each of the following stars, decide whether you think it is possible that it could harbor an advanced civilization. Explain your reasoning in one or two paragraphs.
    a. $10M_{Sun}$ main-sequence star
    b. $1.5M_{Sun}$ main-sequence star
    c. $1.5M_{Sun}$ red giant
    d. $1M_{Sun}$ helium core-fusion star
    e. Red supergiant

42. *Rare Elements.* Lithium, beryllium, and boron are elements with atomic numbers 3, 4, and 5, respectively. Despite their being three of the five simplest elements, Figure 13.19 shows that they are rare compared to many heavier elements. Suggest a reason for their rarity. (*Hint:* Consider the process by which helium fuses into carbon.)

43. *Future Skies.* As a red giant, the Sun will have an angular size in Earth's sky of about 30°. What will sunset and sunrise be like? Do you think the color of the sky will be different from what it is today? Explain.

44. *Research: Historical Supernovae.* Historical accounts exist of supernovae in the years 1006, 1054, 1572, and 1604. Choose one of these supernovae and learn more about historical records of the event. Did the supernova influence human history in any way? Write a two- to three-page summary of your research findings.

Quantitative Problems

Be sure to show all calculations clearly and state your final answers in complete sentences.

45. *An Isolated Star-Forming Cloud.* Isolated molecular clouds can have a temperature as low as 10 K and a particle density as great as 100,000 particles per cubic centimeter. What is the minimum mass that a cloud with these properties needs in order to form a star?

46. *Density of a Red Giant.* Near the end of its life, the Sun's radius will extend nearly to Earth's orbit. Estimate the volume of the Sun at that time using the formula for the volume of a sphere ($V = 4\pi r^3/3$). Using that result, estimate the average matter density of the Sun at that time. How does that density compare with the density of water (1 g/cm³)? How does it compare with the density of Earth's atmosphere at sea level (about 10^{-3} g/cm³)?

47. *Supernova Betelgeuse.* The distance from Earth of the red supergiant Betelgeuse is approximately 643 light-years. If it were to explode as a supernova, it would be one of the brightest stars in the sky. Right now, the brightest star other than the Sun is Sirius, with a luminosity of $26L_{Sun}$ and a distance of 8.6 light-years. How much brighter in our sky than Sirius would the Betelgeuse supernova be if it reached a maximum luminosity of $10^{10}L_{Sun}$?

48. *Construction of Elements.* Using the periodic table in Appendix D, determine which elements are made by the following nuclear fusion reactions. (You can assume the total number of protons in the reaction remains constant.)
    a. Fusion of a carbon nucleus with another carbon nucleus
    b. Fusion of a carbon nucleus with a neon nucleus
    c. Fusion of an iron nucleus with a helium nucleus

49. *Algol's Orbital Separation.* The Algol binary system consists of a $3.7M_{Sun}$ star and a $0.8M_{Sun}$ star with an orbital period of 2.87 days. Use Newton's version of Kepler's third law to calculate the orbital separation of the system. How does that separation compare with the typical size of a red giant star?

Discussion Questions

50. *Connections to the Stars.* In ancient times, many people believed that our lives were somehow influenced by the patterns of the stars in the sky. Modern science has not found any evidence to support this belief, but instead has found that we have a connection to the stars on a much deeper level: We are "star stuff." Discuss in some detail our real connections to the stars as established by modern astronomy. Do you think these connections have any philosophical implications in terms of how we view our lives and our civilization? Explain.

51. *Humanity in A.D. 5,000,000,000.* Do you think it is likely that humanity will survive until the Sun begins to expand into a red giant 5 billion years from now? Why or why not?

Web Projects

52. *Star Birth and the Spitzer Telescope.* The Spitzer Space Telescope is one of astronomers' best tools for learning about star birth. Visit the Spitzer website and look for the latest findings about star formation. Summarize your research in a one- to two-page report.

53. *Picturing Star Birth and Death.* Photographs of stellar birthplaces (molecular clouds) and death places (planetary nebulae and supernova remnants) can be strikingly beautiful, but only a few such photographs are included in this chapter. Search the Internet for additional images. Put the photographs you find into a personal online journal, along with a one-paragraph description of what each one shows. Include at least 20 images.

This image from the Chandra X-Ray Observatory shows the neutron star amidst the hot gas at the center of the Crab Nebula, a supernova remnant.

LEARNING GOALS

14.1 White Dwarfs
- ◆ What is a white dwarf?
- ◆ What can happen to a white dwarf in a close binary system?

14.2 Neutron Stars
- ◆ What is a neutron star?
- ◆ How were neutron stars discovered?
- ◆ What can happen to a neutron star in a close binary system?

14.3 Black Holes: Gravity's Ultimate Victory
- ◆ What is a black hole?
- ◆ What would it be like to visit a black hole?
- ◆ Do black holes really exist?

14.4 Extreme Events
- ◆ What causes gamma-ray bursts?
- ◆ What happens when black holes merge?

ESSENTIAL PREPARATION

1. What determines the strength of gravity? [Section 4.4]

2. How do gravity and energy allow us to understand orbits? [Section 4.4]

3. How massive are newborn stars? [Section 13.1]

4. How does a low-mass star die? [Section 13.2]

5. How does a high-mass star die? [Section 13.3]

Welcome to the afterworld of stars, the fascinating domain of white dwarfs, neutron stars, and black holes. To scientists, these dead stars are ideal laboratories for testing the most extreme predictions of general relativity and quantum theory. To most other people, the eccentric behavior of stellar corpses demonstrates that the universe is stranger than they ever imagined.

Dead stars behave in unusual and unexpected ways that challenge our minds and stretch the boundaries of what we believe is possible. Stars that have finished nuclear fusion have only one hope of staving off the crushing power of gravity: the quantum mechanical effect of degeneracy pressure. But even this strange pressure cannot save the most massive stellar cores. In this chapter, we will study the bizarre properties and occasional catastrophes of the eerie inhabitants of the stellar graveyard.

14.1 White Dwarfs

In the previous chapter, we saw that stars of different masses leave different types of stellar corpses. Low-mass stars like the Sun leave behind white dwarfs when they die. High-mass stars die in the titanic explosions known as supernovae, leaving behind neutron stars and black holes. Let's begin our study of stellar corpses with white dwarfs.

◆ What is a white dwarf?

As we discussed in Chapters 12 and 13, a white dwarf is essentially the exposed core of a star that has died and shed its outer layers in a planetary nebula [Section 13.2]. It is quite hot when it first forms, because it was recently the inside of a star, but it slowly cools with time. White dwarfs are stellar in mass but small in size (radius), which is why they are generally quite dim compared to stars like the Sun [Section 12.2]. The hottest white dwarfs can shine quite brightly in high-energy light such as ultraviolet and X rays (Figure 14.1).

A white dwarf's large mass and small size make gravity very strong near its surface. If gravity were unopposed, it would crush the white dwarf to an even smaller size, so some sort of pressure must be pushing back equally hard to keep the white dwarf stable. Because there is no fusion to maintain heat and pressure inside a white dwarf, the pressure that opposes gravity must come from some other source. The source is *degeneracy pressure*—the same type of pressure that supports the "failed stars" that we call brown dwarfs and that arises when subatomic particles are packed as closely as the laws of quantum mechanics allow [Section 13.1]. More specifically, the degeneracy pressure in white dwarfs arises from closely packed *electrons,* so we call it **electron degeneracy pressure.** A white dwarf exists in a state of balance because the outward push of electron degeneracy pressure matches the inward crush of gravity.

A white dwarf is the corpse of a low-mass star, supported against the crush of gravity by electron degeneracy pressure.

White Dwarf Composition, Density, and Size Because a white dwarf is the core left over after a star has ceased nuclear fusion, its composition reflects the products of the star's final fusion stage. The white dwarf left

Sirius A is the brightest star in the night sky in visible and infrared light.

However, Sirius A is relatively dim in ultraviolet and X rays . . .

The hot white dwarf Sirius B is much less bright in visible and infrared light.

. . . while Sirius B outshines Sirius A in ultraviolet and X rays.

a Sirius as seen in infrared light by the Hubble Space Telescope. **b** Sirius as seen by the Chandra X-Ray Telescope.

▲ FIGURE 14.1

Sirius, the brightest star in the night sky, is actually a binary system consisting of a main-sequence star (Sirius A) and a white dwarf (Sirius B). The main-sequence star is much brighter in infrared and visible light, but the hot white dwarf shines more brightly in high-energy light. (The spikes in the images are artifacts of the telescope optics.)

behind by a $1M_{Sun}$ star like our Sun consists mostly of carbon, since stars like the Sun fuse helium into carbon in their final stage of life.

Despite its ordinary-sounding composition, a scoop of matter from a white dwarf would be unlike anything ever seen on Earth. A typical white dwarf has the mass of the Sun ($1M_{Sun}$) compressed into a volume the size of Earth. If you recall that Earth is smaller than a typical sunspot, you'll realize that packing the entire mass of the Sun into the volume of Earth is no small feat. The density of a white dwarf is so high that a teaspoon of its material would weigh several tons—as much as a small truck—if you could bring it to Earth.

A teaspoon of white dwarf matter would weigh several tons.

More massive white dwarfs are actually smaller in size than less massive ones. For example, a $1.3M_{Sun}$ white dwarf is half the diameter of a $1.0M_{Sun}$ white dwarf (Figure 14.2). The more massive white dwarf is smaller because its greater gravity compresses matter to a much greater density. According to the laws of quantum mechanics, the electrons in a white dwarf respond to this compression by moving faster, which makes the degeneracy pressure strong enough to resist the greater force of gravity. The most massive white dwarfs are therefore the smallest.

The White Dwarf Limit
The fact that electron speeds are higher in more massive white dwarfs leads to a fundamental limit on the maximum mass of a white dwarf. Theoretical calculations show that the electron speeds would reach the speed of light in a white dwarf with a mass of about 1.4 times the mass of the Sun ($1.4M_{Sun}$). Because neither electrons nor anything else can travel faster than the speed of light (see the Special Topic on the next page), no white dwarf can have a mass greater than $1.4M_{Sun}$, which is known as the **white dwarf limit** (also called the *Chandrasekhar limit,* after its discoverer).

A white dwarf cannot have a mass greater than 1.4 times the mass of the Sun.

Strong observational evidence supports this theoretical limit on the mass of a white dwarf. Many known white dwarfs are members of binary systems, and hence we can measure their masses [Section 12.1]. In every observed case, the white dwarfs have masses below $1.4M_{Sun}$, just as our theory predicts.

◆ What can happen to a white dwarf in a close binary system?

Left to itself, a white dwarf will never again shine as brightly as the star it once was. With no source of fuel for fusion, it will simply cool with time into a cold "black dwarf." Its size will never change, because its electron degeneracy pressure will forever keep it stable against the crush of gravity. However, the situation can be quite different for a white dwarf in a close binary system.

Accretion Disks
A white dwarf in a close binary system can gradually gain mass if its companion is a main-sequence or giant star (Figure 14.3). When a clump of mass first spills over from the companion to the white dwarf, it has some small orbital velocity. The law of conservation of angular momentum dictates that the clump must orbit faster and faster as it falls toward the white dwarf's surface. The infalling matter therefore forms a whirlpool-like disk around the white dwarf. Because the process in which material falls onto another body is called *accretion,* this rapidly rotating disk is called an **accretion disk.**

In a close binary system, gas from a companion star can spill toward a white dwarf, forming a swirling accretion disk around it.

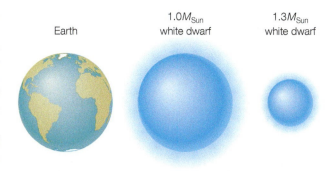

Earth 1.0M_{Sun} white dwarf 1.3M_{Sun} white dwarf

▲ **FIGURE 14.2**
Contrary to what you might expect, more massive white dwarfs are actually smaller (and thus denser) than less massive white dwarfs. Earth is shown for scale.

Just as in the disks that form from material accreting onto protostars [Section 13.1], gas particles in a white dwarf's accretion disk move on orbits that obey Kepler's laws. Gas in the inner region of the disk therefore orbits faster than gas in the outer region. The differences in gas particle speeds lead to friction that removes orbital energy from the inner region, and the loss of energy means that gas gradually spirals inward and eventually settles onto the white dwarf.

special topic   Relativity and the Cosmic Speed Limit

The idea that nothing can travel faster than the speed of light is one of several mind-boggling consequences of Einstein's *special theory of relativity*, published in 1905. Although relativity is often portrayed as being difficult, its basic ideas are easy to understand.

To understand what's "relative" about relativity, imagine a supersonic plane trip at a speed of 1670 km/hr from Nairobi, Kenya, to Quito, Ecuador. How fast is the plane going? At first, this question sounds trivial—we have just said that the plane is going 1670 km/hr. But wait. Nairobi and Quito are both nearly on Earth's equator, and the equatorial speed of Earth's rotation is the same 1670 km/hr at which the plane is flying, but in the opposite direction (see figure). If you lived on the Moon, the plane would appear to remain stationary *while Earth rotated beneath it.*

Einstein's theory tells us that both viewpoints—the plane is going 1670 km/hr for people on Earth but is stationary as viewed from the Moon—are equally valid. That is, questions like "Who is really moving?" and "How fast are you going?" have no absolute answers, and everyone can agree only on the fact that the plane is traveling at 1670 km/hr *relative to* the surface of Earth. The theory of relativity gets its name from the fact that measurements of motion (and of time and space) make sense only when we describe whom or what they are being measured relative to.

Note that while Einstein's theory says that motion is relative, it does *not* say that "everything" is relative. In fact, the theory states that two things in the universe are absolute:

1. The laws of nature are the same for everyone.
2. The speed of light is the same for everyone.

The first absolute, that the laws of nature are the same for everyone, is a more general version of the idea that all viewpoints on motion are equally valid. If they weren't, different observers would disagree about the laws of physics. The second absolute, that the speed of light is the same for everyone, is much more surprising. Ordinarily, we expect speeds to add and subtract. If you watch someone throw a ball forward from a moving car, you see the ball traveling at the speed at which it is thrown *plus* the speed of the car. But if a person shines a light beam from a moving car, you see it moving at precisely the speed of light (about 300,000 km/s), no matter how fast the car is going. This strange fact has been experimentally verified countless times.

The cosmic speed limit follows directly from this fact about the speed of light. To see why, imagine that you have just built the most incredible rocket possible, and you are taking it on a test ride. You

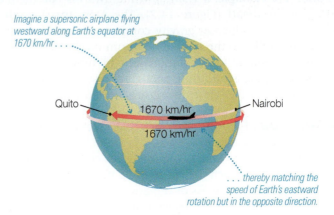

Imagine a supersonic airplane flying westward along Earth's equator at 1670 km/hr...

Quito   1670 km/hr   Nairobi

1670 km/hr

...thereby matching the speed of Earth's eastward rotation but in the opposite direction.

A plane flying at 1670 km/hr from Nairobi to Quito travels precisely opposite Earth's rotation.

push the acceleration button and keep going faster and faster and faster. With enough fuel, you might expect that you'd eventually be moving faster than the speed of light. But we must ask: What is your speed being measured relative to? Remember that everyone always measures light to be traveling at the speed of light, 300,000 km/s, so *you* will find that the light from your rocket's headlights is racing away from you at 300,000 km/s. This shouldn't be too surprising, as it's just another way of saying that you can't catch up with your own light. However, because *everyone* always measures the same speed of light, people back on Earth (or anyplace else) must also say that your headlight beams travel at a speed of 300,000 km/s. Because we already know that you can't keep up with your headlight beams, observers on Earth must find that you are traveling *slower* than the headlight beams, which means slower than the speed of light.

These facts about the relativity of motion and the absoluteness of the speed of light also lead to several other famous consequences of Einstein's theory. For example, they tell us that if you observed a person moving past you at a speed close to the speed of light, you'd see her time running slower than yours, you'd measure her size to be compressed (in the direction of motion) from what you'd measure if she were stationary (relative to you), and you'd conclude that her mass was greater than it would be if she were stationary. The theory also predicts that mass and energy should be equivalent, as stated by Einstein's famous formula $E = mc^2$. All these predictions of relativity have been experimentally tested and verified to high precision.

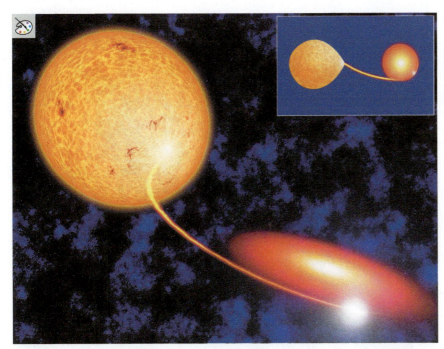

◄ FIGURE 14.3
This artist's conception shows how mass spilling from a companion star (left) toward a white dwarf (right) forms an accretion disk. The white dwarf itself is in the center of the accretion disk—too small to be seen on this scale. Matter streaming onto the disk creates a hot spot where the stream joins the disk. The inset shows how the system looks from above rather than from the side.

Accretion can provide a "dead" white dwarf with a new energy source. The inward-spiraling gas in the accretion disk becomes quite hot as its gravitational potential energy is converted into thermal energy [Section 4.3], allowing it to shine with intense ultraviolet or X-ray radiation. More dramatic events can occur as fresh hydrogen gas from the companion star accumulates on the surface of a white dwarf.

Novae The hydrogen spilling toward the white dwarf from the companion gradually spirals inward through the accretion disk and eventually falls onto the surface of the white dwarf. The white dwarf's strong gravity compresses this hydrogen gas into a thin surface layer. Both the pressure and the temperature rise as the layer builds up with more accreting gas. When the temperature at the bottom of the layer reaches about 10 million K, hydrogen fusion suddenly ignites.

The white dwarf blazes back to life while fusion in its hydrogen layer supplies energy. This thermonuclear flash causes the binary system to shine for a few glorious weeks as a **nova** (Figure 14.4a). A nova is far less luminous than a supernova, but still can shine as brightly as 100,000 Suns. It generates heat and pressure, ejecting most of the material that has accreted onto the white dwarf. This material expands outward, creating a *nova remnant* that sometimes remains visible years after the nova explosion (Figure 14.4b).

A nova is caused by hydrogen fusion on the surface of a white dwarf in a binary star system.

Accretion resumes after a nova explosion subsides, so the entire process can repeat itself. The time between successive novae in a particular system depends on the rate at which hydrogen accretes onto the white dwarf surface and on how highly compressed this hydrogen becomes. The compression of hydrogen is greatest for the most massive white dwarfs, which have the strongest surface gravities. In some cases, novae have been observed to repeat after just a few decades. More commonly, thousands of years may pass between nova outbursts.

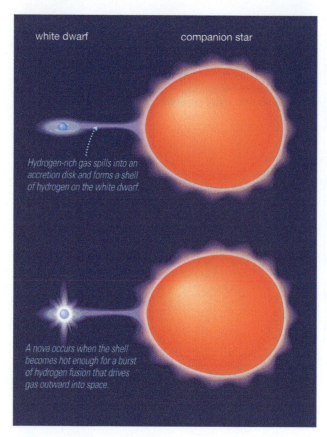

white dwarf   companion star

Hydrogen-rich gas spills into an accretion disk and forms a shell of hydrogen on the white dwarf.

A nova occurs when the shell becomes hot enough for a burst of hydrogen fusion that drives gas outward into space.

a Diagram of the nova process.

▲ **FIGURE 14.4**

A nova occurs when hydrogen fusion ignites on the surface of a white dwarf in a binary star system.

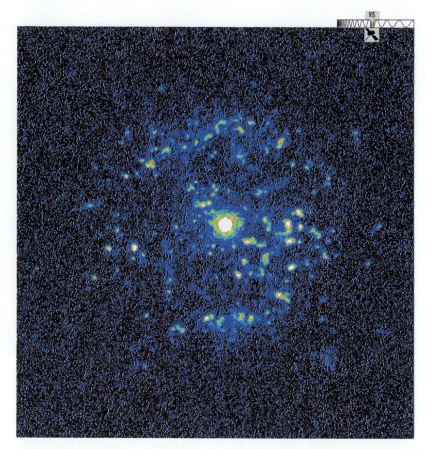

b Hubble Space Telescope image showing blobs of gas ejected from the nova T Pyxidis. The bright spot at the center of the blobs is the binary star system that generated the nova.

White Dwarf Supernovae Each time a nova occurs, the white dwarf ejects some of its mass. Each time a nova subsides, the white dwarf begins to accrete matter again. Theoretical models cannot yet tell us whether the net result should be a gradual increase or decrease in the white dwarf's mass, but if it is an increase, a white dwarf's mass might gradually approach the white dwarf limit. The day it reaches that limit will be the white dwarf's last.

Recall that white dwarfs are made largely of carbon. As a white dwarf's mass approaches $1.4M_{Sun}$, its temperature rises enough to allow carbon fusion to begin. Carbon fusion ignites almost instantly throughout the white dwarf, much like the sudden helium flash in low-mass red giants

A white dwarf that reaches the $1.4M_{Sun}$ white dwarf limit will explode completely in a white dwarf supernova.

[Section 13.2] but releasing far more energy. The white dwarf explodes completely in what we will call a **white dwarf supernova.**

Alternatively, a white dwarf may reach the white dwarf limit by merging with its binary-star companion. Ordinarily, two objects in orbit around each other will remain in those orbits unless a third object (or surrounding gas) interacts with them. However, two white dwarfs orbiting especially close together emit *gravitational waves*

Two very closely orbiting objects will emit gravitational waves, causing them to spiral toward each other and ultimately merge.

(see Special Topic, page 373), which carry energy and angular momentum away from the binary system. The two white dwarfs therefore gradually spiral inward toward each other, ultimately merging together.

If the total mass of the two white dwarfs exceeds $1.4 M_{Sun}$, the result of this merger is a white dwarf supernova.

think about it According to our understanding of novae and white dwarf supernovae, can either of these events ever occur with a white dwarf that is *not* a member of a binary star system? Explain.

Regardless of which process leads to a white dwarf supernova, the resulting carbon explosion is quite different from the iron catastrophe that leads to a supernova ending the life of a high-mass star [Section 13.3], which we will call a **massive star supernova.*** Astronomers can distinguish between the two types of supernova by studying their light. Both types shine brilliantly, with peak luminosities about 10 billion times that of the Sun ($10^{10} L_{Sun}$), but the luminosities of white dwarf supernovae fade steadily, while the decline in brightness of a massive star supernova is often more complicated (Figure 14.5). In addition, spectra of white dwarf supernovae lack hydrogen lines (because white dwarfs contain very little hydrogen), while these lines are prominent in the spectra of most massive star supernovae.

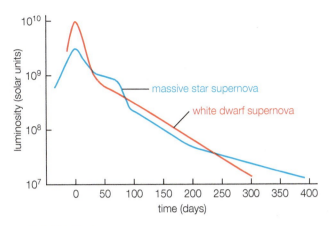

▲ **FIGURE 14.5**

The curves on this graph show how the luminosities of the two types of supernovae fade with time. The white dwarf supernova fades quickly at first and then more gradually a few weeks after the peak, while the massive star supernova fades in a more complicated pattern.

14.2 Neutron Stars

White dwarfs with densities of several tons per teaspoon may seem incredible, but neutron stars are stranger still. The possibility that neutron stars might exist was first proposed in the 1930s, but many astronomers thought it preposterous that nature could make anything so bizarre. Nevertheless, strong evidence now makes it clear that neutron stars really exist.

◆ What is a neutron star?

A **neutron star** is the ball of neutrons created by the collapse of the iron core in a massive star supernova (Figure 14.6). Typically just 10 kilometers in radius yet more massive than the Sun, neutron stars are essentially giant atomic nuclei made almost entirely of neutrons and held together by gravity. Like white dwarfs, neutron stars resist the crush of gravity with degeneracy pressure that arises when particles are packed as closely as nature allows. In the case of neutron stars, however, it is neutrons rather than electrons that are closely packed, so we say that **neutron degeneracy pressure** supports them against the crush of gravity.

A neutron star is a ball of neutrons just a few kilometers in radius but with a mass like that of the Sun.

The force of gravity at the surface of a neutron star is awe-inspiring. Escape velocity is about half the speed of light. If you foolishly chose to visit a neutron star's surface, your body would be squashed immediately into a microscopically thin pancake of subatomic particles.

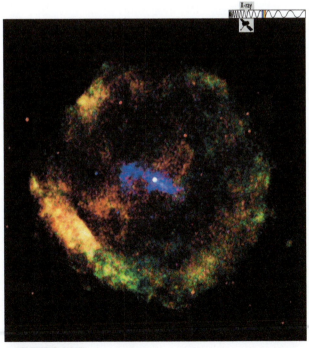

▲ **FIGURE 14.6**

This X-ray image from the Chandra X-Ray Observatory shows the supernova remnant G11.2–03, the remains of a supernova observed by Chinese astronomers in A.D. 386. The white dot at the center is the neutron star left behind by the supernova. The different colors correspond to emission of X rays in different wavelength bands. The region pictured is about 23 light-years across.

---

* Observationally, astronomers classify supernovae as *Type II* if their spectra show hydrogen lines, and *Type I* otherwise. All Type II supernovae are assumed to be massive star supernovae. However, a Type I supernova can be either a white dwarf supernova or a massive star supernova in which the star blew away all its hydrogen before exploding. Type I supernovae fall into three classes, called *Type Ia, Type Ib,* and *Type Ic.* Only Type Ia supernovae are thought to be white dwarf supernovae.

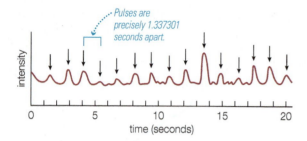

▲ FIGURE 14.7

About 20 seconds of data from the first pulsar discovered by Jocelyn Bell in 1967.

Things would be only slightly less troubling if a bit of neutron star could somehow come to visit you. A paper clip with the density of neutron star material would outweigh Mount Everest. If such a paper clip magically appeared in your hand, you could not prevent it from falling. Down it would plunge, passing through the Earth like a rock falling through air. It would gain speed until it reached Earth's center, and its momentum would carry it onward until it slowed to a stop on the other side of our planet. Then it would fall back down again. If it came in from space, each plunge of the neutron star material would drill a different hole through the rotating Earth. In the words of astronomer Carl Sagan, the inside of the Earth would "look briefly like Swiss cheese" (until the melted rock flowed to fill in the holes) by the time friction finally brought the piece of neutron star to rest at the center of the Earth.

In the unfortunate event that an *entire* neutron star came to visit you, it would not fall at all. Because it would be only about 10 kilometers in radius, the neutron star would probably fit in your hometown.

A neutron star could fit in your hometown, but its gravity would quickly destroy our planet.

Remember, however, that it would be 300,000 times as massive as Earth. As a result, the neutron star's immense surface gravity would quickly destroy your hometown and the rest of civilization. By the time the dust settled, the former Earth would have been squashed into a shell no thicker than your thumb on the surface of the neutron star.

◆ How were neutron stars discovered?

The first observational evidence for neutron stars came in 1967, when a 24-year-old graduate student named Jocelyn Bell discovered a strange source of radio waves. The radio waves pulsed on and off at intervals of precisely 1.337301 seconds (Figure 14.7). At first, astronomers could not think of any natural process that could pulse in such a clockwork way, and they only half-jokingly called the new radio source LGM, for Little Green Men. Today we refer to such rapidly pulsing radio sources as **pulsars.**

The mystery of pulsars was soon solved. By the end of 1968, astronomers had found two smoking guns: Pulsars sat at the centers of two supernova remnants, the Vela Nebula and the Crab Nebula (Figure 14.8). The pulsars are neutron stars left behind by the supernova explosions.

The pulsations arise because the neutron star is spinning rapidly as a result of the conservation of angular momentum [Section 4.3]: As an iron core collapses into a neutron star, its rotation rate must increase as it shrinks in size. The collapse also bunches the magnetic field lines running through the core far more tightly, greatly amplifying the strength of the magnetic field. These intense magnetic fields direct beams of

▶ FIGURE 14.8

This time-lapse image of the pulsar at the center of the Crab Nebula, a supernova remnant, shows its main pulse recurring every 0.033 second. The fainter pulses are thought to come from the pulsar's other lighthouse-like beam. (Photo from the Very Large Telescope of the European Southern Observatory.)

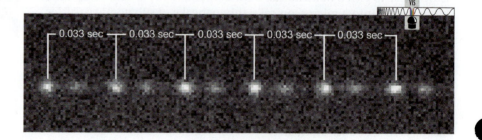

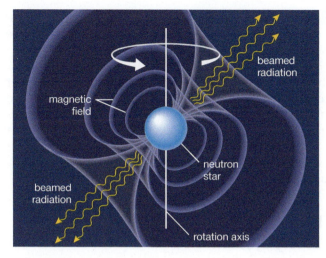

a A pulsar is a rotating neutron star that beams radiation along its magnetic axis.

b If the magnetic axis is not aligned with the rotation axis, the pulsar's beams sweep through space like lighthouse beams. Each time one of the pulsar's beams sweeps across Earth, we see a pulse of radiation.

▲ **FIGURE 14.9**

Radiation from a rotating neutron star can appear to pulse like the beams from a lighthouse.

Neutron stars can spin rapidly and emit beams of radiation along their magnetic poles, which we detect as pulses of radiation if the beams sweep by Earth. radiation out along the magnetic poles, although we do not yet know exactly how they produce that radiation. If a neutron star's magnetic poles are not aligned with its rotation axis, the beams of radiation sweep round and round (Figure 14.9). Like lighthouses, neutron stars emit fairly steady beams of light, but we see a pulse of light each time one of those beams sweeps past Earth.

Pulsars are not quite perfect clocks. The continual twirling of a pulsar's magnetic field generates electromagnetic radiation that carries away energy and angular momentum, causing the neutron star's rotation rate to slow gradually. The pulsar in the Crab Nebula, for example, currently spins about 30 times per second. Two thousand years from now, it will spin less than half as fast. Eventually, a pulsar's spin slows so much and its magnetic field becomes so weak that we can no longer detect it. In addition, some spinning neutron stars may be oriented so that their beams do not sweep past our location. We therefore have the following rule: *All pulsars are neutron stars, but not all neutron stars are pulsars.*

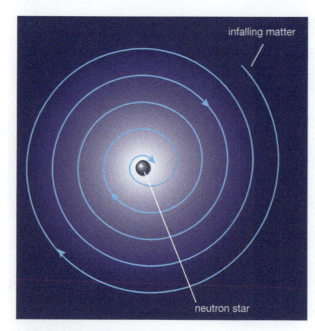

▲ FIGURE 14.10

Matter accreting onto a neutron star adds angular momentum, increasing the neutron star's rate of spin. The spin increases because infalling matter just above the neutron star's surface is orbiting faster than the neutron star is spinning. When this matter hits the neutron star's surface, it boosts the star's spin rate.

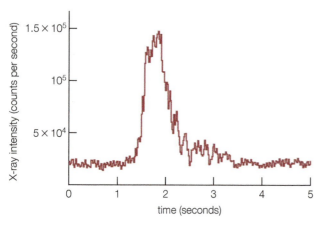

▲ FIGURE 14.11

Light curve of an X-ray burst. In this particular burst, the X-ray luminosity of the neutron star spiked to over six times its usual intensity in a matter of seconds.

think about it Suppose we do not see pulses from a particular neutron star and hence do not call it a pulsar. Is it possible that a civilization living in some other star system would see this neutron star as a pulsar? Explain.

We know that pulsars must be neutron stars because no other massive object could spin so fast. A white dwarf, for example, can spin no faster than about once per second without being torn apart, because a faster spin would mean its surface would be rotating faster than the escape velocity. Pulsars have been discovered that rotate as fast as 716 times per second. Only an object as small and dense as a neutron star could spin so fast without breaking apart.

◆ What can happen to a neutron star in a close binary system?

Like a white dwarf, a neutron star in a close binary system can brilliantly burst back to life as gas overflowing from a companion star creates a hot, swirling accretion disk around it. However, the much stronger gravity of the neutron star makes its accretion disk much hotter and denser than an accretion disk around a white dwarf.

The high temperatures in the inner regions of the accretion disk around a neutron star make it radiate powerfully in X rays. Some close binaries with neutron stars emit 100,000 times as much energy in X rays as our Sun emits in all wavelengths of light combined. Because of this intense X-ray emission, close binaries that contain accreting neutron stars are often called **X-ray binaries.** We have identified hundreds of X-ray binaries in the disk of the Milky Way Galaxy.

The hot accretion disks around neutron stars in close binary systems shine brightly with X rays.

The emission from most X-ray binaries pulsates rapidly as the neutron star spins. However, while other pulsars tend to slow down with time, the pulsation rates of X-ray binaries tend to accelerate, presumably because matter accreting onto the neutron star adds angular momentum (Figure 14.10). Some of these neutron stars rotate so fast that they pulsate every few thousandths of a second (and are called *millisecond pulsars*).

X-Ray Bursts Like accreting white dwarfs that occasionally erupt as novae, accreting neutron stars sporadically erupt with enormous luminosities. Hydrogen-rich material from the companion star builds up on the surface of the neutron star, forming a layer about a meter thick. Pressures at the bottom of this hydrogen layer are high enough to maintain steady fusion, which produces a layer of helium beneath the hydrogen. Helium fusion suddenly ignites when the temperature in this layer builds to 100 million K. The helium fuses rapidly to make carbon and heavier elements, generating a burst of energy that flows from the neutron star in the form of X rays. These **X-ray bursters** typically flare every few hours to every few days (Figure 14.11). Each burst lasts only a few seconds, but during those seconds the system can radiate 100,000 times as much power as the Sun, all in X rays. Within a minute after a burst, the X-ray burster cools back down and accretion resumes.

Neutron Star Mergers Just as we discussed for two closely orbiting white dwarfs, two neutron stars orbiting close together will emit gravitational waves, ultimately causing the two neutron stars to spiral together and merge

(Figure 14.12). The total energy released during this cataclysmic merger can be even greater than that of a massive star supernova, and it happens in an environment rich in neutrons. As a result, neutron star mergers can produce

Gold and other rare elements may have been produced in explosions resulting from neutron star mergers.

a blend of rare elements that differ from those made in a massive star supernova. Most of the neutron star matter remains gravitationally bound to the system, but some of it is ejected into interstellar space. Models of these mergers indicate that they produce most of the gold, platinum, and rare earth elements present in the universe. In other words, the gold in your jewelry was probably made by neutron star mergers.

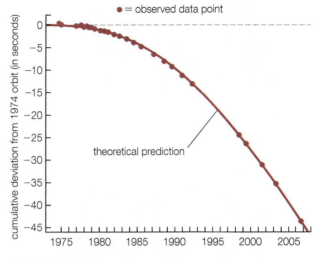

▲ FIGURE 14.12

This diagram shows how the orbital period of a close binary system consisting of two neutron stars has changed over the last several decades. The observed changes precisely match what we expect from a system with an orbital distance that is changing because of gravitational-wave emission. Projecting the trend into the future shows that the two neutrons stars will collide and merge about 85 million years from now. (Data courtesy of Joel Weisberg and Joseph Taylor.)

14.3 Black Holes: Gravity's Ultimate Victory

The story of stellar corpses would be strange enough if it ended with white dwarfs and neutron stars, but it does not. Sometimes, the gravity in a stellar corpse becomes so strong that nothing can prevent the corpse from collapsing under its own weight. The stellar corpse collapses without end, crushing itself out of existence and forming perhaps the most bizarre type of object in the universe: a *black hole*.

◆ What is a black hole?

The "black" in the name **black hole** comes from the fact that nothing—not even light—can escape from a black hole. The escape velocity of any object depends on the strength of its gravity, which depends on its mass and size [Section 4.4]. Decreasing the size of an object of a particular mass makes its gravity stronger and increases its escape velocity. A black hole is so compact that it has an escape velocity greater than the speed of light. Because nothing can travel faster than the speed of light, neither light nor anything else can escape from within a black hole.

The "hole" part of the name *black hole* tells an even stranger story. Einstein discovered that space and time are not distinct, as we usually think of them, but instead are bound up together as four-dimensional

A black hole is a place where gravity is so strong that nothing—not even light—can escape from within it.

spacetime. Moreover, in his general theory of relativity, Einstein showed that what we perceive as gravity arises from *curvature of space-time* (see Special Topic, page 373). Curved spacetime is not easy to visualize, because we can see only three dimensions at once. However, we can understand the idea with a two-dimensional analogy.

Figure 14.13 uses a rubber sheet to represent two-dimensional slices through spacetime. In this analogy, the sheet is flat—corresponding to weak gravity—in a region far from any mass (Figure 14.13a). The sheet becomes curved near a massive object, corresponding to strong gravity (Figure 14.13b), and greater curvature means stronger gravity. In this representation, the curvature of spacetime near a black hole is so great that the rubber sheet forms a bottomless pit (Figure 14.13c). Keep in mind that the illustration is only an analogy, and black holes actually are spherical, not funnel-shaped. Nevertheless, it captures the key idea that a black hole is like a hole in the observable universe in the following sense: If you enter a black hole, you leave our observable universe and can never return.

a A two-dimensional representation of "flat" spacetime. The distances between adjacent circles are the same.

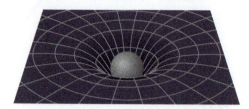

b Gravity arises from curvature of spacetime, represented here by a mass pushing down on the rubber sheet. Notice how the circles become more widely separated near the mass, showing that the curvature is greater as we approach the mass on the sheet.

event horizon

c The curvature of spacetime becomes greater and greater as we approach a black hole, and a black hole itself is a bottomless pit in spacetime.

▲ **FIGURE 14.13**

We can use two-dimensional rubber sheets to show an analogy to curvature in four-dimensional spacetime.

see it for yourself Curved space has different rules of geometry than "flat" space, a difference you can demonstrate with a plastic ball and a marker. Draw a straight line on the ball going one-quarter of the way around it, then make a right-angle turn and draw a line one-quarter of the way around the ball in this new direction, and finish the triangle with a straight line back to where you started. Measure all three angles. How does the sum of the three angles of the triangle on the ball compare to the 180° you would find for that sum in a triangle on a flat surface?

The Event Horizon The boundary between the inside of the black hole and the universe outside is called the **event horizon.** The event horizon essentially marks the point of no return for objects entering a black hole: It is the boundary around a black hole at which the escape velocity equals the speed of light. Nothing that passes within this boundary can ever escape. The event horizon gets its name because we have no hope of learning about any events that occur within it.

We usually think of the "size" of a black hole as the size of its event horizon. Our everyday understanding of "size" is hard to apply inside the event horizon because space and time are so distorted there. However, for someone outside the black hole, the event horizon is shaped like a sphere, and the size of the sphere defines what we call the **Schwarzschild radius** of the black hole (after Karl Schwarzschild, who first computed it with Einstein's general theory of relativity). The Schwarzschild radius depends only on the black hole's mass. A black hole with the mass of the Sun has a Schwarzschild radius of about 3 kilometers—only a little smaller than the radius of a neutron star of the same mass. More massive black holes have larger Schwarzschild radii. For example, a black hole with 10 times the mass of the Sun has a Schwarzschild radius of about 30 kilometers.

In essence, a collapsing stellar core becomes a black hole at the moment it shrinks to a size smaller than its Schwarzschild radius. At that moment, the core disappears from view within its own event horizon. The black hole still contains all the mass and has the gravity associated with that mass, but its outward appearance tells us nothing about what fell in.

Singularity and the Limits to Knowledge We can't observe what happens to a stellar core once it collapses into a black hole, because no information can emerge from within the event horizon. Nevertheless, we can use our understanding of the laws of physics to predict what should occur inside a black hole. Because nothing can stop the crush of gravity in a black hole, all the matter that forms a black hole should ultimately be crushed to an infinitely tiny and dense point in the black hole's center. We call this point a **singularity.**

Unfortunately, the idea of a singularity pushes up against the limits of scientific knowledge today. The problem is that two very successful theories make different predictions about the nature of a singularity. Einstein's general theory of relativity, which successfully explains how gravity works throughout the universe, predicts that spacetime should grow infinitely curved as it enters the pointlike singularity. Quantum physics, which successfully explains the nature of atoms and the spectra of light, predicts that spacetime should fluctuate chaotically near the singularity. These are clearly different claims, and no theory that can reconcile them has yet been found.

◆ **What would it be like to visit a black hole?**

Imagine that you are a pioneer of the future, making the first visit to a black hole. Your target is a black hole with a mass of $10M_{Sun}$ and a Schwarzschild radius of 30 kilometers. As your spaceship approaches the black hole, you fire its engines to put the ship on a circular orbit a few thousand kilometers above the event horizon. This orbit will be perfectly stable—there is no need to worry about getting "sucked in."

Your first task is to test Einstein's general theory of relativity. This theory predicts that time should run more slowly as the force of gravity grows stronger. It also predicts that light coming out of a strong gravitational field should show a redshift, called a *gravitational redshift*, that is due to gravity rather than to the Doppler effect. You test these predictions with the aid of two identical clocks whose numerals glow with blue light. You keep one clock aboard the ship and push the other one, with a small rocket attached, directly toward the black hole (Figure 14.14). The small rocket automatically fires its engines just enough so that the clock falls

---

special topic | General Relativity and Curvature of Spacetime

Einstein's special theory of relativity (see the Special Topic on page 364) is called *special* because it applies only to the special case in which we can ignore gravity. After its publication in 1905, Einstein therefore sought to generalize the theory to include gravity.

In 1907, Einstein hit upon what he later called "the happiest thought of my life." He realized that effects of gravity and effects of acceleration are indistinguishable on small size scales, leading him to conclude that they must be equivalent; we call this idea the *equivalence principle*. To clarify its meaning, imagine that you are sitting in a room with no windows or doors when the room is magically removed from Earth and sent hurtling through space with the acceleration of gravity on Earth, or 9.8 m/s² [Section 4.1]. According to the equivalence principle, you would have no way of knowing that you'd left Earth. Any experiment you performed, such as dropping balls of different weights, would yield the same results you'd get on Earth. Likewise, experiments performed in a freely falling elevator would yield the same results as those performed in an elevator drifting at constant velocity through empty space.

Einstein's new point of view on motion, acceleration, and gravity brought about a radical revision of how we think about space and time. Instead of thinking about the three dimensions of space and the one dimension of time as separate, we learned to think of them as a seamless, four-dimensional entity known as *spacetime*. Einstein showed that a person would feel weightless, as though drifting through empty space, as long as the person's path through four-dimensional spacetime was as straight as possible.

So why do astronauts feel weightless even though they orbit Earth on a curved path? According to general relativity, they are still following the *straightest possible* path through four-dimensional spacetime. The path is curved only because spacetime itself is curved near the Earth. In other words, while Newton would have attributed the curved orbital path of the astronauts to the force of gravity, Einstein attributes it to curvature of spacetime. That is, *gravity arises from curvature of spacetime*.

The curvature of spacetime is caused by mass, and stronger gravity just means greater spacetime curvature. For example, the Sun's gravity curves spacetime more than Earth's gravity, and the strong gravity on the surface of a white dwarf curves spacetime more than the gravity on the surface of the Sun. Although we cannot visualize spacetime curvature, we can visualize two-dimensional analogies to it with rubber sheet diagrams like those shown in Figure 14.13. From this new perspective, planets orbit the Sun because of the way

space is curved by the Sun: Each planet is moving as straight as it can, but space is curved in a way that keeps it going round and round somewhat like a marble in a salad bowl (see figure).

Given that we cannot actually perceive all four dimensions of spacetime at once, you may wonder why scientists think spacetime curvature is real. The answer is that Einstein's theory predicts measurable effects from this curvature. For example, if gravity really does arise from curvature of spacetime, then light paths ought to bend when they pass near large masses. Scientists have measured such bending of starlight as it passes near the Sun, and the amount of bending is precisely what general relativity predicts it should be. We have also observed this bending of light by distant galaxies, a phenomenon called *gravitational lensing* [Section 18.2]. Other verified predictions of general relativity include *gravitational time dilation*, which is the idea that time runs slower in regions of stronger gravity (GPS navigation makes use of this fact), and the idea that closely orbiting massive objects—such as binary systems containing white dwarfs, neutron stars, or black holes—will emit *gravitational waves*. These waves travel at the speed of light and carry information about changes in the structure of spacetime in much the same way that electromagnetic waves [Section 5.2] carry information about changes in electric or magnetic fields. Strange as it may seem, we live in a four-dimensional universe in which space and time are intertwined and can never be disentangled.

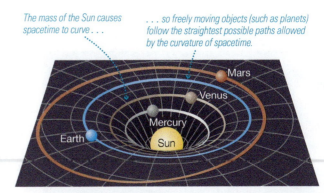

The mass of the Sun causes spacetime to curve . . .

. . . so freely moving objects (such as planets) follow the straightest possible paths allowed by the curvature of spacetime.

Mars · Venus · Mercury · Sun · Earth

According to general relativity, planets orbit the Sun for much the same reason that you can make a marble go around in a salad bowl: The planet is going as straight as it can, but the curvature of spacetime causes its path through space to curve.

The Schwarzschild Radius

The Schwarzschild radius (R_S) of a black hole is given by a simple formula:

$$\text{Schwarzschild radius} = R_S = \frac{2GM}{c^2}$$

where M is the black hole's mass, the gravitational constant $G = 6.67 \times 10^{-11}$ m³/(kg × s²), and the speed of light is $c = 3 \times 10^8$ m/s. With a bit of calculation, this formula can also be written as

$$\text{Schwarzschild radius} = R_S = 3.0 \times \frac{M}{M_{Sun}}\ \text{km}$$

Example: What is the Schwarzschild radius of a black hole with a mass of $10 M_{Sun}$?

Solution: We are given the black hole's mass in solar units, so we use the second formula above and set $M = 10 M_{Sun}$:

$$R_S = 3.0 \times \frac{10 M_{Sun}}{M_{Sun}}\ \text{km} = 30\ \text{km}$$

The Schwarzschild radius of a $10 M_{Sun}$ black hole is about 30 km—roughly the radius of a large city on Earth, but containing 10 times as much mass as the Sun.

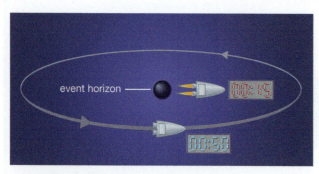

▲ **FIGURE 14.14**

Time runs more slowly on the clock nearer to the black hole, and gravitational redshift makes its glowing blue numerals appear red from your orbiting spaceship.

gradually toward the event horizon. Sure enough, the clock on the rocket ticks more slowly as it heads toward the black hole, and its light becomes increasingly redshifted. When the clock reaches a distance of about 10 kilometers above the event horizon, you see it ticking only half as fast as the clock on your spaceship, and its numerals are red instead of blue.

The rocket has to expend fuel rapidly to keep the clock hovering in the strong gravitational field, and the fuel soon runs out. The clock plunges toward the black hole. From your safe vantage point inside your spaceship, you see the clock ticking more and more slowly as it falls. However, you soon need a radio telescope to "see" it, as the light from the clock face shifts from the red part of the visible spectrum to the infrared, and on into the radio. Finally, its light becomes so redshifted that no conceivable telescope could detect it. Just as the clock vanishes from view, you see that the time on its face has frozen.

Curiosity overwhelms the better judgment of one of your colleagues. He hurriedly climbs into a space suit, grabs the other clock, resets it, and jumps out of the air lock on a trajectory aimed straight for the black hole. Down he falls, clock in hand. He watches the clock, but because he and the clock are traveling together, its time seems to run normally and its numerals stay blue. From his point of view, time seems to neither speed up nor slow down. When his clock reads, say, 00:30, he and the clock pass through the event horizon. There is no barrier, no wall, no hard surface. The event horizon is a mathematical boundary, not a physical one. From his point of view, the clock keeps ticking. He is inside the event horizon, the first human being ever to vanish into a black hole.

Back on the spaceship, you watch in horror as your overly curious friend plunges to his death. Yet, from your point of view, he will *never* cross the event horizon. You'll see time come to a stop for him and his clock just as he vanishes from view because of the huge gravitational redshift of light. When you return home, you can play a video for the judges at your trial, proving that your friend is still outside the black hole. Strange as it may seem, all this is true according to Einstein's theory. From your point of view, your friend takes *forever* to cross the event horizon (even though he vanishes from view because of his ever-increasing redshift). From his point of view, it is but a moment's plunge before he passes into oblivion.

If you fell toward a black hole, you would rapidly accelerate and soon cross the event horizon. But to someone watching from afar, your fall would appear to take forever.

The truly sad part of this story is that your friend did not live to experience the crossing of the event horizon. The force of gravity grew so quickly as he approached the black hole that it pulled much harder on his feet than on his head, simultaneously stretching him lengthwise and squeezing him from side to side (Figure 14.15). In essence, your friend was stretched in the same way the oceans are stretched by the tides, except that the *tidal force* near the black hole is trillions of times stronger than the tidal force of the Moon on Earth [Section 4.4]. No human could survive it.

If he had thought ahead, your friend might have waited to make his jump until you visited a much larger black hole, like one of the *supermassive black holes* thought to reside in the centers of many galaxies [Section 16.4]. A 1 billion solar mass ($10^9 M_{Sun}$) black hole has a Schwarzschild radius of 3 billion kilometers—about the distance from our Sun to Uranus. Although the difficulty of escape from the event horizon of any black hole is equally great, the larger size of a supermassive black hole makes its tidal forces much weaker and hence nonlethal. Your friend could safely plunge through the event horizon. Unfortunately, anything he saw or learned on his continuing plunge toward oblivion would be known to him alone, because there would be no way for him to send information back to you on the outside.

◆ Do black holes really exist?

As was the case for neutron stars, at first most astronomers who contemplated the idea of black holes thought them too strange to be true. Today, our understanding of physics gives us reason to think that black holes ought to be fairly common, and observational evidence strongly suggests that black holes really exist.

The Formation of a Black Hole

The idea that black holes ought to exist comes from considering how they might form. Recall that white dwarfs cannot exceed $1.4 M_{Sun}$, because gravity overcomes electron degeneracy pressure above that mass. Calculations show that the mass of a neutron star has a similar limit that lies somewhere between about 2 and 3 solar masses. Above this mass, neutron degeneracy pressure cannot hold off the crush of gravity in a collapsing stellar core.

A massive star supernova occurs when the electron degeneracy pressure supporting the iron core of a massive star succumbs to gravity, causing the core to collapse catastrophically into a ball of neutrons [Section 13.3]. That is why most supernovae leave neutron stars behind. However, theoretical models show that very massive stars might not succeed in blowing away all their upper layers. If enough matter falls back onto the neutron core, its mass may rise above the neutron star limit. Likewise, a merger between two neutron stars can produce an object that exceeds the neutron-star mass limit.

As soon as the core exceeds the neutron star limit, gravity overcomes the neutron degeneracy pressure and the core collapses once again. This time, no known force can keep the core from collapsing into oblivion as a black hole. Moreover, another consequence of Einstein's theory of relativity makes it highly unlikely that any as-yet-unknown force could intervene either.

Recall that Einstein's theory tells us that energy is equivalent to mass ($E = mc^2$) [Section 4.3], implying that energy, like mass, must exert some gravitational attraction. The gravity of pure energy usually is negligible, but not in a stellar core collapsing beyond the neutron star limit.

According to the known laws of physics, nothing can stop the collapse of a stellar corpse with a mass greater than about 3 solar masses.

Inside a star that collapses beyond the point where it could have been a neutron star, the energy associated with the enhanced internal temperature and pressure acts like additional mass, making the crushing power of gravity even stronger. The more the star collapses, the stronger gravity gets. To the best of our understanding, *nothing* can halt the crush of gravity at this point, so the star must inevitably collapse to form a black hole.

Observational Evidence for Black Holes

The fact that black holes emit no light might make it seem as if they should be impossible to detect. However, a black hole's gravity can influence its surroundings in a way that reveals its presence. Astronomers have discovered many objects that show the telltale signs of an unseen gravitational influence with a large enough mass to suggest a black hole.

Strong observational evidence for black holes formed by supernovae comes from studies of X-ray binaries. Recall that the accretion disk around a neutron star in a close binary system can emit strong X-ray

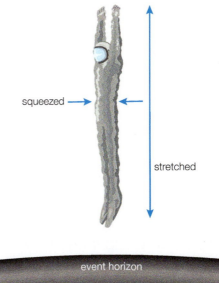

squeezed →

← stretched

event horizon

▲ **FIGURE 14.15**
Tidal forces would be lethal near a black hole formed by the collapse of a star. The black hole's gravity would pull more strongly on the astronaut's feet than on his head, stretching him lengthwise and squeezing him from side to side.

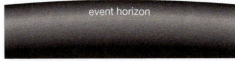

common misconceptions

Black Holes Don't Suck

What would happen if our Sun suddenly became a black hole? For some reason, the idea that Earth and the other planets would inevitably be "sucked in" by the black hole has become part of our popular culture. But it is not true. Although the sudden disappearance of the Sun's light and heat would make our planet cold and dark, Earth's orbit would not change.

Newton's law of gravity tells us that the allowed orbits in a gravitational field are ellipses, hyperbolas, and parabolas [Section 4.4]. Note that "sucking" is not on the list! A spaceship would get into trouble only if it came so close to a black hole—within about three times the Schwarzschild radius—that the force of gravity would deviate significantly from what Newton's law predicts. Otherwise, a spaceship passing near a black hole would simply swing around it on an ordinary orbit (ellipse, parabola, or hyperbola). In fact, because most black holes are so small—typical Schwarzschild radii are far smaller than those of any star or planet—a black hole is actually one of the most difficult things in the universe to fall into by accident.

Some X-ray binaries probably contain accreting black holes rather than accreting neutron stars.

radiation, resulting in an X-ray binary. The accretion disk forms because the neutron star's strong gravity pulls mass from its companion star. Because a black hole has even stronger gravity than a neutron star, a black hole in a close binary system should also be surrounded by a hot, X ray–emitting accretion disk. Some X-ray binaries may therefore contain black holes rather than neutron stars.

One of the most promising black hole candidates is in an X-ray binary called Cygnus X-1 (Figure 14.16). This system contains an extremely luminous star with an estimated mass of $19M_{Sun}$. Based on Doppler shifts of its spectral lines, astronomers have concluded that this star orbits a compact, unseen companion with a mass of about $15M_{Sun}$. Although there is some uncertainty in these mass estimates, the mass of the invisible accreting object clearly exceeds the $3M_{Sun}$ neutron star limit. It is therefore too massive to be a neutron star, so according to current knowledge it must be a black hole. A few dozen other X-ray binaries offer similar evidence for black holes formed

extraordinary claims Neutron Stars and Black Holes Are Real

Subrahmanyan Chandrasekhar

Theoretical calculations predicted the existence of neutron stars and black holes long before their observational discovery, but many astronomers considered these theoretical results too strange to be true.

Subrahmanyan Chandrasekhar, an astrophysicist from India, was only 19 when he completed the calculations showing that there is a white dwarf limit of $1.4M_{Sun}$, and he boldly predicted that a more massive white dwarf would collapse under the force of gravity. He did this work in 1931 while traveling by ship to England, where he hoped to impress the eminent British astrophysicist Sir Arthur Stanley Eddington. However, Eddington ridiculed Chandrasekhar for believing that white dwarfs could collapse. Neutrons had not yet been discovered, little was known about fusion, and no one had any idea what supernovae were. The idea of gravity achieving an ultimate victory seemed nonsensical to Eddington, who speculated that some type of force must prevent gravity from crushing any object.

A few more radical thinkers took collapsing stars more seriously. A Russian physicist, Lev Davidovich Landau, independently computed the white dwarf limit

Sir Arthur Stanley Eddington

Lev Davidovich Landau

in 1932. Neutrons were discovered just a few months later, and Landau speculated that stellar corpses above the white dwarf limit might collapse until neutron degeneracy pressure halted the crush of gravity. While most astronomers found the idea of neutron stars to be unacceptably weird, two European scientists who had emigrated to California, Fritz Zwicky and Walter Baade, were not so skeptical. Without knowing of Landau's ideas, they also independently

Robert Oppenheimer

concluded that neutron stars were possible. In 1934, they suggested that a supernova might result when a stellar core collapses and forms a neutron star—an extraordinarily insightful guess. By 1938, physicist Robert Oppenheimer, working at Berkeley, was contemplating whether neutron stars had a limiting mass of their own. He and his coworkers concluded that the answer was yes and that neutron degeneracy pressure could not resist the crush of gravity when the mass rose above a few solar masses. Because no known force could keep such a star from collapsing indefinitely, Oppenheimer speculated that gravity would achieve ultimate victory, crushing the star into a black hole.

Astronomers gradually came to accept Chandrasekhar's $1.4M_{Sun}$ white dwarf limit, because observations found no white dwarfs more massive than this. However, most astronomers held to a belief that high-mass stars would inevitably shed enough mass late in life to prevent the formation of a more massive collapsed object. Jocelyn Bell's 1967 discovery of pulsars shattered this belief. Within a few months, Thomas Gold of Cornell University correctly suggested that the pulsars were spinning neutron stars. These discoveries forced astronomers to acknowledge that nature was far stranger than they had expected. The verification that neutron stars really exist made the prospect of the still stranger black holes much less difficult to accept.

Jocelyn Bell

Chandrasekhar, who had long since moved to the University of Chicago, was awarded a Nobel Prize in 1984 for his lifelong contributions to astronomy. Landau won a Nobel Prize in 1962 for his work on condensed states of matter. Oppenheimer went on to lead the Manhattan Project, which developed the atomic bomb in 1945. Eddington died in 1944, still convinced that white dwarf stars could not collapse.

Verdict: Strong evidence supports the reality of neutron stars and black holes.

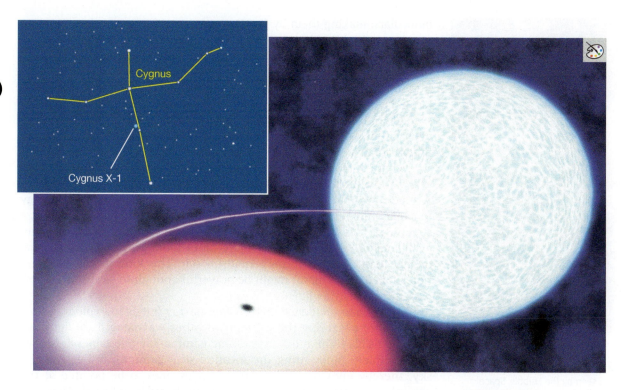

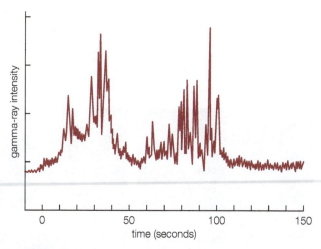

from the collapse of massive stellar cores. As we'll discuss in coming chapters, an even greater body of evidence suggests the existence of *supermassive black holes*—some with masses millions or billions of times that of our Sun—residing at the centers of many galaxies, including our own Milky Way.

think about it Recall that some X-ray binaries that contain neutron stars emit frequent X-ray bursts and are called *X-ray bursters*. Could an X-ray binary that contains a black hole exhibit the same type of X-ray bursts? Why or why not? (*Hint:* Where do the X-ray bursts occur in an X-ray binary with a neutron star?)

14.4 Extreme Events

Supernovae that produce neutron stars are rare and enormously energetic events. However, the events accompanying the formation and growth of black holes are rarer and even more energetic. To identify such events, astronomers watch for extreme outbursts of energy. Two kinds of events are receiving particularly close attention: gamma-ray bursts and gravitational-wave pulses.

◆ What causes gamma-ray bursts?

In the early 1960s, the United States began launching a series of top-secret satellites designed to look for gamma rays emitted by nuclear bomb tests. The satellites soon started detecting occasional bursts of gamma rays, typically lasting a few seconds (Figure 14.17). It took several years for military scientists to become convinced that these **gamma-ray bursts** were coming from space, not from some sinister human activity, and the military then made the discovery public. We have since learned that gamma-ray bursts represent explosions of almost unimaginable power.

Gamma-ray bursts originating billions of light-years from Earth can produce afterglows of visible light (Figure 14.18) that can be seen through

▲ **FIGURE 14.16**

Artist's conception of the Cygnus X-1 system, so named because it is the brightest X-ray source in the constellation Cygnus. The X rays come from the high-temperature gas in the accretion disk surrounding the black hole. The inset shows the location of Cygnus X-1 in the sky.

▲ **FIGURE 14.17**

The intensity of a gamma-ray burst can fluctuate dramatically over time periods of just a few seconds.

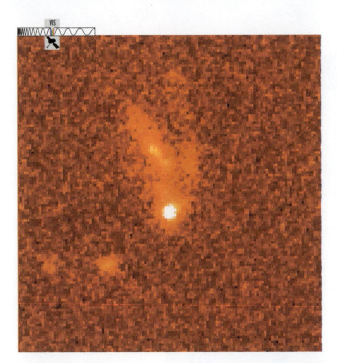

▲ FIGURE 14.18

The bright dot near the center of this photo is the visible-light afterglow of a gamma-ray burst, as seen by the Hubble Space Telescope. The elongated blob extending above the dot is the distant galaxy in which the burst occurred.

binoculars, making them by far the most powerful bursts of energy that we observe in the universe. If a gamma-ray burst emitted light equally in all directions, like a light bulb, then its total luminosity would briefly exceed the combined luminosity of a million galaxies like our Milky Way. Because such a high luminosity is very difficult to explain, scientists think it's likely that gamma-ray bursts channel their energy into narrow searchlight beams, like pulsars, so only some are visible from Earth. But even in this case, for a brief moment, a gamma-ray burst is far more luminous than anything else in the universe.

What could cause such massive outbursts of energy? At least some gamma-ray bursts come from extremely powerful supernova explosions. A supernova that forms a neutron star does not release enough energy to explain the luminosity of the brightest gamma-ray bursts. However, a supernova that forms a black hole crushes even more matter into an even smaller radius, releasing many times more gravitational potential energy than one that forms a neutron star. This kind of event (sometimes called a *hypernova*) might be powerful enough to explain the most extreme gamma-ray bursts.

The primary evidence linking gamma-ray bursts to exploding stars comes from observing these events in other wavelengths of light, including visible light and X rays. Some gamma-ray telescopes are capable of rapidly pinpointing the location of a gamma-ray burst in the sky, thereby allowing other types of telescopes to be pointed at the gamma-ray source almost as soon as the burst is detected. Observations like these have shown that at least some gamma-ray bursts coincide with powerful supernova explosions (Figure 14.19). In a number of other cases, gamma-ray bursts have been observed to come from distant galaxies that are actively forming new stars. Those observations also support the idea that gamma-ray bursts are produced in the explosions of extremely massive stars, because such stars are very short-lived and therefore should be found only in places where stars are actively forming.

Supernovae forming black holes probably account for most gamma-ray bursts, but there is another type whose origin has been harder to verify. Bursts of this second type last only a few seconds, making them more difficult to catch in action with telescopes for other wavelengths. Follow-up observations of these short bursts (which represent about 30% of all gamma-ray bursts) have shown that they do *not* come from supernovae. What could be causing them? The leading hypothesis is a collision in a binary system containing either two neutron stars or a neutron star and a black hole that results in the formation of a more massive black hole. Scientists are actively studying these

▼ FIGURE 14.19

The bright object marked "source of gamma-ray burst" in the first panel of this sequence is the visible-light afterglow of a gamma-ray burst in the outskirts of a distant galaxy (marked "host galaxy"). The light detected from this object once the initial afterglow faded showed that it had the light curve of a massive star supernova.

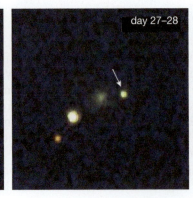

◀ **FIGURE 14.20**
These three panels illustrate gravitational waves coming from a black hole binary. They are based on a supercomputer simulation of the final moments before the black holes merge, and time progresses from left to right. Gravitational waves are depicted in red, orange, and yellow, in order of increasing intensity. The sequence shows that the gravitational-wave intensity from a black hole binary system rapidly increases as the black holes spiral together.

short gamma-ray bursts in hopes of confirming, or contradicting, this hypothesis.

What happens when black holes merge?

We have already seen that the orbits of both binary white dwarf systems and binary neutron star systems decay because gravitational waves carry away some of their orbital energy and angular momentum. The same basic physics also applies to binary systems that consist of two black holes orbiting each other, but the greater masses of the black holes (compared to white dwarfs and neutron stars) lead to much stronger emission of gravitational waves (Figure 14.20). As with the white dwarf and neutron star systems, the two black holes can ultimately merge to make a single, more massive black hole.

Models predict that gravitational waves from a black hole binary become particularly intense during the final death spiral preceding a merger. The resulting pulses of gravitational radiation are so powerful that we can detect black hole mergers that happen billions of light-years away. The Advanced LIGO experiment (see Figure 5.24) first detected such gravitational waves in 2015 (Figure 14.21).

Scientists have detected gravitational waves from merging black holes.

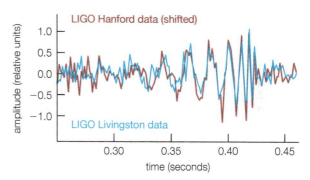

▲ **FIGURE 14.21**
Gravitational-wave signal observed on September 14, 2015 by both detectors of the Advanced LIGO experiment, which are located in Hanford, Washington and Livingston, Louisiana. In order to measure the signal, the LIGO detectors needed to be sensitive enough to record vibrations (strain) of less than one part in 10^{21}. Analysis of the signal shows that it came from the merger of two black holes, each about 30 solar masses, at a distance of 1.3 billion light-years.

---

the big picture | Putting Chapter 14 into Perspective

We have now seen the mind-bending consequences of stellar death. As you think about the bizarre objects of the stellar graveyard described in this chapter, try to keep in mind these "big picture" ideas:

- Despite the strange nature of stellar corpses, clear evidence exists for white dwarfs and neutron stars, and the case for black holes is very strong.

- White dwarfs, neutron stars, and black holes can all have close stellar companions from which they accrete

matter. These binary systems produce some of the most spectacular events in the universe, including novae, white dwarf supernovae, and X-ray bursters.

- Black holes are holes in the observable universe that strongly warp space and time around them. The nature of black hole singularities remains beyond the frontier of current scientific understanding.

my cosmic perspective The idea of a black hole—something with a pull so strong that nothing can escape—is a common metaphor in popular culture, but we humans have little to fear from real black holes. They are at such great distances that their powerful gravitational forces have little chance of ever affecting us.

14.1 White Dwarfs

◆ **What is a white dwarf?**

A white dwarf is the core left over from a low-mass star, supported against the crush of gravity by **electron degeneracy pressure.** A white dwarf typically has the mass of the Sun compressed into a size no larger than Earth.

◆ **What can happen to a white dwarf in a close binary system?**

A white dwarf in a close binary system can acquire hydrogen from its companion through an **accretion disk** that swirls toward the white dwarf's surface. As hydrogen builds up on the white dwarf's surface, it may begin nuclear fusion and cause a **nova** that, for a few weeks, may shine as brightly as 100,000 Suns. In extreme cases, accretion or a merger of two white dwarfs may cause a white dwarf's mass to reach the **white dwarf limit** of $1.4M_{Sun}$, at which point the white dwarf will explode as a **white dwarf supernova.**

14.2 Neutron Stars

◆ **What is a neutron star?**

A neutron star is the ball of neutrons created by the collapse of the iron core in a massive star supernova. It resembles a giant atomic nucleus 10 kilometers in radius and with more mass than the Sun.

◆ **How were neutron stars discovered?**

Neutron stars spin rapidly when they are born, and their strong magnetic fields can direct beams of radiation that sweep through space as the neutron star spins. We detect such neutron stars as **pulsars,** and these pulsars provided the first direct evidence for the existence of neutron stars.

◆ **What can happen to a neutron star in a close binary system?**

Neutron stars in close binary systems can accrete hydrogen from their companions, forming dense, hot accretion disks. The hot gas emits strongly in X rays, so we see these systems as **X-ray binaries.** In some of these systems, frequent bursts of helium fusion occur on the neutron star's surface, causing **X-ray bursts.** Neutron stars can also merge together, and such mergers may be the primary sources of gold and various other rare elements.

14.3 Black Holes: Gravity's Ultimate Victory

◆ **What is a black hole?**

A black hole is a place where gravity has crushed matter into oblivion, creating a hole in the universe from which nothing can escape, not even light. The **event horizon** marks the boundary between our observable universe and the inside of the black hole; the size of a black hole is characterized by its **Schwarzschild radius.**

◆ **What would it be like to visit a black hole?**

You could orbit a black hole just as you could any other object of the same mass. However, you'd see some strange effects if you watched an object fall toward the black hole: Time would seem to run more slowly for the object, and light from it would appear increasingly redshifted as the object fell closer to the black hole. The object would never appear to pass through the event horizon but instead would disappear from view as its light became too redshifted for any instrument to detect it.

◆ **Do black holes really exist?**

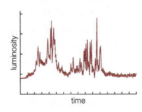

No known force can stop the collapse of a stellar corpse with a mass above the neutron star limit of 2 to 3 solar masses, and theoretical studies of supernovae suggest that such objects should sometimes form. Observational evidence supports this idea: Some X-ray binaries include compact objects far too massive to be neutron stars, making it likely that they are black holes.

14.4 Extreme Events

◆ **What causes gamma-ray bursts?**

Gamma-ray bursts occur in distant galaxies but shine so brightly in the sky that they must be the most powerful explosions we ever observe in the universe. Most gamma-ray bursts appear to come from unusually powerful supernova explosions that may create black holes. Less common short gamma-ray bursts may come from mergers of neutron stars in close binary systems.

◆ **What happens when black holes merge?**

Close binary systems consisting of two black holes emit gravitational waves that causes the average orbital distance to decrease with time until the black holes merge. Gravitational waves from such systems have recently been detected and represent some of the strongest evidence we have for black holes.

visual skills check

Check your understanding of some of the many types of visual information used in astronomy. For additional practice, try the Chapter 14 Visual Quiz at MasteringAstronomy®.

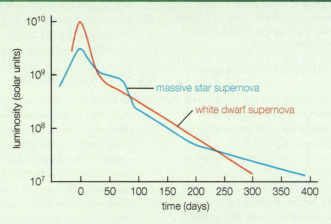

Figure 14.5, repeated above, shows how the luminosities of supernovae change with time. Answer the following questions, using the information provided in the figure.

1. At peak brightness, the white dwarf supernova is approximately _____ times as luminous as the massive star supernova at peak brightness.
   a. 1.5   b. 3
   c. 10   d. 100

2. The luminosity of the white dwarf supernova 175 days after it reaches peak brightness is about _____ of its luminosity at peak brightness.
   a. 30%   b. 10%
   c. 3%   d. 1%

3. Approximately how many days does it take for a white dwarf supernova to decline to 10% of its peak brightness?
   a. 3 days   b. 30 days
   c. 170 days   d. 300 days

4. Approximately how many days does it take for a massive star supernova to decline to 10% of its peak brightness?
   a. 10 days   b. 30 days
   c. 100 days   d. 300 days

5. Approximately how many days does it take for a massive star supernova to decline to 1% of its peak brightness?
   a. 3 days   b. 30 days
   c. 170 days   d. 300 days

exercises and problems

MasteringAstronomy®   For instructor-assigned homework and other learning materials, go to MasteringAstronomy®.

Review Questions

1. What is degeneracy pressure, and how is it important to white dwarfs and neutron stars? What is the difference between *electron degeneracy pressure* and *neutron degeneracy pressure*?

2. Describe the mass, size, and density of a typical white dwarf. How does the size of a white dwarf depend on its mass?

3. What happens to the electron speeds in a more massive white dwarf, and how does this idea lead to the *white dwarf limit* for mass?

4. What is an *accretion disk*? Describe how an accretion disk can provide a white dwarf with a new source of energy.

5. What is a *nova*? Describe the process that creates a nova and what a nova looks like.

6. What processes may cause a *white dwarf supernova*? Observationally, how do we distinguish white dwarf and massive star supernovae?

7. Describe the mass, size, and density of a typical neutron star. What would happen if a neutron star came to your hometown?

8. How do we know that *pulsars* are neutron stars? Are all neutron stars also pulsars? Explain.

9. Explain how the presence of a neutron star can make a close binary star system appear to us as an *X-ray binary*. Why do some of these systems appear to us as *X-ray bursters*?

10. In what sense is a black hole like a hole in the observable universe? Define the *event horizon* and *Schwarzschild radius*.

11. What do we mean by the *singularity* of a black hole? How do we know that our current theories are inadequate to explain what happens at the singularity?

12. Suppose you are falling into a black hole. How will you perceive the passage of your own time? How will outside observers see time passing for you? Briefly explain why your trip is likely to be lethal.

13. Why do we think that black holes should sometimes be formed by supernovae? What observational evidence supports the existence of black holes?

14. What are *gamma-ray bursts*, and how do we think they are produced?

15. Why can emission of *gravitational waves* lead to mergers of white dwarfs, neutron stars, and black holes? What can result from such mergers? How and when was a black hole merger first detected?

Test Your Understanding

⟳ Does It Make Sense?

Decide whether the statement makes sense (or is clearly true) or does not make sense (or is clearly false). Explain clearly; not all these have definitive answers, so your explanation is more important than your chosen answer.

16. The white dwarf at the center of the Helix Nebula has a mass three times the mass of our Sun.

17. I observed a white dwarf supernova occurring at the location of an isolated white dwarf (not a member of a binary system).

18. If you want to find a pulsar, you should look near the remnant of a supernova described by ancient Chinese astronomers.

19. Scientists have just learned that there is a $10M_{Sun}$ black hole lurking near Pluto's orbit.

20. If your spaceship flew within a few thousand kilometers of a black hole, you and your ship would be rapidly sucked into it.

21. We can detect black holes with X-ray telescopes because matter falling into a black hole emits X rays after it smashes into the event horizon.

22. From your point of view, an object falling toward a black hole will *never* cross the event horizon.

23. The best way to search for black holes is to look for small black circles in the sky.

24. Gamma-ray bursts are more likely to be observed in galaxies that are rapidly forming new stars than in galaxies containing only old stars.

25. Gravitational waves are best observed with the Hubble Space Telescope.

Quick Quiz

Choose the best answer to each of the following. Explain your reasoning with one or more complete sentences.

26. Which of these objects has the *smallest* radius? (a) a $1.2M_{Sun}$ white dwarf (b) a $0.6M_{Sun}$ white dwarf (c) Jupiter

27. Which of these objects has the *largest* radius? (a) a $1.2M_{Sun}$ white dwarf (b) a $1.5M_{Sun}$ neutron star (c) a $3.0M_{Sun}$ black hole

28. If we see a nova, we know that we are observing (a) a rapidly rotating neutron star. (b) a gamma ray–emitting supernova. (c) a white dwarf in a binary system.

29. What would happen if the Sun suddenly became a black hole without changing its mass? (a) The black hole would quickly suck in Earth. (b) Earth would gradually spiral into the black hole. (c) Earth would remain in the same orbit.

30. What would happen to a neutron star with an accretion disk orbiting in a direction opposite to the neutron star's spin? (a) Its spin would speed up. (b) Its spin would slow down. (c) Its spin would stay the same.

31. Why do some pairs of neutron stars collide and merge? (a) Occasionally a neutron star moving through space will collide head-on with another neutron star. (b) Gravitational waves from close neutron star binary systems carry away orbital energy and angular momentum. (c) Electromagnetic waves from pulsars carry away angular momentum.

32. Which of these binary systems is most likely to contain a black hole? (a) an X-ray binary containing an O star and another object of equal mass (b) a binary with an X-ray burster (c) an X-ray binary containing a G star and another object of equal mass

33. Viewed from a distance, how would a flashing red light appear as it fell into a black hole? (a) It would appear to flash more quickly. (b) Its flashes would appear bluer. (c) Its flashes would shift to the infrared part of the spectrum.

34. Which of these black holes exerts the *weakest* tidal forces on an object near its event horizon? (a) a $10M_{Sun}$ black hole (b) a $100M_{Sun}$ black hole (c) a 10^6M_{Sun} black hole

35. Where do gamma-ray bursts tend to come from? (a) neutron stars in our galaxy (b) binary systems that also emit X-ray bursts (c) extremely distant galaxies

⟳ Process of Science

36. *Do Black Holes Really Exist?* It is difficult to prove beyond a doubt that black holes really exist because they emit no light. Thus, we cannot see them directly but instead must infer their existence from their gravitational effects on nearby objects. Review the evidence for the existence of black holes presented in this chapter, decide whether you find it convincing or unconvincing, and then defend your position in a one- to two-page essay.

37. *Unanswered Questions.* You have seen in this chapter that current theoretical models make numerous predictions about the nature of black holes but leave many questions unanswered. Briefly describe one important but unanswered question related to black holes. If you think it will be possible to answer this question in the future, describe how we could find an answer, being as specific as possible about the evidence needed. If you think the question will never be answered, explain why you think it is impossible to answer.

Group Work Exercise

38. *A Closer Look at Cygnus X-1.* **Roles:** *Scribe* (takes notes on the group's activities), *Proposer* (proposes explanations to the group), *Skeptic* (points out weaknesses in proposed explanations), *Moderator* (leads group discussion and makes sure everyone contributes). **Activity:** The close binary system Cygnus X-1 contains a supergiant of spectral type O and a black hole with a mass of about $15M_{Sun}$. Consider this system and the artist's conception of it in Figure 14.16 as you do the following.

    a. The *Scribe* and *Proposer* should work together to estimate the radius of the supergiant star using the H-R diagram in Figure 12.10.

    b. The *Skeptic* and *Moderator* should work together to estimate the Schwarzschild radius of the black hole using Cosmic Calculations 14.1.

    c. As a team, compare the results from parts a and b to determine the relative sizes of the O star and the black hole. How much larger is the O star?

    d. The supergiant star in Figure 14.16 is drawn about 5 cm across. Using the team's estimates for the radii of the actual objects in this system, the *Moderator* should determine the approximate size the black hole would have in this figure if it were drawn to scale. How far off is the figure from this approximate size?

    e. Discuss whether it is easy or difficult for matter to fall directly into a black hole of this size. The *Scribe* should briefly summarize the team's discussion.

    f. Notice that the stream of matter from the supergiant star in Figure 14.16 is flowing off to the side of the black hole and onto an accretion disk, rather than straight into the black hole. The *Proposer* should suggest a reason why the stream goes off to the side.

    g. The *Skeptic* should then question the *Proposer's* reasoning, either suggesting an alternative reason or explaining why the matter should instead be flowing straight into the black hole.

    h. Led by the *Moderator*, discuss whether or not the rendering of this stream is an accurate portrayal of how matter flows from the supergiant star toward the black hole. The *Scribe* should briefly summarize the team's discussion.

Investigate Further

Short-Answer/Essay Questions

Life Stories of Stars. Write a one- to two-page life story for the scenarios in 39–42. Each story should be detailed and scientifically correct but also creative. That is, it should be entertaining while at the same time showing that you understand stellar evolution. Be sure to state whether "you" are a member of a binary system.

39. You are a white dwarf of $0.8M_{Sun}$.
40. You are a neutron star of $1.5M_{Sun}$.
41. You are a black hole of $10M_{Sun}$.
42. You are a white dwarf in a close binary system and are accreting matter from your companion star.
43. *Census of Stellar Corpses.* Which kind of object do you think is most common in our galaxy: white dwarfs, neutron stars, or black holes? Explain your reasoning.
44. *Fate of an X-Ray Binary.* The X-ray bursts that happen on the surface of an accreting neutron star are not powerful enough to accelerate the exploding material to escape velocity. Predict what will happen in an X-ray binary system in which the companion star eventually feeds over 3 solar masses of matter into the neutron star's accretion disk.
45. *Why Black Holes Are Safe.* Explain why the principle of conservation of angular momentum makes it very difficult to fall into a black hole.
46. *Surviving the Plunge.* The tidal forces near a black hole with a mass similar to that of a star would tear a person apart before that person could fall through the event horizon. Black hole researchers have pointed out that a fanciful "black hole life preserver" could help counteract those tidal forces. The life preserver would need to have a mass similar to that of an asteroid and would need to be shaped like a flattened hoop and placed around the person's waist. In what direction would the gravitational force from the hoop pull on the person's head? In what direction would it pull on the person's feet? Based on your answers, explain in general terms how the gravitational forces from the "life preserver" would help to counteract the black hole's tidal forces.

Quantitative Problems

Be sure to show all calculations clearly and state your final answers in complete sentences.

47. *Schwarzschild Radii.* Calculate the Schwarzschild radius (in kilometers) for each of the following.
    a. $10^8 M_{Sun}$ black hole in the center of a quasar
    b. $5M_{Sun}$ black hole that formed in the supernova of a massive star
    c. A mini–black hole with the mass of the Moon
    d. A mini–black hole formed when a superadvanced civilization decides to punish you (unfairly) by squeezing you until you become so small that you disappear inside your own event horizon
48. *The Crab Nebula Pulsar Winds Down.* Theoretical models of the slowing of pulsars predict that the age of a pulsar is approximately equal to $p/(2r)$, where p is the pulsar's current period and r is the rate at which the period is slowing with time. Observations of the pulsar in the Crab Nebula show that it pulses 30 times per second, so $p = 0.0333$ second, but the time interval between pulses is growing longer by 4.2×10^{-13} second with each passing second, so $r = 4.2 \times 10^{-13}$ second per second. Using that information, estimate the age of the Crab Nebula pulsar. How does your estimate compare with the true age of the pulsar, which was born in the supernova observed in A.D. 1054?
49. *A Water Black Hole.* A clump of matter does not need to be extraordinarily dense in order to have an escape velocity greater than the speed of light, as long as its mass is large enough. You can use the formula for the Schwarzschild radius R_S to calculate the volume $\frac{4}{3}\pi R_S^3$ inside the event horizon of a black hole of mass M. What does the mass of a black hole need to be in order for its mass divided by its volume to be equal to the density of water (1 g/cm^3)?

50. *Energy of a Supernova.* In a massive star supernova explosion, a stellar core collapses and forms a neutron star roughly 10 kilometers in radius. The gravitational potential energy released in such a collapse is approximately equal to GM^2/r, where M is the mass of the neutron star, r is its radius, and the gravitational constant is $G = 6.67 \times 10^{-11}$ m^3/(kg $\times$ s^2). Using this formula, estimate the amount of gravitational potential energy released in a massive star supernova explosion. How does it compare to the amount of energy released by the Sun during its entire main-sequence lifetime?

51. *Energy of a Black Hole Merger.* Models indicate that the gravitational-wave signal shown in Figure 14.21 came from the merger of two black holes with masses of $29M_{Sun}$ and $36M_{Sun}$ (respectively), and resulted in a single black hole with a mass of $62M_{Sun}$. The difference in total mass between the start and the finish of the merger corresponds to the amount of energy carried away in the form of gravitational waves. Use Einstein's formula $E = mc^2$ to calculate this amount of energy. Be sure to put the mass in units of kilograms and the speed of light in meters per second, so that the resulting energy has units of joules. Compare your answer to the energy released by a supernova (from Problem 50) and to the energy released by the Sun during its entire main-sequence life.

Discussion Questions

52. *Too Strange to Be True?* Despite strong theoretical arguments for the existence of neutron stars and black holes, many scientists rejected the possibility that such objects could really exist until they were confronted with very strong observational evidence. Some people claim that this type of scientific skepticism demonstrates an unwillingness on the part of scientists to give up their deeply held scientific beliefs. Others claim that this type of skepticism is necessary for scientific advancement. What do you think? Defend your opinion.
53. *Black Holes in Popular Culture.* Phrases such as "it disappeared into a black hole" are now common in popular culture. Give a few examples of uses of the term *black hole* in popular culture that are not meant to be taken literally. In what ways are these analogies to real black holes accurate? In what ways are they inaccurate? Why do you think a scientific idea as esoteric as that of a black hole has so strongly captured the public imagination?

Web Projects

54. *Gamma-Ray Bursts.* Go to the website for a mission (such as *Swift* or *Fermi*) studying gamma-ray bursts and find the latest information about these bursts. Write a one- to two-page essay on recent discoveries and how they may shed light on the origin of gamma-ray bursts.
55. *Gravitational Wave Detection.* Go to the website for the Advanced Laser Interferometer Gravitational Observatory. Look up the detections of gravitational waves they have made. How many detections have there been so far? What kinds of objects have made the detected waves? What are the approximate distances of those objects? Write a short report on what you learn.
56. *Black Holes.* Andrew Hamilton, a professor at the University of Colorado, maintains a website with a great deal of information about black holes and what it would be like to visit one. Visit his site and investigate some aspect of black holes that you find particularly interesting. Write a short report on what you learn.

We can understand the entire life cycle of a star in terms of the changing balance between pressure and gravity. This illustration shows how that balance changes over time and why those changes depend on a star's birth mass. *(Stars not to scale.)*

Key

pressure ➡
gravity ⬅

(2) Thermal pressure comes into steady balance with gravity when the core becomes hot enough for hydrogen fusion to replace the thermal energy the star radiates from its surface [Section 11.1].

(3) The balance tips in favor of gravity after the core runs out of hydrogen. Fusion of hydrogen into helium temporarily stops supplying thermal energy in the core. The core again contracts and heats up. Hydrogen fusion begins in a shell around the core. The outer layers expand and cool, and the star becomes redder [Section 13.2].

Hydrogen shell-fusion star

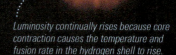

Luminosity continually rises because core contraction causes the temperature and fusion rate in the hydrogen shell to rise.

Main-sequence star

Pressure balances gravity at every point within a main-sequence star.

> 0.08M_{Sun}

(1) Gravity overcomes pressure inside a protostar, causing the core to contract and heat up. A protostar cannot achieve steady balance between pressure and gravity because nuclear fusion is not replacing the thermal energy it radiates into space [Section 13.1].

The balance between pressure and gravity acts as a thermostat to regulate the core temperature:

A drop in core temperature decreases fusion rate, which lowers core pressure, causing the core to contract and heat up.

A rise in core temperature increases fusion rate, which raises core pressure, causing the core to expand and cool down.

Solar Thermostat

The balancing point between pressure and gravity depends on a star's mass:

Balance between pressure and gravity in high-mass stars results in a higher core temperature, a higher fusion rate, greater luminosity, and a shorter lifetime.

Protostar

Contraction converts gravitational potential energy into thermal energy.

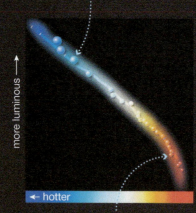

more luminous

← hotter

Balance between pressure and gravity in low-mass stars results in a cooler core temperature, a slower fusion rate, less luminosity, and a longer lifetime.

Degeneracy pressure balances gravity in objects of less than 0.08M_{Sun} before their cores become hot enough for steady fusion. These objects never become stars and end up as brown dwarfs.

> < 0.08M_{Sun}

(4) Balance between thermal pressure and gravity is restored when the core temperature rises enough for helium fusion into carbon, which can once more replace the thermal energy radiated from the core [Section 13.2].

(5) Gravity again gains the upper hand over pressure after the core helium is gone. Just as before, fusion stops replacing the thermal energy leaving the core. The core therefore resumes contracting and heating up, and helium fusion begins in a shell around the carbon core [Section 13.2].

(6) In high-mass stars, core contraction continues, leading to multiple shell fusion that terminates with iron and a supernova explosion [Section 13.3].

(7) At the end of a star's life, either degeneracy pressure has come into permanent balance with gravity or the star has become a black hole. The nature of the end state depends on the mass of the remaining core [Chapter 14].

Helium shell-fusion star

Double shell-fusion star

HIGH-MASS

Multiple shell-fusion star

Black hole

Degeneracy pressure cannot balance gravity in a black hole.

Neutron star ($M < 3M_{Sun}$)

Neutron degeneracy pressure can balance gravity in a stellar corpse with less than about 2–3M_{Sun}.

Luminosity remains steady because helium core fusion restores balance.

White dwarf ($M < 1.4M_{Sun}$)

LOW-MASS

Electron degeneracy pressure balances gravity in the core of a low-mass star before it gets hot enough to fuse heavier elements. The star ejects its outer layers and ends up as a white dwarf.

Electron degeneracy pressure can balance gravity in a stellar corpse of mass < 1.4M_{Sun}.

Brown dwarf ($M < 0.08M_{Sun}$)

Degeneracy pressure keeps a brown dwarf stable in size even as it cools steadily with time.

15 Our Galaxy

This photo from the Spitzer Space Telescope shows infrared light from stars and dusty gas clouds within the central 600 light-years of the Milky Way Galaxy.

LEARNING GOALS

15.1 The Milky Way Revealed
- ◆ What does our galaxy look like?
- ◆ How do stars orbit in our galaxy?

15.2 Galactic Recycling
- ◆ How is gas recycled in our galaxy?
- ◆ Where do stars tend to form in our galaxy?

15.3 The History of the Milky Way
- ◆ What do halo stars tell us about our galaxy's history?
- ◆ How did our galaxy form?

15.4 The Galactic Center
- ◆ What is the evidence for a black hole at our galaxy's center?

n previous chapters, we saw how stars forge new elements and expel them into space. This is not the end of the story, however, because those elements ultimately mix back in with other interstellar gas, making them available to be recycled into new generations of stars. The birth of our solar system and the evolution of life on Earth would not have been possible without this "galactic ecosystem," which acts much like a living ecosystem that gradually recycles material from dead organisms into new life.

In this chapter, we will study our galaxy and the processes that maintain its ecosystem, along with current understanding of galactic history and the evidence for an extremely massive black hole at the galaxy's center. Through it all, we will see that we are not only "star stuff" but also "galaxy stuff"—the product of eons of complex recycling and reprocessing of matter and energy in the Milky Way Galaxy.

ESSENTIAL PREPARATION

1. How does Newton's law of gravity extend Kepler's laws? [Section 4.4]

2. What is light? [Section 5.1]

3. What features of our solar system provide clues to how it formed? [Section 6.2]

4. How do stars form? [Section 13.1]

5. How does a high-mass star die? [Section 13.3]

15.1 The Milky Way Revealed

On a dark night, you can see a faint band of light slicing across the sky through several constellations, including Sagittarius, Cygnus, Perseus, and Orion. This band of light looked like a flowing ribbon of milk to the ancient Greeks, so we now call it the *Milky Way* (see Figure 2.1). In the early 17th century, Galileo used his telescope to prove that the light of the Milky Way comes from myriad individual stars. Together these stars make up the kind of stellar system we call a *galaxy*, echoing the Greek word for "milk," *galactos*.

The true size and shape of our Milky Way Galaxy are hard to guess from how it looks in our night sky. Because we live *inside* the galaxy, trying to determine its structure is somewhat like trying to draw a picture of your house without ever leaving your bedroom. The fact that much of our galaxy's visible light is hidden from our view makes the task even more difficult. Nevertheless, by carefully observing our galaxy and comparing it to others that we see from the outside, we now have a good understanding of the processes that shape it. In this section, we'll begin our exploration of the galaxy by investigating its basic structure and orbital motion.

◆ **What does our galaxy look like?**

Our galaxy consists of a flat disk with spiral arms, a central bulge, and a roughly spherical halo surrounding everything.

Our Milky Way Galaxy holds more than 100 billion stars and is just one among some 100 billion galaxies in the observable universe [Section 1.1]. It is a vast **spiral galaxy,** so named because of the spectacular **spiral arms** illustrated in Figure 15.1a. If we viewed it from the side, as shown in Figure 15.1b, we'd see that the spiral arms are part of a fairly flat **disk** of stars with a bright central **bulge.** The entire disk is surrounded by a dimmer, rounder **halo.** Most of the galaxy's bright stars reside in its disk. The most prominent stars in the halo are found in about 200 *globular clusters* of stars [Section 12.3].

The entire galaxy is about 100,000 light-years in diameter, but the disk is only about 1000 light-years thick. Our Sun is located in the disk about 27,000 light-years from the galactic center—a little more than halfway out

a Artist's conception of the Milky Way viewed from the outside.

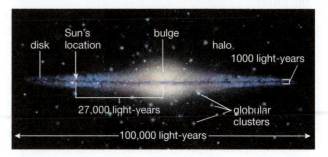

b Edge-on schematic view of the Milky Way.

▲ **FIGURE 15.1**
The Milky Way Galaxy.

from the center to the edge of the disk. Remember that this distance is incredibly huge [Section 1.1]: The few thousand stars visible to the naked eye together fill only a tiny dot in a picture like the one in Figure 15.1.

It took us a long time to learn these facts about the Milky Way's size and shape. Clouds of interstellar gas and dust known collectively as the **interstellar medium** fill the galactic disk, obscuring our view of most of the galaxy when we observe just the visible light. This fact long fooled astronomers into believing that we lived near our galaxy's center, a view refuted only around 1920, when astronomer Harlow Shapley showed that the Milky Way's globular clusters are centered on a point tens of thousands of light-years from our Sun (see Special Topic below).

We now know that the Milky Way is a relatively large galaxy. Within our Local Group of galaxies (see the foldout in the front of the book), only the Andromeda Galaxy is comparable in size. The Milky Way's strong gravity influences smaller galaxies in its vicinity. For example,

The Milky Way is a large galaxy, and several smaller galaxies orbit it.

two small galaxies known as the *Large Magellanic Cloud* and the *Small Magellanic Cloud*—both visible to the naked eye from the Southern Hemisphere—orbit the Milky Way at distances of some 160,000 and 200,000 light-years, respectively. Both Magellanic Clouds are visible to the naked eye from the Southern Hemisphere. Several smaller galaxies lie even closer to the Milky Way. Two of them—known as the Sagittarius Dwarf and Canis Major Dwarf—are each in the process of colliding with the Milky Way's disk. Our galaxy's tidal forces will ultimately rip both of these small galaxies apart.

---

special topic   How Did We Learn the Structure of the Milky Way?

For most of human history, we knew the Milky Way only as an indistinct river of light in the sky. In 1610, Galileo used his telescope to discover that the Milky Way is made up of innumerable stars, but we still did not know the size or shape of our galaxy.

In the late 18th century, British astronomers William and Caroline Herschel (brother and sister) tried to determine the shape of the Milky Way more accurately by counting how many stars lie in each direction. Their approach suggested that the Milky Way's width was five times its thickness. In the early 20th century, Dutch astronomer Jacobus Kapteyn and his colleagues used a more sophisticated star-counting method to gauge the size and shape of the Milky Way. Their results seemed to confirm the Herschels' findings, and suggested that the Sun lies near the center of the galaxy.

Kapteyn's results made astronomers with a sense of history nervous. Only four centuries earlier, before Copernicus challenged the Ptolemaic system, astronomers had believed Earth was the center of the universe. Kapteyn's placement of the Sun near the Milky Way's center seemed to give Earth a central place again. Kapteyn knew that obscuring material could deceive us by hiding the rest of the galaxy like some kind of interstellar fog, but he found no evidence for such a fog.

While Kapteyn was counting stars, American astronomer Harlow Shapley was studying globular clusters. He found that these clusters appeared to be centered on a point tens of thousands of

light-years from the Sun. Shapley concluded that this point marked the true center of our galaxy and that Kapteyn must be wrong.

Today we know that Shapley was right. The Milky Way's interstellar medium is the "fog" that misled Kapteyn. Robert Trumpler, who worked at California's Lick Observatory in the 1920s, established the existence of this dusty gas by studying open clusters of stars. By assuming that all open clusters had about the same diameter, he estimated their distances from their apparent sizes in the sky, much as you might estimate the distances of cars at night from the apparent separation of their headlights. He found that stars in distant clusters appeared dimmer than expected based on their estimated distance, just as a car's headlights might appear in foggy weather. Trumpler concluded that light-absorbing material fills the spaces between the stars, partially obscuring the distant clusters and making them appear fainter than they would otherwise. We thereby learned that interstellar material had misled earlier astronomers, and that the stars visible in the night sky occupy a minuscule portion of the observable universe.

Understanding the effects of interstellar dust allowed astronomers to take it into account in their observations, helping us learn the true size and shape of our galaxy. Then, beginning in the 1950s, careful studies of the motions of stars and gas clouds (made with radio observations of the 21-centimeter line from atomic hydrogen gas) gradually uncovered the detailed structure of the galactic disk, showing us the locations and motions of the spiral arms.

Keep in mind that while the Magellanic Clouds are quite small as galaxies go, they are still vast objects, each containing perhaps 1 billion to a few billion stars. This makes them 1000 or more times the size of typical globular clusters, so they are small only in comparison to an enormous structure like our Milky Way Galaxy.

◆ How do stars orbit in our galaxy?

A spiral galaxy like ours may look like a giant pinwheel, but each individual star follows its own orbital path around the center of the galaxy. The nature of this orbital path depends primarily on whether the star resides in the disk, the halo, or the bulge (Figure 15.2).

Orbits of Disk Stars Stars in the disk orbit the galactic center in roughly circular paths that all go in the same direction in nearly the same plane. However, if you could stand outside the Milky Way and watch it for a few billion years, the disk would resemble a huge merry-go-round. Like horses on a merry-go-round, individual stars bob up and down through the disk as they orbit. The general orbit of a star around the galaxy arises from its gravitational attraction toward the galactic center,

Disk stars orbit the galaxy's center in orderly circles that all go in the same direction, bobbing slightly up and down as they orbit.

while the bobbing arises from the localized pull of gravity within the disk itself. A star that is "too far" above the disk is pulled back into the disk by gravity. Because the density of interstellar gas is too low to slow the star, it flies through the disk until it is "too far" *below* the disk on the other side. Gravity then pulls it back in the other direction. This ongoing process produces the bobbing of the stars.

The up-and-down motions of the disk stars give the disk its thickness of about 1000 light-years—a great distance by human standards, but only about 1% of the disk's 100,000-light-year diameter. In the vicinity of our Sun, each star's orbit takes more than 200 million years, and each up-and-down "bob" takes a few tens of millions of years.

The galaxy's rotation is unlike that of a merry-go-round in one important respect: On a merry-go-round, horses near the edge move much faster than those near the center. But in our galaxy's disk, the orbital velocities of stars near the edge and those near the center are about the same. Stars closer to the center therefore complete each orbit in less time than stars farther out.

Orbits of Halo Stars The orbits of stars in the halo are much less organized. Individual halo stars also orbit the galactic center, but the orientations of their paths are relatively random, and neighboring stars may circle the galactic center in opposite directions. Halo stars swoop from high above the disk to far

Halo stars swoop high above and below the disk on randomly oriented orbits.

below it and back up again, plunging through the disk at velocities so high that the disk's gravity hardly alters their trajectories.

These swooping orbits explain why the halo is much puffier than the disk. Halo stars soar to heights above the disk far greater than the heights achieved by disk stars as they bob up and down. As we'll discuss later, the differences between orbits in the disk and orbits in the halo provide an important clue to how our galaxy formed.

Halo stars travel high above and far below the disk on orbits with random orientations.

Bulge stars also have orbits with random orientations.

Disk stars orbit in circles with the same orientation; except for a little up-and-down motion.

▲ **FIGURE 15.2**
Characteristic orbits of disk stars (yellow), bulge stars (red), and halo stars (green) around the galactic center. (The yellow path exaggerates the up-and-down motion of the disk-star orbits.)

think about it Is there much danger that a halo star swooping through the disk of the galaxy will someday hit the Sun or Earth? Why or why not? (*Hint:* Consider the typical distances between stars on the 1-to-10-billion scale introduced in Chapter 1.)

Orbits of Bulge Stars

The orbits of stars in the bulge are more difficult to measure than those of disk and halo stars, because the bulge is much farther away from us than the nearest disk and halo stars. Bulge-star orbits were once thought to be similar to halo-star orbits because the bulge is puffier than the disk. However, recent observations have shown that the bulge is more complex, with a range of orbital properties among its stars. Some bulge stars have orbits with random orientations like halo stars, while others orbit in the same general direction as disk stars but with more elongated orbits. These orbital patterns give the bulge a cigar-like shape that would look like a bar stretching across the center of the galaxy if we could view it from the outside [Section 16.1].

Some bulge stars orbit like halo stars while others orbit like disk stars.

Stellar Orbits and the Mass of the Galaxy

The Sun's orbital path is fairly typical for a disk star. By measuring the speeds of globular clusters relative to the Sun, astronomers have determined that the Sun and its neighbors orbit the center of the Milky Way at a speed of about 220 kilometers per second (800,000 km/hr [Section 1.3]). Even at this speed, it takes the Sun about 230 million years to complete one orbit around the galactic center. Early dinosaurs were just emerging on Earth when our Sun last visited this side of the galaxy.

We can calculate the mass of the galaxy within the Sun's orbit using the Sun's orbital properties and Newton's version of Kepler's third law.

The orbital motion of the Sun and other stars gives us a way to determine the mass of the galaxy. Recall that Newton's law of gravity determines how quickly objects orbit one another. This fact, embodied in Newton's version of Kepler's third law [Section 4.4], allows us to determine the mass of a relatively large object when we know the period and average distance of a much smaller object in orbit around it.

---

special topic   How Do We Determine Stellar Orbits?

Astronomers learned how stars orbit in the Milky Way by measuring the motions of many different stars relative to the Sun. Although these measurements are easy in principle, they can be difficult in practice.

Determining a star's precise motion relative to the Sun requires knowing the star's true velocity through space. However, our primary means of measuring speeds in the universe—the Doppler effect—can tell us only a star's *radial velocity,* the component of its velocity directed toward or away from us (see Figure 5.15). If we want to know the star's true velocity, we must also measure its *tangential velocity,* the component of its velocity directed across our line of sight.

Tangential velocity is difficult to measure because of the vast distances to stars. Over tens of thousands of years, the tangential velocities of stars cause their apparent positions in our sky to change, which alters the shapes of the constellations. These changes are far too small for human eyes to notice. However, we can measure tangential velocities for many stars by comparing telescopic photographs taken years or decades apart. For example, if photographs taken 10 years apart show that a star has moved across our sky by an angle of 1 arcsecond, we know the star is moving at an angular rate of 0.1 arcsecond per year. We can then convert this angular rate of motion—often called the star's *proper motion*—to a tangential velocity if we also know the star's distance. Because the proper motions of stars at large distances are extremely small, to date we know precise stellar orbits only for relatively nearby stars. However, the European Space Agency's *GAIA* mission is in the process of measuring orbits for up to a billion stars located throughout our galaxy.

For example, we can use the Sun's orbital velocity and its distance from the galactic center to determine the mass of our galaxy lying *within* the Sun's orbit. To understand why the Sun's orbital motion allows us to calculate only the mass within the Sun's orbit, rather than the total mass of the galaxy, we need to consider the difference between the gravitational effects of mass within the Sun's orbit and of mass beyond its orbit. Every part of the galaxy exerts gravitational forces on the Sun as it orbits, but the net force from matter outside the Sun's orbit is relatively small because the pulls from opposite sides of the galaxy virtually cancel one another. In contrast, the net gravitational forces from mass within the Sun's orbit all pull the Sun in the same direction—toward the galactic center. The Sun's orbital velocity therefore responds almost exclusively to the gravitational pull of matter inside its orbit. By using the Sun's 27,000-light-year orbital distance and 220-km/s orbital velocity in Newton's version of Kepler's third law, we find that the total amount of mass within the Sun's orbit is about 2×10^{41} kg, or about 100 billion times the mass of the Sun.

Similar calculations based on the orbits of more distant stars in the Milky Way have revealed one of the greatest mysteries in astronomy—one that we first encountered in Chapter 1 (see Figure 1.14). Photographs of spiral galaxies make it appear as if most of their mass were concentrated near their centers. However, orbital motions tell us that just the opposite is true. If most of the mass were concentrated near the galaxy's center, the orbital speeds of more distant stars would be slower, just as the orbital speeds of the planets decline with distance from the Sun. Instead, we find that orbital speeds remain about the same out to great distances from the galactic center, telling us that most of the galaxy's mass resides far from the center and is distributed throughout the halo. Because we see few stars and virtually no gas or dust in the halo, we conclude that most of the galaxy's mass must not give off any light that we can detect, and hence we refer to it as *dark matter*. We will discuss dark matter in more detail in Chapter 18.

The orbits of the Milky Way's stars reveal that most of the galaxy's mass consists of invisible dark matter in the halo.

15.2 Galactic Recycling

The Milky Way is our home galaxy, but its importance to our existence runs much deeper. The birth of our solar system could not have occurred without the recycling that takes place within the disk of the galaxy and its interstellar medium.

Galactic recycling makes new generations of stars possible and gradually changes the chemical composition of the interstellar medium. Recall that the early universe contained only the two lightest elements, hydrogen and helium. Virtually all other elements (which astronomers tend to refer to as "heavy elements") have been produced and released into space by stars. These elements mix with the existing interstellar gas, raising its proportion of heavy elements, and the galaxy then recycles this gas into new generations of stars. Today, thanks to more than 10 billion years of galactic recycling, elements heavier than helium constitute about 2% of the galaxy's gaseous content by mass. The remaining 98% still consists of hydrogen (about 71%) and helium (about 27%).

cosmic **calculations 15.1**

The Orbital Velocity Formula

Newton's laws of motion and gravity can be used to derive a formula that allows us to calculate the amount of mass contained *within* any object's orbit. For an object on a circular orbit, this *orbital velocity formula* is

$$M_r = \frac{r \times v^2}{G}$$

where M_r is the amount of mass contained within its orbit, r is its orbital radius, v is the object's orbital velocity, and $G = 6.67 \times 10^{-11} \frac{m^3}{kg \times s^2}$ is the gravitational constant. Notice that for a given radius, a larger orbital velocity indicates a larger amount of mass, reflecting the fact that a stronger gravitational pull is necessary to hold a faster-moving object in orbit.

Example: Calculate the mass of the Milky Way Galaxy within the Sun's orbit using the orbital velocity formula.

Solution: We can use the orbital velocity formula with the Sun's orbital velocity of 220 km/s and distance of 27,000 light-years. For consistency with the given units of G, we convert distances to meters, so $v = 2.2 \times 10^5$ m/s and $r = 2.6 \times 10^{20}$ m (because 1 light-year = 9.46×10^{15} meters). Substituting these values for v and r into the orbital velocity formula, we obtain

$$M_r = \frac{r \times v^2}{G}$$
$$= \frac{(2.6 \times 10^{20}\,\text{m}) \times (2.2 \times 10^5 \frac{m}{s})^2}{6.67 \times 10^{-11} \frac{m^3}{kg \times s^2}}$$
$$= 1.9 \times 10^{41}\,\text{kg}$$

The mass of the Milky Way Galaxy within the Sun's orbit is about 2×10^{41} kg. Dividing this mass by the Sun's mass of 2×10^{30} kg, we find that the mass of the galaxy within the Sun's orbit is about 10^{11} (100 billion) solar masses.

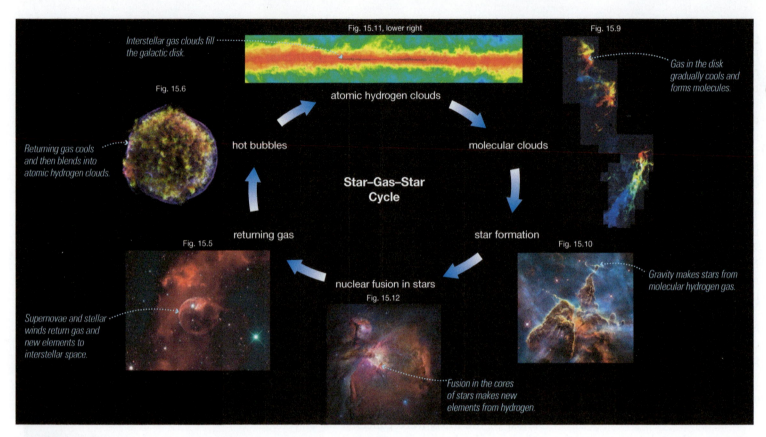

Fig. 15.11, lower right

Interstellar gas clouds fill the galactic disk.

Fig. 15.9

Gas in the disk gradually cools and forms molecules.

atomic hydrogen clouds

Fig. 15.6

hot bubbles

molecular clouds

Returning gas cools and then blends into atomic hydrogen clouds.

Star–Gas–Star Cycle

returning gas

star formation

Fig. 15.5

Fig. 15.10

Gravity makes stars from molecular hydrogen gas.

nuclear fusion in stars

Supernovae and stellar winds return gas and new elements to interstellar space.

Fig. 15.12

Fusion in the cores of stars makes new elements from hydrogen.

▲ **FIGURE 15.3**

A pictorial representation of the star–gas–star cycle. These photos appear individually later in the chapter. Their figure numbers are indicated.

▲ **FIGURE 15.4**

A dying low-mass star returns gas to the interstellar medium in a planetary nebula. This Hubble Space Telescope image shows a planetary nebula known as the Cat's Eye Nebula; it is about 1.2 light-years in diameter.

think about it Recall from Chapter 12 that stars in our galaxy's globular clusters are all extremely old, while stars in open clusters are relatively young. Based on this fact, which stars would you expect to contain a higher proportion of heavy elements: stars in globular clusters or stars in open clusters? Explain.

The galactic recycling process may sound simple enough, but it raises at least one key question: The supernova explosions that scatter most heavy elements into space send debris flying out at speeds of several thousand kilometers per second—much faster than the escape velocity from the galaxy. So how have the chemical riches produced by stars managed to remain in our galaxy? The answer turns out to depend on interactions between the matter expelled by supernovae and the interstellar medium that fills the galactic disk. In this section, we'll examine these interactions in more detail and learn why our existence owes as much to galactic recycling as it does to the manufacturing of heavy elements by stars.

◆ How is gas recycled in our galaxy?

The galactic recycling process proceeds in several stages, making up what we will call the **star–gas–star cycle** that is summarized in Figure 15.3. We have already discussed the ideas shown in the lower three frames of Figure 15.3: Stars are born when gravity causes the collapse of molecular clouds, they shine for millions or billions of years with energy produced by nuclear fusion, and in their deaths they ultimately return much of their material back to the interstellar medium. Let's now complete the loop and look at the remainder of the cycle, starting with the gas ejected by dying stars and finishing back at star birth.

Gas from Dying Stars All stars return much of their original mass to interstellar space in two basic ways: through stellar winds that blow throughout their lives and through "death events" of planetary nebulae (for low-mass stars) or supernovae (for high-mass stars). Low-mass stars generally have weak stellar winds while they are on the main sequence. Their winds grow stronger and carry more material into space when they become red giants. By the time a low-mass star like the Sun ends its life with the ejection of a planetary nebula [Section 13.2], it has returned almost half its original mass to the interstellar medium (Figure 15.4).

High-mass stars lose mass much more dynamically and explosively. The powerful winds from supergiants and massive O and B stars return large amounts of matter into the galaxy. At the ends of their lives, these

Supernovae and high-speed stellar winds can produce hot bubbles of gas in the interstellar medium.

stars explode as supernovae. The high-speed gas ejected into space by supernovae or powerful stellar winds sweeps up surrounding interstellar material, creating a **bubble** of hot, ionized gas (Figure 15.5). These hot bubbles are quite common in the disk of the galaxy, but they are not always easy to detect. While some emit strongly in visible light and others are hot enough to emit large amounts of X rays, many bubbles are evident only through their emission of radio waves.

The bubbles created by supernovae can have even more dramatic effects on the interstellar medium than those created by fast-moving stellar winds. Supernovae generate *shock fronts*—abrupt, high-gas-pressure "walls" that move faster than the speed at which sound waves can travel through interstellar space. A shock front sweeps up surrounding gas as it travels, creating a wall of fast-moving gas on its leading edge. When we observe a *supernova remnant* [Section 13.3], we are seeing walls of gas that have been compressed, heated, and ionized by the outward-moving shock front.

Figure 15.6 shows a young supernova remnant whose shocked gas is hot enough to emit X rays. In contrast, the older supernova remnant shown in Figure 15.7 is cooler because its shock front has swept up more material and has distributed its energy more widely. Eventually, the shocked gas will radiate away most of its original energy, and the expanding wall of gas will slow to subsonic speeds and merge with the surrounding interstellar medium.

In the regions of the galactic disk where many supernovae have recently exploded, we find giant, elongated bubbles extending from young clusters of stars to distances of 3000 light-years or more above or below the disk. These

Giant bubbles of hot gas can erupt out of the disk, spreading their contents over a large region of the galaxy.

probably are places where the bubbles from many individual supernovae have merged to make a giant bubble so large that it cannot be contained within the Milky Way's disk. Once the top of a giant bubble breaks out of the disk, where nearly all of the Milky Way's gas resides, nothing remains to slow its expansion except gravity. The result is a *blowout* that is similar to a volcanic eruption, but on a galactic scale: Hot gas erupts from the disk, spreading out as it shoots upward into the galactic halo (Figure 15.8). The gravity of the galactic disk slows the rise of the gas, eventually pulling it back down. Near the top of its trajectory, the ejected gas starts to cool and form clouds. These clouds then rain back down into the disk, where their contents mix with the gas in a large region of the galaxy.

In addition to producing new elements and churning up the interstellar medium, supernovae may also affect life on Earth by generating **cosmic rays** that can cause genetic mutations in living organisms. Cosmic rays are made of electrons, protons, and atomic nuclei that zip through interstellar space at close to the speed of light. Some cosmic rays penetrate Earth's atmosphere

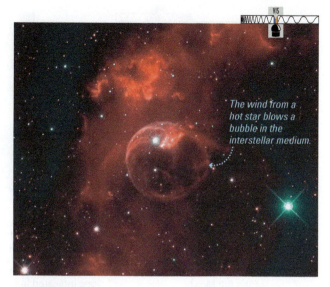

▲ **FIGURE 15.5**

This photo shows a bubble of hot, ionized gas blown by a wind from the hot star near its center. Although it looks much like a soap bubble, it is actually an expanding shell of hot gas about 10 light-years in diameter. It glows where gas piles up as the bubble sweeps outward through the interstellar medium.

The wind from a hot star blows a bubble in the interstellar medium.

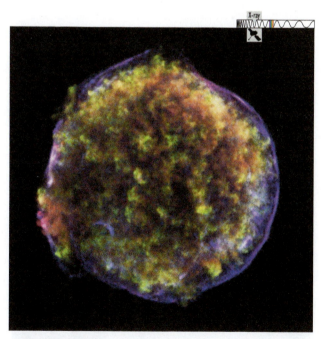

▲ **FIGURE 15.6**

This image shows X-ray emission from hot gas in a young supernova remnant. The most energetic X rays (blue) come from 20-million-degree gas just behind the expanding shock front. Less energetic X rays (green and red) come from the 10-million-degree debris ejected by the exploded star. The remnant is about 20 light-years across.

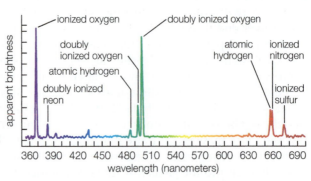

a This visible-light image shows the entire supernova remnant, which is about 130 light-years across and spans an angular width in our sky six times that of the full Moon.

b This Hubble Space Telescope image shows fine filamentary structure in a small piece of the remnant. The colors come from emission lines of the atoms and ions indicated in part c.

c A visible-light spectrum from the Cygnus Loop shows the strong emission lines that account for the distinct colors in the Hubble Space Telescope image.

▲ **FIGURE 15.7**

Emission of visible light from an older supernova remnant, the Cygnus Loop.

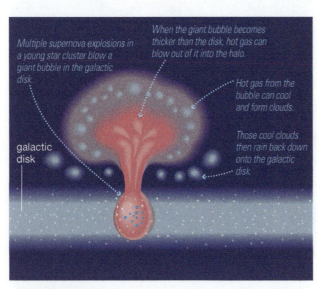

▲ **FIGURE 15.8**

Hot gas erupting from a giant bubble out into the galactic halo eventually cools into gas clouds that rain back down on the disk. This process may be an important part of the galaxy-wide recycling system that incorporates the products of supernova explosions into new generations of stars and planets.

and reach Earth's surface. On average, about one cosmic-ray particle strikes your body each second. The cosmic-ray bombardment rate is 100 times greater at the high altitudes at which jet planes fly, and even more cosmic rays funnel along magnetic field lines to Earth's magnetic poles.

Cooling and Cloud Formation The hot, ionized gas in bubbles heated by supernovae is dynamic and widespread but makes up a relatively small fraction of the gas in the Milky Way. Most of the gas is much cooler—cool enough that hydrogen atoms remain neutral rather than being ionized. We therefore refer to this gas as **atomic hydrogen gas,** although the hydrogen is mixed with neutral atoms of helium and heavier elements in the usual proportions for the galaxy (by mass about 71% hydrogen, 27% helium, and 2% heavier elements). After the gas that forms bubbles cools, it becomes part of the atomic hydrogen gas in the galaxy.

We can map the distribution of atomic hydrogen gas in the Milky Way with radio observations. Atomic hydrogen emits a spectral line with a wavelength of 21 centimeters, which lies in the radio portion of the electromagnetic spectrum (see Figure 5.2). We detect the radio emission of this **21-centimeter line** coming from all directions, telling us that atomic hydrogen gas is distributed throughout the galactic disk. Based on the overall strength of the 21-centimeter emission, the total amount of atomic hydrogen gas in our galaxy is about 5 billion solar masses, which is less than 10% of the galaxy's mass in the form of stars.

Matter remains in the atomic hydrogen stage of the star–gas–star cycle for millions of years. Gravity slowly draws blobs of this gas together into tighter clumps, which radiate energy more efficiently as they grow denser. The blobs therefore cool and contract, forming clouds of cooler and denser gas. This process takes a much longer time than the other steps in the cycle from star death to star birth, which is why so much of the Milky Way's gas is in the atomic hydrogen stage of the star–gas–star cycle.

Clouds of atomic hydrogen also contain a small amount of interstellar dust. Interstellar **dust grains** are tiny, solid flecks of carbon and silicon minerals that resemble particles of smoke and form in the winds of red giant stars [Section 13.2]. Once formed, dust grains remain in the interstellar medium until they are heated and destroyed by a passing shock wave or incorporated into a protostar. Dust grains make up only about 1% of the mass of atomic hydrogen clouds, but they are responsible for the absorption of visible light that prevents us from seeing through the disk of the galaxy.

As the temperature drops further in the center of a cool cloud of atomic hydrogen, hydrogen atoms combine into molecules, making a *molecular cloud* [Section 13.1]. Molecular clouds are the coldest, densest collections of gas in the interstellar medium. They often congregate into *giant molecular clouds* that contain up to a million solar masses of gas. The total mass of molecular clouds in the Milky Way is somewhat uncertain, but it is probably about the same as the total mass of atomic hydrogen gas—about 5 billion solar masses. Temperatures throughout much of this molecular gas hover only a few tens of degrees above absolute zero.

Gas heated by supernovae first cools into atomic hydrogen clouds, then cools further into molecular clouds.

Molecular hydrogen (H_2) is by far the most abundant molecule in molecular clouds, but it is difficult to detect because temperatures are usually too cold for hydrogen molecules to produce emission lines. Most of what we know about molecular clouds comes from observing spectral lines of molecules that make up only a tiny fraction of a cloud's mass but can produce emission lines at lower temperatures. Carbon monoxide (CO) is the most abundant of these molecules. It produces strong emission lines in the radio portion of the spectrum at the 10–30 K temperatures of molecular clouds (Figure 15.9). More than 125 other molecules have been identified in molecular clouds by their radio emission lines, including water (H_2O), ammonia, and ethyl alcohol.

Because molecular clouds are heavy and dense compared to the rest of the interstellar gas, they tend to settle toward the central layers of the Milky Way's disk. This tendency creates a phenomenon you can see with your own eyes: the dark lanes running through the luminous band of light in our sky that we call the Milky Way (see Figure 2.1).

Completing the Cycle   A large molecular cloud gives birth to a cluster of stars. Once a few stars form in a newborn cluster, their radiation begins to erode the surrounding gas in the molecular cloud. Ultraviolet photons from high-mass stars heat and ionize the gas, and winds and radiation pressure push the ionized gas away. This kind of feedback prevents much of the gas in a molecular cloud from turning into stars.

Ultraviolet radiation from newly forming stars can erode the molecular clouds that gave birth to them.

The process of molecular cloud erosion is vividly illustrated in the Carina Nebula, a complex of clouds where new stars are forming (Figure 15.10). The dark, lumpy structures are molecular clouds. Outside the upper edge of the picture, newly formed massive stars glow with ultraviolet radiation. This radiation sears the surface of the molecular clouds, destroying molecules and stripping electrons from atoms. As a result, matter "evaporates" from the molecular clouds and joins the hotter, ionized gas encircling them.

We have arrived back where we started in the star–gas–star cycle. The most massive stars now forming in the Carina Nebula will explode within a few million years, filling the region with bubbles of hot gas and newly formed heavy elements. The expanding bubbles will slow and cool as their gas merges with the widespread atomic hydrogen gas in the galaxy. Eventually, this gas will cool further and coalesce into molecular clouds, forming new stars, new planets, and perhaps even new civilizations.

Despite the recycling of matter from one generation of stars to the next, the star–gas–star cycle cannot go on forever. With each new generation of stars, some of the galaxy's gas becomes permanently locked away in brown dwarfs that never return material to space and in the corpses left behind when stars die. The interstellar medium is slowly running out of gas, and the rate of star formation will gradually taper off until star formation ceases, perhaps some 50 billion years from now.

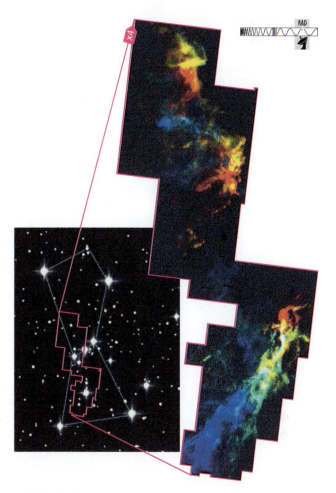

▲ **FIGURE 15.9**

This image shows the complex structure of a molecular cloud in the constellation Orion. The picture was made by measuring Doppler shifts of emission lines from carbon monoxide molecules, and the colors indicate gas motions: Bluer parts are moving toward us and redder parts are moving away from us (relative to the cloud as a whole). This enormous cloud is about 1340 light-years away and several hundred light-years across.

common misconceptions

The Sound of Space

In many science fiction movies, a thunderous sound accompanies the demolition of a spaceship. If the moviemakers wanted to be more realistic, they would silence the explosion. On Earth, we perceive sound when sound waves—which are waves of alternately rising and falling pressure—cause trillions of gas atoms to push our eardrums back and forth. Although sound waves can and do travel through interstellar gas, the extremely low density of this gas means that only a handful of atoms per second would collide with something the size of a human eardrum. It would therefore be impossible for a human ear (or a similar-size microphone) to register any sound, which is why the real sound of space is silence.

▶ **FIGURE 15.10**

A portion of the Carina Nebula, as seen by the Hubble Space Telescope. The dark blobs of gas are molecular clouds, and stars are currently forming in the densest parts of these clouds. The region pictured here is about 3 light-years across.

Radiation from nearby stars is eroding the surfaces of these clouds and causing them to glow . . .

. . . but the densest knots of gas resist that erosion and continue to form stars.

TABLE 15.1 **Typical States of Gas in the Interstellar Medium**

	State of Gas		
	Hot Bubbles	**Atomic Hydrogen Clouds**	**Molecular Clouds**
Primary Constituent	Ionized hydrogen	Atomic hydrogen	Molecular hydrogen
Approximate Temperature	1,000,000 K	100–10,000 K	30 K
Approximate Density (atoms per cm³)	0.01	1–100	300
Description	Pockets of gas heated by stellar winds or supernovae	The most common form of gas, filling much of the galactic disk	Regions of star formation

Putting It All Together: The Distribution of Gas in the Milky Way Different regions of the galaxy are in different stages of the star–gas–star cycle. Because the cycle proceeds over such a long period of time compared to a human lifetime, each stage appears to us as a snapshot. We therefore see the interstellar medium in a wide variety of manifestations, ranging from the million-degree gas of bubbles to the cold, dense gas of molecular clouds. Table 15.1 summarizes the different states in which we observe interstellar gas in the galactic disk.

We can see how these different states of gas are arranged in our galaxy by observing in different wavelengths of light. Figure 15.11 shows seven views of the entire sky. Each view represents the Milky Way Galaxy in a particular wavelength band, made by photographing the sky in every direction from Earth. The infrared image at the top shows what the Milky Way would look like if dusty gas clouds were not blocking our visible-light view. Notice that the dark patches along the disk in the visible-light view correspond to bright regions of long-wavelength infrared emission from dust, as well as to radio emission from molecular and atomic gas clouds. In X rays, the halo is brighter than the disk, because it is filled with X ray–emitting hot gas. In contrast, the disk is brighter in gamma rays than the halo, because collisions between cosmic rays and the nuclei of gas atoms in cold, dense gas clouds produce most of those gamma rays.

To get a complete picture of the Milky Way's star–gas–star cycle, we need to observe its gas in many different wavelengths.

think about it Carefully compare the X-ray image in Figure 15.11 with the radio image of CO (carbon monoxide) molecules. What is the relationship between the dark patches in the X-ray image and the bright patches in the radio image? Explain why this relationship exists.

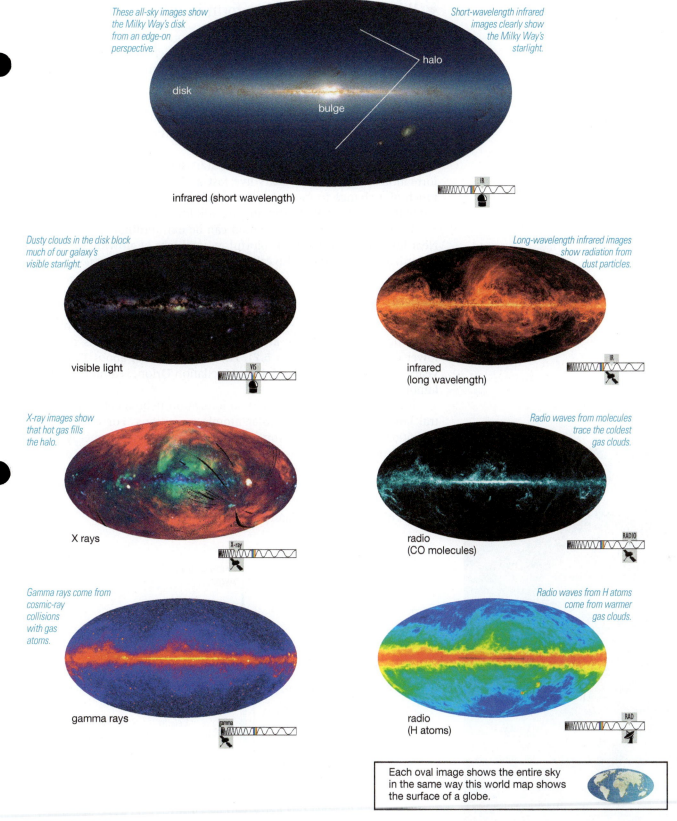

These all-sky images show the Milky Way's disk from an edge-on perspective.

Short-wavelength infrared images clearly show the Milky Way's starlight.

halo

disk

bulge

infrared (short wavelength)

IR

Dusty clouds in the disk block much of our galaxy's visible starlight.

Long-wavelength infrared images show radiation from dust particles.

visible light

VIS

infrared (long wavelength)

IR

X-ray images show that hot gas fills the halo.

Radio waves from molecules trace the coldest gas clouds.

X rays

X-ray

radio (CO molecules)

RADIO

Gamma rays come from cosmic-ray collisions with gas atoms.

Radio waves from H atoms come from warmer gas clouds.

gamma rays

gamma

radio (H atoms)

RAD

Each oval image shows the entire sky in the same way this world map shows the surface of a globe.

▲ **FIGURE 15.11**

All-sky views of the Milky Way Galaxy as it appears in different portions of the spectrum. Observing our galaxy in different wavelengths of light shows us the relationships between its stars, gas clouds, and dust particles. In each image, the brightness corresponds to the intensity of light with that wavelength.

◆ Where do stars tend to form in our galaxy?

The star–gas–star cycle has operated continuously since the Milky Way's birth, yet new stars are not spread evenly across the galaxy. Some regions seem much more fertile than others. Regions rich in molecular clouds tend to spawn new stars easily, while gas-poor regions do not. However, molecular clouds are dark and hard to see. Other signatures of star formation are much more obvious. A quick tour of some star-forming galactic environments will help you spot where the action is.

Star-Forming Regions Hot, massive stars signal a region of active star formation. Because these stars live fast and die young, they don't get much of a chance to move far from their birthplaces and generally die while their lower-mass birthmates are still forming.

Regions of active star formation can be extraordinarily picturesque. Near hot stars we often find colorful, wispy blobs of glowing gas known as **ionization nebulae** (also called *emission nebulae* or *H II regions*). These nebulae glow because electrons in their atoms are raised to high energy levels or ionized when they absorb ultraviolet photons from the hot stars, so the atoms emit light as the electrons return to lower energy levels [Section 5.2]. The Orion Nebula, about 1350 light-years away in the "sword" of the constellation Orion, is among the most famous (Figure 15.12).

Hot, massive stars and ionization nebulae are found only near clouds that are actively forming stars.

Most of the striking colors in an ionization nebula come from spectral lines produced by particular atomic transitions. For example, the transition in which an electron falls from energy level 3 to energy level 2 in a hydrogen atom generates a red photon with a wavelength of 656.3 nanometers (see Figure 5.10). Ionization nebulae appear predominantly red in photographs because of all the red photons released by this particular transition. Transitions in other elements produce other spectral lines of different colors (Figure 15.13).

▲ **FIGURE 15.12**

A Hubble Space Telescope photo of the Orion Nebula, an ionization nebula energized by ultraviolet photons from hot stars.

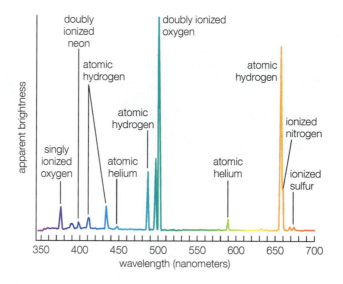

▲ **FIGURE 15.13**

A spectrum of the Orion Nebula. The prominent emission lines reveal the atoms and ions that emit most of the light. Through careful study of these lines, we can determine the nebula's chemical composition.

▲ FIGURE 15.14

The blue tints in this nebula in the constellation Ophiuchus are produced by blue starlight reflecting off dust particles.

▲ FIGURE 15.15

A photo of the Trifid Nebula and its surroundings. (The region pictured is about 50 light-years across.)

The blue and black tints in some star-forming regions have a different origin. Starlight reflected from dust grains produces the blue colors, because interstellar dust grains scatter blue light more easily than red light (Figure 15.14). These *reflection nebulae* are always bluer in color than the stars supplying the light. (The effect is similar to the scattering of sunlight in our atmosphere that makes the sky blue [Section 7.1].) The black regions of nebulae are dark, dusty gas clouds that block our view of the stars beyond them. Figure 15.15 shows a multicolored nebula characteristic of a hot-star neighborhood.

think about it In Figure 15.15, identify the red ionized regions, the blue reflecting regions, and the dark obscuring regions. Briefly explain the origin of the colors in each region.

Spiral Arms Taking a broader view of our galaxy, we can see that the spiral arms must be full of newly forming stars because they bear all the hallmarks of star formation. They are home to both molecular clouds and numerous clusters of young, bright, blue stars surrounded by ionization nebulae. Detailed images of other spiral galaxies show these characteristics more clearly (Figure 15.16). Hot blue stars and ionization nebulae trace out the arms, while the stars between the arms are generally redder and older. We also see enhanced amounts of molecular and atomic gas in the spiral arms, and streaks of interstellar

Dark patches on inner edge of spiral arm show where gas clouds are packing together . . .

. . . and compression of these clouds triggers star formation in the arm.

Blue specks are young stars that formed in the spiral arm.

Red patches are ionization nebulae around the hottest, youngest stars.

Flow of gas and stars through spiral arm

▲ **FIGURE 15.16**

Spiral arms in Galaxy M51. This photo from the Hubble Space Telescope shows M51's two magnificent spiral arms along with a smaller galaxy (at upper right) that is currently interacting with one of those arms. Notice that the spiral arms are much bluer in color than the central bulge. Because massive blue stars live only for a few million years, the relative blueness of the spiral arms tells us that stars must be forming more actively within them than elsewhere in the galaxy. (The large image shows a region roughly 90,000 light-years across.)

dust often obscure the inner sides of the arms themselves. Spiral arms therefore contain both young stars and the material necessary to make new stars.

At first glance, spiral arms look as if they ought to move with the stars, like the fins of a giant pinwheel in space. However, observations indicate that stars throughout most of the disk orbit at approximately the same speed. Stars near the center of the galaxy—which travel in circles with smaller radii—therefore complete an orbit in much less time than stars far from the center. As a result, if the spiral arms moved along with the stars, the central parts of the arms would wind around the galaxy many more times than the outer parts. After a few galactic rotations the arms would become wound into a tight coil. We do not see such tightly wound coils of arms in galaxies, so we conclude that spiral arms must be more like swirling ripples in a whirlpool than like the fins of a giant pinwheel.

More specifically, spiral arms exhibit unusually high levels of star formation. Theoretical models suggest that the spiral pattern of star formation is caused by disturbances called **spiral density waves** that propagate through the gaseous disk of a spiral galaxy like the Milky Way. As a spiral density wave moves through the disk, the changing gravitational forces among stars in the wave alternately cause stars and gas clouds to pack more densely together and then move farther apart. The density changes have little effect on stars, because they always remain too widely separated to collide with each other. However, when a spiral density wave packs gas clouds closer together, gravity within the clouds becomes strong enough to trigger the formation of new star clusters (Figure 15.17). Supernova explosions from massive stars in these clusters can then compress the surrounding clouds further, triggering even more star formation. As this long-lasting pattern of increased density and star formation propagates through the gas and stars of the galaxy's disk, the disk's rotation gradually stretches it into a spiral arm.

Spiral arms are waves of star formation that spread through our galaxy's disk.

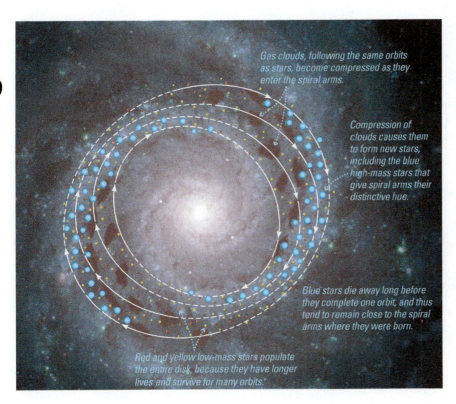

Schematic illustration of star formation produced by spiral density waves. Stars and gas clouds pass through spiral arms as they orbit the galaxy. Each arm is a self-sustaining pattern because gravity slows down stars and gas clouds as they pass through an arm, causing them to be more densely packed there.

Gas clouds, following the same orbits as stars, become compressed as they enter the spiral arms.

Compression of clouds causes them to form new stars, including the blue high-mass stars that give spiral arms their distinctive hue.

Blue stars die away long before they complete one orbit, and thus tend to remain close to the spiral arms where they were born.

Red and yellow low-mass stars populate the entire disk, because they have longer lives and survive for many orbits.

15.3 The History of the Milky Way

Having discussed the basic properties of the Milky Way, we are now ready to turn our attention to the history of our galaxy. As we will see, the differences between disk stars and halo stars provide crucial clues about their origins, which have helped astronomers develop a basic model of how our galaxy formed.

◆ What do halo stars tell us about our galaxy's history?

We have already seen how the disorderly orbits of halo stars differ from the organized orbits of disk stars. Two other key differences distinguish halo stars from disk stars. First, the main-sequence turnoff points for globular clusters [Section 12.3] indicate that stars in the halo are generally very old (at least 12 billion years), while disk stars have many different ages. Second, spectra show that halo stars contain smaller proportions of heavy elements than disk stars. Based on these differences, astronomers divide most of the Milky Way's stars into two distinct populations:

> Halo stars are all old with a very low proportion of heavy elements, while disk stars come in all ages and contain a higher proportion of heavy elements.

1. The **disk population** (sometimes called *Population I*) consists of stars that follow the orderly orbital patterns of the disk. This population includes both young stars and old stars, all of which have heavy-element proportions near 2%, like that of our Sun.

2. The **halo population** (or *Population II*) consists of stars that orbit the center of the galaxy with many different inclinations,

A protogalactic cloud contains pure hydrogen and helium gas amid a larger amount of dark matter.

Halo stars begin to form as gravity pulls the gas toward the center of the cloud.

Conservation of angular momentum ensures that the remaining gas flattens into a spinning disk.

Billions of years later, the star–gas–star cycle supports ongoing star formation within the disk. The lack of cold gas in the halo precludes star formation outside the disk.

▲ **FIGURE 15.18**

This four-picture sequence illustrates a simple schematic model of galaxy formation, showing how a spiral galaxy might develop.

so that they cross through the disk during their orbits. Stars in this population are all old and therefore low in mass. Halo stars can have heavy-element proportions as low as 0.02%—meaning that heavy elements are about 100 times rarer in these stars than in the Sun.

Stars in the bulge exhibit a mixture of disk-star and halo-star characteristics, with a wide range of ages and heavy-element proportions.

We can understand why halo stars differ from disk stars by looking at how the Milky Way's gas is distributed. The halo does not contain the cold, dense molecular clouds required for star formation. Instead, the gas is far hotter and much more spread out. Because star-forming molecular clouds are found only in the disk, new stars can be born only in the disk and not in the halo.

The relative lack of heavy elements in halo stars indicates that they must have formed early in the galaxy's history—before many supernovae had exploded and added heavy elements to star-forming clouds. We therefore conclude that the halo has lacked the gas needed for star formation for a very long time. Apparently, all the Milky Way's cool gas settles into the disk. The only stars that still survive in the halo are long-lived, low-mass stars.

think about it How does the halo of our galaxy resemble the distant future fate of the galactic disk? Explain.

◆ How did our galaxy form?

Any model of our galaxy's formation must account for the differences between disk stars and halo stars. The simplest such model, shown in Figure 15.18, has our galaxy beginning as a giant **protogalactic cloud** of hydrogen and helium gas, along with an even larger amount of dark matter.

Early on, the gravity associated with this protogalactic cloud would have drawn in matter from all directions, creating a cloud that was blobby in shape with little or no organized rotation. Gravity would then have caused localized regions within the cloud to contract and fragment, just as in present-day star-forming clouds [Section 13.1]. As a result, the orbits of stars that formed early on could have had any orientation, accounting for the randomly oriented orbits of stars in the halo. This model also explains the old ages of halo stars, because it tells us that they were the first stars to form in the galaxy. The low proportions of heavy elements found in halo stars are then simply a consequence of early formation, before multiple generations of stars added heavy elements to star-forming clouds.

As time passed, the remaining gas continued to contract under the force of gravity. The law of conservation of angular momentum ensured that the gas in the collapsing proto-galactic cloud (but not the dark matter) ultimately settled into the shape of a flattened, spinning disk, following the same basic processes that occurred on a much smaller scale during the formation of our solar system [Section 6.2]. Collisions among gas particles tended to average out their random motions, leading them to acquire orbits in the same direction and in the same plane. Stars that formed within this spinning disk

Halo stars formed when our galaxy's protogalactic cloud was still large and blobby, and disk stars formed after the gas had settled into a spinning disk.

therefore share the organized motion of the disk. The ages of disk stars, which range from newly born to more than 10 billion years, are explained by the fact that the star–gas–star cycle has supported ongoing star formation ever since the disk formed.

Although the model of Figure 15.18 accounts for the basic differences between halo and disk stars, more detailed studies of heavy-element proportions suggest that this model is a bit too simplistic. If the Milky Way had formed from a single protogalactic cloud, it would have steadily accumulated heavy elements during its inward collapse as stars formed and exploded within it. In that case, the outermost stars in the halo would be the oldest and would have the smallest proportions of heavy elements, and the heavy-element proportions would steadily rise as we looked at halo stars orbiting closer to the disk and bulge. But that is not the pattern we observe.

Instead, we find variations in heavy-element proportions, suggesting that the Milky Way's oldest stars formed in relatively small protogalactic clouds, each with a few globular clusters. These clouds later collided and combined to create a single galaxy, the Milky Way (Figure 15.19). Similar processes may still be under way.

Our galaxy's halo stars may have formed in several small protogalactic clouds that later merged to form one large protogalactic cloud.

Recall that the Sagittarius and Canis Major dwarf galaxies are currently being torn apart as they crash through the Milky Way's disk. A billion years from now, their stars will be indistinguishable from halo stars, because they will all be circling the Milky Way on orbits that carry them high above the disk. Evidence indicates that this process has also occurred in the past: Some halo stars move in organized streams that are probably the remnants of dwarf galaxies torn apart long ago by the Milky Way's gravity.

think about it If the preceding scenario is true, then the Milky Way suffered several collisions early in its history. Explain why we think that galaxy collisions (or collisions between protogalactic clouds) were common in the distant past. (*Hint:* How did the average separations of galaxies in the past compare to their average separations today?)

The distribution of elements also suggests that when the disk finally formed, it began with only about 10% of the amount of heavy elements that it has today, though with a greater concentration of heavy elements toward the bulge. As time passed, the star–gas–star cycle gradually increased the disk's heavy-element content, which is why young stars tend to have higher proportions of heavy elements than older stars.

15.4 The Galactic Center

The center of the Milky Way Galaxy lies in the direction of the constellation Sagittarius and does not look particularly special to our unaided eyes. However, if we could remove the interstellar dust that obscures our view, the galaxy's central bulge would be one of the night sky's most spectacular sights. And deep within the bulge, at the very center of our galaxy, evidence indicates the presence of an extremely massive black hole. Figure 15.20 zooms us into the center, summarizing the evidence for the black hole.

▲ **FIGURE 15.19**

This painting shows a model of how the Milky Way's halo may have formed. The characteristics of stars in the Milky Way's halo suggest that several smaller gas clouds, already bearing some stars and globular clusters, may have merged to form the Milky Way's protogalactic cloud. These stars and star clusters remained in the halo while the gas settled into the Milky Way's disk.

cosmic context FIGURE 15.20 **The Galactic Center**

Zooming into the galactic center in different wavelengths of light reveals strong evidence that the central object, called Sgr A*, is a 4-million-solar-mass black hole.

1 Dusty gas clouds in our galaxy's disk prevent us from observing visible light from the galactic center.

2 Infrared light passes more easily through dusty clouds than visible light, allowing us to observe the stars (white) and gas clouds (red) at our galaxy's center.

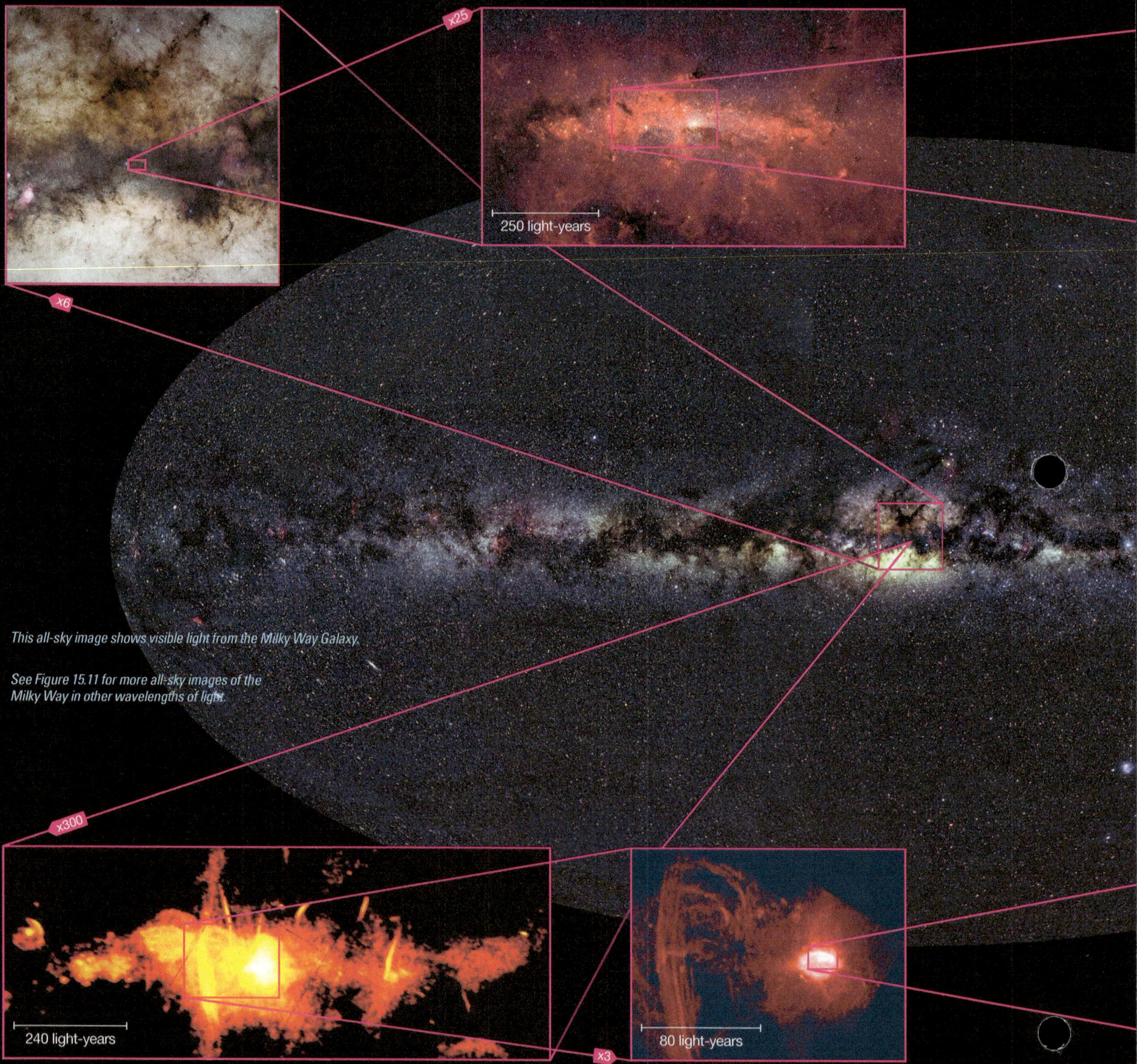

x25
250 light-years
x6
This all-sky image shows visible light from the Milky Way Galaxy.
See Figure 15.11 for more all-sky images of the Milky Way in other wavelengths of light.
x300
240 light-years
x3
80 light-years

7 Radio waves, which also pass through dusty gas clouds, provide further evidence for a central black hole. Here, we see the strong radio emission from the central few hundred light-years.

8 In this radio-wave image, vast threads of emission trace magnetic field lines near the galactic center.

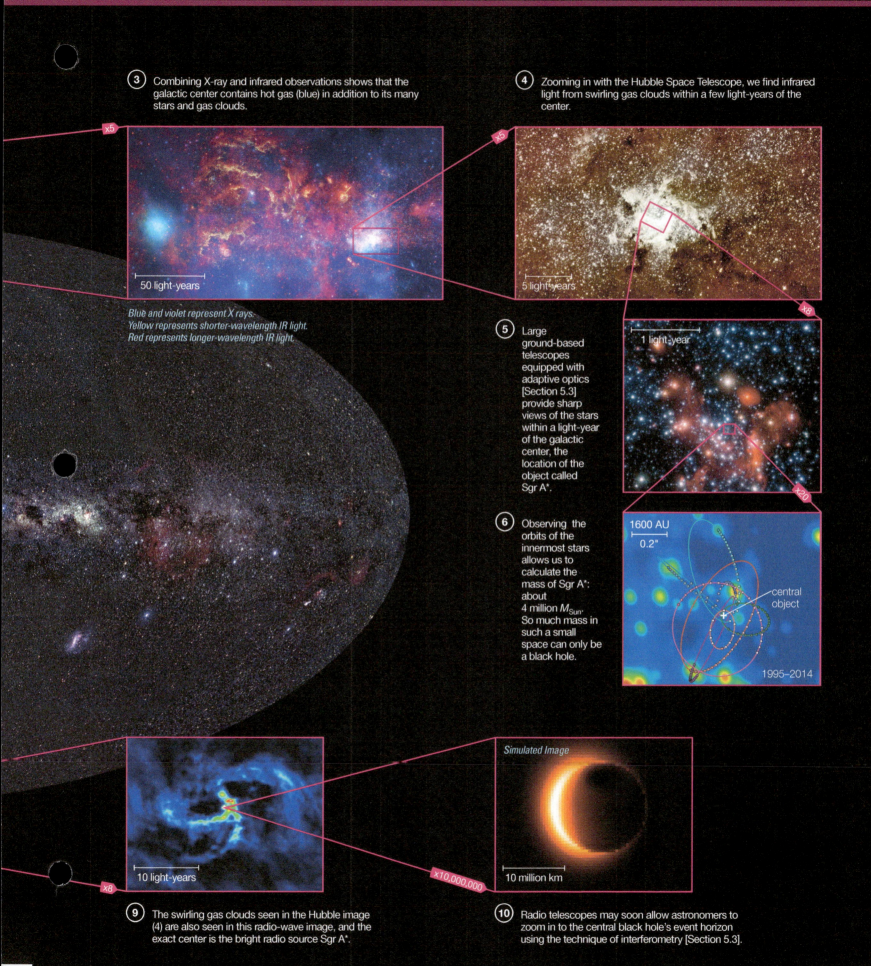

③ Combining X-ray and infrared observations shows that the galactic center contains hot gas (blue) in addition to its many stars and gas clouds.

④ Zooming in with the Hubble Space Telescope, we find infrared light from swirling gas clouds within a few light-years of the center.

50 light-years

Blue and violet represent X rays.
Yellow represents shorter-wavelength IR light.
Red represents longer-wavelength IR light.

5 light-years

⑤ Large ground-based telescopes equipped with adaptive optics [Section 5.3] provide sharp views of the stars within a light-year of the galactic center, the location of the object called Sgr A*.

1 light-year

⑥ Observing the orbits of the innermost stars allows us to calculate the mass of Sgr A*: about 4 million M_{Sun}. So much mass in such a small space can only be a black hole.

1600 AU
0.2"
central object
1995–2014

10 light-years

Simulated Image

10 million km

⑨ The swirling gas clouds seen in the Hubble image (4) are also seen in this radio-wave image, and the exact center is the bright radio source Sgr A*.

⑩ Radio telescopes may soon allow astronomers to zoom in to the central black hole's event horizon using the technique of interferometry [Section 5.3].

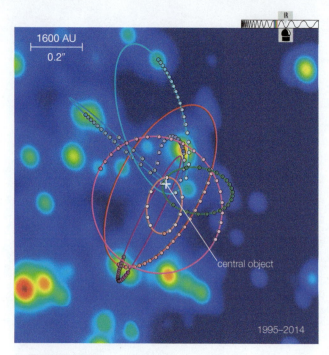

FIGURE 15.21

Evidence for a black hole at the center of our galaxy. Each set of colored dots shows the positions of a particular star at 1-year intervals as observed with the Keck Telescope; calculated orbits are also shown. By applying Newton's version of Kepler's third law, we infer that the central object has a mass about 4 million times that of our Sun, packed into a space no larger than our solar system. (The 1600 AU shown on the scale bar is equivalent to about 9 light-days.)

▼ FIGURE 15.22

This X-ray image from the Chandra X-Ray Observatory shows the central 60 light-years of our galaxy and an X-ray flare from the massive black hole thought to reside there.

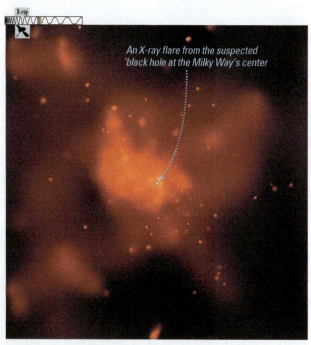

An X-ray flare from the suspected black hole at the Milky Way's center

◆ What is the evidence for a black hole at our galaxy's center?

Although the Milky Way's clouds of gas and dust prevent us from seeing visible light from the center of the galaxy, we can peer into the heart of our galaxy with radio, infrared, and X-ray telescopes. As you follow through the image sequence in Figure 15.20, notice the increasing evidence for something highly energetic occurring in the center of the galaxy. For example, within a few hundred light-years of the center, infrared observations reveal swirling clouds of gas and millions of stars. Bright radio emissions trace out the magnetic fields that thread this turbulent region. Then, as we zoom down toward the exact center, we find a source of radio emission named Sagittarius A* (pronounced "Sagittarius A-star"), or Sgr A* for short, that is quite unlike any other radio source in our galaxy.

see it for yourself Use the star charts in Appendix I to find the constellation Sagittarius, which is easily visible in the evening sky from July through September and looks like a teapot with a handle on the left and a spout on the right (when viewed from the Northern Hemisphere). The center of the Milky Way Galaxy is near the tip of the spout. Can you see the Milky Way's faint band of light passing through the constellations Sagittarius, Cygnus, and Cassiopeia?

Several hundred stars crowd the region within about 1 light-year of Sgr A*, and their motions indicate the presence of an extremely massive object. Figure 15.21 (a larger version of step 6 in Figure 15.20)

Stars quite close to our galaxy's center orbit a compact object about 4 million times as massive as our Sun—almost certainly a huge black hole.

shows the orbital paths that have been observed for some of these stars. Applying Newton's version of Kepler's third law to these orbits shows that the central object has a mass of about 4 million solar masses, all packed into a region of space just a little larger than our solar system. An object that massive within such a small space is almost certainly a black hole.

Some aspects of Sgr A* remain puzzling. Most suspected black holes are thought to accumulate matter through accretion disks that radiate brightly in X rays. These include black holes in binary star systems like Cygnus X-1 [Section 14.3] and some giant black holes at the centers of other galaxies [Section 16.4]. If the black hole at the center of our galaxy had an accretion disk, its X-ray light would easily penetrate the dusty gas of our galaxy and appear fairly bright to our X-ray telescopes. Yet at most times, the X-ray emission from Sgr A* has been relatively faint.

Observations of Sgr A* made in other wavelengths of light are helping us understand this surprising behavior. For example, enormous X-ray flares have been observed coming from the location of the suspected black hole (Figure 15.22). These sudden changes in X-ray brightness probably come from energy released by comet-sized lumps of matter torn apart by the black hole's tidal forces just before disappearing beneath its event horizon. If we continue to observe similar X-ray flares from Sgr A*, then the explanation for its generally low X-ray brightness may be that matter falls into it in big chunks instead of from a smooth, swirling accretion disk. Until we better understand our galaxy's central black hole, it is sure to remain a favorite target for observation.

the big picture — Putting Chapter 15 into Perspective

In this chapter, we have explored the structure, motion, and history of our galaxy, along with the galactic recycling that has made our existence possible. As you review, keep in mind these "big picture" ideas:

- The inability of visible light to pass through interstellar gas and dust concealed the true nature of our galaxy until recent times. Modern astronomical instruments reveal the Milky Way Galaxy to be a dynamic system of stars and gas that continually gives birth to new stars and planetary systems.

- Stellar winds and explosions make interstellar space a violent place. Hot gas tears through the atomic hydrogen that fills much of the galactic disk, leaving expanding bubbles and fast-moving clouds in its wake. All this violence might seem dangerous, but it performs the great service of mixing new heavy elements into the gas of the Milky Way.

- Orbital motions at our galaxy's center indicate that it contains a black hole with 4 million times the mass of our Sun. As we will see in the next chapter, central black holes appear to be a common feature of galaxies.

my cosmic perspective Although the elements from which we are made were forged in stars, we could not exist if stars were not organized into galaxies. The Milky Way Galaxy acts as a giant recycling plant, converting gas expelled from each generation of stars into the next generation and allowing some heavy elements to solidify into planets like our own.

summary of key concepts

15.1 The Milky Way Revealed

◆ **What does our galaxy look like?**

The Milky Way Galaxy is a **spiral galaxy** consisting of a thin **disk** about 100,000 light-years in diameter with a central **bulge** and a spherical **halo** that surrounds the disk. The disk contains an **interstellar medium** of gas and dust, while the halo contains only a small amount of hot gas and virtually no cold gas.

◆ **How do stars orbit in our galaxy?**

Stars in the disk all orbit the galactic center in about the same plane and in the same direction. Halo stars also orbit the center of the galaxy, but their orbits are randomly inclined to the disk of the galaxy. Some bulge stars orbit like halo stars, while others orbit more like disk stars. Orbital motions of stars allow us to determine the distribution of mass in our galaxy.

15.2 Galactic Recycling

◆ **How is gas recycled in our galaxy?**

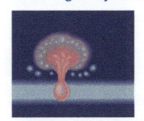

Stars are born from the gravitational collapse of gas clumps in **molecular clouds.** Massive stars explode as supernovae when they die, creating hot **bubbles** in the interstellar medium that contain the new elements made by these stars. This gas cools and mixes with the interstellar medium, forming clouds of **atomic hydrogen gas.** This gas can cool further to make molecular clouds in which new stars are born, completing the **star–gas–star cycle.**

◆ **Where do stars tend to form in our galaxy?**

Active star-forming regions, marked by the presence of hot, massive stars and **ionization nebulae,** are found mostly in **spiral arms.** The spiral arms represent regions where a **spiral density wave** has compressed gas clouds to make star formation more likely.

15.3 The History of the Milky Way

◆ **What do halo stars tell us about our galaxy's history?**

The stars of the **halo population** are old, low-mass stars that have a much smaller proportion of heavy elements than the stars of the **disk population**. Halo stars therefore must have formed early in the galaxy's history, before the gas settled into a disk.

◆ **How did our galaxy form?**

Halo stars probably formed in several different **protogalactic clouds** of hydrogen and helium gas. Gravity pulled those clouds together to form a single larger one. The collapse of this cloud continued until it formed a spinning disk around the galactic center. Stars have formed continuously in the disk since that time, but stars no longer form in the halo.

15.4 The Galactic Center

◆ **What is the evidence for a black hole at our galaxy's center?**

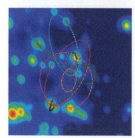

Orbits of stars near the center of our galaxy indicate that it contains a black hole about 4 million times as massive as the Sun. The black hole appears to be powering a bright source of radio emission known as Sgr A*.

---

visual skills check   Check your understanding of some of the many types of visual information used in astronomy. For additional practice, try the Chapter 15 Visual Quiz at MasteringAstronomy®.

Visible light from stars

Infrared light from stars

Radio emission from molecules

X-ray emission from hot gas

These images show the central region of our galaxy in different wavelengths of light. Use these images to answer the following questions.

1. The image of radio emission uses different colors of light to represent different levels of brightness. Which color represents the brightest radio emission? Which color represents the lowest level of brightness?

2. The image of X-ray emission uses different colors of light to represent different levels of brightness. The dark blue color represents the least bright X-ray emission. Which color represents the brightest X-ray emission?

3. How do regions showing strong radio emission from molecules look in the visible-light image? Are they bright or are they dark?

4. How do regions showing strong radio emission from molecules look in the infrared-light image? Are they bright or are they dark?

5. Based on a comparison of the radio, infrared, and visible-light images, we can conclude that gas clouds containing molecules _____.

   a. absorb roughly equal amounts of infrared starlight and visible starlight.
   b. absorb substantial amounts of visible starlight but no infrared starlight.
   c. absorb infrared starlight less effectively than they absorb visible starlight.

6. Compare the radio and X-ray images. Can you conclude that gas clouds containing molecules absorb X rays?

MasteringAstronomy® For instructor-assigned homework and other learning materials, go to MasteringAstronomy®.

Review Questions

1. Draw simple sketches of our galaxy as it would appear face-on and edge-on. Identify the *disk, bulge, halo,* and *spiral arms,* and indicate the galaxy's approximate dimensions.
2. What are the Large and Small Magellanic Clouds, and the Sagittarius and Canis Major Dwarfs?
3. Describe and contrast stellar orbits in the disk, halo, and bulge of our galaxy.
4. How can we use orbital properties to learn about the mass of the galaxy? What have we learned?
5. Summarize the stages of the star–gas–star cycle in Figure 15.3.
6. What creates a bubble of hot, ionized gas? What happens to the gas in the bubble over time?
7. What are *cosmic rays,* and where are they thought to come from?
8. What do we mean by *atomic hydrogen gas*? How common is it, and how do we map its distribution in the galaxy?
9. Briefly summarize the different types of gas present in the disk of the galaxy, and describe how they appear when we view the galaxy in different wavelengths of light.
10. What are *ionization nebulae,* and why are they found near hot, massive stars?
11. How do we know that spiral arms do not rotate like giant pinwheels? What makes spiral arms bright?
12. What triggers star formation within a spiral arm? How do we think spiral arms are maintained?
13. Briefly describe the characteristics that distinguish *disk stars* from *halo stars*.
14. What evidence suggests that the Milky Way formed from the merger of several smaller protogalactic clouds?
15. What is Sgr A*? What evidence suggests that it contains a massive black hole?

Test Your Understanding

🔵 Does It Make Sense?

Decide whether the statement makes sense (or is clearly true) or does not make sense (or is clearly false). Explain clearly; not all of these have definitive answers, so your explanation is more important than your chosen answer.

16. We did not understand the true size and shape of our galaxy until NASA launched satellites into the galactic halo, enabling us to see what the Milky Way looks like from the outside.
17. Planets like Earth probably didn't form around the first stars because there were so few heavy elements back then.
18. If I could see infrared light, the galactic center would look much more impressive.
19. Many spectacular ionization nebulae are seen throughout the Milky Way's halo.
20. The carbon in my diamond ring was once part of interstellar dust grains.
21. The Sun's velocity around the Milky Way tells us that most of our galaxy's dark matter lies near the center of the galactic disk.
22. We know that a black hole lies at our galaxy's center because numerous stars near it have vanished over the past several years, telling us that they've been sucked in.

23. If we could watch a time-lapse movie of a spiral galaxy over millions of years, we'd see many stars being born and dying within the spiral arms.
24. The star–gas–star cycle will keep the Milky Way looking just as bright in 100 billion years as it looks now.
25. Halo stars orbit the center of our galaxy much faster than the disk stars.

Quick Quiz

Choose the best answer to each of the following. Explain your reasoning with one or more complete sentences.

26. What is the shape of the Milky Way's halo? (a) round like a ball (b) flat like a disk (c) flat like a disk but with a hole in the center
27. Where are most of the Milky Way's globular clusters found? (a) in the disk (b) in the bulge (c) in the halo
28. Why do disk stars bob up and down as they orbit the galaxy? (a) because the gravity of other disk stars always pulls them toward the disk (b) because of friction with the interstellar medium (c) because the halo stars keep knocking them back into the disk
29. How do we determine the Milky Way's mass outside the Sun's orbit? (a) from the Sun's orbital velocity and its distance from the center of our galaxy (b) from the orbits of halo stars near the Sun (c) from the orbits of stars and gas clouds orbiting the galactic center at greater distances than the Sun
30. Which part of the galaxy contains the coldest gas? (a) the disk (b) the halo (c) the bulge
31. What is the typical percentage (by mass) of elements other than hydrogen and helium in stars that are forming right now in the vicinity of the Sun? (a) 20% (b) 2% (c) 0.02%
32. Which of these forms of radiation passes most easily through the disk of the Milky Way? (a) red light (b) blue light (c) infrared light
33. Where would you be most likely to find an ionization nebula? (a) in the halo (b) in the bulge (c) in the disk
34. Which kind of star is most likely to be found in the halo? (a) an O star (b) an A star (c) an M star
35. The best measurements of the mass of the black hole at the galactic center come from (a) the orbits of stars in the galactic center. (b) the orbits of gas clouds in the galactic center. (c) the amount of radiation coming from the galactic center.

🔵 Process of Science

36. *Discovering the Structure of the Milky Way.* The story of how we came to learn the structure of the Milky Way (see Special Topic, page 388) is an excellent demonstration of how science progresses. What features of the Milky Way's appearance in our sky led scientists to conclude that its width is much larger than its thickness? Why did they originally believe that the Sun was near the Milky Way's center? What key observations forced scientists to change their views about the location of the Sun within the Milky Way?
37. *Formation of the Milky Way.* Figure 15.18 outlines a basic model that accounts for some but not all of the features of the Milky Way.

What observational evidence indicates that the Milky Way's protogalactic cloud contained virtually no elements other than hydrogen and helium? What evidence suggests that the halo stars formed first and disk stars formed later? What features of the Milky Way are not explained by this basic model?

Group Work Exercise

38. *Star Clusters and Milky Way Structure.* **Roles**: *Scribe* (takes notes on the group's activities), *Proposer* (proposes explanations to the group), *Skeptic* (points out weaknesses in proposed explanations), *Moderator* (leads group discussion and makes sure everyone contributes).
    Activity: Use the all-sky diagram in Appendix H to do the following.
    a. Globular clusters are found in the following constellations: Aquarius, Aquila, Canes Venatici, Capricornus, Carina, Centaurus, Columba, Coma Berenices, Hercules, Hydra, Lepus, Lynx, Lyra, Musca, Ophiuchus, Pegasus, Puppis, Sagittarius, Scorpius, Sculptor, Serpens, Tucana, and Vela. The *Scribe* and *Proposer* should work together to find these constellations and draw conclusions about the locations of globular clusters relative to the Milky Way's disk and relative to the galactic center, which is in the constellation Sagittarius.
    b. Open clusters with ages of 100 million years or less are found in the following constellations: Canis Major, Carina, Cassiopeia, Crux, Cygnus, Monoceros, Norma, Ophiuchus, Perseus, Puppis, Sagittarius, Scorpius, Scutum, Serpens, Taurus, and Vela. The *Moderator* and *Skeptic* should work together to find these constellations and draw conclusions about the locations of young open clusters relative to the Milky Way's disk and relative to the galactic center.
    c. As a team, compare the locations of the two sets of clusters and describe the differences you find.
    d. The *Proposer* should offer an explanation for these differences.
    e. The *Skeptic* should propose an alternative explanation.
    f. The *Moderator* and *Scribe* should discuss these explanations and list some observations that astronomers could perform to determine which explanation is better.

Investigate Further

Short-Answer/Essay Questions

39. *Unenriched Stars.* Suppose you discovered a star made purely of hydrogen and helium. How old do you think it would be? Explain.
40. *Enrichment of Star Clusters.* The gravitational pull of an isolated globular cluster is rather weak—a single supernova explosion can blow all the interstellar gas out of a globular cluster. How might this fact relate to observations indicating that stars ceased to form in globular clusters long ago? How might it relate to the fact that globular clusters are deficient in elements heavier than hydrogen and helium? Summarize your answers in one or two paragraphs.
41. *High-Velocity Star.* The average speed of stars in the solar neighborhood relative to the Sun is about 20 km/s. Suppose you discover a star in the solar neighborhood that is moving at a much higher speed relative to the Sun, say, 200 km/s. What kind of orbit does this star probably have around the Milky Way? In what part of the galaxy does it spend most of its time? Explain.
42. *Future of the Milky Way.* Describe how the Milky Way would look from the outside if you could watch it for the next 100 billion years. How would its appearance change?
43. *Orbits at the Galactic Center.* Using the information in Figure 15.21, identify which two stars reach the highest orbital speeds. Explain how the orbits of those two stars illustrate Kepler's first two laws.

44. *A Nonspinning Galaxy.* How would the development of the Milky Way Galaxy have been different if it had collapsed from protogalactic clouds that had no net angular momentum? Describe how you think our galaxy would look today and explain your reasoning.
45. *Gas Distribution in the Milky Way.* Make a sketch of the gas distribution in the plane of the Milky Way, based on the photographs in Figure 15.11. In your sketch, map out where you would find molecular clouds, atomic hydrogen clouds, and bubbles of hot gas. Explain why each of those components of the interstellar medium is found in the location where you have drawn it.

Quantitative Problems

Be sure to show all calculations clearly and state your final answers in complete sentences.

46. *Mass of the Milky Way's Halo.* The Large Magellanic Cloud orbits the Milky Way at a distance of roughly 160,000 light-years from the galactic center and a velocity of about 300 km/s. Use these values in the orbital velocity formula to estimate the Milky Way's mass within 160,000 light-years from the center.
47. *Mass of the Central Black Hole.* Suppose you observed a star orbiting the galactic center at a speed of 1000 km/s in a circular orbit with a radius of 20 light-days. Calculate the mass of the object that the star was orbiting.
48. *Mass of a Globular Cluster.* Stars in the outskirts of a globular cluster are typically about 50 light-years from the cluster's center, which they orbit at speeds of about 10 km/s. Use these data to calculate the mass of a typical globular cluster.
49. *Mass of Saturn.* The innermost rings of Saturn orbit in a circle with a radius of 67,000 kilometers at a speed of 23.8 km/s. Use the orbital velocity formula to compute the mass contained within the orbit of those rings. Compare your answer with the mass of Saturn listed in Appendix E.

Discussion Questions

50. *Galactic Ecosystem.* We have likened the star–gas–star cycle in our Milky Way to the ecosystem that sustains life on Earth. Here on our planet, water molecules cycle from the sea to the sky to the ground and back to the sea. Our bodies convert atmospheric oxygen molecules into carbon dioxide, and plants convert carbon dioxide back into oxygen molecules. How are the cycles of matter on Earth similar to the cycles of matter in the galaxy? How do they differ? Do you think the term ecosystem is appropriate in discussions of the galaxy?
51. *Galaxy Stuff.* In the chapters on stars, we learned why we are "star stuff." Explain why we are also "galaxy stuff." Does the fact that the entire galaxy was involved in bringing forth life on Earth change your perspective on Earth or on life? If so, how? If not, why not?

Web Projects

52. *Images of the Star–Gas–Star Cycle.* Find pictures on the Internet of ionization nebulae and other forms of interstellar gas in different stages of the star–gas–star cycle. Assemble the pictures into a sequence that tells the story of interstellar recycling, with a one-paragraph explanation of each image.
53. *The Galactic Center.* Search the Internet for recent images of the galactic center, along with information about the massive black hole thought to reside there. Write a two- to three-page report, with pictures, giving an update on current knowledge.

.16

A Universe of Galaxies

This spiral galaxy, known as Messier 106, contains an active galactic nucleus powered by matter accreting onto a black hole with a mass 40 million times that of our Sun.

411

ESSENTIAL PREPARATION

1. How do galaxies move within the universe? [Section 1.3]

2. How do we measure stellar luminosities? [Section 12.1]

3. Where do stars tend to form in our galaxy? [Section 15.2]

4. How did our galaxy form? [Section 15.3]

5. What is the evidence for a black hole at our galaxy's center? [Section 15.4]

Barely a century ago, no one knew for certain whether the universe extended beyond the bounds of our own Milky Way. Today, we know that our observable universe contains some 100 billion galaxies, and that these galaxies come in a variety of shapes and sizes. We also know that our universe is expanding, so nearly all galaxies are being carried away from one another with time.

In this chapter, we will explore galaxies as we observe them throughout the universe. We'll begin by getting acquainted with the patterns observed among galaxies, and then discuss how we measure galactic distances and how those measurements help us learn the expansion rate and age of the universe. Finally, we will study the evolution of galaxies and what it tells us about the history of our universe as a whole.

16.1 Islands of Stars

Figure 16.1 shows an amazing Hubble Space Telescope image, known as the Hubble eXtreme Deep Field, obtained by combining images in which the telescope focused on single, tiny patch of the sky for a total of 23 days. Almost every one of the thousands of blobs of light in this image is a galaxy—an island of stars bound together by gravity, much like our own Milky Way.

Photos like this one allow us to estimate the total number of galaxies in the observable universe, because this patch of sky is quite average and representative of the sky as a whole. We can therefore simply count the number of galaxies in the photo and multiply by the number of such photos it would take to capture the entire sky. The results tell us that the observable universe contains well over 100 billion galaxies. Even without information about the distances of the individual galaxies shown, the images make clear that galaxies come in many sizes, colors, and shapes. Let's begin our study of galaxies by exploring the patterns among their properties.

◆ What patterns do we find among the properties of galaxies?

Astronomers classify galaxies into three main types based on their appearances:

- **Spiral galaxies,** such as our own Milky Way, look like flat white disks with yellowish bulges at their centers. The disks are filled with cool gas and dust, interspersed with hotter ionized gas, and usually display beautiful spiral arms.

- **Elliptical galaxies** are redder, rounder, and often elongated like a football. Compared with spiral galaxies, elliptical galaxies contain very little cool gas and dust, though they often contain very hot ionized gas.

- **Irregular galaxies** appear neither disk-like nor rounded.

The differing colors of galaxies arise from the different kinds of stars that populate them: Spiral and irregular galaxies look white because they contain stars of all different colors and ages, while elliptical galaxies look redder because old, reddish stars produce most of their light. Galaxies also come in a wide range of sizes, from *dwarf galaxies* containing less than 1 million (10^6) stars to *giant galaxies* with more than 1 trillion (10^{12}) stars.

Galaxies come in three major types: spiral, elliptical, and irregular.

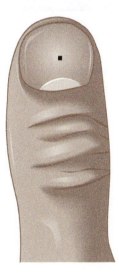

▲ FIGURE 16.1

The Hubble eXtreme Deep Field (XDF) required 23 days of exposure time with the Hubble Space Telescope and shows thousands of galaxies in a typical patch of the sky. To visualize the size of this patch on the celestial sphere, picture on your own thumb the tiny version of the XDF on the thumbnail to the right, then hold your thumb at arm's length.

think about it Take a moment to try to classify the larger galaxies in Figure 16.1. How many appear spiral? Elliptical? Irregular? Do the colors of galaxies seem related to their shapes?

Spiral Galaxies Like the Milky Way, other spiral galaxies also have a thin *disk* and a central *bulge* (Figure 16.2). The bulge merges smoothly into a nearly invisible *halo* that can extend to a radius of more than 100,000 light-years. A spiral galaxy's halo is considerably more difficult to see than its bulge and disk because halo stars are generally dim and spread over a large volume of space.

Recall that the disk and the halo of the Milky Way represent two distinct populations of stars [Section 15.3]: *Disk stars* orbit in the same plane and in-

Spiral galaxies have a disk, bulge, and halo like the Milky Way.

clude stars of all ages and masses, while *halo stars* have randomly oriented orbits and are all old and low in mass. The Milky Way's bulge stars, on the other hand, exhibit a mixture of disk and halo properties. Other spiral galaxies share these characteristics, but the bulge-star orbits in spiral galaxies with particularly large bulges have more in common with the orbits of halo stars than with those of disk stars. We will therefore simplify our discussion of galaxies by dividing their stars into two primary components:

- The **disk component** consists of stars and dusty gas clouds that follow orderly, nearly circular orbits around the galactic center. Galaxy disks always contain a gaseous *interstellar* medium, but the amounts and proportions of molecular, atomic, and ionized gas differ from one galaxy to the next.

▲ FIGURE 16.2

The giant spiral galaxy M101. It is about 170,000 light-years in diameter.

▲ **FIGURE 16.3**

NGC 4594 (the Sombrero Galaxy) is a spiral galaxy with a large bulge and a dusty disk that we see almost edge-on. A much larger but nearly invisible halo surrounds the entire galaxy and merges smoothly into the central bulge. The region shown in this image is about 82,000 light-years across.

▲ **FIGURE 16.4**

NGC 1300, a barred spiral galaxy about 110,000 light-years in diameter.

- The **halo component** (sometimes called the *spheroidal* or *ellipsoidal* component), which we will take to include both the halo and the central bulge, generally has a rounded or elliptical shape. The halo component usually contains little cool gas and dust, and its stars have orbits with many different inclinations. The disk is embedded within it.

All spiral galaxies have both a disk component and a halo component, but there are variations on the general theme. Bulge sizes can vary dramatically; for example, Figure 16.3 shows a spiral galaxy with an unusually large bulge that illustrates the general shape of the halo component. Other spiral galaxies—known as *barred spirals*—appear to have a straight bar of stars cutting across the center, with spiral arms curling away from the ends of the bar (Figure 16.4). Astronomers suspect that the Milky Way itself is a barred spiral galaxy, because of the elongation of our galaxy's bulge.

Some galaxies have disk and halo components like spiral galaxies but appear to lack spiral arms. These *lenticular galaxies* (*lenticular* means "lens-shaped") are sometimes considered an intermediate class between spirals and ellipticals, because they tend to have less cool gas than normal spirals but more than ellipticals. Among large galaxies in the universe, most (75–85%) are spiral or lenticular.

Elliptical Galaxies   Elliptical galaxies differ from spiral galaxies primarily in that they have *only* a halo component and lack a significant disk component. That is, elliptical galaxies look much like the bulge and halo of a spiral galaxy without a disk. Relatively rare *giant elliptical galaxies* are among the most massive galaxies in the universe (Figure 16.5), while small *dwarf elliptical galaxies* are much more common.

Elliptical galaxies usually contain very little dust or cool gas. As a result, they generally have little or no ongoing star formation. Elliptical galaxies therefore look red or yellow in color because they lack the hot, young, blue stars found in the disks of spiral galaxies. However, large elliptical galaxies sometimes contain substantial amounts of very hot gas that emits X rays, much like the gas in the hot bubbles created by supernovae and powerful stellar winds in the Milky Way [Section 15.2].

Elliptical galaxies differ from spiral galaxies in that they do not have significant disks.

The most numerous galaxies in our Local Group lie at the other end of the galaxy size scale. They are round and diskless like large elliptical

▲ **FIGURE 16.5**

M87, a giant elliptical galaxy in the Virgo Cluster, is one of the most massive galaxies in the universe. The region shown is more than 300,000 light-years across.

galaxies, but they are particularly small and much less bright. Because of these distinct differences, astronomers assign these galaxies to a special subtype (known as *dwarf spheroidal galaxies*). Their lack of brightness makes them almost impossible to detect at distances greater than a few million light-years. Nevertheless, astronomers suspect this galaxy subtype to be the most common in the universe, based on the large proportion of them in the Local Group.

Irregular Galaxies   Some of the galaxies we see nearby fall into neither of the two major categories. This class of *irregular galaxies* is a miscellaneous class, encompassing small galaxies such as the Magellanic Clouds (Figure 16.6) and larger "peculiar" galaxies that appear to be in disarray. These blobby star systems are usually white and dusty, like the disks of spirals, and contain many young, massive stars.

Irregular galaxies represent only a small fraction of the nearby large galaxies, but telescopic observations probing deeper into the universe show that irregular shapes are more common among distant galaxies. Because the light of more distant galaxies has taken longer to reach us, these observations tell us that irregular galaxies were more common when the universe was younger.

Irregular galaxies appear to be in disarray.

Hubble's Galaxy Classes   Edwin Hubble devised a system for classifying galaxies that organizes the galaxy types into a diagram shaped like a tuning fork (Figure 16.7). Elliptical galaxies appear on the "handle" at the left, designated by the letter *E* and a number. The larger the number, the flatter the elliptical galaxy: An E0 galaxy is a sphere, and the numbers increase to the highly elongated type E7. The two forks show spiral

▲ **FIGURE 16.6**

The Large Magellanic Cloud, an irregular galaxy that is a small companion to the Milky Way. It is about 30,000 light-years across.

▼ **FIGURE 16.7**

This "tuning fork" diagram illustrates Hubble's galaxy classes.

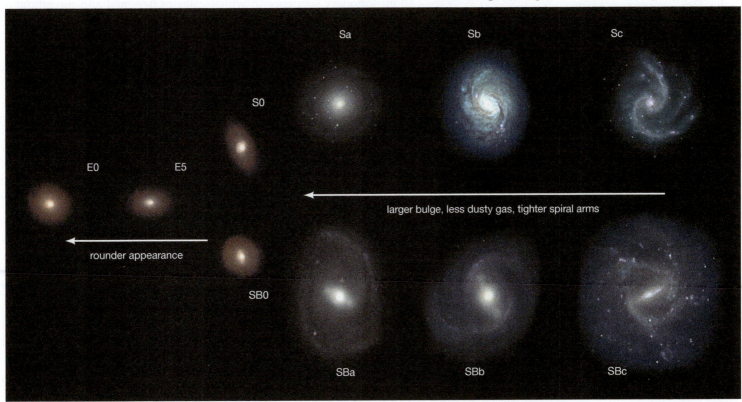

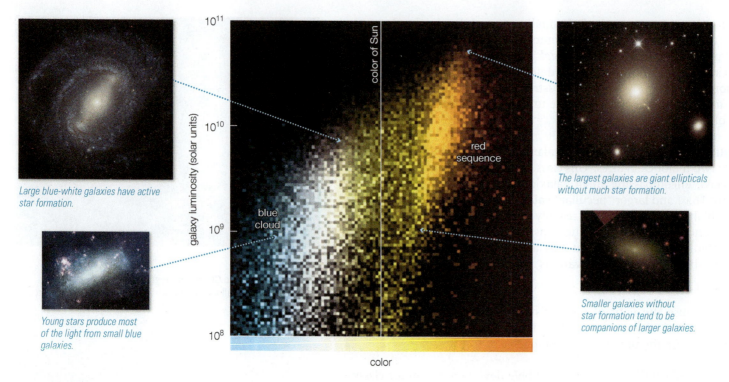

Large blue-white galaxies have active star formation.

Young stars produce most of the light from small blue galaxies.

The largest galaxies are giant ellipticals without much star formation.

Smaller galaxies without star formation tend to be companions of larger galaxies.

▲ **FIGURE 16.8**

A graph plotting galaxy luminosities against colors shows that galaxies fall into two main groups: the *blue cloud*, consisting mostly of spiral and irregular galaxies that are blue to white in color, and the *red sequence*, consisting mostly of elliptical galaxies that are redder in color. (Data from the Sloan Digital Sky Survey.)

▲ **FIGURE 16.9**

Hickson Compact Group 87, a small group of galaxies consisting of a large edge-on spiral galaxy (in the center), two smaller spiral galaxies (above it), and an elliptical galaxy (to its right). The whole group is about 170,000 light-years in diameter. (The other objects in this photograph are foreground stars in our own galaxy.)

galaxies, designated by the letter *S* for ordinary spirals and *SB* for barred spirals, followed by a lowercase *a, b,* or *c:* The bulge size decreases from *a* to *c,* while the amount of dusty gas increases. Lenticular galaxies are designated S0, and irregular galaxies are designated Irr.

Astronomers once suspected that Hubble's tuning fork represented an evolutionary sequence in which galaxies flattened and spread out as they aged, but we now know that is not the case. Instead, other ways of looking at patterns among galaxies have yielded deeper insights.

Patterns in Galaxy Color and Luminosity Astronomers have now measured the properties of millions of galaxies, and can learn more about how galaxies evolve by plotting those properties on a diagram much like the H-R diagram for stars, with galaxy color along the horizontal axis and galaxy luminosity along the vertical axis. Figure 16.8 shows one such diagram. Notice that most galaxies fall into one of two major groups, distinguished primarily by their levels of ongoing star formation:

Plotting the relationship between galaxy color and luminosity reveals patterns produced by galaxy evolution.

- Galaxies in the **blue cloud** are blue because they contain numerous hot, young stars that signify active star formation. This star formation is possible because these galaxies tend to be spiral or irregular galaxies with cold, star-forming gas clouds.

- Galaxies in the **red sequence** usually contain only older stars, which is why they are redder in color. Most galaxies in the red sequence are elliptical in shape, and the ones at the top of the diagram are the most luminous galaxies in the universe.

We will see in Section 16.3 how modern ideas of galaxy evolution explain these patterns among galaxy color and luminosity.

Patterns in Galaxy Groups and Clusters Most of the galaxies in the universe are gravitationally bound together with neighboring galaxies. Spiral galaxies are often found in loose collections of up to a few dozen galaxies, called *groups*. Our Local Group is one example (see Figure 1.1). Figure 16.9 shows another galaxy group.

Larger collections are called galaxy *clusters* and may contain hundreds or thousands of individual galaxies concentrated in a region that can be 10 million or more light-years across (Figure 16.10). Elliptical galaxies make up about half the large galaxies in the central regions of clusters, even though they represent only a small minority (about 15%) of the large galaxies found outside clusters.

Spiral galaxies tend to congregate in small groups, while elliptical galaxies are primarily found in large clusters.

16.2 Distances of Galaxies

To learn more about galaxies than just their shape, color, and type, we need to know how far away they are. Measuring the distances to galaxies is one of the most challenging tasks we face when trying to understand galaxies and the universe as a whole, but the payoff is enormous. Besides telling us where galaxies are located, such measurements also reveal the size and age of the observable universe. In other words, measurements of galactic distances lie at the foundation of what we call **cosmology**—the study of the overall structure and evolution of the universe.

◆ How do we measure the distances to galaxies?

Our determinations of astronomical distances depend on a chain of methods in which each step allows us to measure greater distances. We have already discussed the use of parallax to measure distances to nearby stars [Section 12.1]. Because parallax is the apparent shift in a star's position as Earth orbits the Sun, measuring distances by parallax requires knowing the precise Sun–Earth distance, or astronomical unit (AU). That is why the *first* link in the distance-measurement chain is **radar ranging,** in which radio waves are transmitted from Earth and bounced off Venus. Radio waves travel at the speed of light, so the round-trip travel time for the radar signals tells us Venus's distance from Earth. We can then use Kepler's laws and a little geometry to calculate the length of an AU. Parallax is then the *second* link in the chain, taking us from distances within the solar system to distances within the Milky Way.

Standard Candles The next links in the distance-measurement chain rely on light sources that can serve as **standard candles**, meaning objects whose luminosity we already know. Such objects are useful because we can use their known luminosity and apparent brightness to calculate their distance with the *inverse square law for light* [Section 12.1].

We can determine distance by measuring the apparent brightness of an object whose luminosity we already know and applying the inverse square law for light.

Figure 16.11 shows how standard candles work. All of the lights in the illustration are streetlamps of identical luminosity, but the lamps at greater distance appear dimmer (by the square of their distance). We can therefore

▲ **FIGURE 16.10**
Central part of the galaxy cluster Abell 1689. Most of the fuzzy objects in this photograph are galaxies belonging to the cluster. Yellowish elliptical galaxies outnumber the whiter spiral galaxies. The region pictured is about 2 million light-years across. (A few stars from our own galaxy appear in the foreground as white dots centered on cross-shaped spikes.)

▲ **FIGURE 16.11**
These streetlights are standard candles because they all put out the same amount of light, allowing us to determine their relative distances from their relative brightnesses.

Standard Candles

The inverse square law for light tells us how an object's apparent brightness depends on its luminosity and distance (see Cosmic Calculations 12.1). With a little algebra, we can rewrite this law in a form that enables us to calculate distance if we know luminosity and apparent brightness, as we do for any object that qualifies as a *standard candle*.

$$\text{distance} = \sqrt{\frac{\text{luminosity}}{4\pi \times \text{(apparent brightness)}}}$$

Example: You measure a star's apparent brightness to be 1.0×10^{-12} watt/m^2. The star has the same spectral type and luminosity class as the Sun. How far away is it?

Solution: The star is essentially a twin of the Sun, so we can assume it has the Sun's luminosity of 3.8×10^{26} watts. We use this luminosity and the star's apparent brightness in the formula

$$\text{distance} = \sqrt{\frac{3.8 \times 10^{26} \text{ watts}}{4\pi \times \left(1.0 \times 10^{-12} \, \frac{\text{watt}}{\text{m}^2}\right)}} \approx 5.5 \times 10^{18} \text{ m}$$

The star is about 5.5×10^{18} meters away. We can convert this to light-years by dividing by 9.5×10^{15} meters/light-year (see Appendix A); the result is about 580 light-years.

determine each lamp's distance from its known luminosity. For example, suppose we know that the lamps all emit a luminosity of 1000 watts of light. Then if one lamp appears only ¼ as bright as another, we know that it must be twice as far away as the other.

Unlike light bulbs, astronomical objects do not come marked with wattage. However, an astronomical object can still serve as a standard candle if we have some other way of knowing its true luminosity. Many astronomical objects meet this requirement. For example, any star that is a twin of our Sun—that is, a main-sequence star with spectral type G2—should have about the same luminosity as the Sun. Therefore, if we measure the apparent brightness of a Sun-like star, we can assume it has the Sun's luminosity (3.8×10^{26} watts) and use the inverse square law for light to estimate its distance.

Beyond the distances to which we can measure parallax (about 1500 light-years as of 2016, but expected to increase to tens of thousands of light-years with the *GAIA* spacecraft), we use standard candles for most cosmic distance measurements. These distance measurements always have some uncertainty, because no astronomical object is a perfect standard candle. The challenge of measuring astronomical distances comes down to the challenge of finding the objects that make the best standard candles. The more confident we are about an object's true luminosity, the more certain we are of its distance.

Cepheid Variables The key to measuring distances beyond the Milky Way is to identify objects that can serve as standard candles and that are also bright enough to be detected at great distances. The most useful such objects for relatively nearby galaxies are a special class of extremely luminous stars known as **Cepheid variable stars,** or **Cepheids** for short. These stars vary in luminosity (and hence also in apparent brightness), alternately becoming dimmer and brighter with periods ranging from a few days to a few months. Figure 16.12 shows the brightness variations of a Cepheid with a period of about 50 days.

Cepheids are valuable as standard candles because they obey a **period–luminosity relation:** The longer the period, the more luminous the star (Figure 16.13). This relation was first discovered by Henrietta Leavitt in 1912, through careful observations of Cepheids in the Large Magellanic Cloud. The distance to this small galaxy was not yet known at the time, so Leavitt was unable to determine the true luminosities of the Cepheids. However, because all the stars in the Large Magellanic Cloud are at approximately the same distance from us, she was able to discover the relationship by comparing the apparent brightnesses of these Cepheids. Today, we know the Cepheid period–luminosity relation quite accurately, because we have measured the distances to nearby Cepheids with parallax and used these distances and their apparent brightnesses to calculate true luminosities. We also now know that the period–luminosity relation holds because Cepheids pulsate in size, growing brighter as they grow larger and then dimming as they shrink back in size. Larger (and hence more luminous) Cepheids take longer to pulsate.

As an example of how we use Cepheids to measure distance, look at the point marked with dashed lines in Figure 16.13. Notice that this star's period of 30 days corresponds to a luminosity of about 10,000 Suns. In other words, the fact that this Cepheid's brightness peaks every

Cepheid variable stars are useful for measuring distances because we can determine a Cepheid's luminosity from the period between its peaks of brightness.

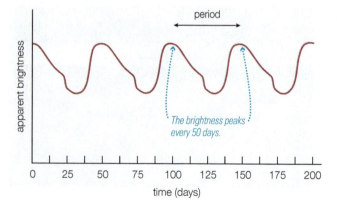

▲ FIGURE 16.12
This graph shows how the brightness of a Cepheid varies with time. The period is the time from one peak of brightness to the next.

30 days means it is effectively screaming out, "Hey, everybody, my luminosity is 10,000 times that of the Sun!" More generally, once we measure a Cepheid's period, we use the period–luminosity relation to learn its luminosity (within about 10%) and can then use the inverse square law for light to determine its distance.

Cepheids have been used for almost a century to measure distances to nearby galaxies, and as we'll discuss shortly, they played a critical role in Edwin Hubble's discoveries. Today, we can use Cepheids to measure distances to galaxies up to 100 million light-years away. This distance may sound large, but it is still quite small compared with the distances of the galaxies in Figure 16.1. To go further, we use the distances determined with Cepheids to learn the luminosities of even brighter standard candles.

Distant Standard Candles Astronomers have identified several types of objects that can be used as standard candles for measuring very large distances, but the most valuable are *white dwarf supernovae*. Recall that white dwarf supernovae are exploding white dwarf stars that have reached the $1.4M_{Sun}$ limit [Section 14.1]. These supernovae all have nearly the same luminosity, probably because they all come from stars of similar mass that explode in nearly the same way. Although white dwarf supernovae are rare in any individual galaxy, numerous examples have been observed within about 50 million light-years of the Milky Way. We can determine the true luminosities of those supernovae by using Cepheids to measure the distances to the galaxies in which they occurred. These measurements not only tell us the true luminosities of white dwarf supernovae, but also confirm that their luminosities are all about the same and therefore that we can use them as standard candles.

Because white dwarf supernovae are so bright—about 10 billion solar luminosities at their peak—we can detect them even when they occur in galaxies billions of light-years away (Figure 16.14). Although the number of galaxies whose distances we can measure with this technique is relatively

White dwarf supernovae are useful for measuring large distances because they are bright and all have about the same peak luminosity.

small (because it works only for galaxies in which a white dwarf supernova has been seen), these galaxies have allowed us to calibrate yet another technique—one that relies on the expansion of the universe.

◆ What is Hubble's law?

The ability to measure distances to galaxies is the key to much of our modern understanding of the size and age of the universe. We can trace the beginning of this understanding directly back to discoveries made by Edwin Hubble, including the law that bears his name.

Hubble and the Andromeda Galaxy Before Hubble's groundbreaking work in the 1920s, no one knew for certain whether the spiral-shaped objects they saw in the sky were merely clouds of gas within the Milky Way—and therefore that the Milky Way represented the entire universe—or distant and distinct galaxies. The opinions of astronomers were split on this issue, which became a subject of great debate.

Hubble put the debate to rest in 1924. Using the new, 100-inch telescope atop southern California's Mount Wilson (Figure 16.15)—the largest telescope in the world at the time—he identified Cepheid variables in

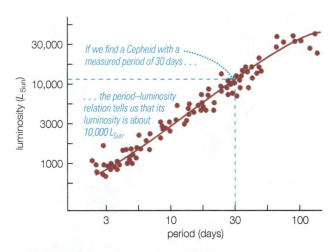

▲ **FIGURE 16.13**

Cepheid period–luminosity relation. The data show that all Cepheids of a particular period have very nearly the same luminosity, making Cepheids excellent standard candles. (Cepheids actually come in two types with two different period–luminosity relations. The relation here is for Cepheids with heavy-element content similar to that of our Sun, or "Type I Cepheids.")

▲ **FIGURE 16.14**

White dwarf supernovae. White arrows in the lower images indicate the supernovae, and the upper images show what these galaxies looked like without supernovae. The first two (from left) are supernovae that exploded when the universe was approximately half its current age; the one to the right exploded about 9 billion years ago.

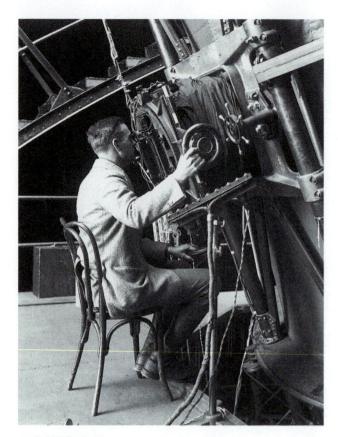

▲ FIGURE 16.15
Edwin Hubble at the Mount Wilson Observatory.

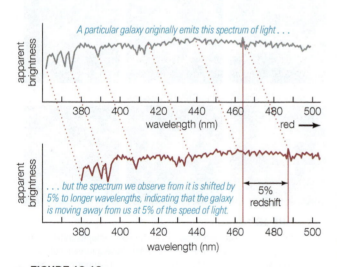

A particular galaxy originally emits this spectrum of light . . .

apparent brightness

380 400 420 440 460 480 500
wavelength (nm)
red ➝

apparent brightness

. . . but the spectrum we observe from it is shifted by 5% to longer wavelengths, indicating that the galaxy is moving away from us at 5% of the speed of light.

5% redshift

380 400 420 440 460 480 500
wavelength (nm)

▲ FIGURE 16.16
Redshifted galaxy spectrum.

the Andromeda Galaxy by comparing photographs of the galaxy taken days apart. Using Henrietta Leavitt's period–luminosity relation, he could then estimate the Cepheid luminosities and calculate the galaxy's distance. He thereby proved that the Andromeda Galaxy resides far beyond the outer reaches of the Milky Way.

This single scientific discovery dramatically changed our view of the universe. Rather than wondering if the Milky Way might be the entire universe, we suddenly knew that it is just one among many galaxies in an enormous universe. The stage was set for an even greater discovery.

Distance and Redshift Astronomers were already aware that the spectra of most spiral galaxies tended to be *redshifted* (Figure 16.16). Recall that redshifts occur when the object emitting the radiation is moving away from us [Section 5.2]. Because Hubble had not yet proved that the spiral galaxies were separate from the Milky Way, no one understood the true significance of their motions.

In the years following his discovery of Cepheids in Andromeda, Hubble sought to estimate more galaxy distances and measure the redshifts of those galaxies. Because even Cepheids were too dim to be seen in most of these galaxies, Hubble needed brighter standard candles for his distance estimates. One of his favorite techniques was to use the brightest object he could see in each galaxy as a standard candle, because he assumed these objects to be very bright stars that would always have about the same luminosity. (We now know that his "brightest objects" were actually star clusters, not individual stars. This means they were not very good standard candles, but they proved good enough for him to draw the correct conclusion.)

In 1929, Hubble announced his conclusion: The more distant a galaxy, the greater its redshift and hence the faster it moves away from us. As we discussed in Chapter 1 (see Figure 1.15), this discovery implies that the entire universe is expanding.

Hubble's Law Hubble's discovery that galaxies at greater distances are moving away from us at greater speeds can be expressed quantitatively with a simple formula called **Hubble's law,** in which v stands for a galaxy's *recession speed*, meaning the speed at which it is moving away from us, and d stands for its distance:

$$v = H_0 \times d$$

The H_0 (pronounced "H-naught") in this equation is a number called **Hubble's constant.**

We usually write Hubble's law in the above form because it expresses the key idea that speeds of galaxies depend on their distances. However, astronomers more often use the law in reverse: They measure a galaxy's speed (from its redshift) and then use Hubble's law to estimate its distance.

Because Hubble's law in principle applies to all distant galaxies for which we can measure a redshift, it is the most useful technique for determining distances to galaxies that are very far away. Nevertheless, we encounter two practical difficulties when we try to use Hubble's law to measure galactic distances:

1. Galaxies do not obey Hubble's law perfectly. Hubble's law gives an exact distance only for a galaxy whose speed is determined solely by the expansion of the universe. In reality, nearly all galaxies experience gravitational tugs from other galaxies, and these tugs alter their speeds from the values predicted by Hubble's law.

2. Even when galaxies obey Hubble's law well, the distances we find with it are only as accurate as our best measurement of Hubble's constant.

The first problem is most serious for nearby galaxies. Within the Local Group, for example, Hubble's law does not work at all: The galaxies in the Local Group are gravitationally bound together with the Milky Way and therefore are *not* moving away from us in accord with Hubble's law. However, Hubble's law works well for more distant galaxies, which have recession velocities so great that any motions caused by the gravitational tugs of neighboring galaxies are tiny in comparison.

The second problem means that, even for distant galaxies, we can know only *relative* distances until we pin down the true value of H_0. For example, Hubble's law tells us that a galaxy moving away from us at 20,000 km/s is twice as far away as one moving at 10,000 km/s, but we can determine the actual distances of the two galaxies only if we know H_0.

One of the main missions of the Hubble Space Telescope was to measure an accurate value of H_0. Astronomers used the telescope to measure distances to Cepheids in galaxies out to about 100 million light-years, and this in turn al-

The quest to measure Hubble's constant was one of the main missions of the Hubble Space Telescope.

lowed them to determine the luminosities of distant standard candles such as white dwarf supernovae. Plotting galactic distances measured with those distant standard candles against the velocities indicated by their redshifts has pinned down the value of H_0 to somewhere between 21 and 23 *kilometers per second per million light-years* (Figure 16.17). That is, a galaxy's

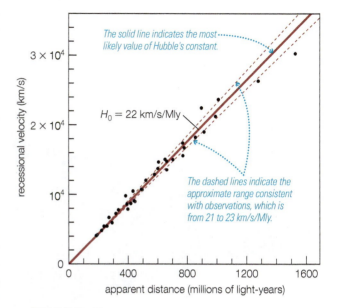

▲ **FIGURE 16.17**

White dwarf supernovae can be used as standard candles to establish Hubble's law out to very large distances. The points on this figure show the apparent distances of white dwarf supernovae and the recession speeds of the galaxies in which they exploded. The fact that these points all fall close to a straight line demonstrates that these supernovae are good standard candles.

---

special topic | Who Discovered the Expanding Universe?

We generally credit Hubble with the discovery of the expanding universe, but like most scientific discoveries, this one was not made in isolation, and it can be difficult to determine exactly how much credit each scientist deserves.

There is no dispute that Hubble and his primary assistant Milton Humason made the distance measurements on which the discovery was based. However, the redshift measurements that Hubble initially used were actually made years earlier by astronomer Vesto Slipher of the Lowell Observatory.

More intriguingly, Belgian priest and astronomer Georges Lemaître (the French pronunciation is approximately "Leh mcht-rch") published a version of what we now call Hubble's law in 1927—two years before Hubble's own publication—and even calculated a value for the rate of universal expansion (Hubble's constant). This fact has led some science historians to suggest that Lemaître deserves more credit than Hubble, but there are at least a

couple of caveats. First, Hubble was unaware of Lemaître's paper, which received little notice because it was published in French in an obscure Belgian journal. Second, Lemaître based his work on Hubble's already published distances to galaxies, along with Slipher's already published redshifts. Moreover, when Lemaître translated his paper into English for a broader audience in 1931, he left out the parts that contained his discovery of universal expansion.* Perhaps Lemaître was simply being modest, but it seems clear that he was willing to cede credit to Hubble. Either way, there's no doubt that Lemaître was the first person to publish the discovery of the expanding universe, but also no doubt that Hubble discovered it independently and was the one who made it famous.

---

*For a while, some people suspected a conspiracy in leaving out these parts, but Mario Livio of the Space Telescope Science Institute recently found letters showing that Lemaître himself made the decision to leave them out.

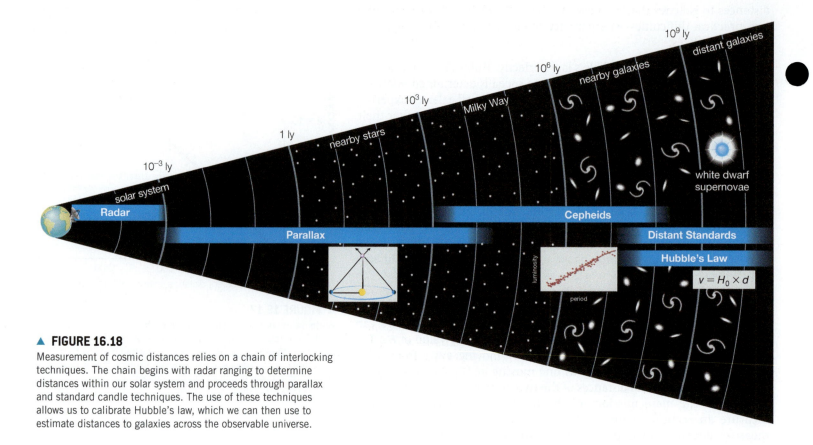

▲ FIGURE 16.18

Measurement of cosmic distances relies on a chain of interlocking techniques. The chain begins with radar ranging to determine distances within our solar system and proceeds through parallax and standard candle techniques. The use of these techniques allows us to calibrate Hubble's law, which we can then use to estimate distances to galaxies across the observable universe.

cosmic calculations 16.2

Hubble's Law

We can use Hubble's law, $v = H_0 \times d$, to measure a galaxy's distance d from its recession speed v. We simply divide both sides of the equation by H_0 to put it in the form

$$d = \frac{v}{H_0}$$

If we want the distance in light-years and know the measured speed in kilometers per second, then we need Hubble's constant in units of *kilometers per second per million light-years* (km/s/Mly).

Example: Estimate the distance to a galaxy whose redshift indicates that it is moving away from us at a speed of 22,000 km/s. Assume that Hubble's constant is $H_0 = 22$ km/s/Mly.

Solution: Putting the given values into our formula, we find

$$d = \frac{v}{H_0} = \frac{22{,}000 \text{ km/s}}{22 \frac{\text{km/s}}{\text{Mly}}} = 1000 \text{ Mly}$$

The galaxy's distance is about 1000 million, or 1 billion, light-years.

speed away from us is between 21 and 23 km/s for every million light-years it is from us. For example, with this range of values for Hubble's constant, Hubble's law predicts that a galaxy located 100 million light-years away should be moving away from us at a speed between 2100 and 2300 km/s.

Distance Chain Summary Figure 16.18 summarizes the techniques we have discussed, showing how they make up a chain of measurements that allows us to determine distances across the universe:

- *Radar ranging:* We measure the Earth–Sun distance by bouncing radio waves off Venus and using some geometry.
- *Parallax:* We measure the distances to nearby stars by observing how their positions appear to change as Earth orbits the Sun. These distances rely on our knowledge of the Earth–Sun distance, determined with radar ranging.
- *Cepheid variables:* We learn the precise period–luminosity relation for Cepheids by using parallax to measure distances to the nearest ones. We can then use this relation to find the luminosities of Cepheids in more distant star clusters or galaxies, and can determine the distances to these Cepheids by applying the inverse square law for light.
- *Distant standards:* By measuring distances to relatively nearby galaxies with Cepheids, we learn the true luminosities of white dwarf supernovae, enabling us to measure great distances throughout the universe.

- *Hubble's law:* Distances measured to galaxies with white dwarf supernovae allow us to measure Hubble's constant, H_0. Once we know H_0, we can use Hubble's law to determine a galaxy's distance from its redshift.

Note that the uncertainty increases with each link in the distance chain. As a result, although we know the Earth–Sun distance at the beginning of the chain extremely accurately, distances to the farthest reaches of the observable universe remain uncertain by about 5%.

◆ How do distance measurements tell us the age of the universe?

Hubble's law is a remarkably powerful tool for understanding the universe. Not only does it tell us that the universe is expanding and give us a way to measure galactic distances, but it also helps us determine the age and size of the observable universe. To see how, we must first consider the expansion of the universe in a little more detail.

The Cosmological Principle   It may be tempting to think of the expanding universe as a ball of galaxies expanding into a void, but this is not the case. To the best of our knowledge, the universe is not expanding *into* anything. As far as we can tell, there is no edge to the distribution of galaxies in the universe. On very large scales, the distribution of galaxies appears to be relatively smooth, meaning that the overall appearance of the universe around you would look more or less the same no matter where you were located. The idea that the universe has no center or edge, and on large scales looks about the same everywhere, is often called the **Cosmological Principle**. Although we cannot prove this principle to be true, it is consistent with all our observations of the universe [Section 18.3].

How can the universe be expanding if it's not expanding *into* anything? In Chapter 1, we likened the expanding universe to a raisin cake baking [Section 1.3], but a cake has a center and edges that grow into empty space as it bakes. A better analogy involves something that can expand but that has no center and no edges. The surface of a balloon can fit the bill, as could an infinite surface such as a sheet of rubber that extends to infinity in all directions.

As far as we can tell, the universe has no center or edge.

Because it's hard to visualize infinity, let's use the surface of a balloon as our analogy to the expanding universe (Figure 16.19). Note that this analogy uses the balloon's two-dimensional *surface* to represent all three dimensions of space. The surface of the balloon represents the entire universe, and the spaces inside and outside the balloon have no meaning in this analogy. Aside from the reduced number of dimensions, the analogy works well because the balloon's spherical surface has no center and no edges, just as no city is the center of Earth's surface and no edges exist where you could walk or sail off of Earth. We can represent galaxies with plastic dots attached to the balloon, and we can make our model universe expand by inflating the balloon.

see it for yourself   Make a one-dimensional model of the expanding universe with a rubber band and some paper clips. Cut the rubber band so that you can stretch it out along a line, then attach four paper clips along it. Pin down the ends of the rubber band and measure the distances from one paper clip to each of the others. Then unpin one of the ends, stretch the rubber band more, pin it down again, and remeasure the distances. How much have they changed? How do your measurements illustrate Hubble's law?

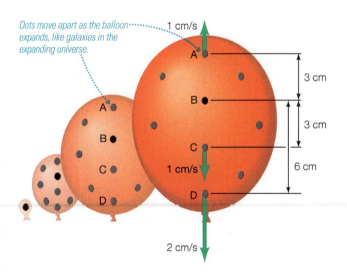

▲ FIGURE 16.19
As the balloon expands, the dots move apart in the same way that galaxies move apart in our expanding universe.

What Is the Universe Expanding Into?

When you first learn about the expansion of the universe, it's natural to assume that the universe must be expanding "into" something. The name "Big Bang" does evoke a mental image of a huge explosion that sent matter flying outward into a vast space that was previously empty, but the scientific view of the Big Bang is very different. According to modern science, the Big Bang filled *all of space* with matter and energy.

How can the Big Bang have started "all of space" expanding without its expanding into some empty void? A complete answer requires Einstein's general theory of relativity, which tells us that space can be "curved" in ways we can measure but cannot visualize (see Special Topic, page 373). But you can understand the key ideas with the balloon analogy. Remember that the balloon's *surface* represents all of space; just as the surface has no center and no edges, space has no center and no edges. In the analogy, the expanding universe is like the expanding surface of the balloon, and the Big Bang was the moment when an extremely tiny balloon first started growing in size. As the balloon expands, any two points on the surface get farther apart because the surface itself is stretching, not because they are moving into a previously empty region above the surface. In the same way, galaxies move apart in the expanding universe because the space between them is stretching, not because they move into a space that was previously empty.

The Age of the Universe   We can now see how Hubble's law leads us to an age for the universe. Imagine that some miniature scientists are living on Dot B in Figure 16.19. Suppose that, 3 seconds after the balloon begins to expand, they measure the following:

Dot A is 3 cm away and moving at 1 cm/s.
Dot C is 3 cm away and moving at 1 cm/s.
Dot D is 6 cm away and moving at 2 cm/s.

These observations are exactly what Hubble's law would tell them to expect: More distant dots are moving away faster. Moreover, because the miniature scientists have measured distances and speeds for multiple dots, they have the information necessary to discover their own version of Hubble's law. Recall that this law can be written $v = H_0 \times d$. For the balloon, v and d would represent any dot's speed and distance, and "Hubble's law" for the balloon would be the following:

$$v = \left(\frac{1}{3\,\text{s}}\right) \times d$$

think about it   Confirm that the formula gives the correct values for the speeds of Dots C and D, as seen from Dot B, 3 seconds after the balloon began expanding. How fast would a dot located 9 cm from Dot B move, according to the scientists on Dot B?

If the miniature scientists think of their balloon as a bubble, they might call the number relating distance to velocity—the term $1/(3\,\text{s})$ in the preceding formula—the "bubble constant." An especially insightful miniature scientist might flip over the "bubble constant" and find that it is exactly equal to the time elapsed since the balloon started expanding. That is, the "bubble constant" $1/(3\,\text{s})$ tells them that the balloon has been expanding for 3 seconds. Perhaps you see where we are heading.

Just as the inverse of the "bubble constant" tells the miniature scientists that their balloon has been expanding for 3 seconds, the inverse of the Hubble constant, or $1/H_0$, tells us something about how long our universe has been expanding. The "bubble constant" for the balloon depends on when it is measured, but it is always equal to 1 divided by the time since the balloon started expanding. Similarly, the Hubble constant actually changes with time but stays roughly equal to 1 divided by the age of the universe. We call it a "constant" because it is the same at all locations in the universe, and because its value does not change noticeably on the time scale of human civilization.

A simple estimate based only on the value of Hubble's constant puts the age of the universe at a little less than 14 billion years. To derive a more precise value, we need to know whether the expansion has been speeding up or slowing down over time, a question we will examine more closely in Chapter 18. If the rate of expansion has slowed with time (because of gravity, for example), then the expansion rate would have been faster in the past. In that case, the universe would have reached its current size faster than we would guess by assuming a constant expansion rate, which means the universe would be *younger* than the age we get from $1/H_0$. If the expansion rate has accelerated with time—and current evidence indicates a small acceleration—then the universe's age would be somewhat more than $1/H_0$. The best available evidence suggests that the universe is very close to 14 billion years old.

The rate at which the universe expands tells us how old it is—about 14 billion years old.

Lookback Time and Cosmological Redshift The expansion of the universe leads to a complication when we try to specify the distance of a faraway galaxy, because its distance is always changing. Here's the problem. Suppose you are trying to determine the distance to a galaxy using observations of a supernova that actually occurred 1 billion years ago. The photons you observe from the supernova and the galaxy it resides in must have traveled a total distance of 1 billion light-years, because they travel at the speed of light and their journey took 1 billion years (Figure 16.20). However, this 1 billion light-year distance is *smaller* than the current distance to the supernova's galaxy and *larger* than its distance at the time the supernova occurred, because the galaxy's distance has been continually increasing during the time it took for the supernova's light to travel to us. So what should we say if someone asks us the "distance" to the galaxy?

Astronomers often find that the clearest way to describe the galaxy's distance is in terms of the time that its light has taken to reach us. We call this time the galaxy's **lookback time,** because it means we are

An object's lookback time is the time it took for the object's light to reach us.

seeing the galaxy as it looked 1 billion years ago. For example, when astronomers say that a galaxy is "1 billion light-years away," they usually mean that the galaxy has a lookback time of 1 billion light-years.

Redshifts of distant galaxies call for a similar shift in thinking. One way to interpret a galaxy's redshift is in terms of the speed at which the galaxy is moving away from us (the *v* in Hubble's law). However, our balloon analogy suggests an alternative interpretation: We can think of galaxies as remaining essentially stationary—like the dots glued to the balloon—while the space between them grows. From this perspective, the expansion of the universe causes photon wavelengths to become longer (redder) with time in much the same way as wavy lines on a rubber sheet stretch out as the sheet expands (Figure 16.21).

In essence, we have a choice when we interpret the redshift of a distant galaxy: We can think of the redshift either as being caused by the Doppler effect as the galaxy moves away from us or as being a **cosmological redshift** that arises from photon stretching in an expanding universe. For distant galaxies, astronomers find it more useful to adopt the latter perspective, in which the expansion of space itself is carrying galaxies along for the ride and stretching the wavelengths of photons. A galaxy's redshift therefore tells us how much space has expanded during the time that the galaxy's light traveled to us, and its lookback time tells us the distance its light has traveled.

The Horizon of the Universe When we began our discussion of the expanding universe, we stressed that the universe as a whole does not seem to have an edge. However, as we discussed in Chapter 1, the age of the universe limits the size of our *observable* universe (see Figure 1.3): We cannot see any object beyond our **cosmological horizon,** which represents the distance at which lookback times equal the 14-billion-year age of the universe.

Note that the cosmological horizon is a boundary in time, not in space. The reason we cannot see past it is *not* because it's any kind of

The size of the observable universe is determined by the age of the universe.

limit on the distances at which galaxies can exist, but rather because anything beyond it has a lookback time that is greater than the age of the universe. Light hasn't had time to reach us from those locations in time and space, so the observable universe ends at the cosmological horizon.

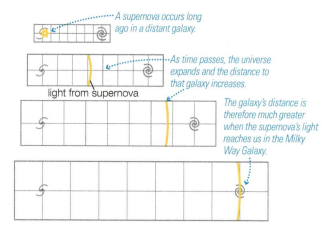

▲ **FIGURE 16.20**

Distances to faraway objects keep changing in an expanding universe. This figure shows how the distance to a galaxy changes during the time it takes for the light from a supernova explosion in that galaxy to reach us. Because the distance changes, the most useful way to describe a galaxy's "distance" is to state its *lookback time*—the time it took for the light to make its journey to Earth.

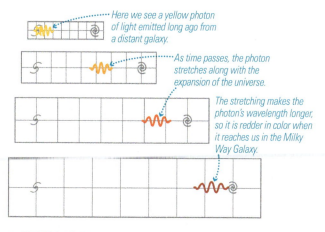

▲ **FIGURE 16.21**

As the universe expands, photon wavelengths stretch. This *cosmological redshift* shifts them toward the red side of the spectrum.

16.3 Galaxy Evolution

The galaxies in Figure 16.1 at the beginning of this chapter extend from relatively nearby almost all the way to the cosmological horizon. Now that we understand how we learn about their lookback times, we are ready to turn our attention to the study of how galaxies form and develop in our expanding universe—a subject known as **galaxy evolution.**

◆ How do we study galaxy evolution?

As in most areas of science, we study galaxy evolution through a combination of observations and theoretical modeling. Observations allow us to observe key aspects of how galaxies evolve through time, while modeling is particularly important for learning about the earliest stages of galaxy formation, which we cannot observe directly.

Observing Galaxies at Different Ages   Powerful telescopes can be used like time machines to observe the life histories of galaxies, because the farther we look into the universe, the further back we look in time [Section 1.1]. The most distant galaxies already had some stars in place by about 13 billion years ago, which means these stars match the ages of the oldest stars in our own galaxy. It therefore seems safe to assume that most galaxies began to form at about this time, in which case most galaxies would be roughly the same age today. A galaxy's lookback time is therefore linked directly to its age. A galaxy with a lookback time of 13 billion years in a 14-billion-year-old universe must be less than a billion years old, while nearby galaxies (with lookback times much less than 1 billion years) can be more than 13 billion years old.

> **think about it**   We've used the term *today* in a very broad sense. For example, a relatively nearby galaxy may be located, say, 20 million light-years away, so we see it as it was 20 million years ago. In what sense is this "today"?

This linkage between lookback time and age gives us a remarkable ability: Simply by photographing galaxies at different distances, we can assemble "family albums" of galaxies in different stages of development.

Most galaxies were born early in the universe's history, so observations of galaxies at different distances also show us galaxies of different ages.

Figure 16.22 shows partial family albums for elliptical, spiral, and irregular galaxies, as well as a set of galaxies seen when they and the universe were very young. Each individual photograph shows a single galaxy at a single stage in its life. Grouping these photographs by galaxy type allows us to see how galaxies of a particular type have changed through time.

Modeling Galaxy Formation   Although current technology allows us to see some galaxies at a time when the universe was only about 1 billion years old, we cannot yet observe the earliest stages of galaxy formation. Scientists hope to see farther back in time with the James Webb Space Telescope, scheduled for launch in 2018 (see Figure 5.29), but we must rely on theoretical models to study the earliest stages of galaxy formation. The most successful models start from two key assumptions, both of which are backed by strong observational evidence [Section 17.2]:

- Hydrogen and helium gas, along with dark matter, filled all of space *almost* uniformly when the universe was very young—say, in the first million years after its birth.

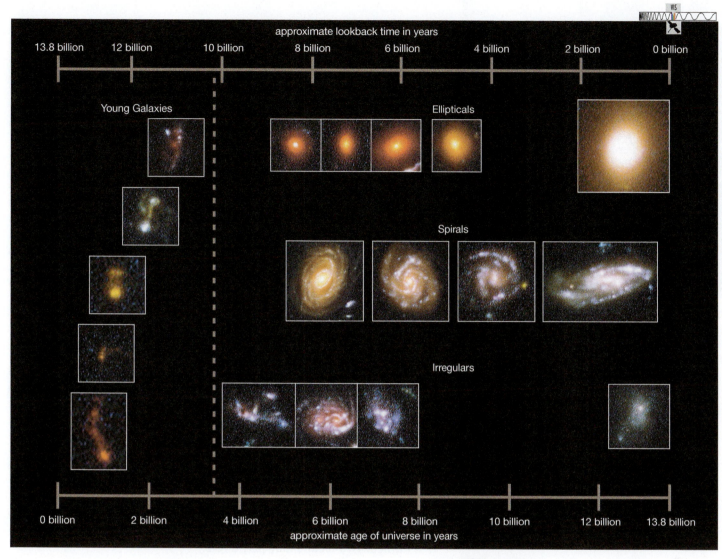

approximate lookback time in years

| 13.8 billion | 12 billion | 10 billion | 8 billion | 6 billion | 4 billion | 2 billion | 0 billion |

Young Galaxies

Ellipticals

Spirals

Irregulars

| 0 billion | 2 billion | 4 billion | 6 billion | 8 billion | 10 billion | 12 billion | 13.8 billion |

approximate age of universe in years

▲ **FIGURE 16.22**

Family albums for elliptical, spiral, and irregular galaxies of different ages, plus some very young galaxies shown on the far left. These photos are all zoomed-in images of galaxies from the Hubble eXtreme Deep Field (see Figure 16.1). We see more distant galaxies as they were when they were younger; the approximate age of the universe is indicated along the bottom horizontal axis, with lookback time shown at the top. (The figure assumes an age of 13.8 billion years for the universe, which is the best current estimate.)

- The distribution of matter in the universe was not *perfectly* uniform—certain regions of the universe started out slightly more dense than others, and these enhanced-density regions served as "seeds" for the formation of galaxies.

Beginning with these assumptions, we can use the laws of physics and computer simulations to model how galaxies formed (Figure 16.23). The models show that the slightly greater pull of gravity around the enhanced-density seeds would have halted the expansion of these regions within about a billion years, even as the universe as a whole continued to expand. The material in these regions then contracted into *protogalactic clouds*, much like those that eventually formed our Milky Way [Section 15.3].

When we discussed the Milky Way, we outlined a basic model for galaxy formation that accounts for the primary features of spiral galaxies

Our most successful models of galaxy formation suggest that protogalactic clouds formed in regions of slightly enhanced density in the early universe.

[Section 15.3]: In that model, the collapsing protogalactic clouds that are destined to form a spiral galaxy radiate away their thermal energy as they contract, which allows them to remain cool. The first generation of stars then grow from the densest, coldest clumps of gas in those clouds, forming

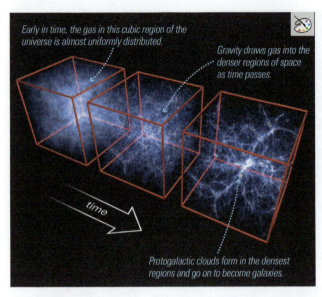

▲ FIGURE 16.23

A computer simulation of the formation of protogalactic clouds. The simulated region of space is about 500 million light-years wide and will go on to form numerous galaxies.

Early in time, the gas in this cubic region of the universe is almost uniformly distributed.

Gravity draws gas into the denser regions of space as time passes.

Protogalactic clouds form in the densest regions and go on to become galaxies.

time

some very massive stars that live and die within just a few million years [Section 13.1]. The supernovae of those massive stars seed the rest of the protogalactic-cloud system with its first sprinkling of heavy elements and generate shock fronts that heat the surrounding interstellar gas. This heating slows the collapse of the protogalactic clouds and the rate at which stars form within them, allowing time for additional gas to settle into a rotating disk as the protogalactic-cloud system merges to form a galaxy.

This basic model accounts for how the Milky Way—and presumably other spiral galaxies—ended up with both disk and halo components and for the differences between disk and halo stars [Section 15.3]. However, it leaves at least two major questions unanswered. First, it assumes that galaxies formed in regions of slightly enhanced density, but does not tell us where these density enhancements came from. The origin of density enhancements in the early universe is one of the major puzzles in astronomy, and we'll revisit it in Chapters 17 and 18. Second, while the basic model explains the origin of spiral galaxies quite well, it does not tell us why some galaxies are elliptical and others irregular.

◆ Why do galaxies differ?

The early stages in the formation and collapse of a protogalactic-cloud system are thought to have been similar for all galaxies. Therefore, to account for the existence of elliptical galaxies, we must ask why some galaxies do not have gas-rich disks. Two general hypotheses have been proposed to explain the absence of gas-rich disks in ellipticals: (1) Elliptical galaxies may end up looking different from spirals because of different birth conditions, or (2) they may begin their lives similarly to spirals but later become elliptical through interactions with other galaxies.

Birth Conditions Hypotheses in the first category trace a galaxy's type back to the characteristics of the protogalactic-cloud system from which it formed. Two factors may play a role:

- **Protogalactic rotation.** A galaxy's type might be determined in part by the rotation of its protogalactic-cloud system. If the original system had a significant amount of angular momentum, it would have rotated quickly as it collapsed. The galaxy it produced would therefore have tended to form a disk, making it a spiral galaxy. If the protogalactic-cloud system had little or no angular momentum, its gas might not have formed a disk at all, so the resulting galaxy would be elliptical.

- **Protogalactic density.** A galaxy's type might also be determined in part by the density of the protogalactic clouds from which it formed. Protogalactic clouds with relatively high gas density would have radiated energy more effectively and cooled more quickly, thereby allowing more rapid star formation. If the star formation proceeded fast enough, all the gas could have been turned into stars before any of it had time to settle into a disk. The resulting galaxy would therefore lack a disk, making it an elliptical galaxy. In contrast, lower-density clouds would have formed stars more slowly, leaving plenty of gas to form the disk of a spiral galaxy.

Evidence for the role of birth conditions comes from a few giant elliptical galaxies at very great distances (Figure 16.24). These galaxies lack blue and white stars, indicating that new stars no longer form

▲ FIGURE 16.24

The light we observe from the distant elliptical galaxy called HUDF-JD2 (circled) left that galaxy when the universe was about 800 million years old. Even though it is very young, the galaxy contains about eight times as many stars as the Milky Way, and its color indicates that few new stars are forming within it.

Elliptical galaxies may have formed from protogalactic clouds with slower rotation or higher density than those that formed spiral galaxies.

within them, even though we are seeing them as they were when the universe was quite young. This finding suggests that all the stars in these elliptical galaxies were born at almost the same time, which is consistent with the idea that all the stars formed before a disk could develop.

Later Interactions Differences in birth conditions probably play important roles in the overall story of why some galaxies have gas-rich disks and others do not. However, birth conditions are probably not the whole story, because they ignore one key fact: Galaxies rarely evolve in perfect isolation.

Think back to our scale model solar system in Chapter 1. On a scale on which the Sun was the size of a grapefruit, the nearest star was like another grapefruit a few thousand kilometers away. Because the average distances between stars are so huge compared to the sizes of stars, collisions between stars are extremely rare. However, if we rescale the universe so that our *galaxy* is the size of a grapefruit, the Andromeda Galaxy is like another grapefruit only about 3 meters away, and a few smaller galaxies lie considerably closer. In other words, the average distances between galaxies are not much larger than the sizes of galaxies, meaning that collisions between galaxies are inevitable.

Galaxy collisions are spectacular events that unfold over hundreds of millions of years (Figure 16.25). During our short lifetimes, we can see only a snapshot of a collision in progress, distorting the shapes of the colliding galaxies. Collisions must have been even more frequent in the distant past, when the universe was smaller and galaxies were closer together. Observations confirm that distorted-looking galaxies—probably galaxy collisions in progress—were more common in the early universe than they are today (see Figure 16.22).

Computer simulations show that a collision between two spiral galaxies can form an elliptical galaxy.

We can learn much more about galactic collisions with the aid of computer simulations that allow us to "watch" collisions

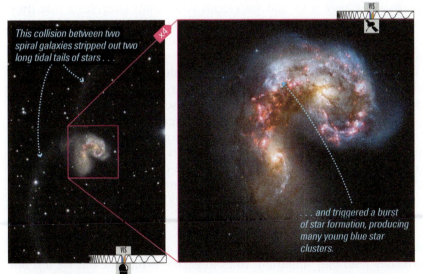

This collision between two spiral galaxies stripped out two long tidal tails of stars . . .

. . . and triggered a burst of star formation, producing many young blue star clusters.

▲ **FIGURE 16.25**

A pair of colliding spiral galaxies known as the Antennae (or NGC 4038/4039). The image taken from the ground (left) reveals their vast tidal tails, while the Hubble Space Telescope image (right) shows the burst of star formation at the center of the collision.

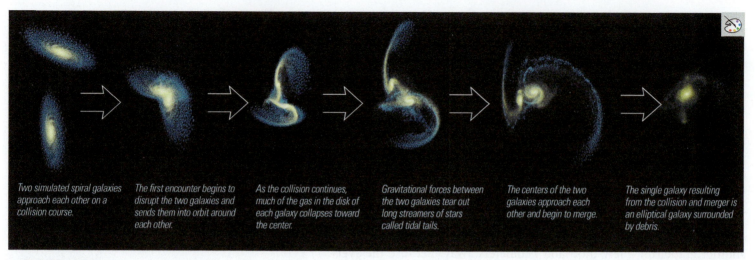

Two simulated spiral galaxies approach each other on a collision course.

The first encounter begins to disrupt the two galaxies and sends them into orbit around each other.

As the collision continues, much of the gas in the disk of each galaxy collapses toward the center.

Gravitational forces between the two galaxies tear out long streamers of stars called tidal tails.

The centers of the two galaxies approach each other and begin to merge.

The single galaxy resulting from the collision and merger is an elliptical galaxy surrounded by debris.

▲ **FIGURE 16.26**

Several stages in a supercomputer simulation of a collision between two spiral galaxies that results in an elliptical galaxy. At least some of the elliptical galaxies in the present-day universe formed in this way. The whole sequence spans about 1.5 billion years.

that in nature take hundreds of millions of years to unfold. These simulations show that a collision between two spiral galaxies can create an elliptical galaxy (Figure 16.26). Tremendous tidal forces between the colliding galaxies tear apart the two disks, randomizing the orbits of their stars. Meanwhile, a large fraction of their gas sinks to the center of the collision and rapidly forms new stars. Supernovae and stellar winds eventually blow away the rest of the gas. When the cataclysm finally settles down, the merger of the two spirals has produced a single elliptical galaxy. Little gas is left for a disk, and the orbits of the stars have random orientations.

Observations support the idea that at least some elliptical galaxies result from collisions and subsequent mergers. Elliptical galaxies dominate the galaxy populations at the cores of dense clusters of galaxies, where collisions should be most frequent.

Observations show that at least some elliptical galaxies have experienced past collisions.

This fact may mean that any spirals once present became ellipticals through collisions. In addition, structural details of elliptical galaxies often attest to a violent past. For example, some elliptical galaxies have stars and gas clouds with orbits suggesting that they are leftover pieces of once-separate galaxies, while others are surrounded by shells of stars that may have formed in gas stripped out of a galaxy by a collision (Figure 16.27).

The most decisive evidence that collisions affect the evolution of elliptical galaxies comes from observations of the **central dominant galaxies** found at the centers of many dense clusters. Central dominant galaxies are giant elliptical galaxies that apparently grew to a huge size by consuming other galaxies through collisions. They frequently contain several tightly bound clumps of stars that probably were the centers of individual galaxies before being swallowed (Figure 16.28). This process of *galactic cannibalism* can create central dominant galaxies more than 10 times as massive as the Milky Way, making them the most massive galaxies in the universe.

Shells are made of stars on orbits that swing back and forth through the galaxy's center.

▲ **FIGURE 16.27**

Elliptical galaxy NGC 474 is surrounded by several distinct shells of stars that were likely produced by collisions with other galaxies.

An Incomplete Answer Birth conditions and subsequent interactions probably both play important roles in galaxy evolution, but we are not yet certain which is more influential. Nevertheless, when we consider both kinds of mechanisms together, they do seem to account for the basic differences between galaxy types. The birth-condition scenarios explain why the vast majority of galaxies are either spiral or elliptical in shape. The interaction scenarios explain why ellipticals are more common in clusters while spirals are more common outside clusters. Even the relatively small fraction of galaxies that are irregular may be explained by these ideas: At least some irregulars probably are galaxies undergoing some sort of disruptive interaction.

These ideas also appear to explain the patterns of galaxy color and galaxy luminosity originally shown in Figure 16.8 and repeated along with explanations in Figure 16.29. All galaxies are thought to begin as star-forming systems that are blue-white in color, making them members of the blue cloud. However, some of them later transition to the red sequence.

All galaxies start as actively star-forming systems, but birth conditions and interactions with other galaxies eventually shut off star formation in many galaxies.

The upper portion of the red sequence forms through mergers that produce large, red, elliptical galaxies. The lower portion of the red sequence consists of smaller galaxies that have used up their cool gas, so star formation has ceased and left them with only cool, red stars. Astronomers are conducting and planning further observations to test these ideas and to complete our understanding of galaxy evolution.

◆ How does gas cycle through galaxies?

Our discussion so far has assumed that all spiral galaxies sustain fairly steady star formation through a star–gas–star cycle like that of the Milky Way [Section 15.2], while elliptical galaxies consist mostly of stars that formed in the distant past. However, observations show that star formation rates can be quite variable. In particular, a small percentage of galaxies—known as **starburst galaxies**—are forming new stars at an astonishingly rapid rate.

Bursts of Star Formation The Milky Way Galaxy produces an average of about one new star per year, but some starburst galaxies form new stars at rates exceeding 100 stars per year. At this rate, a starburst would consume *all* of a galaxy's available gas in less than 100 million years. Because most galaxies have been around for billions of years, we conclude that a starburst must be a relatively short-term stage in a galaxy's life. After the burst of star formation ends, a starburst galaxy presumably returns to being an "ordinary" spiral, elliptical, or irregular galaxy.

The high rate of star formation leads to a high rate of supernova explosions from short-lived, massive stars. The bubbles of hot gas blown out by the many individual supernovae merge into a gigantic bubble so large that it erupts outward into the space around the galaxy, creating a **galactic wind** (Figure 16.30). These ejections of gas can have a dramatic impact on subsequent star formation, especially when they occur in small galaxies. If enough gas is blown out, star formation may shut down for billions of years, resuming only if and when some of the ejected gas cools and falls back into the galaxy.

Starburst galaxies form stars so rapidly that they would run out of gas for star formation in less than 100 million years.

Regulation of the Gas Supply The existence of starbursts with galactic winds suggests that similar processes may regulate the gas supply for star formation in many or all spiral and elliptical galaxies. If so, then a galaxy's star–gas–star cycle may extend over a region many times larger than the galaxy's visible disk, and this large region may actually contain

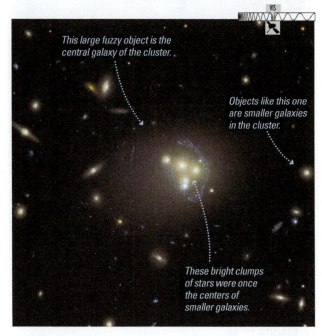

▲ **FIGURE 16.28**

This image shows the central dominant galaxy of the cluster Abell 3827, which has apparently grown by consuming smaller galaxies that have collided with it. Notice that the center of this galaxy contains multiple clumps of stars that probably were once the centers of galaxies. (The blue streaks encircling the central galaxy represent light from a much more distant galaxy that has been distorted by the phenomenon of gravitational lensing [Section 18.2].)

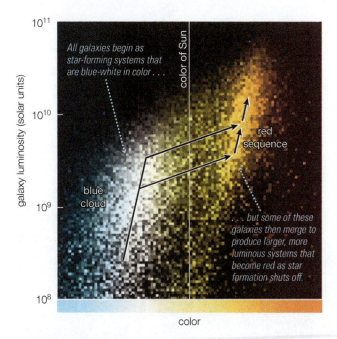

▲ **FIGURE 16.29**

This figure shows the data from Figure 16.8 and explains how the patterns arise according to our current understanding of galaxy evolution. All galaxies are thought to begin as star-forming systems, putting them in the *blue cloud* at the left. Mergers of these galaxies can later produce larger galaxies, some of which cease forming new stars. The schematic black lines show how mergers cause the observed properties of galaxies to shift up and to the right, placing the galaxies on the *red sequence*.

a This visible-light photograph (from the Hubble Space Telescope) shows violently disturbed gas (red) blowing out both above and below the disk.

▲ **FIGURE 16.30**

Visible-light and X-ray views of a starburst galaxy called M82 and its galactic wind. Both images show the same region, which is about 16,000 light-years across.

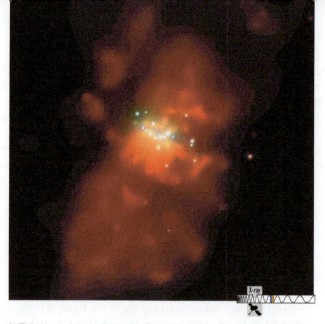

b This X-ray image from the Chandra X-Ray Observatory shows the same region as the visible-light photograph in (a). The reddish region represents X-ray emission from very hot gas blowing out of the disk. The bright dots in the galactic disk probably represent X-ray emission from accretion disks around black holes or neutron stars produced by recent supernovae.

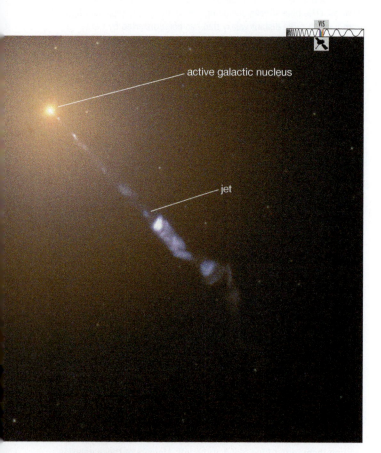

active galactic nucleus

jet

▲ **FIGURE 16.31**

The active galactic nucleus in the elliptical galaxy M87. The bright yellow spot is the active nucleus, and the blue streak is a jet of particles shooting outward from the nucleus at nearly the speed of light.

A galaxy's star–gas–star cycle may extend over a region much larger than its disk.

the majority of the galaxy's gas. This gas would be very hot (because it was ejected by supernovae) and very spread out, making detection of the X-ray and ultraviolet radiation that it would produce difficult. However, models suggest that this gas would represent a hot atmosphere around the visible galaxy, in which gas could cool and condense to make a light rain of cold clouds falling back toward the disk. In other words, the long-term evolution of a galaxy would depend on the cycling of gas in and out of its disk. And as we'll see in the next section, central black holes may play a critical role in this process.

16.4 The Role of Supermassive Black Holes

Galaxy collisions and starbursts may be spectacular, but even more awe-inspiring events can occur deep in the centers of galaxies. A small percentage of galaxies emit truly extreme amounts of radiation from a tiny central region, sometimes accompanied by powerful jets of material shooting outward at nearly the speed of light (Figure 16.31). What could possibly release so much energy from within such a small region of space? In this section, we'll investigate the evidence that has led scientists to conclude that the culprits are **supermassive black holes,** much like the one at the Milky Way's center [Section 15.4] but in many cases far more massive.

◆ What is the evidence for supermassive black holes at the centers of galaxies?

Galaxies with unusually bright centers are called *active galaxies,* and the bright central regions themselves are known as **active galactic nuclei.** The existence of active galactic nuclei was first recognized in the 1940s, when astronomers noticed that about 1% of present-day galaxies (sometimes called *Seyfert galaxies*) have unusually large luminosities coming from small regions near their centers.

The mystery posed by active galactic nuclei became deeper after the discovery in the early 1960s of **quasars**, which we now know to be the most luminous examples of active galactic nuclei (Figure 16.32). The cosmological redshifts of quasars indicated that they were generally billions of light-years from Earth, which implied luminosities that were in some cases more than 1000 times that of the entire Milky Way. These results were so astonishing that, for a while, some scientists feared the distances of quasars had somehow been grossly overestimated. However, observations eventually verified quasar distances and proved that quasars are the bright centers of very distant galaxies.

The fact that most quasars are at very large distances also means that quasars were much more common early in the history of the universe than they are now. Apparently, the engines responsible for supplying the awesome luminosities of quasars somehow become dormant as galaxies age. As we will see, this fact helped astronomers realize that activity in the center of a galaxy is linked to the overall evolution of galaxies.

Sizes of Active Galactic Nuclei

Quasars and other active galactic nuclei puzzled astronomers for decades, not only because of their large luminosities, but also because those luminosities are generated in remarkably small regions of space. Our best visible-light images show these active galactic nuclei to be smaller than 100 light-years across, and higher-resolution radio observations limit their size to no more than about 3 light-years across. Rapid changes in luminosity show them to be even smaller than that.

To understand how variations in luminosity tell us about an object's size, imagine that you are a master of the universe and you want to signal one of your fellow masters a billion light-years away. An active galactic nucleus would make an excellent signal beacon, because it is so bright. However, suppose the smallest nucleus you can find is 1 light-year across. Each time you flash it on, the photons from the front end of the source reach your fellow master a full year before the photons from the back end. If you flash it on and off more than once a year, your signal will be smeared out. Similarly, if you find a source that is 1 light-day across, you can transmit signals that flash on and off no more than once a day. If you want to send signals just a few hours apart, you need a source no more than a few light-hours across.

Occasionally, the luminosity of an active galactic nucleus doubles in a matter of hours, telling us that the source of all that light must be less

Light from an active galactic nucleus comes from a region no larger than our solar system.

than a few light-hours across, not much bigger than our solar system. Explaining how such a tiny part of a galaxy could produce all that light has been a challenge, but not nearly as great a challenge as explaining the immense light outputs of quasars.

The Black Hole Hypothesis

The only known way to account for the extreme luminosities of active galactic nuclei is with a model in which the central object is a supermassive black hole surrounded by a swirling accretion disk of very hot gas (Figure 16.33). This idea is much like the idea we used earlier to explain the emission from X-ray binary star systems [Section 14.2]. The gravitational potential energy of matter falling toward a black hole is converted into kinetic energy, and collisions between infalling particles convert kinetic energy into thermal energy. The

▲ **FIGURE 16.32**
This Hubble Space Telescope photo shows quasar 3C273, the first object to be identified as a quasar (in 1963). Its luminosity is more than 1 trillion times that of the Sun. (The photo shows a region 275,000 light-years across at the distance of the quasar.)

▲ **FIGURE 16.33**
Artist's conception of an accretion disk surrounding a supermassive black hole.

resulting heat causes this matter to emit the intense radiation we observe. As in X-ray binaries, we expect that the infalling matter swirls through an accretion disk before it disappears within the event horizon of the black hole.

Matter falling into a black hole can generate awesome amounts of energy. As a chunk of matter falls to the event horizon of a black hole,

> Active galactic nuclei are thought to be powered by matter falling into supermassive black holes.

as much as 10–40% of its mass-energy ($E = mc^2$) can be converted into thermal energy and ultimately to radiation; notice that this is far more efficient than nuclear fusion, which converts less than 1% of mass-energy into photons. (The precise amount within the 10–40% range depends on the black hole's spin.) This mechanism is so efficient that it can account for the enormous luminosity of a quasar if we assume that the black hole swallows the equivalent of about one star per year.

The supermassive black hole model also accounts for the small sizes implied by quasar variability. A black hole accreting a solar mass of matter per year could grow to contain a billion solar masses if accretion continued at that rate for a billion years. But the event horizon of such a massive black hole would have a radius of only about 3 light-hours, which would still fit within the orbit of Neptune.

Evidence from Orbits Demonstrating that monster black holes really do inhabit the centers of other galaxies has been difficult, largely because black holes themselves do not emit any light. We therefore need to infer their existence from the ways in which they alter their surroundings. In the vicinity of a black hole, matter should be orbiting at high speed around something invisible. We have already seen how the orbits of stars at the center of the Milky Way indicate the existence of a supermassive black hole there [Section 15.4]. What about other galaxies, whose centers are far more distant from us and are therefore much harder to observe?

Detailed observations of matter orbiting at the centers of nearby galaxies, performed during the last two decades, show that supermassive black holes are quite common. In fact, it is possible that *all* galaxies contain super-

> Orbital motions of stars and gas clouds around active galactic nuclei indicate that they do indeed harbor supermassive black holes.

massive black holes at their centers. One prominent example is the relatively nearby galaxy M87, which features a bright active galactic nucleus and a jet that emits both radio and visible light (see Figure 16.31). Images and spectra of M87's central regions reveal that gas some 60 light-years from the center is orbiting an unseen object at a speed of hundreds of kilometers per second (Figure 16.34). This high-speed orbital motion indicates that the central object has a mass at least 2 to 3 billion times that of our Sun.

Astronomers have found similar evidence for central black holes in many other galaxies with active galactic nuclei (including galaxy M106, in the picture on the chapter opener). In each case, a supermassive black hole seems to be the only explanation for the enormous orbital speeds. We may never be 100% certain that these objects are indeed giant black holes. The best we can do is rule out all other possibilities. However, a supermassive black hole is the only thing we know of that could be so massive while remaining unseen. As a result of these observations, the supermassive black hole model is now considered a confirmed explanation of the phenomena associated with active galactic nuclei.

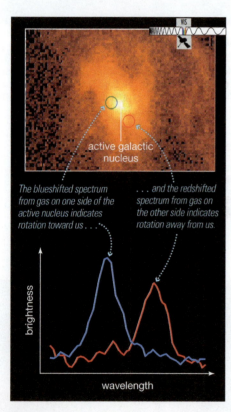

▲ **FIGURE 16.34**
This Hubble Space Telescope photo shows gas near the center of the galaxy M87, and the graph shows Doppler shifts of spectra from gas 60 light-years from the center on opposite sides (the circled regions in the photo). The Doppler shifts tell us that gas is orbiting the galactic center at about 800 km/s. From this speed and the orbital velocity formula, the central object must have a mass at least 2 to 3 billion times that of the Sun.

◆ Do supermassive black holes regulate galaxy evolution?

Even though we see most quasars at great distances—when the universe and the galaxies containing the quasars were young—the black holes responsible for the energy output of quasars must still be present at the centers of some of today's galaxies. The difference is that supermassive black holes in the present-day universe tend to be consuming less material and therefore radiate less energy into space. This realization is what led astronomers to search for evidence of supermassive black holes in nearby galaxies, and these observations have also revealed the relationship shown in Figure 16.35: The mass of a galaxy's central bulge is directly related to the mass of its central black hole.

The consistency of this relationship is striking and somewhat surprising. After all, a galaxy's bulge occupies an enormous volume of space (typically thousands of light-years across) compared to the central black hole, making it hard to see why the properties of one would be so closely linked to those of the other. However, our studies of starbursts suggest a possible connection: The bulge consists of stars that formed in gas clouds, and the central black hole can produce radiation and other forms of activity that can in principle disrupt gas clouds and eject gas from the galaxy. Astronomers do not yet fully understand this linkage, but mounting evidence indicates that central black holes may play a critical role in galaxy evolution.

Radio Galaxies One clue to the relationship between supermassive black holes and galaxy evolution comes from **radio galaxies**, which get their name from their unusually strong radio-wave emission. Modern arrays of radio telescopes can resolve the structure of that emission in vivid detail, showing that most of it comes from pairs of huge *radio lobes* on either side of the galaxy (Figure 16.36).

These radio images show clearly that the plasma in a radio galaxy's lobes has been shot out from the galaxy's center. Apparently, activity around the central supermassive black hole drives two huge jets of matter in opposite directions into space, and observations show that some of the plasma blobs within the jets move outward at speeds close to the speed of light. Calculations show that the jets carry more than enough energy to have a dramatic impact on the hot gas surrounding a galaxy.

Disruption of the Star–Gas–Star Cycle The fact that the jets end in radio lobes, rather than continuing outward as narrow streams, tells us that the jets must be running into something, and that something is presumably the hot gas thought to surround galaxies. In some cases we even see radio galaxies with bent jets and swept-back lobes, indicating that those galaxies are traveling through hot intergalactic gas.

The energy of the plasma jets adds further heat to the hot gas around galaxies, which can prevent this gas from cooling. Although it is usually very difficult to detect the hot gas surrounding galaxies, the jets sometimes make it hot enough to emit detectable X rays. Observations with X-ray telescopes confirm that the jets are indeed interacting strongly with the surrounding gas and depositing large amounts of heat into it (Figure 16.37).

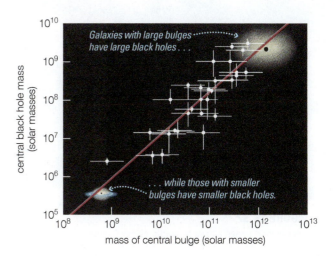

▲ **FIGURE 16.35**

The relationship between the mass of a galaxy's bulge and the mass of its supermassive black hole.

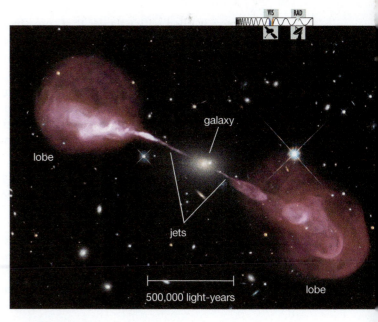

▲ **FIGURE 16.36**

This composite image of radio galaxy Hercules A shows visible light in full color and radio waves in red.

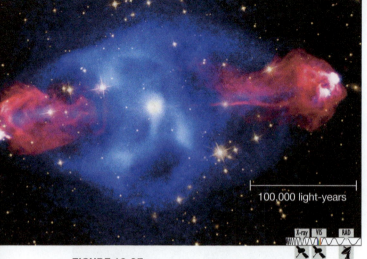

Black Holes and Galaxy Evolution   Studies of both radio galaxies and galactic winds are showing that energetic activity within a galaxy can regulate a galaxy's gas supply at distances that can exceed 10 times the extent of the visible stars. In small galaxies the energy input from starbursts can strongly limit the amount of star formation that happens, but in large galaxies the energy input from black holes may be even more important. Today, the leading hypothesis for why the universe's largest galaxies end up at the top of the red sequence in Figure 16.8 starts by suggesting that a central black hole forms very early in a galaxy's history. Sporadic outbursts of energy, fueled by accretion of gas onto the central black hole, can periodically disrupt the galaxy's supply of star-forming gas and eventually shut the supply off.

This hypothesis can successfully explain the relationship shown in Figure 16.35 between bulge mass and black hole mass, at least in larger galaxies. Furthermore, recent observations suggest that these energetic outbursts may be the critical factor that prevents the largest galaxies from forming even more stars than they already contain. If this interpretation of the observations is correct, large central black holes are *required* in large galaxies, in order to explain why current rates of star formation within them are relatively low.

▲ **FIGURE 16.37**

This composite image shows the radio galaxy Cygnus A in visible light (yellow), radio waves (red), and X rays (blue). The bright spot at the center is visible light from the galaxy itself. Two long jets of energetic particles revealed by the radio-wave image are shooting in opposite directions from the active galactic nucleus into the hot gas surrounding the galaxy. The X-ray image of the hot gas shows the disturbances the jets are creating as they burrow through it.

the big picture   Putting Chapter 16 into Perspective

The picture could hardly get any bigger than it has in this chapter. Looking back through both space and time, we have seen a wide variety of galaxies extending nearly to the limits of the observable universe. As you look back, keep sight of these "big picture" ideas:

- The universe is filled with galaxies that come in a variety of shapes and sizes. In order to learn the histories of these galaxies, we must account for how the universe itself has evolved through time.

- Much of our current understanding of the structure and evolution of the universe is based on measurements of distances to faraway galaxies. These measurements rely on a chain of techniques in which each link in the chain builds upon the links that come before it.

- It has been less than a century since Hubble first proved that the Milky Way is only one of billions of galaxies in the universe. This discovery, and his subsequent

discovery of the universe's expansion, provided the foundation on which modern cosmology has been built. Measurements of the rate of expansion tell us that the universe began about 14 billion years ago.

- Although we do not yet know the complete story of galaxy evolution, we are rapidly learning more. Galaxies probably all began as protogalactic clouds, but they do not always evolve peacefully. Some galaxies collide with their neighbors, often with dramatic results.

- The tremendous energy outputs of quasars and other active galactic nuclei, including those of radio galaxies, are probably powered by gas accreting onto supermassive black holes. The centers of many present-day galaxies must still contain the supermassive black holes that once enabled them to shine as quasars.

my cosmic perspective   Measurements of galaxy distances by Edwin Hubble nearly 100 years ago led to a profound insight into our origins: The universe is *not* eternal and unchanging. Instead, it has been expanding since its birth about 14 billion years ago. Interactions during that time between stars, gas, and even central black holes explain how galaxies like our Milky Way became capable of producing life-sustaining planetary systems like our own.

16.1 Islands of Stars

What patterns do we find among the properties of galaxies?

   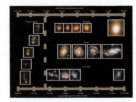

Galaxies come in three major types. **Spiral galaxies** have prominent disks and spiral arms. **Elliptical galaxies** are rounder and redder than spiral galaxies and contain less cool gas and dust. **Irregular galaxies** are neither disklike nor rounded in appearance. Grouping them according to luminosity and color shows that spiral and irregular galaxies tend to belong to a star-forming **blue cloud** in a diagram like the H-R diagram for stars, while elliptical galaxies tend to fall along a non-star-forming **red sequence**. Galaxies are often found in groups, with spiral galaxies tending to reside in relatively small groups and elliptical galaxies usually found in large clusters.

16.2 Distances of Galaxies

How do we measure the distances to galaxies?

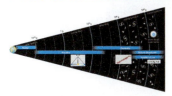 The chain of methods by which we measure galaxy distances begins with **radar ranging** in our own solar system and parallax measurements of distances to nearby stars, then relies on **standard candles** to measure greater distances. Important standard candles include **Cepheid variable stars,** which obey a **period–luminosity relation** that allows us to determine their luminosities and distances, and **white dwarf supernovae** that can be seen even at enormous distances.

What is Hubble's law?

Hubble's law tells us that more distant galaxies are moving away faster: $v = H_0 \times d$, where H_0 is **Hubble's constant.** It allows us to determine a galaxy's distance from the speed at which it is moving away from us, which we can measure from the redshift of its spectrum.

How do distance measurements tell us the age of the universe?

The inverse of Hubble's constant tells us how long it would have taken the universe to reach its present size *if* the expansion rate had never changed and gives an estimated age for the universe of about 14 billion years. A **lookback time** equal to this age therefore marks our **cosmological horizon,** beyond which we cannot see.

16.3 Galaxy Evolution

How do we study galaxy evolution?

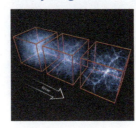 Observationally, we can study galaxies of different ages by using the fact that most galaxies were born early in the universe's history. Therefore, we see more distant galaxies at younger ages than nearer galaxies. Theoretical modeling allows us to study early stages of galaxy formation. The most successful models assume that gravity pulled together regions of the universe that were slightly denser than their surroundings. Gas collected in protogalactic clouds, and stars began to form as the gas cooled.

Why do galaxies differ?

Differences between present-day galaxies probably arise both from conditions in their protogalactic-cloud systems and from collisions with other galaxies. Slowly rotating or high-density systems of protogalactic clouds may form elliptical rather than spiral galaxies. Ellipticals may also form through the collision and merger of two spiral galaxies.

How does gas cycle through galaxies?

Starburst galaxies, which can have new stars forming at more than 100 times the rate of the Milky Way, show that star formation is not always steady. A strong starburst can lead to a supernova-driven **galactic wind** that ejects much of a galaxy's gas into its halo. This process can regulate the star–gas–star cycle in galaxies by creating around a galaxy an extended atmosphere of hot gas, in which cooling clouds rain back down to the galaxy's disk.

16.4 The Role of Supermassive Black Holes

What is the evidence for supermassive black holes at the centers of galaxies?

Some galaxies have unusually bright centers known as **active galactic nuclei,** of which the most luminous are called **quasars.** Quasars are generally found at very great distances, telling us that they were much more common early in the history of the universe. Quasars and other active galactic nuclei are thought to be powered by **supermassive black holes.** As matter falls into a supermassive black hole through an accretion disk, its gravitational potential energy is efficiently transformed into thermal energy and then into light. Observations of orbiting stars and gas clouds in the nuclei of galaxies suggest that *all* galaxies may harbor supermassive black holes at their centers.

◆ **Do supermassive black holes regulate galaxy evolution?**

The masses we measure for central black holes are closely related to the properties of the galaxy around them, suggesting that the growth of these black holes is closely tied to the process of *galaxy evolution*. Observations of **radio galaxies** show that active galactic nuclei can produce powerful jets of matter that heat the surrounding gas. These outbursts may regulate star formation in galaxies by limiting the rate at which gas in the halo can cool.

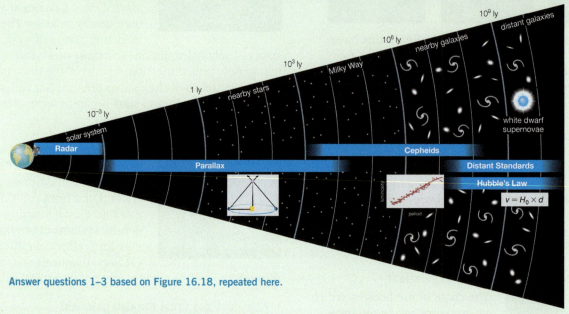

Answer questions 1–3 based on Figure 16.18, repeated here.

1. Which distance measurement technique is most suited to measuring objects at a distance of 10 million light-years?

2. Which distance measurement technique is most suited to measuring objects at a distance of 10 light-years?

3. Which standard-candle technique is best for measuring the distances of very distant galaxies?

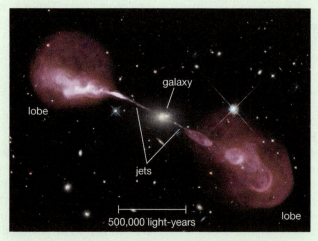

Answer questions 4–7 based on Figure 16.36, repeated here.

4. This image combines observations made with visible light and radio telescopes. Which color in the image represents the radio emission?

5. What is the approximate distance from the far edge of one lobe to the far edge of the other?

6. What is the approximate size of the visible galaxy producing the jets?

MasteringAstronomy® For instructor-assigned homework and other learning materials, go to MasteringAstronomy®.

Review Questions

1. What are the three major types of galaxies, and how do their appearances differ?
2. Distinguish between spiral and elliptical galaxies in terms of the presence or absence of a *disk component* and *halo component*. How does this difference explain the lack of hot, young stars in elliptical galaxies?
3. Where are the *blue cloud* and the *red sequence* on a plot of the relationship between galaxy color and luminosity? On which part of the plot would you tend to find the most luminous galaxies?
4. How do galaxy types in clusters of galaxies differ from those in smaller groups and those of isolated galaxies?
5. What do we mean by a *standard candle*? Explain how we can use standard candles to measure distances.
6. Summarize each of the major links in the distance chain. Why are *Cepheid variable stars* so important? Why are white dwarf supernovae so useful, even though they are quite rare?
7. Explain how Hubble used Cepheid variable stars to prove that the Andromeda Galaxy lies beyond the bounds of the Milky Way.
8. What is *Hubble's law*? Explain what we mean when we say that *Hubble's constant* is between 21 and 23 kilometers per second per million light-years.
9. What is the *Cosmological Principle*, and how is it important to our understanding of the universe?
10. How is the expansion of the surface of an inflating balloon similar to the expansion of the universe? Use the balloon analogy to explain why Hubble's constant is related to the age of the universe.
11. What do we mean by the *lookback time* to a distant galaxy? Briefly explain why lookback times are less ambiguous than distances for discussing objects very far away.
12. What do we mean by a *cosmological redshift*? How does our interpretation of a distant galaxy's redshift differ if we think of it as a cosmological redshift rather than as a Doppler shift?
13. What is the *cosmological horizon*, and what determines how far away it lies?
14. What do we mean by *galaxy evolution*? How do telescopic observations allow us to study galaxy evolution?
15. What are the starting assumptions for models of galaxy formation? Describe how a spiral galaxy is thought to form.
16. Describe two ways in which conditions in a protogalactic cloud might lead to the birth of an elliptical rather than a spiral galaxy.
17. Describe some of the consequences of galaxy collisions. Why were collisions more common in the past?
18. What is a *starburst galaxy*, and why would it tend to have a *galactic wind*? Explain how this idea leads to a model in which spiral and irregular galaxies are surrounded by an extended atmosphere of hot gas.
19. Define *active galactic nuclei* and *quasars*. How can we use variations in luminosity to set limits on the sizes of their emitting regions?
20. What evidence suggests that supermassive black holes really exist?
21. What is a *radio galaxy*? How can radio galaxies affect the gas surrounding them?
22. What evidence suggests that central black holes are connected to the overall evolution of galaxies, and how is the connection thought to work?

Test Your Understanding

Does It Make Sense?

Decide whether the statement makes sense (or is clearly true) or does not make sense (or is clearly false). Explain clearly; not all of these have definitive answers, so your explanation is more important than your chosen answer.

23. If you want to find a lot of elliptical galaxies, you'll have better luck looking in clusters of galaxies than elsewhere in the universe.
24. Cepheids make good standard candles because they all have exactly the same luminosity.
25. After measuring the galaxy's redshift, I used Hubble's law to estimate its distance.
26. The center of the universe is more crowded with galaxies than any other place in the universe.
27. I'd love to live in one of the galaxies near our cosmological horizon, because then I could see the black void into which the universe is expanding.
28. If someone in a galaxy with a lookback time of 4½ billion years had a superpowerful telescope, that person could see our solar system in the process of its formation.
29. Galaxies that are more than 10 billion years old are too far away to see even with our most powerful telescopes.
30. If the Andromeda Galaxy someday collides and merges with the Milky Way, the resulting galaxy may be elliptical.
31. NGC 9645 is a starburst galaxy that has been forming stars at the same furious pace for some 10 billion years.
32. Astronomers proved that quasar 3C473 contains a supermassive black hole when they discovered that its center is dark.

Quick Quiz

Choose the best answer to each of the following. Explain your reasoning with one or more complete sentences.

33. Which of these galaxies do we see at its oldest age? (a) a galaxy in the Local Group (b) a galaxy observed at a distance of 5 billion light-years (c) a galaxy observed at a distance of 10 billion light-years.
34. Which of these galaxies would you most likely find at the *center* of a large cluster of galaxies? (a) a large spiral galaxy (b) a giant elliptical galaxy (c) a small irregular galaxy.
35. We determine the distance of a Cepheid in another galaxy by (a) measuring its parallax. (b) determining its luminosity from the period–luminosity relation and then applying the inverse square law for light. (c) knowing that all Cepheids have about the same luminosity and then applying the inverse square law for light.

36. Which kind of object is the best standard candle for measuring distances to extremely distant galaxies? (a) a white dwarf (b) a Cepheid variable star (c) a white dwarf supernova.

37. Why do virtually all the galaxies in the universe appear to be moving away from our own? (a) We are located near where the Big Bang happened. (b) We are located near the center of the universe. (c) Expansion causes all galaxies to move away from nearly all others.

38. When we observe a distant galaxy whose photons have traveled for 10 billion years before reaching Earth, we are seeing that galaxy as it was when the universe was about (a) 10 billion years old. (b) 7 billion years old. (c) 4 billion years old.

39. Which of these statements expresses a key assumption in our most successful models for galaxy formation? (a) The distribution of matter was perfectly uniform early in time. (b) Some regions of the universe were slightly denser than others. (c) Galaxies formed around supermassive black holes.

40. The luminosity of a quasar is generated in a region the size of (a) the Milky Way. (b) a star cluster. (c) the solar system.

41. The primary source of a quasar's energy is (a) chemical energy. (b) nuclear energy. (c) gravitational potential energy.

42. Observations indicate that galaxies with more massive central black holes tend to also have (a) a greater mass of stars in their central bulges. (b) a greater overall luminosity. (c) a more elliptical shape.

◯ Process of Science

43. *Deviations from Hubble's Law.* Suppose you are measuring distances and velocities of galaxies in order to test Hubble's law. You find that 90% of the galaxies have velocities that are within 200 kilometers per second of the predictions of Hubble's law but 10% have velocities that deviate from the predictions by up to 1000 kilometers per second. Propose a hypothesis that would explain these deviations from Hubble's law, and outline a set of observations that could test your hypothesis.

44. *Black Holes and Galaxy Evolution.* Observations show that the mass of a galaxy's supermassive black hole is closely related to the total mass of stars in its bulge, but astronomers do not yet know the cause of this relationship. Section 16.4 discusses one hypothesis for the cause, which proposes that energetic outbursts from the black hole periodically disrupt a galaxy's star–gas–star cycle. Suggest some observations that could be made to test whether black-hole outbursts can disrupt star formation in galaxies.

45. *Unanswered Questions.* Briefly describe one important but unanswered question related to galaxy evolution. If you think it will be possible to answer that question in the future, describe how we might find an answer, being as specific as possible about the evidence necessary to answer the question. If you think the question will never be answered, explain why you think it is impossible to answer.

Group Work Exercise

46. *Counting Galaxies.* **Roles:** *Scribe* (takes notes on the group's activities), *Proposer* (proposes explanations to the group), *Skeptic* (points out weaknesses in proposed explanations), *Moderator* (leads group discussion and makes sure everyone contributes). **Activity:** Study the Hubble eXtreme Deep Field (Figure 16.1) to complete the following.

a. Each team member should individually estimate the number of galaxies in the image.

b. Compare your estimates, and explain your estimation methods to each other. The *Scribe* should record each team member's estimate.

c. The *Moderator* should lead a discussion of the team's results, with the goal of determining why the different methods may have led to different results.

d. The *Proposer* should suggest a new estimation method that incorporates the best features of the team members' individual methods, and the *Skeptic* should point out potential problems and suggest improvements to the new method. The *Moderator* then leads a discussion, recorded by the *Scribe*, in which the team chooses the best method. The *Scribe* should write down the final method.

e. The team works together to apply the new method, with the *Scribe* recording the result.

Investigate Further

Short-Answer/Essay Questions

47. *Going Deep with Hubble.* The Hubble eXtreme Deep Field in Figure 16.1 was chosen for observation in large part because it is a completely ordinary part of the sky. Why do you think astronomers would want to devote so much precious telescope time to observing totally ordinary regions of the sky in such great detail? Explain your reasoning.

48. *Supernovae in Other Galaxies.* In which type of galaxy would you be most likely to observe a massive star supernova, a giant elliptical galaxy, or a large spiral galaxy? Explain your reasoning.

49. *Hubble's Galaxy Types.* How would you classify the following galaxies using the system illustrated in Figure 16.7? Justify your answers.
   a. Galaxy M101 (Figure 16.2)
   b. Galaxy NGC 4594 (Figure 16.3)
   c. Galaxy NGC 1300 (Figure 16.4)
   d. Galaxy M87 (Figure 16.5)

50. *Color and Luminosity.* The luminosity of our Milky Way galaxy is about $1.5 \times 10^{10} L_{Sun}$. Describe how the Milky Way's light output compares with those of the most luminous galaxies in the blue cloud, based on the information in Figure 16.8. Approximately how much more luminous than the Milky Way are the most luminous galaxies in the red sequence?

51. *Cepheids as Standard Candles.* Suppose you are observing Cepheids in a nearby galaxy. You observe one Cepheid with a period of 8 days between peaks in brightness and another with a period of 35 days. Use Figure 16.13 to estimate the luminosity of each star.

52. *Galaxies at Great Distances.* The most distant galaxies that astronomers have observed are much easier to see in infrared light than in visible light. Explain why that is the case.

53. *Universe on a Balloon.* In what ways is the surface of a balloon a good analogy for the universe? In what ways is this analogy limited? Explain why a miniature scientist living in a polka dot on the balloon would observe all other dots to be moving away, with more distant dots moving away faster.

54. *Life Story of a Spiral.* Imagine that you are a spiral galaxy. Describe your life history from birth to the present day. Your story should be detailed and scientifically consistent, but also creative. That is, it should be entertaining while at the same time incorporating current scientific ideas about the formation of spiral galaxies.

55. *Life Story of an Elliptical.* Imagine that you are an elliptical galaxy. Describe your life history from birth to the present. There are several possible scenarios for the formation of elliptical galaxies, so choose one and stick to it. Be creative while also incorporating scientific ideas that demonstrate your understanding.

56. *Orbits Around Supermassive Black Holes.* The data in Figure 16.34 show the Doppler shifts of emission lines from gas at a distance of 60 light-years from the center of the galaxy M87. Suppose you observed emission lines from gas 30 light-years from the center. How would you expect the Doppler shifts of those lines to be different, assuming that the gas really is orbiting a supermassive black hole? What about gas at 120 light-years from the center?

Quantitative Problems

Be sure to show all calculations clearly and state your final answers in complete sentences.

57. *Counting Galaxies.* Estimate how many galaxies are pictured in Figure 16.1. Explain the method you used to arrive at this estimate. This picture shows about 1/30,000,000 of the sky, so multiply your estimate by 30,000,000 to obtain an estimate of how many galaxies like these fill the entire sky.

58. *Cepheids in M100.* Scientists using the Hubble Space Telescope have observed Cepheids in the galaxy M100. Here are the actual data for three Cepheids in M100:
    - Cepheid 1: luminosity $= 3.9 \times 10^{30}$ watts, brightness 9.3×10^{-19} watt/m^2
    - Cepheid 2: luminosity $= 1.2 \times 10^{30}$ watts, brightness 3.8×10^{-19} watt/m^2
    - Cepheid 3: luminosity $= 2.5 \times 10^{30}$ watts, brightness 8.7×10^{-19} watt/m^2

    Compute the distance to M100 with data from each of the three Cepheids. Do all three distance computations agree? Based on your results, estimate the uncertainty in the distance you have found.

59. *Distances from Hubble's Law.* Imagine that you have obtained spectra for several galaxies and have measured the redshift of each galaxy to determine its speed away from us. Here are your results:
    - Galaxy 1: Speed away from us is 15,000 km/s.
    - Galaxy 2: Speed away from us is 20,000 km/s.
    - Galaxy 3: Speed away from us is 25,000 km/s.

    Estimate the distance to each galaxy from Hubble's law. Assume that $H_0 = 22$ km/s/Mly.

60. *Your Last Hurrah.* Suppose you fell into an accretion disk that swept you into a supermassive black hole. On your way down, the disk radiates 10% of your mass-energy, $E = mc^2$.

a. What is your mass in kilograms? (Recall that 1 kg = 2.2 pounds.) Calculate how much radiative energy will be produced by the accretion disk as a result of your fall into the black hole.

b. Calculate approximately how long a 100-watt light bulb would have to burn to radiate this same amount of energy.

Discussion Questions

61. *Cosmology and Philosophy.* One hundred years ago, many scientists believed that the universe was infinite and eternal, with no beginning and no end. When Einstein first developed his general theory of relativity, he found that it predicted that the universe should be either expanding or contracting. He believed so strongly in an eternal and unchanging universe that he modified the theory, a modification he would later call his "greatest blunder." Why do you think Einstein and others assumed that the universe had no beginning? Do you think that a universe with a definite beginning in time, some 14 billion or so years ago, has any important philosophical implications? Explain.

62. *The Case for Supermassive Black Holes.* The evidence for supermassive black holes at the center of galaxies is strong. However, it is very difficult to prove absolutely that they exist because the black holes themselves emit no light. We can only infer their existence from their powerful gravitational influences on surrounding matter. How compelling is the evidence? Do you think astronomers have proved the case for black holes beyond a reasonable doubt? Defend your opinion.

Web Projects

63. *Galaxy Gallery.* Many fine images of galaxies are available on the Internet. Collect several images of each major type and build a galaxy gallery of your own. Supply a descriptive paragraph about each galaxy.

64. *Greatest Lookback Time.* Look for recent discoveries of objects with the largest lookback times (or redshifts). What is the most distant known object to date? What kind of object is it?

65. *Future Observatories.* The subject of galaxy evolution is a very active area of research. Look for information on current and future observatories involved in investigating galaxy evolution (such as the James Webb Space Telescope). How big are the planned telescopes? What wavelengths will they look at? When will they be built? Write a short summary of a proposed mission.

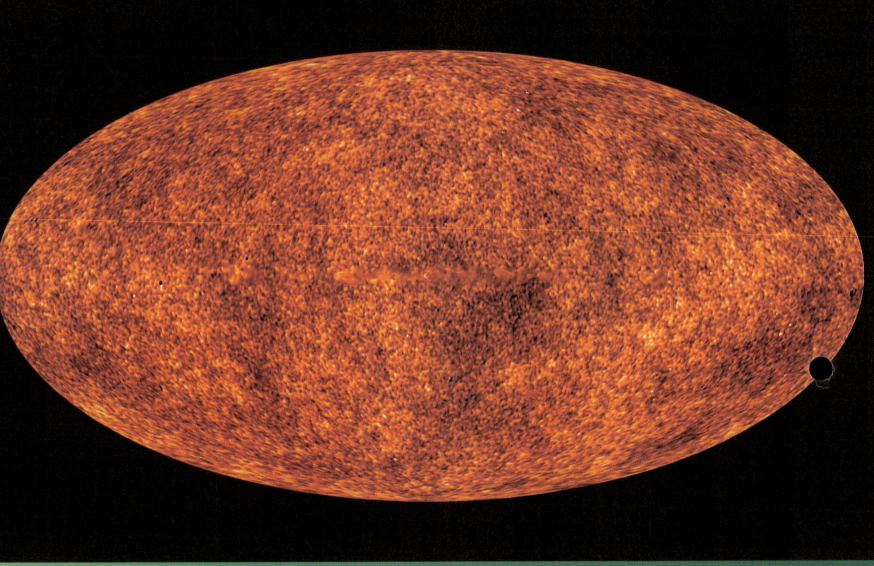

The Big Bang theory is supported by several key lines of evidence, perhaps the most important of which is the existence and characteristics of the cosmic microwave background, shown here in this all-sky image from the *Planck* spacecraft.

LEARNING GOALS

17.1 The Big Bang Theory
- ◆ What were conditions like in the early universe?
- ◆ How did the early universe change with time?

17.2 Evidence for the Big Bang
- ◆ How do observations of the cosmic microwave background support the Big Bang theory?
- ◆ How do the abundances of elements support the Big Bang theory?

17.3 The Big Bang and Inflation
- ◆ What key features of the universe are explained by inflation?
- ◆ Did inflation really occur?

17.4 Observing the Big Bang for Yourself
- ◆ Why is the darkness of the night sky evidence for the Big Bang?

Throughout this book, we have studied how the matter produced in the early universe gradually assembled to make galaxies, stars, and planets. However, we have not yet answered one big question: Where did matter itself come from?

To answer this question, we must go beyond the most distant galaxies and even beyond what we can see near the horizon of the universe. We must go back not just to the origins of matter and energy but all the way back to the beginning of time itself. As we will see, while many questions about the birth of the universe remain unanswered, we now seem to have some understanding of events that must have unfolded as far back as the first fraction of a second after the Big Bang.

17.1 The Big Bang Theory

Is it really possible to study the origin of the entire universe? A century ago, this topic was considered unfit for scientific study. Scientific attitudes began to change with Hubble's discovery that the universe is expanding, which led to the insight that all things very likely sprang into being at a single moment in time, in an event that we have come to call the *Big Bang*. Today, powerful telescopes allow us to view how galaxies have changed over the past 14 billion years, and at great distances we see young galaxies still in the process of forming [Section 16.3]. These observations confirm that the universe has been gradually aging.

Unfortunately, we cannot see back to the very beginning of time. Light from the most distant galaxies shows us what the universe looked like when it was a few hundred million years old. Light from earlier times is more difficult to observe because we are looking back to a time before stars existed. Ultimately, however, we face an even more fundamental problem. The universe is filled with a faint glow of radiation that appears to be the remnant heat of the Big Bang. This faint glow is light that has traveled freely through space since the universe was about 380,000 years old, which is when the universe first became transparent to light. Before that time, light could not pass freely through the universe, so there is no possibility of seeing light from earlier times. Just as we must rely on theoretical modeling to determine what the Sun is like on the inside, we must also use modeling to investigate what the universe was like during its earliest moments.

◆ What were conditions like in the early universe?

The scientific theory that describes what the universe was like early in time is called the **Big Bang theory.** It is based on applying known and tested laws of physics to the idea that everything we see today began as an incredibly hot and dense collection of matter and radiation. The Big Bang theory successfully describes how expansion and cooling of this unimaginably intense mixture of particles and photons could have led to the present universe of stars and galaxies, and it explains key aspects of today's universe with impressive accuracy. Our main goal in this chapter is to understand the evidence supporting the Big Bang theory, but first we must explore what the theory tells us about the early universe.

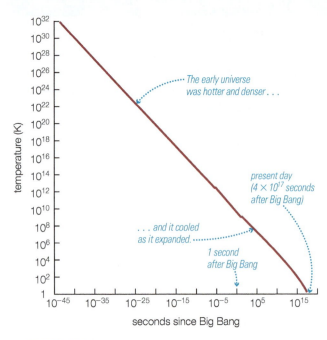

▲ FIGURE 17.1

The universe cools as it expands. This graph shows results of calculations that tell us how the temperature has changed with time. Notice that both axis scales use powers of 10; therefore, even though most of the graph shows temperatures during the first second of the Big Bang, the far right part of the graph actually extends to the present (14 billion years ≈ 4×10^{17} s). The kinks correspond to periods of matter–antimatter annihilation.

The early universe was hotter and denser . . .

. . . and it cooled as it expanded.

present day (4×10^{17} seconds after Big Bang)

1 second after Big Bang

▲ FIGURE 17.2

Aerial photograph of the Large Hadron Collider. The large circle traces the path of the main particle acceleration ring, which lies underground and has a circumference of 27 kilometers.

Observations demonstrate that the universe is cooling with time as it expands, implying that it must have been hotter and denser in the past. Calculating exactly how hot and dense the universe must have been when it was more compressed is much like calculating how the temperature and density of gas in a balloon change when you squeeze it, except that the conditions become much more extreme. Figure 17.1 shows how the temperature of the universe has changed with time, according to such calculations. Note that for most of the universe's history, even back to times just minutes after the Big Bang, conditions were no more extreme than those found in many places in the universe today, such as in the interiors of stars, and therefore can be understood with the same laws of physics that we've applied throughout most of this book. However, at very early times, temperatures were so high that different processes came into play.

The Big Bang theory is a detailed scientific model that describes conditions in the early universe and how they changed with time.

Particle Creation and Annihilation

The universe was so hot during the first few seconds that photons could transform themselves into matter, and vice versa, in accordance with Einstein's formula $E = mc^2$ [Section 4.3]. Reactions that create and destroy matter are now relatively rare in the universe at large, but physicists can reproduce many such reactions with particle accelerators such as the Large Hadron Collider (Figure 17.2).

One such reaction is the creation or destruction of an *electron–antielectron pair* (Figure 17.3). When two photons collide with a total energy greater than twice the mass-energy of an electron (the electron's mass times c^2), they can create two brand-new particles: a negatively charged electron and its positively charged twin, the *antielectron* (also known as a *positron*). The electron is a particle of **matter,** and the antielectron is a particle of **antimatter.** The reaction that creates an electron–antielectron pair also runs in reverse. When an electron and an antielectron meet, they *annihilate* each other totally, transforming all their mass-energy back into photon energy.

Similar reactions can produce or destroy any particle–antiparticle pair, such as a proton and antiproton or a neutron and antineutron. The early universe therefore was filled with an extremely hot and dense blend of photons, matter, and antimatter, converting furiously back and forth. Despite all these vigorous reactions, describing conditions in the early universe is straightforward, at least in principle. We simply need to use the laws of physics to calculate the proportions of the various forms of radiation and matter at each moment in the universe's early history. The only difficulty is our incomplete understanding of the laws of physics.

The very early universe was so hot that energy could be transformed into matter and vice versa.

To date, physicists have observed the behavior of matter and energy at temperatures as high as those that existed in the universe just *one ten-billionth* (10^{-10}) of a second after the Big Bang, giving us confidence that we actually understand what was happening at that early time. Our understanding of physics under the more extreme conditions that prevailed even earlier is less certain, but we have some ideas about what the universe was like when it was a mere 10^{-38} second old, and perhaps a glimmer of what it was like at the age of just 10^{-43} second. These tiny fractions of a second are so small that, for all practical purposes, we are studying the very moment of creation—the Big Bang itself.

Fundamental Forces To understand the changes that occurred in the early universe, it helps to think in terms of *forces*. Everything that happens in the universe today is governed by four distinct forces: *gravity, electromagnetism,* the *strong force,* and the *weak force.* We have already encountered examples of each of these forces in action.

Gravity is the most familiar of the four forces, providing the "glue" that holds planets, stars, and galaxies together. The electromagnetic force, which depends on the electrical charge of a particle instead of its mass, is far stronger than gravity. It is therefore the dominant force between particles in atoms and molecules, responsible for all chemical and biological reactions. However, the existence of both positive and negative electrical charges causes the electromagnetic force to lose out to gravity on large scales, even though both forces decline with distance by an inverse square law. Most large astronomical objects (such as planets and stars) are electrically neutral overall, making the electromagnetic force unimportant on that scale. Gravity therefore becomes the dominant force for such objects, because more mass always means more gravity.

The strong and weak forces operate only over extremely short distances, making them important within atomic nuclei but not on larger scales. The strong force binds protons and neutrons together in atomic nuclei [Section 11.2]. The weak force plays a crucial role in nuclear reactions such as fission and fusion, and it is the only force besides gravity that affects weakly interacting particles such as neutrinos.

Although the four forces behave quite differently from one another, current models of fundamental physics predict that they are just different aspects of a smaller number of more fundamental forces (Figure 17.4). These models predict that at the high temperatures that prevailed in the early universe, the four forces would not have been as distinct as they are today.

As an analogy, think about ice, liquid water, and water vapor. These three substances are quite different from one another in appearance and behavior, yet they are just different phases of the single substance H_2O. In a similar way, experiments have shown that the electromagnetic and weak forces

> The four forces that operate in the universe today may have been unified early in time, becoming distinct as the universe expanded and cooled.

lose their separate identities under conditions of very high temperature or energy and merge together into a single **electroweak force.** At even higher temperatures and energies, the electroweak force may merge with the strong force and ultimately with gravity. Models that predict the merger of the electroweak and strong forces are called **grand unified theories,** or **GUTs** for short.* The merger of the strong, weak, and electromagnetic forces is therefore often called the *GUT force.* Many physicists suspect that at even higher energies, the GUT force and gravity merge into a single "super force" that governs the behavior of everything. (Current ideas for linking all four forces go by names that include *supersymmetry, superstrings, supergravity,* and "theory of everything.")

If these ideas are correct, then the universe was governed solely by the super force in the first instant after the Big Bang. As the universe expanded and cooled, the super force split into gravity and the GUT

---

* The grand unified "theories" are not yet well tested and therefore do not meet our usual definition for being *scientific* theories; however, they are mathematically rigorous and qualify as mathematical theories.

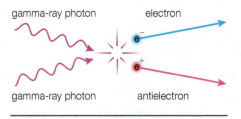

Particle creation

gamma-ray photon · electron

gamma-ray photon · antielectron

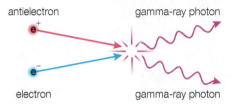

Particle annihilation

antielectron · gamma-ray photon

electron · gamma-ray photon

▲ **FIGURE 17.3**

Electron–antielectron creation and annihilation. Reactions like these constantly converted photons to particles, and vice versa, in the early universe.

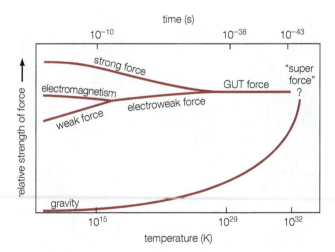

▲ **FIGURE 17.4**

The four forces are distinct at low temperatures but may merge at very high temperatures, such as those that prevailed during the first fraction of a second after the Big Bang.

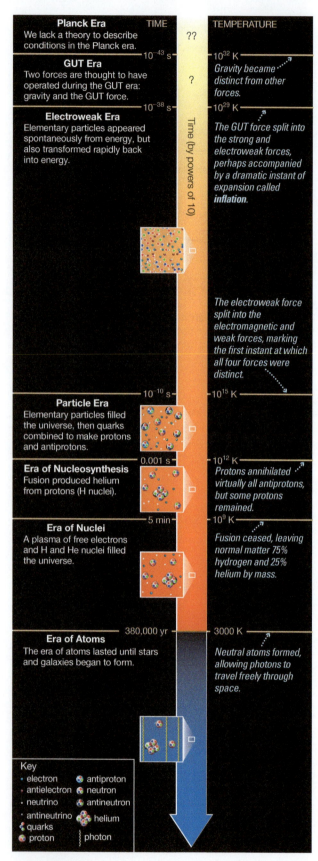

Planck Era
We lack a theory to describe conditions in the Planck era.

GUT Era
Two forces are thought to have operated during the GUT era: gravity and the GUT force.

Electroweak Era
Elementary particles appeared spontaneously from energy, but also transformed rapidly back into energy.

Particle Era
Elementary particles filled the universe, then quarks combined to make protons and antiprotons.

Era of Nucleosynthesis
Fusion produced helium from protons (H nuclei).

Era of Nuclei
A plasma of free electrons and H and He nuclei filled the universe.

Era of Atoms
The era of atoms lasted until stars and galaxies began to form.

TIME
10^{-43} s
10^{-38} s
10^{-10} s
0.001 s
5 min
380,000 yr

Time (by powers of 10)

TEMPERATURE
10^{32} K — Gravity became distinct from other forces.
10^{29} K — The GUT force split into the strong and electroweak forces, perhaps accompanied by a dramatic instant of expansion called **inflation**.

The electroweak force split into the electromagnetic and weak forces, marking the first instant at which all four forces were distinct.

10^{15} K
10^{12} K — Protons annihilated virtually all antiprotons, but some protons remained.
10^9 K — Fusion ceased, leaving normal matter 75% hydrogen and 25% helium by mass.

3000 K — Neutral atoms formed, allowing photons to travel freely through space.

Key
- electron
- antielectron
- neutrino
- antineutrino
- quarks
- proton
- antiproton
- neutron
- antineutron
- helium
- photon

▲ **FIGURE 17.5**

A timeline for the eras of the early universe. The only era not shown is the era of galaxies, which began with the birth of stars and galaxies when the universe was a few hundred million years old.

force, which then split further into the strong and electroweak forces. Ultimately, all four forces became distinct. As we'll see shortly, these changes in the fundamental forces probably occurred before the universe was one ten-billionth of a second old.

◆ How did the early universe change with time?

The Big Bang theory uses scientific understanding of particles and forces to reconstruct the history of the universe. Here we will outline this history as a series of *eras,* or time periods. Each era is distinguished from the next by some major change in physical conditions as the universe cools. You'll find it useful to refer to the timeline shown in Figure 17.5 as you read along. Notice that the time scale in Figure 17.5 runs by powers of 10, which means that early eras were very brief, even though they appear spread out on the figure. It will take you longer to read this chapter than it took the universe to progress through the first five eras we will discuss, by which point the chemical composition of the early universe had already been determined.

The Planck Era   The first era after the Big Bang is called the **Planck era,** named for physicist Max Planck (1858–1947); it represents times before the universe was 10^{-43} second old. According to the laws of quantum mechanics, there should have been substantial energy fluctuations from point to point during this very early time. Because energy and mass are equivalent, Einstein's theory of general relativity tells us that these energy fluctuations should have generated a rapidly changing gravitational field that would have randomly warped space and time. Unfortunately, these fluctuations are predicted to have been so large that our current understanding of physics is inadequate to describe what might have been happening. The main problem is that we do not yet have a theory that links quantum mechanics (our successful theory of the very small) and general relativity (our successful theory of the very big). Perhaps someday we will be able to merge these theories of the very small and the very big into a single "theory of everything." Until that happens, science cannot describe the universe during the Planck era.

We do not yet understand the physics of the universe well enough to describe what it was like during the Planck era.

Nevertheless, we have at least some idea of how the Planck era ended. If you look back at Figure 17.4, you'll see that all four forces are thought to merge into the single, unified super force at temperatures above 10^{32} K—the temperatures that prevailed during the Planck era. In that case, the Planck era would have been a time of ultimate simplicity, when just a single force operated in nature, and it came to an end when the temperature dropped low enough for gravity to become distinct from the other three forces, which were still merged as the GUT force. By analogy to the way ice crystals form as a liquid cools, we say that gravity "froze out" at the end of the Planck era.

The GUT Era   The next era is called the **GUT era,** named for the grand unified theories (GUTs) that predict the merger of the strong, weak, and electromagnetic forces into a single GUT force at temperatures above 10^{29} K (see Figure 17.4). Although different versions of grand unified theories disagree in many details, they all predict that the GUT era was a time during which two forces—gravity and the GUT force—operated in the universe. It came to an end when the GUT force split into the strong and electroweak forces, which happened when the universe was a mere 10^{-38} second old; note that this is less than a trillion-trillion-trillionth of a second.

Our current understanding of physics allows us to say only slightly more about the GUT era than about the Planck era, and none of our ideas about the GUT era have been sufficiently tested to give us great confidence about what occurred during that time. However, if the grand unified theories are correct, the freezing out of the strong and electroweak forces may have released an enormous amount of energy, causing a sudden and dramatic expansion of the universe that we call **inflation.** In a mere 10^{-36} second, pieces of the universe the size of an atomic nucleus may have grown to the size of our solar system. Inflation sounds bizarre, but as we will discuss later, it explains several important features of today's universe.

Energy released near the end of the GUT era may have caused a dramatic expansion of the universe known as inflation.

The Electroweak Era The splitting of the GUT force marked the beginning of an era during which three distinct forces operated: gravity, the strong force, and the electroweak force. We call this time the **electroweak era,** indicating that the electromagnetic and weak forces were still merged together. Intense radiation continued to fill all of space, as it had since the Planck era, spontaneously producing matter and antimatter particles that almost immediately annihilated each other and turned back into photons.

The universe continued to expand and cool throughout the electroweak era, dropping to a temperature of 10^{15} K when it reached an age of 10^{-10} second. This temperature is still 100 million times hotter than the temperature in the core of the Sun today, but it was low enough for the electromagnetic and weak forces to freeze out from the electroweak force. After this instant (10^{-10} second), all four forces were forever distinct in the universe.

The end of the electroweak era marks an important transition not only in the physical universe, but also in human understanding of the universe. The theory that unified the weak and electromagnetic forces, which was developed in the 1970s, predicted the emergence of new types of particles (called the W and Z bosons, or *weak bosons*) at temperatures above the 10^{15} K temperature that pervaded the universe when it was 10^{-10} second old. In 1983, particle-accelerator experiments reached energies equivalent to such high temperatures for the first time. The new particles showed up just as predicted, produced from the extremely high energy in accord with $E = mc^2$. We therefore have direct experimental evidence concerning the conditions in the universe at the end of the electroweak era. We do *not* have any direct experimental evidence of conditions before that time. Our theories concerning the earlier parts of the electroweak era and the GUT era consequently are much more speculative than our theories describing the universe from the end of the electroweak era to the present.

Experimental evidence supports the current theory about the universe since it was 10^{-10} second old, at the end of the electroweak era, but we have no direct evidence about earlier times.

The Particle Era

As long as the universe was hot enough for the spontaneous creation and annihilation of particles to continue, the total number of particles was roughly in balance with the total number of photons. Once it became too cool for this spontaneous exchange of matter and energy to continue, photons became the dominant form of energy in the universe. We refer to the time between the end of the electroweak era and the moment when spontaneous particle production ceased as the **particle era,** to emphasize the importance of subatomic particles during this period.

During the early parts of the particle era (and earlier eras), photons turned into all sorts of exotic particles that we no longer find existing freely in the universe today, including *quarks*—the building blocks of protons and neutrons. By the end of the particle era, all quarks had combined into protons and neutrons, which shared the universe with other particles such as electrons and neutrinos. The particle era came to an end when the universe reached an age of 1 millisecond (0.001 second) and the temperature had fallen to 10^{12} K. At this point, it was no longer hot enough to produce protons and antiprotons spontaneously from pure energy.

If the universe had contained equal numbers of protons and antiprotons (or neutrons and antineutrons) at the end of the particle era, all of the pairs would have annihilated each other, creating photons and leaving essentially no matter in the universe. From the obvious fact that the universe contains matter, we conclude that protons must have slightly outnumbered antiprotons at the end of the particle era.

Protons must have slightly outnumbered antiprotons during the particle era, or we would not be here today.

We can estimate the ratio of matter to antimatter by comparing the present numbers of protons and photons in the universe. These two numbers should have been similar in the very early universe, but today photons outnumber protons by about a billion to one. This ratio indicates that for every billion antiprotons in the early universe, there must have been about a billion and one protons. That is, for each 1 billion protons and antiprotons that annihilated each other at the end of the particle era, a single proton was left over. This seemingly slight excess of matter over antimatter makes up all the ordinary matter in the present-day universe. Some of those protons (and neutrons) left over from when the universe was 0.001 second old are the very ones that make up our bodies.

The Era of Nucleosynthesis

The eras we have discussed so far all occurred within the first 0.001 second of the universe's existence—less time than it takes you to blink an eye. At this point, the protons and neutrons left over after the annihilation of antimatter began to fuse into heavier nuclei. However, the temperature of the universe remained so high that gamma rays blasted apart most of those nuclei as fast as they formed. This dance of fusion and demolition marked the **era of nucleosynthesis,** which ended when the universe was about 5 minutes old. By this time, the density in the expanding universe had dropped so much that fusion no longer occurred, even though the temperature was still about a billion kelvins (10^9 K)—much hotter than the temperature of the Sun's core.

Most of the helium in the universe was made during the first 5 minutes.

When fusion ceased at the end of the era of nucleosynthesis, the chemical content of the universe had become (by mass) about 75% hydrogen and 25% helium, along with trace amounts of deuterium (hydrogen with a neutron) and lithium (the next heaviest element after hydrogen and helium). Except for the small proportion of matter that stars later forged into heavier elements, the chemical composition of the universe remains the same today.

The Era of Nuclei   After fusion ceased, the universe consisted of a very hot plasma of hydrogen nuclei, helium nuclei, and free electrons. This basic picture held for about the next 380,000 years as the universe continued to expand and cool. The fully ionized nuclei moved independently of electrons (rather than being bound with electrons in neutral atoms) during this period, which we call the **era of nuclei.** Throughout this era, photons bounced rapidly from one electron to the next, just as they do deep inside the Sun today [Section 11.2], never managing to travel far between collisions. Any time a nucleus managed to capture an electron to form a complete atom, one of the photons quickly ionized it.

The era of nuclei came to an end when the expanding universe was about 380,000 years old. At this point the temperature had fallen to about 3000 K—roughly half the temperature of the Sun's surface today.

Photons began to travel freely through the universe about 380,000 years after the Big Bang, when electrons first combined with nuclei to make atoms.

Hydrogen and helium nuclei finally captured electrons for good, forming stable, neutral atoms for the first time. With electrons now bound into atoms, the universe became transparent, as if a thick fog had suddenly lifted. Photons, formerly trapped among the electrons, began to stream freely across the universe. We still see these photons today as the *cosmic microwave background,* which we will discuss shortly.

The Eras of Atoms and Galaxies   We've already discussed the rest of the universe's history in earlier chapters. The end of the era of nuclei marked the beginning of the **era of atoms,** when the universe consisted of a mixture of neutral atoms and plasma (ions and electrons), along with a large number of photons. Because the density of matter in the

The first galaxies had formed by the time the universe was a billion years old.

universe differed slightly from place to place, gravity slowly drew atoms and plasma into the higher-density regions, which assembled into protogalactic clouds [Section 16.3]. Stars formed in these clouds, and the clouds subsequently merged to form galaxies.

The first full-fledged galaxies had formed by the time the universe was about 1 billion years old, beginning what we call the **era of galaxies,** which continues to this day. Generation after generation of star formation in galaxies steadily builds elements heavier than helium and incorporates them into new star systems. Some of these star systems develop planets, and on at least one of these planets life burst into being a few billion years ago. Now here we are, thinking about it all.

Early Universe Summary   Figure 17.6 summarizes the major ideas from our brief overview of the history of the early universe as it is described by the Big Bang theory. In the rest of this chapter, we will discuss the evidence that supports this theory. Before you read on, be sure to study the visual summary presented in Figure 17.6.

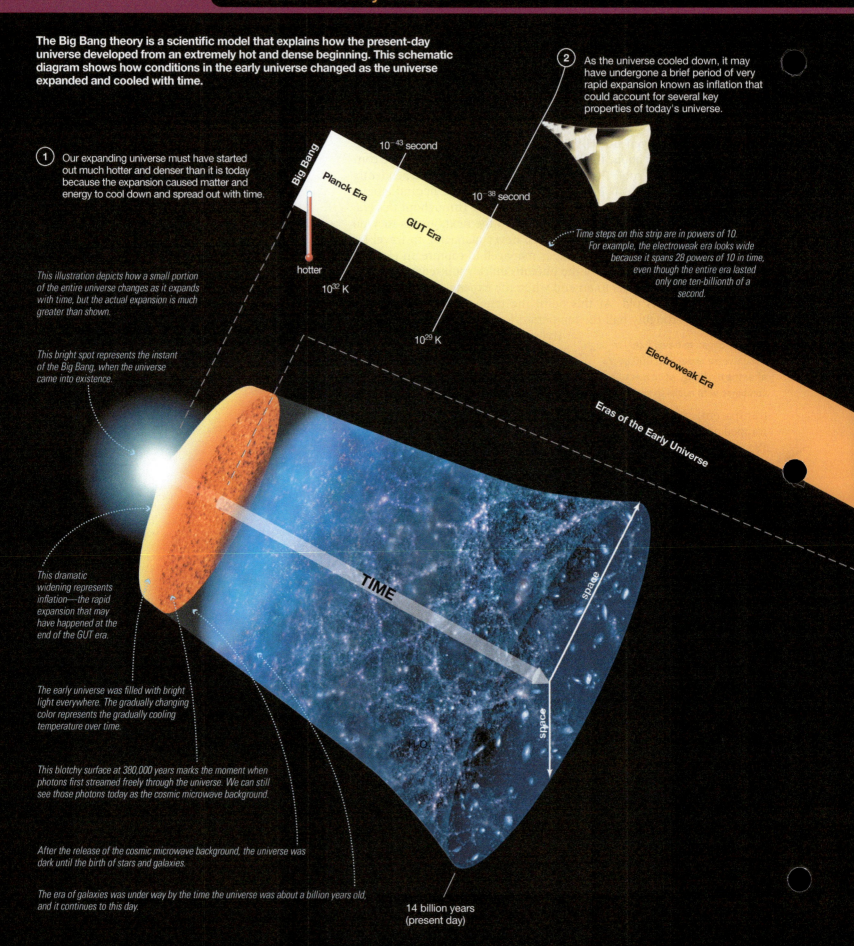

The Big Bang theory is a scientific model that explains how the present-day universe developed from an extremely hot and dense beginning. This schematic diagram shows how conditions in the early universe changed as the universe expanded and cooled with time.

1 Our expanding universe must have started out much hotter and denser than it is today because the expansion caused matter and energy to cool down and spread out with time.

2 As the universe cooled down, it may have undergone a brief period of very rapid expansion known as inflation that could account for several key properties of today's universe.

This illustration depicts how a small portion of the entire universe changes as it expands with time, but the actual expansion is much greater than shown.

This bright spot represents the instant of the Big Bang, when the universe came into existence.

This dramatic widening represents inflation—the rapid expansion that may have happened at the end of the GUT era.

The early universe was filled with bright light everywhere. The gradually changing color represents the gradually cooling temperature over time.

This blotchy surface at 380,000 years marks the moment when photons first streamed freely through the universe. We can still see those photons today as the cosmic microwave background.

After the release of the cosmic microwave background, the universe was dark until the birth of stars and galaxies.

The era of galaxies was under way by the time the universe was about a billion years old, and it continues to this day.

Time steps on this strip are in powers of 10. For example, the electroweak era looks wide because it spans 28 powers of 10 in time, even though the entire era lasted only one ten-billionth of a second.

Big Bang

Planck Era

10^{-43} second

GUT Era

10^{-38} second

Electroweak Era

Eras of the Early Universe

hotter

10^{32} K

10^{29} K

TIME

space

space

14 billion years
(present day)

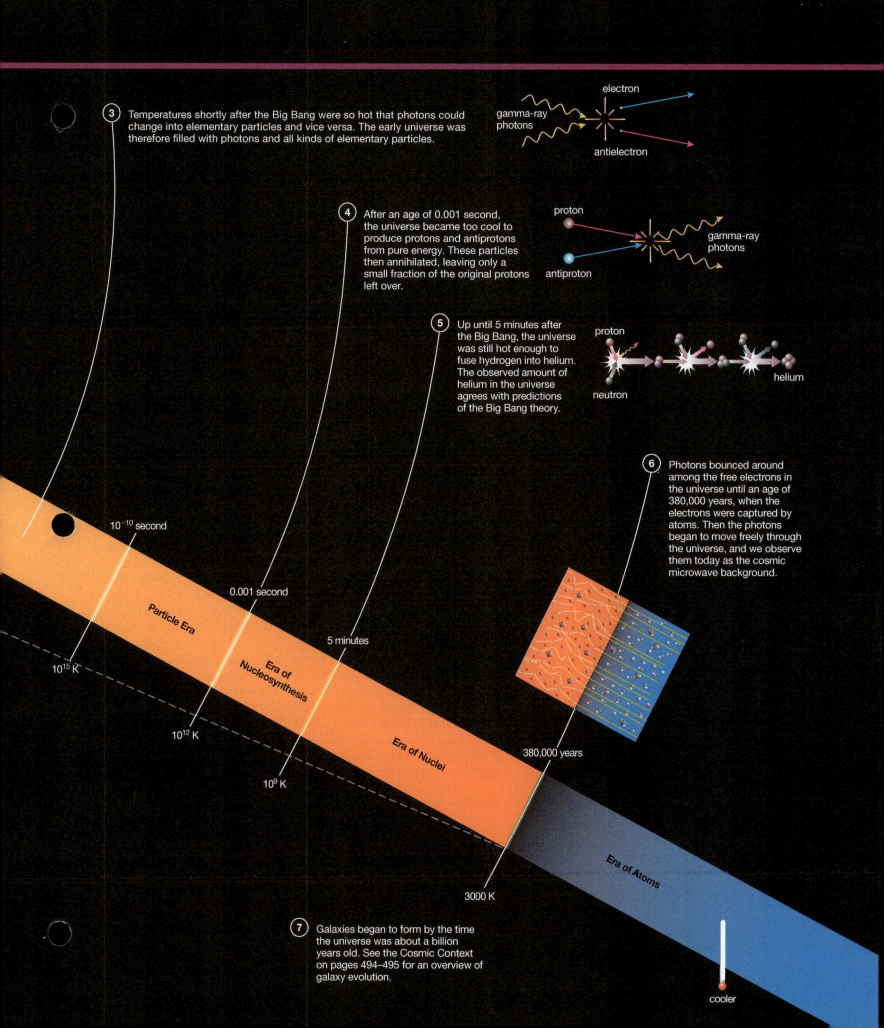

3 Temperatures shortly after the Big Bang were so hot that photons could change into elementary particles and vice versa. The early universe was therefore filled with photons and all kinds of elementary particles.

electron

gamma-ray photons

antielectron

4 After an age of 0.001 second, the universe became too cool to produce protons and antiprotons from pure energy. These particles then annihilated, leaving only a small fraction of the original protons left over.

proton

antiproton

gamma-ray photons

5 Up until 5 minutes after the Big Bang, the universe was still hot enough to fuse hydrogen into helium. The observed amount of helium in the universe agrees with predictions of the Big Bang theory.

proton

neutron

helium

6 Photons bounced around among the free electrons in the universe until an age of 380,000 years, when the electrons were captured by atoms. Then the photons began to move freely through the universe, and we observe them today as the cosmic microwave background.

10^{-10} second

Particle Era

0.001 second

5 minutes

Era of Nucleosynthesis

Era of Nuclei

10^{15} K

10^{12} K

10^9 K

380,000 years

3000 K

Era of Atoms

cooler

7 Galaxies began to form by the time the universe was about a billion years old. See the Cosmic Context on pages 494–495 for an overview of galaxy evolution.

17.2 Evidence for the Big Bang

Like any scientific theory, the Big Bang theory is a model of nature designed to explain a set of observations. The model was inspired by Edwin Hubble's observation that the universe is expanding, which implies that the universe must have been much denser and hotter in the past. However, like any scientific model, it had to make testable predictions about other observable features of the universe, and it has gained wide scientific acceptance because of two major predictions that have been verified:

- The Big Bang theory predicts that the radiation that began to stream across the universe at the end of the era of nuclei should still be present today. Sure enough, we find that the universe is filled with what we call the **cosmic microwave background.** Its characteristics precisely match what the theory predicts.

- The Big Bang theory predicts that some of the original hydrogen in the universe should have fused into helium during the era of nucleosynthesis. Observations of the actual helium content of the universe closely match the predicted amount of helium.

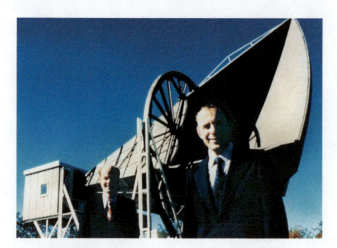

▲ **FIGURE 17.7**
Arno Penzias and Robert Wilson, discoverers of the cosmic microwave background, with the Bell Labs microwave antenna.

◆ How do observations of the cosmic microwave background support the Big Bang theory?

The discovery of the cosmic microwave background was announced in 1965. Arno Penzias and Robert Wilson, two physicists working at Bell Laboratories in New Jersey, were calibrating a sensitive microwave antenna designed for satellite communications (Figure 17.7). (*Microwaves* fall within the radio portion of the electromagnetic spectrum; see Figure 5.2.) Much to their chagrin, they kept finding unexpected "noise" in every measurement they made. The noise was the same no matter where they pointed the antenna, indicating that it came from all directions in the sky and ruling out any possibility that it came from any particular astronomical object or any place on Earth.

Meanwhile, physicists at nearby Princeton University were busy calculating the expected characteristics of the radiation left over from the heat of the Big Bang, which was first predicted by George Gamow and his colleagues in the 1940s. They concluded that, if the Big Bang had really occurred, this radiation should be permeating the entire universe and should be detectable with a microwave antenna. The Princeton group soon met with Penzias and Wilson to compare notes, and both teams realized that the "noise" from the Bell Labs antenna was the predicted cosmic microwave background—the first strong evidence that the Big Bang had really happened. Penzias and Wilson received the 1978 Nobel Prize in physics for their discovery.

The cosmic microwave background consists of microwave photons that have traveled through space since the end of the era of nuclei, when most of the electrons in the universe joined with nuclei to make neutral atoms, which interact less strongly with photons. With very few free electrons left to block them, most of the photons from that time have traveled unobstructed through the universe ever since (Figure 17.8). When we observe the cosmic microwave background, we essentially are seeing back to the end of the era of nuclei, when the universe was only 380,000 years old.

> The cosmic microwave background is radiation left over from the Big Bang.

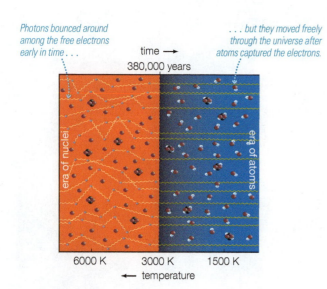

Photons bounced around among the free electrons early in time . . .

time →

380,000 years

. . . but they moved freely through the universe after atoms captured the electrons.

era of nuclei

era of atoms

6000 K   3000 K   1500 K
← temperature

▲ **FIGURE 17.8**
Photons (yellow squiggles) frequently collided with free electrons during the era of nuclei and thus could travel freely only after electrons became bound into atoms. This transition was something like the transition from a dense fog to clear air. The photons released at the end of the era of nuclei, when the universe was about 380,000 years old, make up the cosmic microwave background. Precise measurements of these microwaves tell us what the universe was like at this moment in time.

The Big Bang theory predicts that the cosmic microwave background should have an essentially perfect thermal radiation spectrum [Section 5.2], because it came from the heat of the universe itself. Because the cosmic microwave background came into existence when the universe had cooled to a temperature of about 3000 K, similar to the surface temperature of a red giant star, its spectrum should have originally peaked at a wavelength of about 1000 nanometers (just like the thermal radiation from a red star). Because the universe has since expanded by a factor of about 1000, the wavelengths of these photons should by now have stretched to about 1000 times their original wavelengths [Section 16.2]. We therefore expect the peak wavelength of the cosmic microwave background now to be about a millimeter, squarely in the microwave portion of the spectrum and corresponding to a temperature of a few degrees above absolute zero.

In the early 1990s, a NASA satellite called the *Cosmic Background Explorer (COBE)* was launched to test these ideas about the cosmic microwave background. The results were a stunning success for the Big Bang theory, and earned the 2006 Nobel Prize in physics for *COBE* team leaders George Smoot and John Mather. As shown in Figure 17.9, the cosmic microwave background does indeed have a perfect thermal radiation spectrum, with a peak corresponding to a temperature of 2.73 K.

> The cosmic microwave background—the heat of the universe itself—has a perfect thermal radiation spectrum corresponding to a temperature of 2.73 K.

think about it Suppose the cosmic microwave background did not really come from the heat of the universe itself but instead came from many individual stars and galaxies. Explain why we would not expect it to have a perfect thermal radiation spectrum in that case. How does the spectrum of the cosmic microwave background lend support to the Big Bang theory?

COBE and its successor missions, the *Wilkinson Microwave Anisotropy Probe (WMAP)* and the European *Planck* satellite, have also mapped the temperature of the cosmic microwave background in all directions (Figure 17.10). The temperature turns out to be extraordinarily uniform throughout the universe—just as the Big Bang theory predicts it should be—with variations from one place to another of only a few parts in 100,000. Moreover, these slight variations also represent a predictive success of the Big Bang theory. Recall that our theory of galaxy formation depends on the assumption that the early universe was *not quite* perfectly uniform; some regions of the universe must have started out slightly denser than other regions, so that they could serve as seeds for galaxy formation [Section 16.3]. The small variations in the temperature of the cosmic microwave background indicate that the density of the early universe really did differ slightly from place to place.

In fact, detailed observations of these small temperature variations are very important to studies of galaxy evolution, because all large structures in the universe are thought to have formed around the regions of slightly enhanced density [Section 18.3]. Measuring the patterns of variations in the cosmic microwave background therefore tells us both about what must have happened at even earlier times to create the variations and about the starting conditions that we should use in models of galaxy evolution.

> Maps of the cosmic microwave background reveal the density enhancements from which galaxies and larger structures later formed.

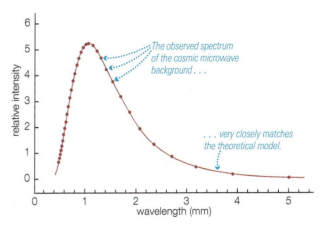

▲ **FIGURE 17.9**

This graph shows the spectrum of the cosmic microwave background recorded by NASA's *COBE* satellite. A theoretically calculated thermal radiation spectrum (smooth curve) for a temperature of 2.73 K perfectly fits the data (dots). (The error bars are smaller than the dots!) This excellent fit is important evidence in favor of the Big Bang theory.

The observed spectrum of the cosmic microwave background . . .

. . . very closely matches the theoretical model.

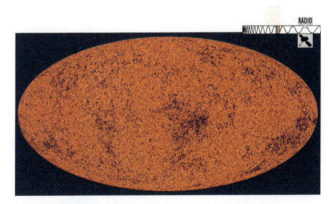

▲ **FIGURE 17.10**

This all-sky map (see larger version on page 442) shows temperature differences in the cosmic microwave background measured by *Planck*. The background temperature is about 2.73 K everywhere, but the brighter regions of this picture are slightly less than 0.0001 K hotter than the darker regions—indicating that the early universe was very slightly lumpy at the end of the era of nuclei. We are essentially seeing what the universe was like at the surface marked "380,000 years" in Figure 17.6. Gravity later drew matter toward the centers of these lumps, forming the structures we see in the universe today.

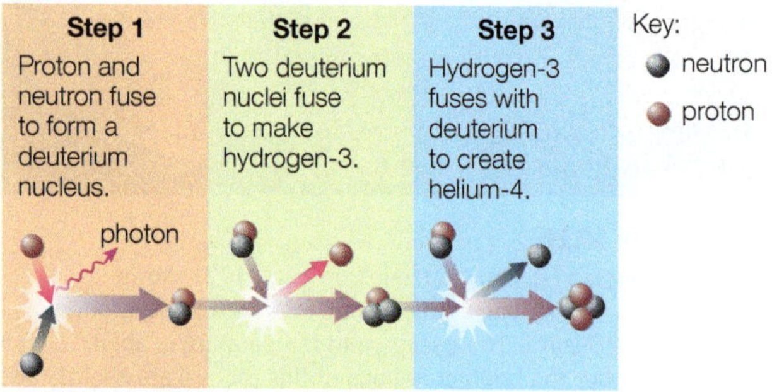

▲ **FIGURE 17.11**

During the 5-minute-long era of nucleosynthesis, virtually all the neutrons in the universe fused with protons to form helium-4. This figure illustrates one of several possible reaction pathways.

How do the abundances of elements support the Big Bang theory?

The Big Bang theory also solves what had previously been another long-standing astronomical problem: the origin of cosmic helium. Everywhere in the universe, about three-quarters of the mass of ordinary matter (not including dark matter) is hydrogen and about one-quarter is helium. The Milky Way's helium fraction is about 28%, and no galaxy has a helium fraction lower than 25%. Although helium is produced by hydrogen fusion in stars, calculations show that this production can account for only a small proportion of the total observed helium. We therefore conclude that the majority of the helium in the universe must already have been present in the protogalactic clouds that preceded the formation of galaxies.

The Big Bang theory makes a specific prediction about the helium abundance. As we discussed earlier, the theory explains the existence of helium as a consequence of fusion that occurred during the era of nucleosynthesis, when the universe itself was hot enough to fuse hydrogen into helium. Combining the current microwave background temperature of 2.73 K with the number of protons we observe in the universe tells us precisely how hot the universe must have been in the distant past, allowing scientists to calculate exactly how much helium should have been made. The result—25% helium—is another impressive success of the Big Bang theory.

Helium Formation in the Early Universe   To see why 25% of ordinary matter became helium, we need to understand what protons and neutrons were doing during the 5-minute era of nucleosynthesis. Early in this era, when the universe's temperature was 10^{11} K, nuclear reactions could convert protons into neutrons, and vice versa. As long as the universe remained hotter than 10^{11} K, these reactions kept the numbers of protons and neutrons nearly equal. But as the universe cooled, neutron–proton conversion reactions began to favor protons.

Neutrons are slightly more massive than protons, and therefore reactions that convert protons to neutrons require energy to proceed (in accordance with $E = mc^2$). As the temperature fell below 10^{11} K, the energy required for neutron production was no longer readily available, so the rate of these reactions slowed. In contrast, reactions that convert neutrons into protons release energy and therefore are unhindered by cooler temperatures. By the time the temperature of the universe fell to 10^{10} K, protons had begun to outnumber neutrons because the conversion reactions ran only in one direction. Neutrons changed into protons, but the protons didn't change back.

For the next few minutes, the universe was still hot and dense enough for nuclear fusion to take place. Protons and neutrons constantly combined to form *deuterium*—the rare form of hydrogen that contains a neutron in addition to a proton in the nucleus—and deuterium nuclei fused to form helium (Figure 17.11). However, during the early part of the era of nucleosynthesis, the helium nuclei were almost immediately blasted apart by one of the many gamma rays that filled the universe.

Fusion began to create long-lasting helium nuclei when the universe was about 1 minute old and had cooled to a temperature at which it contained few destructive gamma rays. Calculations show that the proton-to-neutron ratio at this time should have been about 7 to 1. Moreover, almost all the available neutrons should have been incorporated into

The Big Bang theory predicts that the universe should have a chemical composition of 75% hydrogen and 25% helium by mass, which agrees with the observed composition.

nuclei of helium-4. Figure 17.12 shows that, based on the 7-to-1 ratio of protons to neutrons, the universe should have had a composition of 75% hydrogen and 25% helium by mass at the end of the era of nucleosynthesis. This match between the predicted and observed helium ratios provides strong support for the Big Bang theory.

think about it Briefly explain why it should not be surprising that some galaxies contain a little more than 25% helium, but why it would be very surprising if some galaxies contained less. (*Hint:* Think about how the relative amounts of hydrogen and helium in the universe are affected by fusion in stars.)

Abundances of Other Light Elements

Why didn't the Big Bang produce heavier elements? By the time stable helium nuclei formed, when the universe was about a minute old, the temperature and density of the rapidly expanding universe had already dropped too far for a process like carbon production (three helium nuclei fusing to make a carbon nucleus [Section 13.3]) to occur. Reactions between protons, deuterium nuclei, and helium were still possible, but most of these reactions led nowhere. In particular, fusing two helium-4 nuclei results in a nucleus that is unstable and falls apart in a fraction of a second, as does fusing a proton to a helium-4 nucleus.

A few reactions involving hydrogen-3 (also known as *tritium*) or helium-3 can create long-lasting nuclei. For example, fusing helium-4 and hydrogen-3 produces lithium-7. However, the contributions of these reactions to the overall composition of the universe were minor because hydrogen-3 and helium-3 were so rare. Models of element production in the early universe show that, before the cooling of the universe shut off fusion entirely, such reactions generated only trace amounts of lithium, the next heavier element after helium. Aside from hydrogen, helium, and lithium, all other elements were forged much later in the nuclear furnaces of stars. (Beryllium and boron, which are heavier than lithium but lighter than carbon, were created later when high-energy particles broke apart heavier nuclei that formed in stars.)

Rapid cooling and expansion of the early universe halted nuclear fusion before the universe could produce elements much heavier than helium.

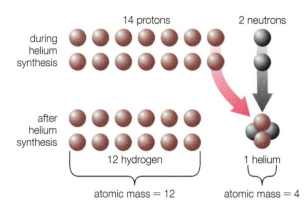

▲ **FIGURE 17.12**
Calculations show that protons outnumbered neutrons 7 to 1, which is the same as 14 to 2, during the era of nucleosynthesis. The result was 12 hydrogen nuclei (individual protons) for each helium nucleus. Therefore, the predicted hydrogen-to-helium mass ratio is 12 to 4, which is the same as 75% to 25%, in agreement with the observed abundance of helium.

---

extraordinary claims   The Universe Doesn't Change with Time

Although the Big Bang theory enjoys wide acceptance among scientists today, alternative ideas have been proposed and considered. One of the cleverest alternatives, developed in the late 1940s, was called the *steady state universe*. This hypothesis accepted the fact that the universe is expanding but rejected the idea of a Big Bang, instead postulating that the universe is infinitely old and always looks about the same on large scales.

This claim might not have seemed that extraordinary before Edwin Hubble discovered that the universe is expanding. However, the claim was made two decades after that discovery and raised the following obvious question: If the universe has been expanding forever, shouldn't every galaxy be infinitely far away from every other galaxy? Proponents of the steady state universe answered by claiming that new galaxies continually form in the gaps that open up as the universe expands, thereby keeping the same average distance between galaxies at all

times. In a sense, the steady state hypothesis said that the creation of the universe is an ongoing and eternal process rather than one that happened all at once with a Big Bang.

Two key discoveries caused the steady state hypothesis to lose favor. First, the 1965 discovery of the cosmic microwave background matched a prediction of the Big Bang theory but was not adequately explained by the steady state hypothesis. Second, a steady state universe should look about the same at all times, but observations made with increasingly powerful telescopes during the last half-century show that galaxies at great distances look younger than nearby galaxies. As a result of these predictive failures, few astronomers still consider the steady state hypothesis to be viable.

Verdict: Rejected.

17.3 The Big Bang and Inflation

When we discussed the eras of the universe earlier in the chapter, we noted that the universe is thought to have undergone a sudden and dramatic expansion, called *inflation*, which may have occurred at the end of the GUT era, when the universe was 10^{-38} second old. (In some models of inflation, the dramatic expansion can happen later, up until the end of the electroweak era.) This idea first emerged in 1981, when physicist Alan Guth was considering the consequences of the separation of the strong force from the GUT force that marked the end of the GUT era. Some models of high-energy physics predict that this separation of forces would have released enormous amounts of energy, and Guth realized that this energy might have caused a short period of inflation. He found that, in a mere 10^{-36} second, inflation could have caused the universe to expand by a factor of 10^{30}. Strange as this idea may sound, it appears to explain several otherwise mysterious features of the present-day universe. Moreover, recent evidence from detailed studies of the cosmic microwave background has provided support for the hypothesis that an early period of inflation really occurred.

◆ What key features of the universe are explained by inflation?

The Big Bang theory has gained wide acceptance because of the strong evidence from the cosmic microwave background and the abundance of helium in the universe. However, the theory in its simplest form leaves several major features of our universe unexplained. The three most pressing questions are the following:

- *Where did the density enhancements that led to galaxies come from?* Recall that successful models of galaxy formation start from the assumption that gravity could collect matter together in regions of the early universe that had slightly enhanced density [Section 16.3]. We know from our observations of variations in the cosmic microwave background that such regions of enhanced density were present in the universe at an age of 380,000 years, but we have not yet explained how these density variations came to exist.

- *Why is the large-scale universe nearly uniform?* Although the slight variations in the cosmic microwave background show that the universe is not *perfectly* uniform on large scales, the fact that it is smooth to within a few parts in 100,000 is remarkable enough that we would not expect it to have occurred by pure chance.

- *Why is the geometry of the universe flat?* Einstein's general theory of relativity tells us that matter and energy produce curvature of spacetime (see Special Topic on page 373). The overall geometry of the universe should therefore have a curvature that depends on the expansion rate and the overall amount of matter and energy the universe contains. However, observational efforts to measure the large-scale geometry of the universe have not yet detected any curvature. As far as we can tell, the large-scale geometry of the universe is flat, meaning that the effects of matter and energy on the overall curvature precisely balance the effects of the expansion rate. This precise balance is another feature that is difficult to attribute to chance.

Taking a scientific approach to the early universe demands that we seek answers to these questions that rely on natural processes, and the

The original Big Bang model left several questions unanswered. hypothesis of inflation provides such answers. That is, if we assume inflation occurred, we find that the density enhancements, large-scale uniformity, and flat geometry are all natural and expected consequences. Note that inflation is a *scientific* hypothesis because it is testable with observations we can perform today, and confidence in the hypothesis is growing because it has passed all the tests it has faced so far.

Density Enhancements: Giant Quantum Fluctuations

To understand how inflation explains the origin of the density enhancements that led to galaxies, we need to recognize a special feature of energy fields. Laboratory-tested principles of quantum mechanics tell us that on very small scales, the energy fields at any point in space are always fluctuating. The distribution of energy through space is therefore very slightly irregular, even in a complete vacuum. The tiny quantum "ripples" that make up the irregularities can be characterized by a wavelength that corresponds roughly to their size. In principle, quantum ripples in the very early universe could have been the seeds for density enhancements that later grew into galaxies. However, the wavelengths of the original ripples were far too small to explain density enhancements like those we see imprinted on the cosmic microwave background.

Inflation would have dramatically increased the wavelengths of these quantum fluctuations. The rapid growth of the universe during the period of inflation would have stretched tiny ripples from a size

Inflation would have stretched tiny quantum ripples to enormous sizes, allowing them to become the density enhancements around which galaxies later formed. smaller than that of an atomic nucleus to the size of our solar system (Figure 17.13), making them large enough to become the density enhancements from which galaxies and larger structures later formed. If that's the case, then the large-scale structure of today's universe started as tiny quantum fluctuations just before the period of inflation.

Uniformity: Equalizing Temperatures and Densities

The remarkable uniformity of the cosmic microwave background might at first seem quite natural, but with further thought it becomes difficult to explain. Imagine observing the cosmic microwave background in a certain part of the sky. You are seeing microwaves that have traveled through the universe since the end of the era of nuclei, just 380,000 years after the Big Bang. You are therefore seeing a region of the universe as it was some 14 billion years ago, when the universe was only 380,000 years old. Now imagine turning around and looking at the background radiation coming from the opposite direction. You are also seeing this region at an age of 380,000 years, and it looks virtually identical in temperature and density. The surprising part is this: The two regions are billions of light-years apart on opposite sides of our observable universe, but we are seeing them as they were when they were only 380,000 years old. They can't possibly have exchanged light or any other information; a signal traveling at the speed of light from one to the other would barely have started its journey. So how did they come to have the same temperature and density?

The inflation hypothesis explains large-scale uniformity by saying that distant regions of our observable universe today were once close enough to exchange radiation. The inflation hypothesis answers this question by saying that even though the two regions cannot have had any contact *since* the time of inflation, they were in contact prior to that time. Before the onset of inflation, when

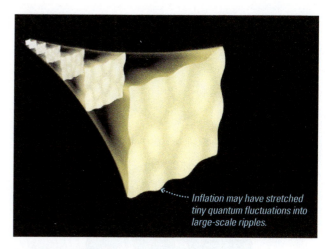

▲ **FIGURE 17.13**
During inflation, ripples in spacetime would have stretched by a factor of perhaps 10^{30}. The peaks of these ripples then would have become the density enhancements that produced all the structure we see in the universe today.

Inflation may have stretched tiny quantum fluctuations into large-scale ripples.

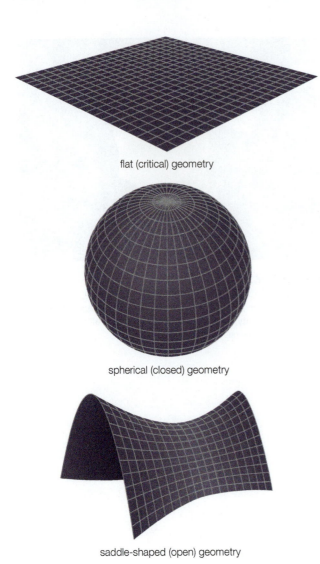

flat (critical) geometry

spherical (closed) geometry

saddle-shaped (open) geometry

▲ **FIGURE 17.14**

The three possible categories of overall geometry for the universe. Keep in mind that the real universe has these "shapes" in more dimensions than we can see.

the universe was 10^{-38} second old, the two regions were less than 10^{-38} light-second away from each other. Radiation traveling at the speed of light would therefore have had time to bounce between the two regions, and this exchange of energy equalized their temperatures and densities. Inflation then pushed these equalized regions to much greater distances, far out of contact with each other. Like criminals getting their stories straight before being locked in separate jail cells, the two regions (and all other parts of the observable universe) came to the same temperature and density before inflation spread them far apart.

Because inflation caused different regions of the universe to separate so far in such a short period of time, many people wonder whether it violates Einstein's theories saying that nothing can move faster than the speed of light. It does not, because nothing actually *moves* through space as a result of inflation or the ongoing expansion of the universe. Recall that the expansion of the universe is the expansion of *space itself.* Objects may in some cases be separating from one another at a speed faster than the speed of light, but no matter or radiation is able to travel between them during that time. In essence, inflation opens up a huge gap in space between objects that were once close together. The objects get very far apart, but nothing ever travels between them at a speed that exceeds the speed of light.

Geometry: Balancing the Universe The third question asks why the overall geometry of the universe is "flat." To understand this idea, we must consider the geometry of the universe in a little more detail.

Recall that Einstein's general theory of relativity tells us that the presence of matter can curve the structure of spacetime (see Special Topic, page 373). We cannot visualize this curvature in all three dimensions of space (or all four dimensions of spacetime), but we can detect its presence by its effects on how light travels through the universe. Although the curvature of the universe can vary from place to place, the universe as a whole must have some overall shape. Almost any shape is possible, but all the possibilities fall into just three general categories (Figure 17.14). Using analogies to objects that we can see in three dimensions, scientists refer to these three categories of shape as *flat* (or critical), *spherical* (or closed), and *saddle shaped* (or open).

According to general relativity, the overall geometry depends on the average density of matter and energy of the universe, and the geometry can be flat only if the combined density of matter plus energy is precisely equal to a value known as the **critical density.** If the universe's average density is less than the critical density, then the overall geometry is saddle-shaped. If its average density is greater than the critical density, then the overall geometry is spherical.

Inflation can explain why the overall geometry is so close to being flat. In terms of Einstein's theory, the effect of inflation on spacetime curvature is similar to the flattening of a balloon's surface when you blow into the balloon (Figure 17.15). The flattening of space caused by inflation would have been so enormous that any curvature the universe might have had previously would be noticeable only on size scales much larger than that of the observable universe. Inflation therefore makes the overall geometry of the universe appear flat, which means that the overall density of matter plus energy must be very close to the critical density.

> Inflation predicts that the overall geometry of the universe should appear flat, and hence that the overall density of matter and energy should equal the critical density.

◆ Did inflation really occur?

We've seen that inflation offers natural answers to our three key questions about the universe, but did it really happen? We cannot directly observe the universe at the very early time when inflation is thought to have occurred. Nevertheless, we can test the idea of inflation through careful studies of the cosmic microwave background. Scientists are only beginning to make observations that test inflation, but the findings to date are consistent with the idea that an early inflationary episode made the universe uniform and flat while planting the seeds of structure formation.

The strongest tests of inflation to date come from studying the patterns of temperature differences in the cosmic microwave background, and in particular the maps made by the *WMAP* and *Planck* satellites (see Figure 17.10). Remember that these maps show tiny temperature differences corresponding to density variations in the universe at the end of the era of nuclei, when the universe was about 380,000 years old.

Observed patterns in the temperature of the cosmic microwave background are consistent with the idea of inflation and reveal details about the universe's geometry, composition, and age.

However, according to models of inflation, these density enhancements were created much earlier, when inflation caused tiny quantum ripples to expand into seeds of structure. Careful observations of the temperature variations in the microwave background can therefore tell us about the structure of the universe at that very early time.

The graph in Figure 17.16 shows how temperature differences between patches of sky in the cosmic microwave background depend on the angular separations of these patches on the celestial sphere. The dots represent data from observations by *Planck* and other microwave telescopes, and the red curve shows the inflation-based model that best fits the observations. Notice the excellent fit between the model predictions and the data. Moreover, the same model also makes specific predictions about other characteristics of our universe, such as its composition and age, and these also agree well with observations. The general agreement between models and observations provides persuasive evidence in favor of the Big Bang in general, and also inspires confidence in the hypothesis of inflation.

The bottom line is that, all things considered, inflation does a remarkable job of explaining features of our universe that are otherwise unaccounted for in the Big Bang theory. Many astronomers and physicists therefore suspect that some process akin to inflation did affect the early universe, but the details of the interaction between high-energy particle physics and the evolving universe remain unclear. If these details can be worked out successfully, we face an amazing prospect—a breakthrough in our understanding of the very smallest particles, achieved by studying the universe on the largest observable scales.

17.4 Observing the Big Bang for Yourself

You might occasionally read an article in a newspaper or a magazine questioning whether the Big Bang really happened. We will never be able to prove with absolute certainty that the Big Bang theory is correct. However, no one has come up with any other model of the universe that so successfully explains so much of what we see. As we have discussed, the Big Bang model makes specific predictions that we have observationally verified, including predictions about the characteristics of the cosmic microwave background and the composition of the universe.

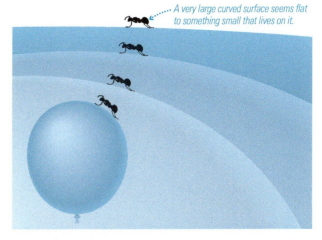

A very large curved surface seems flat to something small that lives on it.

▲ **FIGURE 17.15**

As a balloon expands, its surface seems increasingly flat to an ant crawling along it. Inflation is thought to have made the universe seem flat in a similar way.

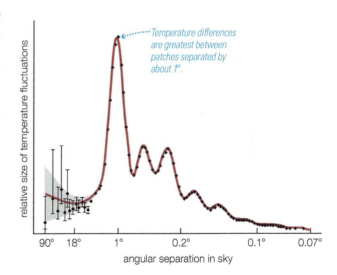

Temperature differences are greatest between patches separated by about 1°.

relative size of temperature fluctuations

angular separation in sky

▲ **FIGURE 17.16**

All-sky maps of the cosmic microwave background like that in Figure 17.10 allow scientists to measure temperature differences between different patches of the sky. This graph shows how the typical sizes of those temperature differences depend on the angular separation of the patches of sky. The data points represent actual measurements of the cosmic microwave background, while the red curve is the prediction of a model that relies on inflation to produce slight variations of temperature and density in the universe. Note the close agreement between the data and the model. (Bars indicate the uncertainty range in the data points.)

It also naturally explains many other features of the universe. So far, we know of nothing that is inconsistent with the Big Bang model.

The Big Bang theory's very success has also made it a target for respected scientists, skeptical nonscientists, and crackpots alike. The nature of scientific work requires that we test established wisdom to make sure it is valid. A sound scientific disproof of the Big Bang theory would be a discovery of great importance. However, stories touted in the news media as disproofs of the Big Bang usually turn out to be disagreements over details rather than fundamental problems that threaten to bring down the whole theory. Yet scientists must keep refining the theory and tracking down disagreements, because once in a while a small disagreement blossoms into a full-blown scientific revolution.

You don't need to accept all you have read without question. The next time you are musing on the universe's origins, try an experiment for yourself. Go outside on a clear night, look at the sky, and ask yourself why it is dark.

see it for yourself How dark is the night sky where you live? Go outside and observe it on a moonless night. Estimate the total number of stars that are visible to you. How many stars like those do you think it would take to completely cover the entire sky?

◆ Why is the darkness of the night sky evidence for the Big Bang?

If the universe were infinite, unchanging, and everywhere the same, then the entire night sky would blaze as brightly as the Sun. Johannes Kepler [Section 3.3] was one of the first people to reach this conclusion, but we now refer to the idea as **Olbers' paradox** after German astronomer Heinrich Olbers (1758–1840).

To better understand Olbers' paradox, imagine that you are in a dense forest on a flat plain. If you look in any direction, you'll likely see a tree. If the forest is small, you might be able to see through some gaps in the trees to the open plains, but larger forests have fewer gaps (Figure 17.17). An infinite forest would have no gaps at all—a tree trunk would block your view along any line of sight.

If the universe were infinite, unchanging, and everywhere the same, then the entire night sky would be bright, because it would be completely covered with stars.

The universe is like a forest of stars in this respect. In an unchanging universe with an infinite number of stars, we would see a star in every direction, making every point in the sky as bright as the Sun's surface. Even the presence of obscuring dust would not change this conclusion. The intense starlight would heat the dust over time until it too glowed like the Sun or evaporated away.

There are only two ways out of this dilemma. Either the universe has a finite number of stars, in which case we would not see a star in every direction, or it changes over time in some way that prevents us from seeing an infinite number of stars. For several centuries after Kepler first recognized the dilemma, astronomers leaned toward the first option. Kepler himself preferred to believe that the universe had a finite number of stars because he thought it had to be finite in space, with some kind of dark wall surrounding everything. Astronomers in the early 20th century preferred to believe that the universe was infinite in space but that we lived inside a finite collection of stars. They thought of the Milky Way as an island floating in a vast black void. However, subsequent observations showed that galaxies fill all of space more or less uniformly. We are therefore left with the second option: The universe changes over time.

a In a large forest, a tree will block your view no matter where you look. Similarly, in an unchanging universe with an infinite number of stars, we would expect to see stars in every direction, making the sky bright even at night.

b In a small forest with a smaller number of trees, you can see open spaces beyond the trees. Because the night sky is dark, the universe must similarly have spaces in which we see nothing beyond the stars, which means either that the number of stars is finite or that the universe changes in a way that prevents us from seeing an infinite number of them.

▲ **FIGURE 17.17**

Olbers' paradox can be understood by thinking of the view through a forest.

The Big Bang theory resolves Olbers' paradox in a particularly simple way. It tells us that we can see only a finite number of stars because the universe began at a particular moment. While the universe may contain an infinite number of stars, we can see only those that lie within the observable universe, inside our cosmological horizon [Section 16.2]. There are other ways in which the universe could change over time and prevent us from seeing an infinite number of stars, so Olbers' paradox does not *prove* that the universe began with a Big Bang. However, we must have some explanation for why the sky is dark at night, and no explanation besides the Big Bang also explains so many other observed properties of the universe so well.

The Big Bang theory is the best explanation we have for why the night sky is dark.

the big picture — Putting Chapter 17 into Context

Our "big picture" now extends all the way back to the earliest moments in time. When you think back on this chapter, keep in mind the following ideas:

- Predicting conditions in the early universe is straightforward, as long as we know how matter and energy behave under such extreme conditions.

- Our current understanding of physics allows us to reconstruct the conditions thought to have prevailed in the universe back to the first 10^{-10} second. Our understanding is less certain back to 10^{-38} second. Beyond 10^{-43} second, we run up against the present limits of human knowledge.

- Although it may sound strange to talk about the universe during its first fraction of a second, our ideas about the Big Bang rest on a solid foundation of observational, experimental, and theoretical evidence. We cannot say with absolute certainty that the Big Bang really happened, but no other model has so successfully explained how our universe came to be as it is.

my cosmic perspective The Big Bang happened long ago, but our connection to it is intimate. All the protons, neutrons, and electrons in your body were made from pure energy during the first few moments of the early universe.

17.1 The Big Bang Theory

◆ What were conditions like in the early universe?

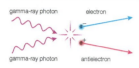

The early universe was filled with radiation and elementary particles. It was so hot and dense that the energy of radiation could turn into particles of **matter** and **antimatter,** which then collided and turned back into radiation.

◆ How did the early universe change with time?

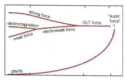

The universe has progressed through a series of eras, each marked by unique physical conditions. We know little about the **Planck era,** when the four fundamental forces may have behaved as one. Gravity became distinct at the start of the **GUT era,** and electromagnetism and the weak force became distinct at the end of the **electroweak era.** Matter particles annihilated all the antimatter particles by the end of the **particle era.** Fusion of protons and neutrons into helium ceased at the end of the **era of nucleosynthesis.** Hydrogen nuclei captured all the free electrons, forming hydrogen atoms at the end of the **era of nuclei.** Galaxies began to form at the end of the **era of atoms.** The **era of galaxies** continues to this day.

17.2 Evidence for the Big Bang

◆ How do observations of the cosmic microwave background support the Big Bang theory?

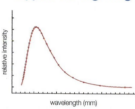

Telescopes that can detect microwaves allow us to observe the **cosmic microwave background**—radiation left over from the Big Bang. Its spectrum matches the characteristics expected of the radiation released at the end of the era of nuclei, confirming a key prediction of the Big Bang theory.

◆ How do the abundances of elements support the Big Bang theory?

The Big Bang theory predicts the ratio of protons to neutrons during the era of nucleosynthesis, and from this predicts that the chemical composition of the universe should be about 75% hydrogen and 25% helium (by mass). The prediction matches observations of the cosmic abundances of elements.

17.3 The Big Bang and Inflation

◆ What key features of the universe are explained by inflation?

The hypothesis that the universe underwent a rapid and dramatic period of **inflation** successfully explains three key features of the universe that are otherwise mysterious: (1) the density enhancements that led to galaxy formation, (2) the smoothness of the cosmic microwave background, and (3) the "flat" geometry of the observable universe.

◆ Did inflation really occur?

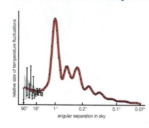

We can test the idea of inflation because it makes specific predictions about the patterns we should observe in the cosmic microwave background. Observations made with microwave telescopes so far match those predictions, lending credence to the idea that inflation (or something much like it) really occurred.

17.4 Observing the Big Bang for Yourself

◆ Why is the darkness of the night sky evidence for the Big Bang?

Olbers' paradox tells us that if the universe were infinite, unchanging, and filled with stars, the sky would be everywhere as bright as the surface of the Sun, and it would not be dark at night. The Big Bang theory resolves this paradox by telling us that the night sky is dark because the universe has a finite age, which means we can see only a finite number of stars in the sky.

visual skills check

Check your understanding of some of the many types of visual information used in astronomy. For additional practice, try the Chapter 17 Visual Quiz at MasteringAstronomy®.

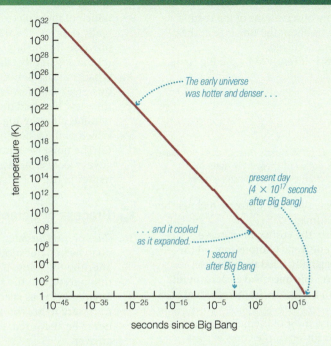

Answer the following questions based on Figure 17.1, which is repeated here.

1. What was the approximate temperature of the universe at an age of 10^{15} s?
   a. about 1 K
   b. about 100 K
   c. about 10^5 K
   d. about 10^{15} K

2. What was the approximate temperature of the universe at an age of 5 minutes?
   a. about 300 K
   b. about 10^6 K
   c. about 10^9 K
   d. about 10^{12} K

3. How much cooler is the universe now (at an age of 4×10^{17} s) than it was at an age of 1 second?
   a. Its current temperature is one hundred-millionth (10^{-8}) the temperature at an age of 1 second.
   b. Its current temperature is one hundred-thousandth (10^{-5}) the temperature at an age of 1 second.
   c. Its current temperature is one hundredth (10^{-2}) the temperature at an age of 1 second.
   d. Its current temperature is one ten-billionth (10^{-10}) the temperature at an age of 1 second.

exercises and problems

MasteringAstronomy® For instructor-assigned homework and other learning materials, go to MasteringAstronomy®.

Review Questions

1. Explain what we mean by the *Big Bang theory.*
2. What is *antimatter*? How were particle–antiparticle pairs created and destroyed in the early universe?
3. What are the four forces that operate in the universe today? Why do we think there were fewer forces operating in the early universe?
4. Make a list of the major eras in the history of the universe, summarizing the important events thought to have occurred during each era.
5. Why can't our current theories describe conditions that existed in the universe during the *Planck era*?
6. What are *grand unified theories*? According to these theories, how many forces operated during the *GUT era*? How are these forces related to the four forces that operate today?

7. What do we mean by *inflation*, and when do we think it occurred?
8. Why do we think there was slightly more matter than antimatter in the early universe? What happened to all the antimatter, and when?
9. How long did the *era of nucleosynthesis* last? Explain why this era was so important in determining the chemical composition of the universe.
10. When we observe the *cosmic microwave background*, at what age are we seeing the universe? How long have the photons in the background been traveling through space? Explain.
11. Briefly describe how the cosmic microwave background was discovered. How do the existence and nature of this radiation support the Big Bang theory?
12. How does the chemical abundance of helium in the universe support the Big Bang theory? Explain.

13. Describe three key questions about the universe that are answered by inflation, and explain how inflation answers each of them.
14. What observational evidence supports the hypothesis of inflation? Be sure to explain how observations of the cosmic microwave background can tell us about the universe at the much earlier time when inflation occurred.
15. What is *Olbers' paradox,* and how is it resolved by the Big Bang theory?

Test Your Understanding

Does It Make Sense?

Decide whether the statement makes sense (or is clearly true) or does not make sense (or is clearly false). Explain clearly; not all of these have definitive answers, so your explanation is more important than your chosen answer.

16. According to the Big Bang theory, the early universe had nearly equal amounts of matter and antimatter.
17. According to the Big Bang theory, the cosmic microwave background was created when energetic photons ionized the neutral hydrogen atoms that originally filled the universe.
18. Observed characteristics of the cosmic microwave background can be explained by assuming that it comes from individual stars and galaxies.
19. According to the Big Bang theory, most of the helium in the universe was created by nuclear fusion in the cores of stars.
20. According to the hypothesis of inflation, large-scale structure in the universe may have originated as tiny quantum fluctuations.
21. According to the hypothesis of inflation, the "flat" geometry of the universe most likely arose by chance.
22. Inflation is a nice idea, but there are no known ways to test whether it really happened.
23. In the distant past, the cosmic microwave background consisted primarily of infrared light.
24. The main reason the night sky is dark is that stars are so far away.
25. Patterns in the cosmic microwave background tell us about conditions in the early universe that ultimately led to galaxy formation.

Quick Quiz

Choose the best answer to each of the following. Explain your reasoning with one or more complete sentences.

26. The current temperature of the universe as a whole is (a) absolute zero. (b) a few K. (c) a few thousand K.
27. The charge of an antiproton is (a) positive. (b) negative. (c) neutral.
28. When a proton and an antiproton collide, they (a) repel each other. (b) fuse together. (c) convert into two photons.
29. Which of the following does *not* provide strong evidence for the Big Bang theory? (a) observations of the cosmic microwave background (b) observations of the amount of hydrogen in the universe (c) observations of the ratio of helium to hydrogen in the universe.
30. When the universe was 380,000 years old, its thermal radiation spectrum consisted mostly of (a) radio and microwave photons. (b) visible and infrared photons. (c) X-ray and ultraviolet photons.
31. Which of the following does inflation help to explain? (a) the uniformity of the cosmic microwave background (b) the amount of helium in the universe (c) the temperature of the cosmic microwave background.

32. Which of the following does inflation help to explain? (a) the origin of hydrogen (b) the origin of galaxies (c) the origin of atomic nuclei.
33. Which of these pieces of evidence supports the idea that inflation really happened? (a) the enormous size of the observable universe (b) the large amount of dark matter in the universe (c) the apparently "flat" geometry of the universe.
34. What is the earliest time from which we observe light in the universe? (a) a few hundred million years after the Big Bang (b) a few hundred thousand years after the Big Bang (c) a few minutes after the Big Bang.
35. Which of the following best explains why the night sky is dark? (a) The universe is not infinite in space. (b) The universe has not always looked the way it looks today. (c) The distribution of matter in the universe is not uniform on very large scales.

Process of Science

36. *Unanswered Questions.* Briefly describe one important but unanswered question about the events that happened shortly after the Big Bang. If you think it will be possible to answer that question in the future, describe how we might find an answer, being as specific as possible about the evidence necessary to answer the question. If you think the question will never be answered, explain why you think it is impossible to answer.
37. *Darkness at Night.* Suppose you are Kepler, pondering the darkness of the night sky without any knowledge of the Big Bang or the expanding universe. Come up with a hypothesis for the darkness of the night sky that would have been plausible in Kepler's time but does not depend on the Big Bang theory. Propose an experiment that scientists might be able to perform today to test that hypothesis.

Group Work Exercise

38. *Testing the Big Bang Theory.* **Roles:** *Scribe* (takes notes on the group's activities), *Advocate* (argues in favor of the Big Bang theory), *Skeptic* (points out weaknesses in the Big Bang theory), *Moderator* (leads group discussion and makes sure everyone contributes). **Activity:** Listed below are five hypothetical observations that are *not* predicted by the Big Bang theory. Your goal for each observation is to decide whether it could be explained with the existing Big Bang theory, could be explained with a revision to the Big Bang theory, or would force us to abandon the Big Bang theory. In each case, begin with the *Advocate* and *Skeptic* discussing the hypothetical discovery, then have the *Scribe* and *Moderator* decide how it fits (or does not fit) with the Big Bang theory. Write down your team's reasoning for each observation.
Hypothetical observation 1: a star cluster with an age of 15 billion years
Hypothetical observation 2: a galaxy with an age of 10 million years
Hypothetical observation 3: a galaxy at a distance of 10 billion light-years whose spectrum is blueshifted
Hypothetical observation 4: a galaxy containing 90% hydrogen and 10% helium
Hypothetical observation 5: evidence for an increase in the cosmic microwave background temperature with time

Investigate Further

Short-Answer/Essay Questions

39. *Life Story of a Proton.* Tell the life story of a proton from its formation shortly after the Big Bang to its presence in the nucleus of an oxygen atom you have just inhaled. Your story should be creative and imaginative, but it should also demonstrate your scientific understanding of as many stages in the proton's life as possible. You can draw on material from the entire book, and your story should be three to five pages long.

40. *Creative History of the Universe.* The story of creation as envisioned by the Big Bang theory is quite dramatic, but it is usually told in a fairly straightforward, scientific way. Write a more dramatic telling of the story, in the form of a short story, play, or poem. Be as creative as you wish, but be sure to remain accurate according to the science as it is understood today.

41. *Re-creating the Big Bang.* Particle accelerators on Earth can push particles to extremely high speeds. When these particles collide, the amount of energy associated with the colliding particles is much greater than the mass-energy the particles have when at rest. As a result, these collisions can produce many other particles out of pure energy. Explain in your own words how the conditions that occur in particle accelerators are similar to the conditions that prevailed shortly after the Big Bang. Also, point out some of the differences between what happens in particle accelerators and what happened in the early universe.

42. *Betting on the Big Bang Theory.* If you had $100, how much money would you wager on the proposition that we have a reasonable scientific understanding of what the universe was like when it was 1 minute old? Explain your bet in terms of the scientific evidence presented in this chapter.

43. *"Observing" the Early Universe.* Explain why we will never be able to observe the era of nucleosynthesis through direct detection of the radiation emitted at that time. How do we learn about this era?

44. *Element Production in the Big Bang.* Nucleosynthesis in the early universe was unable to produce more than trace amounts of elements heavier than helium. Using the information in Figure 13.18, which shows the mass per nuclear particle for many different elements, explain why producing elements like lithium (3 protons), boron (4 protons), and beryllium (5 protons) was so difficult.

45. *Evidence for the Big Bang.* Make a list of at least seven observed features of the universe that are satisfactorily explained by the Big Bang theory (including the idea of inflation).

Quantitative Problems

Be sure to show all calculations clearly and state your final answers in complete sentences.

46. *Energy from Antimatter.* The total annual U.S. power consumption is about 2×10^{20} joules. Suppose you could supply that energy by combining pure matter with pure antimatter. Estimate the total mass of matter–antimatter fuel you would need to supply the United States with energy for 1 year. How does that mass compare with the amount of matter in your car's gas tank? (A gallon of gas has a mass of about 4 kilograms.)

47. *Temperature of the Universe.* What will the temperature of the cosmic microwave background be when the average distances between galaxies are twice as large as they are today? (*Hint*: The peak wavelength of photons in the background will then also be twice as large as it is today.)

48. *Uniformity of the Cosmic Microwave Background.* The temperature of the cosmic microwave background differs by only a few parts in 100,000 across the sky. Compare that level of uniformity to the uniformity of the surface of a table that is 1 meter in size. How big would the largest bumps on that table be if its surface were smooth to one part in 100,000? Could you see bumps of that size on the table's surface?

49. *Daytime at "Night."* According to Olbers' paradox, the entire sky would be as bright as the surface of a typical star if the universe were infinite in space, unchanging in time, and the same everywhere. However, conditions would not need to be quite that extreme for the "nighttime" sky to be as bright as the daytime sky.
    a. Using the inverse square law for light from Cosmic Calculations 12.1, determine the apparent brightness of the Sun in our sky.
    b. Using the inverse square law for light, determine the apparent brightness our Sun would have if it were at a distance of 10 billion light-years.
    c. From your answers to parts a and b, estimate how many stars like the Sun would need to exist at a distance of 10 billion light-years for their total apparent brightness to equal that of our Sun.
    d. Compare your answer to part c with the estimate of 10^{22} stars in our observable universe from Section 1.1. Use your answer to explain why the night sky is much darker than the daytime sky. How much larger would the total number of stars need to be for "night" to be as bright as day?

Discussion Questions

50. *The Moment of Creation.* You've probably noticed that, in our discussion of the Big Bang theory, we never talk about the first instant. Even our most speculative theories take us back only to within 10^{-43} second of creation. Do you think it will *ever* be possible for science to consider the moment of creation itself? Will science ever be able to answer questions such as *why* the Big Bang happened? Defend your opinions.

51. *The Big Bang.* How convincing do you find the evidence for the Big Bang theory? What are its strengths? What does it fail to explain? Do *you* think the Big Bang really happened? Defend your opinion.

Web Projects

52. *Tests of the Big Bang Theory.* The satellites *COBE* and *WMAP* have provided striking confirmation of several predictions of the Big Bang theory. The more recent *Planck* mission was designed to test the Big Bang theory further. Use the Internet to gather pictures and information about *COBE, WMAP,* and *Planck.* Write a one- to two-page report about the strength of the evidence compiled by these satellite missions.

53. *New Ideas in Inflation.* The idea of inflation solves many of the puzzles associated with the standard Big Bang theory, but we are still a long way from confirming that inflation really occurred. Find recent articles that discuss some ideas about inflation and how we might test these ideas. Write a two- to three-page summary of your findings.

18

Dark Matter, Dark Energy, and the Fate of the Universe

This composite image of the Bullet Cluster of galaxies shows that the X ray–emitting hot gas (red) is in a different location from the majority of the mass revealed by gravitational lensing (blue).

LEARNING GOALS

18.1 Unseen Influences in the Cosmos
- ◆ What do we mean by dark matter and dark energy?

18.2 Evidence for Dark Matter
- ◆ What is the evidence for dark matter in galaxies?
- ◆ What is the evidence for dark matter in clusters of galaxies?
- ◆ Does dark matter really exist?
- ◆ What might dark matter be made of?

18.3 Structure Formation
- ◆ What is the role of dark matter in galaxy formation?
- ◆ What are the largest structures in the universe?

18.4 Dark Energy and the Fate of the Universe
- ◆ What is the evidence for an accelerating expansion?
- ◆ Why is flat geometry evidence for dark energy?
- ◆ What is the fate of the universe?

O ver the past several chapters, we have discussed the strong evidence supporting our scientific understanding of the history of the universe. This evidence indicates that our universe was born about 14 billion years ago in the Big Bang and has been expanding ever since. However, in regions that began with slightly enhanced density, gravity took hold and built galaxies, within which some of the hydrogen and helium atoms produced in the Big Bang assembled into stars. We exist today because galactic recycling has incorporated heavier elements made by early generations of stars into new star systems containing planets like Earth.

Scientists broadly agree on this basic outline of universal history, but many details are not yet well understood and a few more significant mysteries still remain. In this chapter, we'll explore two of the greatest remaining mysteries, which concern the nature of the so-called *dark matter* and *dark energy* that appear to make up most of the content of our universe. As we will see, these mysteries likely hold the key not only to a more complete understanding of how the universe has evolved to date, but also to the eventual fate of the universe.

18.1 Unseen Influences in the Cosmos

What is the universe made of? Ask an astronomer this seemingly simple question, and you might see a professional scientist blush with embarrassment. Based on all the evidence available today, the answer to this simple question is "We do not know."

It might seem incredible that we still do not know the composition of most of the universe, but you might also wonder why we should be so clueless. After all, astronomers can measure the chemical composition of distant stars and galaxies from their spectra, so we know that stars and gas clouds are made almost entirely of hydrogen and helium, with small amounts of heavier elements mixed in. But notice the key words "chemical composition." When we say these words, we are talking about the composition of material built from atoms of elements such as hydrogen, helium, carbon, and iron.

While it is true that all familiar objects—including people, planets, and stars—are built from atoms, the same may not be true of the universe as a whole. In fact, we now have good reason to think that the universe is *not* composed primarily of atoms. Instead, observations indicate that the universe consists largely of a mysterious form of mass known as *dark matter* and a mysterious form of energy known as *dark energy*.

◆ What do we mean by dark matter and dark energy?

It's easy for scientists to talk about dark matter and dark energy, but what do these terms really mean? They are nothing more than names given to unseen influences in the cosmos. In both cases observational evidence leads us to think that there is something out there, but we do not yet know exactly what the "something" is.

We might naively think that the major source of gravity that holds galaxies together should be the same gas that makes up their stars. However, observations suggest otherwise. By carefully observing gravitational effects on matter that we can see, such as stars or glowing clouds of gas, we've learned that there must be far more matter than meets the eye. Because this matter gives off little or no light, we call it **dark matter**. In other words, dark matter is simply a name we give to whatever unseen influence is causing the observed gravitational effects. We've already discussed dark matter briefly in Chapters 1 and 15, noting that studies of the Milky Way's rotation suggest that most of our galaxy's mass is distributed throughout its halo while most of the galaxy's light comes from stars and gas clouds in the thin galactic disk (see Figure 1.14).

Dark matter **is the name given to mass that emits no detectable radiation; we infer its existence from its gravitational effects.**

We infer the existence of the second unseen influence from careful studies of the expansion of the universe. After Edwin Hubble first discovered the expansion, it was generally assumed that gravity must slow the expansion with time. However, evidence collected during the last two decades indicates that the expansion of the universe is actually accelerating, implying that some mysterious force counteracts the effects of gravity on very large scales. **Dark energy** is the name most commonly given to the source of this mysterious force, though you may occasionally hear the same unseen influence attributed to *quintessence* or to a *cosmological constant*. Note that while dark matter really is "dark" compared to ordinary matter (because it gives off no light), there's nothing unusually "dark" about dark energy—after all, we don't expect to see light from the mere presence of a force or energy field.

Dark energy **is the name given to the unseen influence that causes the expansion of the universe to accelerate with time.**

Before we continue, it's important to think about dark matter and dark energy in the context of science. Strange as these ideas may seem, they have emerged from careful scientific study conducted in accordance with the hallmarks of science discussed in Chapter 3 (see Figure 3.21). Dark matter and dark energy were each proposed to exist because they seem the simplest ways to explain observed motions in the universe. They've each gained credibility because models of the universe that assume their existence make testable predictions and, at least so far, further observations have borne out some of those predictions. Even if we someday conclude that we were wrong to infer the existence of dark matter or dark energy, we will still need alternative explanations for the observations made to date. One way or the other, what we learn as we explore the mysteries of these unseen influences will forever change our view of the universe.

18.2 Evidence for Dark Matter

Scientific evidence for dark matter has been building for decades and is now at the point where dark matter seems almost indispensable to explaining the current structure of the universe. For that reason, we will devote most of this chapter to dark matter and its presumed role as the dominant source of gravity in our universe, saving further discussion of dark energy for the final section of the chapter.

◆ What is the evidence for dark matter in galaxies?

Several distinct lines of evidence point to the existence of dark matter, including observations of our own galaxy, of other galaxies, and of clusters of galaxies. Let's start with individual galaxies and then proceed on to clusters.

Dark Matter in the Milky Way In Chapter 15, we saw how the Sun's motion around the galaxy reveals the total amount of mass within its orbit. We can similarly use the orbital motion of any other star to measure the mass of the Milky Way within that star's orbit. In principle, we could determine the complete distribution of mass in the Milky Way by doing the same thing with the orbits of stars at every different distance from the galactic center.

In practice, interstellar dust obscures our view of disk stars more than a few thousand light-years away from us, making it very difficult to measure stellar velocities. However, radio waves penetrate this dust, and clouds of atomic hydrogen gas emit a spectral line at the radio wavelength of 21 centimeters [Section 15.2]. Measuring the Doppler shift of this 21-centimeter line tells us a cloud's velocity toward or away from us. With a little geometry, we can then determine the cloud's orbital speed.

We can summarize the results of these measurements with a diagram that plots the orbital speeds of objects in the galaxy against their orbital distances. As a simple example of how we construct such a diagram, sometimes called a *rotation curve,* consider how the rotation speed of a merry-go-round depends on the distance from its center. Every object on a merry-go-round goes around the center in the same amount of time (the rotation period of the merry-go-round). But because objects farther from the center move in larger circles, they must move at faster speeds. The speed is proportional to distance from the center, so the graph illustrating the relationship between speed and distance is a steadily rising straight line (Figure 18.1a).

In contrast, orbital speeds in our solar system *decrease* with distance from the Sun (Figure 18.1b). This drop-off in speed with distance occurs because virtually all the mass of the solar system is concentrated in the Sun. The gravitational force holding a planet in its orbit therefore decreases with distance from the Sun, and a smaller force means a lower orbital speed. Orbital speeds must drop similarly with distance in any other astronomical system that has its mass concentrated at its center.

▼ **FIGURE 18.1**

These graphs show how orbital speed depends on distance from the center in three different systems.

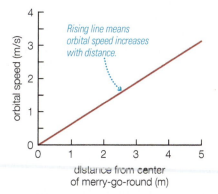

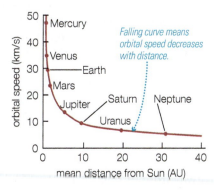

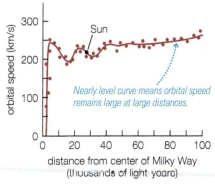

a A rotation curve for a merry-go-round is a rising straight line.

b The rotation curve for the planets in our solar system.

c The rotation curve for the Milky Way Galaxy. Dots represent actual data points for stars or gas clouds.

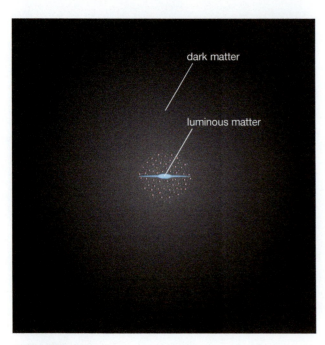

▲ FIGURE 18.2

The dark matter associated with the Milky Way occupies a much larger volume than the galaxy's luminous matter. The radius of this dark-matter halo may be 10 times as large as that of the galaxy's halo of stars.

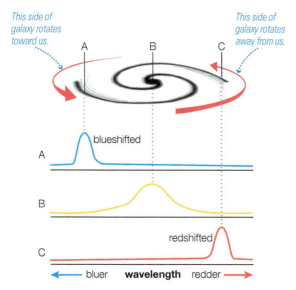

▲ FIGURE 18.3

Measuring the orbital speeds of gas in a spiral galaxy with the 21-centimeter line of atomic hydrogen. Blueshifted lines on the left side of the disk show how fast that side is moving toward us. Redshifted lines on the right side show how fast that side is moving away from us. (This diagram assumes that we first subtract a galaxy's average redshift, so that we can see the shifts that remain due to rotation.)

Figure 18.1c shows how orbital speed depends on distance in the Milky Way Galaxy. Each individual dot represents the orbital speed and distance from the galactic center of a particular star or gas cloud, and the curve running through the dots represents a best fit to the data. Notice that the orbital speeds remain approximately constant beyond the inner few thousand light-years, so most of the curve is relatively flat. Because this is so different from the declining orbital speeds found in our solar system, we conclude that most of the Milky Way's mass must *not* be concentrated at its center. Instead, the orbits of progressively more distant gas clouds must encircle more and more mass. The Sun's orbit encompasses about 100 billion solar masses, but a circle twice as large surrounds twice as much mass, and a larger circle surrounds even more mass.

Orbital speeds in the Milky Way remain high even very far from the center, indicating that a large amount of dark matter lies beyond our galaxy's visible regions.

To summarize, orbital speeds in the Milky Way imply that most of our galaxy's mass lies well beyond the orbit of our Sun. A more detailed analysis suggests that most of this mass is distributed throughout the spherical halo that surrounds the disk of our galaxy, extending to distances well beyond those at which we observe globular clusters and other halo stars. Moreover, the total amount of this mass is more than *10 times* the total mass of all the stars in the disk. Because we have detected very little radiation coming from this enormous amount of mass, it qualifies as dark matter. If we are interpreting the evidence correctly, the luminous part of the Milky Way's disk must be rather like the tip of an iceberg, marking only the center of a much larger clump of mass (Figure 18.2).

think about it Suppose we made a graph of orbital speeds and distances for the moons orbiting Jupiter. Which graph in Figure 18.1 would it most resemble? Why?

Dark Matter in Other Galaxies Other galaxies also seem to contain vast quantities of dark matter. We can determine the amount of dark matter in a galaxy by comparing the galaxy's mass to its luminosity. (More formally, astronomers calculate the galaxy's *mass-to-light ratio;* see Cosmic Calculations 18.1.) First, we use the galaxy's luminosity to estimate the amount of mass that the galaxy contains in the form of stars. Next, we determine the galaxy's total mass by applying the law of gravity to observations of the orbital velocities of stars and gas clouds. If this total mass is larger than the mass that we can attribute to stars, then we infer that the excess mass must be dark matter.

We can measure a galaxy's luminosity as long as we can determine its distance with one of the techniques discussed in Chapter 16. We simply point a telescope at the galaxy in question, measure its apparent brightness, and calculate its luminosity from its distance and the inverse square law for light [Section 12.1]. Measuring the galaxy's total mass requires measuring orbital speeds as far from the galaxy's center as possible. For spiral galaxies, this is usually done with radio observations of the 21-centimeter line from atomic hydrogen gas clouds. Doppler shifts of the 21-centimeter line tell us how fast a cloud is moving toward us or away from us (Figure 18.3).

Once we've measured orbital speeds and distances, we can make a graph similar to Figure 18.1c for any spiral galaxy. Figure 18.4 shows a few examples illustrating that, like the Milky Way, most other spiral galaxies also have orbital speeds that remain high even at great distances from their centers. Detailed analysis tells us that most spiral galaxies also have at least 10 times as much mass in dark matter as they do in stars, with most of this mass far out in their halos.

Elliptical galaxies contain very little atomic hydrogen gas and hence do not produce detectable 21-centimeter radio waves, so we generally measure masses of elliptical galaxies by observing the motions of their stars.

Orbital speeds in the outer regions of other galaxies indicate that they, too, harbor lots of dark matter.

When we compare spectral lines from different regions of an elliptical galaxy, we find that the speeds of the stars remain fairly constant as we look farther from the galaxy's center. Just as in spirals, we conclude that most of the matter in elliptical galaxies must lie beyond the distance where the light trails off and hence must be dark matter. The evidence for dark matter is even more convincing for cases in which we can measure the speeds of globular star clusters orbiting at large distances from the center of an elliptical galaxy. These measurements suggest that elliptical galaxies, like spirals, contain 10 or more times as much mass in dark matter as they do in the form of stars.

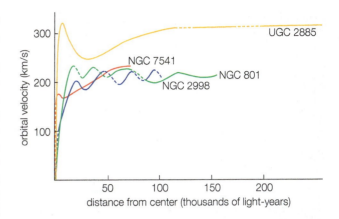

▲ **FIGURE 18.4**

Graphs of orbital speed versus distance for four spiral galaxies. In each galaxy, the orbital speeds remain nearly constant over a wide range of distances from the center, indicating that dark matter is common in spiral galaxies.

◆ What is the evidence for dark matter in clusters of galaxies?

Observations of galaxy clusters suggest that they contain 40 or more times as much mass in dark matter as in stars, which is a greater proportion than is found in individual galaxies. The evidence for dark matter in clusters comes from three different ways of measuring cluster masses: measuring the speeds of galaxies orbiting the center of the cluster, studying the X-ray emission from hot gas between the cluster's galaxies, and observing how the clusters bend light as *gravitational lenses*.

Orbits of Galaxies in Clusters The idea of dark matter is not particularly new. In the 1930s, astronomer Fritz Zwicky was already arguing that clusters of galaxies held enormous amounts of this mysterious stuff (Figure 18.5). Few of his colleagues paid attention, but later observations supported Zwicky's claims.

Zwicky was one of the first astronomers to think of galaxy clusters as huge swarms of galaxies bound together by gravity. It seemed natural to him that galaxies clumped closely together in space should all be orbiting one another, just like the stars in a star cluster. He therefore assumed that he could measure cluster masses by observing galaxy motions and applying Newton's laws of motion and gravitation.

Armed with a spectrograph, Zwicky measured the redshifts of the galaxies in a particular cluster and used these redshifts to calculate the speeds at which the individual galaxies are moving away from us. He determined the *recession speed* of the cluster as a whole—that is, the speed at which the expansion of the universe carries it away from us—by averaging the speeds of its individual galaxies. He then estimated the orbital speed of a galaxy around the cluster by subtracting this average velocity from the individual galaxy's velocity. Finally, he used these orbital speeds to estimate the cluster's mass and compared this mass to the cluster's luminosity.

To his surprise, Zwicky found that clusters of galaxies have much greater masses than their luminosities would suggest. That is, when he estimated the total mass of stars necessary to account for the overall luminosity of a cluster, he found that it was far less than the mass he measured by studying galaxy speeds. He concluded that most of the

▲ FIGURE 18.5

Fritz Zwicky, discoverer of dark matter in clusters of galaxies. Zwicky had an eccentric personality, but some of his ideas that seemed strange in the 1930s proved correct many decades later.

The orbits of galaxies in clusters tell us that galaxy clusters contain huge amounts of dark matter.

matter within these clusters must not be in the form of stars and instead must be almost entirely dark. Many astronomers disregarded Zwicky's result, believing that he must have done something wrong to arrive at such a strange result. However, more sophisticated measurements ultimately confirmed Zwicky's original finding.

Hot Gas in Clusters   A second method for measuring a cluster's mass relies on observing X rays from the hot gas that fills the space between its galaxies (Figure 18.6). This gas (sometimes called the *intracluster medium*) is so hot—typically tens of millions of degrees on the Kelvin scale—that it emits primarily X rays and does not appear in visible-light images. Measurements show that for large clusters, this hot gas can represent up to 7 times as much mass as is found in stars.

The hot gas can tell us about dark matter because its temperature depends on the total mass of the cluster. The gas in most clusters is nearly in a state of *gravitational equilibrium*—that is, the outward gas pressure balances gravity's inward pull [Section 11.1]. In this state of balance, the average kinetic energies of the gas particles are determined primarily by the strength of gravity and hence by the amount of mass within the cluster. Because the temperature of a gas reflects the average kinetic energies of its particles, the gas temperatures we measure with X-ray telescopes tell us the average speeds of the X ray–emitting particles. We can then use these particle speeds to determine the cluster's total mass.

Temperature measurements of hot gas also tell us the amount of dark matter in clusters, and give results that agree with those we infer from galaxy velocities.

The results obtained with this method agree well with the results found by studying the orbital motions of the cluster's galaxies. Even after we account for the mass of the hot gas, we find that the amount of dark matter in clusters of galaxies is at least 40 times the combined mass of the stars in

---

extraordinary claims   **Most of the Universe's Matter Is Dark**

Scientists always take a risk when they publish what they think are groundbreaking results. If their results turn out to be in error, their reputations may suffer. When it came to dark matter, the pioneers in its discovery risked their entire careers. A case in point is Fritz Zwicky and his extraordinary proclamations in the 1930s about dark matter in clusters of galaxies. Most of his colleagues considered him an eccentric who leapt to premature conclusions.

Another pioneer in the discovery of dark matter was Vera Rubin, an astronomer at the Carnegie Institution. Working in the 1960s, she became the first woman to observe under her own name at California's Palomar Observatory, then the largest telescope in the world. (Margaret Burbidge was permitted to observe at Palomar earlier but was required to apply for time under the name of her husband, also an astronomer.) Rubin first saw the gravitational signature of dark matter in spectra that she recorded of stars in the Andromeda Galaxy. She noticed that stars in the outskirts of Andromeda moved at surprisingly high speeds, suggesting a stronger gravitational attraction than the mass of the galaxy's stars alone could explain.

Working with a colleague, Kent Ford, Rubin went on to measure orbital speeds of hydrogen gas clouds in many other spiral galaxies (by studying Doppler shifts in the spectra of hydrogen gas) and discovered that the behavior seen in Andromeda is common. Although Rubin and Ford did not immediately recognize the significance of the results, they were soon arguing that the universe must contain substantial quantities of dark matter.

For a while, many other astronomers had trouble believing the results. Some astronomers suspected that the bright galaxies studied by Rubin and Ford were unusual for some reason. So Rubin and Ford went back to work, obtaining orbital measurements for fainter galaxies. By the 1980s, the evidence that Rubin, Ford, and other astronomers had compiled was so overwhelming that even the critics came around. The orbital speeds could not be explained by the visible mass acting under the known theory of gravity, so either something is amiss in the theory of gravity or vast amounts of dark matter must be present in spiral galaxies. The risks of the pioneers had paid off in a groundbreaking discovery.

Verdict: Strongly supported, though there is still some chance that the observations could indicate a problem with our theory of gravity rather than the existence of dark matter.

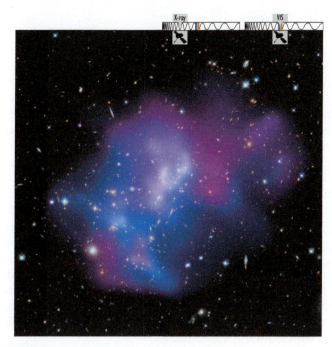

▲ FIGURE 18.6

A distant cluster of galaxies in both visible light and X-ray light. The visible-light photo shows the individual galaxies. The blue-violet overlay shows the X-ray emission from extremely hot gas in the cluster, with blue representing the hottest gas and violet representing cooler gas. Evidence for dark matter comes both from the observed motions of the visible galaxies and from the temperature of the hot gas. (The region shown is about 8 million light-years across.)

▲ FIGURE 18.7

This Hubble Space Telescope photo shows a galaxy cluster (known by the name CL0024+1654) acting as a gravitational lens. The yellow elliptical galaxies are cluster members. The small blue ovals (such as those indicated by the arrows) are multiple images of a single galaxy that lies almost directly behind the cluster's center. (The picture shows a region about 1.4 million light-years across.)

the cluster's galaxies. In other words, the gravity of dark matter seems to be binding the galaxies of a cluster together in much the same way gravity helps bind matter within individual galaxies.

think about it What would happen to a cluster of galaxies if you instantly removed all the dark matter without changing the velocities of the galaxies?

Gravitational Lensing The methods of measuring galaxy and cluster masses that we've discussed so far all ultimately rely on Newton's laws, including his universal law of gravitation. But can we trust these laws on such large size scales? One way to check is to measure masses in a different way. Today, astronomers can do this with observations of *gravitational lensing*.

Gravitational lensing occurs because masses distort spacetime—the "fabric" of the universe [Section 14.3]. Massive objects can therefore act as **gravitational lenses** that bend light beams passing nearby. This prediction of Einstein's general theory of relativity was first verified in 1919 during an eclipse of the Sun. Because the light-bending angle of a gravitational lens depends on the mass of the object doing the bending, we can measure the masses of objects by observing how strongly they distort light paths.

Figure 18.7 shows a striking example of how a cluster of galaxies can act as a gravitational lens. Notice the multiple blue ovals found at several positions around the central clump of yellow galaxies. These ovals all represent gravitationally lensed images of a single blue galaxy that lies almost directly behind the center of the cluster, at a much greater distance. We see multiple images because photons from the more distant galaxy do not follow straight paths to Earth. Instead, the cluster's gravity bends the photon paths, allowing light from the

Gravity's light-bending effects distort the images of galaxies lying behind a cluster, enabling us to measure the cluster's mass without relying on Newton's laws.

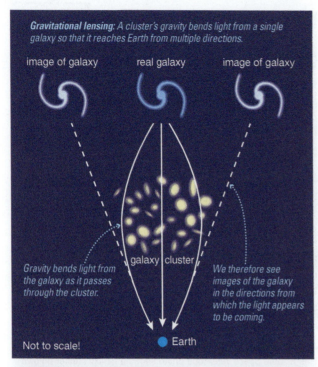

Gravitational lensing: A cluster's gravity bends light from a single galaxy so that it reaches Earth from multiple directions.

image of galaxy   real galaxy   image of galaxy

Gravity bends light from the galaxy as it passes through the cluster.

galaxy cluster

We therefore see images of the galaxy in the directions from which the light appears to be coming.

Not to scale!   ● Earth

Result: Through a telescope on Earth, we see multiple images of what is really a single galaxy.

▲ **FIGURE 18.8**

A cluster's powerful gravity bends light paths from background galaxies to Earth. If light arrives from several different directions, we see multiple images of the same galaxy.

galaxy to arrive at Earth from a few slightly different directions (Figure 18.8). Each alternative path produces a separate, distorted image of the blue galaxy.

Multiple images of a gravitationally lensed galaxy are rare. They occur only when a distant galaxy lies directly behind the lensing cluster. However, single distorted images of gravitationally lensed galaxies are quite common. Figure 18.9 shows a typical example. This picture shows numerous normal-looking galaxies and several arc-shaped galaxies. The oddly curved galaxies are not members of the cluster, nor are they really curved. They are normal galaxies lying far beyond the cluster whose images have been distorted by the cluster's gravity.

Careful analyses of the distorted images created by clusters enable us to measure cluster masses without using Newton's laws. Instead, Einstein's general theory of relativity tells us how massive these clusters must be to generate the observed distortions. Cluster masses derived in this way generally agree with those derived from galaxy velocities and X-ray temperatures. It is reassuring that the three different methods all indicate that clusters of galaxies hold substantial amounts of dark matter.

Cluster masses measured through gravitational lensing agree with those measured from galaxy velocities and gas temperatures.

◆ Does dark matter really exist?

Astronomers have made a strong case for the existence of dark matter, but is it possible that there's a completely different explanation for the observations we've discussed? Addressing this question gives us a chance to see how science progresses.

All the evidence for dark matter rests on our understanding of gravity. For individual galaxies, the case for dark matter rests primarily on applying Newton's laws of motion and gravity to observations of the orbital speeds of stars and gas clouds. We've used the same laws to make the case for dark matter in clusters, along with additional evidence based on gravitational lensing predicted by Einstein's general theory of relativity. It therefore seems that one of the following must be true:

1. Dark matter really exists, and we are observing the effects of its gravitational attraction.

2. There is something wrong with our understanding of gravity that is causing us to mistakenly infer the existence of dark matter.

We cannot completely rule out the second possibility, but most astronomers consider it very unlikely. Newton's laws of motion and gravity are among the most trustworthy tools in science. We have used them time and again to measure masses of celestial objects from their orbital properties. We found the masses of Earth and the Sun by applying Newton's version of Kepler's third law to objects that orbit them [Section 4.4]. We used this same law to calculate the masses of stars in binary star systems, revealing the general relationships between the masses of stars and their outward appearances. Newton's laws have also told us the masses of things we can't see directly, such as the masses of orbiting neutron stars in X-ray binaries and of black holes in active galactic nuclei. Einstein's general theory of relativity likewise stands on solid ground, having been repeatedly tested and verified to high precision in many observations and experiments. We therefore have good reason to trust our current understanding of gravity.

Moreover, many scientists have made valiant efforts to come up with alternative models of gravity that could account for the observations without invoking dark matter. So far, no one has succeeded in doing so in a way that can also explain the many other observations accounted for by our current theories of gravity. Meanwhile, astronomers keep making observations that are difficult to explain without dark matter. For example, in observations of colliding galaxy clusters, most of the mass detected by gravitational lensing is *not* in the same place as the hot gas, even though the hot gas is several times more massive than the cluster's stars (Figure 18.10). This finding is at odds with alternative models of gravity, which predict that the hot gas should be doing most of the gravitational lensing.

Either dark matter exists or our current understanding of gravity is incorrect.

In essence, our high level of confidence in our current understanding of gravity, combined with observations that seem consistent with dark matter but not with alternative hypotheses, gives us high confidence that dark matter really exists. While we should always keep an open mind about the possibility of future changes in our understanding, we will proceed for now under the assumption that dark matter is real.

think about it Should the fact that we have three different ways of measuring cluster masses give us greater confidence that we really do understand gravity and that dark matter really does exist? Why or why not?

◆ What might dark matter be made of?

Although we do not yet know exactly what dark matter is, there would seem to be two basic possibilities: (1) It could be made of *ordinary matter* (also called *baryonic matter*), meaning the familiar type of matter built from protons, neutrons, and electrons, but in forms too dark for us to detect with current technology, or (2) it could be made of *exotic matter,* meaning particle types that are different from those in ordinary atoms and that are dark because they do not interact with light at all.

A first step in distinguishing between the two possibilities is to measure how much dark matter is out there. When discussing the matter content of the universe as a whole, astronomers usually focus on density rather than mass. That is, they take the total amount of some type of matter (such as stars, gas, or dark matter) found in a large but typical volume of space and divide by the volume to determine the average density of this type of matter. These densities are then stated as percentages of the *critical density*—the density of mass-energy needed to make the geometry of the universe flat [Section 17.3]. The critical density is quite small, equivalent to only

▲ **FIGURE 18.9**

Hubble Space Telescope photo of the cluster Abell 383. The thin, elongated galaxies are images of background galaxies distorted by the cluster's gravity. By measuring these distortions, astronomers can determine the total amount of mass in the cluster. (The region pictured is about 1 million light-years across.)

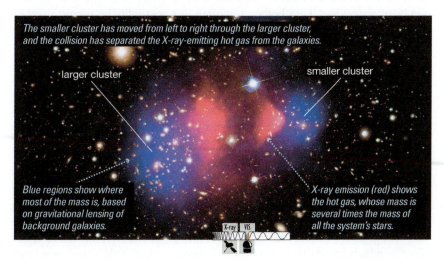

The smaller cluster has moved from left to right through the larger cluster, and the collision has separated the X-ray-emitting hot gas from the galaxies.

larger cluster

smaller cluster

Blue regions show where most of the mass is, based on gravitational lensing of background galaxies.

X-ray emission (red) shows the hot gas, whose mass is several times the mass of all the system's stars.

◀ **FIGURE 18.10**

Observations of the Bullet Cluster show strong evidence for dark matter. The Bullet Cluster actually consists of two galaxy clusters—the smaller one is emerging from a high-speed collision with the larger one. A map of the system's overall mass (blue) made from gravitational lensing observations does *not* line up with X-ray observations (red) showing the location of the system's hot gas. This fact is difficult to explain without dark matter because the gas contains several times as much mass as all the cluster's stars combined. However, it is easy to explain if dark matter exists: The collision has simply stripped the hot gas away from the dark matter on which it was previously centered.

10^{-29} gram per cubic centimeter—roughly what we'd find with an average of just a few hydrogen atoms in a volume the size of a closet. Still, the critical density is much larger than the observed density of matter in the universe.

The observations of galaxies and clusters that we have discussed indicate that the total amount of matter in stars is only about 0.5% of the critical density, while dark matter represents about one-quarter of the critical density. Clearly, there is a lot of dark matter that needs to be accounted for.

Ordinary Matter: Not Enough
Could all this dark matter simply be ordinary matter in some hard-to-observe form? After all, matter doesn't necessarily need to be exotic to be dark. Astronomers consider matter to be "dark" as long as it is too dim for us to see at the great distances of the halo of our galaxy or beyond. Your body is dark matter, because our telescopes could not detect you if you were somehow flung into the halo of our galaxy. Planets, the "failed stars" known as brown dwarfs [Section 13.1], and even some faint red main-sequence stars also qualify as dark matter, because they are too dim for current telescopes to see in the halo.

However, calculations made with the Big Bang model allow scientists to place limits on the total amount of ordinary matter in the universe. Recall that, during the era of nucleosynthesis, protons and neutrons first fused to make deuterium nuclei (nuclei consisting of one proton and one neutron) and these then fused into helium [Section 17.2]. The fact that some deuterium nuclei still exist in the universe indicates that this process stopped before all the deuterium nuclei were used up. The amount of deuterium in the universe today therefore tells us about the density of protons and neutrons (ordinary matter) during the era of nucleosynthesis: The higher the density, the more efficiently fusion would have proceeded. A higher density in the early universe would have therefore left less deuterium in the universe today, and a lower density would have left more deuterium.

> Calculations based on the Big Bang model indicate that ordinary matter cannot account for most of the dark matter.

Calculations based on the observed deuterium abundance indicate that the overall density of ordinary matter in the universe is around 5% of the critical density (Figure 18.11). Detailed studies of temperature patterns in the cosmic microwave background (see Figures 17.9 and 17.16) give the same result. Because this 5% is much too small to account for the one-quarter of the critical density that is dark, we conclude that most of the dark matter cannot be made from ordinary matter.

Exotic Matter: The Leading Hypothesis
We are left with the idea that dark matter is made of exotic particles, and probably of a type that has not yet been discovered. Let's begin to explore this possibility by taking another look at a type of exotic particle that we first encountered in connection with nuclear fusion in the Sun: neutrinos [Section 11.2]. Neutrinos are dark by nature because they have no electrical charge and cannot emit electromagnetic radiation of any kind. Moreover, they are never bound together with charged particles in the way that neutrons are bound in atomic nuclei, so their presence cannot be revealed by associated light-emitting particles. In fact, neutrinos interact with other forms of matter through only two of the four forces: gravity and the *weak force* [Section 17.1]. For this reason, neutrinos are said to be *weakly interacting particles*.

The dark matter in galaxies cannot be made of neutrinos, because these very-low-mass particles travel through the universe at enormous speeds and can easily escape a galaxy's gravitational pull. But what if other weakly interacting particles exist that are similar to neutrinos but considerably heavier? They, too, would evade direct detection, but they would move more slowly, which means that their mutual gravity could hold together a large

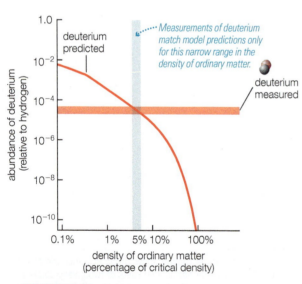

▲ **FIGURE 18.11**

This graph shows how the measured abundance of deuterium leads to the conclusion that the density of ordinary matter is about 5% of the critical density. The horizontal band shows the measured abundance of deuterium. The curve shows predictions of the deuterium abundance based on the Big Bang theory, and how it depends on the density of ordinary matter in the universe. Notice that the prediction matches up with the measurements only in the gray vertical strip, which represents a density of about 5% of the critical density.

collection of them. Such hypothetical particles are called **weakly interacting massive particles,** or **WIMPs** for short. Note that they are subatomic particles, so the "massive" in their name is relative—they are massive only in comparison to lightweight particles like neutrinos. Such particles could make up most of the mass of a galaxy or a cluster of galaxies, but they would be completely invisible in all wavelengths of light. Most astronomers now consider it likely that WIMPs make up the majority of dark matter, and hence the majority of all matter in the universe.

Scientists suspect that dark matter consists of weakly interacting particles that are like neutrinos but more massive.

This hypothesis would also explain why dark matter seems to be distributed throughout spiral galaxy halos rather than concentrated in flattened disks like the visible matter. Recall that galaxies are thought to have formed as gravity pulled together matter in regions of slightly enhanced density in the early universe [Section 16.3]. This matter would have consisted mostly of dark matter mixed with some ordinary hydrogen and helium gas. The ordinary gas could collapse to form a rotating disk because individual gas particles could lose orbital energy: Collisions among many gas particles can convert some of their orbital energy into radiative energy that escapes from the galaxy in the form of photons. In contrast, WIMPs cannot produce photons, and they rarely interact and exchange energy with other particles. As the gas collapsed to form a disk, WIMPs would therefore have remained stuck in orbits far out in the galactic halo—just where most dark matter seems to be located.

Searching for Dark Matter Particles   The case for the existence of WIMPs seems fairly strong but is still circumstantial. Detecting the particles directly would be much more convincing, and physicists are currently searching for them in two different ways. The first and most direct way is with detectors that can potentially capture WIMPs from space. Because these particles are thought to interact only very weakly, the search requires building large, sensitive detectors deep underground, where they are shielded from other particles from space. As of 2016, these detectors have not yet provided confirmed, direct detections of dark matter.

The search for exotic particles of dark matter is currently under way, and scientists are optimistic that a discovery will soon be made.

The second way scientists are currently searching for dark matter particles is with particle accelerators, and in particular with the Large Hadron Collider (see Figure 17.2). None of the particles found as of 2016 has the characteristics of a WIMP, but scientists are optimistic that a discovery may be made within the next few years.

18.3 Structure Formation

The nature of dark matter remains enigmatic, but we are rapidly learning more about its role in the universe. We've already seen that dark matter appears to be the primary source of gravity in galaxies and galaxy clusters. We therefore suspect that the gravitational attraction of dark matter is what formed structures like galaxies and clusters of galaxies in the first place.

◆ What is the role of dark matter in galaxy formation?

Stars, galaxies, and clusters of galaxies are all *gravitationally bound systems*—their gravity is strong enough to hold them together. In most of the gravitationally bound systems we have discussed so far, gravity has completely overwhelmed the expansion of the universe. That is, while the universe as a whole is expanding, space is *not* expanding within our solar system, our galaxy, or our Local Group of galaxies.

Recall that our best model of how galaxies formed, outlined in Section 16.3, envisions them growing from slight density enhancements that were present in the very early universe. During the first few million years after the Big Bang, the universe expanded everywhere. Gradually, the stronger gravity in regions of enhanced density pulled in matter until these regions stopped expanding and became protogalactic clouds, even as the universe as a whole continued to expand. As we've already discussed, dark matter consisting of WIMPs would have remained in the galactic halos because of its inability to radiate away orbital energy, while ordinary gas collapsed to form stars and galactic disks.

think about it State whether each of the following is a gravitationally bound system, and state your reasoning: (a) Earth; (b) a hurricane on Earth; (c) the Orion Nebula; (d) a supernova.

Galaxy clusters probably formed similarly. Early on, all the galaxies that would eventually constitute a cluster were flying apart with the expansion of the universe. The gravity of the dark matter associated with the cluster eventually reversed their trajectories, so the galaxies ultimately fell back inward and began orbiting each other with random orientations, much like the stars in the halo of our galaxy.

The gravity due to dark matter was probably the main force that caused protogalactic clouds to become galaxies and galaxies to group into clusters.

Some clusters apparently have not yet finished forming, because their immense gravity is still drawing in new galaxies. For example, the relatively nearby Virgo Cluster of galaxies (about 60 million light-years away) appears to be tugging on the Milky Way and other galaxies of the Local Group. This gravitational tug may eventually reverse the motion of the Local Group away from the Virgo Cluster, and our galaxy and its companions will join the cluster. On even larger scales, clusters themselves seem to be tugging on one another, hinting that they might be parts of even bigger gravitationally bound systems, called **superclusters,** that are still in the early stages of formation (Figure 18.12). Clearly, the gravity of dark matter is having an enormous and ongoing influence on large structures in the universe.

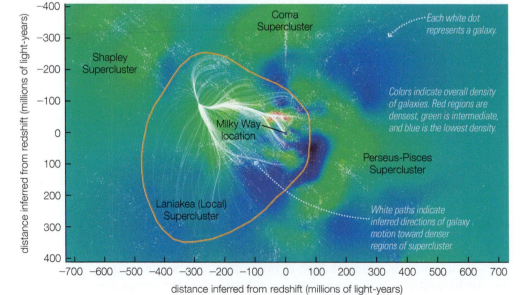

► **FIGURE 18.12**

This diagram shows how galaxies move within the Local Supercluster (or Laniakea), which is the region within the orange line. White dots represent locations of individual galaxies, with distances from the Milky Way inferred from Hubble's law. Galaxies flow toward denser regions of the supercluster along the white lines, which have been inferred from a combination of observations and modeling.

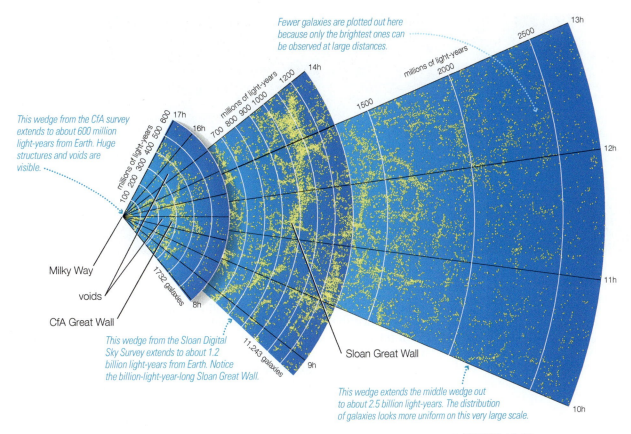

Fewer galaxies are plotted out here because only the brightest ones can be observed at large distances.

This wedge from the CfA survey extends to about 600 million light-years from Earth. Huge structures and voids are visible.

Milky Way

voids

CfA Great Wall

1732 galaxies

This wedge from the Sloan Digital Sky Survey extends to about 1.2 billion light-years from Earth. Notice the billion-light-year-long Sloan Great Wall.

11,243 galaxies

Sloan Great Wall

This wedge extends the middle wedge out to about 2.5 billion light-years. The distribution of galaxies looks more uniform on this very large scale.

▲ **FIGURE 18.13**

Each of these three wedges shows a "slice" of the universe extending outward from our own Milky Way Galaxy. The dots represent galaxies, shown at their measured distances from Earth. We see that galaxies trace out long chains and sheets surrounded by huge voids containing very few galaxies. (The wedges are shown flat but actually are a few angular degrees in thickness; the CfA wedge at left does not actually line up with the two Sloan wedges.)

◆ What are the largest structures in the universe?

Over the past few decades, astronomers have measured the redshifts of millions of galaxies. These redshifts can then be converted to distances with Hubble's law [Section 16.2], thereby allowing astronomers to make three-dimensional maps of the distribution of galaxies in space. Such maps have revealed the existence of **large-scale structures** much vaster than clusters or superclusters of galaxies.

Figure 18.13 shows the distribution of galaxies in three slices of the universe, each extending farther out in distance. Our Milky Way Galaxy is located at the vertex at the far left, and each dot represents an entire galaxy of stars. The slice at the left comes from one of the first surveys of large-scale structure, performed at the Harvard-Smithsonian Center for Astrophysics (CfA) in the 1980s. This map showed that galaxies are not scattered randomly through space but are instead arranged in huge chains and sheets that span many millions of light-years. Clusters of galaxies are located at the intersections of these chains. Between these chains and sheets of galaxies lie giant empty regions called **voids.** The other two slices show data from the Sloan Digital Sky Survey, which measured redshifts for more than a million galaxies spread across about one-fourth of the sky.

Galaxies appear to be arranged in immense structures hundreds of millions of light-years across.

Some of the structures in these pictures are amazingly large. The so-called Sloan Great Wall, clearly visible in the center slice, extends more than 1 billion light-years from end to end. Immense structures such as these apparently have not yet collapsed into randomly orbiting, gravitationally bound systems. The universe may still be growing structures on

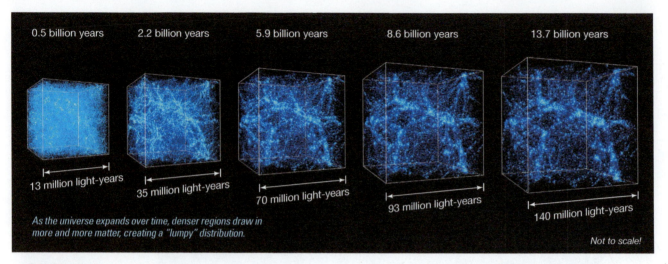

| 0.5 billion years | 2.2 billion years | 5.9 billion years | 8.6 billion years | 13.7 billion years |

13 million light-years

35 million light-years

70 million light-years

93 million light-years

140 million light-years

As the universe expands over time, denser regions draw in more and more matter, creating a "lumpy" distribution.

Not to scale!

▲ **FIGURE 18.14**

Frames from a supercomputer simulation of structure formation. The five boxes depict the development of a cubical region that is now 140 million light-years across. The labels above the boxes give the age of the universe, and the labels below give the size of the box as it expands with time. Notice that the distribution of matter is only slightly lumpy when the universe is young (left frame). Structures grow more pronounced with time as the densest lumps draw in more and more matter.

these very large scales. However, there seems to be a limit to the size of the largest structures. If you look closely at the rightmost slice in Figure 18.13, you'll notice that the overall distribution of galaxies appears nearly uniform on scales larger than about a billion light-years. In other words, on very large scales the universe looks much the same everywhere, in agreement with what we expect from the *Cosmological Principle* [Section 16.2].

Why is gravity collecting matter on such enormous scales? Just as we suspect that galaxies formed from regions of slightly enhanced density in the early universe, we suspect that these larger structures were also regions of enhanced density. Galaxies, clusters, superclusters, and the Sloan Great Wall probably all started as mildly high-density regions of different sizes. The voids in the distribution of galaxies probably started as mildly low-density regions.

According to this model of structure formation, the structures we see in today's universe should mirror the original distribution of dark matter very early in time. Super-computer models of structure formation in the universe can now simulate the growth of gal-

The structure we see in today's universe probably mirrors the distribution of dark matter when the universe was very young.

axies, clusters, and larger structures from tiny density enhancements as the universe evolves (Figure 18.14). The results of these models look remarkably similar to the slices of the universe in Figure 18.13, bolstering our confidence in this scenario. Moreover, the patterns of mass distribution are consistent with the patterns of density enhancements revealed in maps of the cosmic microwave background. Overall, we now have a basic picture of how galaxies and large-scale structures formed in the universe, perhaps starting from quantum fluctuations that occurred when the universe was a tiny fraction of a second old [Section 17.3].

18.4 Dark Energy and the Fate of the Universe

Two competing processes have governed the large-scale development of the universe: (1) the ongoing expansion that began in the Big Bang, which tends to drive galaxies apart from one another, and (2) the gravitational attraction of matter in the universe, which assembles galaxies and larger-scale structures around the density enhancements that emerged from the Big Bang.

These ideas naturally lead us to one of the ultimate questions in astronomy: How will the universe end? After Edwin Hubble discovered the expansion of the universe, astronomers generally assumed that the end would be like one of the two fates in Robert Frost's poem. If gravity were strong enough, the expansion would someday halt and reverse; the universe would then begin collapsing and heating back up, eventually ending in a fiery and cataclysmic crunch. Alternatively, if the total strength of gravity were too weak, gravity would never slow the expansion enough for it to halt and reverse, leading to an icy end in which the universe would grow ever colder as its galaxies moved ever farther apart. Much to astronomers' surprise, more than two decades of observation now show that the expansion is not slowing at all, but instead is accelerating with time. As we've discussed, this implies the existence of some repulsive force produced by the mysterious form of energy that we have come to call *dark energy*.

Some say the world will end in fire,
Some say in ice.
From what I've tasted of desire
I hold with those who favor fire.
But if it had to perish twice,
I think I know enough of hate
To say that for destruction ice
Is also great
And would suffice.

—Robert Frost, "Fire and Ice"

◆ What is the evidence for an accelerating expansion?

Astronomers can learn how the expansion rate of the universe has changed with time by carefully measuring the recession speeds and distances of very distant galaxies. Because those measurements show how fast galaxies were moving when the universe was much younger, careful analysis of their speeds can tell us whether galaxy speeds have slowed down, held steady, or sped up over time.

Four Expansion Models To understand the evidence that has led astronomers to conclude that the expansion is accelerating, let's consider how different amounts of gravity and repulsion cause the expansion rate to change with time according to the following four general models (Figure 18.15):

- A **recollapsing universe.** In the case of extremely strong gravitational attraction and no repulsive force, the expansion would

▼ **FIGURE 18.15**

Four general models for how the universal expansion rate might change with time. Each diagram shows how the size of a circular slice of the universe changes with time in a particular model. The slices are the same size at the present time, marked by the red line, but the models make different predictions about the sizes of the slices in the past and future.

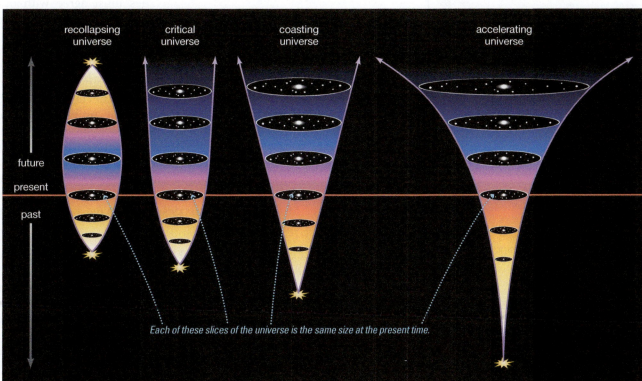

recollapsing universe   critical universe   coasting universe   accelerating universe

future
present
past

Each of these slices of the universe is the same size at the present time.

continually slow down with time and eventually would stop entirely and then reverse. Galaxies would come crashing back together, and the universe would end in a fiery "Big Crunch." We call this a *recollapsing* universe, because the final state, with all matter collapsed together, would look much like the state in which the universe began in the Big Bang.

- A **critical universe.** In the case of gravitational attraction that was not quite strong enough to reverse the expansion in the absence of a repulsive force, the expansion would decelerate forever, leading to a universe that would never collapse but would expand ever more slowly as time progressed. We call this a *critical* universe, because calculations show that it is what we would expect if the total density of the universe were the critical density and only matter (and not dark energy) contributed to this density.

- A **coasting universe.** In the case of weak gravitational attraction and no repulsive force, galaxies would always move apart at approximately the speeds they have today. We call this a *coasting* universe, because it is what we would find if no forces acted to change the expansion rate, much as a spaceship can coast through space at constant speed if no forces act to slow it down or speed it up.

- An **accelerating universe.** In the case of a repulsive force strong enough to overpower gravity, the expansion would *accelerate* with time, causing galaxies to recede from one another with ever-increasing speed.

Note that each general model leads to a different age for the universe today. In all four models, the size of a particular region of space and the expansion rate of the universe are the same for the present (indicated by the horizontal red line in Figure 18.15), because those values must agree with our measurements for the average distance between galaxies today and for Hubble's constant today.

> The expansion rate of the universe depends on the balance between gravity, which acts to slow the expansion, and dark energy, which acts to accelerate it.

However, the four models each extend different lengths into the past. The coasting model assumes that the expansion rate never changes, and its starting point therefore indicates the age of the universe that we would infer from Hubble's constant alone. The recollapsing and critical models both give younger ages for the universe, because these models assume that galaxy speeds would have been faster in the past, so galaxies would have reached their current distances in less total time. The accelerating model leads to an older age, because galaxy speeds would have been slower in the past, so galaxies would have required more time to reach their current distances.

Evidence for Acceleration Figure 18.16 shows the key evidence that has led astronomers to favor the accelerating model. The four solid curves show how the four general models predict that average distance between galaxies should have changed with time. The curves for the accelerating, coasting, and critical universes always continue upward as time increases, because in these cases the universe is always expanding. The steeper the slope, the faster the expansion. In the recollapsing case, the curve begins on an upward slope but eventually turns around and declines as the universe contracts. All the curves

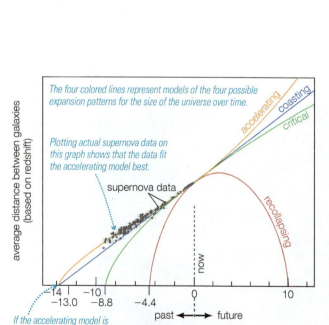

The four colored lines represent models of the four possible expansion patterns for the size of the universe over time.

Plotting actual supernova data on this graph shows that the data fit the accelerating model best.

supernova data

accelerating

coasting

critical

recollapsing

average distance between galaxies (based on redshift)

now

−14   −10   0   10
  −13.0 −8.8   −4.4

past ←——→ future

If the accelerating model is correct, then the universe must be nearly 14 billion years old.

time in billions of years (lookback times for supernovae based on apparent brightness)

▲ **FIGURE 18.16**

Data from white dwarf supernovae are shown, along with four possible models for the expansion of the universe. Each curve shows how the average distance between galaxies changes with time for a particular model. A rising curve means that the universe is expanding, and a falling curve means that the universe is contracting. Notice that the supernova data fit the accelerating universe better than the other models.

pass through the same point and have the same slope at the moment labeled "now," because the current separation between galaxies and the current expansion rate in each case must agree with observations of the present-day universe.

see it for yourself Toss a ball in the air, and observe how it rises and falls. Then make a graph to illustrate your observations, with time on the horizontal axis and height on the vertical axis. Which universe model does your graph most resemble? What is the reason for that resemblance? How would your graph look different if Earth's gravity were not as strong? Would the time for the ball to rise and fall be longer or shorter?

The black dots in Figure 18.16 show actual data, which are based on distance measurements using white dwarf supernovae as standard candles [Section 16.2]. (The horizontal line through each dot indicates the range of uncertainty in the measured lookback time.) Although there is some scatter in the data points, they clearly fit the curve for the accelerating model better than any of the other models. In other words, the observations agree best with a model of the universe in which the expansion is now accelerating with time.

Distances measured to faraway white dwarf supernovae indicate that the expansion of the universe is speeding up.

The Nature of Dark Energy The acceleration of the expansion clearly implies the existence of some force that acts to push galaxies apart, and the source of this force has been dubbed *dark energy*. Keep in mind, however, that scientists have little idea of what the nature of dark energy might actually be. None of the four known forces in nature could provide a force to oppose gravity, and while some theories of fundamental physics suggest ways in which energy could fit the bill, no known type of energy produces the right amount of acceleration.

Continued observations of distant supernovae have the potential to tell us exactly how large an effect dark energy has had throughout cosmic history and whether the strength of this effect has changed with time. Already there are some intriguing hints. For example, it appears that the expansion was *not* accelerating immediately after the Big Bang, indicating that gravity was strong enough to slow the expansion during the universe's first few billion years.

We still do not know what dark energy is, but observations of how it changes with time might provide clues.

Apparently, it took a few billion years for dark energy to become dominant enough to cause acceleration. (The curve for the accelerating model in Figure 18.16 shows this scenario.) Interestingly, this type of behavior is consistent with an idea that Einstein once introduced but later disavowed in his general theory of relativity, leading some scientists to suggest that dark energy might successfully be described by a term in Einstein's equations that describe gravity (see the Special Topic, page 484). Nevertheless, even if this idea turns out to be correct, we remain a long way from an actual understanding of dark energy's nature.

◆ Why is flat geometry evidence for dark energy?

The evidence for the existence of dark energy provided by observations of an accelerating expansion seems quite strong, but it is important to remember that it is based on measurements of distances to white dwarf supernovae. While we have good reason to think that these supernovae make reliable standard candles, having just a single source of evidence would be cause for at least some concern. Fortunately, a second and independent line of evidence also supports the existence of dark energy.

Flatness and Dark Energy   Recall that Einstein's general theory of relativity tells us that the overall geometry of the universe can take one of three general forms—spherical, flat, or saddle shaped (see Figure 17.14)—and that careful observations of the cosmic microwave background provide strong evidence that the actual geometry is flat (see Figure 17.16). As we discussed in Chapter 17, a flat geometry implies that the total density of *matter plus energy* in the universe is exactly equal to the critical density.

Now, recall that observations of dark matter indicate that the total density of matter amounts to only about 30% of the critical density. In

> The universe's flat geometry implies a total density equal to the critical density. Matter alone falls far short of this density, so most of the universe must be composed of energy.

that case, the remaining 70% of the universe's density must be in the form of energy. Tellingly, the amount of dark energy required to explain the observed acceleration of the expansion also is about 70% of the critical density. The startling conclusion: About 70% of the total mass-energy of the universe takes the form of dark energy.

Inventory of the Universe   We began this chapter by noting that astronomers today must admit the embarrassing fact that we do not yet know what most of the universe is made of. It appears to be made of things we call dark matter and dark energy, but we do not yet know the true nature of either one. Nevertheless, the observations we have discussed allow us to make quantitative statements about our ignorance. According to the model that best explains the observed temperature patterns in the cosmic microwave background, the total density of matter plus energy in the universe is equal to the critical density, and the universe's mass-energy has the following components:

---

special topic | **Einstein's Greatest Blunder**

Shortly after Einstein completed his general theory of relativity in 1915, he found that it predicted that the universe could not be standing still: The mutual gravitational attraction of all the matter would make the universe collapse. Because Einstein thought at the time that the universe should be eternal and static, he decided to alter his equations. In essence, he inserted a "fudge factor" called the *cosmological constant* that acted as a repulsive force to counteract the attractive force of gravity.

Had he not been so convinced that the universe should be standing still, Einstein might instead have come up with the correct explanation for why the universe is not collapsing: because it is still expanding from the event of its birth. After Hubble discovered universal expansion, Einstein supposedly called his invention of the cosmological constant "the greatest blunder" of his career.

Now that observations of very distant galaxies (using white dwarf supernovae as standard candles) have shown that the universe's expansion is accelerating, Einstein's idea of a universal repulsive force doesn't seem so far-fetched. In fact, observations to date are consistent with the idea that dark energy has properties virtually identical to those that Einstein originally proposed for the cosmological constant. In particular, the amount of dark energy in each volume of space seems to remain unchanged while the universe expands, as if the vacuum of space itself was constantly rippling with energy—which is just what the cosmological constant does in Einstein's equations. We'll need more measurements to know for sure, but it is beginning to seem that Einstein's greatest blunder may not have been a blunder after all.

- Ordinary matter (made up of protons, neutrons, electrons) makes up about 5% of the total mass-energy of the universe. Note that this model prediction agrees with what we find from observations of deuterium in the universe. Some of this matter is in the form of stars (about 0.5% of the universe's mass-energy). The rest is presumed to be in the form of intergalactic gas, such as the hot gas found in galaxy clusters.

- Some form of exotic dark matter—most likely weakly interacting massive particles (WIMPs)—makes up about 27% of the mass-energy of the universe, in close agreement with what we infer from measurements of the masses of clusters of galaxies.

- Dark energy makes up the remaining 68% of the mass-energy of the universe, accounting both for the observed acceleration of the expansion and for the pattern of temperatures in the cosmic microwave background.

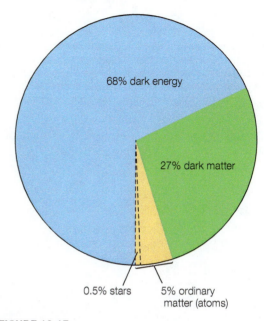

▲ **FIGURE 18.17**
This pie chart shows the proportion of each of the major components of matter and energy in the universe, based on current evidence.

The composition of the universe: 68% dark energy, 27% exotic dark matter, 5% ordinary matter.

Figure 18.17 summarizes this inventory of the universe with a pie chart, and Figure 18.18 summarizes the evidence we have discussed for the existence of dark matter and dark energy. We may not yet know what either dark matter or dark energy actually is, but our measurements of how much matter and energy may be out there are becoming more precise.

The Age of the Universe Models that explain the temperature variations in the cosmic microwave background not only give us an inventory of the universe but also make precise predictions about the age of the universe. According to the model that gives the best agreement to the data (the same model used for the inventory above), the age of the universe is about 13.8 billion years, with an uncertainty of about 0.1 billion years (100 million years). That is why, throughout this book, we have said that the universe is "about 14 billion years old." Note that this age is in good agreement with what we infer from Hubble's constant and observed changes in the expansion, and also agrees well with the fact that the oldest stars in the universe appear to be about 13 billion years old.

◆ What is the fate of the universe?

We are now ready to return to the question of the fate of the universe. If we think in terms of Robert Frost's poetry at the beginning of this section, the recollapsing universe is the only one of our four possible expansion models that has an end in fire, and the data do not fit that model. Therefore, it seems that the universe is doomed to expand forever, its galaxies receding ever more quickly into an icy, empty future. The end, it would seem, is more likely to be like that described in the poetry excerpt from astronomer Rebecca Elson (1960–1999) to the right.

Into its own beyond,
Till it exhausts itself and lies down cold,
Its last star going out.

—Rebecca Elson, from "Let There Always Be Light"

think about it Do you think that one of the possible fates (fire or ice) is preferable to the other? Why or why not?

The Next 10^{100} Years What will happen in the distant future if the universe continues to expand forever? We can use our current understanding of physics to hypothesize about the answer.

Scientists suspect that most of the matter in the universe is *dark matter* we cannot see, and that the expansion of the universe is accelerating because of a *dark energy* we cannot directly detect. Both dark matter and dark energy have been proposed to exist because they bring our models of the universe into better agreement with observations, in accordance with the process of science. This figure presents some of the evidence supporting the existence of dark matter and dark energy.

1 Dark Matter in Galaxies: Applying Newton's laws of gravity and motion to the orbital speeds of stars and gas clouds suggests that galaxies contain much more matter than we observe in the form of stars and glowing gas.

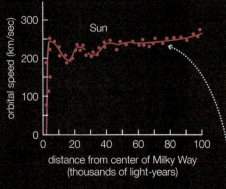

Orbital speeds of stars and gas clouds remain high even quite far from our galaxy's center. . .

dark matter

luminous matter

. . . indicating that the visible portion of our galaxy lies at the center of a much larger volume of dark matter.

HALLMARK OF SCIENCE   **A scientific model must seek explanations for observed phenomena that rely solely on natural causes.** Orbital motions within galaxies demand a natural explanation, which is why scientists proposed the existence of dark matter.

2 Dark Matter in Clusters: Further evidence for dark matter comes from studying galaxy clusters. Observations of galaxy motions, hot gas, and gravitational lensing all suggest that galaxy clusters contain far more matter than we can directly observe in the form of stars and gas.

This cluster of galaxies acts as a gravitational lens to bend light from a single galaxy behind it into the multiple blue shapes in this photo. The amount of bending allows astronomers to calculate the total amount of matter in the cluster.

HALLMARK OF SCIENCE   **Science progresses through creation and testing of models of nature that explain the observations as simply as possible.** Dark matter accounts for our observations of galaxy clusters more simply than alternative hypotheses.

(3) **Structure Formation:** If dark matter really is the dominant source of gravity in the universe, then its gravitational force must have been what assembled galaxies and galaxy clusters in the first place. We can test this prediction using supercomputers to model the formation of large-scale structures both with and without dark matter. Models with dark matter provide a better match to what we observe in the real universe.

(4) **Universal Expansion and Dark Energy:** The expansion of a universe consisting primarily of dark matter would slow down over time because of gravity, but observations have shown that the expansion is actually speeding up. Scientists hypothesize that a mysterious *dark energy* is causing the expansion to accelerate. Models that include both dark matter and dark energy agree more closely with observations of distant supernovae and the cosmic microwave background than models containing dark matter alone.

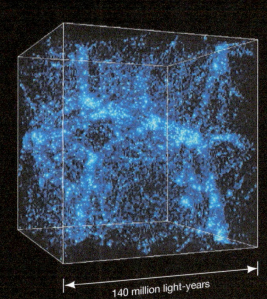

140 million light-years

Supercomputer models in which dark matter is the dominant source of gravity show galaxies organized into strings and sheets similar in size and shape to those we observe in the real universe.

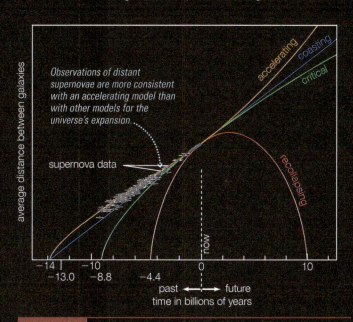

average distance between galaxies

Observations of distant supernovae are more consistent with an accelerating model than with other models for the universe's expansion.

supernova data

accelerating

coasting

critical

recollapsing

now

| −14 | −10 | −4.4 | 0 | 10 |
| −13.0 | −8.8 | | | |

past ←→ future
time in billions of years

HALLMARK OF SCIENCE **A scientific model makes testable predictions about natural phenomena. If predictions do not agree with observations, the model must be revised or abandoned.** Observations of the universe's expansion have forced us to modify our models of the universe to include dark energy along with dark matter.

First, the answer obviously depends on how much the expansion of the universe accelerates in the future. Some scientists speculate that the

Based on current data, the universe seems destined to expand forever, ending in darkness.

repulsive force due to dark energy might strengthen with time. In that case, perhaps in a few tens of billions of years, the growing repulsive force would tear apart our galaxy, our solar system, and even matter itself in a catastrophic event sometimes called the "Big Rip." However, evidence for this type of growing repulsion is very weak, and it seems more likely that the expansion will continue to accelerate more gradually.

If the universe continues to expand in this way, galaxies and galaxy clusters will remain gravitationally bound far into the future. Galaxies will not always look the same, however, because the star–gas–star cycle [Section 15.2] cannot continue forever. With each generation of stars, more mass becomes locked up in planets, brown dwarfs, white dwarfs, neutron stars, and black holes. Eventually, about a trillion years from now, even the longest-lived stars will burn out, and the galaxies will fade into darkness.

At this point, the only new action in the universe will occur on the rare occasions when two objects—such as two brown dwarfs or two white dwarfs—collide within a galaxy. The vast distances separating star systems in galaxies make such collisions extremely rare. For example, the probability of our Sun (or the white dwarf that it will become) colliding with another star is so small that it would be expected to happen only once in a quadrillion (10^{15}) years. However, given a long enough period of time, even low-probability events will eventually happen many times. If a star system experiences a collision once in a quadrillion years, it will experience about 100 collisions in 100 quadrillion (10^{17}) years. By the time the universe reaches an age of 10^{20} years, star systems will have suffered an average of 100,000 collisions each, making a time-lapse history of any galaxy look like a cosmic game of billiards.

These multiple collisions will severely disrupt galaxies. As in any gravitational encounter, some objects lose energy in such collisions and some gain energy. Objects that gain enough energy will be flung into intergalactic space, to be carried away from their home galaxies with the expansion of the universe. Objects that lose energy will eventually fall to the galactic center, leaving behind a single supermassive black hole where a galaxy used to be. The remains of the universe will consist of widely separated black holes with masses as great as a trillion solar masses, and widely scattered planets, brown dwarfs, and stellar corpses. If Earth somehow survives, it will be a frozen chunk of rock in the darkness of the expanding universe, billions of light-years away from any other solid object.

If grand unified theories [Section 17.1] are correct, Earth still cannot last forever. These models predict that protons will eventually fall apart. The predicted lifetime of protons is extremely long: a half-life of at least 10^{33} years. However, if protons really do decay, then by the time the universe is 10^{40} years old, Earth and all other atomic matter will have disintegrated into radiation and subatomic particles.

The final phase may come through a mechanism proposed by physicist Stephen Hawking. He has predicted that black holes must eventually "evaporate," turning their mass-energy into radiation (now called *Hawking radiation*). The process is so slow that we do not expect to be able to see it from any existing black holes, but if it really occurs, then black holes in the distant future will disappear in brilliant bursts of radiation. The largest black holes will last the longest, but even trillion-solar-mass black holes will evaporate sometime after the universe reaches an age of 10^{100} years. From then on, the universe will consist only of individual photons and subatomic particles, each separated by enormous distances from the others. Nothing new will ever happen, and no events will ever occur that would allow an omniscient observer to distinguish past from future. In a sense, the universe will finally have reached the end of time.

Forever Is a Long Time Lest any of this sound depressing, keep in mind that we are talking about incredibly long times. Remember that 10^{11} years is already nearly 10 times the current age of the universe (because 14 billion years is the same as 1.4×10^{10} years), 10^{12} years is another 10 times that, and so on. A time of 10^{100} years is so long that we can scarcely describe it, but one way to think about it (thanks to the late Carl Sagan) is to imagine that you wanted to write on a piece of paper a number that consisted of a 1 followed by 10^{100} zeros (that is, the number $10^{10^{100}}$). It sounds easy, but a piece of paper large enough to hold all those zeros *would not fit in the observable universe* today. If that still does not alleviate your concerns, you may be glad to know that a few creative thinkers are already speculating about ways in which the universe might avoid an icy fate or undergo rebirth, even after the end of time.

Perhaps of greater significance, speculating about the future of the universe means speculating about forever, and forever leaves us with a very long time in which to make new discoveries. After all, it is only in the past century that we learned that we live in an expanding universe, and only in the past couple of decades that we were surprised to learn that the expansion is accelerating. The universe may yet hold other surprises that might force us to rethink what might happen between now and the end of time.

the big picture Putting Chapter 18 into Context

We have found that there may be much more to the universe than meets the eye. Dark matter too dim for us to see seems to far outweigh the stars, and a mysterious dark energy may be even more prevalent. Together, dark matter and dark energy have probably been the dominant agents of change in the overall history of the universe. Here are some key "big picture" points to remember about this chapter:

- Dark matter and dark energy sound very similar, but they are hypothesized to explain different observations. Dark matter is thought to exist because we detect its gravitational influence. Dark energy is a term given to the source of the force that is accelerating the expansion of the universe.

- Either dark matter exists or we do not understand how gravity operates across galaxy-size distances. There are many reasons to be confident about our understanding of gravity, leading most astronomers to conclude that dark matter is real.

- Dark matter seems to be by far the most abundant form of mass in the universe, and therefore the primary source of the gravity that formed galaxies and larger-scale structures from tiny density enhancements that existed in the early universe. We still do not know what dark matter is, but we suspect it is largely made up of as-yet-undiscovered subatomic particles.

- The existence of dark energy is supported by evidence from observations both of the expansion rate through time and of temperature variations in the cosmic microwave background. Together, these observations have led to a model of the universe that gives us precise values for the inventory of its contents and its age.

- The fate of the universe seems to depend on whether its expansion continues forever, and the acceleration of the expansion suggests that it will. Nevertheless, forever is a long time, and only time will tell whether new discoveries will alter our speculations about the distant future.

my cosmic perspective Dark matter may seem far removed from our lives here on Earth, but our galaxy and the stars within it would not have formed without the assistance of its gravitational attraction.

18.1 Unseen Influences in the Cosmos

◆ What do we mean by dark matter and dark energy?

Dark matter and dark energy have never been directly observed, but each has been proposed to exist because it seems the simplest way to explain a set of observed motions in the universe. **Dark matter** is the name given to the unseen mass whose gravity governs the observed motions of stars and gas clouds. **Dark energy** is the name given to the form of energy thought to be causing the expansion of the universe to accelerate.

18.2 Evidence for Dark Matter

◆ What is the evidence for dark matter in galaxies?

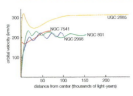 The orbital velocities of stars and gas clouds in galaxies do not change much with distance from the center of the galaxy. Applying Newton's laws of gravitation and motion to these orbits leads to the conclusion that the total mass of a galaxy is far larger than the mass of its stars. Because no detectable visible light is coming from this additional mass, we call it dark matter.

◆ What is the evidence for dark matter in clusters of galaxies?

 We have three different ways of measuring the amount of dark matter in clusters of galaxies: from galaxy orbits, from the temperature of the hot gas in clusters, and from the **gravitational lensing** predicted by Einstein. All these methods are in agreement, indicating that the total mass of dark matter in a galaxy cluster is at least 40 times the mass of its stars.

◆ Does dark matter really exist?

We infer that dark matter exists from its gravitational influence on the matter we can see, leaving two possibilities: Either dark matter exists or there is something wrong with our understanding of gravity. We cannot rule out the latter possibility, but we have good reason to be confident about our current understanding of gravity and the idea that dark matter is real.

◆ What might dark matter be made of?

Some of the dark matter could be ordinary (baryonic) matter in the form of dim stars or planetlike objects, but the amount of deuterium left over from the Big Bang and the patterns in the cosmic microwave background both indicate that ordinary matter adds up to only about one-sixth of the total amount of matter. The rest of the matter is hypothesized to be exotic (nonbaryonic) dark matter consisting of as-yet-undiscovered subatomic particles called **WIMPs.**

18.3 Structure Formation

◆ What is the role of dark matter in galaxy formation?

Because most of a galaxy's mass is in the form of dark matter, the gravity due to that dark matter is probably what formed protogalactic clouds and then galaxies from slight density enhancements in the early universe.

◆ What are the largest structures in the universe?

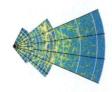 Galaxies appear to be distributed in gigantic chains and sheets that surround great **voids.** These **large-scale structures** trace their origin directly back to regions of slightly enhanced density early in time.

18.4 Dark Energy and the Fate of the Universe

◆ What is the evidence for an accelerating expansion?

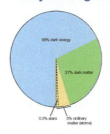 Observations of distant supernovae show that the expansion of the universe has been speeding up for the last several billion years. No one knows the nature of the mysterious force that could be causing this acceleration. However, its characteristics are consistent with models in which the force is produced by a form of dark energy that pervades the universe.

◆ Why is flat geometry evidence for dark energy?

 Observations of the cosmic microwave background also support the existence of dark energy because they demonstrate that the overall geometry of the universe is nearly flat. According to Einstein's general theory of relativity, the universe can be flat only if the total amount of mass-energy it contains is equal to the critical density, but measurements of the total amount of matter show that it represents only about 30% of the critical density. We therefore infer that about 70% of the total mass-energy is in the form of dark energy—the same amount implied by the supernova observations.

◆ What is the fate of the universe?

If dark energy is indeed what's driving the acceleration of the universe's expansion, then we expect the expansion to continue accelerating into the future, as long as the effects of dark energy do not change with time and there are no other factors that affect the fate of the universe.

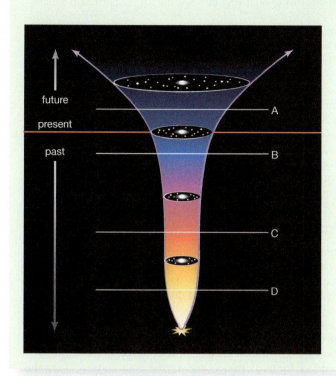

The schematic figure to the left shows a more complicated expansion history than the four idealized models shown in Figure 18.15. Answer the following questions, using the information given in this figure.

1. At time A, is the expansion of the universe accelerating, coasting, or decelerating?

2. At time B, is the expansion of the universe accelerating, coasting, or decelerating?

3. At time C, is the expansion of the universe accelerating, coasting, or decelerating?

4. At time D, is the expansion of the universe accelerating, coasting, or decelerating?

exercises and problems

MasteringAstronomy®   For instructor-assigned homework and other learning materials, go to MasteringAstronomy®.

Review Questions

1. Define *dark matter* and *dark energy,* and clearly distinguish between them. What types of observations have led scientists to propose the existence of each of these unseen influences?

2. Describe how orbital speeds in the Milky Way depend on distance from the galactic center. How does this relationship indicate the presence of large amounts of dark matter?

3. How do orbital speeds depend on distance from the galactic center in other spiral galaxies, and what does this tell us about dark matter in spiral galaxies?

4. How do we measure the masses of elliptical galaxies? What do these masses lead us to conclude about dark matter in elliptical galaxies?

5. Briefly describe the three different ways of measuring the mass of a cluster of galaxies. Do the results from the different methods agree? What do they tell us about dark matter in galaxy clusters?

6. What is *gravitational lensing*? Why does it occur? How can we use it to estimate the masses of lensing objects?

7. Briefly explain why the conclusion that dark matter exists rests on assuming that we understand gravity correctly. Is it possible that our understanding of gravity is not correct? Explain.

8. In what sense is dark matter "dark"? Briefly explain why objects like you, planets, and even dim stars qualify as dark matter.

9. What evidence indicates that most of the matter in the universe cannot be ordinary (baryonic) matter?

10. Explain what we mean when we say that a neutrino is a *weakly interacting particle*. Why can't the dark matter in galaxies be made of neutrinos?

11. What do we mean by *WIMPs*? Why does it seem likely that dark matter consists of these particles, even though we do not yet know what they are?

12. Briefly describe the various *large-scale structures* of the universe, the role of dark matter in their formation, and why the largest structures probably reflect the density patterns of the early universe.

13. Describe and compare the four general patterns for the expansion of the universe: *recollapsing, critical, coasting,* and *accelerating*. What evidence supports the accelerating model and the existence of dark energy?

14. How do observations of the cosmic microwave background provide evidence for dark energy?

15. Based on current evidence, what is the overall inventory of the mass-energy content of the universe?

16. What implications does the evidence for dark energy have for the fate of the universe?

Test Your Understanding

Decide whether the statement makes sense (or is clearly true) or does not make sense (or is clearly false). Explain clearly; not all of these have definitive answers, so your explanation is more important than your chosen answer.

17. Strange as it may sound, most of both the mass and the energy in the universe may take forms that we are unable to detect directly.
18. A cluster of galaxies is held together by the mutual gravitational attraction of all the stars in its galaxies.
19. We can estimate the total mass of a cluster of galaxies by studying the distorted images of galaxies whose light passes through the cluster.
20. Clusters of galaxies are the largest structures that we have so far detected in the universe.
21. The primary evidence for an accelerating universe comes from observations of young stars in the Milky Way.
22. There is no doubt remaining among astronomers that the fate of the universe is to expand forever.
23. Dark matter is called "dark" because it blocks light from traveling between the stars.
24. Dark energy is the energy associated with the motion of particles of dark matter.
25. Evidence that the expansion of the universe is accelerating comes from observations showing that the average distance between galaxies is increasing faster now than it was 5 billion years ago.
26. If dark matter consists of WIMPs, then we should be able to observe photons produced by collisions between these particles.

Quick Quiz

Choose the best answer to each of the following. Explain your reasoning with one or more complete sentences.

27. Dark matter is inferred to exist because (a) we see lots of dark patches in the sky. (b) it explains how the expansion of the universe can be accelerating. (c) we can observe its gravitational influence on visible matter.
28. Dark energy has been hypothesized to exist in order to explain (a) observations suggesting that the expansion of the universe is accelerating. (b) the high orbital speeds of stars far from the center of our galaxy. (c) explosions that seem to create giant voids between galaxies.
29. Measurements of how orbital speeds depend on distance from the center of our galaxy tell us that stars in the outskirts of the galaxy (a) orbit the galactic center just as fast as stars closer to the center. (b) rotate rapidly on their axes. (c) travel in straight, flat lines rather than elliptical orbits.
30. Strong evidence for the existence of dark matter comes from observations of (a) our solar system. (b) the center of the Milky Way. (c) clusters of galaxies.
31. A photograph of a cluster of galaxies shows distorted images of galaxies that lie behind it at greater distances. This is an example of what astronomers call (a) dark energy. (b) spiral density waves. (c) gravitational lensing.
32. Based on the observational evidence, is it possible that dark matter doesn't really exist? (a) No, the evidence for dark matter is too strong for us to think it could be in error. (b) Yes, but only if there is something wrong with our current understanding of how gravity should work on large scales. (c) Yes, but only if all the observations themselves are in error.

33. Based on current evidence, which of the following is considered a likely candidate for the majority of the dark matter in galaxies? (a) subatomic particles that we have not yet been able to detect (b) swarms of relatively dim red stars (c) supermassive black holes
34. Which region of the early universe was most likely to become a galaxy? (a) a region whose matter density was lower than average (b) a region whose matter density was higher than average (c) a region with an unusual concentration of dark energy
35. The major evidence for the idea that the expansion of the universe is accelerating comes from observations of (a) white dwarf supernovae. (b) the orbital speeds of stars within galaxies. (c) the evolution of quasars.
36. Which of the following possible types of universe would *not* expand forever? (a) a critical universe (b) an accelerating universe (c) a recollapsing universe

Process of Science

37. *Dark Matter.* Overall, how convincing do you consider the case for the existence of dark matter? Write a short essay in which you explain what we mean by dark matter, describe the evidence for its existence, and discuss your opinion about the strength of the evidence.
38. *Dark Energy.* Overall, how convincing do you consider the case for the existence of dark energy? Write a short essay in which you explain what we mean by dark energy, describe the evidence for its existence, and discuss your opinion about the strength of the evidence.
39. *Alternative Gravity.* Suppose someone proposed a new theory of gravity that claimed to explain observations of motion in galaxies and clusters of galaxies without the need for dark matter. Briefly describe at least one other test that you would expect the new theory to be able to pass if it was, in fact, a better theory of gravity than general relativity, which is currently our best explanation of how gravity works.

Group Work Exercise

40. *Dark Matter and Distorted Galaxies.* **Roles:** *Scribe* (takes notes on the group's activities), *Proposer* (proposes explanations to the group), *Skeptic* (points out weaknesses in proposed explanations), *Moderator* (leads group discussion and makes sure everyone contributes). **Activity:**
    a. Study the gravitational lensing diagram in Figure 18.8 and notice how gravitational lensing causes the image of a background galaxy to shift to a position *farther* from the center of the cluster. The *Proposer* should explain how this shift affects the lensed image of a galaxy and predict how the lensed image of a spherical galaxy would look. The *Skeptic* should then decide whether she or he agrees with the *Proposer*'s reasoning and, if not, should offer an alternative prediction.
    b. On a large piece of paper, the *Scribe* should draw a diagram like the one that follows, using a straightedge to make sure the lines are straight. They should all intersect at the same place, and the circle should be close to the point of intersection. (Note that the point at which the lines intersect represents the center of a galaxy cluster, and the circle represents the true position and shape of a spherical galaxy at a much greater distance from Earth.)

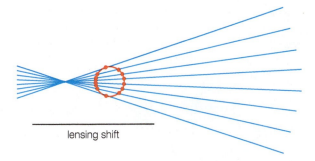

lensing shift

c. The *Moderator* should then determine the effect of the cluster's lensing shift on the galaxy's image as follows. For each dot on the circle, draw another dot farther to the right along the same line, so that the distance between the two dots is equal to the length of the line labeled "lensing shift." Then connect the new dots to see the shape of the lensed image. Does it agree with the *Proposer's* prediction? Was the *Skeptic's* prediction better? How does it compare with the lensed galaxy images in Figure 18.9? Discuss the possible reasons for any discrepancies you find.

Investigate Further

Short-Answer/Essay Questions

41. *The Future Universe.* Based on current evidence concerning the growth of structure in the universe, briefly describe what you would expect large-scale structures in the universe to look like about 10 billion years from now.
42. *Dark Matter and Life.* State and explain at least two reasons one might use to argue that dark matter is (or was) essential for life to exist on Earth.
43. *Orbital Speed vs. Radius.* Draw graphs showing how orbital speed depends on distance from the galactic center for each of the following three hypothetical galaxies. Make sure the horizontal axis has approximate distances labeled.
    a. a galaxy with all its mass concentrated at its center
    b. a galaxy with constant mass density within 20,000 light-years of its center, and zero density beyond that distance
    c. a galaxy with constant mass density within 20,000 light-years of its center, and beyond that an enclosed mass that increases proportionally to the distance from the center
44. *Dark Energy and Supernova Brightness.* When astronomers began measuring the brightnesses and redshifts of distant white dwarf supernovae, they expected to find that the expansion of the universe was slowing down. Instead they found that it was speeding up. Were the distant supernovae brighter or fainter than expected? Explain why. (*Hint:* In Figure 18.16, the position of a supernova point on the vertical axis depends on its redshift. Its position on the horizontal axis depends on its brightness—supernovae seen further back in time are not as bright as those seen closer in time.)
45. *What Is Dark Matter?* Describe at least three possible constituents of dark matter. Explain how we would expect each to interact with light, and how we might go about detecting its existence.
46. *Alternative Gravity.* How would gravity have to be different in order to explain the rotation curves of galaxies without the need for dark matter? Would gravity need to be stronger or weaker than expected at very large distances? Explain.

Quantitative Problems

Be sure to show all calculations clearly and state your final answers in complete sentences.

47. *White Dwarf M/L.* What is the mass-to-light ratio of a $1M_{Sun}$ white dwarf with a luminosity of $0.001L_{Sun}$?
48. *Supergiant M/L.* What is the mass-to-light ratio of a $30M_{Sun}$ supergiant star with a luminosity of $300,000L_{Sun}$?
49. *Solar System M/L.* What is the mass-to-light ratio of the solar system?
50. *Mass from Orbital Velocities.* Study the graph of orbital speeds for the spiral galaxy NGC 7541, which is shown in Figure 18.4.
    a. Use the orbital velocity formula (see Cosmic Calculations 15.1) to determine the mass (in solar masses) of NGC 7541 enclosed within a radius of 30,000 light-years from its center. (*Hint:* 1 light-year = 9.461×10^{15} m.)
    b. Use the orbital velocity formula to determine the mass of NGC 7541 enclosed within a radius of 60,000 light-years from its center.
    c. Based on your answers to parts a and b, what can you conclude about the distribution of mass in this galaxy?
51. *Weighing a Cluster.* A cluster of galaxies has a radius of about 5.1 million light-years (4.8×10^{22} m) and an intracluster medium with a temperature of 6×10^{7} K. The relationship between the equivalent orbital speeds of protons in the gas and the gas temperature is approximately $v = (140 \text{ m/s}) \times T$. Estimate the mass of the cluster using the orbital velocity formula (Cosmic Calculations 15.1). Give your answer in both kilograms and solar masses. Suppose that the combined luminosity of all the stars in the cluster is $8 \times 10^{12}L_{Sun}$. What is the cluster's mass-to-light ratio?

Discussion Questions

52. *Dark Matter or Revised Gravity.* One possible explanation for the evidence we find for dark matter is that we are currently using the wrong law of gravity to measure the masses of very large objects. If we really do misunderstand gravity, then many fundamental theories of physics, including Einstein's theory of general relativity, will need to be revised. Which explanation for our observations do you find more appealing: dark matter or revised gravity? Explain why. Why do you suppose most astronomers find dark matter more appealing?
53. *Our Fate.* Scientists, philosophers, and poets alike have speculated about the fate of the universe. How would you prefer the universe as we know it to end: in a "Big Crunch" or through eternal expansion? Explain the reasons behind your preference.

Web Projects

54. *Gravitational Lenses.* Gravitational lensing occurs in numerous astronomical situations. Compile a catalog of examples from the Internet with photos of lensed stars, quasars, and galaxies. Give a one-paragraph explanation of what is shown in each photo.
55. *Accelerating Universe.* Search for the most recent information about the acceleration of the expansion of the universe. Write a one- to three-page report on your findings.
56. *The Nature of Dark Matter.* Find and study recent reports on the possible nature of dark matter. Write a one- to three-page report that summarizes the latest ideas about what dark matter is made of.

All galaxies, including our Milky Way, developed as gravity pulled together matter in regions of the universe that started out slightly denser than surrounding regions. The central illustration depicts how galaxies formed over time, starting from the Big Bang in the upper left and proceeding to the present day in the lower right, as space gradually expanded according to Hubble's law.

380,000 years

1 billion years

Big Bang

TIME

① Dramatic inflation early in time is thought to have produced large-scale ripples in the density of the universe. All the structure we see today formed as gravity drew additional matter into the peaks of these ripples [Section 17.3].

Inflation may have stretched tiny quantum fluctuations into large-scale ripples.

② Observations of the cosmic microwave background show us what the regions of enhanced density were like about 380,000 years after the Big Bang [Section 17.2].

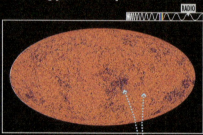

Image from *Planck*

Variations in the cosmic microwave background show that regions of the universe differed in density by only a few parts in 100,000.

③ Large-scale surveys of the universe show that gravity has gradually shaped early regions of enhanced density into a web-like structure, with galaxies arranged in huge chains and sheets [Section 18.3].

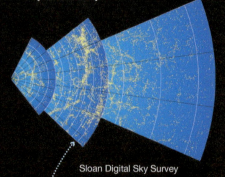

Sloan Digital Sky Survey

The web-like patterns of structure observed in large-scale galaxy surveys agree with those seen in large-scale computer simulations of structure formation.

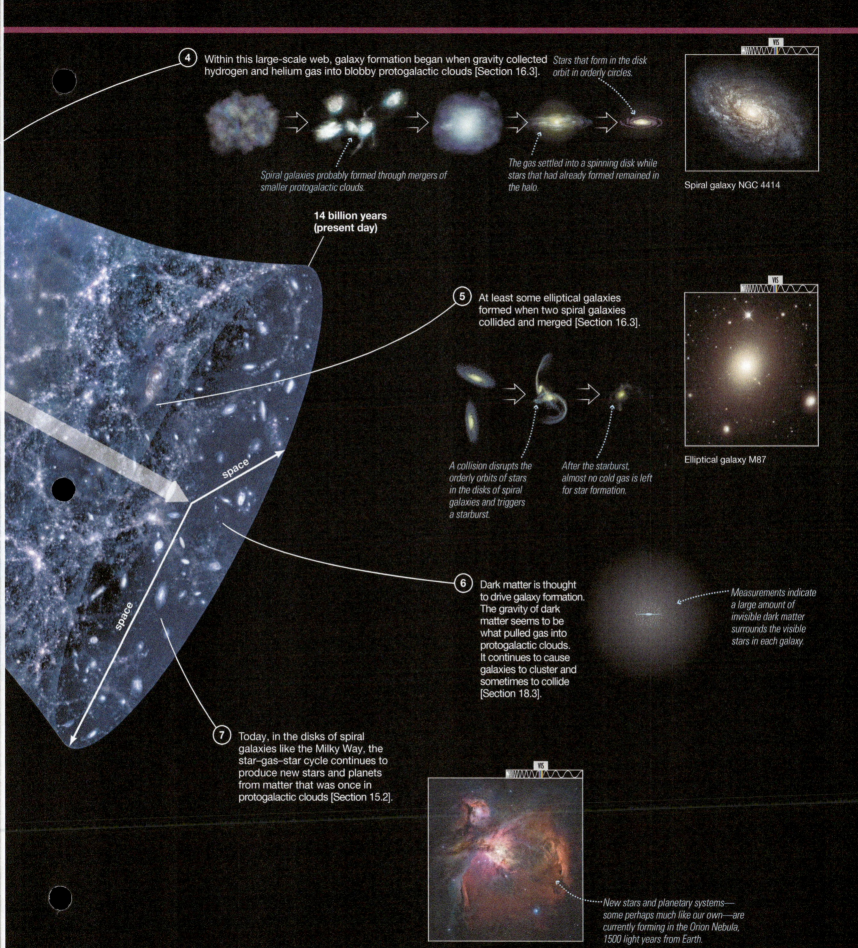

④ Within this large-scale web, galaxy formation began when gravity collected hydrogen and helium gas into blobby protogalactic clouds [Section 16.3].

Stars that form in the disk orbit in orderly circles.

Spiral galaxies probably formed through mergers of smaller protogalactic clouds.

The gas settled into a spinning disk while stars that had already formed remained in the halo.

Spiral galaxy NGC 4414

14 billion years (present day)

⑤ At least some elliptical galaxies formed when two spiral galaxies collided and merged [Section 16.3].

space

A collision disrupts the orderly orbits of stars in the disks of spiral galaxies and triggers a starburst.

After the starburst, almost no cold gas is left for star formation.

Elliptical galaxy M87

space

⑥ Dark matter is thought to drive galaxy formation. The gravity of dark matter seems to be what pulled gas into protogalactic clouds. It continues to cause galaxies to cluster and sometimes to collide [Section 18.3].

Measurements indicate a large amount of invisible dark matter surrounds the visible stars in each galaxy.

⑦ Today, in the disks of spiral galaxies like the Milky Way, the star–gas–star cycle continues to produce new stars and planets from matter that was once in protogalactic clouds [Section 15.2].

New stars and planetary systems—some perhaps much like our own—are currently forming in the Orion Nebula, 1500 light years from Earth.

The Orion Nebula

19 Life in the Universe

The Allen Telescope Array (Hat Creek, California) is used in the search for extraterrestrial intelligence (SETI).

LEARNING GOALS

19.1 Life on Earth
- When did life arise on Earth?
- How did life arise on Earth?
- What are the necessities of life?

19.2 Life in the Solar System
- Could there be life on Mars?
- Could there be life in the outer solar system?

19.3 Life Around Other Stars
- What are the requirements for surface life?
- What kinds of extrasolar worlds might be habitable?
- How could we detect life on extrasolar planets?

19.4 The Search for Extraterrestrial Intelligence
- How many civilizations are out there?
- How does SETI work?

19.5 Interstellar Travel and Its Implications for Civilization
- How difficult is interstellar travel?
- Where are the aliens?

Throughout this book, we have explored the evidence supporting the modern scientific story of the universe. We have seen very strong evidence that we live in an expanding universe born about 14 billion years ago in the Big Bang, and in which life has been made possible by the recycling of gas by stars within our galaxy. We've also seen how the processes that give birth to stars also give birth to planets, and explored how our own planet came to be. But we have not yet explored the most profound question of all: Are we alone?

The universe contains worlds beyond imagination—more than 100 billion star systems in our galaxy alone, and some 100 billion galaxies in the observable universe. But we do not yet know whether any world besides Earth has ever been home to life. In this chapter, we will discuss current understanding of the origin of life on Earth, and how we are using this understanding to consider the possibility of finding life elsewhere. In the process, we will see that the search for extraterrestrial life may have astonishing implications for our own future.

ESSENTIAL PREPARATION

1. How do we detect planets around other stars? [Section 10.1]

2. What unique features of Earth are important for life? [Section 7.5]

3. Why are Jupiter's Galilean moons geologically active? [Section 8.2]

4. What geological activity do we see on Titan and other moons? [Section 8.2]

5. What is the significance of the main sequence? [Section 12.2]

19.1 Life on Earth

It may seem that aliens are everywhere. Aliens abound in television shows and movies, and it's not hard to find websites claiming alien atrocities or a government conspiracy to hide alien corpses in "Area 51." Most scientists are deeply skeptical of such reports (see Extraordinary Claims, page 522), but scientific interest in the possibility of alien life is quite real. The scientific search for life in the universe even has its own name: **astrobiology.**

Most research in astrobiology focuses in one of the following three major areas: (1) seeking to understand the origin and evolution of life on Earth, so that we might learn the conditions under which life could arise and evolve elsewhere; (2) searching for worlds, both in our own solar system and beyond, that might have conditions suitable for life; and (3) searching for evidence of actual life on other worlds. In this section, we'll focus our attention on the first area of astrobiology research.

◆ When did life arise on Earth?

An important step in learning about the origin and evolution of life on Earth is finding out when life first arose. We cannot draw definitive conclusions from the single example of Earth, but if life on Earth arose relatively rapidly, perhaps it's easy to get life on a world with Earth-like conditions. On the other hand, if life took a long time to arise on Earth, perhaps it's difficult to get life, which would make life on other worlds seem less likely. The weight of evidence today supports the idea that life arose quickly and easily on Earth.

Evidence indicates that life arose early in Earth's history, offering at least a hint that life might arise with similar ease on other Earth-like worlds.

We learn about the history of life on Earth through the study of **fossils,** relics of organisms that lived and died long ago. Most fossils form when dead organisms fall

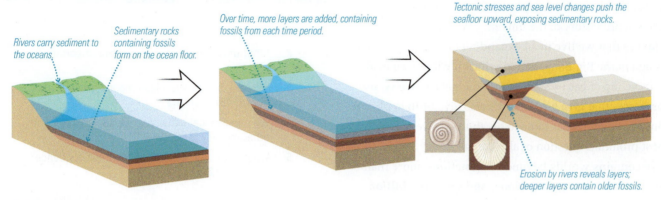

Rivers carry sediment to the oceans.

Sedimentary rocks containing fossils form on the ocean floor.

Over time, more layers are added, containing fossils from each time period.

Tectonic stresses and sea level changes push the seafloor upward, exposing sedimentary rocks.

Erosion by rivers reveals layers; deeper layers contain older fossils.

▲ **FIGURE 19.1**

Formation of sedimentary rock. Each layer represents a particular time and place in Earth's history and is characterized by fossils of organisms that lived in that time and place.

to the bottom of a sea (or other body of water) and are gradually buried by layers of sediments. The sediments are produced by erosion on land and carried by rivers to the sea. Over millions of years, sediments pile up on the seafloor, and the weight of the upper layers compresses underlying layers into rock. Erosion or tectonic activity can later expose the fossils (Figure 19.1). In some places, such as the Grand Canyon, the sedimentary layers record hundreds of millions of years of Earth's history (Figure 19.2).

The key to reconstructing the history of life is to determine the dates at which fossil organisms lived. The *relative* ages of fossils found in different layers are easy to determine: Deeper layers formed earlier and contain more ancient fossils. Radiometric dating [Section 6.4] confirms these relative ages and gives us fairly precise absolute ages for fossils. Based on the layering of rocks and fossils, geologists divide Earth's 4½-billion-year history into a set of distinct intervals that make up what we call the **geological time scale.** Figure 19.3 shows the names of the various intervals on a timeline, along with numerous important events in Earth's history.

think about it Based on Figure 19.3, how does the length of time during which animals and plants have lived on land compare to the length of time during which life has existed? How does the length of time during which humans have existed compare to the length of time since mammals and dinosaurs first arose?

You might wonder why the geological time scale shows so much more detail for the last few hundred million years than it does for earlier times. The answer is that fossils become increasingly difficult to find as we look deeper into Earth's history, for three major reasons. First, older rocks are much rarer than younger rocks, because most of Earth's surface is geologically young. Second, even when we find very old rocks, they often have been subject to transformations (caused by heat and pressure) that would have destroyed any fossil evidence they may have contained. Third, nearly all life prior to a few hundred million years ago was microscopic, and microscopic fossils are much more difficult to identify.

Despite these difficulties, geologists have found a few very old rocks that suggest life was already thriving on Earth 3.5 billion years

▲ **FIGURE 19.2**

The rock layers of the Grand Canyon record more than 500 million years of Earth's history.

Fossil evidence suggests that life on Earth was already thriving by 3.5 billion years ago.

ago, and possibly for hundreds of millions of years before that. One strong line of evidence comes from rocks called *stromatolites,* which are strikingly similar in size, shape, and interior structure to sections of

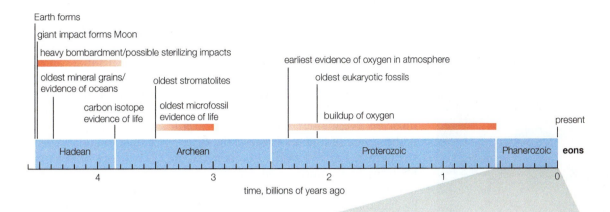

Earth forms

giant impact forms Moon

heavy bombardment/possible sterilizing impacts

oldest mineral grains/evidence of oceans

oldest stromatolites

earliest evidence of oxygen in atmosphere

oldest eukaryotic fossils

carbon isotope evidence of life

oldest microfossil evidence of life

buildup of oxygen

present

| Hadean | Archean | Proterozoic | Phanerozoic | **eons** |

4   3   2   1   0

time, billions of years ago

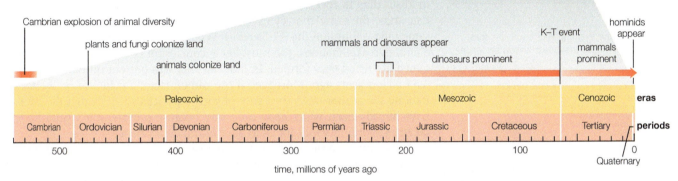

Cambrian explosion of animal diversity

plants and fungi colonize land

animals colonize land

mammals and dinosaurs appear

dinosaurs prominent

K–T event

hominids appear

mammals prominent

| Paleozoic | Mesozoic | Cenozoic | **eras** |

| Cambrian | Ordovician | Silurian | Devonian | Carboniferous | Permian | Triassic | Jurassic | Cretaceous | Tertiary | **periods** |

500   400   300   200   100   0
Quaternary

time, millions of years ago

modern-day, mat-shaped formations known as "living stromatolites" (Figure 19.4). Chemical analysis further supports the idea that the ancient stromatolites are fossils of ancient microbes. Moreover, if the microbes that made the ancient stromatolites are like the microbes in the living stromatolites today, then at least some ancient microbes produced energy by photosynthesis. Because photosynthesis is a fairly sophisticated metabolic process, we presume that it must have taken at least a moderately long time for this process to evolve in living organisms. In other words, the oldest stromatolites suggest that photosynthetic life already existed some 3.5 billion years ago, in which case more primitive life must have existed even earlier.

It's more difficult to search for evidence of life from more than 3.5 billion years ago, because rocks of such ages have usually undergone too much transformation to preserve intact fossils. However, another

▲ **FIGURE 19.3**

The geological time scale. Notice that the lower timeline is an expanded view of the last portion of the upper timeline. The eons, eras, and periods are defined by changes observed in the fossil record. The absolute ages come from radiometric dating. (The *K–T event* is the geological term for the impact linked to the mass extinction of the dinosaurs [Section 9.5].)

▼ **FIGURE 19.4**

Rocks called *stromatolites* offer evidence of microbial life existing as early as 3.5 billion years ago.

a These knee-high mats at Shark Bay, Western Australia, are colonies of microbes known as "living stromatolites."

b The banded structure in this section from one of the Shark Bay stromatolites is formed by layers of sediment attaching to the microbial mats.

c This section of a 3.5-billion-year-old stromatolite (found in the Strelley Pool Formation in Western Australia) shows the same type of structure found in living stromatolites. The black layers are organic deposits that are the remains of ancient microbial mats. (The ruler is marked in centimeters.)

▲ **FIGURE 19.5**

This ancient rock formation on the island of Akilia (off the coast of southern West Greenland) may hold the oldest known evidence of life.

line of evidence suggests that life was already present at least 3.85 billion years ago. Some of the most ancient rocks in the world are found on the island of Akilia near Greenland (Figure 19.5). Because these rocks are sedimentary (which means they are a mix of sediments that can contain mineral grains of many different ages), we cannot determine exactly when they formed. However, they are cut through by volcanic rock that radiometric dating shows to be 3.85 billion years old, so the sediments must be even older. Intriguingly, careful analysis of carbon isotopes within these sediments suggests that they once held living organisms.

Carbon has two stable isotopes: carbon-12, with six protons and six neutrons in its nucleus, and carbon-13, which has one extra neutron (see Figure 5.6). Living organisms incorporate carbon-12 slightly more easily than carbon-13. As a result, the fraction of carbon-13 is always a bit lower in fossils than in rock samples that lack fossils. All life and all fossils tested to date show the same characteristic ratio of the two carbon isotopes, and this same ratio is found in the Akilia rocks. While we cannot completely rule out alternative explanations, the most likely conclusion is that these rocks contained life when they formed. In that case, life was present on Earth at least 3.85 billion years ago. Moreover, if the carbon isotope evidence really does indicate life at that time, the rarity of such old rocks suggests that life must have already been widespread on Earth, because otherwise it would have taken extraordinary luck for us to have found the evidence.

How much earlier might life have arisen? Evidence from ancient mineral grains indicates that Earth may already have had oceans by about 4.3 billion years ago, suggesting at least the possibility that life could have existed at that time. However, recall that the first few hundred million years of the solar system's history was the period of the *heavy bombardment* [Section 6.3], during which Earth should have been struck numerous times by large asteroids or comets. Based on the sizes of lunar craters from this time period, some impacts of the heavy bombardment should have been large enough to vaporize much or all of the oceans, and perhaps to melt portions of Earth's crust. These impacts are therefore often called "sterilizing impacts," because they probably would have extinguished any life that might already have existed. Moreover, even if some life managed to survive these impacts, the surviving microbes would almost certainly have been in deep ocean or underground environments, making it unlikely that any kind of evidence would be preserved in rocks we find today.

To summarize, it would have been difficult for life to survive if it arose more than 3.9 billion years ago, and even if it did we probably could not find evidence of it. When we combine this fact with the fossil evidence indicating widespread life by 3.5 billion years ago and the carbon isotope evidence that pushes this date back to more than 3.85 billion years ago, we are led to a remarkable conclusion: We have evidence of life on Earth going nearly as far back in time as such evidence could exist, and perhaps nearly as far back as life could have taken hold. The implication is that life on Earth arose relatively easily, suggesting that the same might have happened on many other worlds.

Carbon isotope evidence suggests that life was present on Earth at least 3.85 billion years ago, which may be almost as early as it was possible for life to take hold.

◆ How did life arise on Earth?

The early origin of life is suggestive of the idea that life arose easily, but we would be more confident of this idea if we could learn *how* life arose on Earth. Scientists explore this question primarily by learning about how life has evolved over time, which tells us what early life may have looked like, and then by doing laboratory experiments to learn how life might have started.

The Theory of Evolution The fossil record clearly shows that life has gone through great changes over time. If we are going to understand how life arose, we must understand what causes these changes, so that we can trace life back to its origin. The unifying theory through which scientists understand life on Earth is the **theory of evolution,** first published by Charles Darwin in 1859.

Evolution simply means "change with time." Even before Darwin, numerous scientists (some going back to ancient times) had recognized evidence for evolution in the fossil record, but no one before Darwin successfully explained how species might undergo change. In essence, the fossil record provides strong evidence that evolution *has* occurred, while Darwin's theory of evolution explains *how* it occurs.

The fossil record shows us how life has changed through time, and the theory of evolution explains how these changes have occurred.

Darwin studied relationships between living species (most famously in the Galápagos Islands) as well as fossils of extinct species. Based on the evidence he collected, he put forth a simple model to explain how evolution occurs. As described by biologist Stephen Jay Gould (1941–2002), Darwin built his model from "two undeniable facts and an inescapable conclusion":

- *Fact 1: overproduction and competition for survival.* Any localized population of a species has the potential to produce far more offspring than the local environment can support with resources such as food and shelter. This overproduction leads to a competition for survival among the individuals of the population.

- *Fact 2: individual variation.* Individuals in a population of any species vary in many heritable traits (traits passed from parents to offspring). No two individuals are exactly alike, and some individuals possess traits that make them better able to compete for food and other vital resources.

- *The inescapable conclusion: unequal reproductive success.* In the struggle for survival, those individuals whose traits best enable them to survive and reproduce will, on average, leave the largest number of offspring that in turn survive to reproduce. Therefore, in any local environment, heritable traits that enhance survival and successful reproduction will become progressively more common in succeeding generations.

It is this unequal reproductive success that Darwin called **natural selection,** meaning that genetic traits that provide a reproductive advantage will naturally win out (be "selected") over less advantageous traits. Over time, natural selection can help individuals of a species become better able to compete for scarce resources. If enough small individual variations accumulate, natural selection can even give rise to an entirely new species. Darwin backed his model with so much evidence that it quickly gained the status of a scientific *theory* [Section 3.4], and ongoing research during the more than 150 years since he published the theory has only given it further support.

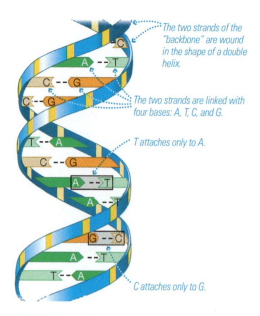

The two strands of the "backbone" are wound in the shape of a double helix.

The two strands are linked with four bases: A, T, C, and G.

T attaches only to A.

C attaches only to G.

▲ **FIGURE 19.6**

This diagram represents a small piece of a DNA molecule, which looks like a zipper twisted into a spiral. Hereditary information is contained in the "teeth" linking the strands. These "teeth" are the DNA bases. Only four DNA bases are used, and they can link up between the two strands only in specific ways: T attaches only to A, and C attaches only to G. (The color coding is arbitrary and is used only to represent different types of chemical groups; in the backbone, blue and yellow represent sugar and phosphate groups, respectively.)

The Mechanism of Evolution Darwin's theory of evolution by natural selection tells us that species adapt and change by passing hereditary traits from one generation to the next. However, Darwin did not know precisely how these traits were passed down, nor did he know why there is variation among individuals or how new traits can appear in a population. In essence, his theory *predicted* that there must be some biochemical mechanism for heredity, one that would preserve most hereditary information from prior generations but still allow for small changes. Today, we know that this mechanism is embodied in the properties of the molecule called **DNA** (short for "*d*eoxyribo*n*ucleic *a*cid"). In that sense, the discovery of DNA was a tremendous predictive success for Darwin's theory.

Living organisms reproduce by copying DNA and passing these copies on to their descendants. A molecule of DNA consists of two long strands—somewhat like the interlocking strands of a zipper—wound together in the spiral shape known as a double helix (Figure 19.6). The instructions for assembling a living organism are written in the precise order of four chemical bases (abbreviated A, T, G, and C for the first letters of their chemical names) that make up the interlocking portions of the DNA "zipper." These bases pair up in a way that ensures that both strands of a DNA molecule contain the same genetic information. By unwinding and allowing new strands (made from chemicals floating around inside a cell) to form alongside the original ones, a single DNA molecule can give rise to two identical copies of itself. That is how genetic material is copied and passed on to future generations.

Evolution occurs because the transfer of genetic information from one generation to the next is not always perfect. An organism's DNA may occasionally be altered by copying errors or by external influences, such as ultraviolet light from the Sun or exposure to toxic or radioactive chemicals. Any change in an organism's DNA is called a **mutation.** Many mutations are lethal, killing the cell in which the mutation occurs.

special topic **What Is Life?**

You may have noticed that while we've been talking about life, we haven't actually defined the term. It's surprisingly difficult to draw a clear boundary between life and nonlife, but one way to start is by listing distinguishing features common to all known life. For example, nearly all living organisms on Earth appear to share the following six key properties:

1. *Order:* Living organisms are not random collections of molecules but rather have molecules arranged in orderly patterns that form cell structures.
2. *Reproduction:* Living organisms are capable of reproducing.
3. *Growth and development:* Living organisms grow and develop in patterns determined at least in part by heredity.
4. *Energy utilization:* Living organisms use energy to fuel their many activities.
5. *Response to the environment:* Living organisms actively respond to changes in their surroundings. For example, organisms may alter their chemistry or movements in the presence of a food source.
6. *Evolutionary adaptation:* Life evolves through natural selection, as organisms pass on traits that make them better adapted to survival in their local environments.

These six properties are all important, but biologists today regard evolution as the most fundamental and unifying of them. Evolutionary adaptation is the only property that can explain the great diversity of life on Earth. Moreover, understanding how evolution works allows us to understand how all the other properties came to be. A simple definition of *life* might therefore be "something that can reproduce and evolve through natural selection."

This definition of life probably suffices for most practical purposes, but some cases may still challenge it. For example, computer scientists have written programs (that is, lines of computer code) that can reproduce themselves (that is, create additional sets of identical lines of code). By adding to the programs instructions that allow random changes, computer scientists can even make "artificial life" that evolves on a computer. Should this "artificial life," which consists of nothing but electronic signals processed by computer chips, be considered alive?

The fact that we have such difficulty distinguishing the living from the nonliving on Earth suggests that we should be very cautious about constraining our search for life elsewhere. No matter what definition of life we choose, there's always the possibility that we'll someday encounter something that challenges it. Nevertheless, the properties of reproduction and evolution seem likely to be shared by most if not all life in the universe, and therefore provide a useful starting point as we consider how to explore the possibility of extraterrestrial life.

Some, however, may improve a cell's ability to survive and reproduce. The cell then passes on this improvement to its offspring.

Our understanding of the molecular mechanism of natural selection has put the theory of evolution on a stronger foundation than ever. While no theory can ever be proved true beyond all doubt, the theory of evolution is as solid as any theory in science, including the theory of gravity and the theory of atoms. Biologists routinely witness evolution occurring before their eyes among laboratory microorganisms, or over periods of just a few decades among plants and animals subjected to environmental stress. Moreover, the theory of evolution has become the underpinning of virtually all modern biology, medicine, and agriculture. For example, agricultural scientists apply the idea of natural selection to develop pest control strategies that reduce the populations of harmful insects without harming the populations of beneficial ones; medical researchers test new drugs on animals that are genetically similar to humans, because the theory of evolution tells us that genetically similar species should have somewhat similar physiological responses; and biologists study the genetic relationships between organisms by comparing their DNA.

The theory of evolution is backed by a tremendous body of evidence, and there is no more scientific doubt about evolution than there is about gravity or atoms.

The First Living Organisms

The basic chemical nature of DNA is virtually identical among all living organisms. This fact, along with other biochemical similarities shared by all living organisms, tells us that all life on Earth today can trace its origins to a common ancestor that lived long ago.

We are unlikely to find fossils of the earliest organisms, but we can learn about early life through careful studies of the DNA of living organisms. Biologists can determine the evolutionary relationships among living species by comparing the sequences of bases in their DNA. For example, two organisms whose DNA sequences differ in five places for a particular gene are probably more distantly related than two organisms whose gene sequences differ in only one place. Many such DNA comparisons suggest that all living organisms are related in a way depicted schematically by the "tree of life" in Figure 19.7. Although details in the structure of this tree remain uncertain (particularly because species can sometimes swap genes), it indicates that life on Earth is divided into three major groupings, or *domains,* called *bacteria, archaea,* and *eukarya,* and that all three domains share a common ancestor. Notice that plants and animals represent only two tiny branches of the domain *eukarya.*

All living organisms today evolved from a common ancestor that lived long ago.

Organisms on branches located closer to the root of the tree of life must contain DNA that is evolutionarily older, suggesting that they more closely resemble the organisms that lived early in Earth's history. Some of the modern-day organisms that appear to be evolutionarily oldest are microbes that live in very hot water around seafloor volcanic vents (Figure 19.8). Unlike most life at Earth's surface, which depends on sunlight, these organisms get energy from chemical reactions in the hot, mineral-rich water around the volcanic vents.

The idea that early organisms might have lived in such "extreme" conditions may seem surprising, but it makes sense when we think about it. The deep ocean environment would have been protected from harmful ultraviolet radiation that bathed Earth's surface before our atmosphere had oxygen or an ozone layer, and may also have

Early organisms probably arose near a source of chemical energy, such as deep-sea volcanic vents.

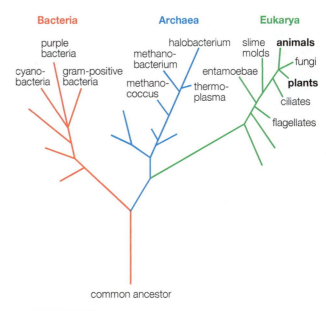

▲ **FIGURE 19.7**

The tree of life, showing evolutionary relationships determined by comparison of DNA sequences in different organisms. Just two small branches represent *all* plant and animal species.

▲ **FIGURE 19.8**

This photograph shows a volcanic vent on the ocean floor that spews out hot, mineral-rich water. DNA studies indicate that the microbes living near these vents are evolutionarily older than most other living organisms, hinting that early life may have arisen in similar environments.

▲ FIGURE 19.9

Stanley Miller poses with a reproduction of the experimental setup he first used in the 1950s to study pathways to the origin of life. (He worked with Harold Urey, so the experiment is called the Miller-Urey experiment.)

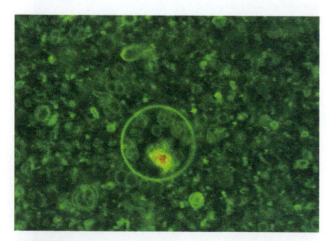

▲ FIGURE 19.10

This microscopic photo made with the aid of fluorescent dyes shows short strands of RNA (red) contained within an enclosed membrane (green circle), both of which formed spontaneously with the aid of clay minerals beneath them. The structure is often called a "pre-cell" because it has many characteristics of a living cell but it is not alive.

offered protection from the effects of impacts. Moreover, the chemical pathways used to extract energy from mineral-rich hot water are simpler than other chemical pathways (such as photosynthesis) used by living organisms to obtain energy. For these reasons, many scientists suspect that life arose near deep-sea vents or in similar environments that provided chemical energy.

The Transition from Chemistry to Biology The theory of evolution explains how the earliest organisms evolved into the great diversity of life on Earth today. But where did the first organisms come from? We may never know for sure, because we are unlikely to find fossils that show the transition from nonlife to life. However, over the past several decades, scientists have conducted laboratory experiments designed to mimic the conditions that existed on the young Earth.

Such experiments were first performed in the 1950s (Figure 19.9), and they have been refined and improved since that time. In essence, the experiments mix chemicals thought to have been present on the early Earth and then "spark" the chemicals with electricity to simulate lightning or other energy sources. The chemical reactions that follow have produced nearly all the major molecules of life, including amino acids and DNA bases. Many of these molecules are also found in meteorites, suggesting that some organic molecules may have arrived from space.

More recent laboratory experiments have shown that naturally occurring clays, which should have been abundant in the early oceans, can catalyze the formation of short strands of RNA, a molecule much

Laboratory experiments show that the molecules of life form easily under conditions that existed early on Earth.

like a single strand of DNA. These same clay minerals can also lead to the formation of microscopic enclosed membranes—often called "pre-cells" because of their resemblance to living cells—sometimes with RNA inside them (Figure 19.10).

Because some RNA molecules (though not those so far produced in the laboratory) are capable of self-replication, we can envision the process summarized in Figure 19.11. Naturally formed chemical building blocks were catalyzed by clay to make self-replicating RNA molecules, some of which became enclosed in microscopic pre-cells. Pre-cells in which RNA replicated faster and more accurately were more likely to spread, leading to a type of positive feedback that would have encouraged even faster and more accurate replication. With these types of chemical reactions occurring all over Earth, it may only have been a matter of time until some of the pre-cells turned into RNA-based life, which eventually evolved into DNA-based life. We may never know whether life really arose in this way, but unless we are missing some major piece of the puzzle, it seems plausible to imagine that life started through a natural sequence of chemical processes.

Could Life Have Migrated to Earth? Our scenario suggests that life could have arisen naturally here on Earth. However, an alternative possibility is that life arose somewhere else first—perhaps on Venus or Mars—and then migrated to Earth on meteorites. Remember that we have collected meteorites that were blasted by impacts from the surfaces of the Moon and Mars [Section 9.1]. Calculations suggest that Venus, Earth, and Mars all should have exchanged many tons of rock, especially in the early days of the solar system when impacts were more common.

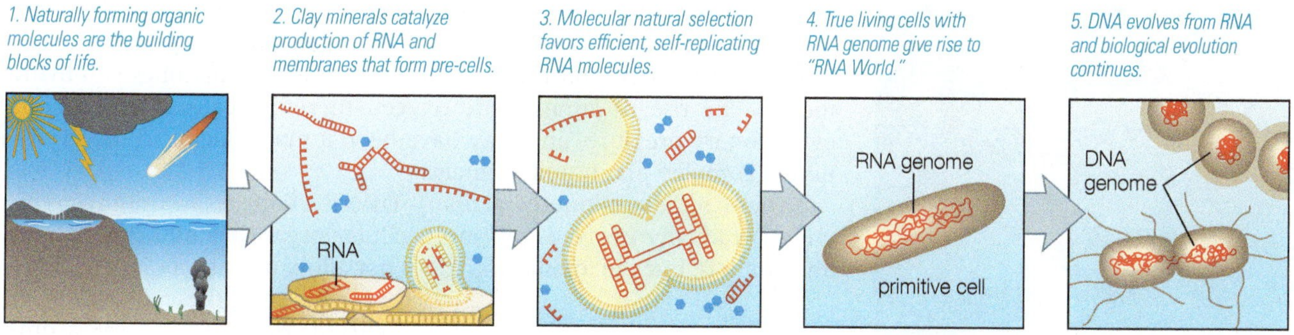

1. Naturally forming organic molecules are the building blocks of life.

2. Clay minerals catalyze production of RNA and membranes that form pre-cells.

3. Molecular natural selection favors efficient, self-replicating RNA molecules.

4. True living cells with RNA genome give rise to "RNA World."

5. DNA evolves from RNA and biological evolution continues.

RNA

RNA genome

primitive cell

DNA genome

▲ **FIGURE 19.11**
A summary of the steps by which chemistry early on Earth may have led to the origin of life.

The idea that life could travel through space to Earth once seemed outlandish. After all, it's hard to imagine a more forbidding environment than that of space, with no air, no water, and constant bombardment by dangerous radiation from the Sun and stars. However, the presence of organic molecules in meteorites and comets tells us that the building blocks of life can survive in space, and tests have shown that some microbes can survive in space for years. There's even a known set of animal species, known as *tardigrades*, that can survive some time in space (Figure 19.12).

In a sense, Earth, Venus, and Mars have been "sneezing" on each other for billions of years. Life could conceivably have originated on any of these three planets and been transported to the others. It's an intriguing thought, but it does not change our basic scenario for the origin of life—it simply moves it from one planet to another.

A Brief History of Life on Earth We are now ready to take a quick look at the history of life on Earth, summarized in the geological timeline of Figure 19.3. Earth formed about 4½ billion years ago, and the giant impact thought to have formed the Moon [Section 6.3] probably happened soon after. Mineral evidence suggests that Earth had oceans by about 4.3 billion years ago, and those early oceans would have been natural laboratories for chemical reactions that could have led to life. If life formed very early, it may have been disrupted by the heavy bombardment. But once life took hold, evolution rapidly diversified it.

Despite the rapid pace of evolution, the most complex organisms remained single-celled for at least a billion years after life first arose. All of these organisms probably lived in the oceans, because the lack of a protective ozone layer made the land inhospitable. Things began to change only when oxygen started building up in Earth's atmosphere.

Nearly all the oxygen in our atmosphere was originally released through photosynthesis by single-celled organisms known as *cyanobacteria* (Figure 19.13). Fossil evidence indicates that cyanobacteria were producing oxygen through photosynthesis by at least 2.7 billion years ago, and possibly for hundreds of millions of years before that. However, oxygen did not immediately begin to accumulate in the atmosphere. For hundreds of millions of years, chemical reactions with surface rocks pulled oxygen back out of the atmosphere nearly as fast as the cyanobacteria could produce it. But these tiny organisms were abundant and persistent, and eventually the surface rock was so saturated with oxygen that the rate of oxygen removal slowed down. Oxygen then began to accumulate in the atmosphere, though it may not have reached a level that we could have breathed until just a few hundred million years ago.

▲ **FIGURE 19.12**
It almost doesn't look real, but this photograph shows a tiny animal called a tardigrade (also called a "water bear") that is about a millimeter long. Tardigrades can survive an incredible range of "extreme" conditions, including at least some time in the near-vacuum of space.

▲ FIGURE 19.13

This photo shows microscopic chains of modern cyanobacteria. The ancestors of these living organisms produced essentially all the oxygen in Earth's atmosphere.

Today, we often think of oxygen as a necessity for life. However, oxygen was probably poisonous to most organisms living before about 2 billion years ago (and remains a poison to many microbes still living today). The rise of atmospheric oxygen therefore caused tremendous evolutionary pressure, and may have been a major factor in the evolution of complex plants and animals.

There were undoubtedly many crucial changes as primitive microbes gradually evolved into multicellular organisms and early plants and animals, but the fossil record does not allow us to pinpoint the times at which all these changes occurred. However, we see a dramatic change in the fossil record during the *Cambrian period,* beginning about 542 million years ago. During this period, animal life evolved from tiny and primitive organisms into all the basic body types (phyla) that we find on Earth today. This remarkable diversification occurred in such a short time relative to Earth's history that it is often called the *Cambrian explosion.*

> Living organisms remained single-celled for most of Earth's history, with larger plants and animals arising only in the past few hundred million years.

Early dinosaurs and mammals arose some 225 to 250 million years ago, but dinosaurs at first proved more successful and dominated for well over 100 million years. Their sudden demise 65 million years ago [Section 9.5] paved the way for the evolution of large mammals—and ultimately to us. The earliest humans appeared on the scene only a few million years ago, or after 99.9% of Earth's history to date had already occurred. Our few centuries of industry and technology have come after 99.99999% of Earth's history.

◆ What are the necessities of life?

Our understanding of the timing and possible origins of life on Earth suggests that life might similarly arise on other worlds, but this would be the case only if they contained the necessities of life. So what are these necessities? While it is possible that life on other worlds could be quite different from life on Earth, it's easiest to begin the search for life by looking for conditions in which organisms from Earth could live.

If we think about ourselves, the requirements for life seem fairly stringent: We need abundant oxygen in an atmosphere that is otherwise not poisonous, we need temperatures in a fairly narrow range of conditions, and we need abundant and varied food sources. However, the discovery of life in "extreme" environments—such as in the hot water near undersea volcanic vents—shows that many microbes (often called *extremophiles*) can survive in a much wider range of conditions.

Organisms living in hot water prove that at least some microbes can survive in much higher temperatures than we would have guessed. Other organisms live in other extremes. In the freezing cold but very dry valleys of Antarctica, scientists have found microbes that live *inside* rocks, surviving on tiny droplets of liquid water and energy from sunlight. Life, including tardigrades (see Figure 19.12), has been found in Antarctic lakes buried so deep beneath the ice that they have not been exposed to open air or sunlight for millions of years. In many places, microscopic life has also been found deep underground in water that fills pores within subterranean rock. We have found life thriving in environments so acidic, alkaline, or salty that humans would be poisoned almost instantly. We have even found microbes that can survive high doses of radiation, making it possible for them to survive for many years in the radiation-filled environment of space.

If we compare all the different forms of life on Earth, we find that life as a whole has only three basic requirements:

- A source of nutrients (atoms and molecules) from which to build living cells
- Energy to fuel the activities of life, whether from sunlight, from chemical reactions, or from the heat of Earth itself
- Liquid water

These requirements give us a basic road map for the search for life elsewhere. If we want to find life on other worlds, it makes sense to start by searching for worlds that offer these basic necessities.

Interestingly, only the third requirement (liquid water) seems to pose much of a constraint. Organic molecules are present almost everywhere—even on meteorites and comets. Many worlds are large enough to retain internal heat that could provide energy for life, and virtually all worlds have sunlight (or starlight) bathing their surfaces, although the inverse square law for light [Section 12.1] means that light provides less energy to worlds farther from their star. Nutrients and energy should therefore be available to some degree on almost every planet and moon. In contrast, liquid water is relatively rare. Therefore, the search for liquid water—or possibly a liquid of some other type— drives the search for life in our universe.

Life on Earth requires nutrients, energy, and liquid water. Liquid water is the only one that is not common on other worlds.

19.2 Life in the Solar System

The liquid water requirement rules out most of the worlds in our solar system as possible homes for life. Mercury and the Moon are barren and dry, Venus is too hot for liquid water on its surface (though its clouds contain droplets of water laced with acid), and most of the small bodies of the outer solar system are too cold. The jovian planets may have droplets of liquid water in some of their clouds (see Figure 8.6), but the strong vertical winds on these planets mean any droplets cannot stay liquid for long, making these planets unlikely homes for life. Nevertheless, several worlds *do* show evidence of liquid water or some other liquid mixture that seems to offer at least some possibility of life. The most notable are Mars and a few of the jovian moons. In this section, we'll consider the possibility of extraterrestrial life in our own solar system.

◆ Could there be life on Mars?

When we search for worlds that have the necessities of life as we know them from life on Earth, we are looking for what we call **habitable worlds.** In other words, a habitable world is one that seems to have all the necessities of life.

think about it Is it possible for a world to be habitable but not actually have life? Is it possible for a world to have life but not be habitable? Explain.

Mars seems to have fit the bill for habitability in the past, because strong evidence indicates that liquid water flowed on Mars during its early history [Section 7.3]. Moreover, while the atmospheric pressure is too low for liquid water to be stable on the surface today, Mars still

A habitable world is one that seems to have all the necessities of life.

has abundant water ice. It therefore seems likely that at least some pockets of liquid water might still exist underground, particularly near sources of volcanic heat. The idea that Mars was once habitable and might still have some habitable environments underground explains why Mars has been the number one target in the search for life in the solar system.

Searching for Life on Mars The most direct way to search for life on Mars is by looking for it with landers and rovers. Unfortunately, a direct search turns out to be more difficult than we first imagined, a lesson learned from the two *Viking* landers that reached Mars in 1976. Each was equipped with a robotic arm for scooping up soil samples, which were fed into several on-board robotically controlled experiments that had been designed to search for evidence of life. For example, one of the experiments mixed scooped-up Martian soil with organic nutrients from Earth, then looked for evidence that the nutrients were being taken up by microscopic organisms in the soil.

Unfortunately, the *Viking* experiments produced confounding results. While some of them at first seemed suggestive of life, scientists later realized that the results could also be explained by chemical reactions that would occur under the conditions found on Mars. Moreover, one of the experiments looked for evidence of organic molecules in the soil and found none at all, which seemingly precluded the existence of carbon-based life. However, some 30 years after the *Viking* experiments were conducted, scientists discovered that they had missed an important part of the soil chemistry on Mars: The soil contains a salt (known as *perchlorate*) that destroys organic molecules when it is heated as it was in the *Viking* experiments. The *Curiosity* rover has since found strong evidence that the Martian soil *does* contain at least some organic molecules (Figure 19.14).

The main lesson from the *Viking* experiments, then, was that searching for life with robotic experiments requires a deeper understanding of

Martian soil chemistry introduces complexities that have hindered the search for life, but scientists remain optimistic that future missions may answer the question of whether Mars has or ever had life.

Martian soil chemistry and other issues than we yet have, which is a major reason why subsequent missions have sought to help us understand these issues, rather than attempting a direct search for life. Scientists are particularly interested in a "sample return mission" in which a robotic spacecraft would collect rocks on Mars and bring them back to Earth, where they could be analyzed in many more ways than would be possible with predesigned robotic experiments sent to Mars. While no such mission has yet been funded, NASA's plans for a rover to be launched in 2020 include instruments that may be able to search for signs of past life and a system to grab and store samples that would be eventually picked up and returned to Earth by future missions.

In the meantime, there is one other way to search for evidence of past life on Mars, and it doesn't even require going there: We can study meteorites whose chemical composition suggests they came from Mars. Martian meteorites have more than once been claimed to show evidence of Martian life, though most scientists have found the claims unconvincing to date. Nevertheless, studies of these meteorites continue, and at minimum should help us understand more about the geological history of Mars.

▲ **FIGURE 19.14**

This is an artist's conception of the *Curiosity* rover's ChemCam firing its laser at a rock on Mars, so that it can spectroscopically analyze the composition of the vaporized rock. ChemCam is one of the instruments on *Curiosity* used to search for organic material on Mars. The inset shows a rock that *Curiosity* first drilled (large hole) and then studied spectroscopically by using its laser to vaporize the dusty debris; the laser made the row of smaller holes.

Methane on Mars A less direct way to investigate the possibility of life on Mars is to look at atmospheric gases that might provide clues. Scientists are particularly intrigued by claims of detection of methane (CH_4) in the Martian atmosphere. Atmospheric methane is quickly destroyed by ultraviolet light from the Sun, so its presence would imply that it is being continually released from the Martian surface.

Claims of methane gas detection were first announced in 2003 and 2004, based on telescopic observations from Earth and the *Mars Express* orbiter at Mars. Subsequent telescopic observations also suggested that the methane level varies with time. However, the signals were weak enough that many scientists doubted the claims. In 2014, the *Curiosity* rover detected methane on the Martian surface, but some scientists argue that, rather than coming from Mars, the methane leaked from the rover itself. *Curiosity* is continuing its measurements, but as this book goes to press in 2016, the debate over whether Mars releases methane is still unresolved.

If the methane is real, its source would almost certainly have to be either geological activity or life, and either possibility would have important implications for the search for life on Mars. If the source is geological, the amount of volcanic heat necessary for methane release would probably also be sufficient to maintain pockets of liquid water underground, which would be a potential habitat for life. And, of course, if the source is biological, it should be only a matter of time until we are able to identify the organisms responsible.

◆ Could there be life in the outer solar system?

Beyond Mars, the cold temperatures of the outer solar system make it unlikely that we could ever find surface liquid water. However, as we discussed in Chapters 8 and 9, numerous worlds may have subsurface lakes or oceans of liquid water, and at least one world—Saturn's moon Titan—has surface lakes of ethane and methane.

Jupiter's Potentially Habitable Moons Recall that three of Jupiter's moons show evidence of having subsurface oceans: Europa, Ganymede, and Callisto. Of these three, Europa is considered the strongest candidate for potential life. The ice and rock from which Europa formed undoubtedly included the necessary chemical ingredients for life, and Europa's internal heating (primarily due to tidal heating) is strong enough to power volcanic vents on its seafloor. If life on Earth arose near such undersea volcanic vents, Europa would seem to have everything needed for life to originate.

If Europa's ocean really exists, it may have seafloor vents much like those in Earth's oceans, where early life may have thrived.

The possibility of life on Europa is especially interesting because, unlike any potential life on Mars, it would not necessarily have to be microscopic, given the enormous depth of the suspected ocean (see Figure 8.17). However, the potential energy sources for life on Europa are far more limited than the energy sources for life on Earth, mainly because sunlight could not fuel photosynthesis in the subsurface ocean. As a result, most scientists suspect that any life that might exist on Europa would probably be microscopic and primitive. Energy considerations also explain why scientists consider life to be less likely on Ganymede and Callisto, because these moons would have even less energy for life in their subsurface oceans than Europa.

Searching for life on Europa will probably be difficult, since the ocean is buried beneath several kilometers of ice. However, as we discussed in Chapter 8, study of Europa's icy surface indicates that at least some water occasionally wells up from below and freezes, so it is possible that we could find frozen microbes or other evidence of life through spectroscopy or direct sampling of the surface. More ambitiously, some scientists are considering the possibility of someday sending a robotic submarine that would melt its way through the ice to search for life in the ocean below.

Titan and Enceladus   Saturn's moons Titan and Enceladus are both considered promising candidates for life. Let's start with Titan.

Titan's surface is far too cold for liquid water, but it has lakes and rivers of liquid methane and ethane [Section 8.2]. Although many biologists think it unlikely, it is possible that this liquid could support life as water does on Earth. There are also at least slim prospects for water-based life on Titan. The ultraviolet light that hits Titan's atmosphere produces a wide range of organic molecules that, over billions of years, should have accumulated as a deep layer of organic sediment on the surface; these sediments may be the material in the dunes found on Titan (Figure 19.15). Occasional impacts by comets or asteroids would provide enough heat locally to melt any water ice and create pockets of warm water that might persist for a thousand years or so, offering at least a chance for some interesting organic chemistry if not the development of life. Titan may also have icy volcanism in which volcanoes occasionally erupt with a "lava" of liquid water from below ground. One final possibility for life on Titan is a subsurface ocean. *Cassini* measurements show small changes in Titan's shape as it orbits Saturn, suggesting that it may have an ocean of liquid water or a colder ammonia–water mixture far beneath its icy crust.

> Other possible homes to life in our solar system include Jupiter's moons Ganymede and Callisto and Saturn's moons Titan and Enceladus.

Enceladus seems at least equally promising, since recent evidence supports the idea that it has a global subsurface ocean [Section 8.2]. Moreover, Enceladus's ice fountains presumably spray out material from the subsurface ocean, which becomes frozen as it emerges. This offers the tantalizing possibility that, if the ocean contains life, the icy fountains might spray out flash-frozen life forms. For this reason, scientists hope to send a mission that could sample the ice spray in search of life on Enceladus. (*Cassini* flew through the ice fountains, but was not equipped with instruments to sample the ice, since the existence of the fountains was not known when *Cassini* was built.)

Beyond Saturn   The discovery of geological activity and of the likely existence of a subsurface ocean on Enceladus has led scientists to realize that potential habitats for life might be more widespread than had previously been imagined. In particular, while we know that Enceladus is

▲ **FIGURE 19.15**

Dunes on Titan. The dark streaks in this radar image are thought to be windblown dunes, possibly made of hydrocarbon sediments.

tidally heated, scientists still do not fully understand why there is enough heating to melt subsurface ice, suggesting that we might find similar surprises elsewhere in the solar system. For example, recall that *Voyager* observations of Neptune's moon Triton (see Figure 8.27) show evidence of at least some geological activity, so it is possible that Triton could hold subsurface liquid water. Similarly, the surprising geological activity revealed in the *New Horizons* flyby of Pluto [Section 9.4] has scientists speculating about subsurface lakes or oceans even on that distant world. We do not yet know whether any world in our solar system besides Earth has ever been home to life, but it is becoming increasingly clear that there are numerous places worth searching for evidence of life.

19.3 Life Around Other Stars

Studies of extrasolar planets suggest that our galaxy contains billions of planetary systems, so the prospects for life elsewhere might seem quite good. But numbers alone don't tell the whole story. In this section, we'll consider the prospects for life on worlds orbiting other stars.

◆ What are the requirements for surface habitability?

When we consider the search for life beyond our solar system, we must distinguish between *surface* life like that on Earth and *subsurface* life like that we envision as a possibility on Mars or Europa. While large telescopes could in principle allow us to discover surface life on extrasolar planets, no foreseeable technology will allow us to find life that is hidden deep underground in other star systems (unless the subsurface life has a noticeable effect on the planet's atmosphere). Therefore, we'll begin our consideration of extrasolar life by focusing on the search for planets with habitable surfaces, meaning surfaces with temperatures and pressures that could allow liquid water to exist.

As we discussed in Chapter 10, current data already suggest that most stars have planets. Our technology is not yet good enough for us to study these planets in enough detail to determine whether any of them have habitable surfaces, but we can make educated guesses if we know the requirements for surface habitability. If we think back on all the ideas we've discussed in this book, we find that there are four major factors that have made Earth so hospitable to the long-term evolution of life:

1. A distance from the star that is great enough to allow water vapor to condense as rain and make oceans, but not so far that all the water freezes

2. Volcanism that can release trapped gases from the interior, including water vapor and carbon dioxide, to make an atmosphere and oceans

3. Plate tectonics that can support a climate-regulating carbon dioxide cycle

4. A planetary magnetic field that protects the atmosphere from the solar wind

Let's briefly examine each of these requirements in more detail.

The Habitable Zone We know from our study of Venus that even if a planet seems to be geologically much like Earth, it will be too hot for life if it is too close to its star. In particular, if we moved Earth inward toward

The approximate habitable zones around our Sun, a star with one-half the mass of our Sun (spectral type K), and a star with one-tenth the mass of our Sun (spectral type M), shown to scale. The habitable zone becomes increasingly smaller and closer in for stars of lower mass and luminosity.

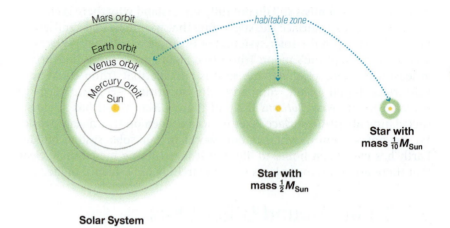

the Sun, there is some location between Earth's orbit and the orbit of Venus at which Earth would suffer a runaway greenhouse effect (see Figure 7.39). This location marks the inner boundary of what we call the Sun's **habitable zone**—the range of distances from the Sun at which a planet like Earth could have oceans and surface life. Similarly, if we moved Earth outward in its orbit, there would be a location beyond which Earth would freeze. This location must be somewhere near the orbit of Mars, because we know Mars had liquid water in the past but is now frozen. (In fact, many scientists suspect that if Mars had been larger, it might still be habitable.) The left diagram in Figure 19.16 shows the approximate boundaries of the Sun's habitable zone.

We can similarly map the habitable zones of other stars. For ordinary (main-sequence) stars, the location and extent of a habitable zone depend only on the star's mass, because mass determines the star's surface temperature and total energy output (luminosity). Because lower-mass stars are cooler and less luminous, their habitable zones are closer in and smaller in total extent. Figure 19.16 compares the habitable zones of two lower-mass stars to that of our Sun. It's worth noting that in many cases habitable zones are also possible in binary and multiple star systems. For example, if two stars in a binary system are widely enough separated, planets could orbit one star without being perturbed much by the other. Alternatively, if two stars orbit closely, there can be a habitable zone that encircles both of them, offering up the prospect of worlds with two suns in the sky (Figure 19.17).

> The habitable zone is the region around a star in which it is *possible* for a planet to have "just right" conditions for surface liquid water.

think about it The habitable zone is sometimes called the "Goldilocks zone." Recall or reread the story "Goldilocks and the Three Bears," then explain how the "Goldilocks zone" gets its name.

Note that in the search for life, we generally don't pay much attention to stars that are significantly more massive than the Sun, for two reasons. First, higher-mass stars are much less common than lower-mass stars (see Figure 13.6). Second, recall that higher-mass stars have shorter lifetimes. The example of Earth suggests that life may not take hold for up to a few hundred million years after a planet first forms, so we would expect to find life only around stars with lifetimes of hundreds of millions of years or more. This constraint rules out stars with more than a few times the mass of our Sun.

Volcanism and Plate Tectonics   Being in a star's habitable zone is not enough to make a world habitable. The Moon, for example, is clearly in our Sun's habitable zone (because it is at the same distance from the Sun as Earth), but it is not habitable. Long-term surface habitability also requires an atmosphere and a relatively stable climate.

Recall that the terrestrial atmospheres came from volcanic outgassing, and that volcanism and tectonics are driven by internal heat [Section 7.1]. We therefore conclude that Earth's habitability required the *combination* of being in the Sun's habitable zone and being large enough to have retained the internal heat needed for substantial volcanic outgassing. The Moon is not habitable despite being in the Sun's habitable zone because of its small size, which did not allow enough volcanic outgassing to produce a substantial atmosphere.

Most scientists suspect that internal heat and outgassing were enough to produce the conditions that led to an origin of life on Earth, and therefore that the same would be true for other worlds of similar size in habitable zones. However, the fact that Earth has remained habitable for some 4 billion years is due to its long-term climate stability, which we trace to the climate self-regulation that comes from the carbon dioxide cycle (see Figure 7.44). The carbon dioxide cycle, in turn, depends on plate tectonics, suggesting that plate tectonics may be necessary for long-term habitability. If so,

> To have a habitable surface, a planet in the habitable zone must also be large enough to have substantial volcanism and, perhaps, plate tectonics.

then a key question about habitability around other stars is whether plate tectonics should be expected on Earth-size planets in habitable zones. The answer to this question is not yet known, because Venus provides an example of a nearly Earth-size planet that does not show evidence of plate tectonics. But recall that one hypothesis ties Venus's lack of plate tectonics to its high surface temperature [Section 7.4], which comes from the runaway greenhouse effect that occurred because Venus is too close to the Sun to be in the habitable zone. If this hypothesis is correct, then Venus might have had plate tectonics if it were in the habitable zone, in which case we might expect plate tectonics on any Earth-size world in a habitable zone. The same might be true of terrestrial worlds larger than Earth, though we do not know.

Overall, internal heat and volcanism appear to be necessary conditions for an origin of life on a planet's surface, while plate tectonics is clearly helpful to long-term habitability. Still, the great diversity among the billions of expected Earth-like planets argues that we should keep an open mind about other possible ways that long-term habitability might occur.

Global Magnetic Field Recall that the leading hypothesis for explaining how Mars changed from being warm and wet to its current frozen condition invokes the loss of its magnetic field (see Figure 7.34). Once the magnetic field was gone, the atmosphere was vulnerable to the solar wind, and over time Mars lost so much atmospheric gas that it could no longer maintain much heat through the greenhouse effect. Earth has been spared a similar fate because its magnetic field protects our atmosphere from solar wind particles. (Venus's lack of magnetic field means it has lost a lot of gas, but it had so much gas to begin with that it still has a thick atmosphere.)

If this hypothesis is correct, then Earth's global magnetic field has been crucial to maintaining conditions for life over a period of some 4 billion years. We might therefore expect a magnetic field to make long-term habitability more likely on other worlds. Recall that the basic requirements for a planet to maintain a global magnetic field are enough internal heat to maintain core convection and fast enough rotation to twist and distort the convection pattern (see Figure 7.5). The first requirement will already be met by any world that is large enough to have ongoing volcanism and tectonics, so only the second requirement—fast enough rotation—is new on our list. However, it probably is not difficult to meet, as many worlds likely either spin relatively fast like Earth or have large enough cores to support convection and magnetic fields even with slower rotation rates. Again, the sheer number and diversity of expected terrestrial planets make it seem reasonable to think that some without magnetic fields might maintain enough of their atmospheres to remain habitable.

A Recipe for Habitability? Having a habitable surface requires being in the habitable zone, and volcanism, plate tectonics, and a magnetic field all require that a planet be at least moderately large (larger than Mars, perhaps as large as or larger than Earth). The question is whether having this size is *sufficient* to have those or other characteristics that could support long-term habitability, or whether Earth has attributes that will prove to be rare even among worlds that seem superficially similar.

Although we do not know for sure, most planetary scientists suspect that Earth-like size is indeed sufficient. In that case, we would have a simple recipe for long-term surface habitability: Any planet that is within a star's habitable zone and that is similar to Earth in size would be expected to have a habitable surface. If this is correct, then current data suggest there should be billions of worlds with habitable surfaces in our galaxy alone.

> Although there is still uncertainty, it may be that the only requirements for a habitable surface are Earth-like size and location in a star's habitable zone.

◆ What kinds of extrasolar worlds might be habitable?

One of the key lessons we've learned in the study of extrasolar planets is that planets come in a wider range of types than we find in our own solar system. This fact makes scientists wonder if habitability might also be broader when we consider extrasolar worlds, particularly when we consider subsurface habitability in addition to surface habitability. Here, we'll explore a few of the intriguing possibilities.

Moons with Habitable Surfaces While Europa, Enceladus, and other moons may have subsurface oceans of liquid water, no moons in our solar system have liquid water on their surfaces. However, we can envision at least two possible ways in which such moons might exist in other star systems.

One possibility is that they might form as a consequence of giant impacts like that thought to have formed our own Moon. For example, while our Moon has a mass only about 1/80 that of Earth, we might imagine that a similar giant impact on a super-Earth could lead to the formation of a moon with an Earth-like mass. To understand the second possibility, recall that many other solar systems have jovian-size planets that have apparently migrated inward [Section 10.3]. Like the jovian planets or our own solar system, we might expect such planets to have been born with numerous large moons, and it is possible that these moons would remain in orbit as the planet migrates inward. If so, and if the planet ends up within the habitable zone, then the moons might be sufficiently large to have habitable surfaces.

Super-Earths and Water Worlds in Extended Habitable Zones The standard way in which habitable zones are calculated—such as the zones shown in Figure 19.16—assumes rocky planets (like Venus, Earth, and Mars) warmed by the greenhouse effect of water vapor and carbon dioxide in their atmospheres. But is it possible that other types of worlds might have habitable surfaces over a broader zone? Some scientists suspect that this might indeed be the case for worlds somewhat larger than Earth, such as super-Earths and water worlds (see Figure 10.12). Because these planets can be several times as massive as Earth, their atmospheres might retain substantial amounts of hydrogen gas captured from a solar nebula during planet formation. Hydrogen can act as a greenhouse gas, which means it could keep these planets warm enough to maintain surface liquid water well beyond the boundaries usually assumed for habitable zones. Some estimates suggest that these planets could have habitable surfaces at distances up to 10 or more times as far from their star as the standard calculations would suggest. For example, a super-Earth or water world with a substantial hydrogen atmosphere might be habitable at Saturn's distance from the Sun, or even beyond. Research into this idea is actively continuing, raising the intriguing possibility that surface habitability may be much more common than we have generally assumed.

> With thick enough atmospheres, super-Earths and water worlds might have habitable surfaces even well outside the standard habitable zone.

Subsurface Habitability Just as we found in our own solar system, where we've identified Mars and several jovian moons as possibly having habitable subsurface regions, it is possible that there could be far more worlds with subsurface than surface habitability. These worlds fall into several categories. First, there are planets like Mars that are too small to have the climate stability that would have allowed them to maintain habitable surfaces for billions of years, but that might have had habitable surfaces for a shorter time and subsurface liquid water for far longer. Second, there are worlds that might be similar to Europa, Enceladus, and perhaps even Pluto, with potential zones of subsurface habitability. Beyond that, super-Earths or water worlds located beyond habitable zones might also offer opportunities for finding subsurface liquid water. For example, because we expect super-Earths to contain much more internal heat than Earth, they might have deep subsurface zones of liquid water even if they are far enough from their star to have completely frozen surfaces.

Orphan Planets There is another category of potentially habitable worlds that we have not yet discussed: **orphan planets** (sometimes called *rogue planets*) that do not orbit a star. These "planets" pose a definitional challenge since we usually *define* a planet as something that orbits a star, but it's easy to understand where the idea comes from. Recall that one mechanism by which jovian planets may migrate inward in some planetary systems is through close gravitational encounters in which one object loses energy and moves inward while the other object gains energy and is flung outward. Given that migration appears to be quite common in planetary systems, it seems reasonable to expect that many planets have been flung outward into interstellar space in this way; it's even possible that planet-size bodies might form independently (without first being born around a star) from fragments of collapsing interstellar clouds. Orphan planets would be extremely difficult to detect directly because they are so dim, but one recent estimate suggested that orphan planets may be twice as numerous as the stars in our galaxy.

You might at first guess that planets without a star would lack energy for life, but plenty of life on Earth lives without sunlight in the deep oceans or underground, and some of this life survives on energy traceable to Earth's internal heat alone. A planet like Earth retains internal heat for billions of years, and therefore might have active volcanism and tectonics—and the prospects of subsurface liquid water—even if it were not orbiting the Sun. In that case, orphan planets might frequently have subsurface habitable zones.

Even "orphan planets" that do not orbit a star seem to offer some prospects for habitability.

Some researchers speculate that some orphan planets might even have *surface* habitability. Much as with the idea of extended habitable zones, this might be possible if the planets have a thick enough hydrogen atmosphere. In that case, internal heat leaking outward might be trapped by the greenhouse effect, and in some places this might raise surface temperature above the freezing point of water. Indeed, with a thick enough atmosphere, even a planet the size of Earth could potentially offer such conditions while floating freely in interstellar space. A key question, then, is whether a thick hydrogen atmosphere is likely around planets similar in size to Earth. No one really knows, but remember that the main reason Earth and the other terrestrial planets did not retain hydrogen from the solar nebula is that their small sizes allowed the lightweight hydrogen to escape. Because the speed of hydrogen atoms depends on the temperature, which would be far lower for planets ejected into interstellar space, it is possible that hydrogen atmospheres might be retained, if they formed in the first place.

The Bottom Line: Wide-Ranging Possibilities Overall, there seems to be great potential for habitability throughout the universe, both on planets that might be Earth-like in character and appearance and on a variety of worlds that might have other forms of habitability. Moreover, we have focused primarily on habitability by life that uses water as its liquid medium, but the case of Titan makes us wonder whether other liquids might also be used. Of course, being habitable does not necessarily mean having life. We are therefore ready to turn our attention to the question of how we might actually discover life beyond the solar system.

How could we detect life on extrasolar planets?

At present, we have no way to search for actual life on any of the many extrasolar worlds that may potentially be habitable. None of our indirect methods for learning about these worlds (astrometric, Doppler, transits) can offer information that would tell us whether they have life, and our direct detection capabilities remain too rudimentary to detect the presence of life. However, we expect our direct observational capabilities to improve dramatically in coming decades, and as a result, scientists are working on strategies that might allow future telescopes to search for life on worlds around other stars. Note that, as we've briefly discussed already, we will probably be limited to searching for evidence of surface life; detecting subsurface life on such distant worlds will be far more difficult, if it is possible at all.

If our telescope technology becomes sufficiently powerful, we can hope to learn about potential life through both images and spectra. For example, large telescopes in space may be able to obtain images with high enough resolution to tell us whether extrasolar planets have continents and oceans, and perhaps even allow us to monitor seasonal changes. Spectra should prove even more useful. Mod-

It seems reasonable to imagine that life is common in the universe, but we'll need much more powerful telescopes to test this idea.

erate-resolution infrared spectra can reveal the presence and abundance of many atmospheric gases (Figure 19.18), and careful analysis of atmospheric makeup might tell us whether a planet has life. On Earth, for example, the large abundance of oxygen (21% of our atmosphere) is a direct result of photosynthetic life. Abundant oxygen in the atmosphere of a distant world might similarly indicate the presence of life. Other evidence might come from any ratio of atmospheric gases that would naturally change unless it were being actively maintained by the respiration of living organisms. Although we cannot do it yet, it seems likely that within a few decades we will have methods for learning whether distant planets have life.

think about it Considering all the factors we have discussed, do *you* believe that life will prove to be rare or common in the universe? Defend your opinion.

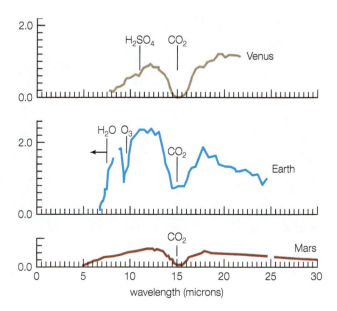

▲ **FIGURE 19.18**
The infrared spectra of Venus, Earth, and Mars as they might be seen from afar, showing absorption features that point to the presence of carbon dioxide (CO_2), ozone (O_3), and sulfuric acid (H_2SO_4) in their atmospheres. While carbon dioxide is present in all three spectra, only our own planet has appreciable oxygen (and hence ozone)—a product of photosynthesis. If we could make similar spectral analyses of distant planets, we might detect atmospheric gases that would indicate life.

19.4 The Search for Extraterrestrial Intelligence

If life turns out to be common, it's natural to wonder whether other worlds might also have intelligent life and civilizations. If such civilizations exist, we might be able to find them simply by listening for signals that they are sending into interstellar space, either in deliberate attempts to contact other civilizations or as a means of communicating among themselves. The search for signals from other civilizations is generally known as the **search for extraterrestrial intelligence,** or **SETI** for short.

◆ How many civilizations are out there?

SETI efforts have a chance to succeed only if other advanced civilizations are broadcasting signals that we could receive. To judge the chances of SETI success, we'd need to know how many civilizations are broadcasting such signals right now.

▲ **FIGURE 19.19**

Astronomer Frank Drake, with the equation he first wrote in 1961. (With Dr. Drake's approval, we use a slightly modified form of his equation in this book.)

Given that we do not even know whether microbial life exists anywhere beyond Earth, we certainly don't know whether other civilizations exist, let alone how many there might be. Nevertheless, for the purposes of planning a search for extraterrestrial intelligence, it is useful to have an organized way of thinking about the number of civilizations that might be out there. To keep our discussion simple, let's consider only the number of potential civilizations in our own galaxy. We can always extend our estimate to the rest of the universe by simply multiplying the result we find for our galaxy by 100 billion, the approximate number of galaxies in our universe.

The Drake Equation In 1961, astronomer Frank Drake wrote a simple equation designed to summarize the factors that would determine the number of civilizations we might contact (Figure 19.19). This equation is now known as the **Drake equation,** and in principle it gives us a simple way to calculate the number of civilizations capable of interstellar communication that are currently sharing the Milky Way Galaxy with us. In a form slightly modified from the original, the Drake equation looks like this:

The Drake equation summarizes the factors that determine the number of civilizations in our galaxy with which we could potentially communicate.

$$\text{Number of civilizations} = N_{HP} \times f_{life} \times f_{civ} \times f_{now}$$

This equation will make sense once you understand the meaning of each factor:

- N_{HP} is the number of habitable planets in the galaxy—that is, the number of planets that could *potentially* have life.

- f_{life} is the fraction of habitable planets that actually have life. For example, $f_{life} = 1$ would mean that all habitable planets have life, and $f_{life} = \frac{1}{1,000,000}$ would mean that only 1 in a million habitable planets has life. Therefore, the product $N_{HP} \times f_{life}$ tells us the number of life-bearing planets in the galaxy.

- f_{civ} is the fraction of the life-bearing planets on which a civilization capable of interstellar communication *has at some time* arisen. For example, $f_{civ} = \frac{1}{1000}$ would mean that such a civilization has existed on 1 out of 1000 planets with life, while the other 999 out of 1000 have not had a species that learned to build radio transmitters, high-powered lasers, or other devices for interstellar conversation. When we multiply this factor by the first two factors to form the product $N_{HP} \times f_{life} \times f_{civ}$, we get the total number of planets on which intelligent beings have evolved and developed a communicating civilization at some time in the galaxy's history.

- f_{now} is the fraction of these civilization-bearing planets that happen to have a civilization *now*, as opposed to, say, millions or billions of years in the past. This factor is important because we can hope to contact only civilizations that are broadcasting signals we could receive at present. (In estimating f_{now}, we assume that the light-travel time for signals from other stars has been taken into account.)

We do not know the values of any of the factors in the Drake equation, but it is still a useful tool for organizing our thinking.

Because the product of the first three factors tells us the total number of civilizations that have *ever* arisen in the galaxy, multiplying by f_{now} tells us how many civilizations we could potentially make contact with today. In other words, the result of the Drake equation should in principle tell us the number of civilizations that we might hope to contact. Unfortunately, we can make a reasonable estimate only for the first term (N_{HP}) in the Drake equation, which means we cannot actually calculate its result. Nevertheless, the equation is a useful way of organizing our thinking, as we can see by considering the potential values for each of its factors.

think about it Try the following sample numbers in the Drake equation. Suppose that there are 1000 habitable planets in our galaxy, that 1 in 10 habitable planets has life, that 1 in 4 planets with life has at some point had an intelligent civilization, and that 1 in 5 civilizations that have ever existed is in existence now. How many civilizations would exist at present? Explain.

The Number of Life-Bearing Planets

Let's begin with the first two factors in the Drake equation, whose product ($N_{HP} \times f_{life}$) tells us the number of life-bearing planets in our galaxy. As we've discussed, statistics from the *Kepler* mission have already shown that Earth-size planets are common, and early data suggest that such planets are also common in habitable zones. Overall, it now seems reasonable to suppose that a significant fraction of all stars have at least one habitable planet, which means there should be many billions of—perhaps more than 100 billion—habitable planets in our galaxy.

The factor f_{life} presents more difficulty, because we do not yet have any reliable way to estimate the fraction of habitable planets on which life actually arose. Still, we've argued that the rapid appearance of life on Earth and the fact that laboratory experiments suggest possible ways in which life could have arisen both make it seem that life should be likely

Our galaxy is likely to have billions of habitable planets, but we cannot yet estimate the fraction of them that actually have life.

on habitable worlds. In that case, we might expect most or all habitable planets to also have life, making the fraction f_{life} close to 1. Of course, until we have solid evidence that life arose anywhere else, such as on Mars, it is also possible that Earth has been very lucky and that f_{life} is so close to zero that life has never arisen on any other planet in our galaxy.

The Question of Intelligence

Even if life-bearing planets are common, civilizations capable of interstellar communication might not be. The fraction of life-bearing planets that at some time have such civilizations, f_{civ}, depends on at least two things. First, a planet would have to have a species evolve with sufficient intelligence to develop interstellar communication. In other words, the planet needs a species at least as smart as we are. Second, that species would have to develop a civilization with technology at least as advanced as ours.

Although we cannot really be sure, most scientists suspect that only the first requirement is difficult to meet. A fundamental assumption in nearly all of science today is that we are not "special" in any particular way. We live on a fairly typical planet orbiting an ordinary star in a normal galaxy, and we assume that living creatures elsewhere—whether they prove to be rare or common—would be subjected to evolutionary pressures similar to those that have operated on Earth. Therefore, if species with intelligence similar to ours have arisen elsewhere, we assume that they would have similar sociological drives that would eventually lead them to develop the technology necessary for interstellar communication.

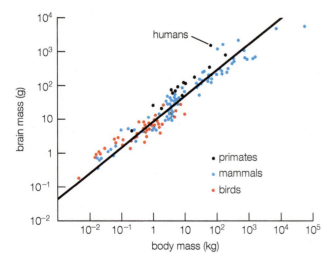

▲ FIGURE 19.20

This graph shows how brain mass compares to body mass for some mammals (including primates) and birds. The straight line represents an average of the ratio of brain mass to body mass, so animals that fall above the line are smarter than average and animals that fall below the line are less smart. Note that the scale uses powers of 10 on both axes. (Data from Harry J. Jerison, 1973.)

If this assumption is correct, then the fraction f_{civ} depends primarily on the question of whether sufficient intelligence is rare or common among life-bearing planets. As with the question of life of any kind, the short answer is that we just don't know, but we can get at least some insight by considering what happened on Earth.

Look again at Figure 19.3. While life arose quite quickly on Earth, nearly all life remained microbial until just a few hundred million years ago, and it took 4½ billion years for humans to arrive on the scene. This slow progress toward intelligence might suggest that producing a civilization is very difficult even when life is present. On the other hand, roughly half the stars in the Milky Way are older than our Sun, so if Earth's case is typical, then plenty of planets have existed long enough for intelligence to arise.

Life on Earth arose quickly, but Earth was 4½ billion years old by the time life became intelligent enough to learn about it.

Another way to address the question is by considering our level of intelligence in comparison to that of other animals on Earth. We can get a rough measure of intelligence by comparing brain mass to total body mass (a measure sometimes called the *encephalization quotient,* or EQ). Figure 19.20 shows the brain weights for a sampling of birds and mammals plotted against their body weights. There is a clear and expected trend in that heavier animals have heavier brains. By drawing a straight line that fits these data, we can define an average value of brain mass for each body mass. Animals whose brain mass falls above the line are smarter than average, while animals whose brain mass falls below the line are less mentally agile. Keep in mind that it is the *vertical* distance above the line that tells us how much smarter a species is than the average, and that the scale goes in powers of 10 on both axes. If you look closely, you'll see that the data point for humans lies significantly farther above the line than the data point for any other species. By this measure of intelligence, we are far smarter than any other species that has ever existed on Earth.

Some people use this fact to argue that even on a planet with complex life, a species as intelligent as we are would be very rare. They say that even if there is an evolutionary drive toward intelligence in general, it takes extreme luck to reach our level of intelligence. After all, while it's evolutionarily useful to have enough intelligence to capture prey and evade other predators, it's not clear why natural selection would lead to brains big enough to build spacecraft. However, the same data can be used to reach an opposite conclusion. The scatter in the levels of intelligence among different animals tells us that some variation should be expected, and statistical analysis shows that we are not unreasonably far above the average. It might therefore be inevitable that some species would develop our level of intelligence on any planet with complex life.

Technological Lifetimes For the sake of argument, let's assume that life and intelligence are reasonably likely, so thousands or millions of planets in our galaxy have at some time given birth to a civilization capable of interstellar communication. In that case, the final factor in the Drake equation, f_{now}, determines the likelihood of there being someone whom we could contact now. The value of this factor depends on how long civilizations survive.

Consider our own example. In the roughly 12 billion years during which our galaxy has existed, we have been capable of interstellar communication via radio for only about 60 years. If we were to destroy ourselves tomorrow, then other civilizations could have received signals from us during only 60 years out of the galaxy's 12-billion-year existence, equivalent to 1 part in 200 million of the galaxy's history. If such a short technological lifetime is typical of civilizations, then f_{now} would be only 1/200,000,000 and some 200 million civilization-bearing planets would need to have existed at one time or another in the Milky Way in order for us to have a decent chance of finding another civilization out there now.

However, we'd expect f_{now} to be so small only if we are on the brink of self-destruction—after all, the fraction will grow larger for as long as our civilization survives. That is, if civilizations are at all common, survivability is the key factor in whether any are out there now. If

Even if civilizations have arisen on many planets, they are unlikely to exist today unless they avoided early self-destruction.

most civilizations self-destruct shortly after achieving the technology for interstellar communication, then we are almost certainly alone in the galaxy at present. But if most survive and thrive for thousands or millions of years, the Milky Way may be brimming with civilizations—most of them far more advanced than our own.

think about it Describe a few reasons why a civilization capable of interstellar communication would also be capable of self-destruction. Overall, do you believe our civilization can survive for thousands or millions of years? Defend your opinion.

◆ How does SETI work?

If there are indeed other civilizations out there, then in principle we ought to be able to make contact with them. Based on our current understanding of physics, it seems likely that even very advanced civilizations would communicate much as we do—by encoding signals in radio waves or other forms of light. Most SETI researchers use large radio telescopes to search for alien radio signals (Figure 19.21). A few researchers are studying other parts of the electromagnetic spectrum as well. For example, some scientists use visible-light telescopes to search for communications encoded as laser pulses. Of course, advanced civilizations may well have invented communication technologies that we cannot even imagine, let alone detect.

A good way to think about our chances of picking up an alien signal is to imagine what aliens would need to do to pick up signals from us.

In principle, aliens within about 60 light-years of Earth could watch past television broadcasts. Our own current SETI efforts could pick up only much stronger signals.

We have been sending relatively high-powered transmissions into space since about the 1950s in the form of television broadcasts. In principle, anyone within about 60 light-years of Earth could watch our old television shows (perhaps a frightening thought). However, in order to detect our broadcasts, they would need far larger and more sensitive radio telescopes than we have today. If their technology were at the same level as ours, they could receive a signal from us only if we deliberately broadcast an unusually high-powered transmission.

To date, humans have made only a few attempts to broadcast our existence in this way. The most famous was a 3-minute transmission made in 1974 with a radar transmitter on the Arecibo radio telescope, in which a simple pictorial message was beamed toward the globular cluster M13 (Figure 19.22). This target was chosen in part because it

▲ **FIGURE 19.21**
The Allen Telescope Array, in Hat Creek, California, is being used to search for radio signals from extraterrestrial civilizations.

▲ **FIGURE 19.22**
In 1974, a short message was broadcast to the globular cluster M13 using the Arecibo radio telescope. The picture shown here was encoded by using two different radio frequencies, one for "on" and one for "off" (the colors shown here are arbitrary). To decode the message, the aliens would need to realize that the bits are meant to be arranged in a rectangular grid as shown, but that should not be difficult: The grid has 73 rows and 23 columns, and aliens would presumably know that these are both prime numbers. The picture represents the Arecibo radio dish, our solar system, a human stick figure, and a schematic of DNA and the eight simple molecules used in its construction.

contains a few hundred thousand stars, seemingly offering a good chance that at least one has a civilization around it. However, M13 is about 25,000 light-years from Earth, so it will take some 25,000 years for our signal to get there and another 25,000 years for any response to make its way back to Earth.

Several SETI projects under way or in development would be capable of detecting signals like the one we broadcast from Arecibo if they came from civilizations within a few hundred light-years. These SETI efforts scan millions of radio frequency bands simultaneously. If anyone nearby is deliberately broadcasting on an ongoing basis, we have a good chance of detecting the signals.

think about it SETI efforts are often controversial because of their cost (which is privately funded) and uncertain chance of success, but supporters say the cost is justified because contact with an extraterrestrial intelligence would be such an important discovery. Do you agree? Defend your opinion.

extraordinary claims — Aliens Are Visiting Earth in UFOs

In this chapter, we discuss aliens as a possibility, not a reality. However, opinion polls suggest that up to half the American public believes that aliens are already visiting us. What can science say about this extraordinary claim?

The bulk of the claimed evidence for alien visitation consists of sightings of UFOs—unidentified flying objects. Many thousands of UFOs are reported each year, and no one doubts that unidentified objects are being seen. The question is whether they are alien spacecraft.

Aliens have long been a staple of science fiction, but modern interest in UFOs began with a widely reported sighting in 1947. While flying a private plane near Mount Rainier in Washington State, businessman Kenneth Arnold saw nine mysterious objects streaking across the sky. He told a reporter that the objects "flew erratic, like a saucer if you skip it across the water." (One possible explanation is that he saw meteors skipping across the atmosphere.) He did *not* say that the objects were saucer-shaped, but the reporter wrote of "flying saucers." The story was front-page news throughout the United States, and "flying saucers" soon invaded popular culture, if not our planet.

The flying saucer reports also interested the U.S. Air Force, largely out of concern that the UFOs might represent new types of aircraft developed by the Soviet Union. For two decades, the Air Force hired teams of academics to study UFO reports. In most cases, these experts were able to specify a plausible identification of the UFO. The explanations included bright stars and planets, aircraft rockets, balloons, birds, meteors, atmospheric phenomena, and the occasional hoax. In a few cases, the investigators could not deduce what was seen, but their overall conclusion was that there was no reason to believe the UFOs were either highly advanced Soviet craft or visitors from other worlds.

Believers discounted the Air Force denials, claiming to have other evidence of alien visitation. So far, none of this evidence has withstood scientific scrutiny. Photographs and film clips are nearly always either fuzzy or obviously faked. Crop circles are easily made by pranksters, and claimed pieces of alien spacecraft turn out to have more mundane origins. Champions of alien visitation generally explain away the lack of clear evidence either as a government cover-up or as a failure of the scientific community to take the subject seriously, but neither explanation is compelling.

It's conceivable that a government might *try* to put the lid on evidence of alien visits, though the motivation for doing so is unclear. The usual explanations are that the public couldn't handle the news or that the government is taking secret advantage of the alien materials to design new military hardware (via "reverse engineering"). However, given that half the population already believes in alien visitors, the shock of discovery would seem unlikely to cause panic. As for reverse-engineering extraterrestrial spacecraft, any society that could routinely cross interstellar distances would be far beyond us technologically. Reverse engineering their spaceships is as unlikely as Neandertals constructing personal computers just because a laptop somehow landed in their cave. In addition, while a government might successfully hide evidence for a short time, over decades the lure of talk show fame and riches would surely cause someone to reveal the conspiracy. Moreover, unless the aliens landed only in the United States, the conspiracy would have to include all governments on Earth, which seems highly unlikely given the world's political conditions.

Claims that the scientific community is uninterested also fall apart on scrutiny. Scientists are constantly competing with one another to be the first to make a great discovery, and clear evidence of alien visitors would rank high on the all-time list. The fact that few scientists are engaged in such study reflects not a lack of interest, but a lack of evidence worthy of study.

Of course, absence of evidence is not evidence of absence. Most scientists are open to the possibility that we might someday find evidence of alien visits, and given what we've learned about the prospects of life in the universe, it is at least plausible to imagine that such visits occur. So far, however, we have no hard evidence to support the belief that aliens are already here.

Verdict: Not established. While it's conceivable that aliens could be visiting Earth, the claimed evidence is insufficient to warrant confidence in such an extraordinary claim.

19.5 Interstellar Travel and Its Implications for Civilization

So far, we have discussed ways of detecting distant civilizations without leaving the comfort of our own planet. Could we ever actually visit other worlds in other star systems? A careful analysis of this question turns out to have profound implications for the future of our civilization. To see why, we first need to consider the prospects for achieving interstellar travel.

◆ How difficult is interstellar travel?

In many science fiction movies, our descendants race around the galaxy in starships, circumventing nature's prohibition on faster-than-light travel by entering hyperspace, wormholes, or warp drive. Unfortunately, we do not have good reason to think that any of these science fiction technologies are really possible. In that case, we will be limited to speeds slower than the speed of light, and today we cannot even begin to approach that speed. Nevertheless, we have already sent out our first emissaries to the stars, and there's no reason to believe that we won't develop better technologies in the future.

The Challenge of Interstellar Travel We have launched five spacecraft that will leave our solar system and eventually travel among the stars: the planetary probes *Pioneer 10, Pioneer 11, Voyager 1, Voyager 2,* and *New Horizons*. These spacecraft are traveling about as fast as anything ever built by humans, but their speeds are still less than 1/10,000 the speed of light. It would take each of them some 100,000 years just to reach the next nearest star system (Alpha Centauri), but their trajectories won't take them anywhere near it. Instead, they will simply continue their journey without passing close to any nearby stars, wandering the Milky Way for millions or even billions of years to come. The *Pioneer* and *Voyager* spacecraft carry greetings from Earth, just in case someone comes across one of them someday (Figure 19.23).

▼ **FIGURE 19.23**

Messages aboard the *Pioneer* and *Voyager* spacecraft, which are bound for the stars.

a The *Pioneer* plaque, about the size of an automobile license plate. The human figures are shown in front of a drawing of the spacecraft to give them a sense of scale. The "prickly" graph to their left shows the Sun's position relative to nearby pulsars, and Earth's location around the Sun is shown below. Binary code indicates the pulsar periods; because pulsars slow with time, the periods will allow someone reading the plaque to determine when the spacecraft was launched.

b *Voyagers 1* and *2* carry a phonograph record—a 12-inch gold-plated copper disk containing music, greetings, and images from Earth.

If we want to make interstellar journeys within human lifetimes, we will need starships that can travel at speeds close to the speed of light. We will need entirely new types of engines to reach such high speeds.

High-speed interstellar travel would require thousands of times as much energy as the entire world currently uses each year.

The energy requirements of interstellar spacecraft may pose an even more daunting challenge. For example, the energy needed to accelerate a single ship the size of *Star Trek*'s *Enterprise* to just half the speed of light would be more than 2000 times the total annual energy use of the world today. Clearly, interstellar travel will require vast new sources of energy. In addition, fast-moving starships will require new types of shielding to protect crew members from instant death. As a starship travels through interstellar gas at near-light speed, ordinary atoms and ions will hit it like a deadly flood of high-energy cosmic rays.

If we succeed in building starships capable of traveling at speeds close to the speed of light, the crews will face significant social challenges. According to well-tested principles of Einstein's theory of relativity (see Special Topic, page 364), time will run much more slowly on a spaceship that travels at high speed to the stars than it does here on Earth. For example, in a ship traveling at an average speed of 99.9% of the speed of light, the 50-light-year round trip to the star Vega would take the travelers aboard only about 2 years—but more than 50 years would pass on Earth while they were gone. The crew would therefore need only 2 years' worth of provisions and would age only 2 years during the voyage, but they would return to a world quite different from the one they left. Family and friends would be older or deceased, new technologies might have made their knowledge and skills obsolete, and many political and social changes may have occurred in their absence. The crew would face a difficult adjustment when they came home to Earth.

Starship Design Despite all the difficulties, some scientists and engineers have already proposed designs that could in principle take us to nearby stars. In the 1960s, a group of scientists proposed *Project Orion*, which envisioned accelerating a spaceship with repeated detonations of relatively small hydrogen bombs. Each explosion would take place a few

▶ **FIGURE 19.24**
Artist's conception of the *Project Orion* starship, showing one of the small hydrogen bomb detonations that would propel it. Debris from the detonation strikes the flat disk, called the pusher plate, at the back of the spaceship. The central sections (enclosed in a lattice) hold the bombs, and the front sections house the crew.

tens of meters behind the spaceship and would propel the ship forward as the vaporized debris impacted a "pusher plate" on the back of the spacecraft (Figure 19.24). Calculations showed that a spaceship accelerated by the rapid-fire detonation of a million H-bombs could reach Alpha Centauri in just over a century. In principle, we could build an *Orion* spacecraft with existing technology, though it would be very expensive and would require an exception to the international treaty banning nuclear detonations in space.

High-speed interstellar travel remains well beyond our current capabilities, but we can envision future technologies that could make it possible.

A century to the nearest star isn't bad, but it still wouldn't make interstellar travel easy. Unfortunately, no available technology could go much faster. The problem is mass: Making a rocket faster requires more fuel, but adding fuel adds mass and makes it more difficult for the rocket to accelerate. Calculations show that even in the best case, rockets carrying nuclear fuel could achieve speeds no more than a few percent of the speed of light. However, we can envision some possible future technologies that might get around this problem.

One idea suggests powering starships with engines that generate energy through matter–antimatter annihilation. While nuclear fusion converts less than 1% of the mass of atomic nuclei into energy, matter–antimatter annihilation [Section 17.1] converts *all* the annihilated mass into energy. Starships with matter–antimatter engines could potentially reach speeds of 90% or more of the speed of light. At these speeds, the slowing of time predicted by relativity becomes noticeable, putting many nearby stars within a few years' journey for the crew members. However, because no natural reservoirs of antimatter exist, we would have to be able to manufacture many tons of antimatter and then store it safely for the trip—capabilities that are far beyond our present means.

An even more speculative and futuristic design, known as an *interstellar ramjet,* would collect interstellar hydrogen with a gigantic scoop, using the collected gas as fuel for its nuclear engines (Figure 19.25). By collecting fuel along the way, the ship would not need to carry the weight of fuel on board. However, because the density of interstellar gas is so low, the scoop would need to be enormous. As astronomer Carl Sagan said, we are talking about "spaceships the size of worlds."

The bottom line is that while we face enormous obstacles to achieving interstellar travel, there's no reason to think it's impossible. If we can avoid self-destruction and if we continue to explore space, our descendants might well journey to the stars.

◆ Where are the aliens?

Imagine that we survive long enough to become interstellar travelers and that we begin to colonize habitable planets around nearby stars. As each colony grows, it may send out explorers to other star systems. Even if our starships traveled at relatively low speeds—say, a few percent of the speed of light—we could have dozens of outposts around nearby stars within a few centuries. In 10,000 years, our descendants would be spread among stars within a few hundred light-years of Earth. In a few million years, we could have outposts throughout the Milky Way Galaxy. We would have become a true galactic civilization.

Now, if we take the idea that *we* could develop a galactic civilization within a few million years and combine it with the reasonable (though

▼ **FIGURE 19.25**

Artist's conception of a spaceship powered by an interstellar ramjet. The giant scoop in the front (left) collects interstellar hydrogen for use as fusion fuel.

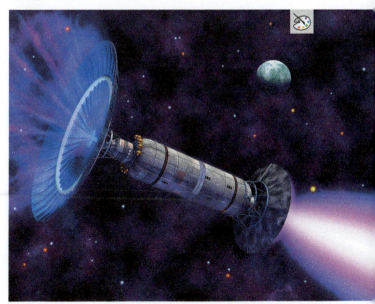

If we can develop interstellar travel in the future, then it would seem that other civilizations should have developed the capability long ago.

unproved) idea that civilizations ought to be common, we are led to an astonishing conclusion: Someone else should already have created a galactic civilization.

To see why, let's take some sample numbers. Suppose the overall odds of a civilization arising around a star are about the same as your odds of winning the lottery, or 1 in a million. Using a low estimate of 100 billion stars in the Milky Way Galaxy, this would mean there are some 100,000 civilizations in our galaxy alone. Moreover, current evidence suggests that stars and planetary systems like our own could have formed for at least 5 billion years before our solar system was even born, in which case the first of these 100,000 civilizations would have arisen at least 5 billion years ago. Others would have arisen, on average, about every 50,000 years. Under these assumptions, we would expect the youngest civilization besides ourselves to be some 50,000 years ahead of us technologically, and most would be millions or billions of years ahead of us.

We thereby encounter a strange paradox: Plausible arguments suggest that a galactic civilization should already exist, yet we have so far found no evidence of such a civilization. This paradox is often called *Fermi's paradox,* after the Nobel Prize–winning physicist Enrico Fermi. During a 1950 conversation with other scientists about the possibility of extraterrestrial intelligence, Fermi responded to speculations by asking, "So where is everybody?"

This paradox has many possible solutions, but broadly speaking we can group them into three categories:

1. *We are alone.* There is no galactic civilization because civilizations are extremely rare—so rare that we are the first to have arisen on the galactic scene, perhaps even the first in the universe.

2. *Civilizations are common, but no one has colonized the galaxy.* There are at least three possible reasons why this might be the case. Perhaps interstellar travel is much harder or more expensive than we have guessed, and civilizations are unable to venture far from their home worlds. Perhaps the desire to explore is unusual, and other societies either never leave their home star systems or stop exploring before they've colonized much of the galaxy. Most ominously, perhaps many civilizations have arisen, but they have all destroyed themselves before achieving the ability to colonize the stars.

3. *There IS a galactic civilization,* but it has not yet revealed its existence to us.

We do not know which, if any, of these explanations is the correct solution to the question "Where are the aliens?" However, each category of solution has astonishing implications for our own species.

Consider the first solution—that we are alone. If this is true, then our civilization is a remarkable achievement. It implies that through all of cosmic evolution, among countless star systems, we are the first piece of our galaxy or the universe ever to know that the rest of the universe exists. Through us, the universe has attained self-awareness. Some philosophers and many religions argue that the ultimate purpose of life is to become truly self-aware. If so, and if we are alone, then the destruction of our civilization and the loss of our scientific knowledge would represent an inglorious end to something that took the universe some 14 billion years to achieve. From this point of view, humanity becomes all the more precious, and the collapse of our civilization would be all the more tragic.

The second category of solutions has much more terrifying implications. If thousands of civilizations before us have all failed to achieve interstellar travel on a large scale, what hope do we have? Unless we somehow think differently than all other civilizations, this solution says that we will never go far in space. Because we have always explored when the opportunity arose, this solution almost inevitably leads to the conclusion that failure will come about because we destroy ourselves. We hope that this answer is wrong.

The third solution is perhaps the most intriguing. It says that we are newcomers on the scene of a galactic civilization that has existed for millions or billions of years before us. Perhaps this civilization is deliberately leaving us alone for the time being and will someday decide the time is right to invite us to join it.

No matter what the answer turns out to be, learning it will surely mark a turning point in the brief history of our species. Moreover, this turning point is likely to be reached within the next few decades or centuries. We already have the ability to destroy our civilization. If we do so, then our fate is sealed. But if we survive long enough to develop technology that can take us to the stars, the possibilities seem almost limitless.

---

the big picture — Putting Chapter 19 into Perspective

Throughout our study of astronomy, we have taken the "big picture" view of trying to understand how we fit into the universe. Here, at last, we have returned to Earth and we have examined the role of our own generation in the big picture of human history. Tens of thousands of past human generations have walked this Earth. Ours is the first generation with the technology to study the far reaches of our universe, to search for life elsewhere, and to travel beyond our home planet. It is up to us to decide whether we will use this technology to advance our species or to destroy it.

Imagine for a moment the grand view, a gaze across the centuries and millennia from this moment forward. Picture our descendants living among the stars, having created or joined a great galactic civilization. They will have the privilege of experiencing ideas, worlds, and discoveries far beyond our wildest imagination. Perhaps, in their history lessons, they will learn of our generation—the generation that history placed at the turning point and that managed to steer its way past the dangers of self-destruction and onto the path to the stars.

---

summary of key concepts

19.1 Life on Earth

◆ When did life arise on Earth?

Fossil evidence puts the origin of life at least 3.5 billion years ago, and carbon isotope evidence pushes this date to more than 3.85 billion years ago. Life therefore arose within a few hundred million years after Earth's birth, and possibly in a much shorter time.

◆ How did life arise on Earth?

Genetic evidence suggests that all life on Earth evolved from a common ancestor, which may have resembled microbes that live today in hot water near undersea volcanic vents. We do not know how this first organism arose, but laboratory experiments suggest that it may have been the result of natural chemical processes on the early Earth. Once life arose, it rapidly diversified and evolved through **natural selection.**

◆ What are the necessities of life?

Life on Earth thrives in a wide range of environments and in general seems to require only three things: a source of nutrients, a source of energy, and a liquid that is most likely water.

19.2 Life in the Solar System

◆ Could there be life on Mars?

Mars once had conditions that may have been conducive to an origin of life. If life arose, it might still survive in pockets of liquid water underground.

◆ Could there be life in the outer solar system?

Numerous worlds in the outer solar system may have subsurface lakes or oceans of liquid water, including Jupiter's moons Europa, Ganymede, and Callisto; Saturn's moons Enceladus and Titan; Neptune's moon Triton; and Pluto. Titan could also have life using methane or ethane as a liquid medium, since it has lakes of those substances on its surface.

19.3 Life Around Other Stars

◆ What are the requirements for surface life?

To have a habitable surface, a planet must reside in its star's **habitable zone.** Earth's long-term habitability has also been made possible by volcanism to create an atmosphere, plate tectonics to help maintain a stable climate, and a magnetic field to protect the atmosphere from the solar wind. It seems likely, but not certain, that other Earth-size worlds in habitable zones would also have habitable conditions on their surfaces.

◆ What kinds of extrasolar worlds might be habitable?

Surface habitability seems possible for planets or moons similar in size and composition to Earth and located within the habitable zone, and the habitable zone may extend farther for super-Earths or water worlds with thick hydrogen atmospheres. Subsurface habitability may be even more common, since it is possible on any world with enough internal heat to keep water liquid beneath the surface. **Orphan planets,** which do not orbit a star, also offer intriguing possibilities for subsurface life, and possibly even for surface life if they have thick enough atmospheres.

◆ How could we detect life on extrasolar planets?

Future telescopes should allow us to obtain crude images or spectra of planets within stellar habitable zones. An image of an extrasolar planet—even if only a few pixels in size—might indicate the presence of continents and oceans, or of seasonal changes. Spectroscopic analysis could tell us much more, and might reveal atmospheric gases that would be evidence of life.

19.4 The Search for Extraterrestrial Intelligence

◆ How many civilizations are out there?

We don't know, but the **Drake equation** gives us a way to organize our thinking about the question. The equation (in a modified form) says that the number of civilizations in the Milky Way Galaxy with which we could potentially communicate is $N_{HP} \times f_{life} \times f_{civ} \times f_{now}$, where N_{HP} is the number of habitable planets in the galaxy, f_{life} is the fraction of habitable planets that actually have life on them, f_{civ} is the fraction of life-bearing planets on which a civilization capable of interstellar communication has at some time arisen, and f_{now} is the fraction of all these civilizations that exist now.

◆ How does SETI work?

SETI, the search for extraterrestrial intelligence, generally refers to efforts to detect signals—such as radio or laser communications—coming from civilizations on other worlds.

19.5 Interstellar Travel and Its Implications for Civilization

◆ How difficult is interstellar travel?

Convenient interstellar travel remains well beyond our technological capabilities because of the technological requirements for engines, the enormous energy needed to accelerate spacecraft to speeds near the speed of light, and the difficulties of shielding the crew from radiation. Nevertheless, people have proposed ways around all these difficulties, and it seems reasonable to think that we will someday achieve interstellar travel if our civilization survives long enough.

◆ Where are the aliens?

A civilization capable of interstellar travel ought to be able to colonize the galaxy in a few million years or less, and the galaxy was around for billions of years before Earth was born. It therefore seems that someone should have colonized the galaxy long ago—yet we have no evidence of other civilizations. Every possible explanation for this surprising fact has astonishing implications for our species and our place in the universe.

visual skills check

Use the following questions to check your understanding of some of the many types of visual information used in astronomy. For additional practice, try the Chapter 19 Visual Quiz at MasteringAstronomy®.

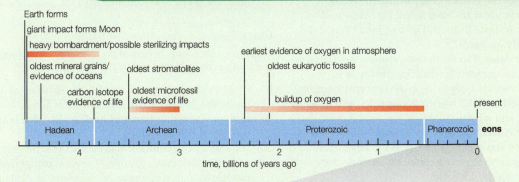

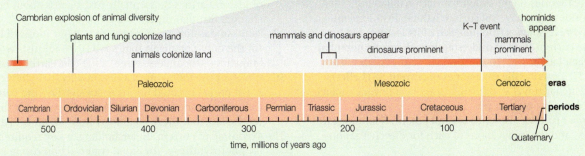

The figure above, which repeats Figure 19.3, shows the geological time scale. Use this figure to answer the following questions.

1. List the following events in the order in which they occurred, from first to last.
   a. earliest humans
   b. earliest animals
   c. impact causes extinction of dinosaurs
   d. earliest mammals
   e. earliest plants living on land
   f. first time there is significant oxygen in Earth's atmosphere
   g. first life on Earth

2. List the following time frames in order by how long they lasted, from the longest one to the shortest one.
   a. Hadean eon   b. Proterozoic eon
   c. Paleozoic era   d. Cretaceous period

3. Which of the following are time frames in which we are living today? More than one may apply.
   a. Quaternary period   b. Tertiary period
   c. Cenozoic era   d. Phanerozoic eon
   e. Paleozoic era

4. How long did the Cambrian explosion last?
   a. less than 1 year   b. about a decade
   c. about 10,000 years   d. about 20 million years
   e. about 500 million years

5. When did the heavy bombardment end?
   a. about 4.5 billion years ago
   b. between about 4.3 and 4.5 billion years ago
   c. between about 3.8 and 4.0 billion years ago
   d. exactly 3.85 billion years ago

6. How long have mammals been present on Earth?
   a. about 1 million years   b. about 65 million years
   c. about 225 million years   d. about 510 million years

exercises and problems

MasteringAstronomy® For instructor-assigned homework and other learning materials, go to MasteringAstronomy®.

Review Questions

1. What is *astrobiology,* and what type of research does it involve?
2. How do we study the history of life on Earth? Describe the *geological time scale* and a few of the major events along it.

3. Summarize the evidence pointing to an early origin of life on Earth. How far back in Earth's history did life exist?
4. Why is the *theory of evolution* so critical to our understanding of the history of life on Earth? Explain how evolution proceeds by *natural selection,* and what happens to DNA that allows species to evolve.

5. How are laboratory experiments helping us study the origin of life on Earth? Explain.

6. Give a brief overview of the history of life on Earth. What evidence points to a common ancestor for all life? How and when did oxygen accumulate in Earth's atmosphere? When did larger animals diversify on Earth?

7. Is it possible that life migrated to Earth from elsewhere? Explain.

8. Describe the range of environments in which life thrives on Earth. What three basic requirements apply to life in all these environments?

9. What is a *habitable world*? Which worlds in our solar system seem potentially habitable, and why?

10. Briefly summarize the current status of the search for life on Mars.

11. What do we mean by a star's *habitable zone*? What key factors have given Earth long-term surface habitability, and do they seem likely on other worlds? Explain.

12. What types of worlds might support surface habitability? What types of worlds might have subsurface habitability? Explain.

13. What is the *Drake equation*? Define each of its factors, and describe the current state of understanding about the potential values of each factor.

14. What is *SETI*? Describe the capabilities of current SETI efforts.

15. Why is interstellar travel so difficult? Describe a few technologies that might someday make it possible.

16. What is *Fermi's paradox*? Describe several potential solutions to the paradox, and the implications of each for our civilization.

Test Your Understanding

◐ Fantasy or Science Fiction?

For each of the following futuristic scenarios, decide whether it is plausible or implausible according to our present understanding of science. Explain clearly; not all of these have definitive answers, so your explanation is more important than your chosen answer.

17. The first human explorers on Mars discover the ruins of an ancient civilization, including remnants of tall buildings and temples.

18. The first human explorers on Mars drill a hole into a Martian volcano to collect a sample of soil from deep underground. On analyzing the soil, they discover that it holds living microbes resembling terrestrial bacteria but with a different biochemistry.

19. In 2040, a spacecraft lands on Europa and melts its way through the ice into the Europan ocean. It finds numerous strange, living microbes, along with a few larger organisms that feed on the microbes.

20. It's the year 2075. A giant telescope on the Moon, consisting of hundreds of small telescopes linked together across a distance of 500 km, has just captured a series of images of a planet around a distant star that clearly show seasonal changes in vegetation.

21. A century from now, after completing a careful study of planets around stars within 100 light-years of Earth, astronomers discover that the most diverse life exists on a planet orbiting a young star that formed just 100 million years ago.

22. In 2040, a brilliant teenager working in her garage builds a coal-powered rocket that can travel at half the speed of light.

23. In the year 2750, we receive a signal from a civilization telling us that the *Voyager 2* spacecraft recently crash-landed on its planet, which orbits a nearby star.

24. Crew members of the matter–antimatter spacecraft *Star Apollo*, which left Earth in the year 2165, return to Earth in the year 2450, looking only a few years older than when they left.

25. Aliens from a distant star system invade Earth with the intent to destroy us and occupy our planet, but we successfully fight them off when their technology proves no match for ours.

26. A single great galactic civilization exists. It originated on a single planet long ago but is now made up of beings from many different planets, assimilated into the galactic culture.

Quick Quiz

Choose the best answer to each of the following. Explain your reasoning with one or more complete sentences.

27. Fossil evidence suggests that life on Earth arose (a) almost immediately after Earth formed. (b) within a few hundred million years after Earth formed. (c) about a billion years before the rise of the dinosaurs.

28. The theory of evolution is (a) a scientific theory, supported by extensive evidence. (b) one of several competing scientific models that all seem equally successful in explaining the nature of life on Earth. (c) essentially just a guess about how life changes through time.

29. Plants and animals are (a) the two major forms of life on Earth. (b) the only organisms that have DNA. (c) just two small branches of the diverse "tree of life" on Earth.

30. Which of the following is a reason why early living organisms on Earth probably could not have survived on the surface? (a) the lack of an ozone layer (b) the lack of oxygen for them to breathe (c) the fact that these organisms were single-celled

31. According to current understanding, a key requirement for life is (a) photosynthesis. (b) liquid water. (c) an ozone layer.

32. Which of the following worlds is *not* considered a candidate for harboring life? (a) Europa (b) Mars (c) the Moon

33. How does the habitable zone around a star of spectral type G compare to that around a star of spectral type M? (a) It is larger. (b) It is hotter. (c) It is closer to its star.

34. In the Drake equation, suppose that the term $f_{life} = \frac{1}{2}$. What would this mean? (a) Half the stars in the Milky Way Galaxy have a planet with life. (b) Half of all life forms in the universe are intelligent. (c) Half of the habitable worlds in the galaxy actually have life, while the other half don't.

35. The amount of energy that would be needed to accelerate a large spaceship to half the speed of light is (a) about 100 times as much energy as is needed to launch the Space Shuttle. (b) more than 2000 times the current annual global energy consumption. (c) more than the amount of energy released by a supernova.

36. According to current scientific understanding, the idea that the Milky Way Galaxy might be home to a civilization millions of years more advanced than ours is (a) a virtual certainty. (b) extremely unlikely. (c) one reasonable answer to Fermi's paradox.

◐ Process of Science

37. *The Science of Astrobiology.* The study of astrobiology is sometimes criticized as being the study of something for which we have no evidence, since we do not yet have evidence of life beyond Earth. Is astrobiology a science or speculation? Defend your opinion.

38. *Unanswered Questions.* In a sense, this entire chapter was about one big, unanswered question: Are we alone in the universe? But as we attempt to answer this "big" question, there are many smaller questions that we might wish to answer along the way. Describe one currently unanswered question about life in the universe that we might be able to answer with new missions or experiments over the next couple of decades. What kinds of evidence will we need to answer the question? How will we know when it is answered?

Group Work Exercise

39. *Habitable Planets?* **Roles:** *Scribe* (takes notes on the group's activities), *Proposer* (proposes explanations to the group), *Skeptic* (points out weaknesses in proposed explanations), *Moderator* (leads group discussion and makes sure everyone contributes). **Activity:** List the hypothetical planets described below in order of most probable to least probable, and explain the reasons for your ranking in each case.
    a. a planet orbiting a star of spectral type B (approximately 10 solar masses) in a circular orbit with an average temperature expected to be 300 K
    b. a planet orbiting a Sun-like star in a circular orbit at a distance two times Earth's orbital distance from the Sun
    c. a planet orbiting a star with a luminosity one-quarter the Sun's luminosity in a circular orbit at a distance one-half Earth's orbital distance from the Sun
    d. a planet orbiting a Sun-like star in an elliptical orbit ranging from Earth's orbital distance to ten times Earth's orbital distance from the Sun

Investigate Further

Short-Answer/Essay Questions

40. *Most Likely to Have Life.* Suppose you were asked to vote in a contest to name the world in our solar system (besides Earth) "most likely to have life." Which world would you cast your vote for? Explain and defend your choice in a one-page essay.

41. *Likely Suns.* Study the stellar data for nearby stars given in Appendix F, Table F.1. Which star on the list would you expect to have the largest habitable zone? Which would have the second-largest habitable zone? If we rule out multiple-star systems, which star would you expect to have the highest probability of having a habitable planet? Explain your answers.

42. *Is Life Common?* Based on what you have learned in this book, do you think life will ultimately prove to be rare, common, or something in between? Write a one- to two-page essay explaining and defending your opinion.

43. *Solution to the Fermi Paradox.* Among the various possible solutions to the question "Where are the aliens?" which do you think is most likely? Write a one- to two-page essay in which you explain why you favor this solution.

44. *What's Wrong with This Picture?* Many science fiction stories have imagined the galaxy divided into a series of empires, each having arisen from a different civilization on a different world, that hold each other at bay because they are all at about the same level of military technology. Is this a realistic scenario? Explain.

45. *Aliens in the Movies.* Choose a science fiction movie (or television show) that involves an alien species. Do you think aliens like this could really exist? Write a one- to two-page critical review of the movie or show, focusing primarily on the question of whether it portrays the aliens in a scientifically reasonable way.

Quantitative Problems

Be sure to show all calculations clearly and state your final answers in complete sentences.

46. *SETI Search.* Suppose there are 10,000 civilizations broadcasting radio signals in the Milky Way Galaxy right now. On average, how many stars would we have to search before we would expect to hear a signal? Assume there are 500 billion stars in the galaxy. How does your answer change if there are only 100 civilizations instead of 10,000?

47. *SETI Signal.* Consider a civilization broadcasting a signal with a power of 10,000 watts. The Arecibo radio telescope, which is about 300 m in diameter, could detect this signal if it were coming from as far away as 100 light-years. Suppose instead that the signal is being broadcast from the other side of the Milky Way Galaxy, about 70,000 light-years away. How large a radio telescope would we need to detect this signal? (*Hint:* Use the inverse square law for light.)

48. *Cruise Ship Energy.* Suppose we have a spaceship about the size of a typical ocean cruise ship today, which means it has a mass of about 100 million kg, and we want to accelerate the ship to a speed of 10% of the speed of light.
    a. How much energy would be required? (*Hint:* You can find the answer simply by calculating the kinetic energy of the ship when it reaches its cruising speed; because 10% of the speed of light is still small compared to the speed of light, you can use this formula: kinetic energy $= \frac{1}{2} \times m \times v^2$.)
    b. How does your answer compare to total worldwide energy use at present, which is about 5×10^{22} joules per year?
    c. The typical cost of energy today is roughly 5¢ per 1 million joules. At this price, how much would it cost to generate the energy needed by this spaceship?

49. *Matter–Antimatter Engine.* Consider the spaceship from Problem 48. Suppose you want to generate the energy to get it to cruising speed using matter–antimatter annihilation. How much antimatter would you need to produce and take on the ship? (*Hint:* When matter and antimatter meet, they turn all their mass into energy equivalent to mc^2.)

Discussion Questions

50. *Funding the Search for Life.* Imagine that you are a member of Congress who decides how much government funding goes to research in different areas of science. How much would you allot to the search for life in the universe compared to the amount allotted to research in other areas of astronomy and planetary science? Why?

51. *Distant Dream or Near-Reality?* Considering all the issues surrounding interstellar flight, when (if ever) do you think we are likely to begin traveling among the stars? Why?

52. *The Turning Point.* Discuss the idea that our generation has acquired a greater responsibility for the future than any previous generation. Do you agree with this assessment? If so, how should we deal with this responsibility? Defend your opinion.

Web Projects

53. *Astrobiology News.* Go to NASA's astrobiology site and read some of the recent news about the search for life in the universe. Choose one article and write a one- to two-page summary of the research.

54. *The Search for Extraterrestrial Intelligence.* Learn more about SETI at the SETI Institute home page, and summarize your findings in one page or less.

55. *Starship Design.* Read about a proposal for starship propulsion or design. How would the proposed starship work? What new technologies would be needed, and what existing technologies could be applied? Write a one- to two-page report on your research.

56. *Advanced Spacecraft Technologies.* NASA supports many efforts to incorporate new technologies into spaceships. Although few of them may be suitable for interstellar colonization, most are innovative and fascinating. Learn about one such NASA project, and write a short overview of your findings.

Throughout this book, we have seen that the history of the universe has proceeded in a way that has made our existence on Earth possible. This figure summarizes some of the key ideas, and leads us to ask: If life arose here, shouldn't it also have arisen on many other worlds? We do not yet know the answer, but scientists are actively seeking to learn whether life is rare or common in the universe.

1. The protons, neutrons, and electrons in the atoms that make up Earth and life were created out of pure energy during the first few moments after the Big Bang, leaving the universe filled with hydrogen and helium gas [Section 17.1].

electron

gamma-ray photons

antielectron

Matter can be created from energy: $E = mc^2$.

2. Ripples in the density of the early universe were necessary for life to form later on. Without those ripples, matter would never have collected into galaxies, stars, and planets [Section 17.3].

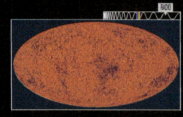

RADIO

We observe the seeds of structure formation in the cosmic microwave background.

3. The attractive force of gravity pulls together the matter that makes galaxies, stars, and planets [Section 4.4].

M_1 $F_g = G \dfrac{M_1 M_2}{d^2}$ M_2

d

Every piece of matter in the universe pulls on every other piece.

4. Our planet and all the life on it is made primarily of elements formed by nuclear fusion in high-mass stars and dispersed into space by supernovae [Section 13.3].

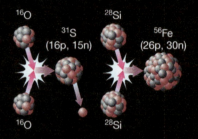

^{16}O

^{31}S (16p, 15n)

^{28}Si

^{56}Fe (26p, 30n)

^{16}O

^{28}Si

High-mass stars have cores hot enough to make elements heavier than carbon.

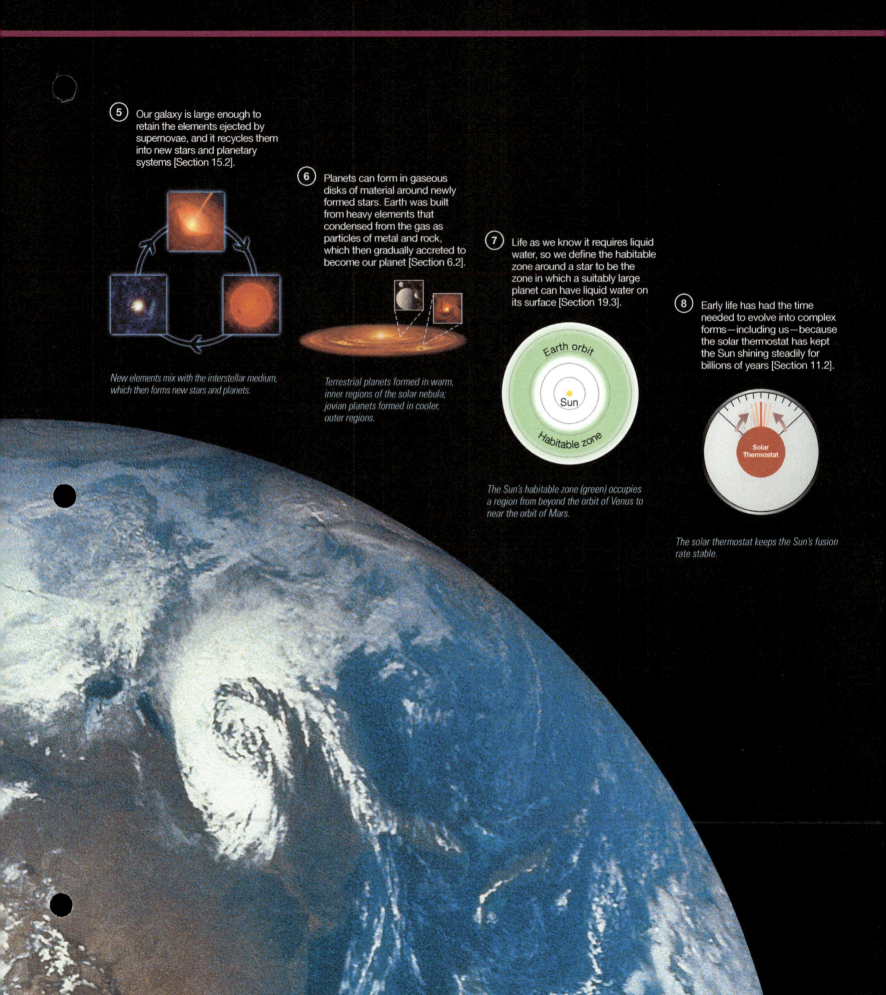

⑤ Our galaxy is large enough to retain the elements ejected by supernovae, and it recycles them into new stars and planetary systems [Section 15.2].

New elements mix with the interstellar medium, which then forms new stars and planets.

⑥ Planets can form in gaseous disks of material around newly formed stars. Earth was built from heavy elements that condensed from the gas as particles of metal and rock, which then gradually accreted to become our planet [Section 6.2].

Terrestrial planets formed in warm, inner regions of the solar nebula; jovian planets formed in cooler, outer regions.

⑦ Life as we know it requires liquid water, so we define the habitable zone around a star to be the zone in which a suitably large planet can have liquid water on its surface [Section 19.3].

Earth orbit

Sun

Habitable zone

The Sun's habitable zone (green) occupies a region from beyond the orbit of Venus to near the orbit of Mars.

⑧ Early life has had the time needed to evolve into complex forms—including us—because the solar thermostat has kept the Sun shining steadily for billions of years [Section 11.2].

Solar Thermostat

The solar thermostat keeps the Sun's fusion rate stable.

A   Useful Numbers

Astronomical Distances

1 AU $\approx 1.496 \times 10^8$ km $= 1.496 \times 10^{11}$ m

1 light-year $\approx 9.46 \times 10^{12}$ km $= 9.46 \times 10^{15}$ m

1 parsec (pc) $\approx 3.09 \times 10^{13}$ km ≈ 3.26 light-years

1 kiloparsec (kpc) $= 1000$ pc $\approx 3.26 \times 10^3$ light-years

1 megaparsec (Mpc) $= 10^6$ pc $\approx 3.26 \times 10^6$ light-years

Astronomical Times

1 solar day (average) $= 24^h$

1 sidereal day $\approx 23^h56^m4.09^s$

1 synodic month (average) ≈ 29.53 solar days

1 sidereal month (average) ≈ 27.32 solar days

1 tropical year ≈ 365.242 solar days

1 sidereal year ≈ 365.256 solar days

Universal Constants

Speed of light: $c = 3.00 \times 10^5$ km/s $= 3 \times 10^8$ m/s

Gravitational constant: $G = 6.67 \times 10^{-11} \dfrac{m^3}{kg \times s^2}$

Planck's constant: $h = 6.63 \times 10^{-34}$ joule $\times$ s

Stefan-Boltzmann constant: $\sigma = 5.67 \times 10^{-8} \dfrac{watt}{m^2 \times K^4}$

Mass of a proton: $m_p = 1.67 \times 10^{-27}$ kg

Mass of an electron: $m_e = 9.11 \times 10^{-31}$ kg

Energy and Power Units

Basic unit of energy: 1 joule $= 1 \dfrac{kg \times m^2}{s^2}$

Basic unit of power: 1 watt $= 1$ joule/s

Electron-volt: 1 eV $= 1.60 \times 10^{-19}$ joule

Useful Sun and Earth Reference Values

Mass of the Sun: $1 M_{Sun} \approx 2 \times 10^{30}$ kg

Radius of the Sun: $1 R_{Sun} \approx 696{,}000$ km

Luminosity of the Sun: $1 L_{Sun} \approx 3.8 \times 10^{26}$ watts

Mass of Earth: $1 M_{Earth} \approx 5.97 \times 10^{24}$ kg

Radius (equatorial) of Earth: $1 R_{Earth} \approx 6378$ km

Acceleration of gravity on Earth: $g = 9.8$ m/s^2

Escape velocity from surface of Earth: $v_{escape} = 11.2$ km/s $= 11{,}200$ m/s

B Useful Formulas

- Universal law of gravitation for the force between objects of mass M_1 and M_2, with distance d between their centers:

$$F = G\frac{M_1 M_2}{d^2}$$

- Newton's version of Kepler's third law, which applies to any pair of orbiting objects, such as a star and planet, a planet and moon, or two stars in a binary system; p is the orbital period, a is the distance between the centers of the orbiting objects, and M_1 and M_2 are the object masses:

$$p^2 = \frac{4\pi^2}{G(M_1 + M_2)}a^3$$

- Escape velocity at distance R from center of object of mass M:

$$v_{escape} = \sqrt{\frac{2GM}{R}}$$

- Relationship between a photon's wavelength (λ), frequency (f), and the speed of light (c):

$$\lambda \times f = c$$

- Energy of a photon of wavelength λ or frequency f:

$$E = hf = \frac{hc}{\lambda}$$

- Stefan-Boltzmann law for thermal radiation at temperature T (on the Kelvin scale):

$$\text{emitted power per unit area} = \sigma T^4$$

- Wien's law for the peak wavelength (λ_{max}) thermal radiation at temperature T (on the Kelvin scale):

$$\lambda_{max} = \frac{2{,}900{,}000}{T}\ \text{nm}$$

- Doppler shift (radial velocity is positive if the object is moving away from us and negative if it is moving toward us):

$$\frac{\text{radial velocity}}{\text{speed of light}} = \frac{\text{shifted wavelength} - \text{rest wavelength}}{\text{rest wavelength}}$$

- Angular separation (α) of two points with an actual separation s, viewed from a distance d (assuming d is much larger than s):

$$\alpha = \frac{s}{2\pi d} \times 360°$$

- Inverse square law for light (d is the distance to the object):

$$\text{apparent brightness} = \frac{\text{luminosity}}{4\pi d^2}$$

- Parallax formula (distance d to a star with parallax angle p in arcseconds):

$$d \text{ (in parsecs)} = \frac{1}{p \text{ (in arcseconds)}}$$

$$\text{or } d \text{ (in light-years)} = 3.26 \times \frac{1}{p \text{ (in arcseconds)}}$$

- The orbital velocity law, to find the mass M_r contained within the circular orbit of radius r for an object moving at speed v:

$$M_r = \frac{r \times v^2}{G}$$

C   A Few Mathematical Skills

This appendix reviews the following mathematical skills: powers of 10, scientific notation, working with units, the metric system, and finding a ratio. You should refer to this appendix as needed while studying the textbook.

C.1   Powers of 10

Powers of 10 indicate how many times to multiply 10 by itself. For example:

$$10^2 = 10 \times 10 = 100$$

$$10^6 = 10 \times 10 \times 10 \times 10 \times 10 \times 10 = 1,000,000$$

Negative powers are the reciprocals of the corresponding positive powers. For example:

$$10^{-2} = \frac{1}{10^2} = \frac{1}{100} = 0.01$$

$$10^{-6} = \frac{1}{10^6} = \frac{1}{1,000,000} = 0.000001$$

Table C.1 lists powers of 10 from 10^{-12} to 10^{12}. Note that powers of 10 follow two basic rules:

1. A positive exponent tells how many zeros follow the 1. For example, 10^0 is a 1 followed by no zeros, and 10^8 is a 1 followed by eight zeros.

2. A negative exponent tells how many places are to the right of the decimal point, including the 1. For example, $10^{-1} = 0.1$ has one place to the right of the decimal point; $10^{-6} = 0.000001$ has six places to the right of the decimal point.

Multiplying and Dividing Powers of 10

Multiplying powers of 10 simply requires adding exponents, as the following examples show:

$$10^4 \times 10^7 = \underbrace{10,000}_{10^4} \times \underbrace{10,000,000}_{10^7}$$

$$= \underbrace{100,000,000,000}_{10^{4+7}\,=\,10^{11}} = 10^{11}$$

$$10^5 \times 10^{-3} = \underbrace{100,000}_{10^5} \times \underbrace{0.001}_{10^{-3}}$$

$$= \underbrace{100}_{10^{5+(-3)}\,=\,10^2} = 10^2$$

$$10^{-8} \times 10^{-5} = \underbrace{0.00000001}_{10^{-8}} \times \underbrace{0.00001}_{10^{-5}}$$

$$= \underbrace{0.0000000000001}_{10^{-8+(-5)}\,=\,10^{-13}} = 10^{-13}$$

Dividing powers of 10 requires subtracting exponents, as in the following examples:

$$\frac{10^5}{10^3} = \underbrace{100,000}_{10^5} \div \underbrace{1000}_{10^3}$$

$$= \underbrace{100}_{10^{5-3}\,=\,10^2} = 10^2$$

$$\frac{10^3}{10^7} = \underbrace{1000}_{10^3} \div \underbrace{10,000,000}_{10^7}$$

TABLE C.1   Powers of 10

Zero and Positive Powers			Negative Powers		
Power	Value	Name	Power	Value	Name
10^0	1	One			
10^1	10	Ten	10^{-1}	0.1	Tenth
10^2	100	Hundred	10^{-2}	0.01	Hundredth
10^3	1000	Thousand	10^{-3}	0.001	Thousandth
10^4	10,000	Ten thousand	10^{-4}	0.0001	Ten-thousandth
10^5	100,000	Hundred thousand	10^{-5}	0.00001	Hundred-thousandth
10^6	1,000,000	Million	10^{-6}	0.000001	Millionth
10^7	10,000,000	Ten million	10^{-7}	0.0000001	Ten-millionth
10^8	100,000,000	Hundred million	10^{-8}	0.00000001	Hundred-millionth
10^9	1,000,000,000	Billion	10^{-9}	0.000000001	Billionth
10^{10}	10,000,000,000	Ten billion	10^{-10}	0.0000000001	Ten-billionth
10^{11}	100,000,000,000	Hundred billion	10^{-11}	0.00000000001	Hundred-billionth
10^{12}	1,000,000,000,000	Trillion	10^{-12}	0.000000000001	Trillionth

$$= \underbrace{\frac{0.0001}{10^{3-7} = 10^{-4}}} = 10^{-4}$$

$$\frac{10^{-4}}{10^{-6}} = \underbrace{0.0001}_{10^{-4}} \div \underbrace{0.000001}_{10^{-6}}$$

$$= \underbrace{100}_{10^{-4-(-6)} = 10^2} = 10^2$$

Powers of Powers of 10

We can use the multiplication and division rules to raise powers of 10 to other powers or to take roots. For example:

$$(10^4)^3 = 10^4 \times 10^4 \times 10^4 = 10^{4+4+4} = 10^{12}$$

Note that we can get the same end result by simply multiplying the two powers:

$$(10^4)^3 = 10^{4 \times 3} = 10^{12}$$

Because taking a root is the same as raising to a fractional power (e.g., the square root is the same as the $\frac{1}{2}$ power, the cube root is the same as the $\frac{1}{3}$ power, etc.), we can use the same procedure for roots, as in the following example:

$$\sqrt{10^4} = (10^4)^{1/2} = 10^{4 \times (1/2)} = 10^2$$

Adding and Subtracting Powers of 10

Unlike multiplying and dividing powers of 10, there is no shortcut for adding or subtracting powers of 10. The values must be written in longhand notation. For example:

$$10^6 + 10^2 = 1,000,000 + 100 = 1,000,100$$

$$10^8 + 10^{-3} = 100,000,000 + 0.001 = 100,000,000.001$$

$$10^7 - 10^3 = 10,000,000 - 1000 = 9,999,000$$

Summary

We can summarize our findings using n and m to represent any numbers:

- To *multiply* powers of 10, *add* exponents:
  $10^n \times 10^m = 10^{n+m}$
- To *divide* powers of 10, *subtract* exponents:
  $\frac{10^n}{10^m} = 10^{n-m}$
- To *raise* powers of 10 to other powers, multiply exponents:
  $(10^n)^m = 10^{n \times m}$
- To add or subtract powers of 10, first write them out longhand.

C.2 Scientific Notation

When we are dealing with large or small numbers, it's generally easier to write them with powers of 10. For example, it's much easier to write the number 6,000,000,000,000 as 6×10^{12}. This format, in which a number *between* 1 and 10 is multiplied by a power of 10, is called **scientific notation.**

Converting a Number to Scientific Notation

We can convert numbers written in ordinary notation to scientific notation with a simple two-step process:

1. Move the decimal point to come after the *first* nonzero digit.
2. The number of places the decimal point moves tells you the power of 10; the power is *positive* if the decimal point moves to the left and *negative* if it moves to the right.

Examples:

$$3042 \xrightarrow[\text{3 places to left}]{\text{decimal needs to move}} 3.042 \times 10^3$$

$$0.00012 \xrightarrow[\text{4 places to right}]{\text{decimal needs to move}} 1.2 \times 10^{-4}$$

$$226 \times 10^2 \xrightarrow[\text{2 places to left}]{\text{decimal needs to move}} (2.26 \times 10^2) \times 10^2 = 2.26 \times 10^4$$

Converting a Number from Scientific Notation

We can convert numbers written in scientific notation to ordinary notation by the reverse process:

1. The power of 10 indicates how many places to move the decimal point; move it to the *right* if the power of 10 is positive and to the *left* if it is negative.
2. If moving the decimal point creates any open places, fill them with zeros.

Examples:

$$4.01 \times 10^2 \xrightarrow[\text{2 places to right}]{\text{move decimal}} 401$$

$$3.6 \times 10^6 \xrightarrow[\text{6 places to right}]{\text{move decimal}} 3,600,000$$

$$5.7 \times 10^{-3} \xrightarrow[\text{3 places to left}]{\text{move decimal}} 0.0057$$

Multiplying or Dividing Numbers in Scientific Notation

Multiplying or dividing numbers in scientific notation simply requires operating on the powers of 10 and the other parts of the number separately.

Examples:

$$(6 \times 10^2) \times (4 \times 10^5) = (6 \times 4) \times (10^2 \times 10^5)$$
$$= 24 \times 10^7 = (2.4 \times 10^1) \times 10^7$$
$$= 2.4 \times 10^8$$

$$\frac{4.2 \times 10^{-2}}{8.4 \times 10^{-5}} = \frac{4.2}{8.4} \times \frac{10^{-2}}{10^{-5}} = 0.5 \times 10^{-2-(-5)} = 0.5 \times 10^3$$
$$= (5 \times 10^{-1}) \times 10^3 = 5 \times 10^2$$

Note that, in both these examples, we first found an answer in which the number multiplied by a power of 10 was *not* between 1 and 10. We therefore followed the procedure for converting the final answer to scientific notation.

Addition and Subtraction with Scientific Notation

In general, we must write numbers in ordinary notation before adding or subtracting.

Examples:

$$(3 \times 10^6) + (5 \times 10^2) = 3{,}000{,}000 + 500$$
$$= 3{,}000{,}500 = 3.0005 \times 10^6$$
$$(4.6 \times 10^9) - (5 \times 10^8) = 4{,}600{,}000{,}000 - 500{,}000{,}000$$
$$= 4{,}100{,}000{,}000 = 4.1 \times 10^9$$

When both numbers have the *same* power of 10, we can factor out the power of 10 first.

Examples:

$$(7 \times 10^{10}) + (4 \times 10^{10}) = (7 + 4) \times 10^{10}$$
$$= 11 \times 10^{10} = 1.1 \times 10^{11}$$
$$(2.3 \times 10^{-22}) - (1.6 \times 10^{-22}) = (2.3 - 1.6) \times 10^{-22}$$
$$= 0.7 \times 10^{-22} = 7.0 \times 10^{-23}$$

C.3 Working with Units

Showing the units of a problem as you solve it usually makes the work much easier and also provides a useful way of checking your work. If an answer does not come out with the units you expect, you probably did something wrong. In general, working with units is very similar to working with numbers, as the following guidelines and examples show.

Five Guidelines for Working with Units

Before you begin any problem, think ahead and identify the units you expect for the final answer. Then operate on the units along with the numbers as you solve the problem. The following five guidelines may be helpful when you are working with units:

1. Mathematically, it doesn't matter whether a unit is singular (e.g., meter) or plural (e.g., meters); we can use the same abbreviation (e.g., m) for both.

2. You cannot add or subtract numbers unless they have the *same* units. For example, 5 apples + 3 apples = 8 apples, but the expression 5 apples + 3 oranges cannot be simplified further.

3. You *can* multiply units, divide units, or raise units to powers. Look for key words that tell you what to do.

   - *Per* suggests division. For example, we write a speed of 100 kilometers per hour as

$$100 \ \frac{km}{hr} \ \text{or} \ 100 \ \frac{km}{1 \ hr}$$

- *Of* suggests multiplication. For example, if you launch a 50-kg space probe at a launch cost *of* $10,000 per kilogram, the total cost is

$$50 \ kg \times \frac{\$10{,}000}{kg} = \$500{,}000$$

- *Square* suggests raising to the second power. For example, we write an area of 75 square meters as 75 m^2.

- *Cube* suggests raising to the third power. For example, we write a volume of 12 cubic centimeters as 12 cm^3.

4. Often the number you are given is not in the units you wish to work with. For example, you may be given that the speed of light is 300,000 km/s but need it in units of m/s for a particular problem. To convert the units, simply multiply the given number by a *conversion factor:* a fraction in which the numerator (top of the fraction) and denominator (bottom of the fraction) are equal, so that the value of the fraction is 1; the number in the denominator must have the units that you wish to change. In the case of changing the speed of light from units of km/s to m/s, you need a conversion factor for kilometers to meters. Thus, the conversion factor is

$$\frac{1000 \ m}{1 \ km}$$

Note that this conversion factor is equal to 1, since 1000 meters and 1 kilometer are equal, and that the units to be changed (km) appear in the denominator. We can now convert the speed of light from units of km/s to m/s simply by multiplying by this conversion factor:

$$\underbrace{300{,}000 \ \frac{km}{s}}_{\substack{\text{speed of light} \\ \text{in km/s}}} \times \underbrace{\frac{1000 \ m}{1 \ km}}_{\substack{\text{conversion from} \\ \text{km to m}}} = \underbrace{3 \times 10^8 \ \frac{m}{s}}_{\substack{\text{speed of light} \\ \text{in m/s}}}$$

Note that the units of km cancel, leaving the answer in units of m/s.

5. It's easier to work with units if you replace division with multiplication by the reciprocal. For example, suppose you want to know how many minutes are represented by 300 seconds. We can find the answer by dividing 300 seconds by 60 seconds per minute:

$$300 \ s \div 60 \ \frac{s}{min}$$

However, it is easier to see the unit cancellations if we rewrite this expression by replacing the division with multiplication by the reciprocal (this process is easy to remember as "invert and multiply"):

$$300 \ s \div 60 \ \frac{s}{min} = 300 \ s \times \underbrace{\frac{1 \ min}{60 \ s}}_{\substack{\text{invert} \\ \text{and multiply}}} = 5 \ min$$

We now see that the units of seconds (s) cancel in the numerator of the first term and the denominator of the second term, leaving the answer in units of minutes.

More Examples of Working with Units

Example 1. How many seconds are there in 1 day?

Solution: We can answer the question by setting up a *chain* of unit conversions in which we start with 1 *day* and end up with *seconds*. We use the facts that there are 24 hours per day (24 hr/day), 60 minutes per hour (60 min/hr), and 60 seconds per minute (60 s/min):

$$1 \text{ day} \times \underbrace{\frac{24 \text{ hr}}{\text{day}}}_{\substack{\text{conversion} \\ \text{from} \\ \text{day to hr}}} \times \underbrace{\frac{60 \text{ min}}{\text{hr}}}_{\substack{\text{conversion} \\ \text{from} \\ \text{hr to min}}} \times \underbrace{\frac{60 \text{ s}}{\text{min}}}_{\substack{\text{conversion} \\ \text{from} \\ \text{min to s}}}$$

(starting value under "1 day")

$$= 86{,}400 \text{ s}$$

Note that all the units cancel except *seconds*, which is what we want for the answer. There are 86,400 seconds in 1 day.

Example 2. Convert a distance of 10^8 cm to km.

Solution: The easiest way to make this conversion is in two steps, since we know that there are 100 centimeters per meter (100 cm/m) and 1000 meters per kilometer (1000 m/km):

$$10^8 \text{ cm} \times \underbrace{\frac{1 \text{ m}}{100 \text{ cm}}}_{\substack{\text{conversion} \\ \text{from} \\ \text{cm to m}}} \times \underbrace{\frac{1 \text{ km}}{1000 \text{ m}}}_{\substack{\text{conversion} \\ \text{from} \\ \text{m to km}}}$$

(starting value under "10^8 cm")

$$= 10^8 \text{ cm} \times \frac{1 \text{ m}}{10^2 \text{ cm}} \times \frac{1 \text{ km}}{10^3 \text{ m}} = 10^3 \text{ km}$$

Alternatively, if we recognize that the number of kilometers should be smaller than the number of centimeters (because kilometers are larger), we might decide to do this conversion by dividing as follows:

$$10^8 \text{ cm} \div \frac{100 \text{ cm}}{\text{m}} \div \frac{1000 \text{ m}}{\text{km}}$$

In this case, before carrying out the calculation, we replace each division with multiplication by the reciprocal:

$$10^8 \text{ cm} \div \frac{100 \text{ cm}}{\text{m}} \div \frac{1000 \text{ m}}{\text{km}}$$

$$= 10^8 \text{ cm} \times \frac{1 \text{ m}}{100 \text{ cm}} \times \frac{1 \text{ km}}{1000 \text{ m}}$$

$$= 10^8 \text{ cm} \times \frac{1 \text{ m}}{10^2 \text{ cm}} \times \frac{1 \text{ km}}{10^3 \text{ m}}$$

$$= 10^3 \text{ km}$$

Note that we again get the answer that 10^8 cm is the same as 10^3 km, or 1000 km.

Example 3. Suppose you accelerate at 9.8 m/s^2 for 4 seconds, starting from rest. How fast will you be going?

Solution: The question asked "how fast?" so we expect to end up with a speed. Therefore, we multiply the acceleration by the amount of time you accelerated:

$$9.8 \, \frac{\text{m}}{\text{s}^2} \times 4 \text{ s} = (9.8 \times 4) \, \frac{\text{m} \times \text{s}}{\text{s}^2} = 39.2 \, \frac{\text{m}}{\text{s}}$$

Note that the units end up as a speed, showing that you will be traveling 39.2 m/s after 4 seconds of acceleration at 9.8 m/s^2.

Example 4. A reservoir is 2 km long and 3 km wide. Calculate its area, in both square kilometers and square meters.

Solution: We find its area by multiplying its length and width:

$$2 \text{ km} \times 3 \text{ km} = 6 \text{ km}^2$$

Next we need to convert this area of 6 km² to square meters, using the fact that there are 1000 meters per kilometer (1000 m/km). Note that we must square the term 1000 m/km when converting from km² to m²:

$$6 \text{ km}^2 \times \left(1000 \, \frac{\text{m}}{\text{km}} \right)^2 = 6 \text{ km}^2 \times 1000^2 \, \frac{\text{m}^2}{\text{km}^2}$$

$$= 6 \text{ km}^2 \times 1{,}000{,}000 \, \frac{\text{m}^2}{\text{km}^2}$$

$$= 6{,}000{,}000 \text{ m}^2$$

The reservoir area is 6 km², which is the same as 6 million m².

C.4 The Metric System (SI)

The modern version of the metric system, known as *Système Internationale d'Unites* (French for "International System of Units") or **SI,** was formally established in 1960. Today, it is the primary measurement system in nearly every country in the world with the exception of the United States. Even in the United States, it is the system of choice for science and international commerce. The basic units of length, mass, and time in the SI are

- The **meter** for length, abbreviated m
- The **kilogram** for mass, abbreviated kg
- The **second** for time, abbreviated s

Multiples of metric units are formed by powers of 10, using a prefix to indicate the power. For example, *kilo* means 10^3 (1000), so a kilometer is 1000 meters; a microgram is 0.000001 gram, because *micro* means 10^{-6}, or one millionth. Some of the more common prefixes are listed in Table C.2.

TABLE C.2 SI (Metric) Prefixes

	Small Values			Large Values	
Prefix	Abbreviation	Value	Prefix	Abbreviation	Value
Deci	d	10^{-1}	Deca	da	10^{1}
Centi	c	10^{-2}	Hecto	h	10^{2}
Milli	m	10^{-3}	Kilo	k	10^{3}
Micro	μ	10^{-6}	Mega	M	10^{6}
Nano	n	10^{-9}	Giga	G	10^{9}
Pico	p	10^{-12}	Tera	T	10^{12}

Metric Conversions

Table C.3 lists conversions between metric units and units used commonly in the United States. Note that the conversions between kilograms and pounds are valid only on Earth, because they depend on the strength of gravity.

TABLE C.3 Metric Conversions

To Metric	From Metric
1 inch = 2.540 cm	1 cm = 0.3937 inch
1 foot = 0.3048 m	1 m = 3.28 feet
1 yard = 0.9144 m	1 m = 1.094 yards
1 mile = 1.6093 km	1 km = 0.6214 mile
1 pound = 0.4536 kg	1 kg = 2.205 pounds

Example 1. International athletic competitions generally use metric distances. Compare the length of a 100-meter race to that of a 100-yard race.

Solution: Table C.3 shows that 1 m = 1.094 yd, so 100 m is 109.4 yd. Note that 100 meters is almost 110 yards; a good "rule of thumb" to remember is that distances in meters are about 10% longer than the corresponding number of yards.

Example 2. How many square kilometers are in 1 square mile?

Solution: We use the square of the miles-to-kilometers conversion factor:

$$(1 \text{ mi}^2) \times \left(\frac{1.6093 \text{ km}}{1 \text{ mi}}\right)^2 = (1 \text{ mi}^2) \times \left(1.6093^2 \frac{\text{km}^2}{\text{mi}^2}\right)$$

$$= 2.5898 \text{ km}^2$$

Therefore, 1 square mile is 2.5898 square kilometers.

C.5 Finding a Ratio

Suppose you want to compare two quantities, such as the average density of Earth and the average density of Jupiter. The way we do such a comparison is by dividing, which tells us the *ratio* of the two quantities. In this case, Earth's average density is 5.52 g/cm³ and Jupiter's average density is 1.33 g/cm³ (see Figure 8.1), so the ratio is

$$\frac{\text{average density of Earth}}{\text{average density of Jupiter}} = \frac{5.52 \text{ g/cm}^3}{1.33 \text{ g/cm}^3} = 4.15$$

Notice how the units cancel on both the top and the bottom of the fraction. We can state our result in two equivalent ways:

- The ratio of Earth's average density to Jupiter's average density is 4.15.
- Earth's average density is 4.15 times Jupiter's average density.

Sometimes, the quantities that you want to compare may each involve an equation. In such cases, you could, of course, find the ratio by first calculating each of the two quantities individually and then dividing. However, it is much easier if you first express the ratio as a fraction, putting the equation for one quantity on top and the other on the bottom. Some of the terms in the equation may then cancel out, making any calculations much easier.

Example 1. Compare the kinetic energy of a car traveling at 100 km/hr to that of the same car traveling at 50 km/hr.

Solution: We do the comparison by finding the ratio of the two kinetic energies, recalling that the formula for kinetic energy is $\frac{1}{2}mv^2$. Since we are not told the mass of the car, you might at first think that we don't have enough information to find the ratio. However, notice what happens when we put the equations for each kinetic energy into the ratio, calling the two speeds v_1 and v_2:

$$\frac{\text{K.E. car at } v_1}{\text{K.E. car at } v_2} = \frac{\frac{1}{2}m_{\text{car}} v_1^2}{\frac{1}{2}m_{\text{car}} v_2^2} = \frac{v_1^2}{v_2^2} = \left(\frac{v_1}{v_2}\right)^2$$

All the terms cancel except those with the two speeds, leaving us with a very simple formula for the ratio. Now we put in 100 km/hr for v_1 and 50 km/hr for v_2:

$$\frac{\text{K.E. car at } 100 \text{ km/hr}}{\text{K.E. car at } 50 \text{ km/hr}} = \left(\frac{100 \text{ km/hr}}{50 \text{ km/hr}}\right)^2 = 2^2 = 4$$

The ratio of the car's kinetic energies at 100 km/hr and 50 km/hr is 4. That is, the car has four times as much kinetic energy at 100 km/hr as it has at 50 km/hr.

Example 2. Compare the strength of gravity between Earth and the Sun to the strength of gravity between Earth and the Moon.

Solution: We do the comparison by taking the ratio of the Earth–Sun gravity to the Earth–Moon gravity. In this case, each quantity is found from the equation of Newton's law of gravity. (See Section 4.4.) Thus, the ratio is

$$\frac{\text{Earth–Sun gravity}}{\text{Earth–Moon gravity}} = \frac{G\frac{M_{\text{Earth}}M_{\text{Sun}}}{(d_{\text{Earth−Sun}})^2}}{G\frac{M_{\text{Earth}}M_{\text{Moon}}}{(d_{\text{Earth−Moon}})^2}}$$

$$= \frac{M_{\text{Sun}}}{(d_{\text{Earth−Sun}})^2} \times \frac{(d_{\text{Earth−Moon}})^2}{M_{\text{Moon}}}$$

Note how all but four of the terms cancel; the last step comes from replacing the division with multiplication by the reciprocal (the "invert and multiply" rule for division). We can simplify the work further by rearranging the terms so that we have the masses and distances together:

$$\frac{\text{Earth–Sun gravity}}{\text{Earth–Moon gravity}} = \frac{M_{\text{Sun}}}{M_{\text{Moon}}} \times \frac{(d_{\text{Earth−Moon}})^2}{(d_{\text{Earth−Sun}})^2}$$

Now it is just a matter of looking up the numbers (see Appendix E) and calculating:

$$\frac{\text{Earth–Sun gravity}}{\text{Earth–Moon gravity}} = \frac{1.99 \times 10^{30} \text{ kg}}{7.35 \times 10^{22} \text{ kg}} \times \frac{(384.4 \times 10^3 \text{ km})^2}{(149.6 \times 10^6 \text{ km})^2}$$

$$= 179$$

In other words, the Earth–Sun gravity is 179 times stronger than the Earth–Moon gravity.

D The Periodic Table of the Elements

Key

12	— Atomic number
Mg	— Element's symbol
Magnesium	— Element's name
24.305	— Atomic mass*

*Atomic masses are fractions because they represent a weighted average of atomic masses of different isotopes—in proportion to the abundance of each isotope on Earth. For elements with no stable isotopes, the atomic mass of the longest-lived isotope is given in parentheses.

1 H Hydrogen 1.00794																	2 He Helium 4.003
3 Li Lithium 6.941	4 Be Beryllium 9.01218											5 B Boron 10.81	6 C Carbon 12.011	7 N Nitrogen 14.007	8 O Oxygen 15.999	9 F Fluorine 18.988	10 Ne Neon 20.179
11 Na Sodium 22.990	12 Mg Magnesium 24.305											13 Al Aluminum 26.98	14 Si Silicon 28.086	15 P Phosphorus 30.974	16 S Sulfur 32.06	17 Cl Chlorine 35.453	18 Ar Argon 39.948
19 K Potassium 39.098	20 Ca Calcium 40.08	21 Sc Scandium 44.956	22 Ti Titanium 47.88	23 V Vanadium 50.94	24 Cr Chromium 51.996	25 Mn Manganese 54.938	26 Fe Iron 55.847	27 Co Cobalt 58.9332	28 Ni Nickel 58.69	29 Cu Copper 63.546	30 Zn Zinc 65.39	31 Ga Gallium 69.72	32 Ge Germanium 72.59	33 As Arsenic 74.922	34 Se Selenium 78.96	35 Br Bromine 79.904	36 Kr Krypton 83.80
37 Rb Rubidium 85.468	38 Sr Strontium 87.62	39 Y Yttrium 88.9059	40 Zr Zirconium 91.224	41 Nb Niobium 92.91	42 Mo Molybdenum 95.94	43 Tc Technetium (98)	44 Ru Ruthenium 101.07	45 Rh Rhodium 102.906	46 Pd Palladium 106.42	47 Ag Silver 107.868	48 Cd Cadmium 112.41	49 In Indium 114.82	50 Sn Tin 118.71	51 Sb Antimony 121.75	52 Te Tellurium 127.60	53 I Iodine 126.905	54 Xe Xenon 131.29
55 Cs Cesium 132.91	56 Ba Barium 137.34		72 Hf Hafnium 178.49	73 Ta Tantalum 180.95	74 W Tungsten 183.85	75 Re Rhenium 186.207	76 Os Osmium 190.2	77 Ir Iridium 192.22	78 Pt Platinum 195.08	79 Au Gold 196.967	80 Hg Mercury 200.59	81 Ti Thallium 204.383	82 Pb Lead 207.2	83 Bi Bismuth 208.98	84 Po Polonium (209)	85 At Astatine (210)	86 Rn Radon (222)
87 Fr Francium (223)	88 Ra Radium 226.0254		104 Rf Rutherfordium (263)	105 Db Dubnium (262)	106 Sg Seaborgium (266)	107 Bh Bohrium (267)	108 Hs Hassium (277)	109 Mt Meitnerium (268)	110 Ds Darmstadtium (281)	111 Rg Roentgenium (272)	112 Cn Copernicium (285)	113 Nh Nihonium (286)	114 Fl Flerovium (289)	115 Mc Moscovium (288)	116 Lv Livermorium (293)	117 Ts Tennessine (294)	118 Og Oganesson (294)

Lanthanide Series

57 La Lanthanum 138.906	58 Ce Cerium 140.12	59 Pr Praseodymium 140.908	60 Nd Neodymium 144.24	61 Pm Promethium (145)	62 Sm Samarium 150.36	63 Eu Europium 151.96	64 Gd Gadolinium 157.25	65 Tb Terbium 158.925	66 Dy Dysprosium 162.50	67 Ho Holmium 164.93	68 Er Erbium 167.26	69 Tm Thulium 168.934	70 Yb Ytterbium 173.04	71 Lu Lutetium 174.967

Actinide Series

89 Ac Actinium 227.028	90 Th Thorium 232.038	91 Pa Protactinium 231.036	92 U Uranium 238.029	93 Np Neptunium 237.048	94 Pu Plutonium (244)	95 Am Americium (243)	96 Cm Curium (247)	97 Bk Berkelium (247)	98 Cf Californium (251)	99 Es Einsteinium (252)	100 Fm Fermium (257)	101 Md Mendelevium (258)	102 No Nobelium (259)	103 Lr Lawrencium (260)

E Planetary Data

TABLE E.3 Satellites of the Solar System (as of 2016)[a]

Planet Satellite	Radius or Dimensions[b] (km)	Distance from Planet (10³ km)	Orbital Period[c] (Earth days)	Mass[d] (kg)	Density[d] (g/cm³)	Notes About the Satellite
Earth						
Moon	1738	384.4	27.322	7.349×10^{22}	3.34	Moon: Probably formed in giant impact.
Mars						
Phobos	$13 \times 11 \times 9$	9.38	0.319	1.3×10^{16}	1.9	Phobos, Deimos: Probable captured asteroids.
Deimos	$8 \times 6 \times 5$	23.5	1.263	1.8×10^{15}	2.2	
Jupiter						
Small inner moons (4 moons)	8–83	128–222	0.295–0.674	—	—	Metis, Adrastea, Amalthea, Thebe: Small moonlets within and near Jupiter's ring system.
Io	1821	421.6	1.769	8.933×10^{22}	3.57	Io: Most volcanically active object in the solar system.
Europa	1565	670.9	3.551	4.797×10^{22}	2.97	Europa: Possible oceans under icy crust.
Ganymede	2634	1070.0	7.155	1.482×10^{23}	1.94	Ganymede: Largest satellite in solar system; unusual ice geology.
Callisto	2403	1883.0	16.689	1.076×10^{23}	1.86	Callisto: Cratered iceball.
Irregular group 1 (7 moons)	4–85	7500–17,000	130–457	—	—	Themisto, Leda, Himalia, Lysithea, Elara, and others: Probable captured moons with inclined orbits.
Irregular group 2 (52 moons)	1–30	17,000–29,000	490–980	—	—	Ananke, Carme, Pasiphae, Sinope, and others: Probable captured moons in inclined backward orbits.
Saturn						
Small inner moons (12)	3–89	117–212	0.5–1.2	—	—	Pan, Atlas, Prometheus, Pandora, Epimetheus, Janus, and others: Small moonlets within and near Saturn's ring system.
Mimas	199	185.52	0.942	3.70×10^{19}	1.17	Mimas, Enceladus, Tethys: Small and medium-size iceballs, many with interesting geology.
Enceladus	249	238.02	1.370	1.2×10^{20}	1.24	
Tethys	530	294.66	1.888	6.17×10^{20}	1.26	
Calypso and Telesto	8–12	294.66	1.888	—	—	Calypso and Telesto: Small moonlets sharing Tethys's orbit.
Dione	559	377.4	2.737	1.08×10^{21}	1.44	Dione: Medium-size iceball, with interesting geology.
Helene and Polydeuces	2–16	377.4	2.737	1.6×10^{16}	—	Helene and Polydeuces: Small moonlets sharing Dione's orbit.
Rhea	764	527.04	4.518	2.31×10^{21}	1.33	Rhea: Medium-size iceball, with interesting geology.
Titan	2575	1221.85	15.945	1.35×10^{23}	1.88	Titan: Dense atmosphere shrouds surface; ongoing geological activity.
Hyperion	$180 \times 140 \times 112$	1481.1	21.277	2.8×10^{19}	—	Hyperion: Only satellite known not to rotate synchronously.
Iapetus	718	3561.3	79.331	1.59×10^{21}	1.21	Iapetus: Bright and dark hemispheres show greatest contrast in the solar system.
Phoebe	110	12,952	−550.4	1×10^{19}	—	Phoebe: Very dark; material ejected from Phoebe may coat one side of Iapetus.
Irregular groups (37 moons)	2–16	11,300–25,200	450–930 −550 to −150	—	—	Probable captured moons with highly inclined and/or backward orbits.

Uranus

Moon	Radius (km)	Distance (10³ km)	Period (days)	Mass (kg)	Density (g/cm³)	Notes
Small inner moons (13 moons)	5–81	49–98	0.3–0.9	—	—	*Cordelia, Ophelia, Bianca, Cressida, Desdemona, Juliet, Portia, Rosalind, Cupid, Belinda, Perdita, Puck, Mab:* Small moonlets within and near Uranus's ring system.
Miranda	236	129.8	1.413	6.6×10^{19}	1.26	*Miranda, Ariel, Umbriel, Titania, Oberon:* Small and medium-size iceballs, with some interesting geology.
Ariel	579	191.2	2.520	1.35×10^{21}	1.65	
Umbriel	584.7	266.0	4.144	1.17×10^{21}	1.44	
Titania	788.9	435.8	8.706	3.52×10^{21}	1.59	
Oberon	761.4	582.6	13.463	3.01×10^{21}	1.50	
Irregular group (9 moons)	5–95	4280–21,000	260–2800	—	—	*Francisco, Caliban, Stephano, Trinculo, Sycorax, Margaret, Prospero, Setebos, Ferdinand:* Probable captured moons; several in backward orbits.

Neptune

Moon	Radius (km)	Distance (10³ km)	Period (days)	Mass (kg)	Density (g/cm³)	Notes
Small inner moons (5 moons)	29–96	48–74	0.30–0.55	—	—	*Naiad, Thalassa, Despina, Galatea, Larissa:* Small moonlets within and near Neptune's ring system.
Proteus	218 × 208 × 201	117.6	1.121	6×10^{19}	—	
Triton	1352.6	354.59	−5.875	2.14×10^{22}	2.0	*Triton:* Probable captured Kuiper belt object—largest captured object in solar system.
Nereid	170	5588.6	360.125	3.1×10^{19}	—	*Nereid:* Small, icy moon; very little known.
Irregular group (6 moons)	12–27	16,600–49,300	1880–9750	—	—	*2002 N1, N2, N3, N4, 2003 N, 2004 N1:* Possible captured moons in inclined or backward orbit.

Pluto

Moon	Radius (km)	Distance (10³ km)	Period (days)	Mass (kg)	Density (g/cm³)	Notes
Charon	606	19.6[e]	6.38	1.59×10^{21}	1.702	*Charon:* Unusually large compared to Pluto; may have formed in giant impact.
Styx	1.8–9.8	42.4	20.2	—	—	*Styx, Nix, Kerberos, Hydra:* Newly discovered moons outside Charon's orbit.
Nix	54 × 41 × 36	48.7	24.9	—	—	
Kerberos	2.6–14	57.8	32.2	—	—	
Hydra	43 × 33[f]	64.7	38.2	—	—	

Eris

Moon	Radius (km)	Distance (10³ km)	Period (days)	Mass (kg)	Density (g/cm³)	Notes
Dysnomia	50	37.4	15.8	—	—	*Dysnomia:* Approximate properties determined in June 2007.

[a]*Note:* Authorities differ substantially on many of the values in this table.

[b]$a \times b \times c$ values for the dimensions are the approximate lengths of the axes (center to edge) for irregular moons.

[c]Negative sign indicates backward orbit.

[d]Masses and densities are most accurate for those satellites visited by a spacecraft on a flyby. Masses for the smallest moons have not been measured but can be estimated from the radius and an assumed density.

[e]Distance to system center of mass is 17.5×10^3 km.

[f]Third dimension not measured.

F Stellar Data

Stars Within 12 Light-Years

Star	Distance (ly)	Spectral Type		RA h	RA m	Dec °	Dec '	Luminosity (L/L_{Sun})
Sun	0.000016	G2	V	—	—	—	—	1.0
Proxima Centauri	4.2	M5.0	V	14	30	−62	41	0.0006
α Centauri A	4.4	G2	V	14	40	−60	50	1.6
α Centauri B	4.4	K0	V	14	40	−60	50	0.53
Barnard's Star	6.0	M4	V	17	58	+04	42	0.005
Wolf 359	7.8	M5.5	V	10	56	+07	01	0.0008
Lalande 21185	8.3	M2	V	11	03	+35	58	0.03
Sirius A	8.6	A1	V	06	45	−16	42	26.0
Sirius B	8.6	DA2	White dwarf	06	45	−16	42	0.002
BL Ceti	8.7	M5.5	V	01	39	−17	57	0.0009
UV Ceti	8.7	M6	V	01	39	−17	57	0.0006
Ross 154	9.7	M3.5	V	18	50	−23	50	0.004
Ross 248	10.3	M5.5	V	23	42	+44	11	0.001
ε Eridani	10.5	K2	V	03	33	−09	28	0.37
Lacaille 9352	10.7	M1.0	V	23	06	−35	51	0.05
Ross 128	10.9	M4	V	11	48	+00	49	0.003
EZ Aquarii A	11.3	M5	V	22	39	−15	18	0.0006
EZ Aquarii B	11.3	—	—	22	39	−15	18	0.0004
EZ Aquarii C	11.3	—	—	22	39	−15	18	0.0003
61 Cygni A	11.4	K5	V	21	07	+38	42	0.17
61 Cygni B	11.4	K7	V	21	07	+38	42	0.10
Procyon A	11.4	F5	IV–V	07	39	+05	14	8.6
Procyon B	11.4	DA	White dwarf	07	39	+05	14	0.0005
Gliese 725 A	11.5	M3	V	18	43	+59	38	0.02
Gliese 725 B	11.5	M3.5	V	18	43	+59	38	0.01
GX Andromedae	11.6	M1.5	V	00	18	+44	01	0.03
GQ Andromedae	11.6	M3.5	V	00	18	+44	01	0.003
ε Indi A	11.8	K5	V	22	03	−56	45	0.30
ε Indi B	11.8	T1.0	Brown dwarf	22	04	−56	46	—
ε Indi C	11.8	T6.0	Brown dwarf	22	04	−56	46	—
DX Cancri	11.8	M6.0	V	08	30	+26	47	0.0003
τ Ceti	11.9	G8.5	V	01	44	−15	57	0.67
GJ 1061	12.0	M5.0	V	03	36	−44	31	0.001

Note: These data were provided by the RECONS project, courtesy of Dr. Todd Henry (January, 2010). The luminosities are all total (bolometric) luminosities. The DA stellar types are white dwarfs. The coordinates are for the year 2000. The bolometric luminosity of the brown dwarfs is primarily in the infrared and has not been measured accurately yet.

Twenty Brightest Stars

Star	Constellation	RA		Dec		Distance	Spectral		Apparent	Luminosity
		h	m	°	′	(ly)	Type		Magnitude	(L/L_{Sun})
Sirius	Canis Major	6	45	−16	42	8.6	A1	V	−1.46	26
Canopus	Carina	6	24	−52	41	313	F0	Ib–II	−0.72	13,000
α Centauri	Centaurus	14	40	−60	50	4.4	G2	V	−0.01	1.6
							K0	V	1.3	0.53
Arcturus	Boötes	14	16	+19	11	37	K2	III	−0.06	170
Vega	Lyra	18	37	+38	47	25	A0	V	0.04	60
Capella	Auriga	5	17	+46	00	42	G0	III	0.75	70
							G8	III	0.85	77
Rigel	Orion	5	15	−08	12	772	B8	Ia	0.14	70,000
Procyon	Canis Minor	7	39	+05	14	11.4	F5	IV–V	0.37	7.4
Betelgeuse	Orion	5	55	+07	24	643	M2	Iab	0.41	120,000
Achernar	Eridanus	1	38	−57	15	144	B5	V	0.51	3600
Hadar	Centaurus	14	04	−60	22	525	B1	III	0.63	100,000
Altair	Aquila	19	51	+08	52	17	A7	IV–V	0.77	10.5
Acrux	Crux	12	27	−63	06	321	B1	IV	1.39	22,000
							B3	V	1.9	7500
Aldebaran	Taurus	4	36	+16	30	65	K5	III	0.86	350
Spica	Virgo	13	25	−11	09	260	B1	V	0.91	23,000
Antares	Scorpio	16	29	−26	26	604	M1	Ib	0.92	38,000
Pollux	Gemini	7	45	+28	01	34	K0	III	1.16	45
Fomalhaut	Piscis Austrinus	22	58	−29	37	25	A3	V	1.19	18
Deneb	Cygnus	20	41	+45	16	2500	A2	Ia	1.26	170,000
β Crucis	Crux	12	48	−59	40	352	B0.5	IV	1.28	37,000

Note: Three of the stars on this list, Capella, α Centauri, and Acrux, are binary systems with members of comparable brightness. They are counted as single stars because that is how they appear to the naked eye. All the luminosities given are total (bolometric) luminosities. The coordinates are for the year 2000.

G Galaxy Data

Galaxy Name	Distance (millions of ly)	Type[a]	RA h	RA m	Dec °	Dec ′	Luminosity (millions of L_{Sun})
Milky Way	—	Sbc	—	—	—	—	15,000
WLM	3.0	Irr	00	02	−15	30	50
IC 10	2.7	dIrr	00	20	+59	18	160
Cetus	2.5	dE	00	26	−11	02	0.72
NGC 147	2.4	dE	00	33	+48	30	131
And III	2.5	dE	00	35	+36	30	1.1
NGC 185	2.0	dE	00	39	+48	20	120
NGC 205	2.7	E	00	40	+41	41	370
And VIII	2.7	dE	00	42	+40	37	240
M32	2.6	E	00	43	+40	52	380
M31	2.5	Sb	00	43	+41	16	21,000
And I	2.6	dE	00	46	+38	00	4.7
SMC	0.19	Irr	00	53	−72	50	230
And IX	2.9	dE	00	52	+43	12	—
Sculptor	0.26	dE	01	00	−33	42	2.2
LGS 3	2.6	dIrr	01	04	+21	53	1.3
IC 1613	2.3	Irr	01	05	+02	08	64
And V	2.9	dE	01	10	+47	38	—
And II	1.7	dE	01	16	+33	26	2.4
M33	2.7	Sc	01	34	+30	40	2800
Phoenix	1.5	dIrr	01	51	−44	27	0.9
Fornax	0.45	dE	02	40	−34	27	15.5
EGB0427 + 63	4.3	dIrr	04	32	+63	36	9.1
LMC	0.16	Irr	05	24	−69	45	1300
Carina	0.33	dE	06	42	−50	58	0.4
Canis Major	0.025	dIrr	07	15	−28	00	—
Leo A	2.2	dIrr	09	59	+30	45	3.0
Sextans B	4.4	dIrr	10	00	+05	20	41
NGC 3109	4.1	Irr	10	03	−26	09	160
Antlia	4.0	dIrr	10	04	−27	19	1.7
Leo I	0.82	dE	10	08	+12	18	4.8
Sextans A	4.7	dIrr	10	11	−04	42	56
Sextans	0.28	dE	10	13	−01	37	0.5
Leo II	0.67	dE	11	13	+22	09	0.6
GR 8	5.2	dIrr	12	59	+14	13	3.4
Ursa Minor	0.22	dE	15	09	+67	13	0.3
Draco	2.7	dE	17	20	+57	55	0.3
Sagittarius	0.08	dE	18	55	−30	29	18
SagDIG	3.5	dIrr	19	30	−17	41	6.8
NGC 6822	1.6	Irr	19	45	−14	48	94
DDO 210	2.6	dIrr	20	47	−12	51	0.8
IC 5152	5.2	dIrr	22	03	−51	18	70
Tucana	2.9	dE	22	42	−64	25	0.5
UKS2323-326	4.3	dE	23	26	−32	23	5.2
And VII	2.6	dE	23	38	+50	35	—
Pegasus	3.1	dIrr	23	29	+14	45	12
And VI	2.8	dE	23	52	+24	36	—

[a] Types beginning with S are spiral galaxies classified according to Hubble's system (see Chapter 16). Type E galaxies are elliptical or spheroidal. Type Irr galaxies are irregular. The prefix d denotes a dwarf galaxy. This list is based on a list originally published by M. Mateo in 1998 and augmented by discoveries of Local Group galaxies made between 1998 and 2005.

Galaxy Name (M / NGC)[c]	RA h	RA m	Dec °	Dec '	RV_{hel}[d]	RV_{gal}[e]	Type[f]	Nickname
M31/NGC 224	00	43	+41	16	−300 ± 4	−122	Spiral	Andromeda
M32/NGC 221	00	43	+40	52	−145 ± 2	32	Elliptical	
M33/NGC 598	01	34	+30	40	−179 ± 3	−44	Spiral	Triangulum
M49/NGC 4472	12	30	+08	00	997 ± 7	929	Elliptical/Lenticular/ Seyfert	
M51/NGC 5194	13	30	+47	12	463 ± 3	550	Spiral/Interacting	Whirlpool
M58/NGC 4579	12	38	+11	49	1519 ± 6	1468	Spiral/Seyfert	
M59/NGC 4621	12	42	+11	39	410 ± 6	361	Elliptical	
M60/NGC 4649	12	44	+11	33	1117 ± 6	1068	Elliptical	
M61/NGC 4303	12	22	+04	28	1566 ± 2	1483	Spiral/Seyfert	
M63/NGC 5055	13	16	+42	02	504 ± 4	570	Spiral	Sunflower
M64/NGC 4826	12	57	+21	41	408 ± 4	400	Spiral/Seyfert	Black Eye
M65/NGC 3623	11	19	+13	06	807 ± 3	723	Spiral	
M66/NGC 3627	11	20	+12	59	727 ± 3	643	Spiral/Seyfert	
M74/NGC 628	01	37	+15	47	657 ± 1	754	Spiral	
M77/NGC 1068	02	43	−00	01	1137 ± 3	1146	Spiral/Seyfert	
M81/NGC 3031	09	56	+69	04	−34 ± 4	73	Spiral/Seyfert	
M82/NGC 3034	09	56	+69	41	203 ± 4	312	Irregular/Starburst	
M83/NGC 5236	13	37	−29	52	516 ± 4	385	Spiral/Starburst	
M84/NGC 4374	12	25	+12	53	1060 ± 6	1005	Elliptical	
M85/NGC 4382	12	25	+18	11	729 ± 2	692	Spiral	
M86/NGC 4406	12	26	+12	57	−244 ± 5	−298	Elliptical/Lenticular	
M87/NGC 4486	12	30	+12	23	1307 ± 7	1254	Elliptical/Central Dominant/Seyfert	Virgo A
M88/NGC 4501	12	32	+14	25	2281 ± 3	2235	Spiral/Seyfert	
M89/NGC 4552	12	36	+12	33	340 ± 4	290	Elliptical	
M90/NGC 4569	12	37	+13	10	−235 ± 4	−282	Spiral/Seyfert	
M91/NGC 4548	12	35	+14	30	486 ± 4	442	Spiral/Seyfert	
M94/NGC 4736	12	51	+41	07	308 ± 1	360	Spiral	
M95/NGC 3351	10	44	+11	42	778 ± 4	677	Spiral/Starburst	
M96/NGC 3368	10	47	+11	49	897 ± 4	797	Spiral/Seyfert	
M98/NGC 4192	12	14	+14	54	−142 ± 4	−195	Spiral/Seyfert	
M99/NGC 4254	12	19	+14	25	2407 ± 3	2354	Spiral	
M100/NGC 4321	12	23	+15	49	1571 ± 1	1525	Spiral	
M101/NGC 5457	14	03	+54	21	241 ± 2	360	Spiral	
M104/NGC 4594	12	40	−11	37	1024 ± 5	904	Spiral/Seyfert	Sombrero
M105/NGC 3379	10	48	+12	35	911 ± 2	814	Elliptical	
M106/NGC 4258	12	19	+47	18	448 ± 3	507	Spiral/Seyfert	
M108/NGC 3556	11	09	+55	57	695 ± 3	765	Spiral	
M109/NGC 3992	11	55	+53	39	1048 ± 4	1121	Spiral	
M110/NGC 205	00	40	+41	41	−241 ± 3	−61	Elliptical	

[a]Galaxies identified in the catalog published by Charles Messier in 1781; these galaxies are relatively easy to observe with small telescopes.

[b]Data obtained from NED: NASA/IPAC Extragalactic Database (http://ned.ipac.caltech.edu). The original Messier list of galaxies was obtained from SED, and the list data were updated to 2001 and M102 was dropped.

[c]The galaxies are identified by their Messier number (M followed by a number) and NGC number, which comes from the *New General Catalog* published in 1888.

[d]Radial velocity in kilometers per second, with respect to the Sun (heliocentric). Positive values mean motion away from the Sun; negative values are toward the Sun.

[e]Radial velocity in kilometers per second, with respect to the Milky Way Galaxy, calculated from the RV_{hel} values with a correction for the Sun's motion around the galactic center.

[f]Galaxies are first listed by their primary type (spiral, elliptical, or irregular) and then by any other special categories that apply (see Chapter 16).

TABLE G.3 Nearby, X-Ray Bright Clusters of Galaxies

Cluster Name	Redshift	Distance[a] (billions of ly)	Temperature of Intracluster Medium (millions of K)	Average Orbital Velocity of Galaxies[b] (km/s)	Cluster Mass[c] ($10^{15} M_{Sun}$)
Abell 2142	0.0907	1.26	101. ± 2	1132 ± 110	1.6
Abell 2029	0.0766	1.07	100. ± 3	1164 ± 98	1.5
Abell 401	0.0737	1.03	95.2 ± 5	1152 ± 86	1.4
Coma	0.0233	0.32	95.1 ± 1	821 ± 49	1.4
Abell 754	0.0539	0.75	93.3 ± 3	662 ± 77	1.4
Abell 2256	0.0589	0.82	87.0 ± 2	1348 ± 86	1.4
Abell 399	0.0718	1.00	81.7 ± 7	1116 ± 89	1.1
Abell 3571	0.0395	0.55	81.1 ± 3	1045 ± 109	1.1
Abell 478	0.0882	1.23	78.9 ± 2	904 ± 281	1.1
Abell 3667	0.0566	0.79	78.5 ± 6	971 ± 62	1.1
Abell 3266	0.0599	0.84	78.2 ± 5	1107 ± 82	1.1
Abell 1651a	0.0846	1.18	73.1 ± 6	685 ± 129	0.96
Abell 85	0.0560	0.78	70.9 ± 2	969 ± 95	0.92
Abell 119	0.0438	0.61	65.6 ± 5	679 ± 106	0.81
Abell 3558	0.0480	0.67	65.3 ± 2	977 ± 39	0.81
Abell 1795	0.0632	0.88	62.9 ± 2	834 ± 85	0.77
Abell 2199	0.0314	0.44	52.7 ± 1	801 ± 92	0.59
Abell 2147	0.0353	0.49	51.1 ± 4	821 ± 68	0.56
Abell 3562	0.0478	0.67	45.7 ± 8	736 ± 49	0.48
Abell 496	0.0325	0.45	45.3 ± 1	687 ± 89	0.47
Centaurus	0.0103	0.14	42.2 ± 1	863 ± 34	0.42
Abell 1367	0.0213	0.30	41.3 ± 2	822 ± 69	0.41
Hydra	0.0126	0.18	38.0 ± 1	610 ± 52	0.36
C0336	0.0349	0.49	37.4 ± 1	650 ± 170	0.35
Virgo	0.0038	0.05	25.7 ± 0.5	632 ± 41	0.20

Note: This table lists the 25 brightest clusters of galaxies in the X-ray sky from a catalog by J. P. Henry (2000).

[a]Cluster distances were computed using a value for Hubble's constant of 21.5 km/s/million light-years.

[b]The average orbital velocities given in this column are the velocity component along our line of sight. This velocity should be multiplied by the square root of 2 to get the average orbital velocity.

[c]This column gives each cluster's mass within the largest radius at which the intracluster medium can be in gravitational equilibrium. Because our estimates of that radius depend on Hubble's constant, these masses are inversely proportional to Hubble's constant, which we have assumed to be 21.5 km/s/million light-years.

The 88 Constellations

Constellation Locations

These two charts each show half of the celestial sphere in projection, so that you can use them to learn the approximate locations of the constellations. The grid lines are marked by the right ascension and declination.

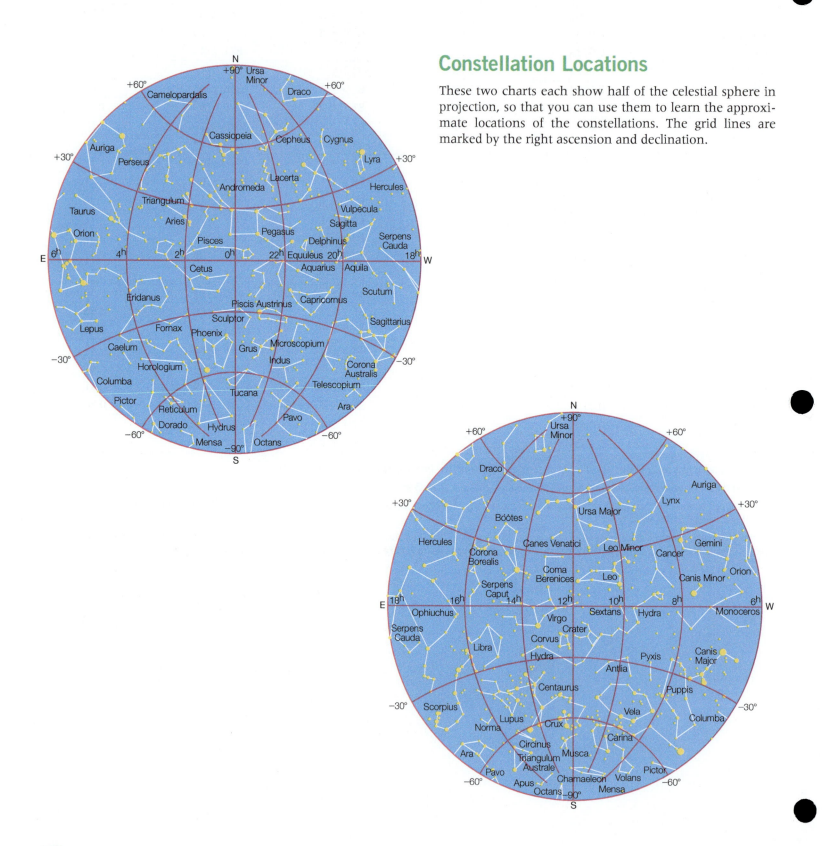

Constellation Names (English Equivalent in Parentheses)

Andromeda (The Chained Princess)
Antlia (The Air Pump)
Apus (The Bird of Paradise)
Aquarius (The Water Bearer)
Aquila (The Eagle)
Ara (The Altar)
Aries (The Ram)
Auriga (The Charioteer)
Boötes (The Herdsman)
Caelum (The Chisel)
Camelopardalis (The Giraffe)
Cancer (The Crab)
Canes Venatici (The Hunting Dogs)
Canis Major (The Great Dog)
Canis Minor (The Little Dog)
Capricornus (The Sea Goat)
Carina (The Keel)
Cassiopeia (The Queen)
Centaurus (The Centaur)
Cepheus (The King)
Cetus (The Whale)
Chamaeleon (The Chameleon)

Circinus (The Drawing Compass)
Columba (The Dove)
Coma Berenices (Berenice's Hair)
Corona Australis (The Southern Crown)
Corona Borealis (The Northern Crown)
Corvus (The Crow)
Crater (The Cup)
Crux (The Southern Cross)
Cygnus (The Swan)
Delphinus (The Dolphin)
Dorado (The Goldfish)
Draco (The Dragon)
Equuleus (The Little Horse)
Eridanus (The River)
Fornax (The Furnace)
Gemini (The Twins)
Grus (The Crane)
Hercules
Horologium (The Clock)
Hydra (The Sea Serpent)

Hydrus (The Water Snake)
Indus (The Indian)
Lacerta (The Lizard)
Leo (The Lion)
Leo Minor (The Little Lion)
Lepus (The Hare)
Libra (The Scales)
Lupus (The Wolf)
Lynx (The Lynx)
Lyra (The Lyre)
Mensa (The Table)
Microscopium (The Microscope)
Monoceros (The Unicorn)
Musca (The Fly)
Norma (The Level)
Octans (The Octant)
Ophiuchus (The Serpent Bearer)
Orion (The Hunter)
Pavo (The Peacock)
Pegasus (The Winged Horse)
Perseus (The Hero)
Phoenix (The Phoenix)
Pictor (The Painter's Easel)
Pisces (The Fish)

Piscis Austrinus (The Southern Fish)
Puppis (The Stern)
Pyxis (The Compass)
Reticulum (The Reticle)
Sagitta (The Arrow)
Sagittarius (The Archer)
Scorpius (The Scorpion)
Sculptor (The Sculptor)
Scutum (The Shield)
Serpens (The Serpent)
Sextans (The Sextant)
Taurus (The Bull)
Telescopium (The Telescope)
Triangulum (The Triangle)
Triangulum Australe (The Southern Triangle)
Tucana (The Toucan)
Ursa Major (The Great Bear)
Ursa Minor (The Little Bear)
Vela (The Sail)
Virgo (The Virgin)
Volans (The Flying Fish)
Vulpecula (The Fox)

All-Sky Constellation Map

This map of the entire sky shows the locations of all the constellations, in much the same way that a world map shows all of the countries on Earth. It does not use the usual celestial coordinate system of right ascension and declination, but instead is oriented so that the Milky Way Galaxy's center is at the center of the map and the Milky Way's disk (shown in shades of lighter blue) stretches from left to right across the map.

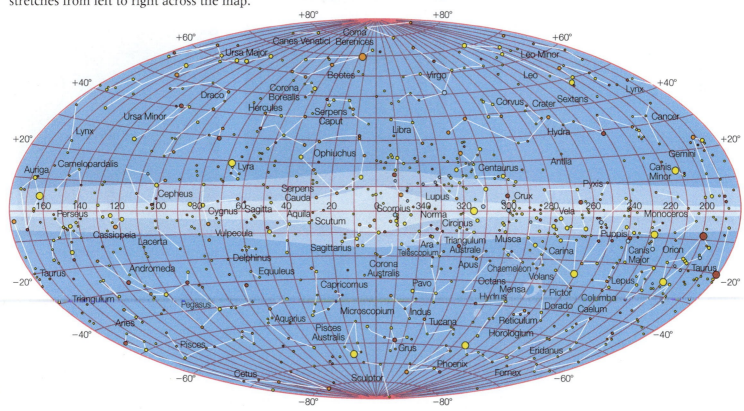

Star Charts

How to use the star charts:

Check the times and dates under each chart to find the best one for you. Take it outdoors within an hour or so of the time listed for your date. Bring a dim flashlight to help you read it.

On each chart, the round outside edge represents the horizon all around you. Compass directions around the horizon are marked in yellow. Turn the chart around so that the edge marked with the direction you're facing (for example, north, southeast) is down. The stars above this horizon now match the stars you are facing. Ignore the rest until you turn to look in a different direction.

The center of the chart represents the sky overhead, so a star plotted on the chart halfway from the edge to the center can be found in the sky halfway from the horizon to straight up.

The charts are drawn for 40°N latitude (for example, Denver, New York, Madrid). If you live far south of there, stars in the southern part of your sky will appear higher than on the chart and stars in the north will be lower. If you live far north of there, the reverse is true.

See pages A-20 through A-23 for full-size star charts.

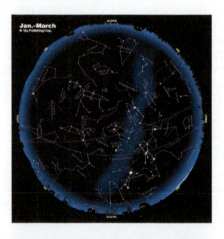

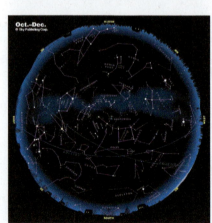

Star charts ©1999 *Sky & Telescope*

Jan.–March
© Sky Publishing Corp.

©1999 *Sky & Telescope*

Use this chart January, February, and March.

Early January—1 A.M.   Early February—11 P.M.   Early March—9 P.M.

Late January—Midnight   Late February—10 P.M.   Late March—Dusk

Use this chart April, May, and June.

Early April—3 A.M.*   Early May—1 A.M.*   Early June—11 P.M.*
Late April—2 A.M.*   Late May—Midnight*   Late June—Dusk

* Daylight Saving Time

Use this chart July, August, and September.

Early July—1 A.M.*   Early August—11 P.M.*   Early September—9 P.M.*
Late July—Midnight*   Late August—10 P.M.*   Late September—Dusk

* Daylight Saving Time

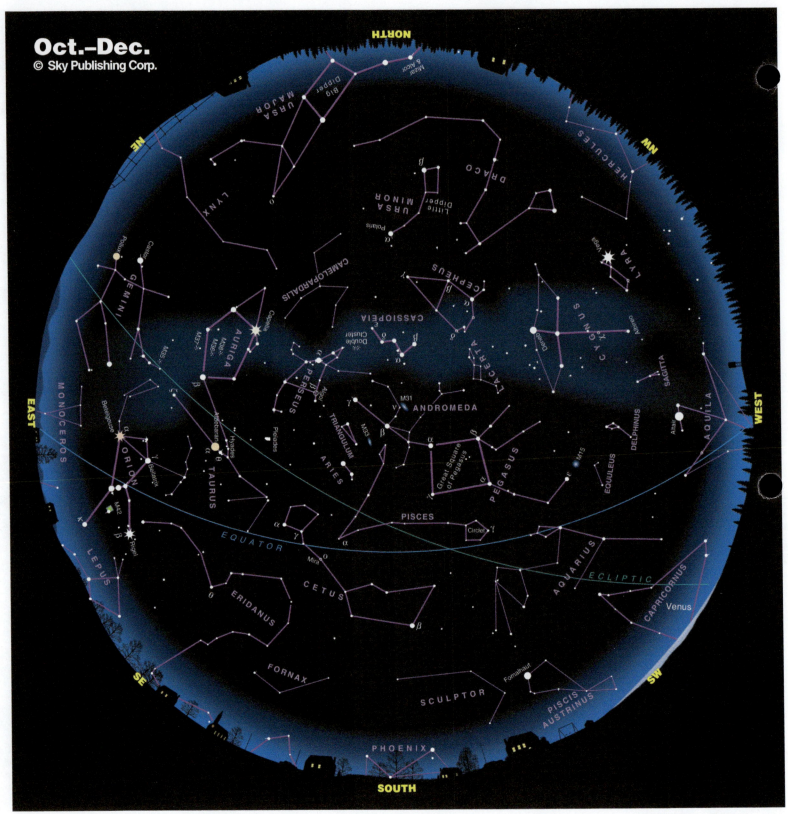

Oct.–Dec.
© Sky Publishing Corp.

©1999 *Sky & Telescope*

Use this chart October, November, and December.

Early October—1 A.M.*   Early November—10 P.M.   Early December—8 P.M.
Late October—Midnight*   Late November—9 P.M.   Late December—7 P.M.

* Daylight Saving Time

J Key to Icons on Figures

You'll see the following icons on figures throughout the book. They are used to indicate the wavelength of light shown in each image, and to identify photo-realistic artworks and images made by computer simulations.

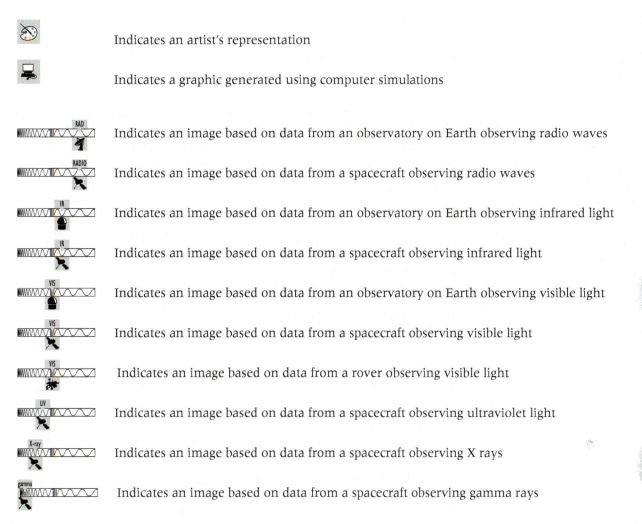

Indicates an artist's representation

Indicates a graphic generated using computer simulations

Indicates an image based on data from an observatory on Earth observing radio waves

Indicates an image based on data from a spacecraft observing radio waves

Indicates an image based on data from an observatory on Earth observing infrared light

Indicates an image based on data from a spacecraft observing infrared light

Indicates an image based on data from an observatory on Earth observing visible light

Indicates an image based on data from a spacecraft observing visible light

Indicates an image based on data from a rover observing visible light

Indicates an image based on data from a spacecraft observing ultraviolet light

Indicates an image based on data from a spacecraft observing X rays

Indicates an image based on data from a spacecraft observing gamma rays

Glossary

absolute magnitude A measure of an object's luminosity; defined to be the apparent magnitude the object would have if it were located exactly 10 parsecs away.

absolute zero The coldest possible temperature, which is 0 K = −273.15°C.

absorption (of light) The process by which matter absorbs radiative energy.

absorption line A dark band (a "line") on an otherwise bright rainbow of light, occurring when light viewed through a diffraction element such as a prism shows a deficit of photons at or near a specific wavelength.

accelerating universe A universe in which a repulsive force (*see* cosmological constant) causes the expansion of the universe to accelerate with time. Its galaxies will recede from one another increasingly faster, and it will become cold and dark more quickly than a coasting universe.

acceleration The rate at which an object's velocity changes. Its standard units are m/s².

acceleration of gravity The acceleration of a falling object. On Earth, the acceleration of gravity, designated by *g*, is 9.8 m/s².

accretion The process by which small objects gather together to make larger objects.

accretion disk A rapidly rotating disk of material that gradually falls inward as it orbits a starlike object (e.g., white dwarf, neutron star, or black hole).

active galactic nuclei The unusually luminous centers of some galaxies, thought to be powered by accretion onto supermassive black holes. Quasars are the brightest type of active galactic nuclei; radio galaxies also contain active galactic nuclei.

active galaxy A term sometimes used to describe a galaxy that contains an *active galactic nucleus*.

adaptive optics A technique in which telescope mirrors flex rapidly to compensate for the bending of starlight caused by atmospheric turbulence.

Algol paradox A paradox concerning the binary star Algol, which contains a subgiant star that is less massive than its main-sequence companion.

altitude (above horizon) The angular distance between the horizon and an object in the sky.

amino acids The building blocks of proteins.

analemma The figure-8 path traced by the Sun over the course of a year when viewed at the same place and the same time each day; it represents the discrepancies between apparent and mean solar time.

Andromeda Galaxy (M31; the Great Galaxy in Andromeda) The nearest large spiral galaxy to the Milky Way.

angular momentum Momentum attributable to rotation or revolution. The angular momentum of an object moving in a circle of radius *r* is the product *m* × *v* × *r*.

angular resolution (of a telescope) The smallest angular separation that two point-like objects can have and still be seen as distinct points of light (rather than as a single point of light).

angular size (or **angular distance**) A measure of the angle formed by extending imaginary lines outward from our eyes to span an object (or the space between two objects).

annihilation *See* matter–antimatter annihilation.

annular solar eclipse A solar eclipse during which the Moon is directly in front of the Sun but its angular size is not large enough to fully block the Sun; thus, a ring (or *annulus*) of sunlight is still visible around the Moon's disk.

Antarctic Circle The circle on Earth with latitude 66.5°S.

antielectron The antimatter equivalent of an electron. It is identical to an electron in virtually all respects, except it has a positive rather than a negative electrical charge.

antimatter Any particle with the same mass as a particle of ordinary matter but whose other basic properties, such as electrical charge, are precisely opposite.

aphelion The point at which an object orbiting the Sun is farthest from the Sun.

apogee The point at which an object orbiting Earth is farthest from Earth.

apparent brightness The amount of light reaching us *per unit area* from a luminous object; often measured in units of watts/m².

apparent magnitude A measure of the apparent brightness of an object in the sky, based on the ancient system developed by Hipparchus.

apparent retrograde motion The apparent motion of a planet, as viewed from Earth, during the period of a few weeks or months when it moves westward relative to the stars in our sky.

apparent solar time Time measured by the actual position of the Sun in the local sky, defined so that noon is when the Sun is *on* the meridian.

arcminute (or **minute of arc**) 1/60 of 1°.

arcsecond (or **second of arc**) 1/60 of an arcminute, or 1/3600 of 1°.

Arctic Circle The circle on Earth with latitude 66.5°N.

asteroid A relatively small and rocky object that orbits a star; asteroids are officially considered part of a category known as "small solar system bodies."

asteroid belt The region of our solar system between the orbits of Mars and Jupiter in which asteroids are heavily concentrated.

astrobiology The study of life on Earth and beyond; it emphasizes research into questions of the origin of life, the conditions under which life can survive, and the search for life beyond Earth.

astrometric method The detection of extrasolar planets through the side-to-side motion of a star caused by gravitational tugs from the planet.

astronomical unit (AU) The average distance (semimajor axis) of Earth from the Sun, which is about 150 million km.

atmosphere A layer of gas that surrounds a planet or moon, usually very thin compared to the size of the object.

atmospheric pressure The surface pressure resulting from the overlying weight of an atmosphere.

atmospheric structure The layering of a planetary atmosphere due to variations in temperature with altitude. For example, Earth's atmospheric structure from the ground up consists of the troposphere, stratosphere, thermosphere, and exosphere.

atomic hydrogen gas Gas composed mostly of hydrogen atoms, though in space it is generally mixed with helium and small amounts of other elements as well; it is the most common form of interstellar gas.

atomic mass number The combined number of protons and neutrons in an atom.

atomic number The number of protons in an atom.

atoms Consist of a nucleus made from protons and neutrons, surrounded by a cloud of electrons.

aurora Dancing lights in the sky caused by charged particles entering our atmosphere; called the *aurora borealis* in the Northern Hemisphere and the *aurora australis* in the Southern Hemisphere.

axis tilt (of a planet in our solar system) The amount by which a planet's axis is tilted with respect to a line perpendicular to the ecliptic plane.

azimuth (usually called **direction** in this book) Direction around the horizon from due north, measured clockwise in degrees. For example, the azimuth of due north is 0°, due east is 90°, due south is 180°, and due west is 270°.

bar The standard unit of pressure, approximately equal to Earth's atmospheric pressure at sea level.

barred spiral galaxies Spiral galaxies that have a straight bar of stars cutting across their centers.

baryonic matter Ordinary matter made from atoms (so called because the nuclei of atoms contain protons and neutrons, which are both baryons).

baryons Particles, including protons and neutrons, that are made from three quarks.

basalt A type of dark, high-density volcanic rock that is rich in iron and magnesium-based silicate minerals; it forms a runny (easily flowing) lava when molten.

belts (on a jovian planet) Dark bands of sinking air that encircle a jovian planet at a particular set of latitudes.

Big Bang The name given to the event thought to mark the birth of the universe.

Big Bang theory The scientific theory of the universe's earliest moments, stating that all the matter in our observable universe came into being at a single moment in time as an extremely hot, dense mixture of subatomic particles and radiation.

Big Crunch The name given to the event that would presumably end the universe if gravity ever reverses the universal expansion and the universe someday begins to collapse.

binary star system A star system that contains two stars.

biosphere The "layer" of life on Earth.

blackbody radiation *See* thermal radiation.

black hole A bottomless pit in spacetime. Nothing can escape from within a black hole, and we can never again detect or observe an object that falls into a black hole.

black smokers Structures around seafloor volcanic vents that support a wide variety of life.

BL Lac objects A class of active galactic nuclei that probably represent the centers of radio galaxies whose jets happen to be pointed directly at us.

blowout Ejection of the hot, gaseous contents of a superbubble when it grows so large that it bursts out of the cooler layer of gas filling the galaxy's disk.

blueshift A Doppler shift in which spectral features are shifted to shorter wavelengths, observed when an object is moving toward the observer.

bosons Particles, such as photons, to which the exclusion principle does not apply.

bound orbits Orbits on which an object travels repeatedly around another object; bound orbits are elliptical in shape.

brown dwarf An object too small to become an ordinary star because electron degeneracy pressure halts its gravitational collapse before fusion becomes self-sustaining; brown dwarfs have mass less than $0.08 M_{Sun}$.

bubble (interstellar) An expanding shell of hot, ionized gas driven by stellar winds or supernovae, with very hot and very low density gas inside.

bulge (of a spiral galaxy) The central portion of a spiral galaxy that is roughly spherical (or football shaped) and bulges above and below the plane of the galactic disk.

Cambrian explosion The dramatic diversification of life on Earth that unfolded over only tens of millions of years, beginning about 542 million years ago.

carbonate rock A carbon-rich rock, such as limestone, that forms underwater from chemical reactions between sediments and carbon dioxide. On Earth, most of the outgassed carbon dioxide currently resides in carbonate rocks.

carbon dioxide cycle (CO_2 cycle) The process that cycles carbon dioxide between Earth's atmosphere and surface rocks.

carbon stars Stars whose atmospheres are especially carbon-rich, thought to be near the ends of their lives; carbon stars are the primary sources of carbon in the universe.

Cassini division A large, dark gap in Saturn's rings, visible through small telescopes on Earth.

CCD (charge coupled device) A type of electronic light detector that has largely replaced photographic film in astronomical research.

celestial coordinates The coordinates of right ascension and declination that fix an object's position on the celestial sphere.

celestial equator (CE) The extension of Earth's equator onto the celestial sphere.

celestial navigation Navigation on the surface of the Earth accomplished by observations of the Sun and stars.

celestial sphere The imaginary sphere on which objects in the sky appear to reside when observed from Earth.

Celsius (temperature scale) The temperature scale commonly used in daily activity internationally, defined so that, on Earth's surface, water freezes at 0°C and boils at 100°C.

center of mass (of orbiting objects) The point at which two or more orbiting objects would balance if they were somehow connected; it is the point around which the orbiting objects actually orbit.

central dominant galaxy A giant elliptical galaxy found at the center of a dense cluster of galaxies, apparently formed by the merger of several individual galaxies.

Cepheid variable star (or Cepheid for short) A particularly luminous type of pulsating variable star that follows a period–luminosity relation and hence is very useful for measuring cosmic distances.

Chandrasekhar limit *See* white dwarf limit.

charged particle belts Zones in which ions and electrons accumulate and encircle a planet.

chemical enrichment The process by which the abundance of heavy elements (heavier than helium) in the interstellar medium gradually increases over time as these elements are produced by stars and released into space.

chemical potential energy Potential energy that can be released through chemical reactions; for example, food contains chemical potential energy that your body can convert to other forms of energy.

chondrites Another name for primitive meteorites. The name comes from the round chondrules within them. *Achondrites*, meaning "without chondrules," is another name for processed meteorites.

chromosphere The layer of the Sun's atmosphere below the corona; most of the Sun's ultraviolet light is emitted from this region, in which the temperature is about 10,000 K.

circulation cells (or Hadley cells) Large-scale cells (similar to convection cells) in a planet's atmosphere that transport heat between the equator and the poles.

circumpolar star A star that always remains above the horizon for a particular latitude.

climate The long-term average of weather.

close binary A binary star system in which the two stars are very close together.

closed universe A universe in which spacetime curves back on itself to the point where its overall shape is analogous to that of the surface of a sphere.

cluster of galaxies A collection of a few dozen or more galaxies bound together by gravity; smaller collections of galaxies are simply called *groups*.

cluster of stars A group of anywhere from several hundred to a million or so stars; star clusters come in two types—open clusters and globular clusters.

CNO cycle The cycle of reactions by which intermediate- and high-mass stars fuse hydrogen into helium.

coasting universe A model of the universe in which the universe expands forever with little change in its rate of expansion; in the absence of a repulsive force (*see* cosmological constant), a coasting universe is one in which the actual mass density is *smaller* than the critical density.

color-coded image An image that represents information or forms of light in any way that makes an object appear different than it would appear if we looked at its true, visible-light colors. Sometimes called a *false-color image*.

coma (of a comet) The dusty atmosphere of a comet, created by sublimation of ices in the nucleus when the comet is near the Sun.

comet A relatively small, icy object that orbits a star. Like asteroids, comets are officially considered part of a category known as "small solar system bodies."

comparative planetology The study of the solar system by examining and understanding the similarities and differences among worlds.

compound (chemical) A substance made from molecules consisting of two or more atoms with different atomic numbers.

condensates Solid or liquid particles that condense from a cloud of gas.

condensation The formation of solid or liquid particles from a cloud of gas.

conduction (of energy) The process by which thermal energy is transferred by direct contact from warm material to cooler material.

conjunction (of a planet with the Sun) An event in which a planet and the Sun line up in our sky.

conservation of angular momentum (law of) The principle that, in the absence of net torque (twisting force), the total angular momentum of a system remains constant.

conservation of energy (law of) The principle that energy (including mass-energy) can be neither created nor destroyed, but can only change from one form to another.

conservation of momentum (law of) The principle that, in the absence of net force, the total momentum of a system remains constant.

constellation A region of the sky; 88 official constellations cover the celestial sphere.

continental crust The thicker lower-density crust that makes up Earth's continents. It is made when remelting of seafloor crust allows lower-density rock to separate and erupt to the surface. Continental crust ranges in age from very young to as old as about 4 billion years (or more).

continuous spectrum A spectrum (of light) that spans a broad range of wavelengths without interruption by emission or absorption lines.

convection The energy transport process in which warm material expands and rises while cooler material contracts and falls.

convection cell An individual small region of convecting material.

convection zone (of a star) A region in which energy is transported outward by convection.

Copernican revolution The dramatic change, initiated by Copernicus, that occurred when we learned that Earth is a planet orbiting the Sun rather than the center of the universe.

core (of a planet) The dense central region of a planet that has undergone differentiation.

core (of a star) The central region of a star, in which nuclear fusion can occur.

Coriolis effect The effect due to rotation that causes air or objects on a rotating surface or planet to deviate from straight-line trajectories.

corona (solar) The tenuous uppermost layer of the Sun's atmosphere; most of the Sun's X rays are emitted from this region, in which the temperature is about 1 million K.

coronal holes Regions of the corona that barely show up in X-ray images because they are nearly devoid of hot coronal gas.

coronal mass ejections Bursts of charged particles from the Sun's corona that travel outward into space.

cosmic microwave background The remnant radiation from the Big Bang, which we detect using radio telescopes sensitive to microwaves (which are short-wavelength radio waves).

cosmic rays Particles such as electrons, protons, and atomic nuclei that zip through interstellar space at close to the speed of light.

cosmological constant The name given to a term in Einstein's equations of general relativity. If it is not zero, then it represents a repulsive force or a type of energy (sometimes called *dark energy* or *quintessence*) that might cause the expansion of the universe to accelerate with time.

cosmological horizon The boundary of our observable universe, which is where the lookback time is equal to the age of the universe. Beyond this boundary in spacetime, we cannot see anything at all.

Cosmological Principle The idea that matter is distributed uniformly throughout the universe on very large scales, meaning that the universe has neither a center nor an edge.

cosmological redshift The redshift we see from distant galaxies, caused by the fact that expansion of the universe stretches all the photons within it to longer, redder wavelengths.

cosmology The study of the overall structure and evolution of the universe.

cosmos An alternative name for the universe.

crescent (phase) The phase of the Moon (or of a planet) in which just a small portion (less than half) of the visible face is illuminated by sunlight.

critical density The precise average density for the entire universe that marks the dividing line between a recollapsing universe and one that will expand forever.

critical universe A model of the universe in which the universe expands more and more slowly as time progresses; in the absence of a repulsive force (*see* cosmological constant), a critical universe is one in which the average mass density *equals* the critical density.

crust (of a planet) The low-density surface layer of a planet that has undergone differentiation.

curvature of spacetime A change in the geometry of space that is produced in the vicinity of a massive object and is responsible for the force we call gravity. The overall geometry of the universe may also be curved, depending on its overall mass-energy content.

cycles per second Units of frequency for a wave; describes the number of peaks (or troughs) of a wave that pass by a given point each second. Equivalent to *hertz*.

dark energy Name sometimes given to energy that could be causing the expansion of the universe to accelerate. *See* cosmological constant.

dark matter Matter that we infer to exist from its gravitational effects but from which we have not detected any light; dark matter apparently dominates the total mass of the universe.

daylight saving time Standard time plus 1 hour, so that the Sun appears on the meridian around 1 p.m. rather than around noon.

decay (radioactive) *See* radioactive decay.

December solstice Both the point on the celestial sphere where the ecliptic is farthest south of the celestial equator and the moment in time when the Sun appears at that point each year (around December 21).

declination (dec) Analogous to latitude, but on the celestial sphere; it is the angular north-south distance between the celestial equator and a location on the celestial sphere.

deferent The large circle upon which a planet follows its circle-upon-circle path around Earth in the (Earth-centered) Ptolemaic model of the universe. *See also* epicycle.

degeneracy pressure A type of pressure unrelated to an object's temperature, which arises when electrons (electron degeneracy pressure) or neutrons (neutron degeneracy pressure) are packed so tightly that the exclusion and uncertainty principles come into play.

degenerate object An object, such as a brown dwarf, white dwarf, or neutron star, in which degeneracy pressure is the primary pressure pushing back against gravity.

density (mass) The amount of mass per unit volume of an object. The average density of any object can be found by dividing its mass by its volume. Standard metric units are kilograms per cubic meter, but density is more commonly stated in units of grams per cubic centimeter.

deuterium A form of hydrogen in which the nucleus contains a proton and a neutron, rather than only a proton (as is the case for most hydrogen nuclei).

differential rotation Rotation in which the equator of an object rotates at a different rate than the poles.

differentiation The process by which gravity separates materials according to density, with high-density materials sinking and low-density materials rising.

diffraction grating A finely etched surface that can split light into a spectrum.

diffraction limit The angular resolution that a telescope could achieve if it were limited only by the interference of light waves; it is smaller (i.e., better angular resolution) for larger telescopes.

dimension (mathematical) Describes the number of independent directions in which movement is possible; for example, the surface of Earth is two-dimensional because only two independent directions of motion are possible (north-south and east-west).

direction (in local sky) One of the two coordinates (the other is altitude) needed to pinpoint an object in the local sky. It is the direction, such as north, south, east, or west, in which you must face to see the object. *See also* azimuth.

disk (of a galaxy) The portion of a spiral galaxy that looks like a disk and contains an interstellar medium with cool gas and dust; stars of many ages are found in the disk.

disk population The stars that orbit within the disk of a spiral galaxy; sometimes called *Population I*.

DNA (deoxyribonucleic acid) The molecule that constitutes the genetic material of life on Earth.

Doppler effect (shift) The effect that shifts the wavelengths of spectral features in objects that are moving toward or away from the observer.

Doppler method The detection of extrasolar planets through the motion of a star toward and away from the observer caused by gravitational tugs from the planet.

double shell–fusion star A star that is fusing helium into carbon in a shell around an inert carbon core and is fusing hydrogen into helium in a shell at the top of the helium layer.

down quark One of the two quark types (the other is the up quark) found in ordinary protons and neutrons. It has a charge of $-\frac{1}{3}$.

Drake equation An equation that lays out the factors that play a role in determining the number of communicating civilizations in our galaxy.

dust (or **dust grains**) Tiny solid flecks of material; in astronomy, we often discuss interplanetary dust (found within a star system) or interstellar dust (found between the stars in a galaxy).

dust tail (of a comet) One of two tails seen when a comet passes near the Sun (the other is the plasma tail). It is composed of small solid particles pushed away from the Sun by the radiation pressure of sunlight.

dwarf elliptical galaxy A small elliptical galaxy with less than about a billion stars.

dwarf galaxies Relatively small galaxies, consisting of less than about 10 billion stars.

dwarf planet An object that orbits the Sun and is massive enough for its gravity to have made it nearly round in shape, but that does not qualify as an official planet because it has not cleared its orbital neighborhood. The dwarf planets of our solar system include the asteroid Ceres and the Kuiper belt objects Pluto, Eris, Haumea, and Makemake.

dwarf spheroidal galaxy A subclass of dwarf elliptical galaxies that are round and diskless like other elliptical galaxies but are particularly small and much less bright. They are the most numerous subtype of galaxy in the Local Group and possibly in the universe.

Earth-orbiters (spacecraft) Spacecraft designed to study Earth or the universe from Earth orbit.

eccentricity A measure of how much an ellipse deviates from a perfect circle; defined as the center-to-focus distance divided by the length of the semimajor axis.

eclipse An event in which one astronomical object casts a shadow on another or crosses our line of sight to the other object.

eclipse seasons Periods during which lunar and solar eclipses can occur because the nodes of the Moon's orbit are aligned with Earth and the Sun.

eclipsing binary A binary star system in which the two stars happen to be orbiting in the plane of our line of sight, so that each star will periodically eclipse the other.

ecliptic The Sun's apparent annual path among the constellations.

ecliptic plane The plane of Earth's orbit around the Sun.

ejecta (from an impact) Debris ejected by the blast of an impact.

electrical charge A fundamental property of matter that is described by its amount and as either positive or negative; more technically, a measure of how a particle responds to the electromagnetic force.

electromagnetic field An abstract concept used to describe how a charged particle would affect other charged particles at a distance.

electromagnetic radiation Another name for light of all types, from radio waves through gamma rays.

electromagnetic spectrum The complete spectrum of light, including radio waves, infrared light, visible light, ultraviolet light, X rays, and gamma rays.

electromagnetic wave A synonym for *light*, which consists of waves of electric and magnetic fields.

electromagnetism (or electromagnetic force) One of the four fundamental forces; it is the force that dominates atomic and molecular interactions.

electron degeneracy pressure Degeneracy pressure exerted by electrons, as in brown dwarfs and white dwarfs.

electrons Fundamental particles with negative electric charge; the distribution of electrons in an atom gives the atom its size.

electron-volt (eV) A unit of energy equivalent to 1.60×10^{-19} joule.

electroweak era The era of the universe during which only three forces operated (gravity, strong force, and electroweak force), lasting from 10^{-38} second to 10^{-10} second after the Big Bang.

electroweak force The force that exists at high energies when the electromagnetic force and the weak force exist as a single force.

element (chemical) A substance made from individual atoms of a particular atomic number.

ellipse A type of oval that happens to be the shape of bound orbits. An ellipse can be drawn by moving a pencil along a string whose ends are tied to two tacks; the locations of the tacks are the *foci* (singular: *focus*) of the ellipse.

elliptical galaxies Galaxies that appear rounded in shape, often longer in one direction, like a football. They have no disks and contain little cool gas and dust compared to spiral galaxies, though they often contain hot, ionized gas.

elongation (greatest) For Mercury or Venus, the point at which it appears farthest from the Sun in our sky.

emission (of light) The process by which matter emits energy in the form of light.

emission line A bright band (a "line") of single color, superimposed on a fainter or completely absent rainbow of light, occurring when light viewed through a diffraction element such as a prism shows an excess of photons at or near a specific wavelength.

emission nebula *See* ionization nebula.

energy Broadly speaking, what can make matter move. The three basic types of energy are kinetic, potential, and radiative.

energy balance (in a star) The balance between the rate at which fusion releases energy in the star's core and the rate at which the star's surface radiates this energy into space.

epicycle The small circle upon which a planet moves while simultaneously going around a larger circle (the *deferent*) around Earth in the (Earth-centered) Ptolemaic model of the universe.

equation of time An equation describing the discrepancies between apparent and mean solar time.

equinox *See* March equinox *and* September equinox.

equivalence principle The fundamental starting point for general relativity, which states that the effects of gravity are exactly equivalent to the effects of acceleration.

era of atoms The era of the universe lasting from about 500,000 years to about 1 billion years after the Big Bang, during which it was cool enough for neutral atoms to form.

era of galaxies The present era of the universe, which began with the formation of galaxies when the universe was about 1 billion years old.

era of nuclei The era of the universe lasting from about 3 minutes to about 380,000 years after the Big Bang, during which matter in the universe was fully ionized and opaque to light. The cosmic background radiation was released at the end of this era.

era of nucleosynthesis The era of the universe lasting from about 0.001 second to about 3 minutes after the Big Bang, by the end of which virtually all of the neutrons and about one-seventh of the protons in the universe had fused into helium.

erosion The wearing down or building up of geological features by wind, water, ice, and other phenomena of planetary weather.

eruption The process of releasing hot lava onto a planet's surface.

escape velocity The speed necessary for an object to completely escape the gravity of a large body such as a moon, planet, or star.

evaporation The process by which atoms or molecules escape into the gas phase from a liquid.

event Any particular point along a worldline; all observers will agree on the reality of an event but may disagree about its time and location.

event horizon The boundary that marks the "point of no return" between a black hole and the outside universe; events that occur within the event horizon can have no influence on our observable universe.

evolution (biological) The gradual change in populations of living organisms responsible for transforming life on Earth from its primitive origins to its great diversity today.

exchange particle A type of subatomic particle that transmits one of the four fundamental forces; according to the standard model of physics, these particles are always exchanged whenever two objects interact through a force.

excited state (of an atom) Any arrangement of electrons in an atom that has more energy than the ground state.

exclusion principle The law of quantum mechanics that states that two fermions cannot occupy the same quantum state at the same time.

exosphere The hot, outer layer of an atmosphere, where the atmosphere "fades away" to space.

expansion (of the universe) The idea that the space between galaxies or clusters of galaxies is growing with time.

exposure time The amount of time during which light is collected to make a single image.

extrasolar planet A planet orbiting a star other than our Sun.

extremophiles Living organisms that are adapted to conditions that are "extreme" by human standards, such as very high or low temperature or a high level of salinity or radiation.

Fahrenheit (temperature scale) The temperature scale commonly used in daily activity in the United States; defined so that, on Earth's surface, water freezes at 32°F and boils at 212°F.

fall equinox *See* September equinox, which is commonly called the fall equinox by people living in the Northern Hemisphere.

false-color image *See* color-coded image.

fault (geological) A place where lithospheric plates slip sideways relative to one another.

feedback processes Processes in which a small change in some property (such as temperature) leads to changes in other properties that either amplify or diminish the original small change.

fermions Particles, such as electrons, neutrons, and protons, that obey the exclusion principle.

Fermi's paradox The question posed by Enrico Fermi about extraterrestrial intelligence—"So where is everybody?"—which asks why we have not observed other civilizations even though simple arguments would suggest that some ought to have spread throughout the galaxy by now.

field An abstract concept used to describe how a particle would interact with a force. For example, the idea of a gravitational field describes how a particle would react to the local strength of gravity, and the idea of an electromagnetic field describes how a charged particle would respond to forces from other charged particles.

filter (for light) A material that transmits only particular wavelengths of light.

fireball A particularly bright meteor.

first-quarter (phase) The phase of the Moon that occurs one-quarter of the way through each cycle of phases, in which precisely half of the visible face is illuminated by sunlight.

fission *See* nuclear fission.

flare star A small, spectral type M star that displays particularly strong flares on its surface.

flat (or **Euclidean**) **geometry** The type of geometry in which the rules of geometry for a flat plane hold, such as that the shortest distance between two points is a straight line and that the sum of the angles in a triangle is 180°.

flat universe A universe in which the overall geometry of spacetime is flat (Euclidean), as would be the case if the density of the universe was equal to the critical density.

flybys (spacecraft) Spacecraft that fly past a target object (such as a planet), usually just once, as opposed to entering a bound orbit of the object.

focal plane The place where an image created by a lens or mirror is in focus.

foci Plural of *focus.*

focus (of a lens or mirror) The point at which rays of light that were initially parallel (such as those from a distant star) converge.

focus (of an ellipse) One of two special points within an ellipse that lie along the major axis; these are the points around which we could stretch a pencil and string to draw an ellipse. When one object orbits a second object, the second object lies at one focus of the orbit.

force Anything that can cause a change in momentum.

formation properties (of planets) In this book, for the purpose of understanding geological processes, planets are defined to be born with four formation properties: size (mass and radius), distance from the Sun, composition, and rotation rate.

fossil Any relic of an organism that lived and died long ago.

frame of reference *See* reference frame.

free-fall The condition in which an object is falling without resistance; objects are weightless when in free-fall.

free-float frame A frame of reference in which all objects are weightless and hence float freely.

frequency The rate at which peaks of a wave pass by a point, measured in units of 1/s, often called *cycles per second* or *hertz.*

frost line The boundary in the solar nebula beyond which ices could condense; only metals and rocks could condense within the frost line.

fundamental forces There are four known fundamental forces in nature: gravity, the electromagnetic force, the strong force, and the weak force.

fundamental particles Subatomic particles that cannot be divided into anything smaller.

fusion *See* nuclear fusion.

galactic cannibalism The term sometimes used to describe the process by which large galaxies merge with other galaxies in collisions. *Central dominant galaxies* are products of galactic cannibalism.

galactic fountain A model for the cycling of gas in the Milky Way Galaxy in which fountains of hot, ionized gas rise from the disk into the halo and then cool and form clouds as they sink back into the disk.

galactic wind A wind of low-density but extremely hot gas flowing out from a starburst galaxy, created by the combined energy of many supernovae.

galaxy A great island of stars in space, containing millions, billions, or even trillions of stars, all held together by gravity and orbiting a common center.

galaxy cluster *See* cluster of galaxies.

galaxy evolution The formation and development of galaxies.

Galilean moons The four moons of Jupiter that were discovered by Galileo: Io, Europa, Ganymede, and Callisto.

gamma-ray burst A sudden burst of gamma rays from deep space; such bursts apparently come from distant galaxies, but their precise mechanism is unknown.

gamma rays Light with very short wavelengths (and hence high frequencies)—shorter than those of X rays.

gap moons Tiny moons located within a gap in a planet's ring system. The gravity of a gap moon helps clear the gap.

gas phase The phase of matter in which atoms or molecules can move essentially independently of one another.

gas pressure The force (per unit area) pushing on any object due to surrounding gas. *See also* pressure.

general theory of relativity Einstein's generalization of his special theory of relativity so that the theory also applies when we consider effects of gravity or acceleration.

genetic code The "language" that living cells use to read the instructions chemically encoded in DNA.

geocentric model Any of the ancient Greek models that were used to predict planetary positions under the assumption that Earth lay in the center of the universe.

geocentric universe The ancient belief that Earth is the center of the entire universe.

geological activity Processes that change a planet's surface long after formation, such as volcanism, tectonics, and erosion.

geological processes The four basic geological processes are impact cratering, volcanism, tectonics, and erosion.

geological time scale The time scale used by scientists to describe major eras in Earth's past.

geology The study of surface features (on a moon, planet, or asteroid) and the processes that create them.

geostationary satellite A satellite that appears to stay stationary in the sky as viewed from Earth's surface, because it orbits in the same time it takes Earth to rotate and orbits in Earth's equatorial plane.

geosynchronous satellite A satellite that orbits Earth in the same time it takes Earth to rotate (one sidereal day).

giant galaxies Galaxies that are unusually large, typically containing a trillion or more stars. Most giant galaxies are elliptical, and many contain multiple nuclei near their centers.

giant impact A collision between a forming planet and a very large planetesimal, such as is thought to have formed our Moon.

giant molecular cloud A very large cloud of cold, dense interstellar gas, typically containing up to a million solar masses worth of material. *See also* molecular clouds.

giants (luminosity class III) Stars that appear just below the supergiants on the H-R diagram because they are somewhat smaller in radius and lower in luminosity.

gibbous (phase) The phase of the Moon (or of a planet) in which more than half but less than all of the visible face is illuminated by sunlight.

global positioning system (GPS) A system of navigation by satellites orbiting Earth.

global warming An expected increase in Earth's global average temperature caused by human input of carbon dioxide and other greenhouse gases into the atmosphere.

global wind patterns (or **global circulation**) Wind patterns that remain fixed on a global scale, determined by the combination of surface heating and the planet's rotation.

globular cluster A spherically shaped cluster of up to a million or more stars; globular clusters are found primarily in the halos of galaxies and contain only very old stars.

gluons The exchange particles for the strong force.

grand unified theory (GUT) A theory that unifies three of the four fundamental forces—the strong force, the weak force, and the electromagnetic force (but not gravity)—in a single model.

granulation (on the Sun) The bubbling pattern visible in the photosphere, produced by the underlying convection.

gravitation (law of) *See* universal law of gravitation.

gravitational constant The experimentally measured constant G that appears in the law of universal gravitation:

$$G = 6.67 \times 10^{-11} \frac{m^3}{kg \times s^2}$$

gravitational contraction The process in which gravity causes an object to contract, thereby converting gravitational potential energy into thermal energy.

gravitational encounter An encounter in which two (or more) objects pass near enough so that each can feel the effects of the other's gravity and they can therefore exchange energy.

gravitational equilibrium A state of balance in which the force of gravity pulling inward is precisely counteracted by pressure pushing outward; also referred to as *hydrostatic equilibrium*.

gravitational lensing The magnification or distortion (into arcs, rings, or multiple images) of an image caused by light bending through a gravitational field, as predicted by Einstein's general theory of relativity.

gravitationally bound system Any system of objects, such as a star system or a galaxy, that is held together by gravity.

gravitational potential energy Energy that an object has by virtue of its position in a gravitational field; an object has more gravitational potential energy when it has a greater distance that it can potentially fall.

gravitational redshift A redshift caused by the fact that time runs slowly in gravitational fields.

gravitational time dilation The slowing of time that occurs in a gravitational field, as predicted by Einstein's general theory of relativity.

gravitational waves Waves, predicted by Einstein's general theory of relativity, that travel at the speed of light and transmit distortions of space through the universe. Although they have not yet been observed directly, we have strong indirect evidence that they exist.

gravitons The exchange particles for the force of gravity.

gravity One of the four fundamental forces; it is the force that dominates on large scales.

grazing incidence (in telescopes) Reflections in which light grazes a mirror surface and is deflected at a small angle; commonly used to focus high-energy ultraviolet light and X rays.

great circle A circle on the surface of a sphere whose center is at the center of the sphere.

greatest elongation *See* elongation (greatest).

Great Red Spot A large, high-pressure storm on Jupiter.

greenhouse effect The process by which greenhouse gases in an atmosphere make a planet's surface temperature warmer than it would be in the absence of an atmosphere.

greenhouse gases Gases, such as carbon dioxide, water vapor, and methane, that are particularly good absorbers of infrared light but are transparent to visible light.

Gregorian calendar Our modern calendar, introduced by Pope Gregory in 1582.

ground state (of an atom) The lowest possible energy state of the electrons in an atom.

group (of galaxies) A few to a few dozen galaxies bound together by gravity. *See also* cluster of galaxies.

GUT era The era of the universe during which only two forces operated (gravity and the grand-unified-theory, or GUT, force), lasting from 10^{-43} second to 10^{-38} second after the Big Bang.

GUT force The proposed force that exists at very high energies when the strong force, the weak force, and the electromagnetic force (but not gravity) all act as one.

H II region *See* ionization nebula.

habitable world A world with environmental conditions under which life could *potentially* arise or survive.

habitable zone The region around a star in which planets could potentially have surface temperatures at which liquid water could exist.

Hadley cells *See* circulation cells.

half-life The time it takes for half of the nuclei in a given quantity of a radioactive substance to decay.

halo (of a galaxy) The spherical region surrounding the disk of a spiral galaxy.

halo component (of a galaxy) The portion of any galaxy that is spherical (or football-like) in shape and contains very little cool gas; it generally contains only very old stars. For spiral galaxies, the halo component includes both the halo and the bulge (but not the disk); elliptical galaxies have only a halo component.

halo population Stars that orbit within the halo component of a galaxy; sometimes called *Population II*. Elliptical galaxies have only a halo population (they lack a disk population), while spiral galaxies have halo population stars in their bulges and halos.

Hawking radiation Radiation predicted to arise from the evaporation of black holes.

heavy bombardment The period in the first few hundred million years after the solar system formed during which the tail end of planetary accretion created most of the craters found on ancient planetary surfaces.

heavy elements In astronomy, generally all elements *except* hydrogen and helium.

helium-capture reactions Fusion reactions that fuse a helium nucleus into some other nucleus; such reactions can fuse carbon into oxygen, oxygen into neon, neon into magnesium, and so on.

helium flash The event that marks the sudden onset of helium fusion in the previously inert helium core of a low-mass star.

helium-fusing star (or **helium core-fusion star**) A star that is currently fusing helium into carbon in its core.

helium fusion The fusion of three helium nuclei to form one carbon nucleus; also called the *triple-alpha reaction*.

hertz (Hz) The standard unit of frequency for light waves; equivalent to units of $1/s$.

Hertzsprung-Russell (H-R) diagram A graph plotting individual stars as points, with stellar luminosity on the vertical axis and spectral type (or surface temperature) on the horizontal axis.

high-mass stars Stars born with masses above about $8M_{Sun}$; these stars will end their lives by exploding as supernovae.

horizon A boundary that divides what we can see from what we cannot see.

horizontal branch The horizontal line of stars that represents helium-fusing stars on an H-R diagram for a cluster of stars.

horoscope A predictive chart made by an astrologer; in scientific studies, horoscopes have never been found to have any validity as predictive tools.

hot Jupiter A class of planet that is Jupiter-like in size but orbits very close to its star, causing it to have a very high surface temperature.

hot spot (geological) A place within a plate of the lithosphere where a localized plume of hot mantle material rises.

hour angle (HA) The angle or time (measured in hours) since an object was last on the meridian in the local sky; defined to be 0 hours for objects that are on the meridian.

Hubble's constant A number that expresses the current rate of expansion of the universe; designated H_0, it is usually stated in units of km/s/Mpc. The reciprocal of Hubble's constant is the age the universe would have *if* the expansion rate had never changed.

Hubble's law Mathematical expression of the idea that more distant galaxies move away from us faster: $v = H_0 \times d$, where v is a galaxy's speed away from us, d is its distance, and H_0 is Hubble's constant.

hydrogen compounds Compounds that contain hydrogen and were common in the solar nebula, such as water (H_2O), ammonia (NH_3), and methane (CH_4).

hydrogen shell fusion Hydrogen fusion that occurs in a shell surrounding a stellar core.

hydrosphere The "layer" of water on Earth consisting of oceans, lakes, rivers, ice caps, and other liquid water and ice.

hydrostatic equilibrium *See* gravitational equilibrium.

hyperbola The precise mathematical shape of one type of unbound orbit (the other is a parabola) allowed under the force of gravity; at great distances from the attracting object, a hyperbolic path looks like a straight line.

hypernova A term sometimes used to describe a supernova (explosion) of a star so massive that it leaves a black hole behind.

hyperspace Any space with more than three dimensions.

hypothesis A tentative model proposed to explain some set of observed facts, but which has not yet been rigorously tested and confirmed.

ice ages Periods of global cooling during which the polar caps, glaciers, and snow cover extend closer to the equator.

ices (in solar system theory) Materials that are solid only at low temperatures, such as the hydrogen compounds water, ammonia, and methane.

ideal gas law The law relating the pressure, temperature, and number density of particles in an ideal gas.

image A picture of an object made by focusing light.

imaging (in astronomical research) The process of obtaining pictures of astronomical objects.

impact The collision of a small body (such as an asteroid or comet) with a larger object (such as a planet or moon).

impact basin A very large impact crater, often filled by a lava flow.

impact crater A bowl-shaped depression left by the impact of an object that strikes a planetary surface (as opposed to burning up in the atmosphere).

impact cratering The excavation of bowl-shaped depressions (*impact craters*) by asteroids or comets striking a planet's surface.

impactor The object responsible for an impact.

inflation (of the universe) A sudden and dramatic expansion of the universe thought to have occurred at the end of the GUT era.

infrared light Light with wavelengths that fall in the portion of the electromagnetic spectrum between radio waves and visible light.

inner solar system Generally considered to encompass the region of our solar system out to about the orbit of Mars.

intensity (of light) A measure of the amount of energy coming from light of specific wavelength in the spectrum of an object.

interferometry A telescopic technique in which two or more telescopes are used in tandem to produce much better angular resolution than the telescopes could achieve individually.

intermediate-mass stars Stars born with masses between about $2M_{Sun}$ and $8M_{Sun}$; these stars end their lives by ejecting a planetary nebula and becoming a white dwarf.

interstellar cloud A cloud of gas and dust between the stars.

interstellar dust grains Tiny solid flecks of carbon and silicon minerals found in cool interstellar clouds; they resemble particles of smoke and form in the winds of red giant stars.

interstellar medium The gas and dust that fills the space between stars in a galaxy.

interstellar ramjet A hypothesized type of spaceship that uses a giant scoop to sweep up interstellar gas for use in a nuclear fusion engine.

interstellar reddening The change in the color of starlight as it passes through dusty gas. The light appears redder because dust grains absorb and scatter blue light more effectively than red light.

intracluster medium Hot, X-ray-emitting gas found between the galaxies within a cluster of galaxies.

inverse square law A law followed by any quantity that decreases with the square of the distance between two objects.

inverse square law for light The law stating that an object's apparent brightness depends on its actual luminosity and the inverse square of its distance from the observer:

$$\text{apparent brightness} = \frac{\text{luminosity}}{4\pi \times (\text{distance})^2}$$

inversion (atmospheric) A local weather condition in which air is colder near the surface than higher up in the troposphere—the opposite of the usual condition, in which the troposphere is warmer at the bottom.

ionization The process of stripping an electron from an atom.

ionization nebula A colorful, wispy cloud of gas that glows because neighboring hot stars irradiate it with ultraviolet photons that can ionize hydrogen atoms; also called an *emission nebula* or *H II region*.

ionosphere A portion of the thermosphere in which ions are particularly common (because of ionization by X rays from the Sun).

ions Atoms with a positive or negative electrical charge.

Io torus A donut-shaped charged-particle belt around Jupiter that approximately traces Io's orbit.

irregular galaxies Galaxies that look neither spiral nor elliptical.

isotopes Forms of an element that have the same number of protons but different numbers of neutrons.

jets High-speed streams of gas ejected from an object into space.

joule The international unit of energy, equivalent to about 1/4000 of a Calorie.

jovian nebulae The clouds of gas that swirled around the jovian planets, from which the moons formed.

jovian planets Giant gaseous planets similar in overall composition to Jupiter.

Julian calendar The calendar introduced in 46 B.C. by Julius Caesar and used until the Gregorian calendar replaced it.

June solstice Both the point on the celestial sphere where the ecliptic is farthest north of the celestial equator and the moment in time when the Sun appears at that point each year (around June 21).

Kelvin (temperature scale) The most commonly used temperature scale in science, defined such that absolute zero is 0 K and water freezes at 273.15 K.

Kepler's first law Law stating that the orbit of each planet about the Sun is an ellipse with the Sun at one focus.

Kepler's laws of planetary motion Three laws discovered by Kepler that describe the motion of the planets around the Sun.

Kepler's second law The principle that, as a planet moves around its orbit, it sweeps out equal areas in equal times. This tells us that a planet moves faster when it is closer to the Sun (near perihelion) than when it is farther from the Sun (near aphelion) in its orbit.

Kepler's third law The principle that the square of a planet's orbital period is proportional to the cube of its average distance from the Sun (semi-major axis), which tells us that more distant planets move more slowly in their orbits; in its original form, written $p^2 = a^3$. *See also* Newton's version of Kepler's third law.

kinetic energy Energy of motion, given by the formula $\frac{1}{2}mv^2$.

Kirchhoff's laws A set of rules that summarizes the conditions under which objects produce thermal, absorption line, or emission line spectra. In brief: (1) An opaque object produces thermal radiation. (2) An absorption line spectrum occurs when thermal radiation passes through a thin gas that is cooler than the object emitting the thermal radiation. (3) An emission line spectrum occurs when we view a cloud of gas that is warmer than any background source of light.

Kirkwood gaps On a plot of asteroid semi-major axes, regions with few asteroids as a result of orbital resonances with Jupiter.

K-T event (or **K-T impact**) The collision of an asteroid or comet 65 million years ago that caused the mass extinction best known for wiping out the dinosaurs. *K* and *T* stand for the geological layers above and below the event.

Kuiper belt The comet-rich region of our solar system that resides between about 30 and 100 AU from the Sun. Kuiper belt comets have orbits that lie fairly close to the plane of planetary orbits and travel around the Sun in the same direction as the planets.

Kuiper belt object Any object orbiting the Sun within the region of the Kuiper belt, although the term is most often used for relatively large objects. For example, Pluto and Eris are considered large Kuiper belt objects.

Large Magellanic Cloud One of two small, irregular galaxies (the other is the Small Magellanic Cloud) located about 150,000 light-years away; it probably orbits the Milky Way Galaxy.

large-scale structure (of the universe) Generally refers to the structure of the universe on size scales larger than that of clusters of galaxies.

latitude The angular north-south distance between Earth's equator and a location on Earth's surface.

leap year A calendar year with 366 rather than 365 days. Our current calendar (the Gregorian calendar) incorporates a leap year every 4 years (by adding February 29) except in century years that are not divisible by 400.

length contraction The effect in which you observe lengths to be shortened in reference frames moving relative to you.

lens (gravitational) *See* gravitational lensing.

lenticular galaxies Galaxies that look lens-shaped when seen edge-on, resembling spiral galaxies without arms. They tend to have less cool gas than normal spiral galaxies but more cool gas than elliptical galaxies.

leptons Fermions *not* made from quarks, such as electrons and neutrinos.

life track A track drawn on an H-R diagram to represent the changes in a star's surface temperature and luminosity during its life; also called an *evolutionary track*.

light-collecting area (of a telescope) The area of the primary mirror or lens that collects light in a telescope.

light curve A graph of an object's intensity against time.

light gases (in solar system theory) Hydrogen and helium, which never condense under solar nebula conditions.

light pollution Human-made light that hinders astronomical observations.

light-year (ly) The distance that light can travel in 1 year, which is 9.46 trillion km.

liquid phase The phase of matter in which atoms or molecules are held together but move relatively freely.

lithosphere The relatively rigid outer layer of a planet; generally encompasses the crust and the uppermost portion of the mantle.

Local Bubble (interstellar) The bubble of hot gas in which our Sun and other nearby stars apparently reside. *See also* bubble (interstellar).

Local Group The group of galaxies to which the Milky Way Galaxy belongs. The Local Group has at least 70 members, most relatively small compared to the Milky Way.

local sidereal time (LST) Sidereal time for a particular location, defined according to the position of the March equinox in the local sky. More formally, the local sidereal time at any moment is defined to be the hour angle of the March equinox.

local sky The sky as viewed from a particular location on Earth (or another solid object). Objects in the local sky are pinpointed by the coordinates of *altitude* and *direction* (or azimuth).

local solar neighborhood The portion of the Milky Way Galaxy that is located relatively close (within a few hundred to a couple thousand light-years) to our Sun.

Local Supercluster The supercluster of galaxies to which the Local Group belongs.

longitude The angular east-west distance between the prime meridian (which passes through Greenwich, England) and a location on Earth's surface.

lookback time The amount of time since the light we see from a distant object was emitted. If an object has a lookback time of 400 million years, we are seeing it as it looked 400 million years ago.

low-mass stars Stars born with masses less than about $2M_{Sun}$; these stars end their lives by ejecting a planetary nebula and becoming a white dwarf.

luminosity The total power output of an object, usually measured in watts or in units of solar luminosities ($L_{Sun} = 3.8 \times 10^{26}$ watts).

luminosity class A category describing the region of the H-R diagram in which a star falls. Luminosity class I represents supergiants, III represents giants, and V represents main-sequence stars; luminosity classes II and IV are intermediate to the others.

lunar eclipse An event that occurs when the Moon passes through Earth's shadow, which can occur only at full moon. A lunar eclipse may be total, partial, or penumbral.

lunar maria The regions of the Moon that look smooth from Earth and actually are impact basins.

lunar month *See* synodic month.

lunar phase *See* phase (of the Moon or a planet).

magma Underground molten rock.

magnetic braking The process by which a star's rotation slows as its magnetic field transfers its angular momentum to the surrounding nebula.

magnetic field The region surrounding a magnet in which it can affect other magnets or charged particles.

magnetic field lines Lines that represent how the needles on a series of compasses would point if they were laid out in a magnetic field.

magnetosphere The region surrounding a planet in which charged particles are trapped by the planet's magnetic field.

magnitude system A system for describing stellar brightness by using numbers, called *magnitudes*, based on an ancient Greek way of describing the brightnesses of stars in the sky. This system uses *apparent magnitude* to describe a star's apparent brightness and *absolute magnitude* to describe a star's luminosity.

main sequence The prominent line of points (representing *main-sequence stars*) running from the upper left to the lower right on an H-R diagram.

main-sequence lifetime The length of time for which a star of a particular mass can shine by fusing hydrogen into helium in its core.

main-sequence stars (luminosity class V) Stars whose temperature and luminosity place them on the main sequence of the H-R diagram. Main-sequence stars release energy by fusing hydrogen into helium in their cores.

main-sequence turnoff point The point on a cluster's H-R diagram where its stars turn off from the main sequence; the age of the cluster is equal to the main-sequence lifetime of stars at the main-sequence turnoff point.

mantle (of a planet) The rocky layer that lies between a planet's core and crust.

March equinox Both the point in Pisces on the celestial sphere where the ecliptic crosses the celestial equator and the moment in time when the Sun appears at that point each year (around March 21).

Martian meteorites Meteorites found on Earth that are thought to have originated on Mars.

mass A measure of the amount of matter in an object.

mass-energy The potential energy of mass, which has an amount $E = mc^2$.

mass exchange (in close binary star systems) The process in which tidal forces cause matter to spill from one star to a companion star in a close binary system.

mass extinction An event in which a large fraction of the species living on Earth go extinct, such as the event in which the dinosaurs died out about 65 million years ago.

mass increase (in relativity) The effect in which an object moving past you seems to have a mass greater than its rest mass.

massive star supernova A supernova that occurs when a massive star dies, initiated by the catastrophic collapse of its iron core; often called a *Type II supernova*.

mass-to-light ratio The mass of an object divided by its luminosity, usually stated in units of solar masses per solar luminosity. Objects with high mass-to-light ratios must contain substantial quantities of dark matter.

matter–antimatter annihilation An event that occurs when a particle of matter and a particle of antimatter meet and convert all of their mass-energy to photons.

mean solar time Time measured by the average position of the Sun in the local sky over the course of the year.

meridian A half-circle extending from your horizon (altitude 0°) due south, through your zenith, to your horizon due north.

metallic hydrogen Hydrogen that is so compressed that the hydrogen atoms all share electrons and thereby take on properties of metals, such as conducting electricity. It occurs only under very high-pressure conditions, such as those found deep within Jupiter.

metals (in solar system theory) Elements, such as nickel, iron, and aluminum, that condense at fairly high temperatures.

meteor A flash of light caused when a particle from space burns up in our atmosphere.

meteorite A rock from space that lands on Earth.

meteor shower A period during which many more meteors than usual can be seen.

Metonic cycle The 19-year period, discovered by the Babylonian astronomer Meton, over which the lunar phases occur on the same dates.

microwaves Light with wavelengths in the range of micrometers to millimeters. Microwaves are generally considered to be a subset of the radio wave portion of the electromagnetic spectrum.

mid-ocean ridges Long ridges of undersea volcanoes on Earth, along which mantle material erupts onto the ocean floor and pushes apart the existing seafloor on either side. These ridges are essentially the source of new seafloor crust, which then makes its way along the ocean bottom for millions of years before returning to the mantle at a subduction zone.

Milankovitch cycles The cyclical changes in Earth's axis tilt and orbit that can change the climate and cause ice ages.

Milky Way Used both as the name of our galaxy and to refer to the band of light we see in the sky when we look into the plane of our galaxy.

millisecond pulsars Pulsars with rotation periods of a few thousandths of a second.

minor planets An alternative name for *asteroids*.

model (scientific) A representation of some aspect of nature that can be used to explain and predict real phenomena without invoking myth, magic, or the supernatural.

molecular bands The tightly bunched lines in an object's spectrum that are produced by molecules.

molecular cloud fragments (or molecular cloud cores) The densest regions of molecular clouds, which usually go on to form stars.

molecular clouds Cool, dense interstellar clouds in which the low temperatures allow hydrogen atoms to pair up into hydrogen molecules (H_2).

molecular dissociation The process by which a molecule splits into its component atoms.

molecule Technically, the smallest unit of a chemical element or compound; in this text, the term refers only to combinations of two or more atoms held together by chemical bonds.

momentum The product of an object's mass and velocity.

moon An object that orbits a planet.

moonlets Very small moons that orbit within the ring systems of jovian planets.

mutations Errors in the copying process when a living cell replicates itself.

natural selection The process by which mutations that make an organism better able to survive get passed on to future generations.

neap tides The lower-than-average tides on Earth that occur at first- and third-quarter moon, when the tidal forces from the Sun and Moon oppose each other.

nebula A cloud of gas in space, usually one that is glowing.

nebular capture The process by which icy planetesimals capture hydrogen and helium gas to form jovian planets.

nebular theory The detailed theory that describes how our solar system formed from a cloud of interstellar gas and dust.

net force The overall force to which an object responds; the net force is equal to the rate of change in the object's momentum, or equivalently to the object's mass × acceleration.

neutrino A type of fundamental particle that has extremely low mass and responds only to the weak force; neutrinos are leptons and come in three types—electron neutrinos, mu neutrinos, and tau neutrinos.

neutron degeneracy pressure Degeneracy pressure exerted by neutrons, as in neutron stars.

neutrons Particles with no electrical charge found in atomic nuclei, built from three quarks.

neutron star The compact corpse of a high-mass star left over after a supernova; it typically has a mass comparable to the mass of the Sun in a volume just a few kilometers in radius.

newton The standard unit of force in the metric system:

$$1 \text{ newton} = 1 \, \frac{\text{kg} \times \text{m}}{\text{s}^2}$$

Newton's first law of motion Principle that, in the absence of a net force, an object moves with constant velocity.

Newton's laws of motion Three basic laws that describe how objects respond to forces.

Newton's second law of motion Law stating how a net force affects an object's motion. Specifically, force = rate of change in momentum, or force = mass × acceleration.

Newton's third law of motion Principle that, for any force, there is always an equal and opposite reaction force.

Newton's universal law of gravitation *See* universal law of gravitation.

Newton's version of Kepler's third law A generalization of Kepler's third law used to calculate the masses of orbiting objects from measurements of orbital period and distance; usually written as

$$p^2 = \frac{4\pi^2}{G(M_1 + M_2)}a^3$$

nodes (of Moon's orbit) The two points in the Moon's orbit where it crosses the ecliptic plane.

nonbaryonic matter Matter that is not part of the normal composition of atoms, such as neutrinos or the hypothetical WIMPs. (More technically, particles that are not made from three quarks.)

nonscience As defined in this book, any way of searching for knowledge that makes no claim to follow the scientific method, such as seeking knowledge through intuition, tradition, or faith.

north celestial pole (NCP) The point on the celestial sphere directly above Earth's North Pole.

nova The dramatic brightening of a star that lasts for a few weeks and then subsides; it occurs when a burst of hydrogen fusion ignites in a shell on the surface of an accreting white dwarf in a binary star system.

nuclear fission The process in which a larger nucleus splits into two (or more) smaller particles.

nuclear fusion The process in which two (or more) smaller nuclei slam together and make one larger nucleus.

nucleus (of a comet) The solid portion of a comet—the only portion that exists when the comet is far from the Sun.

nucleus (of an atom) The compact center of an atom made from protons and neutrons.

observable universe The portion of the entire universe that, at least in principle, can be seen from Earth.

Occam's razor A principle often used in science, holding that scientists should prefer the simpler of two models that agree equally well with observations; named after the medieval scholar William of Occam (1285–1349).

Olbers' paradox A paradox pointing out that if the universe were infinite in both age and size (with stars found throughout the universe), then the sky would not be dark at night.

Oort cloud A huge, spherical region centered on the Sun, extending perhaps halfway to the nearest stars, in which trillions of comets orbit the Sun with random inclinations, orbital directions, and eccentricities.

opacity A measure of how much light a material absorbs compared to how much it transmits; materials with higher opacity absorb more light.

opaque Describes a material that absorbs light.

open cluster A cluster of up to several thousand stars; open clusters are found only in the disks of galaxies and often contain young stars.

open universe A universe in which spacetime has an overall shape analogous to the surface of a saddle.

opposition The point at which a planet appears opposite the Sun in our sky.

optical quality The ability of a lens, mirror, or telescope to obtain clear and properly focused images.

orbit The path followed by a celestial body because of gravity; an orbit may be *bound* (elliptical) or *unbound* (parabolic or hyperbolic).

orbital energy The sum of an orbiting object's kinetic and gravitational potential energies.

orbital resonance A situation in which one object's orbital period is a simple ratio of another object's period, such as $1/2$, $1/4$, or $5/3$. In such cases, the two objects periodically line up with each other, and the extra gravitational attractions at these times can affect the objects' orbits.

orbital velocity law A variation on Newton's version of Kepler's third law that allows us to use a star's orbital speed and distance from the galactic center to determine the total mass of the galaxy contained *within* the star's orbit; mathematically,

$$M_r = \frac{r \times v^2}{G}$$

where M_r is the mass contained within the star's orbit, r is the star's distance from the galactic center, v is the star's orbital velocity, and G is the gravitational constant.

orbiters (of other worlds) Spacecraft that go into orbit of another world for long-term study.

outer solar system Generally considered to encompass the region of our solar system beginning at about the orbit of Jupiter.

outgassing The process of releasing gases from a planetary interior, usually through volcanic eruptions.

oxidation Chemical reactions, often with rocks on the surface of a planet, that remove oxygen from the atmosphere.

ozone The molecule O_3, which is a particularly good absorber of ultraviolet light.

ozone depletion The decline in levels of atmospheric ozone found worldwide on Earth, especially in Antarctica, in recent years.

ozone hole A place where the concentration of ozone in the stratosphere is dramatically lower than is the norm.

pair production The process in which a concentration of energy spontaneously turns into a particle and its antiparticle.

parabola The precise mathematical shape of a special type of unbound orbit allowed under the force of gravity. If an object in a parabolic orbit loses only a tiny amount of energy, it will become bound.

paradigm (in science) A general pattern of thought that tends to shape scientific study during a particular time period.

paradox A situation that, at least at first, seems to violate common sense or contradict itself. Resolving paradoxes often leads to deeper understanding.

parallax The apparent shifting of an object against the background, due to viewing it from different positions. *See also* stellar parallax.

parallax angle Half of a star's annual back-and-forth shift due to stellar parallax; related to the star's distance according to the formula

$$\text{distance in parsecs} = \frac{1}{p}$$

where p is the parallax angle in arcseconds.

parsec (pc) The distance to an object with a parallax angle of 1 arcsecond; approximately equal to 3.26 light-years.

partial lunar eclipse A lunar eclipse during which the Moon becomes only partially covered by Earth's umbral shadow.

partial solar eclipse A solar eclipse during which the Sun becomes only partially blocked by the disk of the Moon.

particle accelerator A machine designed to accelerate subatomic particles to high speeds in order to create new particles or to test fundamental theories of physics.

particle era The era of the universe lasting from 10^{-10} second to 0.001 second after the Big Bang, during which subatomic particles were continually created and destroyed, and ending when matter annihilated antimatter.

peculiar velocity (of a galaxy) The component of a galaxy's velocity relative to the Milky Way that deviates from the velocity expected by Hubble's law.

penumbra The lighter, outlying regions of a shadow.

penumbral lunar eclipse A lunar eclipse during which the Moon passes only within Earth's penumbral shadow and does not fall within the umbra.

perigee The point at which an object orbiting Earth is nearest to Earth.

perihelion The point at which an object orbiting the Sun is closest to the Sun.

period–luminosity relation The relation that describes how the luminosity of a Cepheid variable star is related to the period between peaks in its brightness; the longer the period, the more luminous the star.

phase (of matter) The state determined by the way in which atoms or molecules are held together; the common phases are solid, liquid, and gas.

phase (of the Moon or a planet) The state determined by the portion of the visible face of the Moon (or of a planet) that is illuminated by sunlight. For the Moon, the phases cycle through new, waxing crescent, first-quarter, waxing gibbous, full, waning gibbous, third-quarter, waning crescent, and back to new.

photon An individual particle of light, characterized by a wavelength and a frequency.

photosphere The visible surface of the Sun, where the temperature averages just under 6000 K.

pixel An individual "picture element" in a digital image.

Planck era The era of the universe prior to the Planck time.

Planck's constant A universal constant, abbreviated h, with a value of $h = 6.626 \times 10^{-34}$ joule $\times$ s.

Planck time The time when the universe was 10^{-43} second old, before which random energy fluctuations were so large that our current theories are powerless to describe what might have been happening.

planet A moderately large object that orbits a star and shines primarily by reflecting light from its star. More precisely, according to a definition approved in 2006, a planet is an object that (1) orbits a star (but is itself neither a star nor a moon); (2) is massive enough for its own gravity to give it a nearly round shape; and (3) has cleared the neighborhood around its orbit. Objects that meet the first two criteria but not the third, including Ceres, Pluto, and Eris, are designated *dwarf planets*.

planetary geology The extension of the study of Earth's surface and interior to apply to other solid bodies in the solar system, such as terrestrial planets and jovian planet moons.

planetary migration A process through which a planet can move from the orbit on which it is born to a different orbit that is closer to or farther from its star.

planetary nebula The glowing cloud of gas ejected from a low-mass star at the end of its life.

planetesimals The building blocks of planets, formed by accretion in the solar nebula.

plasma A gas consisting of ions and electrons.

plasma tail (of a comet) One of two tails seen when a comet passes near the Sun (the other is the dust tail). It is composed of ionized gas blown away from the Sun by the solar wind.

plates (on a planet) Pieces of a lithosphere that apparently float upon the denser mantle below.

plate tectonics The geological process in which plates are moved around by stresses in a planet's mantle.

polarization (of light) The property of light describing how the electric and magnetic fields of light waves are aligned; light is said to be *polarized* when all of the photons have their electric and magnetic fields aligned in some particular way.

Population I *See* disk population.

Population II *See* halo population.

positron *See* antielectron.

potential energy Energy stored for later conversion into kinetic energy; includes gravitational potential energy, electrical potential energy, and chemical potential energy.

power The rate of energy usage, usually measured in watts (1 watt = 1 joule/s).

precession The gradual wobble of the axis of a rotating object around a vertical line.

precipitation Condensed atmospheric gases that fall to the surface in the form of rain, snow, or hail.

pressure The force (per unit area) pushing on an object. In astronomy, we are generally interested in pressure applied by surrounding gas (or plasma). Ordinarily, such pressure is related to the temperature of the gas (*see* thermal pressure). In objects such as white dwarfs and neutron stars, pressure may arise from a quantum effect (*see* degeneracy pressure). Light can also exert pressure (*see* radiation pressure).

primary mirror The large, light-collecting mirror of a reflecting telescope.

prime focus (of a reflecting telescope) The first point at which light focuses after bouncing off the primary mirror; located in front of the primary mirror.

prime meridian The meridian of longitude that passes through Greenwich, England; defined to be longitude 0°.

primitive meteorites Meteorites that formed at the same time as the solar system itself, about 4.6 billion years ago. Primitive meteorites from the inner asteroid belt are usually stony, and those from the outer belt are usually carbon-rich.

processed meteorites Meteorites that apparently once were part of a larger object that "processed" the original material of the solar nebula into another form. Processed meteorites can be rocky if chipped from the surface or mantle, or metallic if blasted from the core.

proper motion The motion of an object in the plane of the sky, perpendicular to our line of sight.

protogalactic cloud A huge, collapsing cloud of intergalactic gas from which an individual galaxy formed.

proton–proton chain The chain of reactions by which low-mass stars (including the Sun) fuse hydrogen into helium.

protons Particles with positive electrical charge found in atomic nuclei, built from three quarks.

protoplanetary disk A disk of material surrounding a young star (or protostar) that may eventually form planets.

protostar A forming star that has not yet reached the point where sustained fusion can occur in its core.

protostellar disk A disk of material surrounding a protostar; essentially the same as a protoplanetary disk, but may not necessarily lead to planet formation.

protostellar wind The relatively strong wind from a protostar.

protosun The central object in the forming solar system that eventually became the Sun.

pseudoscience Something that purports to be science or may appear to be scientific but that does not adhere to the testing and verification requirements of the scientific method.

Ptolemaic model The geocentric model of the universe developed by Ptolemy in about 150 A.D.

pulsar A neutron star from which we observe rapid pulses of radiation as it rotates.

pulsating variable stars Stars that grow alternately brighter and dimmer as their outer layers expand and contract in size.

quantum laws The laws that describe the behavior of particles on a very small scale; *see also* quantum mechanics.

quantum mechanics The branch of physics that deals with the very small, including molecules, atoms, and fundamental particles.

quantum state The complete description of the state of a subatomic particle, including its location, momentum, orbital angular momentum, and spin, to the extent allowed by the uncertainty principle.

quantum tunneling The process in which, thanks to the uncertainty principle, an electron or other subatomic particle appears on the other side of a barrier that it does not have the energy to overcome in a normal way.

quarks The building blocks of protons and neutrons; quarks are one of the two basic types of fermions (leptons are the other).

quasar The brightest type of active galactic nucleus.

radar mapping Imaging of a planet by bouncing radar waves off its surface, especially important for Venus and Titan, where thick clouds mask the surface.

radar ranging A method of measuring distances within the solar system by bouncing radio waves off planets.

radial motion The component of an object's motion directed toward or away from us.

radial velocity The portion of any object's total velocity that is directed toward or away from us. This part of the velocity is the only part that we can measure with the Doppler effect.

radiation pressure Pressure exerted by photons of light.

radiation zone (of a star) A region of the interior in which energy is transported primarily by radiative diffusion.

radiative diffusion The process by which photons gradually migrate from a hot region (such as the solar core) to a cooler region (such as the solar surface).

radiative energy Energy carried by light; the energy of a photon is Planck's constant times its frequency, or $h \times f$.

radioactive decay The spontaneous change of an atom into a different element, in which its nucleus breaks apart or a proton turns into an electron. This decay releases heat in a planet's interior.

radioactive element (or radioactive isotope) A substance whose nucleus tends to fall apart spontaneously.

radio galaxy A galaxy that emits unusually large quantities of radio waves; thought to contain an active galactic nucleus powered by a supermassive black hole.

radio lobes The huge regions of radio emission found on either side of radio galaxies. The lobes apparently contain plasma ejected by powerful jets from the galactic center.

radiometric dating The process of determining the age of a rock (i.e., the time since it solidified) by comparing the present amount of a radioactive substance to the amount of its decay product.

radio waves Light with very long wavelengths (and hence low frequencies)—longer than those of infrared light.

random walk A type of haphazard movement in which a particle or photon moves through a series of bounces, with each bounce sending it in a random direction.

recession velocity (of a galaxy) The speed at which a distant galaxy is moving away from us because of the expansion of the universe.

recollapsing universe A model of the universe in which the collective gravity of all its matter eventually halts and reverses the expansion, causing the galaxies to come crashing back together and the universe to end in a fiery Big Crunch.

red giant A giant star that is red in color.

red-giant winds The relatively dense but slow winds from red giant stars.

redshift (Doppler) A Doppler shift in which spectral features are shifted to longer wavelengths, observed when an object is moving away from the observer.

reference frame (or frame of reference) In the theory of relativity, what two people (or objects) share if they are *not* moving relative to one another.

reflecting telescope A telescope that uses mirrors to focus light.

reflection (of light) The process by which matter changes the direction of light.

reflection nebula A nebula that we see as a result of starlight reflected from interstellar dust grains. Reflection nebulae tend to have blue and black tints.

refracting telescope A telescope that uses lenses to focus light.

resonance *See* orbital resonance.

rest wavelength The wavelength of a spectral feature in the absence of any Doppler shift or gravitational redshift.

retrograde motion Motion that is backward compared to the norm. For example, we see Mars in apparent retrograde motion during the periods of time when it moves westward, rather than the more common eastward, relative to the stars.

revolution The orbital motion of one object around another.

right ascension (RA) Analogous to longitude, but on the celestial sphere; the angular east-west distance between the March equinox and a location on the celestial sphere.

rings (planetary) The collections of numerous small particles orbiting a planet within its Roche tidal zone.

Roche tidal zone The region within two to three planetary radii (of any planet) in which the tidal forces tugging an object apart become comparable to the gravitational forces holding it together; planetary rings are always found within the Roche tidal zone.

rocks (in solar system theory) Materials common on the surface of Earth, such as silicon-based minerals, that are solid at temperatures and pressures found on Earth but typically melt or vaporize at temperatures of 500–1300 K.

rotation The spinning of an object around its axis.

rotation curve A graph that plots rotational (or orbital) velocity against distance from the center for any object or set of objects.

runaway greenhouse effect A positive feedback cycle in which heating caused by the greenhouse effect causes more greenhouse gases to enter the atmosphere, which further enhances the greenhouse effect.

saddle-shaped (or hyperbolic) geometry The type of geometry in which the rules—such as that two lines that begin parallel eventually diverge—are most easily visualized on a saddle-shaped surface.

Sagittarius Dwarf A small dwarf elliptical galaxy that is currently passing through the disk of the Milky Way Galaxy.

saros cycle The period over which the basic pattern of eclipses repeats, which is about 18 years $11\frac{1}{3}$ days.

satellite Any object orbiting another object.

scattered light Light that is reflected into random directions.

Schwarzschild radius A measure of the size of the event horizon of a black hole.

science The search for knowledge that can be used to explain or predict natural phenomena in a way that can be confirmed by rigorous observations or experiments.

scientific method An organized approach to explaining observed facts through science.

scientific theory A model of some aspect of nature that has been rigorously tested and has passed all tests to date.

seafloor crust On Earth, the thin, dense crust of basalt created by seafloor spreading.

seafloor spreading On Earth, the creation of new seafloor crust at mid-ocean ridges.

search for extraterrestrial intelligence (SETI) The name given to observing projects designed to search for signs of intelligent life beyond Earth.

secondary mirror A small mirror in a reflecting telescope, used to reflect light gathered by the primary mirror toward an eyepiece or instrument.

sedimentary rock A rock that formed from sediments created and deposited by erosional processes. The sediments tend to build up in distinct layers, or *strata*.

seismic waves Earthquake-induced vibrations that propagate through a planet.

selection effect (or selection bias) A type of bias that arises from the way in which objects of study are selected and that can lead to incorrect conclusions. For example, when you are counting animals in a jungle it is easiest to see brightly colored animals, which could mislead you into thinking that these animals are the most common.

semimajor axis Half the distance across the long axis of an ellipse; in this text, it is usually referred to as the *average* distance of an orbiting object, abbreviated *a* in the formula for Kepler's third law.

September equinox Both the point in Virgo on the celestial sphere where the ecliptic crosses the celestial equator and the moment in time when the Sun appears at that point each year (around September 21).

Seyfert galaxies The name given to a class of galaxies that are found relatively nearby and that have nuclei much like those of quasars, except that they are less luminous.

shepherd moons Tiny moons within a planet's ring system that help force particles into a narrow ring; a variation on *gap moons*.

shield volcano A shallow-sloped volcano made from the flow of low-viscosity basaltic lava.

shock wave A wave of pressure generated by gas moving faster than the speed of sound.

sidereal day The time of 23 hours 56 minutes 4.09 seconds between successive appearances of any particular star on the meridian; essentially, the true rotation period of Earth.

sidereal month The time required for the Moon to orbit Earth once (as measured against the stars); about $27\frac{1}{4}$ days.

sidereal period (of a planet) A planet's actual orbital period around the Sun.

sidereal time Time measured according to the position of stars in the sky rather than the position of the Sun in the sky. *See also* local sidereal time.

sidereal year The time required for Earth to complete exactly one orbit as measured against the stars; about 20 minutes longer than the tropical year on which our calendar is based.

silicate rock A silicon-rich rock.

singularity The place at the center of a black hole where, in principle, gravity crushes all matter to an infinitely tiny and dense point.

Small Magellanic Cloud One of two small, irregular galaxies (the other is the Large Magellanic Cloud) located about 150,000 light-years away; it probably orbits the Milky Way Galaxy.

small solar system body An asteroid, comet, or other object that orbits a star but is too small to qualify as a planet or dwarf planet.

snowball Earth Name given to a hypothesis suggesting that, some 600–700 million years ago, Earth experienced a period in which it became cold enough for glaciers to exist worldwide, even in equatorial regions.

solar activity Short-lived phenomena on the Sun, including the emergence and disappearance of individual sunspots, prominences, and flares; sometimes called *solar weather*.

solar circle The Sun's orbital path around the galaxy, which has a radius of about 28,000 light-years.

solar day 24 hours, which is the average time between appearances of the Sun on the meridian.

solar eclipse An event that occurs when the Moon's shadow falls on Earth, which can occur only at new moon. A solar eclipse may be total, partial, or annular.

solar flares Huge and sudden releases of energy on the solar surface, probably caused when energy stored in magnetic fields is suddenly released.

solar luminosity The luminosity of the Sun, which is approximately 4×10^{26} watts.

solar maximum The time during each sunspot cycle at which the number of sunspots is the greatest.

solar minimum The time during each sunspot cycle at which the number of sunspots is the smallest.

solar nebula The piece of interstellar cloud from which our own solar system formed.

solar neutrino problem A now-solved problem in which, during the latter decades of the 20th century, there appeared to be disagreement between the predicted and observed number of neutrinos coming from the Sun.

solar prominences Vaulted loops of hot gas that rise above the Sun's surface and follow magnetic field lines.

solar sail A large, highly reflective (and thin, to minimize mass) piece of material that can "sail" through space using pressure exerted by sunlight.

solar system (or star system) A star (sometimes more than one star) and all the objects that orbit it.

solar thermostat *See* stellar thermostat; the solar thermostat is the same idea applied to the Sun.

solar wind A stream of charged particles ejected from the Sun.

solid phase The phase of matter in which atoms or molecules are held rigidly in place.

solstice *See* December solstice *and* June solstice.

sound wave A wave of alternately rising and falling pressure.

south celestial pole (SCP) The point on the celestial sphere directly above Earth's South Pole.

spacetime The inseparable, four-dimensional combination of space and time.

spacetime diagram A graph that plots a spatial dimension on one axis and time on another axis.

special theory of relativity Einstein's theory that describes the relativity of time and space based on the fact that the laws of nature are the same for everyone and that everyone always measures the same speed of light.

spectral lines Bright or dark lines that appear in an object's spectrum, which we can see when we pass the object's light through a prismlike device that spreads out the light like a rainbow.

spectral resolution The degree of detail that can be seen in a spectrum; the higher the spectral resolution, the more detail we can see.

spectral type A way of classifying a star by the lines that appear in its spectrum; it is related to surface temperature. The basic spectral types are designated by letters (OBAFGKM, with O for the hottest stars and M for the coolest) and are subdivided with numbers from 0 through 9.

spectrograph An instrument used to record spectra.

spectroscopic binary A binary star system whose binary nature is revealed because we detect the spectral lines of one or both stars alternately becoming blueshifted and redshifted as the stars orbit each other.

spectroscopy (in astronomical research) The process of obtaining spectra from astronomical objects.

spectrum (of light) *See* electromagnetic spectrum.

speed The rate at which an object moves. Its units are distance divided by time, such as m/s or km/hr.

speed of light The speed at which light travels, which is about 300,000 km/s.

spherical geometry The type of geometry in which the rules—such as that lines that begin parallel eventually meet—are those that hold on the surface of a sphere.

spheroidal component (of a galaxy) *See* halo component.

spheroidal galaxy Another name for an *elliptical galaxy*. Also see dwarf spheroidal galaxy.

spheroidal population *See* halo population.

spin (quantum) *See* spin angular momentum.

spin angular momentum The inherent angular momentum of a fundamental particle; often simply called *spin*.

spiral arms The bright, prominent arms, usually in a spiral pattern, found in most spiral galaxies.

spiral density waves Gravitationally driven waves of enhanced density that move through a spiral galaxy and are responsible for maintaining its spiral arms.

spiral galaxies Galaxies that look like flat white disks with yellowish bulges at their centers. The disks are filled with cool gas and dust, interspersed with hotter ionized gas, and usually display beautiful spiral arms.

spreading centers (geological) Places where hot mantle material rises upward between plates and then spreads sideways, creating new seafloor crust.

spring equinox *See* March equinox, which is commonly called the spring equinox by people living in the Northern Hemisphere.

spring tides The higher-than-average tides on Earth that occur at new and full moon, when the tidal forces from the Sun and Moon both act along the same line.

standard candle An object for which we have some means of knowing its true luminosity, so that we can use its apparent brightness to determine its distance with the luminosity–distance formula.

standard model (of physics) The current theoretical model that describes the fundamental particles and forces in nature.

standard time Time measured according to the internationally recognized time zones.

star A large, glowing ball of gas that generates energy through nuclear fusion in its core. The term *star* is sometimes applied to objects that are in the process of becoming true stars (e.g., protostars) and to the remains of stars that have died (e.g., neutron stars).

starburst galaxy A galaxy in which stars are forming at an unusually high rate.

star cluster *See* cluster of stars.

star–gas–star cycle The process of galactic recycling in which stars expel gas into space, where it mixes with the interstellar medium and eventually forms new stars.

star system *See* solar system.

state (quantum) *See* quantum state.

steady state theory A now-discredited theory that held that the universe had no beginning and looks about the same at all times.

Stefan–Boltzmann constant A constant that appears in the laws of thermal radiation, with value

$$\sigma = 5.7 \times 10^{-8} \frac{\text{watt}}{\text{m}^2 \times \text{K}^4}$$

stellar evolution The formation and development of stars.

stellar parallax The apparent shift in the position of a nearby star (relative to distant objects) that occurs as we view the star from different positions in Earth's orbit of the Sun each year.

stellar thermostat The regulation of a star's core temperature that comes about when a star is in both energy balance (the rate at which fusion releases energy in the star's core is balanced with the rate at which the star's surface radiates energy into space) and gravitational equilibrium.

stellar wind A stream of charged particles ejected from the surface of a star.

stratosphere An intermediate-altitude layer of Earth's atmosphere that is warmed by the absorption of ultraviolet light from the Sun.

stratovolcano A steep-sided volcano made from viscous lavas that can't flow very far before solidifying.

string theory New ideas, not yet well-tested, that attempt to explain all of physics in a much simpler way than current theories.

stromatolites Rocks thought to be fossils of ancient microbial colonies.

strong force One of the four fundamental forces; it is the force that holds atomic nuclei together.

subduction (of tectonic plates) The process in which one plate slides under another.

subduction zones Places where one plate slides under another.

subgiant A star that is between being a main-sequence star and being a giant; subgiants have inert helium cores and hydrogen-fusing shells.

sublimation The process by which atoms or molecules escape into the gas phase from the solid phase.

summer solstice *See* June solstice, which is commonly called the summer solstice by people living in the Northern Hemisphere.

sunspot cycle The period of about 11 years over which the number of sunspots on the Sun rises and falls.

sunspots Blotches on the surface of the Sun that appear darker than surrounding regions.

superbubble Essentially a giant interstellar bubble, formed when the shock waves of many individual bubbles merge to form a single giant shock wave.

superclusters The largest known structures in the universe, consisting of many clusters of galaxies, groups of galaxies, and individual galaxies.

super-Earth An extrasolar planet whose size and density suggest it has an Earth-like composition, made largely of rock and metal.

supergiants The very large and very bright stars (luminosity class I) that appear at the top of an H-R diagram.

supermassive black holes Giant black holes, with masses millions to billions of times that of our Sun, thought to reside in the centers of many galaxies and to power active galactic nuclei.

supernova The explosion of a star.

Supernova 1987A A supernova witnessed on Earth in 1987; it was the nearest supernova seen in nearly 400 years and helped astronomers refine theories of supernovae.

supernova remnant A glowing, expanding cloud of debris from a supernova explosion.

surface area–to–volume ratio The ratio defined by an object's surface area divided by its volume; this ratio is larger for smaller objects (and vice versa).

synchronous rotation The rotation of an object that always shows the same face to an object that it is orbiting because its rotation period and orbital period are equal.

synchrotron radiation A type of radio emission that occurs when electrons moving at nearly the speed of light spiral around magnetic field lines.

synodic month (or **lunar month**) The time required for a complete cycle of lunar phases, which averages about $29\frac{1}{2}$ days.

synodic period (of a planet) The time between successive alignments of a planet and the Sun in our sky; measured from opposition to opposition for a planet beyond Earth's orbit, or from superior conjunction to superior conjunction for Mercury and Venus.

tangential motion The component of an object's motion directed across our line of sight.

tangential velocity The portion of any object's total velocity that is directed across (perpendicular to) our line of sight. This part of the velocity cannot be measured with the Doppler effect. It can be measured only by observing the object's gradual motion across our sky.

tectonics The disruption of a planet's surface by internal stresses.

temperature A measure of the average kinetic energy of particles in a substance.

terrestrial planets Rocky planets similar in overall composition to Earth.

theories of relativity (special and general) Einstein's theories that describe the nature of space, time, and gravity.

theory (in science) *See* scientific theory.

theory of evolution The theory, first advanced by Charles Darwin, that explains how evolution occurs through the process of natural selection.

thermal emitter An object that produces a thermal radiation spectrum; sometimes called a *blackbody*.

thermal energy The collective kinetic energy, as measured by temperature, of the many individual particles moving within a substance.

thermal escape The process in which atoms or molecules in a planet's exosphere move fast enough to escape into space.

thermal pressure The ordinary pressure in a gas arising from motions of particles that can be attributed to the object's temperature.

thermal pulses The predicted upward spikes in the rate of helium fusion, occurring every few thousand years, that occur near the end of a low-mass star's life.

thermal radiation The spectrum of radiation produced by an opaque object that depends only on the object's temperature; sometimes called *blackbody radiation*.

thermosphere A high, hot, X-ray-absorbing layer of an atmosphere, just below the exosphere.

third-quarter (phase) The phase of the Moon that occurs three-quarters of the way through each cycle of phases, in which precisely half of the visible face is illuminated by sunlight.

tidal force A force that occurs when the gravity pulling on one side of an object is larger than that on the other side, causing the object to stretch.

tidal friction Friction within an object that is caused by a tidal force.

tidal heating A source of internal heating created by tidal friction. It is particularly important for satellites with eccentric orbits such as Io and Europa.

time dilation The effect in which you observe time running more slowly in reference frames moving relative to you.

time monitoring (in astronomical research) The process of tracking how the light intensity from an astronomical object varies with time.

torque A twisting force that can cause a change in an object's angular momentum.

total apparent brightness *See* apparent brightness. The word "total" is sometimes added to make clear that we are talking about light across all wavelengths, not just visible light.

totality (eclipse) The portion of a total lunar eclipse during which the Moon is fully within Earth's umbral shadow or a total solar eclipse during which the Sun's disk is fully blocked by the Moon.

total luminosity *See* luminosity. The word "total" is sometimes added to make clear that we are talking about light across all wavelengths, not just visible light.

total lunar eclipse A lunar eclipse in which the Moon becomes fully covered by Earth's umbral shadow.

total solar eclipse A solar eclipse during which the Sun becomes fully blocked by the disk of the Moon.

transit An event in which a planet passes in front of a star (or the Sun) as seen from Earth. Only Mercury and Venus can be seen in transit of our Sun. The search for transits of extrasolar planets is an important planet detection strategy.

transmission (of light) The process in which light passes through matter without being absorbed.

transparent Describes a material that transmits light.

tree of life (evolutionary) A diagram that shows relationships between different species as inferred from genetic comparisons.

triple-alpha reaction *See* helium fusion.

Trojan asteroids Asteroids found within two stable zones that share Jupiter's orbit but lie 60° ahead of and behind Jupiter.

tropical year The time from one March equinox to the next, on which our calendar is based.

Tropic of Cancer The circle on Earth with latitude 23.5°N, which marks the northernmost latitude at which the Sun ever passes directly overhead (which it does at noon on the June solstice).

Tropic of Capricorn The circle on Earth with latitude 23.5°S, which marks the southernmost latitude at which the Sun ever passes directly overhead (which it does at noon on the December solstice).

tropics The region on Earth surrounding the equator and extending from the Tropic of Capricorn (latitude 23.5°S) to the Tropic of Cancer (latitude 23.5°N).

troposphere The lowest atmospheric layer, in which convection and weather occur.

turbulence Rapid and random motion.

21-cm line A spectral line from atomic hydrogen with wavelength 21 cm (in the radio portion of the spectrum).

ultraviolet light Light with wavelengths that fall in the portion of the electromagnetic spectrum between visible light and X rays.

umbra The dark central region of a shadow.

unbound orbits Orbits on which an object comes in toward a large body only once, never to return; unbound orbits may be parabolic or hyperbolic in shape.

uncertainty principle The law of quantum mechanics that states that we can never know both a particle's position and its momentum, or both its energy and the time it has the energy, with absolute precision.

universal law of gravitation The law expressing the force of gravity (F_g) between two objects, given by the formula

$$F_g = G \frac{M_1 M_2}{d^2}$$

$$\left(\text{where } G = 6.67 \times 10^{-11} \frac{m^3}{kg \times s^2} \right)$$

universal time (UT) Standard time in Greenwich, England (or anywhere on the prime meridian).

universe The sum total of all matter and energy.

up quark One of the two quark types (the other is the down quark) found in ordinary protons and neutrons; has a charge of $+\frac{2}{3}$.

vaporization The process by which atoms or molecules escape into the gas phase from the liquid or solid phase; more technically, vaporization from a liquid is called *evaporation* and vaporization from a solid is called *sublimation*.

velocity The combination of speed and direction of motion; it can be stated as a speed in a particular direction, such as 100 km/hr due north.

virtual particles Particles that "pop" in and out of existence so rapidly that, according to the uncertainty principle, they cannot be directly detected.

viscosity The thickness of a liquid described in terms of how rapidly it flows; low-viscosity liquids flow quickly (e.g., water), while high-viscosity liquids flow slowly (e.g., molasses).

visible light The light our eyes can see, ranging in wavelength from about 400 to 700 nm.

visual binary A binary star system in which both stars can be resolved through a telescope.

voids Huge volumes of space between superclusters that appear to contain very little matter.

volatiles Substances, such as water, carbon dioxide, and methane, that are usually found as gases, liquids, or surface ices on the terrestrial worlds.

volcanic plains Vast, relatively smooth areas created by the eruption of very runny lava.

volcanism The eruption of molten rock, or lava, from a planet's interior onto its surface.

waning (phases) The set of phases in which less and less of the visible face of the Moon is illuminated; the phases that come after full moon but before new moon.

water world An extrasolar planet whose size and density suggest it is made largely of water, without a substantial amount of hydrogen or helium gas.

watt The standard unit of power in science; defined as 1 watt = 1 joule/s.

wavelength The distance between adjacent peaks (or troughs) of a wave.

waxing (phases) The set of phases in which more and more of the visible face of the Moon is becoming illuminated; the phases that come after new moon but before full moon.

weak bosons The exchange particles for the weak force.

weak force One of the four fundamental forces; it is the force that mediates nuclear reactions, and it is the only force besides gravity felt by weakly interacting particles.

weakly interacting particles Particles, such as neutrinos and WIMPs, that respond only to the weak force and gravity; that is, they do not feel the strong force or the electromagnetic force.

weather The ever-varying combination of winds, clouds, temperature, and pressure in a planet's troposphere.

weight The net force that an object applies to its surroundings; in the case of a stationary body on the surface of Earth, it equals mass × acceleration of gravity.

weightlessness A weight of zero, as occurs during free-fall.

white dwarf limit (or **Chandrasekhar limit**) The maximum possible mass for a white dwarf, which is about $1.4 M_{Sun}$.

white dwarfs The hot, compact corpses of low-mass stars, typically with a mass similar to that of the Sun compressed to a volume the size of Earth.

white dwarf supernova A supernova that occurs when an accreting white dwarf reaches the white-dwarf limit, ignites runaway carbon fusion, and explodes like a bomb; often called a *Type Ia supernova*.

WIMPs A possible form of dark matter consisting of subatomic particles that are dark because they do not respond to the electromagnetic force; stands for *weakly interacting massive particles*.

winter solstice *See* December solstice, which is commonly called the winter solstice by people living in the Northern Hemisphere.

worldline A line that represents an object on a spacetime diagram.

wormholes The name given to hypothetical tunnels through hyperspace that might connect two distant places in the universe.

X-ray binary A binary star system that emits substantial amounts of X rays, thought to be from an accretion disk around a neutron star or black hole.

X-ray burster An object that emits a burst of X rays every few hours to every few days; each burst lasts a few seconds and is thought to be caused by helium fusion on the surface of an accreting neutron star in a binary system.

X-ray bursts Bursts of X rays coming from sudden ignition of fusion on the surface of an accreting neutron star in an X-ray binary system.

X rays Light with wavelengths that fall in the portion of the electromagnetic spectrum between ultraviolet light and gamma rays.

Zeeman effect The splitting of spectral lines by a magnetic field.

zenith The point directly overhead, which has an altitude of 90°.

zodiac The constellations on the celestial sphere through which the ecliptic passes.

zones (on a jovian planet) Bright bands of rising air that encircle a jovian planet at a particular set of latitudes.

Credits

Foldout

Front NASA/WMAP Science Team, NASA Earth Observing System, Stephen J Mojzsis **Back** Russell Shively/Shutterstock, Pearson Education, Inc., Dorling Kindersley Limited, Dave King/Dorling Kindersley Limited, Linda Whitwam/Dorling Kindersley Limited, Pius Lee/Shutterstock, NASA

Frontmatter

p. viii Jeffrey Bennett, Megan Donahue **p. ix** Nicholas Schneider, Mark Voit **p. xxii** Copyright Neil deGrasse Tyson **p. xxiv** Jeffrey Bennett

Chapter 1

Chapter Opener NASA, ESA, S. Beckwith (STScI) and the HUDF Team **1.1** (Milky Way Galaxy) Michael Carroll **1.2** Jerry Lodriguss/Science Source **1.4** Jeffrey Bennett **1.6** NASA **1.7** (bottom) Akira Fujii **1.8** Jeffrey Bennett **1.10** (photos, left to right) Blakeley Kim/Pearson Education, Inc.; Corel Corporation; NASA Earth Observing System; Steve Vidler/Alamy Stock Photo **1.14** NASA Earth Observing System; Pearson Education, Inc. **p. 21** NASA Earth Observing System

Chapter 2

Chapter Opener D. Nunuk/Science Source **2.1** Wally Pacholka **2.17** Frank Zullo/Science Source **2.18** Arnulf Husmo/Stone/Getty Images **2.20** (bottom) NASA/JPL-Caltech/University of Arizona **2.25** (top and bottom) Akira Fujii; (center) EPA/Stephen Morrison/Alamy **2.26** NASA **2.27** Akira Fujii **2.29** Tunc Tezel

Chapter 3

Chapter Opener NASA **3.2** Karl Kost/Alamy Stock Photo **3.3** Timo Kohlbacher/Shutterstock **3.4** SuperStock **3.5** William E Woolam **3.6** Lawrence Rigby Latin America/Alamy Stock Photo **3.7** Amy Nichole Harris/Shutterstock **3.8** Walter Meayers Edwards/National Geographic **3.9a** Sheila Terry/Science Source **3.9b** Patrick Pihl/Interpix/Alamy Stock Photo **p. 62** (top) World History Archive/SuperStock (bottom) SuperStock **3.12** Gianni Tortoli/Science Source **p. 63** (middle) quote by Johannes Kepler (bottom right photo) Science Source **p. 66** Fine Art Images/SuperStock **3.17** Stuart J. Robbins **3.18** Galileo Galilei **3.22** (top left) Tunc Tezel **p. 80** Timo Kohlbacher/Shutterstock; Scala/Art Resource **p. 81** Yerkes Observatory; Huntington Library/SuperStock; NASA

Chapter 4

Chapter Opener Debra Meloy Elmegreen (Vassar College) et al., & the Hubble Heritage Team (AURA/STScI/NASA) http://apod.nasa.gov/apod/ap991109.html **p. 88** Sir Godfrey Kneller **4.5** (left to right) Goddard Institute for Space Studies/NASA; Bruce Kluckhohn/Getty Images; Goddard Institute for Space Studies/NASA **4.8** (clockwise from top) Stockbyte/Getty Images; Eric Gevaert/Alamy Stock Photo; Don Hammond/Getty Images **4.13** U.S. Department of Energy **4.20** (left to right) Jon Arnold/Travel Pix Collection/SuperStock; Photononstop/SuperStock

Chapter 5

Chapter Opener N.A. Sharp, NOAO/NSO/Kitt Peak FTS/AURA/NSF **5.1** Richard Megna/Fundamental Photographs **5.17** Yerkes Observatory **5.18** National Optical Astronomy Observatories **5.19** Richard Wainscoat/Alamy Stock Photo (inset) Russ Underwood/W.M. Keck Observatory **5.20** Epa european pressphoto agency b.v./Alamy Stock Photo **5.21** Steve Vidler/Alamy Stock Photo **5.22** ESO/B. Tafreshi (twanight.org) **5.23** CXC/SAO **5.24** Handout/MCT/Newscom **5.25** Earth Observatory/NOAA NGDC **5.26** Richard Wainscoat/Alamy Stock Photo **5.27** NASA **5.29** NASA/Chris Gunn **5.30** Canada France Hawaii Telescope

Chapter 6

Chapter Opener ALMA (ESO/NAOJ/NRAO) **6.2a** NSO/AURA/NSF **6.2b** NASA/Marshall Space Flight Center **6.3** Johns Hopkins University Applied Physics Laboratory/Carnegie Institution of Washington/NASA **6.4** (left to right) Marshall Space Flight Center/NASA; Science Source **6.5a** ARC Science Simulations **6.5b** Earth Observing System/NASA **6.6** (left to right) USGS; NASA **6.7** Jeffrey Bennett/ARC Science Simulations **6.8** NASA/Jet Propulsion Laboratory **6.9** ARC Science Simulations **6.10** ARC Science Simulations **6.11** NEAR Project, JHU APL, NASA **6.12** NASA Goddard Institute for Space Studies **6.13** Photography and image processing by Miloslav Druckmüller **6.15** M. Robberto/ESA/Space Telescope Science Institute/Hubble Space Telescope Orion Treasury Project Team/NASA **6.17a** NASA, ESA, J. Debes (STScI), H. Jang-Condell (University of Wyoming), A. Weinberger (Carnegie Institution of Washington), A. Roberge (Goddard Space Flight Center), and G. Schneider (University of Arizona/Steward Observatory) **6.17b** ALMA (ESO/NAOJ/NRAO), ESA/Hubble and NASA. Acknowledgement: Judy Schmidt **6.20** Nicholas M. Schneider **6.21** NASA **6.22** NASA Earth Observing System **6.23** NASA Earth Observing System

Chapter 7

Chapter Opener NASA Goddard Space Flight Center Image by Reto Stöckli (land surface, shallow water, clouds). Enhancements by Robert Simmon (ocean color, compositing, 3D globes, animation). Data and technical support: MODIS Land Group; MODIS Science Data Support Team; MODIS Atmosphere Group; MODIS Ocean Group. Additional data: USGS EROS Data Center (topography); USGS Terrestrial Remote Sensing Flagstaff Field Center (Antarctica); Defense Meteorological Satellite Program (city lights) **7.1** (Mercury, Earth, Mars) Jet Propulsion Laboratory/NASA (Venus) NASA **7.3** Richard Megna/Fundamental Photographs **7.5** NASA **7.6b** Doug Duncan **7.7** Don Davis **7.8** Brad Snowder, Western Washington University **7.9** Paul Chesley/Stone/Getty Images **7.10** U.S. Geological Survey **7.12a** Gene Ahrens/Photoshot **7.12b** Joachim Messerschmidt/Photoshot **7.12c** Mytho/Fotolia **7.12d** Florian Werner/Look/AGE Fotostock **7.16** (left) Alan Dyer/Stocktrek Images, Inc./Alamy Stock Photo (right) Johns Hopkins University Applied Physics Laboratory/Carnegie Institution of Washington/NASA **7.17** (left) Frank Barrett (right) Malcolm Park/Oxford Scientific/Getty Images **7.18** Pearson Education, Inc. **7.19a** Goddard Institute for Space Studies/NASA **7.19b** Goddard Space Flight Center/NASA **7.20** Johns Hopkins University Applied Physics Laboratory/Carnegie Institution of Washington/NASA **7.21** Jet Propulsion Laboratory/NASA **7.23** Goddard Institute for Space Studies/NASA **7.24** Goddard Institute for Space Studies/NASA (inset) Jet Propulsion Laboratory/NASA **7.25** Goddard Institute for Space Studies/NASA (inset) European Space Agency **7.26** NASA/Jet Propulsion Laboratory **7.27a** EROS Data Center **7.27b** National Air and Space Museum **7.27c** JPL/NASA **7.28a** Jet Propulsion Laboratory/NASA **7.28b** Dr. Marjorie A Chan **7.29** NASA/JPL-Caltech/MSSS and PSI **7.30** JPL-Caltech/MSSS/NASA **7.31** JPL-Caltech/Univ. of Arizona/NASA **7.32** JPL-Caltech/Univ. of Arizona/NASA **7.33** NASA/GSFC **7.35** University of Colorado/NASA **7.36** Jet Propulsion Laboratory/NASA **7.37** European Space Agency/NASA/JPL-Caltech/Univ. of Arizona **7.38** Ted Stryk/NASA **7.39** (top) ARC Science Simulations (bottom) NASA Marshall Space Flight Center **7.41** NOAA National Geophysical Data Center **7.46** National Oceanic & Atmospheric Administration **p. 211** NASA.

Chapter 8

Chapter Opener NASA/JPL/Space Science Institute **8.1** NASA/Jet Propulsion Laboratory **8.5** (right) NASA **8.6** (left) NASA/Jet Propulsion Laboratory **8.8** NASA **8.9** NASA/JPL **8.10a,b,d** NASA/Jet Propulsion Laboratory **8.10c** Lawrence Sromovsky, Univ. of Wisconsin-Madison/Keck Observatory **8.11** NASA/Jet Propulsion Laboratory **8.12a,c,f** NASA Earth Observing System **8.12b,d,e** NASA/Jet Propulsion Laboratory **8.13** NASA/Jet Propulsion Laboratory **8.14a** NASA/JPL **8.14b** NASA/JHU/APL **8.16** (left and right) NASA/Jet Propulsion Laboratory **8.17** (center) Britney Schmidt/Dead Pixel FX/Univ. of Texas at Austin (right) NASA/Jet Propulsion Laboratory **8.18** NASA/Jet Propulsion Laboratory **8.19** NASA/Jet Propulsion Laboratory **8.20** NASA/Jet Propulsion Laboratory (inset) NASA Earth Observing System **8.21** (left and right) NASA/Jet Propulsion Laboratory (center) NASA **8.22** NASA/Jet Propulsion Laboratory **8.23** NASA/Jet Propulsion Laboratory **8.24** NASA/Jet Propulsion Laboratory **8.25** NASA **8.26** NASA/Jet Propulsion Laboratory **8.27** NASA/Jet Propulsion Laboratory **8.28** (left) NASA/JPL/Lunar & Planetary Laboratory (right) NASA/Jet Propulsion Laboratory **8.29a** Lunar and Planetary Laboratory **8.29b,c** NASA/Jet Propulsion Laboratory **8.30** NASA/Jet Propulsion Laboratory **8.31** (top) Imke de Pater (other images) NASA/Jet Propulsion Laboratory **p. 237** NASA/JPL

Chapter 9

Chapter Opener (left to right) Lowell Observatory; NASA, ESA, and M. Buie (Southwest Research Institute); NASA/Johns Hopkins University Applied Physics Laboratory/Southwest Research Institute **9.1** Eleanor F. Helin/JPL/NASA **9.2a** Peter Ceravolo **9.2b** Niescja Turner and Carter Emmart **9.3** European Southern Observatory **9.4** NASA **9.5** Jonathan Blair/Getty Images **9.6** NASA/JPL-Caltech/UCLA/MPS/DLR/IDA **9.7** JPL-Caltech/UCLA/MBS/DLR/IDANASA **9.8a** JPL/NASA **9.8b,c** JHU/APL/NASA **9.8d** JAXA **9.9** Nick Schneider **9.12b** BiStar Astronomical Observatory (inset) JPL-Caltech/Univ. of Arizona/ESA/NASA **9.13a,c** ESA/Rosetta/NAVCAM **9.13b** NASA/Jet Propulsion

Laboratory **9.14b** ICSTARS Inc. **9.17** NASA (insets) NASA/JHUAPL/SwRI; NASA/JHUAPL/SwRI; NASA/Johns Hopkins University Applied Physics Laboratory/Southwest Research Institute; NASA/JHUAPL/SwRI **9.18** NASA **9.19** NASA/JHUAPL/SwRI **9.20** NASA/Johns Hopkins University Applied Physics Laboratory/Southwest Research Institute **9.21a** STScI/NASA **9.21b** Bergeron, Joe **9.21c** Mount Stromlo and Siding Spring Observatories, ANU/Science Source **9.21d** Goddard Space Flight Center/NASA **9.22** Kirk R. Johnson **9.23** Sharpton, Virgil L. **9.24** Pearson Education, Inc. **9.25** ©World History Archive/age fotostock **9.26** AP Images **p. 262** NASA/Jet Propulsion Laboratory

Chapter 10

Chapter Opener A.-L. Maire/LBTO **10.6** (part 3) A.-L. Maire/LBTO **10.13** Frederic Masset **p. 286** (Mercury, Earth, Mars) NASA/JPL-Caltech; (Venus) Galileo Spacecraft, JPL, NASA, Copyright Calvin J. Hamilton; (Moon) Akira Fujii **p. 287** (Jupiter, Earth's Moon) NASA/JPL-Caltech (Europa) NASA/Jet Propulsion Laboratory (Earth) NASA Goddard Space Flight Center, Image by Reto Stöckli (land surface, shallow water, clouds). Enhancements by Robert Simmon (ocean color, compositing, 3D globes, animation); (impact) Pearson Education, Inc.

Chapter 11

Chapter Opener NASA/SDO/AIA **11.1** Peter Muller/Cultura/Getty Images **11.10b** The Institute for Solar Physics **11.11** Courtesy of SOHO/MDI consortium. SOHO is a project of international cooperation between ESA and NASA **11.12** Kamioka Observatory, ICRR, University of Tokyo **11.13** INFN—Gran Sasso National Laboratories **11.14a** Royal Swedish Academy of Sciences **11.14b** National Solar Observatory **11.16b** NASA/Stanford-Lockheed Institute for Space Research's TRACE Team **11.17** NASA **11.18** Yohkoh Soft X-ray Telescope, ISAS, Japan **11.19** NASA/SDO/AIA/GSFC **11.20 and p. 306** NASA

Chapter 12

Chapter Opener Alan Dyer/Stocktrek Images/Getty Images **12.4** NASA, ESA, W. Clarkson (Indiana University and UCLA), and K. Sahu (STScI) **12.5** Image Courtesy of Harvard College Observatory **12.9** (center) David Nash (davesastro.co.uk) **12.14** Stocktrek Images/Getty Images **12.15** NASA/Jet Propulsion Laboratory

Chapter 13

Chapter Opener X-ray: NASA/CXC/Rutgers/G. Cassam-Chenaï, J.Hughes et al.; Radio: NRAO/AUI/NSF/GBT/VLA/Dyer, Maddalena & Cornwell; Optical: Middlebury College/F.Winkler, NOAO/AURA/NSF/CTIO Schmidt & DSS **13.1** Stocktrek Images/Robert Gendler/Glow Images **13.2** Prof. Matthew Bate, University of Exeter, U.K. **13.3a** NASA, ESA, and the Hubble Heritage Team (STScI/AURA) **13.3b** Far-infrared: ESA/Herschel/PACS/SPIRE/Hill, Motte, HOBYS Key Programme Consortium; X-ray: ESA/XMM-Newton/EPIC/XMM-Newton-SOC/Boulanger **13.5** NASA/JPL/Caltech **13.7a** Dr. Robert Hurt, Caltech IPAC/NASA **13.7b** European Southern Observatory **13.12a** NASA, WIYN, NOAO, ESA, Hubble Helix Nebula Team, M. Meixner (STScI), & T. A. Rector (NRAO) **13.12b** NASA/ESA/Hubble **13.21** NASA, ESA, J. Hester, A. Loll (ASU) **13.22** European Southern Observatory **13.23** NASA/Jet Propulsion Laboratory

Chapter 14

Chapter Opener NASA/HST/CXC/ASU/J. Hester et al. **14.1a** NASA, ESA, H. Bond (STScI) and M.

Barstow (University of Leicester) **14.1b** NASA/SAO/CXC **14.3** Pearson Education, Inc. **14.4b** Mike Shara, Bob Williams, and David Zurek (Space Telescope Science Institute); Roberto Gilmozzi (European Southern Observatory); Dina Prialnik (Tel Aviv University); and NASA **14.6** NASA/Jet Propulsion Laboratory **14.8** European Southern Observatory **p. 376** (Chandrasekhar) Bettmann/Contributor/Getty Images (Eddington) Science Source (Landau) Interfoto/Personalities/Alamy Stock Photo (Oppenheimer) Everett Collection Inc/Alamy Stock Photo (Bell) PA Images/Alamy Stock Photo **14.18** Andrew Fruchter (STScI) and NASA/ESA **14.19** S. Kulkarni, J. Bloom, P. Price, Caltech-NRAO GRB Collaboration **14.20** NASA Goddard Space Flight Center

Chapter 15

Chapter Opener NASA **15.1a** Mike Carroll **15.3a,f** NASA/Jet Propulsion Laboratory **15.3b** John Bally, University of Colorado **15.3d** NASA **15.3c** NASA, ESA, and The Hubble Heritage Team (STScI/AURA) **15.3e** Robert J. Vanderbei **15.4** NASA, ESA, HEIC, and The Hubble Heritage Team (STScI/AURA) **15.5** Robert J. Vanderbei **15.6** NASA/Jet Propulsion Laboratory **15.7a** Stocktrek Images/SuperStock **15.7b** NASA, ESA, and the Hubble Heritage Team (STScI/AURA)-ESA/Hubble **15.9** John Bally, University of Colorado **15.10** NASA, ESA, and M. Livio and the Hubble 20th Anniversary Team (STScI) **15.11** (top) 2MASS/J. Carpenter, T. H. Jarrett, & R. Hurt (second row, left) Axel Mellinger (second row, right) ESA/NASA/JPL-Caltech (third row, left) Max-Planck-Institut fur extraterrestrische Physik (MPE) and S. L. Snowden/ROSAT Mission (third row, right) ESA/NASA/Planck Collaboration (fourth row, left) NASA/DOE/Fermi LAT Collaboration (fourth row) LAMBDA/Astrophysics Science Division at NASA/GSFC **15.12** NASA, ESA, M. Robberto (STScI/ESA) et al. **15.14** ESO/Igor Chekalin **15.15** ESO **15.16** NASA, Hubble Heritage Team (STScI/AURA), ESA, S. Beckwith (STScI) **15.17** NASA, Hubble Heritage Team (STScI/AURA), ESA, S. Beckwith (STScI) **15.20a** Axel Mellinger **15.20b,c** S. Kulkarni, J. Bloom, P. Price, Caltech-NRAO GRB Collaboration **15.20d,e,f,h,i,j,k** European Southern Observatory **15.20g** UCLA Galactic Center Group **15.21** Keck/UCLA Galactic Center Group **15.22** NASA/Jet Propulsion Laboratory **p. 408** (top to bottom) Axel Mellinger; NASA/Jet Propulsion Laboratory; NASA Goddard Space Flight Center, Max-Planck-Institut fur Astrophysik (MPA); NASA/JPL

Chapter 16

Chapter Opener NASA **16.1** R. Williams (STScI), the Hubble Deep Field Team and NASA **16.2** NASA, ESA, K. Kuntz (JHU), F. Bresolin (University of Hawaii), J. Trauger (Jet Propulsion Lab), J. Mould (NOAO), Y.-H. Chu (University of Illinois, Urbana), and STScI **16.3** Hubble Heritage Team (AURA/STScI/NASA) **16.4** Hubble Heritage Team, ESA, NASA **16.5** Robert Gendler/Science Source **16.6** Anglo-Australian Observatory/Royal Observatory Edinburgh **16.7** National Optical Astronomy Observatories **16.8** (top left) ESA/Hubble & NASA. Acknowledgement: Judy Schmidt (bottom left) Royal Observatory, Edinburgh/Anglo-Australian Telescope Board/Science Source (center) NOAO Gemini Science Center (top right) Robert Gendler (bottom right) Sloan Digital Sky Survey (SDSS) Collaboration, www.sdss.org **16.9** GMOS-S Commissioning Team, Gemini Observatory/Association of Universities for Research in Astronomy/NOAO/NSF **16.10** NASA/Jet Propulsion Laboratory **16.14** Space Telescope Science Institute **16.15** Huntington Library/SuperStock **16.22** NASA **16.23** Max Planck Institute

for Astrophysics **16.24** NASA, ESA and R. Hurt (Spitzer Science Center) **16.25** NASA, ESA, HEIC, and The Hubble Heritage Team (STScI/AURA) **16.26** Visualization by Frank Summers, Space Telescope Science Institute, Simulation by Chris Mihos, Case Western Reserve University, and Lars Hernquist, Harvard University **16.27** Duc/Cuillandre, NCRS/CEA/CFHT **16.28** European Southern Observatory **16.29** NASA/Jet Propulsion Laboratory **16.30a** NASA, ESA, The Hubble Heritage Team (STScI/AURA) **16.30b** NASA/CXC/SAO/PSU/CMU **16.31** J. A. Biretta et al., Hubble Team (STScI/AURA) **16.32** NASA/ESA and J. Bahcall (IAS) **16.33** NASA **16.34** NASA/Jet Propulsion Laboratory **16.36** NASA **16.37** NASA **p. 438** (bottom) NASA

Chapter 17

Chapter Opener ESA and the Planck Collaboration **17.2** CERN-European Organization for Nuclear Research **17.7** Reprinted with permission of Nokia Corporation **17.10** ESA and the Planck Collaboration **17.17a** Greg Thow/Moment/Getty Images **17.17b** WDG Photo/Shutterstock

Chapter 18

Chapter Opener X-ray: NASA/CXC/CfA/M.Markevitch et al.; Lensing Map: NASA/STScI; ESO WFI; Magellan/U.Arizona/D.Clowe et al. Optical: NASA/STScI; Magellan/U.Arizona/D.Clowe et al. **18.5** California Institute of Technology Archives **18.6** X-ray: NASA/CXC/IfA/C. Ma et al.; Optical: NASA/STScI/IfA/C. Ma et al. **18.7** NASA, ESA, and M.J. Jee (Johns Hopkins University) **18.9** NASA, ESA, J. Richard (CRAL) and J.-P. Kneib (LAM) Acknowledgement: Marc Postman (STScI) **18.10** X-ray: NASA/CXC/CfA/M. Markevitch et al.; Lensing Map: NASA/STScI; ESO WFI; Magellan/U.Arizona/D.Clowe et al. Optical: NASA/STScI; Magellan/U.Arizona/D.Clowe et al. **18.12** R. Brent Tully **18.14** Anatoly Klypin (NMSU) and Andrey Kravtsov (University of Chicago)/NCSA **p. 481** Lines of poem "Fire and Ice" by Robert Frost **p. 485** Rebecca Elson "Let There Always Be Light" from *A Responsibility to Awe*. 2002 Carcanet Press Limited. **18.18a** NASA, ESA, J. Richard (Center for Astronomical Research/Observatory of Lyon, France), and J.-P. Kneib (Astrophysical Laboratory of Marseille, France) **18.18b** NASA Earth Observing System **18.18c** Anatoly Klypin (NMSU) and Andrey Kravtsov (University of Chicago)/NCSA **p. 494** ESA and the Planck Collaboration **p. 495** (top to bottom) NASA, ESA, K. Kuntz (JHU), F. Bresolin (University of Hawaii), J. Trauger (Jet Propulsion Lab), J. Mould (NOAO), Y.-H. Chu (University of Illinois, Urbana), and STScI; Robert Gendler; NASA, ESA, M. Robberto (Space Telescope Science Institute/ESA) and the Hubble Space Telescope Orion Treasury Project Team

Chapter 19

Chapter Opener Seth Shostak **19.2** Darlene Cutshall/Shutterstock **19.4a** Jane Gould/Alamy Stock Photo **19.4b** B Christopher/Alamy Stock Photo **19.4c** NASA/Jet Propulsion Laboratory **19.5** Stephen J. Mojzsis **19.8** Photo Copyright Woods Hole Oceanographic Institution **19.9** Reuters **19.10** Martin M. Hanczyc **19.12** Eye of Science/Science Source **19.13** M I (Spike) Walker/Alamy Stock Photo **19.14** NASA/JPL/Jean-Luc Lacour (inset) NASA/JPL-Caltech/MSSS **19.15** NASA/Jet Propulsion Laboratory **19.17** Carroll, Mike **19.19** Seth Shostak **19.21** Seth Shostak **19.22** National Astronomy and Ionosphere Center **19.23** NASA/Jet Propulsion Laboratory **19.24** NASA/Jet Propulsion Laboratory **19.25** NASA/Jet Propulsion Laboratory **p. 532** ESA and the Planck Collaboration

Index

Big Dipper
    pointer stars of, 28, 31
    stars in handle of, 318
Big Rip, 488
binary star systems, 317. *See also* close
        binary systems
    formation of, 337
    lives of, 353, 356
    mass exchange in, 356
    measuring stellar masses in,
        317–318
    types of, 317–318
    X-ray, 370, 375–376
binoculars, 123
blackbody radiation, 116. *See also*
        thermal radiation
black holes, 157, 351, 371–372, 375–376
    active galactic nuclei and, 433–434
    at center of Milky Way Galaxy, 403–406
    evaporation of, 488–489
    event horizon of, 372
    formation of, 375
    galaxy evolution and, 435–436
    galaxy formation and, 436
    hypothesis on, 433–434
    merging of, 379
    observational evidence for, 375–376
    singularity and, 372
    spacetime curvature and, 371
    supermassive, 375–376, 432–435
    from supernovae, 351, 355, 378
    visiting, 372–375
blowout, 393
"blueberries," 190
blue cloud, 416, 431
blueness, of sky, 181
blueshift, 118–119, 121, 268, 272, 317,
        318, 470
Boltzmann's constant, 294
bombardment. *See* heavy bombardment
Borexino neutrino detector, 298
boron, formation of, 455
bosons, 447
bound orbits, 96, 97
Brahe, Tycho, 62–63, 65, 66, 69, 71
brain mass vs. body mass, 520
brightness, apparent, 311, 314–315,
        317–318
brown dwarf, 315
    as dark matter, 476
    degeneracy pressure and, 339–340,
        343, 385
    formation of, 339–340
    planet orbiting, 339
bubble, gas, 392, 393, 394, t396
bulge, of Milky Way, 17, 387, 413
bulge stars, 389, 390
Bullet Cluster, 466, 475
Burbidge, Margaret, 472
Butterfly Nebula, 345

calculus, 88
calendars, 14–15, 57
Callisto (moon of Jupiter), 99, 145,
        222, 223, 226–227
    possibility of life on, 510
    properties of, A-10
Calories, 92
Calypso (moon of Saturn), 223, A-10
Cambrian explosion, 506
Cambrian period, 506
Canada-France-Hawaii Telescope, 129
Cancer, 32
Canis Major Dwarf galaxy, 26, 388, 403
Canis Minor constellation, 26
Cannon, Annie Jump, 316
cantaloupe terrain, 231
Capricornus, 32
captured moons, 161–162, 231

carbon, 11
    fusion of, 344, 349, 366
    isotopes of, 110, 500
carbonate rocks, 197, 201
carbon dating, 500
carbon dioxide, 142, 143, 144
    global warming and, 202–203
    on Mars, 191–192, 517
    on Venus, 196–197, 517
    from volcanic outgassing, 179
carbon dioxide cycle (CO$_2$ cycle),
        201–202, 209, 513
Carina Nebula, 395, 396
Cassini division, 232, 233
Cassini spacecraft, 146, 214, 223, 227,
        228–229, 230, 231, 233, 234, 510
catastrophism, 258
Catholic Church, Galileo and, 67
Cat's Eye Nebula, 392
celestial equator, 26, 29, 30
celestial pole, 26, 29, 30, 31
celestial sphere
    constellations on, 26, A-17–A-18
    Milky Way on, 27
    movement of stars on, 28–29
    overview of, 26
Celsius scale, 93
Centaurus constellation, 9
center of mass, 267
central dominant galaxies, 430
Cepheid variable stars (Cepheids),
        418–419, 420, 421, 422
Ceres, 7, 65, 149, 243, 244–245, 247
C-G, Comet (67P/Churyumov-
        Gerasimenko), 249
Chandrasekhar, Subrahmanyan, 376
Chandrasekhar limit, 363, 377
Chandra X-Ray Observatory, 125, 126,
        128, 361, 362, 367, 406, 432
    design of, 125–126
    observations of Sagittarius A* by, 406
    starburst galaxy image from, 432
Chandrayaan-1 spacecraft, 184
charged particle belts, 177
Charon (moon of Pluto), 99, 243, 253, 254
    formation of, 163, 252
    properties of, 254, A-11
    synchronous rotation of, 99, 149
Chelyabinsk, Russia, 258, 259
chemical elements, 110, 115–116
chemical potential energy, 92, 95
CHEOPS, 270
Chicxulub crater, 258
China, astronomy in ancient, 57, 60
chromosphere, 292, 300–301
Churyumov-Gerasimenko, Comet, 249
circumpolar stars, 28, 29, 30
civilizations, calculated number of in
        Milky Way Galaxy, 518
climate
    on Earth, 199, 201–205, 303–304, 513
    on Mars, 191–193
clocks, sundials as, 56
close binary systems, 337. *See also*
        binary star systems
    black holes in, 376
    formation of, 337
    lives of, 353, 356
    mass exchange in, 356
    neutron stars in, 370–371
    tidal forces in, 356
    white dwarfs in, 363–367
closed (recollapsing) universe, 458,
        481–482, 485
close encounter hypothesis, 153
clouds. *See also* interstellar clouds; Ma-
        gellanic Clouds; molecular clouds;
        protogalactic clouds
    on jovian planets, 219–220

clusters. *See* galaxy clusters
CNO cycle, 348
coasting universe, 481, 482, 487
COBE (Cosmic Background Explorer)
        spacecraft, 453
cold dark matter. *See* WIMPs
collisions. *See also* giant impact(s)
    in asteroid belt, 247, 251
    expansion of universe and, 485,
        487, 488–489
    of galaxies, 429–430
    mass extinctions from, 256–257
    momentum and force in, 84–85
    planetary rings from, 233
    in planet formation, 161
color(s)
    of Earth's sky, 180–181
    of galaxies, 412, 416, 428, 431
    of jovian planets, 219–220
    of Pluto, 254
    of pre-explosion supernovae, 394
    of stars, 314–315
    in visible spectrum, 105, 120
Columbus, and knowledge of Earth's
        shape, 61
coma, in comets, 248–249
comets, 151, 152, 241–242, 248–252,
        287. *See also* names of specific
        comets
    current risk of impacts from,
        257–258, 259
    definition of, 4, 5, 152, 241–242
    dwarf planets as, 248
    gravitational encounters of, 97
    impact of with Jupiter, 255–256
    influence of jovian planets on, 255,
        259–260
    and Kuiper belt objects, 242–243,
        251–252
    largest known, 152
    observation of by Tycho, 62
    origin of, 160–161, 251–252
    structure and composition of,
        248–249
    tails of, 248–251
Compte, Auguste, 122
condensation
    and frost line, 158
    in solar system formation, 157–158,
        164
    temperatures of in solar nebula, t157
conservation laws, 90–95
    of angular momentum, 90, 91–92,
        134–135, 155, 363, 402
    of energy, 91–92, 94–95, 97,
        134–135, 155
    of momentum, 90–91
Constantinople, astronomy in, 60
constant state of free-fall, 87
constant velocity, 84, 88
constellations, 25–26
    names and locations of, A-17–A-18
    precession and, 36, 38–39
    variations in, 29–31
continuous spectrum, 112, 113, 120–121
convection, internal heat of planet and,
        176, 208
convection zone, in Sun, 292, 296, 300
conversion factors, A-4, A-7
Copernican revolution, 61–67, 68, 69,
        70–71, 88
Copernicus, Nicholas, 61–62, 65, 69, 71
core
    of Europa, 226
    of jovian planets, 218, 279
    of low-mass star, 341–342
    of Sun, 290, 292–293, 341–345
    of terrestrial planet, 173–174, 185
core-bounce process, 351–352

Coriolis effect, 221
corona(e)
    of Sun, 44, 117, 292, 300–301
    on Venus, 194
coronal holes, 301
coronal mass ejections, 301
COROT 7b, 278, 279
COROT 14b, 277, 278, 279
cosmic address, 2–3
Cosmic Background Explorer (COBE)
        spacecraft, 453
cosmic calendar, 14–15
cosmic microwave background, 442,
        452–453, 513
    inflation and, 456–459
    photons from era of nuclei in, 449,
        452
    spectrum of, 453
    temperature of, 453, 454–455, 457
    uniformity of, 453, 457–458
cosmic rays, 126, 393–394, 396
cosmological constant, 468, 484
cosmological horizon, 425
Cosmological Principle, 423, 480
cosmological redshift, 425
cosmology, 417
cosmos. *See* universe
Crab Nebula, 352, 361, 368, 369
crescent moon, 43, 55
critical density, 458, 475
    closeness of density of universe to,
        458, 475–476
    dark energy and, 484
    dark matter and, 475–476, 484
    as dividing line between expansion
        and collapse of universe, 458
critical universe, 481, 482
crop circles, 522
crust
    oceanic, 200–201
    planetary, 174
Curiosity rover, 144, 190, 191, 508, 509
curvature of spacetime. *See* spacetime
cyanobacteria, 505, 506
cycles per second, 107
Cygnus A radio galaxy, 436
Cygnus Loop, 394
Cygnus X-1 system, 377, 406

dark energy, 17, 468, 480–482, 483,
        485, 487
    and acceleration of expansion of
        universe, 482–483
    cosmological constant and, 468, 484
    critical density and, 484
    definition of, 468, 483
    fate of universe and, 480–489
    flatness and, 484
dark matter, 17, 391, 468, 469–471
    composition of, 475–477
    critical density and, 475–476, 484
    definition of, 468
    density of, 475–476
    distribution of in early universe,
        478–480
    and distribution of mass in Milky
        Way Galaxy, 17, 391, 469–470
    in elliptical galaxies, 471
    evidence for, 468–477
    evidence against neutrinos as, 476
    exotic, 485
    in galaxy clusters, 472–474, 486
    gravitational lensing and, 473–474,
        475, 486
    role of in galaxy formation,
        477–478, 495
    in spiral galaxies, 470–471, 477
    and temperature of hot gas in clus-
        ters, 472–473

meter showers, 243, 250–251, t251
meter, A-6, A-7
methane, 182, 206
  on jovian planets, 146, 147, 148, 152, 215, 220
  on Mars, 509
  in solar nebula, t157, 157
  on Titan, 146, 227–229
Metonic cycle, 57
metric system, A-6–A-7
microlensing, 270
micrometeorites, 183–184
Micronesian stick chart, 58
microwaves, 107, 108. See also cosmic microwave background
mid-ocean ridges, 200
"midsummer," 37
Milky Way Galaxy, 3, 4, 386, 387
  all-sky views of, 397, 404–405
  on celestial sphere, 26–27
  center of, 403, 406
  dark lanes in, 27, 395
  dark matter in, 391, 469–470
  diameter of, 387
  disk vs. spheroidal populations of, 413–414
  element distribution in, 402–403
  formation of, 402–403, 494–495
  gas distribution in, 391, 394
  location of, 3, 4
  mass of, 17, 390–391, 469–470
  motion of stars within, 16–17, 389–391
  movement of, 17–18, 19
  in night sky, 25, 27, 67, 387
  number of stars in, 9, 387
  rotation of, 16–17, 19
  rotation curve for, 469–470
  size of, 9
  spiral arms of, 387, 399–400, 401
  star-forming regions of, 398–399
  structure of, 27, 326, 387–388
  and Virgo Cluster, 478
Miller, Stanley, 504
Miller–Urey experiment, 504
millimeter waves, 108
millisecond pulsars, 370
Mimas (moon of Saturn), 222, 229
  impact of on Saturn's rings, 234
  properties of, A-10
minerals, t157, 157
Miranda (moon of Uranus), 222, 231, A-11
mirrors, of telescopes, 123
Mizar, 318
models, scientific, 59–60, t74, 444
molecular clouds, 395
  giant, 395
  in Milky Way, 395–396
  and spiral density waves, 400
  and star formation, 334–336, 395–396, 402–403
  in star–gas–star cycle, 392, 395–396
  temperature/density of, t396
molecules, 110, 116
momentum, 84–85. See also angular momentum
Monoceros constellation, 26
month
  origin of word for, 40
  synodic, A-1
Mont-Saint-Michel, France, 100
Moon (of Earth), 39–46, 143, 161, 162–163
  age of, 166
  angular size of, 27, 28, 29
  calendars and, 57
  "dark side" of, 42
  in daytime, 41
  diameter of, 29
  distance of from Earth, 5

and Earth tides, 98–100
eclipses of, 42–46, t45, 59
as fifth terrestrial world, 152
formation of, 162–163
Galileo's observations of, 66
geological character of, 183–184
and habitability, 515
illusion of size of, 28
interior structure of, 174
mass on, 86
movement of in night sky, 28, 40, 46
orbit of, 40, 41–43
phases of, 40–41
properties of, A-10
rocks from, 162, 166
shadows on, 40
surface of, 66, 173, 178, 183–184
synchronous rotation of, 41–42, 99
moonlets, 234–235
moons, 5. See also jovian moons; names of specific moons
  captured, 161–162, 222, 231
  gap, 234
  habitability of, 515
  orbits of, 151, 153
  shepherd, 234
  of solar system, t150, 161–162, A-10–A-11
  tidal forces and, 99, 100
motion, 83–85. See also planetary motion
  mass vs. weight and, 85–86
  Newton's laws of, 66, 87–90, 96, 473
  proper, 390
  radial vs. tangential components of, 119
Mount Wilson Observatory, 126, 419, 420
MRO (Mars Reconnaissance Orbiter), 191, 192
muon neutrinos, 298
Muslim calendar, 57
mutation, 502

nanometer, 108, 116
NASA, 228, 267, 270, 288, 453, 508. See also spacecraft and missions
natural selection, 501, 503
navigation, by ancient Polynesians, 58
Nazca, Peru, 57
neap tide, 100
Near-Earth Asteroid Rendezvous (NEAR) spacecraft, 152, 246
nebula, 153, 399. See also solar nebula
nebular hypothesis, 153
nebular theory of solar system formation, 151–154, 158–160, 279–280, 281
neon, fusion of, 349, 350, 351
Neptune, 148, 252, 278
  atmosphere of, 221
  composition of, 215, 216
  formation of, 215–216
  Great Dark Spot of, 221
  interior structure of, 218
  moons of, 222, 231, 255, 511, A-11
  properties of, t150, 152, A-9
  rings of, 148, 234
Nereid (moon of Neptune), 222, A-11
net force, 85, 88, 89
neutrinos, 126, 294, 476
  in particle era, 448
  solar, 297–298
  from supernovae, 351
  types of, 298
  weak force and, 439, 476
neutron degeneracy pressure, 351, 367, 375–376
neutrons
  in atomic structure, 109–110

in era of nucleosynthesis, 448–449, 454–455
mass of, A-1
in particle era, 448
neutron stars, 111, 351, 355
  black holes from, 351, 375–376
  discovery of, 368, 377
  mergers of, 370–371
  properties of, 367–368
  and pulsars, 368–370, 377
  from supernovae, 351, 352, 355
New Horizons spacecraft, 8, 97, 149, 224, 240, 252, 253–254, 255, 511, 523
new moon, 40, 41, 42, 43, 100
Newton, Isaac, 65, 66, 68, 74–75, 83, 87–88, 95, 96, 373
  experiments with light by, 106
Newton's laws of motion, 66, 87–90, 96, 471, 473, 474, 486
Newton's universal law of gravitation, 95–96, 98, 390, 473, 474, 486, A-2
Newton's version of Kepler's third law, 96, 253, 271, t275, 291, 318, 390, 391, 406, 474, A-2
NGC 474, 430
NGC 1300, 414
NGC 4038/4039, 429
NGC 4594 spiral galaxy, 414
nickel, t157, 157, 173
night sky
  angular size and distances in, 27–28, 29
  and celestial sphere, 26, 27, 28–29
  constellations in, 29–32, A-17–A-18
  darkness of, as evidence of Big Bang, 460–461
  local, 27, 29, 30
  Milky Way in, 25, 27, 67, 386
  Moon movement in, 39–43
  patterns in, 25–32
  planets in, 46–48
  star charts of, A-19–A-23
  star movement in, 28–32
  variations in, 29–32
nitrogen, 11, 179, 227–228, 254
Nix (moon of Pluto), 243, A-11
nodes, of Moon's orbit, 43, 45
nonscience, vs. science, 69–73
north celestial pole, 26, 28, 29, 30
Northern Hemisphere, 28, 29, 31, 33
North Star, 16, 31, 38
nova (plural: novae), 62, 365
nova remnant, 365
nuclear "burning," in Sun, 292
nuclear fission, 293, 350, 445
nuclear fusion, 11, 350, 445
  of carbon, 344, 349, 366
  in era of nucleosynthesis, 448–449, 454
  of helium, 343–344, 348, 349, 370, 385
  and helium-capture reactions, 349, 454–455
  in high-mass stars, 347–348
  of hydrogen, 292, 337, 341–342, 348, 365, 384
  iron and, 349–350
  in low-mass stars, 341–342
  of magnesium, 351
  of neon, 351
  neutrinos from, 294, 297–298
  in nova, 365
  vs. nuclear fission, 293, 350
  of oxygen, 349, 350
  in red giant stage, 341–342
  of silicon, 351
  in star formation, 334
  in Sun, 290, 293–298, 341

nuclear weapons, 258
nuclei, era of, 446, 449, 457
nucleosynthesis, era of, 446, 448–449, 454, 455
nucleus
  atomic, 109–110
  of comet, 248
number density, in ideal gas law, 294

OBAFGKM sequence, of spectral types, t315, 315–317, 319–323, 327, 328
obelisk, Egyptian, 56
Oberon (moon of Uranus), 222, A-11
objectivity, in science, 72–73
observable universe, 5, 6, 10, 25–28, 425
observations, in science, 55–56, 58, 69, 70–71
observatories. See also Hubble Space Telescope; space telescopes
  ALMA, 125, 136, 156
  LICK, 388
  Mauna Kea (Hawaii), 124, 127
  Mount Wilson, 126, 419, 420
  Palomar, 472
  Paranal (Chile), 127
  Solar Dynamics, 301
  Super-Kamiokande neutrino, 298
  Yerkes (Chicago), 123
Occam's razor, 69, 72
oceanic crust, 200–201
Olbers, Heinrich, 460
Olbers' paradox, 460–461
Olympus Mons (Mars), 188
Oort cloud, 139, 152, 160, 242, 251–252, 259, 287
  Jupiter and, 251, 259
  location of, 139, 152, 251
opaque materials, 111
open clusters, 326
open (coasting) universe, 481, 482
Ophiuchus, 31, 32, 399
Oppenheimer, Robert, 376
Opportunity rover, 189–190
orbit(s), 5, 7
  of asteroids, 247
  and atmospheric drag, 98
  backward, 148, 231, 255
  belief in circular, 63, 66, 71
  bound vs. unbound, 96, 97
  around center of mass, 267
  of comet, 248–250, 251–252
  of Earth, 15–16, A-9
  elliptical, 63–64, 66, 71
  of Eris, A-9
  and escape velocity, 98
  of extrasolar planets, 270–271
  of galaxies in clusters, 471–472
  of Io, 224–225
  in Kepler's laws of planetary motion, 64–65, 95, 390
  of Mars, A-9
  of Mercury, 141, A-9
  of Moon, 39–42
  and orbital velocity formula, 391, A-2
  parabolic, 96
  of planets in solar system, 138–139, A-9
  of Pluto, 252, A-9
  solar system formation and, 138–139, 151
  of stars, 389–391, 400
  of Sun, 267–268, 390–391, A-9
  of Triton, 148
orbital angular momentum, 91
orbital eccentricity, 271, A-9
orbital energy, 97, 99, 161
orbital inclination, A-9
orbital period, 62, 65, 96, 99, t150, 247, 270–271, A-9